Where in the World with Astro*Carto*Graphy?

David Meadows

ISBN: 0-86690-486-7

First Printing: 1998

Published by:
American Federation of Astrologers, Inc.
PO Box 22040
6535 S. Rural Road
Tempe, AZ 85285-2040

Printed in the United States of America

Dedication

This book is dedicated to Jim Lewis who introduced me to Astro*Carto*Graphy and taught me everything he knew about it. My knowledge of Astro*Carto*Graphy has completely changed my life which I have Jim Lewis to thank for. Without Jim Lewis this book on Astro*Carto*Graphy would never have been possible.

Acknowledgements

My special thanks to Greg Howe at Astro Numeric Service who supplied me with all the Astro*Carto*Graphy and Cyclo*Carto*Graphy maps that are featured in this book. Usually, Cyclo*Carto*Graphy is drawn on a transparent overlay, but Greg Howe reprogrammed his computer so that the lines could be drawn on a world map.

My thanks to Roy Gillett at Astrolabe UK for helping me to transfer the paran printouts and Astro Mapping listings from Nova to a text processing program. As Nova is a DOS astrological computer program I couldn't copy what I needed into a windows text processing program, but Roy Gillett solved this problem very quickly.

Finally, I would like to thank my wife, Martina, for her love, help and support during the time I was writing this book, and especially for what we have experienced together with Astro*Carto*Graphy.

Contents

Illustrations

Introduction

Astro*Carto*Graphy is an amazingly accurate astrological technique that only a small number of astrologers understand in great depth. Yet, by using all the techniques involved in Astro*Carto*Graphy one has the opportunity to change one's life for the better.

Whenever I try out a new astrological technique I always start by using it in my own natal horoscope, and if the results are favorable I then apply the technique to the charts of people I know. But if the technique does not work in my own natal horoscope I do not look at it any further.

When I first became aware of the Astro*Carto*Graphy technique I decided to try it out myself. I purchased my own individual Astro*Carto*Graphy map, and I could not believe how accurate it was. Having seen the effectiveness of Astro*Carto*Graphy in my own life I decided I would learn as much as I could about this astrological technique. This was 1983, but apart from the material provided with an Astro*Carto*Graphy map there was nothing available to learn about this technique.

In March 1986 the inventor of Astro*Carto*Graphy, American astrologer Jim Lewis, held a weekend seminar in London on Astro*Carto*Graphy. What I learnt that weekend proved to be the turning point of my life. I have applied the various Astro*Carto*Graphy techniques in my own life, and as a result my life has changed completely.

Since attending this seminar in 1986 I have specialized in Astro*Carto*Graphy. My involvement with Astro*Carto*Graphy was more significant when Jim Lewis asked me if I would like to become his U.K. representative. I accepted his offer without hesitation, and as a result I developed my interest in Astro*Carto*Graphy.

I have lectured on Astro*Carto*Graphy at over 35 different astrology groups in Great Britain. My lecture was always well received by astrologers, and as a result hundreds of astrologers decided to purchase their own individual Astro*Carto*Graphy map. But the one question that different astrologers kept asking was how could they learn about Astro*Carto*Graphy. It was a question that I couldn't really answer because no book has been written that explains how to use all the different Astro*Carto*Graphy techniques. But I knew how these astrologers felt as I had the same problem myself.

The first book to feature Astro*Carto*Graphy in detail was published in 1989. *The Astro*Carto*Graphy Book of Maps* by Jim Lewis and Ariel Guttman shows how 136 famous people found their place. For each famous person there is a natal horoscope, an Astro*Carto*Graphy map and a brief biography. The last chapter shows how Astro*Carto*Graphy works for events and nations. This is an excellent book to learn more about Astro*Carto*Graphy.

One year later a book called *Working with Astrology* by Michael Harding and Charles Harvey was published. This book has a chapter on Astro*Carto*Graphy containing 25 pages. It is a very interesting introduction on Astro*Carto*Graphy which shows how important this astrological technique is.

In 1991 a book called *Navigating by the Stars* by Edith Hathaway was published. This book contains seventy pages on Astro*Carto*Graphy showing examples how it works in the lives of well-known people.

The subject I hated most at school was history which I found really boring. I can remember receiving a result of four percent in my history exam, and that was only because I correctly answered that the Olympic Games are held every four years!! But since I started using Astro*Carto*Graphy I'm really interested in history and world events! It was very easy to see that Astro*Carto*Graphy is a very powerful technique in the field of mundane astrology.

Jim Lewis has written many articles on Astro*Carto*Graphy giving examples of how effective it is for American presidents and events that take place in the world. He has also written articles how Astro*Carto*Graphy works in a country's national horoscope.

There are two books available that explain how Astro*Carto*Graphy works in mundane astrology. In a book called *The Astrology of the Macrocosm* edited by Joan McEvers there is a chapter by Jim Lewis consisting of thirty pages. The chapter entitled "Bringing Mundane Astrology Down to Earth" shows how to use Astro*Carto*Graphy in mundane astrology. It is written very clearly and is easy to understand. The other book, called *Mundane Astrology*, by Baigent, Campion and Harvey has a chapter on Astro*Carto*Graphy consisting of sixteen pages. This is another good introduction to Astro*Carto*Graphy with interesting examples.

There is certainly much more material available on Astro*Carto*Graphy than there was when I wanted to learn more about this technique in 1983. Although one is able to get a good understanding of what Astro*Carto*Graphy is and how it works from the material that is available, there is a lot of ground that hasn't been covered. No book has been written that explains all the different techniques that are used in Astro*Carto*Graphy. It is also very important to take Cyclo*Carto*Graphy into consideration. One can also use an Astro*Carto*Graphy solar return map and

Astro*Carto*Graphy solar arc progressions, but very little has been written about these techniques.

It is not possible to change one's natal horoscope, so one has to learn to live with the positive and negative parts of it. What does change are the temporary influences in life which are shown by the progressions and transits to the natal chart. There are periods in life when one seems to experience nothing but problems. Other times one feels really happy in life as everything is going well.

The most important decision a person makes is **where** to live. At first this may not seem very important as some people have no choice where to live. This is because of family reasons, job or other responsibilities that people have at a location where they live. But, by relocating, a person has the possibility to change his or her life. Deciding **where** to relocate, and the best time **when** to do so, is extremely difficult when only the natal horoscope is used. Surely, there must be an easier way to find out this information. Yes there is, and that is by using Astro*Carto*Graphy.

As no book is available that explains how to use all the different Astro*Carto*Graphy techniques I decided it was time to change this situation by writing a book myself. At first Astro*Carto*Graphy looks to be quite complicated, but once the basic principles are understood it is very easy to put into practice.

I will explain briefly how I have written this book by describing the contents of each chapter.

Chapter 1: This short chapter explains what Astro*Carto*Graphy is, how it started and what it can be used for.

Chapter 2: When I lectured at astrology groups in Great Britain the hardest part for astrologers to understand was how an Astro*Carto*Graphy map is drawn. Astrologers could see that a computer calculates and draws an Astro*Carto*Graphy map, but they found it difficult to understand how the lines appear on the map. Without some knowledge of astronomy it is difficult to see how an Astro*Carto*Graphy map is related to a natal horoscope. In this chapter I have gone into a lot of detail to explain how an Astro*Carto*Graphy map is drawn. I have tried to explain everything in simple language and with the help of many illustrations as it is the most difficult part in understanding how Astro*Carto*Graphy works.

Chapter 3: Once an Astro*Carto*Graphy map is drawn the interpretation of it can begin. At first the interpretation of an Astro*Carto*Graphy map looks to be very complicated, but once the two basic rules are understood the interpretation of a map is very easy.

Chapter 4: Here examples show how an Astro*Carto*Graphy map is interpreted. I have used the Astro*Carto*Graphy maps of some well-known people to explain the step-by-step interpretation. But I have also included the Astro*Carto*Graphy map of a client as the experiences at different locations are really amazing. I am very thankful that this client gave me the permission for the Astro*Carto*Graphy map to be featured in this book.

Chapter 5: Astro*Carto*Graphy shows **where** in the world planetary energies will manifest. If a person is planning to move, it is not enough just to look at their Astro*Carto*Graphy map. This person would need to know **when** is the best time to move, and the basic technique in answering this question is by using Cyclo*Carto*Graphy. My own experience has shown that the use of Cyclo*Carto*Graphy is very important, and when it is not considered a move could turn out to be a big disaster.

Chapter 6: Cyclo*Carto*Graphy is the most important technique to see how an Astro*Carto*Graphy map is temporarily influenced. But using the yearly Astro*Carto*Graphy solar return map can provide additional information, and very often reiterates the Cyclo*Carto*Graphy temporary influences.

Chapter 7: Another technique for determining the temporary influences is Astro*Carto*Graphy solar arc progressions. This technique is very easy to use, and the information obtained frequently reiterates what is found by Cyclo*Carto*Graphy and the Astro*Carto*Graphy solar return map.

Chapter 8: This chapter takes a look at how Astro*Carto*Graphy works for two people together in a relationship. If a married couple wish to relocate, then more than one Astro*Carto*Graphy map needs to be considered. In natal astrology an astrologer can use the birth charts of the married couple, their composite chart, relationship chart and wedding chart to find out how they will get on with each other. I have looked at all these techniques using Astro*Carto*Graphy to find out which ones can be used.

Chapter 9: Astro*Carto*Graphy is a very effective technique when it is used in mundane astrology. I have examined significant world events during the 1980s and 1990s to show how Astro*Carto*Graphy works in mundane astrology. Some of the techniques are very advanced and only used in mundane astrology, and although they are difficult to understand, very accurate results can be achieved.

Chapter 10: Without an accurate time of birth an Astro*Carto*Graphy map cannot be drawn. In this chapter I have shown how Astro*Carto*Graphy can be used to rectify a time of birth.

Chapter 11: In this chapter I explain how I have changed my life by using Astro*Carto*Graphy. One of the events in my life turned out to be disastrous, but the use of Astro*Carto*Graphy explained why. All the various Astro*Carto*Graphy techniques have been used to show how they give the complete picture. When I lecture at astrology groups astrologers are always amazed about my personal experiences with Astro*Carto*Graphy, so I am including them in this book.

Chapter 12: How does one use Astro*Carto*Graphy for clients? Having done a great many Astro*Carto*Graphy personal interpretations for clients I have learnt from experience, the most effective way in approaching this kind of work. This chapter shows how I interpret Astro*Carto*Graphy for clients, and I have tried to make it as easy as possible to understand.

If you are unhappy at your present location, then the information in this book could well change your life. Even if you are happy where you live, Astro*Carto*Graphy will explain how different locations affect you.

As I was writing this book I received the shocking news that Jim Lewis only had a few months to live. Doctors had diagnosed cancer, but they couldn't do anything to help him. Although I had prepared myself for the inevitable, it was a very sad time for me when the news arrived that he had died.

Just before Jim Lewis died he started to write a book called *The Psychology of Astro*Carto*Graphy*. Kenneth Irving agreed to finish writing this book for Jim Lewis so that it could still be published. This new book contains detailed interpretations of all the planetary lines and crossings used in Astro*Carto*Graphy. It also gives detailed interpretations for the planetary lines on a Cyclo*Carto*Graphy overlay. These interpretations are from the Astro*Carto*Graphy E*plained computer program which Jim Lewis helped to develop.

This book is excellent for the interpretation of an Astro*Carto*Graphy map and Cyclo*Carto*Graphy overlay. But for learning about the many different ways that Astro*Carto*Graphy can be used there is still no book available that explains the various techniques.

In this book I have concentrated more on the different techniques that are used in Astro*Carto*Graphy as detailed interpretations for the planetary lines are featured in other books. Astrologers do not realize that there is more to Astro*Carto*Graphy than planetary lines drawn on a world map. I hope that this book will show that Astro*Carto*Graphy is a complete system of relocation astrology which works most effectively.

Without the knowledge I learnt from Jim Lewis this book would never have been written. It is my hope that this book will show just what an important contribution Jim Lewis has made to astrology.

Basic Techniques

Chapter 1/What Is Astro*Carto*Graphy?

In the nineteenth century people didn't have access to the modern methods of transport and communication that we now take for granted. As a result it took weeks for people to travel long distances. For most people it wasn't possible to visit many different places around the world. In those days there weren't such things as satellites or cable television. Sometimes it took days, or even weeks, for people to find out what was happening in different countries. The majority of people hardly moved from where they were born and only travelled short distances. At the time, locations around the world were not important to people; therefore, astrologers had no need to examine how planets influenced people at different locations.

The twentieth century has seen revolutionary developments connected with travel and communication. Now it is possible for people to visit many different countries, many of which only involve a few hours travelling time. At every travel agency it is possible to book a holiday in a short space of time, and the next day be there. Travel agents have a large number of brochures available to cater for every type of holiday. When we watch television it is possible to see worldwide events live via satellite. When the Berlin Wall was opened, people around the world switched on their televisions to see this unbelievable event taking place live in their living rooms. When Nelson Mandella was released from prison the same was true. We are citizens of the world, either through travel or communication.

As opportunities started to become available for people to visit different parts of the world, astrologers began to examine how horoscopes were influenced at different locations.

Astro*Carto*Graphy is an easy to use technique that shows the astrological influences at every location in the world. The planetary lines are drawn on a map of the world. Even people with no knowledge of astrology are able to use Astro*Carto*Graphy as the map is supplied with an interpretation book. Names and addresses of where an Astro*Carto*Graphy map can be purchased are listed in Appendix 1.

Maybe you are thinking how the lines on an Astro*Carto*Graphy map are calculated. The simplest way of explaining this is that the position of the planets in a horoscope are then transferred on to a map of the world. In Chapter 2, I will explain in more detail the connection between a horoscope and an Astro*Carto*Graphy map.

How Astro*Carto*Graphy Started

American astrologer Jim Lewis had been doing "routine charts" for five years, improving his interpretation skills in the process. Then a psychic friend told him that his "guides" had advised him he was to correct some prior misconception in astrology before moving on.

Jim Lewis had always been interested in visiting different places around the world. He visited more than forty countries, and his experiences allowed him to gain knowledge of himself in various circumstances.

In 1975, Jim Lewis started working on relocation horoscopes. He came to the conclusion that his own personal experiences at many different locations could be explained astrologically. The usual way to do relocation astrology at this time was to draw a new chart for each location. This new chart was calculated as if a person were born there, instead of at the actual birthplace. After a while it occurred to Jim Lewis that the information obtained from these relocated horoscopes could be better shown as lines on a world map. This would then save having to calculate a new chart for each location.

The first map he drew was his own. Suddenly, locations that he had visited took on a new meaning for him, confirming what he had experienced at different locations.

At first, it was only his intention to draw these time-consuming maps for his small number of travel-minded clients. Then friends offered to help him. One friend introduced him to a computer programmer. Three attempts were made over the next year before a successful program was developed to draw the planetary lines on a world map. Another friend offered Jim Lewis the use of his printing press. Other friends volunteered to typeset and design his interpretation book. Some friends offered to do the artwork for the map and book. The result of all this help given to Jim Lewis was Astro*Carto*Graphy. Since 1976 these unique maps have been purchased by people from all around the world.

The Advantage of Astro*Carto*Graphy Over the Normal Natal Horoscope

A natal horoscope shows the personality of a person, character traits and the kind of life that will be experienced. It shows the strengths and weaknesses that a person has in life. The houses of a horoscope pinpoint the area of life where these energies will manifest.

What a people can't do is change their horoscope. Once a person is born he or she keeps the horoscope for the rest of his or her life, so the positive and negative character traits have to be accepted. What does change is time as we grow older. There are times in life when nothing seems to go right for us, experiencing one problem after the other. Then, at other times, Lady Luck seems to be on our side and everything we do is a success. These changes in our lives are shown by progressions and transits to the natal horoscope. These are temporary influences, and the length of time they last depends on the progressed or transiting planet involved.

The one thing that a person is free to choose is **where** to live. Some people feel limited as to where they can live because of job, family responsibilities or other commitments at their present location.

Although you can't change your horoscope, it is possible to change your life by moving to a different location. But the natal horoscope is only valid for the place where you were born. It is not possible to use a natal horoscope on its own to find a good place to move to.

Admittedly, it's possible to calculate new horoscopes for different locations as is the norm in relocation astrology. But this would mean calculating innumerable horoscopes to find suitable locations. Even then there is no guarantee that the relocated horoscope would provide the correct information! In the next chapter I will explain in more detail why the relocated horoscope is not 100 percent accurate.

These problems do not occur with Astro*Carto*Graphy. When the lines are drawn on an Astro*Carto*Graphy map the **true** position of the planets are used. Once the Astro*Carto*Graphy map is drawn, no further calculations are needed as all the information required is on the map. In Chapter 3 I will explain how to interpret an Astro*Carto*Graphy map.

Uses of Astro*Carto*Graphy

There are two ways in which Astro*Carto*Graphy can be used. First, you can look at specific locations and see what planetary influences are there. A good way to start is by examining locations that you've already been to—either where you have lived or visited. Then you determine how each location is influenced and what experiences you would have there. Sometimes you visit a place where you really feel at home and didn't want to leave. Other times a holiday resort is not what you expected and you couldn't wait to go home. By examining your Astro*Carto*Graphy map you will be able to understand why certain experiences occurred there.

With Astro*Carto*Graphy it is possible to choose where to go for holidays. You will know in advance what to expect at any location in the world, thus eliminating places that are very unfavorable for a holiday. That is the second way in which Astro*Carto*Graphy can be used. You look at the map to find locations that are suitable for a specific area in life. For example, a British businessman would like to sell his products internationally. His Astro*Carto*Graphy map would show him where in the world it would be best for him to expand his business.

The most important way in which Astro*Carto*Graphy can be used is for relocation. There are many people who are very unhappy at the location where they live. Some decide to emigrate, but most do not know what to expect at the new location. Other people move to a foreign country for job reasons, but usually they have no choice in the location. There are many reasons why people relocate, but for these people it is a move into the unknown.

If a person decided to relocate and used Astro*Carto*Graphy, then this person would know in advance what to expect at the new location. For people who want to relocate, but don't know where, Astro*Carto*Graphy can help them like nothing else. Americans wanting to relocate would be able to do so without any problem as America is a big country. The expanding European community gives people more opportunity to relocate within Europe.

Astrologers who are interested in mundane astrology (the astrology of world events) can use Astro*Carto*Graphy to good effect. Since 1979 the *Astro*Carto*Graphy Source Book of Mundane Maps* has been published. This book contains maps for eclipses, lunations, solstices and equinoxes that take place during the year. In Chapter 9 I will explain how Astro*Carto*Graphy can be used in mundane astrology.

Astro*Carto*Graphy has revolutionized astrology. The rest of this book explains the various techniques that are used in Astro*Carto*Graphy and how to use them in your own life.

Chapter 2/The Astro*Carto*Graphy Map

The most difficult part about Astro*Carto*Graphy is understanding how a map is drawn. For many people this is not important because a computer draws the map. It is not essential to understand how the planetary lines appear on an Astro*Carto*Graphy map, but for completeness I will explain it. Some knowledge of astronomy is needed to understand how an Astro*Carto*Graphy map is drawn, and once you understand how the planetary lines appear on a world map you will have no difficulty using Astro*Carto*Graphy.

What Is Not Used with Astro*Carto*Graphy

The main interpretation factors of a horoscope are the planets, aspects, signs and houses. Some astrologers include the Moon's Nodes, Chiron, Part of Fortune, fixed stars, the Uranian planets and the asteroids. Other astrologers use additional techniques with a horoscope, such as midpoints, harmonic charts, heliocentric, Hindu charts, Arabian parts, Sabian symbols, etc. What it all means is that a horoscope is cluttered, and it's no wonder that newcomers are sometimes confused about what techniques to use. None of these additional horoscope factors and techniques are used in Astro*Carto*Graphy.

That leaves the planets, signs and houses of a horoscope. When an Astro*Carto*Graphy map is drawn it's like a horoscope being transferred onto a world map.

Horoscopic houses are not used in Astro*Carto*Graphy. Even among astrologers there is disagreement about the houses because of the numerous house systems that can be used. Placidus houses seem to be the most popular among astrologers as tables have been available for many years. Next in popularity is the Koch house system, followed by equal house. Many astrologers who specialize in horary astrology use the Regiomontanus house system.

But as Robert Hand points out in *Essays on Astrology,* there are problems with house systems in extreme latitudes. Some systems, like Placidus and Koch, break down when calculating horoscopes in extreme latitudes. Some astrologers argue that this is not so important because not many people live in those regions. What these astrologers don't realize is that more than one million people live above the Arctic Circle. At certain times of the day Placidus and Koch house systems cannot be used for places on or **above** the Arctic Circle. By not using houses, Astro*Carto* Graphy resolves this problem.

Also not used in Astro*Carto*Graphy are signs of the zodiac. Astrologers have always used Aries, Taurus, Gemini, Cancer, etc. to interpret a horoscope. Then, it depends on what zodiac is used. The tropical zodiac is most used and is based on the seasons of the year. But in Hindu astrology the sidereal zodiac, which is based on the movement of the stars, is used. Again, Astro*Carto*Graphy omits an astrological problem.

Aspects between planets are used by every astrologer when interpreting a horoscope. Even different astrological techniques such as midpoints and harmonic charts use aspects between the planets, e.g. when the Sun is at twenty degrees Cancer and Jupiter is at twenty-one degrees Scorpio, then Sun and Jupiter are in a trine aspect. Astro*Carto*Graphy does not use such planetary aspects!

This means that the following aspects are not used in Astro*Carto*Graphy:

Degrees	Aspects	Degrees	Aspects
30	Semisextile	36	Semiquintile
40	Novile	45	Semisquare
51.26	Septile	60	Sextile
72	Quintile	90	Square
102.51	Bi-septile	120	Trine
135	Sesquiquadrate	144	Quintile
150	Quincunx		

You may have noticed that the conjunction (zero degrees) and opposition (180 degrees) aspects are not included in this list. These two planetary aspects are to some extent used on the Astro*Carto*Graphy map, but with consideration of the true body aspects. This will be explained later in this chapter.

What Is Used in Astro*Carto*Graphy

Having eliminated most of the interpretation factors in a horoscope, you are probably wondering what is left. I have wiped out most of a horoscope in a short space of time, yet there is a connection between an Astro*Carto*Graphy map and a horoscope!

Well, the planets are still there. You may think that not much can be done with planets on their own. Surely, there must be some kind of planetary aspects used in Astro*Carto*Graphy?

Planetary aspects are used in Astro*Carto*Graphy, but not in a way that can be easily found in a horoscope. These kind of planetary aspects are called **parans,** which are very easy to see on an Astro*Carto*Graphy map.

The other kind of aspect that is used in Astro*Carto*Graphy is the conjunction to the horoscope angles. When a planet is conjunct the Midheaven, Imum Coeli, Ascendant or Descendant, it is considered to be a very powerful planet in a horoscope. Research carried out by the Gauquelins showed this to be the case; for example, Mars was prominent at the Ascendant and Midheaven in the horoscopes of well-known sportsmen

The usual ten planets are used in

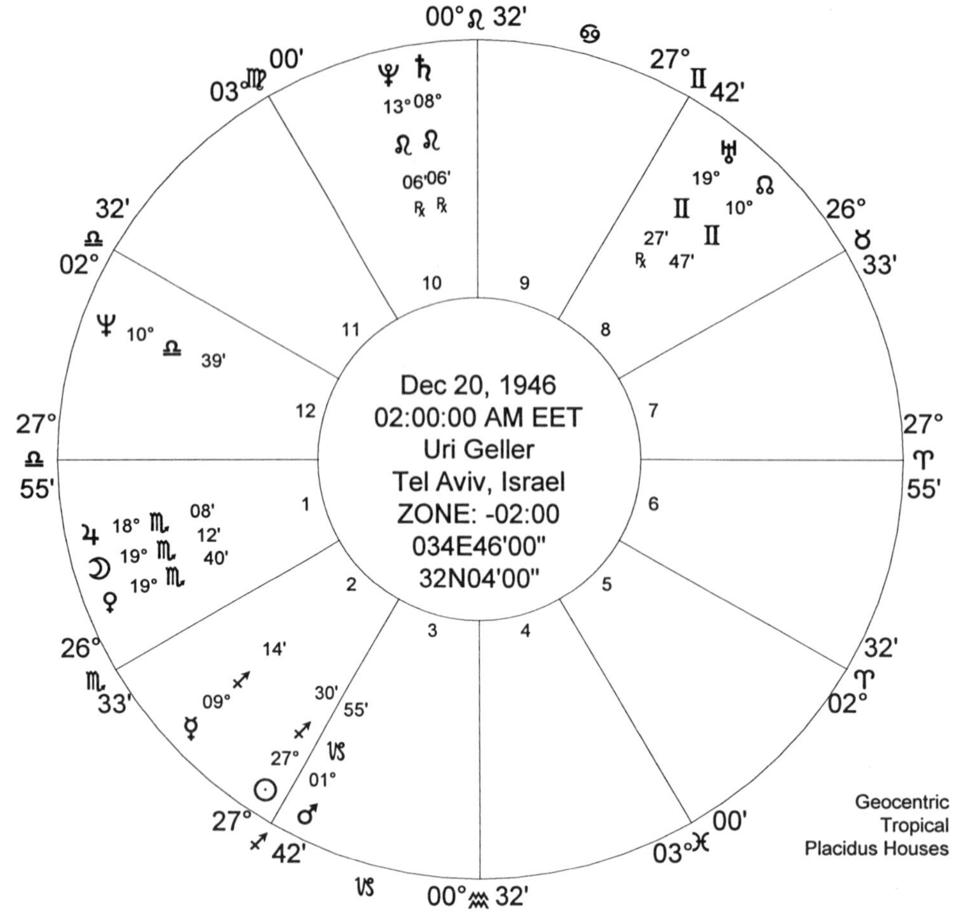

Astro*Carto*Graphy—Sun, Moon, Mercury, Venus, Mars, Jupiter, Saturn, Uranus, Neptune and Pluto. The Astro*Carto*Graphy map shows where all the planets were exactly angular when a person was born. Instead of drawing a horoscope for date, time and place of birth, the same data is used to draw the planetary angular lines on a world map.

Differences between Astro*Carto*Graphy and Relocation Astrology

Before showing the differences between the two systems I will explain how relocation astrology works through the use of Uri Geller's horoscope.

The natal horoscope is calculated from three items of information:

a. Date of birth
b. Time of birth
c. Place of birth

Uri Geller is an Israeli who has extraordinary powers—mind reading, spoon bending, etc. But magicians are not convinced, saying they can do the same, so it remains a mystery whether or not he has genuine psychic powers.

Data for Uri Geller's natal horoscope:

a. Date of birth: December 20, 1946
b. Time of birth: 2:00 a.m. EET
c. Place of birth: Tel Aviv, Israel

Planets that are conjunct the angles (Ascendant, Descendant, Midheaven and Imum Coeli) are considered powerful. The orb that is generally used for planets conjunct the angles is eight degrees.

Uri Geller's natal horoscope has Saturn conjunct the Midheaven (orb = 7°34"). Uri Geller's Astro*Carto*Graphy map has a SA MH line near Tel Aviv, which is the same as Saturn conjunct the Midheaven in his natal horoscope. This explains how the natal horoscope and Astro*Carto*Graphy map are connected.

Suppose a person moves to a new place. Astrologers can then draw a relocation horoscope for this new location and see how the person will do there. A relocation horoscope is calculated from the following information:

a. Date of birth
b. Time of birth
c. New location

A relocation horoscope is calculated as if the person were **born** at the new location. The positions of the planets remain in the same part of the zodiac. What changes is the positions of the four angles, so different planets will be angular.

Uri Geller's psychic powers were tested in London. A relocation horoscope shows which planets he had angular in that city. The relocation horoscope is calculated as follows:

a. Date of birth: December 20, 1946
b. Time of birth: 2:00 a.m. EET
c. New location: London, 51N30, 0W10

The relocation horoscope for London puts the angles at these positions:

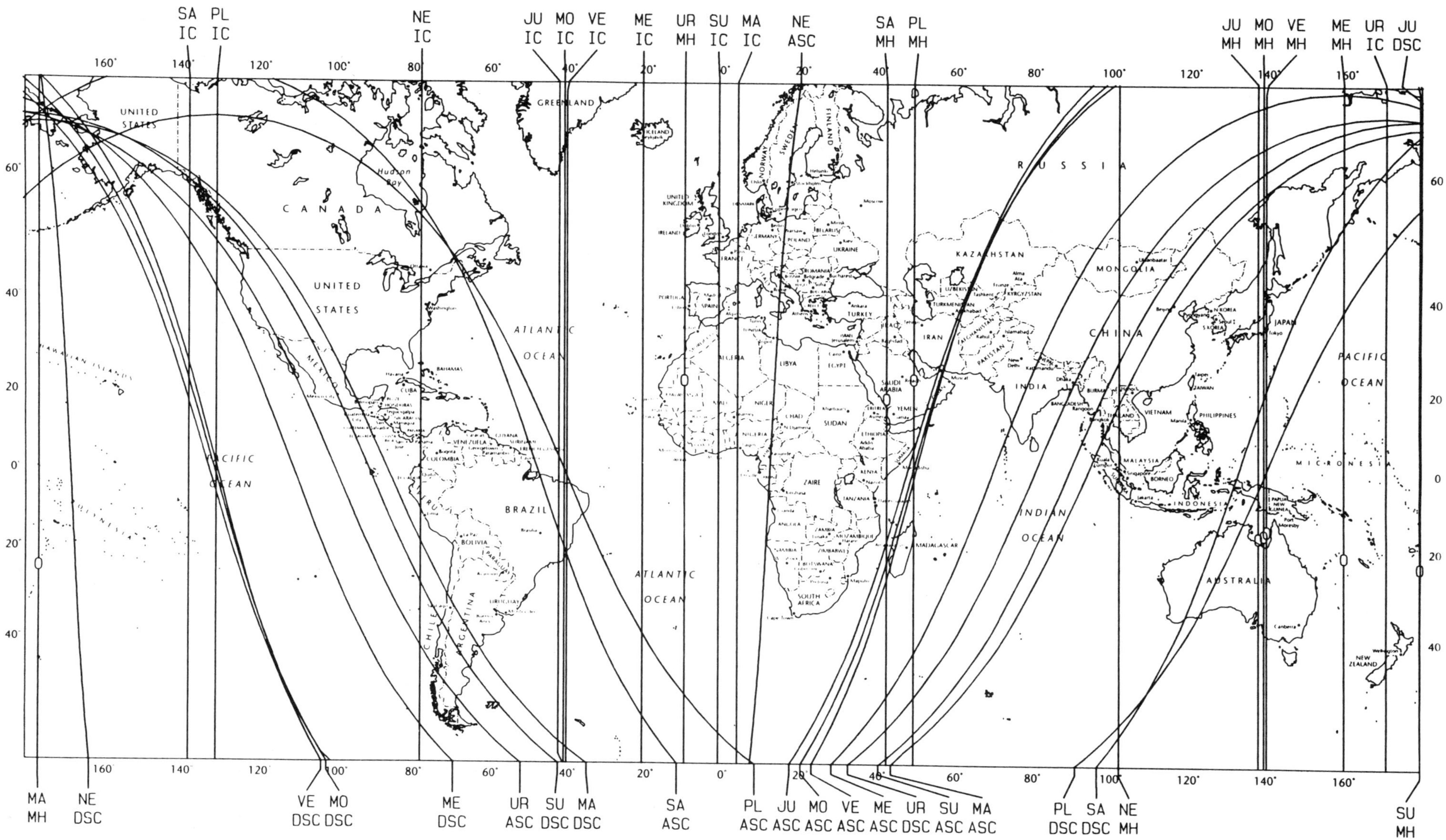

Uri Geller

December 20, 1946 2:00 AM Time Zone -2:00

04/16/96

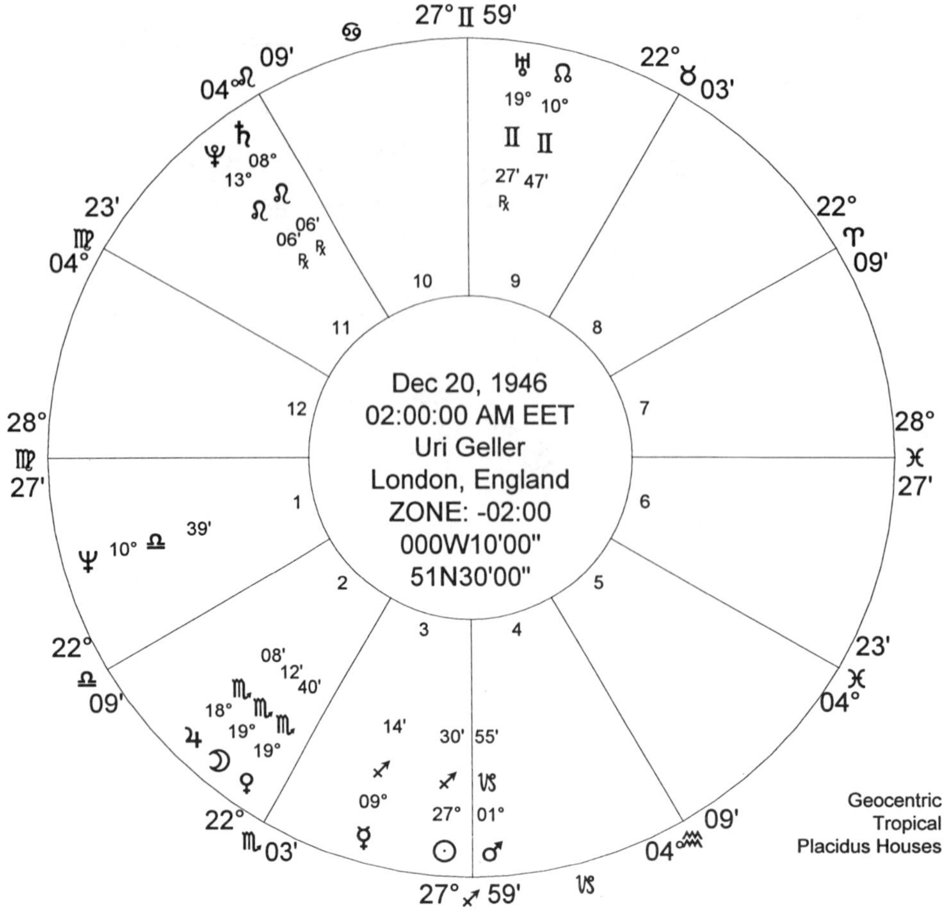
Dec 20, 1946
02:00:00 AM EET
Uri Geller
London, England
ZONE: -02:00
000W10'00"
51N30'00"
Geocentric
Tropical
Placidus Houses

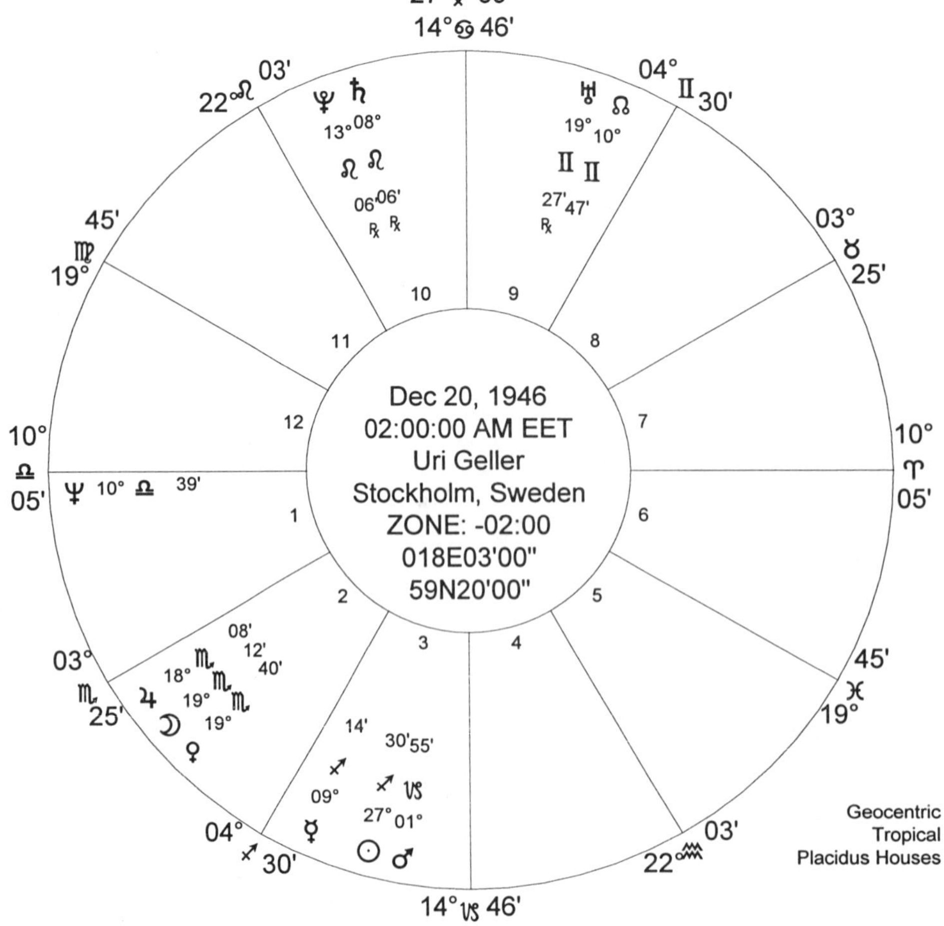
Dec 20, 1946
02:00:00 AM EET
Uri Geller
Stockholm, Sweden
ZONE: -02:00
018E03'00"
59N20'00"
Geocentric
Tropical
Placidus Houses

ASC	28 VIR 27
DSC	28 PIS 27
MH	27 GEM 59
IC	27 SAG 59

Uri Geller's relocation horoscope shows that he has Sun (27 SAG 30) and Mars (1 CAP 55) both conjunct the IC. When we examine his Astro*Carto*Graphy map we see SU IC and MA IC lines near to London. So the connection between a relocation horoscope and the Astro*Carto*Graphy map is easy to understand.

Suppose that Uri Geller wanted to move to Stockholm, Sweden. We could then calculate a relocation horoscope and see which planets he has angular there.

a. Date of birth: December 20, 1946

b. Time of birth: 2:00 a.m. EET

c. New location: Stockholm, 59N20, 18E03

The relocation horoscope for Stockholm shows the angles at:

ASC	10 LIB 05
DSC	10 ARI 05
MH	14 CAN 46
IC	14 CAP 46

At Stockholm, Uri Geller has Neptune (10 LIB 39) conjunct the ASC. On his Astro*Carto*Graphy map NE ASC is exact at Stockholm. This agrees with NE exactly conjunct the ASC in the relocation horoscope (orb thirty-four minutes).

These three locations for Uri Geller show that the angular planets are drawn on a map of the world. This means that an Astro*Carto*Graphy map shows where all the planets were angular around the world at a time a person was born.

Now that a connection between a relocation horoscope and an Astro*Carto*Graphy map is understood, take the test below to make sure. The test is in three parts.

Test 1 (Simon):

a. Date of birth: May 4, 1978
b. Time of birth: 6:34 a.m. BST
c. Place of birth: Leicester, England, 52N39, 1W09
Question 1: Which two planetary lines are closest to Leicester on the Astro*Carto*Graphy map?

Question 2: Which planetary line is nearest Leicester on the Astro*Carto*Graphy map?

Test 2 (Wendy):

a. Date of birth: June 11, 1954
b. Time of birth: 3:50 a.m. CET
c. New location: Kemijärvi, Finland, 66N40, 27E25
Question 1: Which three planetary lines are nearest Kemijärvi on the Astro*Carto*Graphy map?

Test 3 (Nigel):

a. Date of birth: January 9, 1946
b. Time of birth: 11:32 a.m. GMT
c. Place of birth: Manchester, England, 53N30, 2W15
Question 1: Which planetary lines are near Manchester on the Astro*Carto*Graphy map?

Question 2: Which planetary line is closest Manchester on the Astro*Carto*Graphy map?

Most astrologers will probably have the following answers:

Test 1, Question 1	VE ASC	MA IC	
Test 1, Question 2	VE ASC		
Test 2, Question 1	PL IC	VE ASC	UR ASC
Test 3, Question 1	JU DSC	VE MH	
Test 3, Question 2	JU DSC		

Astrologers with these results will be shocked to learn they're wrong!

When I examined Tel Aviv, London and Stockholm for Uri Geller, it was easy to see the connection between the relocation horoscope and the Astro*Carto*Graphy map. Using an orb of eight degrees for planets conjunct the angles, the test questions are analyzed below.

Test 1—Question 1:

Angle		**Planet**		**Orb**
ASC	6 GEM 58	VE	8 GEM 24	1°26"
IC	1 LEO 51	MA	9 LEO 25	7°34"

Therefore, **VE ASC** and **MA IC** are closest Leicester on the Astro*Carto*Graphy map.

Test 1—Question 2:

From the previous question it is obvious that **VE ASC** is the closest planetary line to Leicester.

Test 2—Question 1:

Angle		**Planet**		**Orb**
IC	26 LEO 36	PL	22 LEO 51	3°45"
ASC	27 CAN 37	VE	22 CAN 10	5°27"
ASC	27 CAN 37	UR	21 CAN 17	6°20"

Therefore the three planetary lines nearest Kemijärvi on the Astro*Carto*Graphy map are **PL IC**, **VE ASC** and **UR ASC**.

Test 3—Question 1:

Angle		**Planet**		**Orb**
DSC	23 LIB 22	JU	25 LIB 43	2°21"
MH	8 CAP 24	VE	13 CAP 05	4°41"

Therefore, the planetary lines nearest Manchester on the Astro*Carto*Graphy map are **JU DSC** and **VE MH**.

Test 3—Question 2:

From the previous question it is clear to see that **JU DSC** is the planetary line closest Manchester.

These are the same answers that most astrologers arrived at themselves, yet all the answers are wrong!

At the beginning of this section I mentioned that there are differences between relocation astrology and Astro*Carto*Graphy. The above test questions were deliberately chosen because they reveal the differences between the two systems.

The example locations used for Uri Geller show **no** differences between relocation astrology and Astro*Carto*Graphy. But when astrologers use **only** the relocation horoscope, then there will be times when the wrong information is obtained.

By now you are probably becoming impatient to learn the correct answers to the test questions, so here they are:

Test 1—Question 1:

The two planetary lines closest to Leicester on the Astro*Carto*Graphy map are **PL DSC** and **VE DSC**.

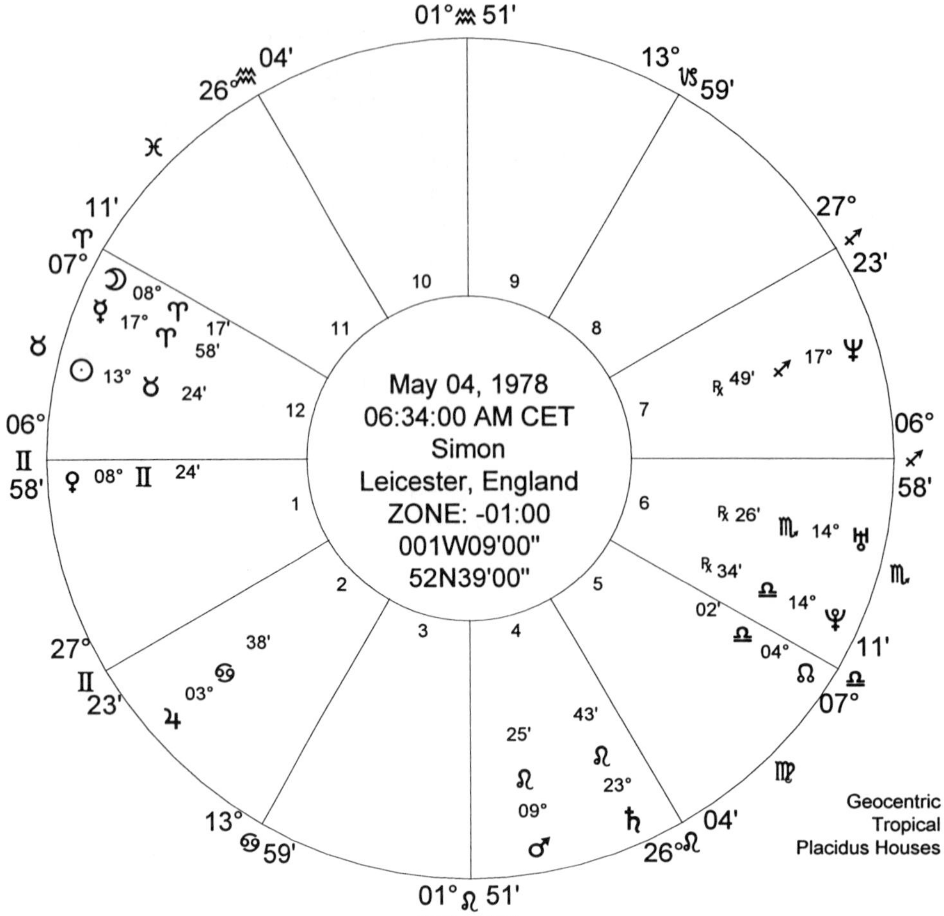
May 04, 1978
06:34:00 AM CET
Simon
Leicester, England
ZONE: -01:00
001W09'00"
52N39'00"
Geocentric
Tropical
Placidus Houses

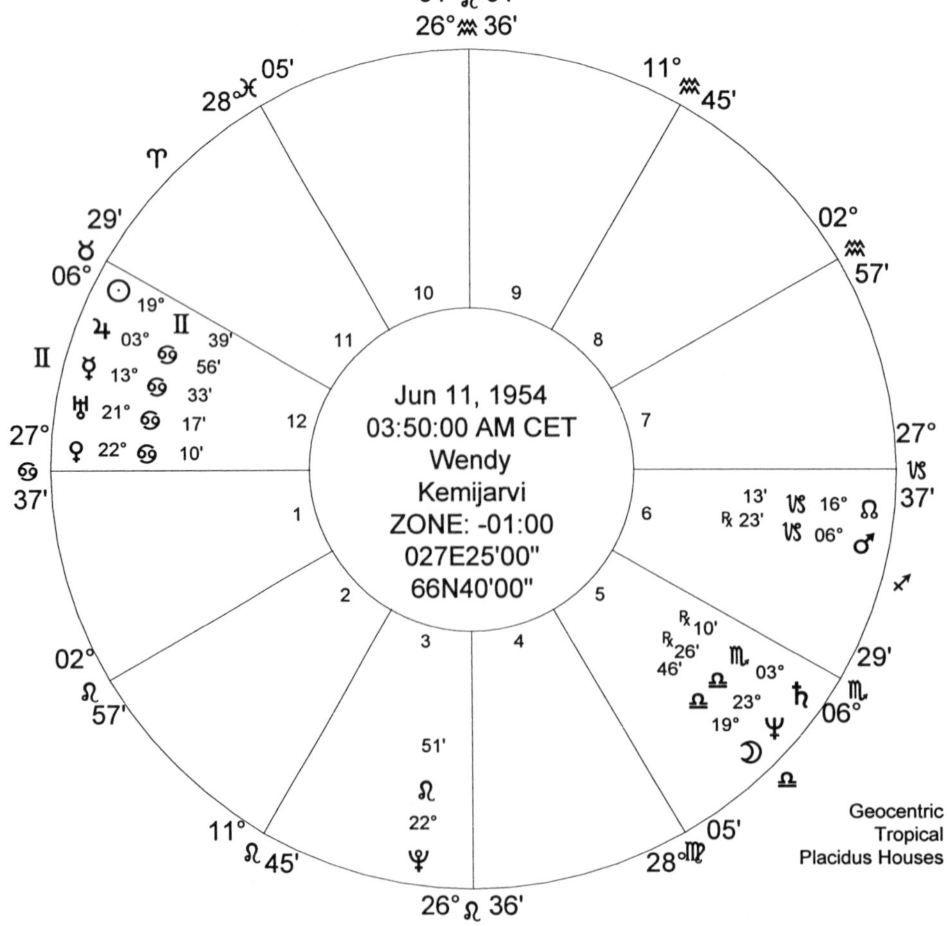
Jun 11, 1954
03:50:00 AM CET
Wendy
Kemijarvi
ZONE: -01:00
027E25'00"
66N40'00"
Geocentric
Tropical
Placidus Houses

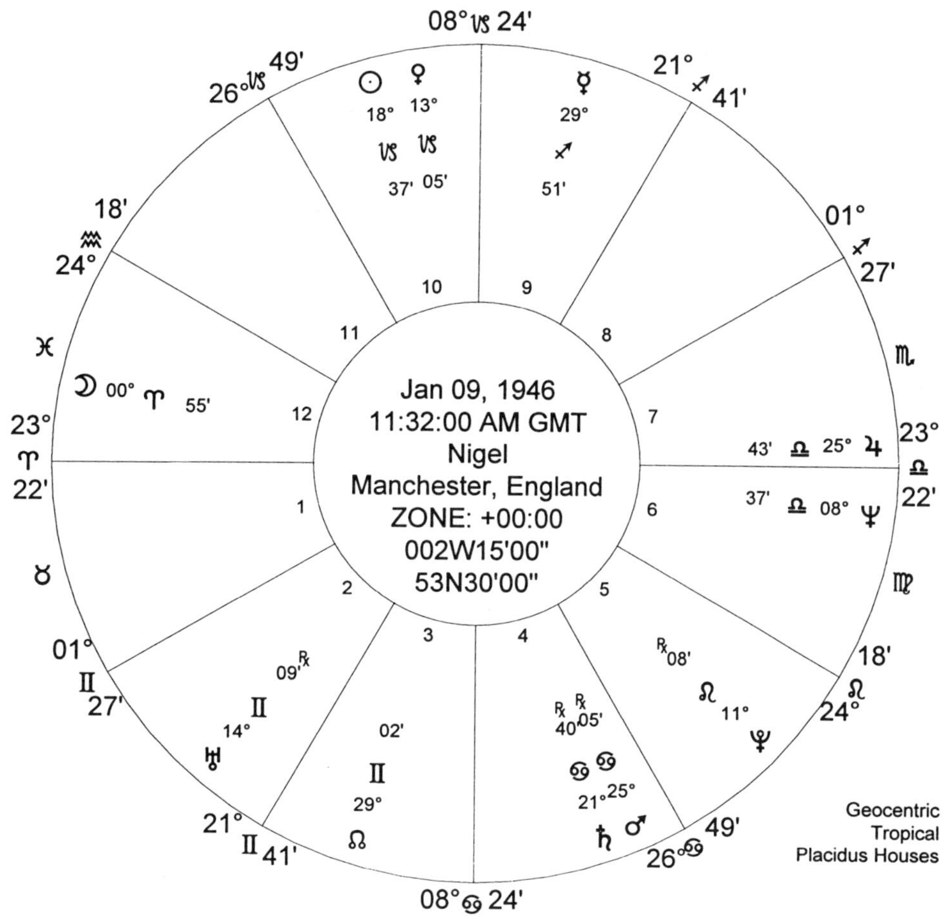

Test 1—Question 2:

PL DSC is the closest planetary line to Leicester on the Astro*Carto*Graphy map (unbelievable, but true).

Test 2—Question 1:

The three planetary lines nearest to Kemijärvi on the Astro*Carto*Graphy map are **PL IC**, **PL ASC** and **PL DSC**.

This was a trick question because I asked which three **planetary lines** are nearest to Kemijärvi. I did not ask which three **planets** are nearest to Kemijärvi, which was easy to see from the relocation horoscope. Yet on the Astro*Carto*Graphy map there are **three** Pluto lines at Kemijärvi.

Test 3—Question 1:

The planetary lines nearest to Manchester on the Astro*Carto*Graphy map are **MO ASC**, **JU DSC**, **NE ASC** and **VE MH**.

The question did not ask for a certain number of planetary lines nearest to Manchester, but **which**. There are in fact **four** planetary lines near to Manchester.

Test 3—Question 2:

MO ASC is the planetary line closest to Manchester on the Astro*Carto*Graphy map.

You will find the Astro*Carto*Graphy maps to these test questions at the end of this chapter. When you examine them, and the relevant relocation horoscopes, you will see the differences between the two systems.

The Importance of Celestial Latitude

Why are there differences between relocation astrology and Astro*Carto*Graphy? When a horoscope is drawn on a piece of paper, certain information is not shown. The heavens are three-dimensional, whereas the usual horoscope is two-dimensional. What information is missing?

The two-dimensional chart does not show the **celestial latitude** of the planets. Most astrologers do not consider the celestial latitude of a planet to be important. Celestial latitude is the distance a planet is north or south of the ecliptic, but a horoscope is drawn as if **all** the planets are **on** the ecliptic. When an Astro*Carto*Graphy map is drawn, the celestial latitude of the planets is taken into consideration.

A horoscope shows the planetary positions in celestial longitude. This is the distance of a planet along the **ecliptic** from zero degrees Aries. This distance is measured eastward in degrees and minutes of arc from where the ecliptic and celestial equator cross at zero degrees Aries. See Diagram 1, Ecliptical Coordinate System.

Some astrologers use parallels of declination when interpreting a horoscope. You would expect the declination of a planet to be the distance north or south of the **ecliptic**. But this is not the case. Declination is the distance north or south of the **celestial equator**. See Diagram 2, Equatorial System.

The equatorial system uses the **right ascension** (distance

Diagram 1: Ecliptical Coordinate System

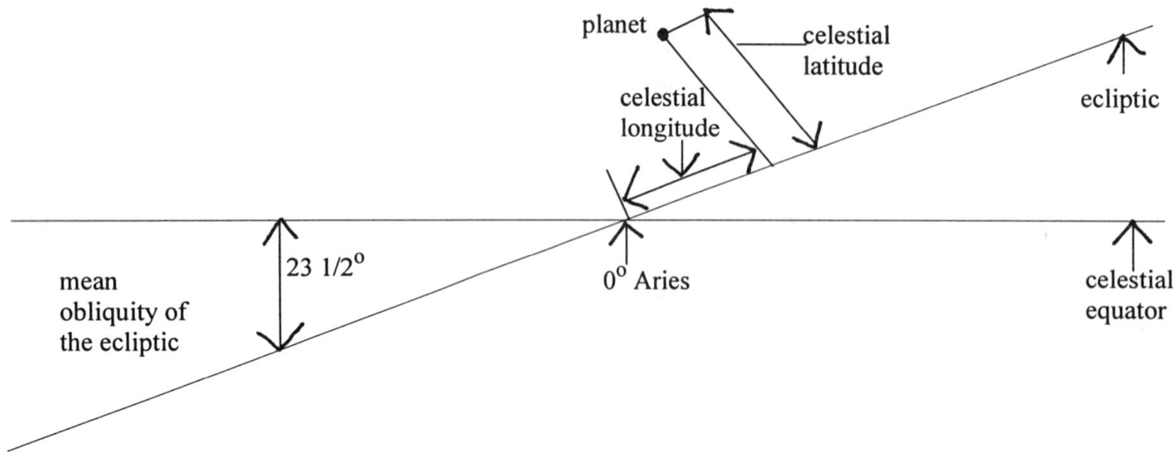

Diagram 2: Equatorial System

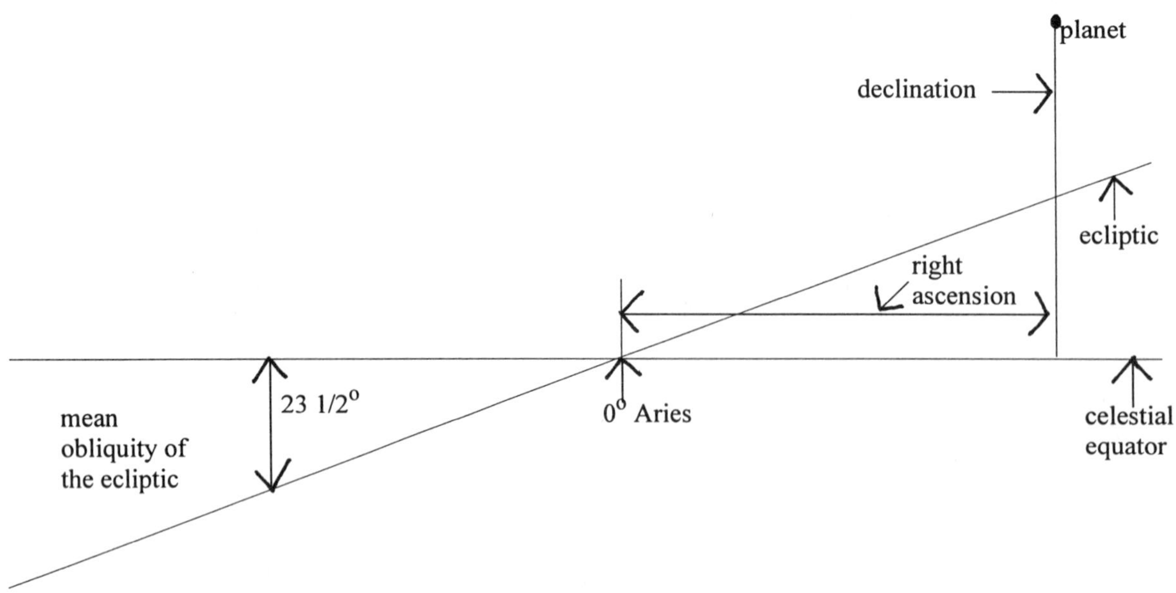

along the **celestial equator** from zero degrees Aries) and **declination** of planets. The ecliptical coordinate system uses the **celestial longitude** and **celestial latitude** of planets.

When a horoscope is interpreted an astrologer uses the **celestial longitude** (ecliptical coordinate system) and **declination** (equatorial system) of the planets. Logically, a horoscope should be interpreted by using **right ascension** and **declination** together. But astrologers use a part of the ecliptical coordinate system and a part of the equatorial system when interpreting a horoscope.

Astro*Carto*Graphy uses **right ascension, declination** and **celestial latitude** of the planets when a map is drawn. This means that an Astro*Carto*Graphy map shows the **true position** of the planets.

Attempts have been made to show the celestial latitude of planets by drawing a three-dimensional chart. In *Recent Advances in Natal Astrology* by Geoffrey Dean (pages 190-192) and *Elements of House Division* by Ralph William Holden (pages 145-147) are found explanations of the 3-D Celestial Birth Map designed by Blundsen. This is very time consuming to draw, but does give more accurate information than the usual horoscope.

The importance of celestial latitude is shown very clearly in the horoscope for John. Here we see that Moon (22 SCO 31) and Pluto (22 SCO 43) are in exact conjunction by **celestial longitude**. But when celestial latitude of both the Moon and Pluto are taken into consideration they are not in conjunction at all! Diagram 3(a) shows that the two planets are conjunct in the natal chart as celestial latitude is **not** taken into consideration. Diagram 3(b) shows the Moon and Pluto conjunction with celestial latitude included. The exact angle between Moon and Pluto is 18 degrees 35 minutes!

In the *A.A. Journal* (April 1986) there is an article by Terry Dwyer about latitude corrected aspects. He includes a type-in computer program to calculate the true aspects between planets. This program calculated that Moon and Pluto are 18°35" apart in John's horoscope.

When we examine John's Astro*Carto*Graphy map the Moon and Pluto are **not** exactly together. The MH and IC Moon and Pluto lines are five degrees apart, still in conjunction. If John's horoscope had been drawn in **right ascension** the Moon would be at 19 SCO 04 and Pluto at 24 SCO 02, a difference of just under five degrees. This shows the clear connection between right ascension and Astro*Carto*Graphy. The Moon and Pluto ASC and DSC lines are in completely different places. Only at geographical latitude 14N36 and 14S36 are Moon and Pluto ASC and DSC lines exactly together. This will be explained in more detail in the next section (declination of the planets).

In Test 1 it seemed impossible that PL DSC was the closest line to Leicester. In Simon's horoscope Pluto is 52°24" **below** the Descendant. In actual fact Pluto has **not** yet set in Simon's horoscope! This is shown very clearly on the **Paran Printout** which is an option available on the NOVA astrological computer program.

The Paran Printout for Simon shows the following information:

	h m s	d m
L.S.T. of chart	20 16 25	304 06
Pluto sets at	20 17 07	304 17
Venus rises at	20 14 48	303 42
Mars anti-culminates at	20 49 42	312 26

This shows that Pluto will exactly set forty-two seconds **later**; hence it is in the **seventh** house of Simon's horoscope. Venus was exactly on the ASC one minute, thirty-seven seconds **previously**, so Venus is in the **twelfth** house. Mars will be on the IC in thirty-three minutes, seventeen seconds. Mars is eight degrees, twenty minutes away from the IC, hence it is out of orb from Leicester on the Astro*Carto*Graphy map.

Diagram 4(a) and 4(b) show why both Pluto and Venus are above the horizon in Simon's natal chart.

For more information about how the Paran Printout is calculated, read *Essays on Astrology* by Robert Hand. The chapters entitled "A New Approach to Transits" and "Astrology's Second Dimension: Declination and Latitude" explain in detail how to find out which planets are **exactly** on the four angles of a chart.

The Sun is always on the ecliptic, so it is not affected by celestial latitude. Pluto can have up to 17°19" celestial latitude, 5°18" for the Moon, 7° for Mercury, 8°30" for Venus and up to 4° for most of the other planets (page 280, *Mundane Astrology* by Baigent, Campion and Harvey). The biggest difference between the celestial longitude of a planet and its true position is with Pluto as we saw in Simon's horoscope. But even for the Moon and Mars there can be a difference up to twenty degrees between its celestial longitude position and where they are exactly on the Ascendant. Saturn can have a difference up to ten degrees, and up to five degrees for Jupiter.

When celestial latitude of the planets is taken into consideration, then it is clear that the usual way of drawing a horoscope is not adequate enough in showing all the necessary information.

Declination of Planets

When interpreting a horoscope some astrologers use parallels of declination. As mentioned earlier, declination of a planet is the distance north or south of the celestial equator.

An orb of one degree is generally used for parallels of declination. A parallel is when two planets have the same declination either north or south. For example, in Simon's horoscope (Test 1), the Sun (15N52) and Saturn (15N03) are in parallel aspect. A contraparallel aspect is when both planets have the same declination, but one planet is north and the other is south of the celestial equator. For example, in Simon's horoscope Saturn (15N03) and Uranus (15S47) are in contraparallel aspect.

In Astro*Carto*Graphy the **declination** of a planet has an effect on the ASC and DSC lines. When you look at an Astro*Carto*Graphy map you will see that the MH and IC lines are always vertical and the ASC and DSC lines are

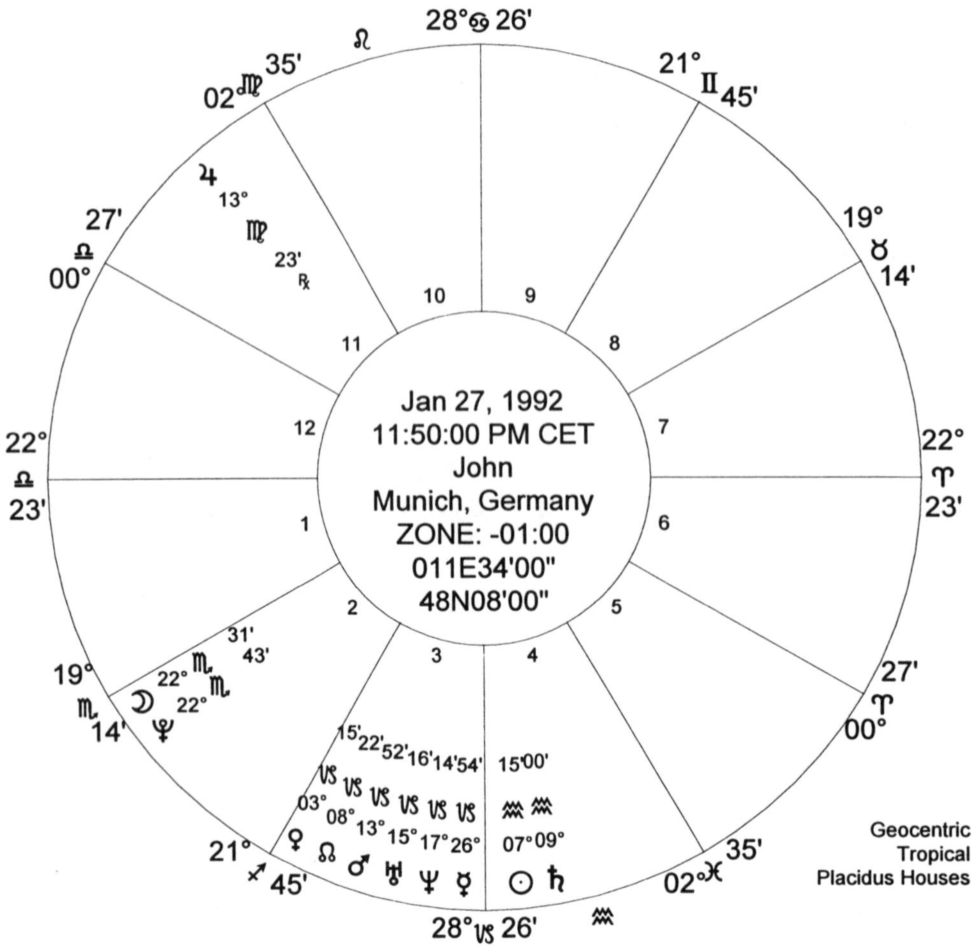

curved. But some ASC and DSC lines are more curved than others. For example, on Uri Geller's Astro*Carto*Graphy map the SU and MA ASC and DSC lines are curved a great deal. Whereas, Uri Geller's NE ASC and DSC lines are hardly curved at all.

The curve of an ASC and DSC line is determined by the planet's declination. The higher the declination of a planet, the more curved the ASC and DSC lines will be. When a planet has zero degrees declination, then the ASC and DSC lines will be vertical on the Astro*Carto*Graphy map. See Diagram 5 for examples of the curving ASC and DSC lines based on a planet's declination.

The third example with Jupiter at twenty-two degrees declination shows that the apex of the ASC and DSC lines is where the JU MH is located. A planet with over twenty degrees declination will show the ASC and DSC lines joining together with the corresponding MH or IC line passing through the apex. If a planet has over twenty degrees **south** declination, then it will be the **MH** line at the apex. When a planet has over twenty degrees **north** declination, then the **IC** line will be at the apex. Planets with less than twenty degrees declination will have ASC and DSC lines making an apex that is not shown on the map. E.g. Uri Geller has SA ASC and DSC lines making an apex **not** shown on his map. The Astro*Carto*Graphy map only shows up to seventy degrees north geographical latitude. This means that Uri Geller's SA ASC and DSC lines apex **above** seventy degrees north geographical latitude.

The geographical latitude where a planet's ASC and DSC lines will apex can be calculated by subtracting its declination from ninety degrees. For example, if a planet has twenty-two degrees declination, then the apex of the ASC and DSC lines will be at sixty-eight degrees geographical latitude.

When a planet has **north** declination, then the ASC curving line will travel **anti-clockwise** from bottom to top on the Astro*Carto*Graphy map. A planet with **south** declination moves in a **clockwise** direction from bottom to top on the Astro*Carto*Graphy map. See Diagram 6.

On an Astro*Carto*Graphy map the declination of a planet can be determined by looking at its MH line. On every MH line there is a small square which represents the zenith of the planet. The geographical latitude where a planet's zenith is located represents its declination.

Most times a planet's declination will be between the Tropic of Cancer (23°27" north of the equator) and the Tropic of Capricorn (23°27" south of the equator). But some planets go above 23°27" minutes in declination. This means there will be times when a planet is **above** the Tropic of Cancer or **below** the Tropic of Capricorn on the Astro*Carto*Graphy map. For example, in Test 2, Wendy's Mars has 26°42" south declination. On her Astro*Carto*Graphy map the zenith of Mars is below the Tropic of Capricorn. See Diagram 7 for an example of

Diagram 3A

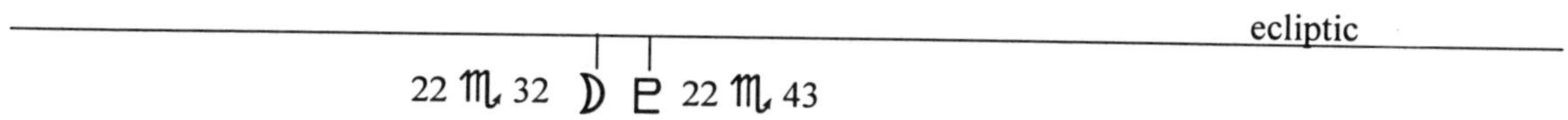

☽ is conjunct ♇, orb 11 minutes

Diagram 3B

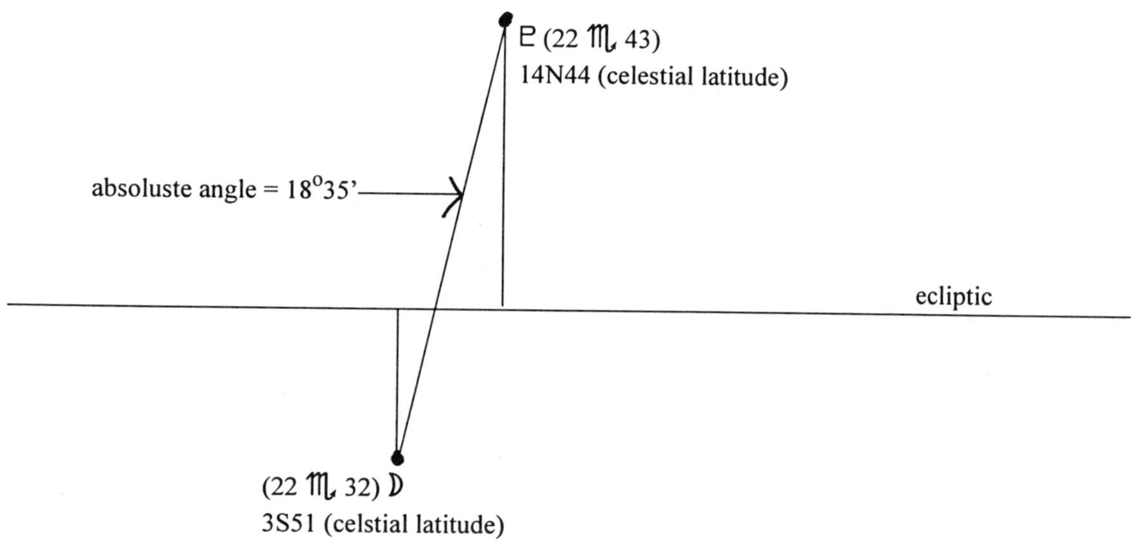

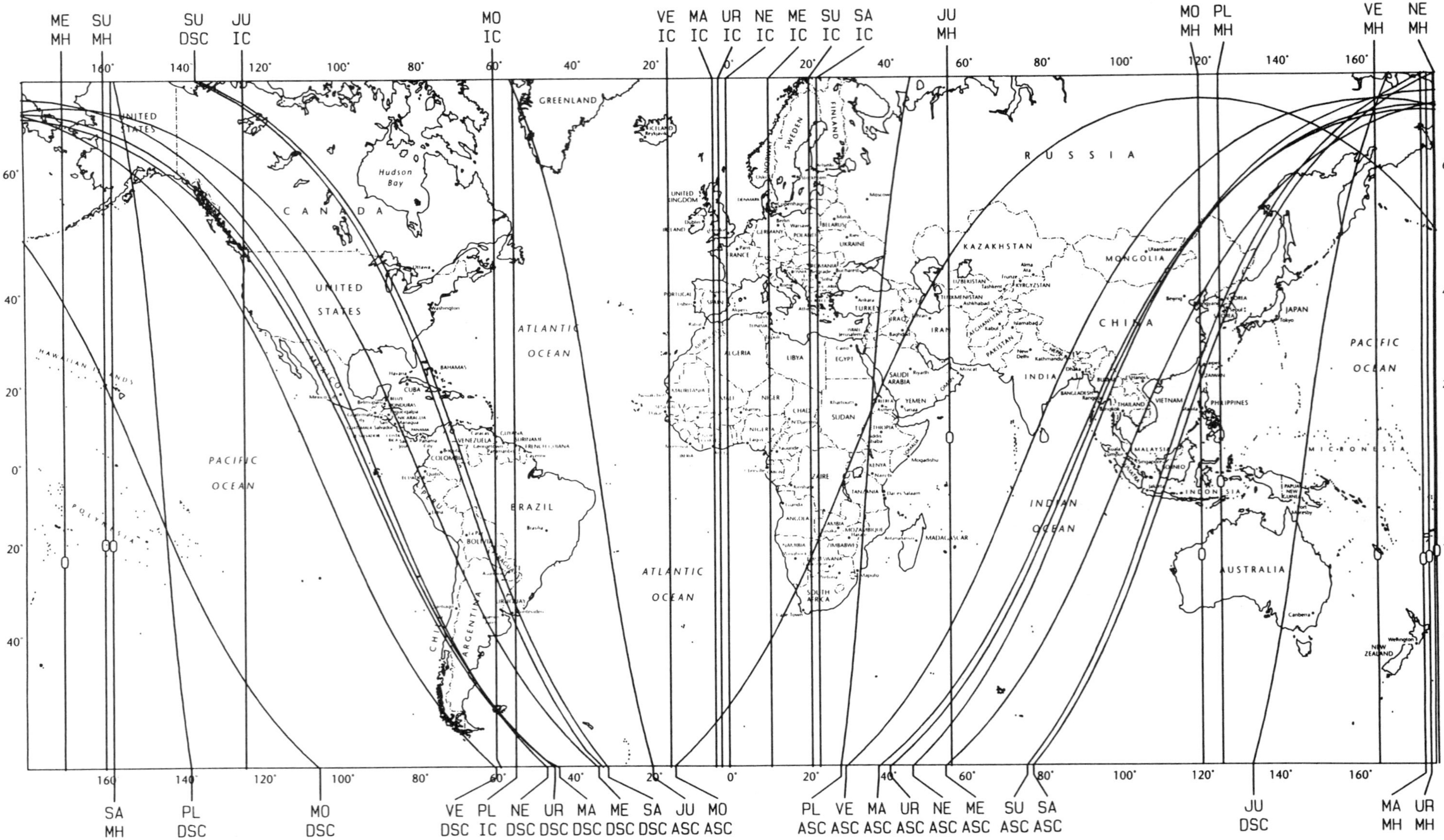

John

January 27, 1992 11:50 PM CET

05/09/96

05/09/96

John
January 27, 1992 11:50 PM CET

ASTRO*CARTO*GRAPHY LATITUDE CROSSINGS

NORTH LATITUDES

69N55 NE/MH * SA/ASC
69N51 MA/MH * SU/ASC
69N38 UR/MH * SA/ASC
69N24 MA/MH * SA/ASC
69N24 PL/IC * JU/ASC
68N23 UR/MH * NE/ASC
68N21 MA/MH * NE/ASC
68N04 SA/ASC * NE/ASC
68N03 ME/MH * NE/DSC
68N01 JU/IC * SA/DSC
67N49 SU/ASC * JU/DSC
67N48 JU/DSC * NE/ASC
67N48 PL/MH * MO/DSC
67N47 SU/ASC * NE/ASC
67N42 VE/MH * NE/ASC
67N42 JU/IC * SU/DSC
67N38 SU/ASC * VE/DSC
67N38 VE/ASC * NE/ASC
67N38 VE/DSC * JU/DSC
67N37 VE/DSC * SA/ASC
67N32 VE/MH * SU/ASC
67N26 ME/DSC * NE/DSC
67N15 SU/MH * ME/DSC
67N13 NE/MH * ME/ASC
67N13 MA/MH * VE/DSC
67N06 SA/MH * ME/DSC
67N06 UR/MH * VE/DSC
67N05 ME/ASC * VE/DSC
67N05 UR/MH * ME/ASC
67N04 ME/ASC * UR/ASC
67N04 VE/DSC * UR/DSC
67N03 MA/MH * UR/ASC
67N03 NE/MH * UR/DSC
67N00 SU/MH * NE/DSC
66N57 MA/MH * ME/ASC
66N56 NE/MH * VE/DSC
66N55 JU/DSC * SA/ASC
66N53 ME/DSC * PL/DSC
66N51 VE/MH * SA/ASC
66N43 SA/MH * NE/DSC
66N40 VE/MH * JU/DSC
66N33 ME/MH * UR/DSC
66N33 UR/MH * MA/DSC
66N31 VE/MH * UR/ASC
66N31 NE/MH * MA/DSC
66N30 JU/DSC * UR/ASC
66N30 ME/ASC * MA/ASC
66N22 SA/ASC * UR/ASC
66N18 VE/DSC * MA/DSC
66N14 NE/DSC * PL/DSC
66N10 MO/DSC * VE/ASC
66N07 VE/MH * MA/ASC
66N04 MA/ASC * JU/DSC
65N57 SU/ASC * UR/ASC
65N54 ME/MH * MA/DSC
65N49 MA/ASC * SA/ASC
65N27 ME/MH * VE/DSC
65N24 SU/ASC * MA/ASC
65N23 VE/MH * ME/ASC
65N18 SU/MH * UR/DSC
65N11 MO/DSC * NE/ASC
65N08 ME/ASC * JU/DSC
64N58 SA/MH * UR/DSC
64N39 MO/DSC * UR/ASC
64N31 SU/MH * MA/DSC
64N30 MO/DSC * MA/ASC
64N14 UR/DSC * PL/DSC
64N09 SA/MH * MA/DSC
63N57 MO/DSC * SU/ASC
63N34 MO/DSC * SA/ASC
63N24 MO/DSC * ME/ASC
63N16 MA/DSC * PL/DSC
63N00 SU/MH * VE/DSC
62N39 ME/ASC * SA/ASC

CENTRAL LATITUDES

62N24 SA/MH * VE/DSC
61N57 PL/MH * VE/ASC
61N46 MO/DSC * JU/DSC
61N13 Anchorage
60N36 VE/DSC * PL/DSC
60N20 VE/MH * MO/DSC
60N07 SU/ASC * ME/ASC
60N03 MO/MH * VE/ASC

59N20 JU/IC * ME/DSC
59N20 Stockholm
59N17 MA/ASC * UR/ASC
55N45 Moscow
55N41 PL/MH * NE/ASC
55N41 Copenhagen
55N22 PL/MH * MA/ASC
55N09 PL/MH * UR/ASC
54N15 JU/IC * NE/DSC
53N55 MA/MH * MO/DSC
53N02 MA/ASC * NE/ASC
52N50 UR/MH * MO/DSC
52N30 Berlin
52N08 MO/MH * MA/ASC
52N01 MO/MH * NE/ASC
51N44 MO/MH * UR/ASC
51N30 London
51N17 NE/MH * MO/DSC
50N56 JU/IC * UR/DSC

49N14 Vancouver
49N01 JU/IC * MA/DSC
48N50 Paris
48N20 UR/ASC * NE/ASC
47N36 Seattle
47N17 JU/MH * MO/ASC
45N32 Montreal
45N26 PL/MH * ME/ASC
44N59 Minneapolis
43N39 Toronto
42N22 Boston
42N20 Detroit
41N54 Rome
41N53 Chicago
40N43 New York

39N46 ME/MH * MO/DSC
39N44 Denver
39N21 MO/MH * ME/ASC
38N53 Washington
38N37 St. Louis
37N47 San Francisco
37N23 JU/IC * VE/DSC
36N40 PL/MH * SU/ASC
36N10 Las Vegas
35N40 Tokyo
34N04 Los Angeles
33N45 Atlanta
33N27 Phoenix
32N47 Dallas
32N38 PL/MH * SA/ASC
30N03 Cairo

29N58 New Orleans
29N46 Houston
28N39 Delhi
26N04 MO/MH * SU/ASC
25N47 Miami
22N34 Calcutta
22N16 Hong Kong
21N52 SU/MH * MO/DSC
21N19 Honolulu
21N05 MO/MH * SA/ASC

19N26 Mexico City
17N41 SA/MH * MO/DSC

SOUTH LATITUDES

14N36 MO/ASC * PL/ASC
14N35 Manila
10N30 Caracas

6N27 Lagos

14S36 MO/DSC * PL/DSC
17S41 SA/IC * MO/ASC

21S05 MO/IC * SA/DSC
21S52 SU/IC * MO/ASC
22S54 Rio De Janeiro
26S04 MO/IC * SU/DSC
26S12 Johannesburg

32S38 PL/IC * SA/DSC
33S55 Sydney
34S36 Buenos Aires
36S40 PL/IC * SU/DSC
36S52 Auckland
37S23 JU/MH * VE/ASC
37S50 Melbourne
39S21 MO/IC * ME/DSC
39S46 ME/IC * MO/ASC

45S26 PL/IC * ME/DSC
47S17 JU/IC * MO/DSC
48S20 UR/DSC * NE/DSC
49S01 JU/MH * MA/ASC

50S56 JU/MH * UR/ASC
51S17 NE/IC * MO/ASC
51S44 MO/IC * UR/DSC
52S01 MO/IC * NE/DSC
52S08 MO/IC * MA/DSC
52S50 UR/IC * MO/ASC
53S02 MA/DSC * NE/DSC
53S55 MA/IC * MO/ASC
54S15 JU/MH * NE/ASC
55S09 PL/IC * UR/DSC
55S22 PL/IC * MA/DSC
55S41 PL/IC * NE/DSC
59S17 MA/DSC * UR/DSC
59S20 JU/MH * ME/ASC

Simon	Natal Chart	Paran Printout
May 4, 1978	Leicester	6:34:0 AM BST
52N39'00"	20:16:25 ST	1W09'00"
Tropical		Geocentric

Pl	Rise			Upper			Set			Lower		
	d	m	h m s	d	m	h m s	d	m	h m s	d	m	h m s
☉	289	05	19:16:21	40	57	2:43:47	152	48	10:11:14	220	57	14:43:47
☽	273	44	18:14:54	7	43	0:30:53	101	43	6:46:52	187	43	12:30:53
☿	281	42	18:46:47	17	37	1:10:30	113	33	7:34:13	197	37	13:10:30
♀	303	42	20:14:48	66	33	4:26:10	189	23	12:37:33	246	33	16:26:10
♂	14	18	0:57:11	132	26	8:49:42	250	33	16:42:13	312	26	20:49:42
♃	329	22	21:57:28	93	57	6:15:50	218	33	14:34:11	273	57	18:15:50
♄	35	56	2:23:44	146	33	9:46:14	257	11	17:08:44	326	33	21:46:14
♅	153	50	10:15:19	222	05	14:48:21	290	21	19:21:22	42	05	2:48:21
♆	197	49	13:11:15	256	54	17:07:37	316	00	21:03:58	76	54	5:07:37
♇	96	11	6:24:46	200	14	13:20:56	304	17	20:17:07	20	14	1:20:56
☊	95	48	6:23:11	183	42	12:14:47	271	36	18:06:23	3	42	0:14:47
As	304	06	20:16:25	65	08	4:20:33	186	10	12:24:40	245	08	16:20:33
MC	242	10	16:08:40	304	06	20:16:25	6	02	0:24:10	124	06	8:16:25
Vx	130	09	8:40:35	205	49	13:43:16	281	30	18:45:58	25	49	1:43:16
EP	285	32	19:02:07	34	06	2:16:25	142	41	9:30:44	214	06	14:16:25
⊗	283	21	18:53:23	29	41	1:58:45	136	02	9:04:07	209	41	13:58:45

☊	lower	3	42	☊	rise	95	48\	☿	lower	197	37\	☽	rise	273	44\
MC	set	6	02	♇	rise	96	11/	♆	rise	197	49/	♃	lower	273	57/
☽	upper	7	43	☽	set	101	43	♇	upper	200	14	Vx	set	281	30\
♂	rise	14	18	☿	set	113	33	Vx	upper	205	49	☿	rise	281	42/
☿	upper	17	37	MC	lower	124	06	⊗	lower	209	41	⊗	rise	283	21
♇	lower	20	14	Vx	rise	130	09	EP	lower	214	06	EP	rise	285	32
Vx	lower	25	49	♂	upper	132	26	♃	set	218	33	☉	rise	289	05
⊗	upper	29	41	⊗	set	136	02	☉	lower	220	57	♅	set	290	21
EP	upper	34	06	EP	set	142	41	♅	upper	222	05	♀	rise	303	42\
♄	rise	35	56	♄	upper	146	33	MC	rise	242	10	MC	upper	304	06/\
☉	upper	40	57	☉	set	152	48	As	lower	245	08	As	rise	304	06/\
♅	lower	42	05	♅	rise	153	50	♀	lower	246	33	♇	set	304	17/
As	upper	65	08	☊	upper	183	42	♂	set	250	33	♂	lower	312	26
♀	upper	66	33	As	set	186	10	♆	upper	256	54\	♆	set	316	00
♆	lower	76	54	☽	lower	187	43	♄	set	257	11/	♄	lower	326	33
♃	upper	93	57	♀	set	189	23	☊	set	271	36	♃	rise	329	22

Diagram 4A

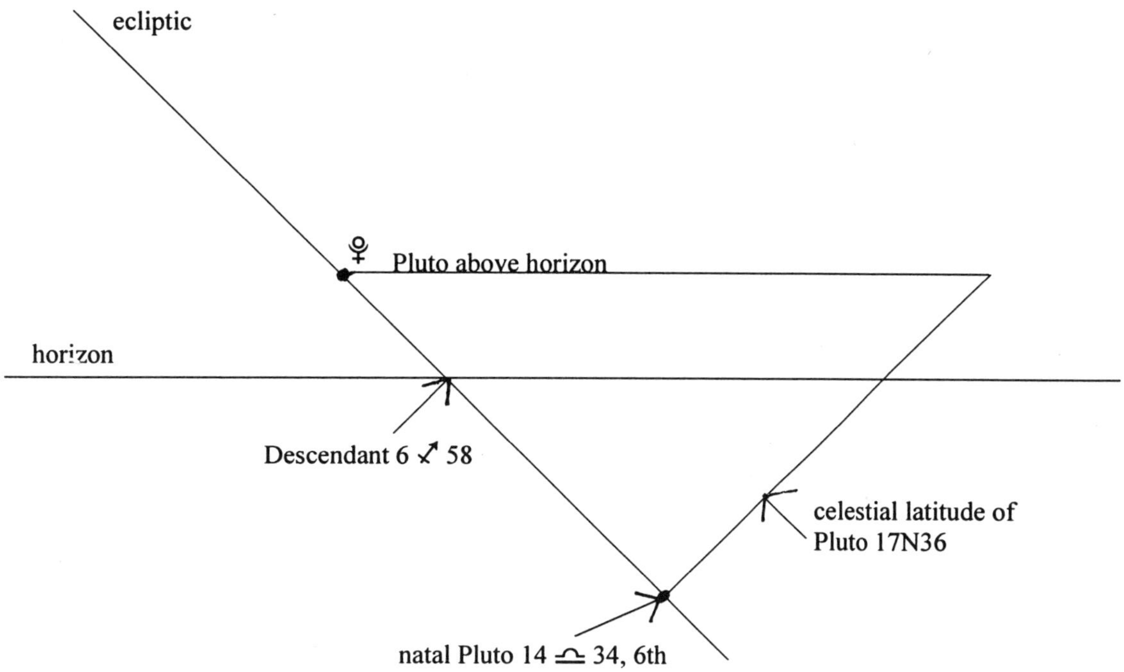

Diagram 4B

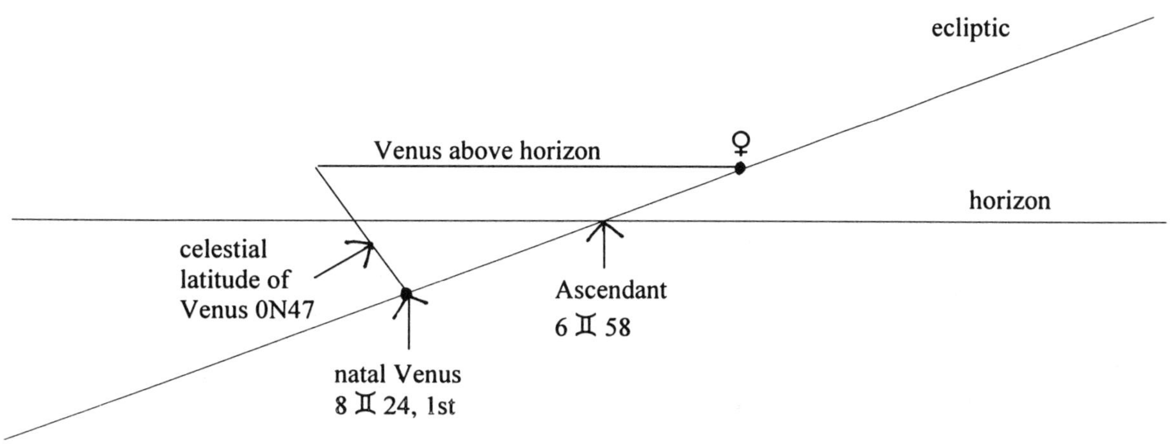

Diagram 5—Declination of Planets

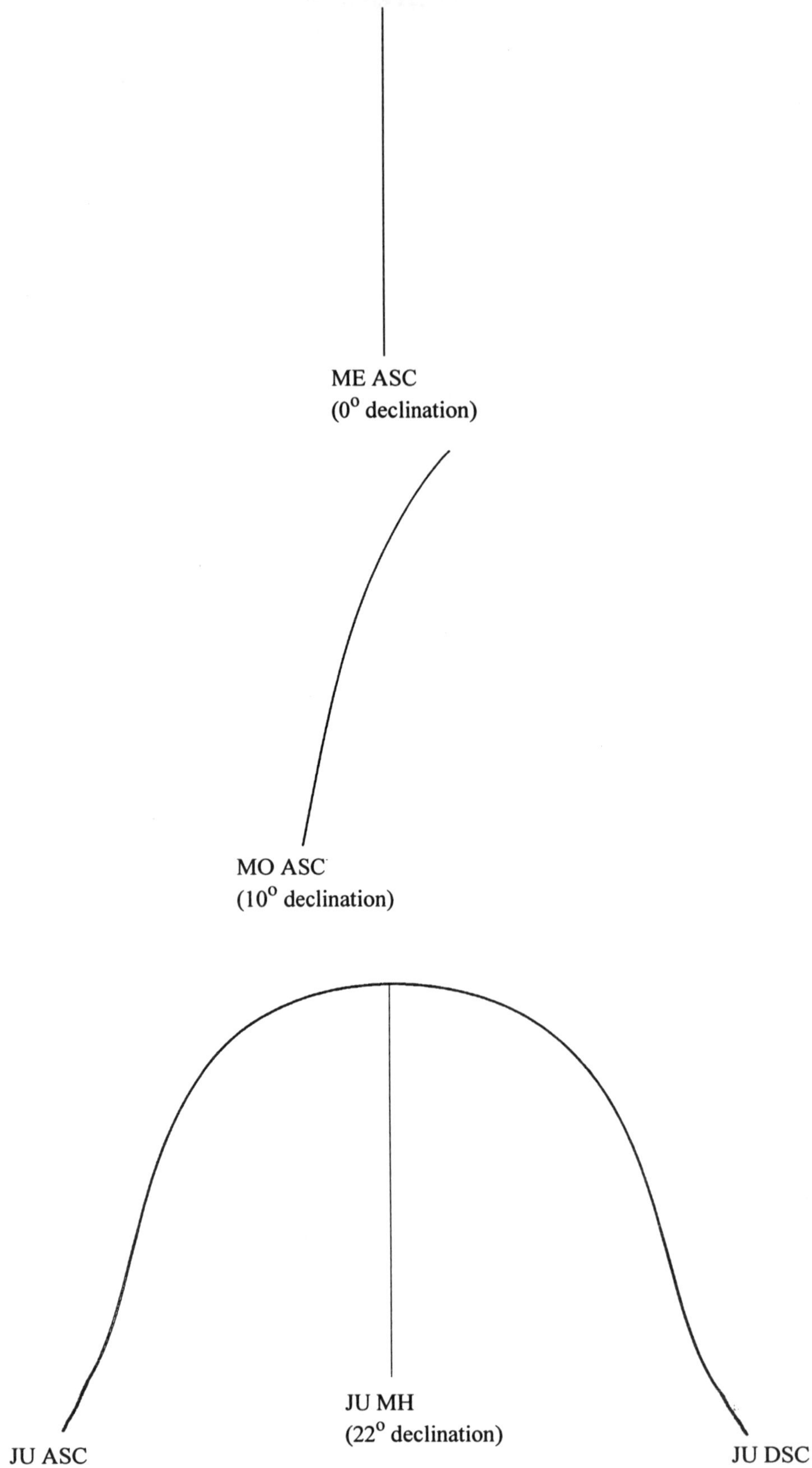

Diagram 6

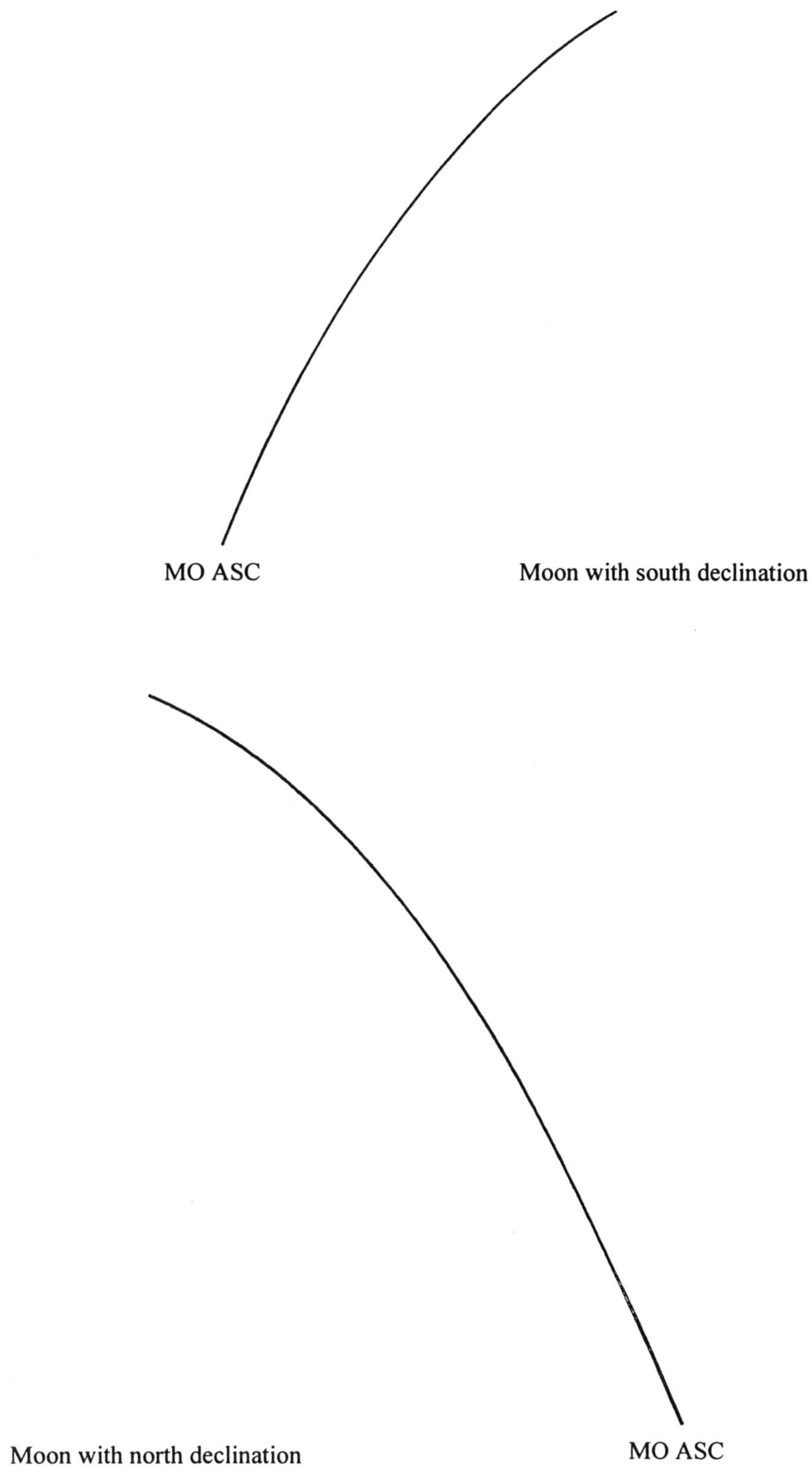
MO ASC
Moon with south declination
Moon with north declination
MO ASC

Diagram 7

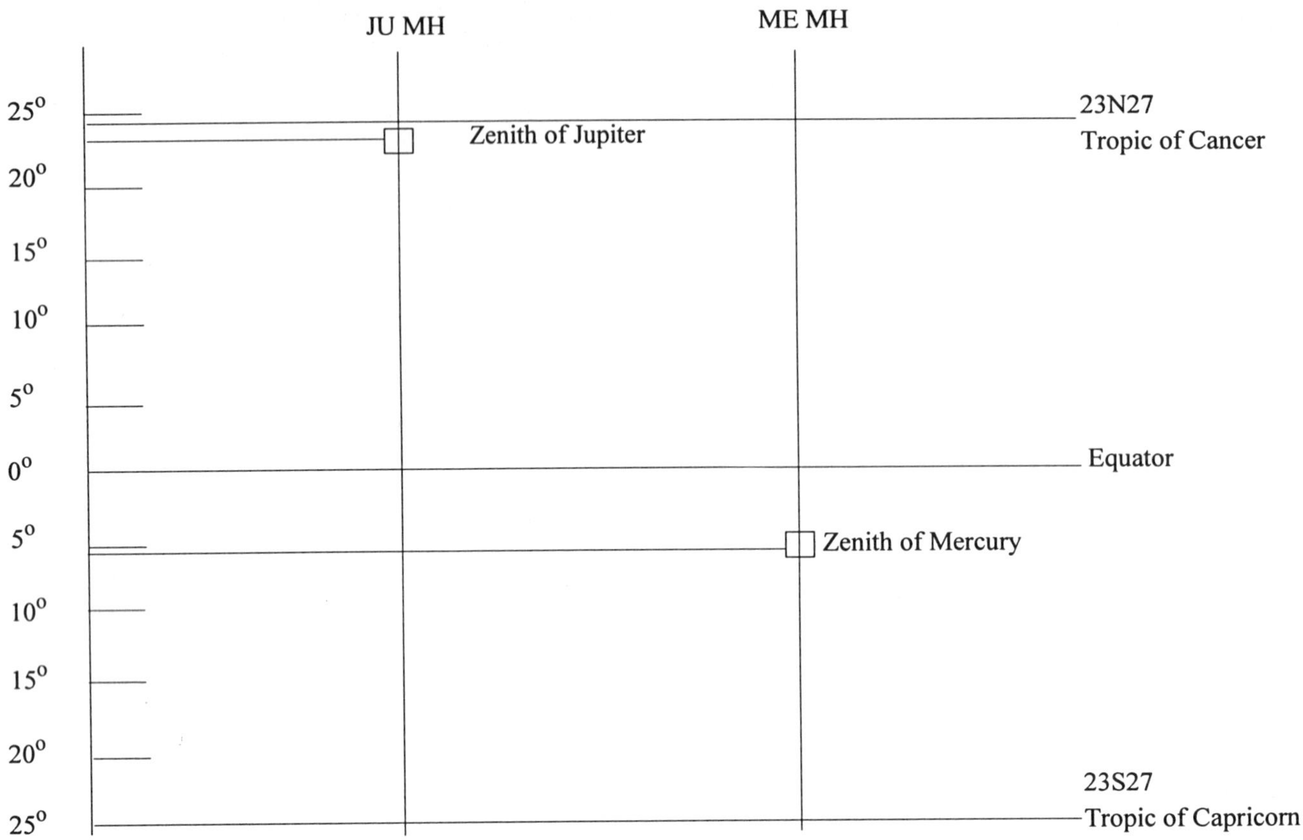

Declination of Jupiter = 22 1/2° N (approx)

Declination of Mercury = 6° S (approx)

Mercury and Jupiter MH lines.

If the ecliptic were drawn on an Astro*Carto*Graphy map, it would then be possible to see the approximate declination and celestial latitude of the planets. See Diagram 8 for examples of Jupiter, Saturn, Mercury and Uranus. Declination is found by looking where the zenith is located on a planet's MH line. From Diagram 8 the approximate declinations are:

Jupiter	4° north
Saturn	22° north
Mercury	2° south
Uranus	18° south

The celestial latitude is the distance north or south of the ecliptic. In Diagram 8 the celestial latitude is shown by the line drawn ninety degrees to the ecliptic. Saturn and Mercury are **above** the ecliptic, so they have north celestial latitude. Jupiter and Uranus are below the ecliptic, hence they have south celestial latitude.

After this explanation about declination, the answer to Test 2 will be clearer. The question was which three planetary lines were nearest to Kemijärvi. The answer is PL IC, PL ASC and PL DSC. On the Paran Printout for Wendy is the following information:

	h m s	**d m**
L.S.T. of chart	21 55 19	328 50
Pluto anti-culminates at	21 54 44	328 41
Pluto rises at	22 23 52	335 58
Pluto sets at	21 25 37	321 24

This doesn't show how near PL ASC and PL DSC are to Kemijärvi. At the latitude for Kemijärvi PL ASC will rise in just over seven degrees. Whereas, PL DSC set just over seven degrees ago.

But a Paran Printout calculated for latitude **67N40** (one degree north of Kemijärvi) shows clearer information about PL ASC and PL DSC.

	h m s	**d m**
L.S.T. of chart	21 55 19	328 50
Pluto anti-culminates at	21 54 44	328 41
Pluto rises at	21 54 44	328 41
Pluto sets at	21 54 44	328 41

This shows that PL IC, PL ASC and PL DSC are all exactly conjunct each other.

We can calculate the exact geographical latitude for the apex of the PL ASC and PL DSC lines. Pluto has 23N10 declination. Subtracting this amount from ninety degrees gives a geographical latitude of 66N50 which is very near to Kemijärvi.

In Wendy's horoscope many planets have twenty-three degrees or more declination. Planets with more than 23N20 declination (ninety degrees minus latitude for Kemijärvi 66N40) will have the same information on the Paran Printouts for 66N40 and 67N40. For Wendy that is Mercury (23N47), Venus (23N26) and Mars (26N42). Jupiter has declination 23N19 which is below 23N20, hence the differences for Jupiter on the Paran Printouts.

In Test 3, the Moon is 22°27" minutes **above** the Ascendant in Nigel's horoscope. But on his Astro*Carto*Graphy map MO ASC is exact at Manchester. This can be explained to a certain extent by the celestial latitude of the Moon being 5S15. More difficult to understand is the position of Neptune. In Nigel's horoscope Neptune is 14°45" below the Descendant but within orb of Manchester on his Astro*Carto*Graphy map. How is this possible when Neptune only has 1N28 celestial latitude? Surely, this wouldn't make such a big difference.

Before explaining why Neptune is within orb of Manchester for Nigel, information about calculating horoscopes for locations above the arctic circle is necessary.

In *Recent Advances in Natal Astrology* by Geoffrey Dean (pages 194-197), he explains the calculation of polar charts. Robert Hand, in *Essays on Astrology*, wrote a chapter on the Ascendant, Midheaven and Vertex in extreme latitudes. The main problem is with house systems, as they break down when a horoscope is calculated for a location above the Arctic Circle.

I thought nothing more about this problem as I had no need to calculate a horoscope where the location was above the Arctic Circle. Like other astrologers, I didn't consider this problem to be all that important, but my attitude changed in 1993.

From July 30 to August 13, 1993, I was on holiday in northern Finland, staying at a small resort called Ruka Loma which is 27 kilometers north of Kuusamo. Ruka Loma is about 72 kilometers (forty-five miles) below the Arctic Circle, and "sunset" was a strange experience. It started to get dark at 11:45 p.m., and it was only "half dark" for a couple of hours. At 2:00 a.m. it was broad daylight again. How does the Sun move around the horoscope for these two hours at locations such as Ruka Loma?

While on holiday I crossed over the Arctic Circle on the way to Kemijärvi, and on the way back I crossed over again at Rovaniemi (where Santa Claus lives!). Kemijärvi, population 12,000, is above the Arctic Circle, and for two months in summer the Sun shines at midnight; in mid-winter it is dark twenty-four hours a day. Astrologically, this raised many interesting questions for which I didn't have answers. Does the Sun suddenly jump from DSC to ASC where there is daylight all day in summer? Do houses work at locations above the Arctic Circle? What happens in winter when the Sun doesn't set? What would an Astro*Carto*Graphy map look at these times?

After returning from Finland, I did some research on horoscopes calculated for Kemijärvi. I began by using June 21, 1993 to calculate horoscopes for Kemijärvi, as this is around the time of the longest day in the Northern Hemisphere. First, I calculated twenty-four horoscopes, one for every hour, but that wasn't enough to answer the questions. Next I calculated half-hourly horoscopes for 12:00 a.m. to 6:00 a.m. and 9:00 p.m. to 11:00 p.m. (see Table 1 for a listing of the ASC and MH for these times).

From 12:00 a.m. to 2:00 a.m. the ASC and MH are at the

Diagram 8

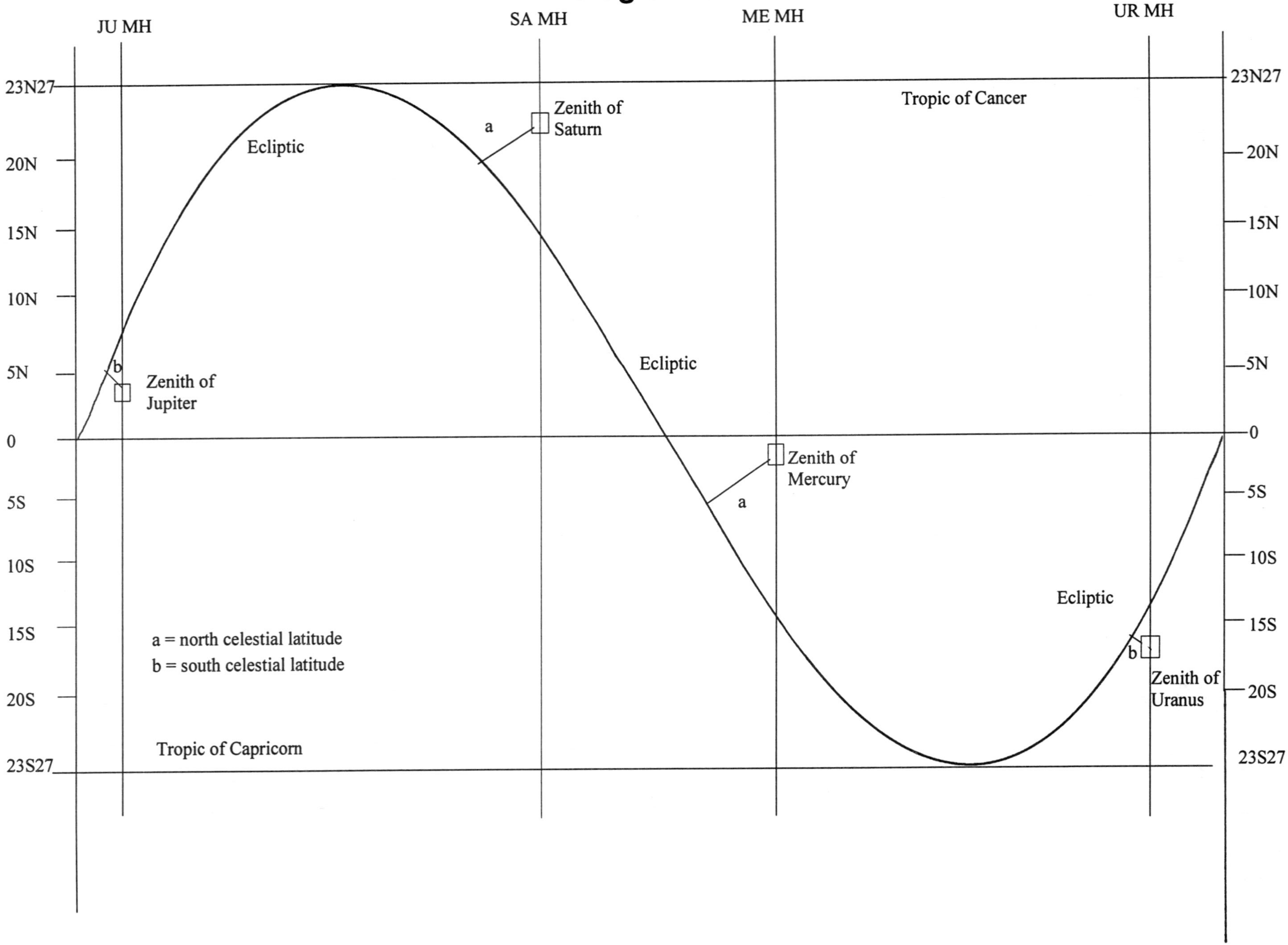
JU MH
SA MH
ME MH
UR MH
23N27
20N
15N
10N
5N
0
5S
10S
15S
20S
23S27
Tropic of Cancer
Tropic of Capricorn
Ecliptic
Ecliptic
Ecliptic
Zenith of Jupiter
Zenith of Saturn
Zenith of Mercury
Zenith of Uranus
a
b
a = north celestial latitude
b = south celestial latitude

Wendy
June 11, 1954
66N40'00"
Tropical

Natal Chart
Kemijarvi
21:55:19 ST

Paran Printout
3:50:0 AM CET
27E25'00"
Geocentric

Pl	Rise			Upper			Set			Lower		
	d	m	h m s	d	m	h m s	d	m	h m s	d	m	h m s
☉	268	20	17:53:19	78	45	5:14:59	249	10	16:36:39	258	45	17:14:59
☽	137	19	9:09:15	196	15	13:05:00	255	11	17:00:45	16	15	1:05:00
☿	284	50	18:59:20	104	50	6:59:20	284	50	18:59:20	284	50	18:59:20
♀	294	16	19:37:06	114	16	7:37:06	294	16	19:37:06	294	16	19:37:06
♂	97	06	6:28:23	277	06	18:28:23	97	06	6:28:23	97	06	6:28:23
♃	276	21	18:25:23	94	17	6:17:10	272	14	18:08:56	274	17	18:17:10
♄	146	17	9:45:08	211	50	14:07:21	277	23	18:29:33	31	50	2:07:21
♅	311	49	20:47:16	113	05	7:32:22	274	22	18:17:27	293	05	19:32:22
♆	130	05	8:40:19	202	20	13:29:19	274	35	18:18:19	22	20	1:29:19
♇	335	58	22:23:52	148	41	9:54:44	321	24	21:25:37	328	41	21:54:44
☊	271	00	18:04:00	287	35	19:10:21	304	11	20:16:42	107	35	7:10:21
As	328	50	21:55:19	119	41	7:58:45	270	33	18:02:11	299	41	19:58:45
MC	270	11	18:00:45	328	50	21:55:19	27	28	1:49:53	148	50	9:55:19
Vx	191	26	12:45:43	230	32	15:22:06	269	37	17:58:30	50	32	3:22:06
EP	269	29	17:57:57	58	50	3:55:19	208	10	13:52:41	238	50	15:55:19
⊗	201	23	13:25:33	235	28	15:41:53	269	33	17:58:12	55	28	3:41:53

☽	lower	16	06	As	upper	119	41	☽	set	255	11	☿	rise	284	50\
♆	lower	22	20	♆	rise	130	05	☉	lower	258	45	☿	set	284	50/\
MC	set	27	28	☽	rise	137	19	☉	rise	268	20	☿	lower	284	50/
♄	lower	31	50	♄	rise	146	17	EP	rise	269	29\	☊	upper	287	35
Vx	lower	50	32	♇	upper	148	41\	⊗	set	269	33/\	♅	lower	293	05
⊗	lower	55	28	MC	lower	148	50/	Vx	set	269	37/\	♀	rise	294	16\
EP	upper	58	50	Vx	rise	191	26	MC	rise	270	11/\	♀	set	294	16/\
☉	upper	78	45	☽	upper	196	15	As	set	270	33/\	♀	lower	294	16/
♃	upper	94	17	⊗	rise	201	23\	☊	rise	271	00/	As	lower	299	41
♂	rise	97	06\	♇	upper	202	20/	♃	set	272	14	☊	set	304	11
♂	set	97	06/\	EP	set	208	10	♃	lower	274	17\	♅	rise	311	49
♂	lower	97	06/	♄	upper	211	50	♅	set	274	22/\	♇	set	321	24
☿	upper	104	50	Vx	upper	230	32	♆	set	274	35/	♇	lower	328	41\
☊	lower	107	35	⊗	upper	235	28	♃	rise	276	21\	As	rise	328	50/\
♅	upper	113	05	EP	lower	238	50	♂	upper	277	06/\	MC	upper	328	50/
♀	upper	114	16	☉	set	249	10	♄	set	277	23/	♇	rise	335	58

Wendy	Natal Chart	Paran Printout
June 11, 1954	Kemijarvi	3:50:0 AM CET
67N40'00"	21:55:19 ST	7E25'00"
Tropical		Geocentric

Pl	Rise			Upper			Set			Lower		
	d	m	h m s	d	m	h m s	d	m	h m s	d	m	h m s
☉	258	45	17:14:56	78	44	5:14:56	258	44	17:14:56	258	44	17:14:56
☽	139	03	9:16:14	196	15	13:05:00	253	27	16:53:47	16	15	1:05:00
☿	284	50	18:59;20	104	50	6:59:20	284	50	18:59:20	284	50	18:59:20
♀	294	16	19:37:06	114	16	7:37:06	294	16	19:37:06	294	16	19:37:06
♂	97	06	6:28:23	277	06	1828:23	97	06	6:28:23	97	06	6:28:23
♃	274	17	18:17:10	94	17	6:17:10	274	17	18:17:10	274	17	18:17:10
♄	147	36	9:50:23	211	50	14:07:21	276	05	18:24:18	31	50	2:07:21
♅	299	08	19:56:33	113	05	7:32:22	287	03	19:08:10	293	05	19:32:22
♆	131	00	8:44:00	202	20	13:29:19	273	40	18:14:38	22	20	1:29:19
♇	328	41	21:54:44	148	41	9:54:44	328	41	21:54:44	328	41	21:54:44
☊	107	35	7:10:21	287	35	19:10:21	107	35	7:10:21	107	35	7:10:21
As	328	50	21:55:19	122	10	8:08:39	275	30	18:22:00	302	10	20:08:39
MC	271	57	18:07:48	328	50	21:55:19	25	42	1:42:50	148	50	9:55:19
Vx	195	52	13:03:30	230	53	15:23:32	265	54	17:43:34	50	53	3:23:32
EP	264	14	17:36:58	58	50	3:55:19	213	25	14:13:40	238	50	15:55:19
⊗	211	23	14:05:33	237	56	15:51:43	264	29	17:37:54	57	56	3:51:43

☽	lower	16	06	♅	upper	113	05	EP	lower	238	50/	☿	set	284	50\
♆	lower	22	20	♀	upper	114	16	☽	set	253	27	☿	lower	284	50/\
MC	set	25	42	As	upper	122	10	☉	set	258	45\	☿	rise	284	50/
♄	lower	31	50	♆	rise	131	00	☉	lower	258	45/\	♅	set	287	03\
Vx	lower	50	53	☽	rise	139	03	☉	rise	258	45/	☊	upper	287	35/
⊗	lower	57	56\	♄	rise	147	36	EP	rise	264	14\	♅	lower	293	05
EP	upper	58	50/	♇	upper	148	41\	⊗	set	264	29/	♀	set	294	16\
☉	upper	78	45	MC	lower	148	50/	Vx	set	265	54	♀	lower	294	16/\
♃	upper	94	17	Vx	rise	195	52\	MC	rise	271	57	♀	rise	294	16/
♂	rise	97	06\	☽	upper	196	15/	♆	set	273	40\	♅	rise	299	08
♂	set	97	06/\	♆	upper	202	20	♃	set	274	17/\	As	lower	302	10/
♂	lower	97	06/	⊗	rise	211	23\	♃	lower	274	17/\	♇	set	328	41\
☿	upper	104	50	♄	upper	211	50/	♄	rise	274	17/	♇	lower	328	41/\
☊	set	107	35\	EP	set	213	25	As	set	275	30\	♇	rise	328	41/\
☊	lower	107	35/\	Vx	upper	230	53	♄	set	276	05/	As	rise	328	50/\
☊	rise	107	35/\	⊗	upper	237	56\	♂	upper	277	06	MC	upper	328	50/

following positions:

	ASC	MH
12:00 a.m.	20 SAG 40	12 SAG 56
12:30 a.m.	23 SAG 33	19 SAG 54
1:00 a.m.	23 SAG 58	26 SAG 49
1:30 a.m.	05 CAN 39	03 CAP 43
2:00 a.m.	06 CAN 39	10 CAP 39

So somewhere between 1:00 a.m. and 1:30 a.m. the ASC changed from SAG to CAN. I decided to calculate horoscopes for every five minutes between these two times which gives the following positions:

	ASC	MH
1:00 a.m.	23 SAG 58	26 SAG 49
1:05 a.m.	22 SAG 03	27 SAG 58
1:10 a.m.	13 SAG 54	29 SAG 07
1:15 a.m.	12 LEO 36	00 CAP 16
1:20 a.m.	10 CAN 34	01 CAP 25
1:25 a.m.	06 CAN 46	02 CAP 34
1:30 a.m.	05 CAN 39	03 CAP 43

The ASC changed from SAG to CAN between 1:10 a.m. and 1:20 a.m., and the ASC at 12 LEO 36 for 1:15 a.m. could be the midway point. So I calculated horoscopes for every minute from 1:10 a.m. to 1:20 a.m., which gives the following positions:

	ASC	MH
1:10 a.m.	13 SAG 54	29 SAG 07
1:11 a.m.	08 SAG 53	29 SAG 21
1:12 a.m.	29 SCO 22	29 SAG 35
1:13 a.m.	07 SCO 32	29 SAG 49
1:14 a.m.	20 VIR 52	00 CAP 02
1:15 a.m.	12 LEO 36	00 CAP 16
1.16 a.m.	26 CAN 32	00 CAP 30
1.17 a.m.	19 CAN 03	00 CAP 44
1.18 a.m.	14 CAN 54	00 CAP 58
1.19 a.m.	12 CAN 19	01 CAP 11
1:20 a.m.	10 CAN 34	01 CAP 25

So within five minutes (1:11 a.m. to 1:16 a.m.) the ASC jumped from SAG to CAN, which answered the question that the Sun goes suddenly from DSC to ASC.

At the same time it showed other interesting facts. On June 21, 1993 the ASC wasn't in CAP, AQU, PIE, ARI, TAU and GEM, which is half the horoscope. This is the result of the difference between signs of long and short ascension.

The listings show that the ASC changed direction twice:

12:30 a.m.	23 SAG 33	1:25 a.m.	6 CAN 46
1:00 a.m.	23 SAG 58	1:30 a.m.	5 CAN 39
1:05 a.m.	22 SAG 03	2:00 a.m.	6 CAN 39

Between 12:30 a.m. and 1:00 a.m. the ASC moved direct-stationary-retrograde, whereas between 1:30 a.m. and 2:00 a.m. the ASC moved retrograde-stationary-direct.

Next I calculated the angles for every five minutes between 12:30 a.m. and 1:00 a.m.:

	ASC	MH
12:30 a.m.	23 SAG 33	19 SAG 54
12:35 a.m.	23 SAG 57	21 SAG 03
12:40 a.m.	24 SAG 18	22 SAG 13
12:45 a.m.	24 SAG 33	23 SAG 22
12:50 a.m.	24 SAG 41	24 SAG 31
12:55 a.m.	24 SAG 34	25 SAG 40
1:00 a.m.	23 SAG 58	26 SAG 49

The ASC changed direction about 12:50 a.m., so I decided to calculate the angles for every minute between 12:50 a.m. and 12:55 a.m.

	ASC	MH
12:50 a.m.	24 SAG 41	24 SAG 31
12:51 a.m.	24 SAG 41	24 SAG 45
12:52 a.m.	24 SAG 41	24 SAG 59
12:53 a.m.	24 SAG 39	25 SAG 12
12:54 a.m.	24 SAG 37	25 SAG 26
12:55 a.m.	24 SAG 34	25 SAG 40

Can you see why the ASC changed direction between 12:50 a.m. and 12:51 a.m.? The ASC also changed direction between 1:30 a.m. and 2:00 a.m., so again I calculated the angles for intervals of every five minutes:

	ASC	MH
1:30 a.m.	05 CAN 39	03 CAP 43
1:35 a.m.	05 CAN 20	04 CAP 52
1:40 a.m.	05 CAN 21	06 CAP 01
1:45 a.m.	05 CAN 33	07 CAP 11
1:50 a.m.	05 CAN 52	08 CAP 20
1:55 a.m.	06 CAN 14	09 CAP 29
2:00 a.m.	06 CAN 39	10 CAP 39

So between 1:35 a.m. and 1:40 a.m. the ASC changed direction; therefore, I calculated the angles for every minute between these times:

	ASC	MH
1:35 a.m.	5 CAN 20	4 CAP 52
1:36 a.m.	5 CAN 19	5 CAP 06
1:37 a.m.	5 CAN 19	5 CAP 20
1:38 a.m.	5 CAN 19	5 CAP 34
1:39 a.m.	5 CAN 20	5 CAP 48
1:40 a.m.	5 CAN 21	6 CAP 01

This shows that the ASC changed direction at about 1:37 a.m. Perhap now you can see the connection between 12:51 a.m. and 1:37 a.m. when the ASC changed direction. If not, look at the MH positions.

The ASC changed direction when the MH was exactly conjunct the ASC or DSC. If a person were born at 12:51 a.m. on June 21, 1993 at Kemijärvi, he or she would have MH exactly conjunct ASC. House systems that use the IC as the fourth cusp and the MH as the tenth cusp cannot be used. Placidus and Koch house systems break down completely for this date, time and place. Perhaps only systems such as equal house can be used for this data. On the other hand, if a person were born at 2:30 p.m. on June 21, 1993 at Kemijärvi, then Placidus or Koch houses could be used without any problem. So the **time** of birth on this date determines whether Placidus or Koch houses can be used.

There is an interval of forty-six minutes between MH CON ASC and MH CON DSC in this example. Where Astro*Carto*Graphy is concerned there is only one time between 12:51 a.m. and 1:37 a.m. when the ASC/DSC and **IC** are all conjunct at Kemijärvi. An A*C*G map calculated for 1:12 a.m. (approximate midpoint in time) shows SU IC exactly conjunct at Kemijärvi. In fact, the A*C*G map shows SU IC, SU ASC and SU DSC **all** conjunct at Kemijärvi. The SU IC line is at the apex of the SU ASC and SU DSC curving line. This A*C*G map shows **eight** planets within orb of Kemijärvi, with only Venus and Pluto having no influence.

At first I thought the occurrence of MH CON ASC and MH CON DSC only applied at certain times of the year, when the Sun didn't set in summer or didn't rise in winter. But charts set for different months still produced the same occurrence of MH CON ASC/DSC axis.

A few examples are the following dates and times for Kemijärvi (exact conjunctions not calculated):

Time	Date	ASC	MH
10:00 p.m.	July 30, 1993	24 SAG 07	21 SAG 35
11:00 p.m.	July 30, 1993	05 CAN 19	05 CAP 24
7:30 p.m.	Sept. 21, 1993	05 CAN 19	05 CAP 01
6:30 a.m.	March 21, 1993	05 CAN 55	03 CAP 19
12:30 p.m.	Dec. 21, 1993	05 CAN 23	04 CAP 31

This suggests that if a person is born in Kemijärvi with the MH around 24-25 SAG and 5-6 CAP, then the MH will be exactly conjunct ASC or DSC in the natal chart. Placidus and Koch houses can't be used when the ASC is about to change direction.

This illustrates how a natal chart looks for a location above the Arctic Circle, and shows how the ASC changes direction twice when it is conjunct the MH/IC axis.

	ASC	MH
12:00 a.m.	20 SAG 40	12 SAG 56
12:30 a.m.	23 SAG 33	19 SAG 54
1:00 a.m.	23 SAG 58	26 SAG 49
1:30 a.m.	05 CAN 39	03 CAP 43
2:00 a.m.	06 CAN 39	10 CAP 39
2:30 a.m.	09 CAN 35	17 CAP 37
3:00 a.m.	12 CAN 49	24 CAP 41
3:30 a.m.	16 CAN 11	01 AQU 53
4:00 a.m.	19 CAN 36	09 AQU 13
4:30 a.m.	23 CAN 05	16 AQU 43
5:00 a.m.	26 CAN 36	24 AQU 24
5:30 a.m.	00 LEO 09	02 PIS 14
6:00 a.m.	03 LEO 45	10 PIS 13
6:30 a.m.	07 LEO 24	18 PIS 19
7:30 a.m.	14 LEO 49	04 ARI 41
8:30 a.m.	22 LEO 25	20 ARI 56
9:30 a.m.	00 VIR 11	06 TAU 43
10:30 a.m.	08 VIR 07	21 TAU 51
11:30 a.m.	16 VIR 10	06 GEM 20
12:30 p.m.	24 VIR 18	20 GEM 21
1:30 p.m.	02 LIB 28	04 CAN 10
2:30 p.m.	10 LIB 38	18 CAN 05
3:30 p.m.	18 LIB 43	02 LEO 22
4:30 p.m.	26 LIB 42	17 LEO 13
5:30 p.m.	04 SCO 32	02 VIR 45
6:30 p.m.	12 SCO 12	18 VIR 51
7:30 p.m.	19 SCO 41	05 LIB 13
8:30 p.m.	27 SCO 00	21 LIB 28
9:00 p.m.	00 SAG 36	29 LIB 25
9:30 p.m.	04 SAG 09	07 SCO 13
10:00 p.m.	07 SAG 39	14 SCO 52
10:30 p.m.	11 SAG 07	22 SCO 20
11:30 p.m.	17 SAG 52	06 SAG 48

Back to Nigel's horoscope. At the latitude for Manchester (53N30) all the signs of the zodiac can be on the Ascendant. As there are twenty-four hours in a day and twelve zodiac signs it would appear that the Ascendant changes sign approximately every two hours. But this is not the case. Signs such as Leo, Virgo, Libra and Scorpio spend approximately three hours each on the Ascendant. That is half a day. Pisces and Aries are only on the Ascendant for about forty-eight minutes each.

The Paran Printout for Nigel shows the following:

	h m s	d m
L.S.T. of chart	18 36 37	279 09
Moon rises at	18 36 01	279 00
Jupiter sets at	18 49 16	282 19
Neptune sets at	18 22 47	275 42
Venus culminates at	18 57 08	284 17

This shows that Neptune set fourteen minutes previously. Fourteen minutes later the Ascendant moved another **eight degrees**. But when the Ascendant is in Leo, Virgo, Libra or Scorpio, then in fourteen minutes the Ascendant only moves two to three degrees. With Nigel having an Aries Ascendant, the Moon and Neptune are nearer to the ASC/DSC axis than what they appear to be.

Right Ascension Natal Horoscope

When an Astro*Carto*Graphy map is drawn, the planetary lines are calculated by using right ascension, declination and celestial latitude. A horoscope is drawn using **celestial longitude** of the planets, but this measurement is **not** used in drawing an Astro*Carto*Graphy map. Logically, it would make more sense to draw a horoscope by using right ascension of the planets. But ephemerides list the celestial longitude of the planets, so astrologers use these measurements. Astrological computer programs usually have an option available to calculate a horoscope in right ascension.

Calculating a horoscope in right ascension means that the positions of the planets are measured along the **celestial equator**. Celestial longitude is calculated with the planets along the **ecliptic**. The difference between a planet measured in right ascension and celestial longitude will usually be within four degrees. The exception is Pluto when it has high celestial latitude; then the difference will be greater.

Compare Simon's normal horoscope to his natal chart drawn in right ascension. The difference between the planets in celestial longitude and right ascension is within or just over three degrees (Sun to Neptune). Pluto has a difference of five and a half degrees because of its high celestial latitude.

Wendy's normal horoscope shows that Pluto is $3^{o}45"$ from the IC. The Paran Printout shows the difference to be only nine minutes. Wendy's right ascension horoscope also shows that Pluto is only nine minutes from the IC. This means that the right ascension horoscope shows the true distance of a planet to the MH/IC axis.

Nigel	Natal Chart	Paran Printout
Jan. 9, 1946	Manchester	11:32:0 AM GMT
53N30'00"	18:36:37 ST	2W15'00"
Tropical		Geocentric

Pl	Rise			Upper			Set			Lower		
	d	m	h m s	d	m	h m s	d	m	h m s	d	m	h m s
☉	233	33	15:34:11	290	10	19:20:40	346	47	23:07:09	110	10	7:20:40
☽	279	00	18:36:01	2	57	0:11:47	86	53	5:47:34	182	57	12:11:47
☿	215	30	14:22:00	269	51	17:59:23	324	12	21:36:47	89	51	5:59:23
♀	229	51	15:19:22	284	17	18:57:08	338	44	22:34:55	104	17	6:57:08
♂	348	23	23:13:31	117	51	7:51:22	247	18	16:29:14	297	51	19:51:22
♃	126	20	8:25:19	204	19	13:37:17	282	19	18:49:16	24	19	1:37:17
♄	351	00	23:24:02	113	24	7:33:35	235	47	15:43:07	293	24	19:33:35
♅	308	41	20:34:45	72	48	4:51:12	196	55	13:07:39	252	48	16:51:12
♆	101	18	6:45:13	188	30	12:34:00	275	42	18:22:47	8	30	0:34:00
♇	9	16	0:37:03	135	31	9:02:03	261	46	17:27:04	315	31	21:02:03
☊	323	04	21:32:16	88	57	5:55:46	214	49	14:19:16	268	57	17:55:46
As	279	09	18:36:37	21	38	1:26:32	124	07	8:16:26	201	38	13:26:32
MC	224	31	14:58:02	279	09	18:36:37	333	48	22:15:12	99	09	6:36:37
Vx	100	59	6:43:58	186	56	12:27:44	272	53	18:11:30	6	56	0:27:44
EP	273	48	18:15:13	9	09	0:36:37	104	30	6:58:01	189	09	12:36:37
⊗	330	33	22:02:11	96	11	6:24:45	221	50	14:47:18	276	11	18:24:45

☽	upper	2	57	EP	set	104	30/	⊗	set	221	50	MC	upper	279	09/
Vx	lower	6	56	☉	lower	110	10	MC	rise	224	31	♃	set	282	19
♆	lower	8	30\	♄	upper	113	24	♀	rise	229	51	♀	upper	284	17
EP	upper	9	09/\	♂	upper	117	51	☉	rise	233	33	☉	upper	290	10
♇	rise	9	16/	As	set	124	07	♄	set	235	47	♄	lower	293	24
As	upper	21	38	♃	rise	126	20	♂	set	247	18	♂	lower	297	51
♃	lower	24	19	♇	upper	135	31	♅	lower	252	48	♅	rise	308	41
♅	upper	72	48	☽	lower	182	57	♇	set	261	46	♇	lower	315	31
☽	set	86	53	Vx	upper	186	56	☊	lower	268	57\	☊	rise	323	04
☊	upper	88	57\	♆	upper	188	30\	☿	upper	269	51/	☿	set	324	12
☿	lower	89	51/	EP	lower	189	09/	Vx	set	272	53\	⊗	rise	330	33
⊗	upper	96	11	♅	set	196	55	EP	rise	273	48/	MC	set	333	48
MC	lower	99	09	As	lower	201	38	♆	set	275	42\	♀	set	338	44
Vx	rise	100	59\	♃	upper	204	19	⊗	lower	276	11/	☉	set	346	47
♆	rise	101	18/	☊	set	214	49\	☽	rise	279	00\	♂	rise	348	23
♀	lower	104	17\	☿	rise	215	30/	As	rise	279	09/\	♄	rise	351	00

Simon
May 04, 1978
Leicester, England
06:34:00 AM GMD
ZONE: -01:00
001W09'00"
52N39'00"

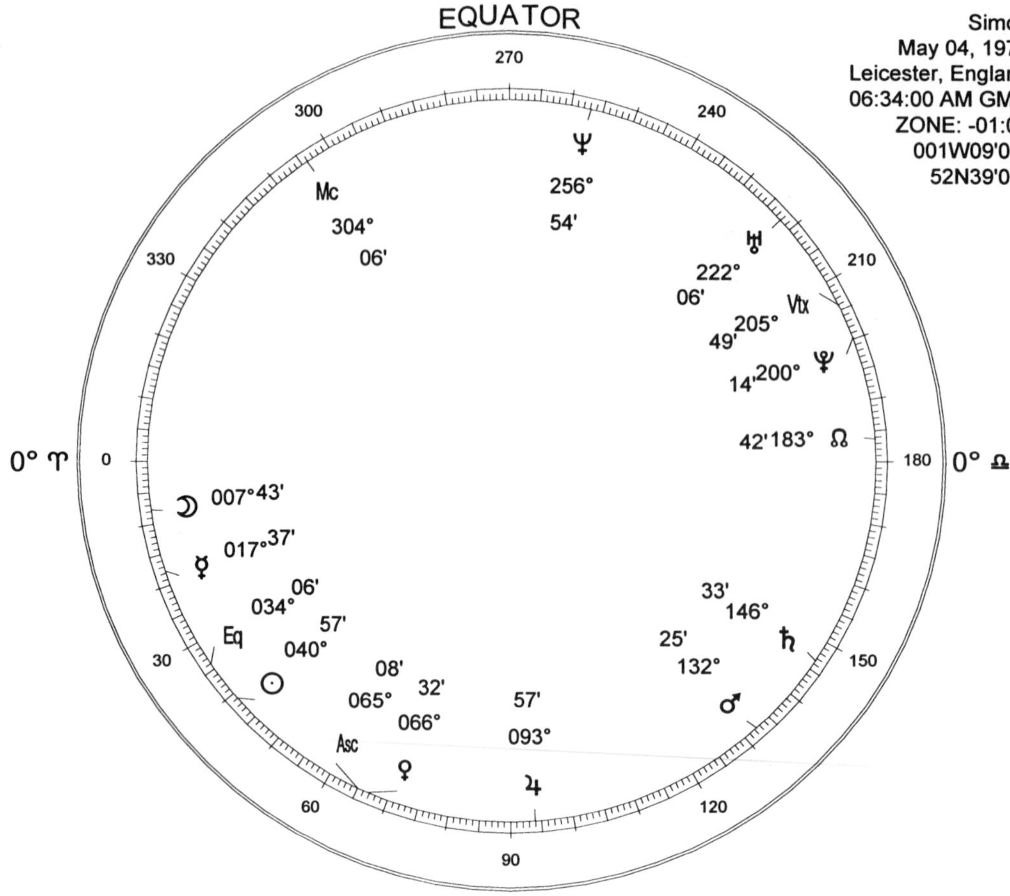

Wendy
Jun 11, 1954
Kemijarvi
03:50:00 AM CET
ZONE: -01:00
027E25'00"
66N40'00"

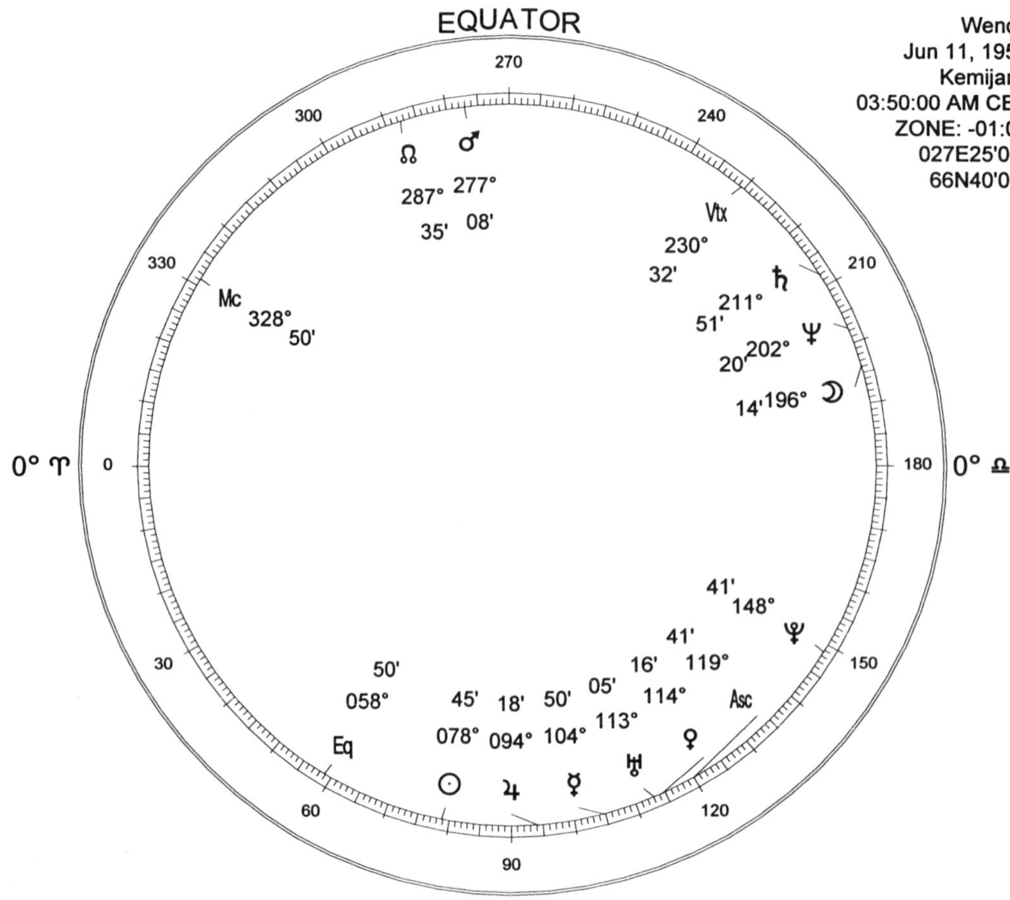

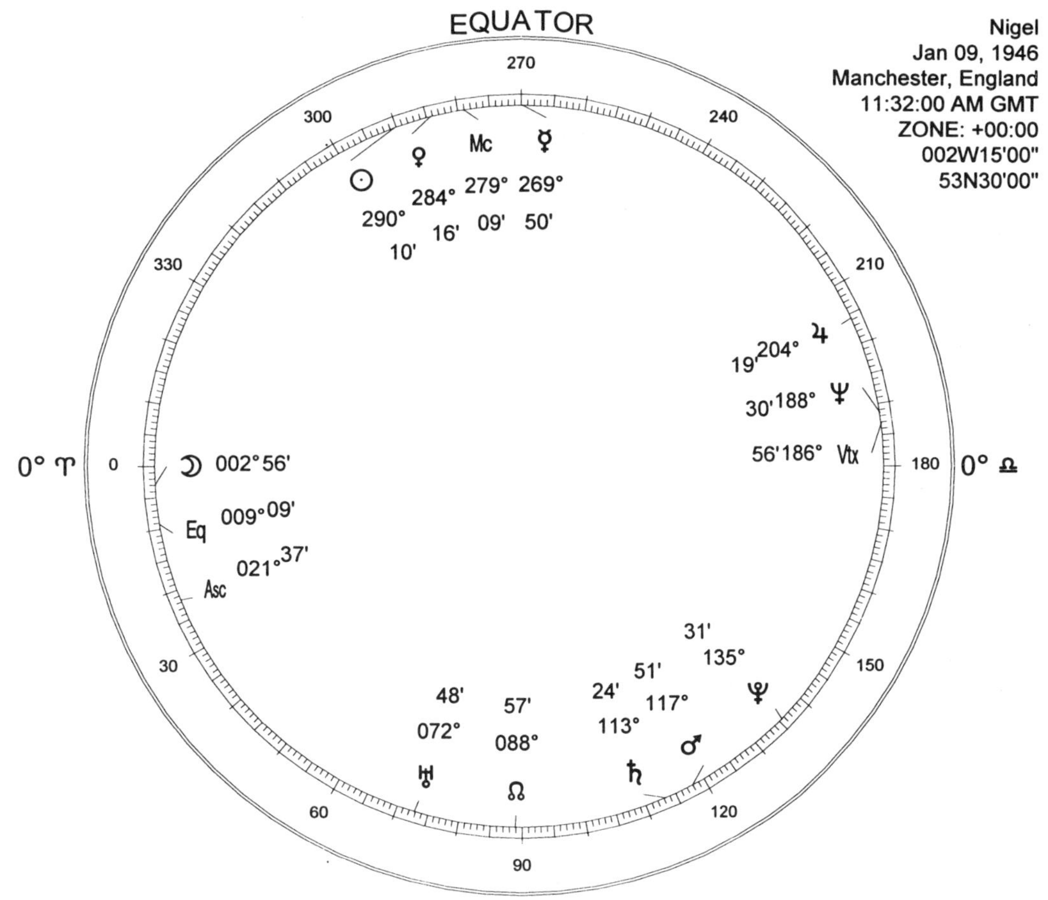

What the right ascension horoscope does not show is the true distance of a planet to the ASC/DSC axis. In Nigel's right ascension horoscope the Moon is nineteen degrees from the Ascendant. But the Paran Printout shows that the Moon is really only nine minutes away from the Ascendant. The reason for this is that more than right ascension and declination are needed to calculate when a planet is **exactly** on the Ascendant or Descendant. More information on this is included in the next section, which explains how the planetary lines are drawn on an Astro*Carto*Graphy map.

The advantage of drawing a horoscope in right ascension is that the true orb of the aspects can be seen. The difference between aspects in right ascension and celestial longitude is more noticeable for conjunctions and oppositions. An example of a conjunction aspect was shown with the Moon and Pluto in John's horoscope.

Right ascension of the planets is used in other Astro*Carto*Graphy techniques. Astro*Carto*Graphy solar arc progressions and Cyclo*Carto*Graphy are easier to understand when the natal horoscope is calculated in right ascension.

Drawing the Astro*Carto*Graphy Map

Appendix 1 shows how the planetary lines are calculated for an Astro*Carto*Graphy map. When these calculations are done by hand, it takes a long time to find out the positions of all the planetary lines, and once the calculations are completed the planetary lines are drawn on a world map. This is what Jim Lewis did before a computer program was designed to do the work.

An Astro*Carto*Graphy map is drawn for geographical longitude from sixty degrees south to seventy degrees north. The MH and IC lines are fairly easy to calculate as they are based on the right ascension of the planets. The ASC and DSC lines are much more complicated to calculate as the geographical latitude is also taken into consideration.

The following list shows the positions of the Sun MH, IC, ASC and DSC for Simon. The SU ASC and SU DSC lines have been calculated for every degree of geographical latitude from sixty degrees south to seventy degrees north. If you had used a scientific calculator for Simon's SU ASC and SU DSC lines, then **242** different calculations would have been made. Even using an orb of five degrees in geographical latitude as shown in Appendix 1 means fifty-two or fifty-four different calculations to find out a planet's ASC and DSC positions.

The NOVA astrology computer program was used to calculate Simon's Astro*Carto*Graphy Sun lines. This program only calculates the positions of the MH, IC, ASC and DSC lines. It would still take some time to draw all the lines on a world map. Appendix 3 examines computer software that draws Astro*Carto*Graphy maps.

Diagram 9(a) shows **oblique ascension** and the **ascensional difference** which are both needed to calculate the Astro*Carto*Graphy ASC lines. Diagram 9(b) is the reverse of Diagram 9(a) to show **oblique descension** and **ascensional difference** for calculating the DSC lines.

Simon	Natal Chart	Astro-Mapping
May 4, 1978	Leicester	6:34:0 AM BST
52N39'00"	20:16:25 ST	1W09'00"
Tropical	Geocentric	

Sun is on the upper Meridian at 95 E 42
is on the Zenith at 15 N 52
is on the lower Meridian at 84 W 18
is on the Nadir at 15 S 52
never sets at 74 N 08
never rises at 74 S 08

At	60 S 00	Su rises at long.	35 E 11	and sets at long. 156 E 12
At	59 S 00	Su rises at long.	33 E 55	and sets at long. 157 E 28
At	58 S 00	Su rises at long.	32 E 45	and sets at long. 158 E 39
At	57 S 00	Su rises at long.	31 E 39	and sets at long. 159 E 45
At	56 S 00	Su rises at long.	30 E 37	and sets at long. 160 E 47
At	55 S 00	Su rises at long.	29 E 38	and sets at long. 161 E 45
At	54 S 00	Su rises at long.	28 E 43	and sets at long. 162 E 40
At	53 S 00	Su rises at long.	27 E 51	and sets at long. 163 E 32
At	52 S 00	Su rises at long.	27 E 01	and sets at long. 164 E 22
At	51 S 00	Su rises at long.	26 E 14	and sets at long. 165 E 09
At	50 S 00	Su rises at long.	25 E 29	and sets at long. 165 E 54
At	49 S 00	Su rises at long.	24 E 46	and sets at long. 166 E 37
At	48 S 00	Su rises at long.	24 E 05	and sets at long. 167 E 18
At	47 S 00	Su rises at long.	23 E 26	and sets at long. 167 E 57
At	46 S 00	Su rises at long.	22 E 48	and sets at long. 168 E 35
At	45 S 00	Su rises at long.	22 E 12	and sets at long. 169 E 11
At	44 S 00	Su rises at long.	21 E 37	and sets at long. 169 E 46
At	43 S 00	Su rises at long.	21 E 04	and sets at long. 170 E 20
At	42 S 00	Su rises at long.	20 E 31	and sets at long. 170 E 52
At	41 S 00	Su rises at long.	20 E 00	and sets at long. 171 E 23
At	40 S 00	Su rises at long.	19 E 29	and sets at long. 171 E 54
At	39 S 00	Su rises at long.	19 E 00	and sets at long. 172 E 23
At	38 S 00	Su rises at long.	18 E 31	and sets at long. 172 E 52
At	37 S 00	Su rises at long.	18 E 03	and sets at long. 173 E 20
At	36 S 00	Su rises at long.	17 E 36	and sets at long. 173 E 47
At	35 S 00	Su rises at long.	17 E 10	and sets at long. 174 E 13
At	34 S 00	Su rises at long.	16 E 45	and sets at long. 174 E 39
At	33 S 00	Su rises at long.	16 E 20	and sets at long. 175 E 03
At	32 S 00	Su rises at long.	15 E 55	and sets at long. 175 E 28
At	31 S 00	Su rises at long.	15 E 31	and sets at long. 175 E 52
At	30 S 00	Su rises at long.	15 E 08	and sets at long. 176 E 15
At	29 S 00	Su rises at long.	14 E 45	and sets at long. 176 E 38
At	28 S 00	Su rises at long.	14 E 23	and sets at long. 177 E 00
At	27 S 00	Su rises at long.	14 E 01	and sets at long. 177 E 22
At	26 S 00	Su rises at long.	13 E 40	and sets at long. 177 E 44
At	25 S 00	Su rises at long.	13 E 18	and sets at long. 178 E 05
At	24 S 00	Su rises at long.	12 E 58	and sets at long. 178 E 25

At	23 S 00	Su rises at long.	12 E 37	and sets at long. 178 E 46
At	22 S 00	Su rises at long.	12 E 17	and sets at long. 179 E 06
At	21 S 00	Su rises at long.	11 E 57	and sets at long. 179 E 26
At	20 S 00	Su rises at long.	11 E 38	and sets at long. 179 E 45
At	19 S 00	Su rises at long.	11 E 18	and sets at long. 179 W 55
At	18 S 00	Su rises at long.	10 E 59	and sets at long. 179 W 36
At	17 S 00	Su rises at long.	10 E 41	and sets at long. 179 W 18
At	16 S 00	Su rises at long.	10 E 22	and sets at long. 178 W 59
At	15 S 00	Su rises at long.	10 E 04	and sets at long. 178 W 40
At	14 S 00	Su rises at long.	9 E 45	and sets at long. 178 W 22
At	13 S 00	Su rises at long.	9 E 27	and sets at long. 178 W 04
At	12 S 00	Su rises at long.	9 E 09	and sets at long. 177 W 46
At	11 S 00	Su rises at long.	8 E 52	and sets at long. 177 W 28
At	10 S 00	Su rises at long.	8 E 34	and sets at long. 177 W 11
At	9 S 00	Su rises at long.	8 E 16	and sets at long. 176 W 53
At	8 S 00	Su rises at long.	7 E 59	and sets at long. 176 W 36
At	7 S 00	Su rises at long.	7 E 42	and sets at long. 176 W 18
At	6 S 00	Su rises at long.	7 E 24	and sets at long. 176 W 01
At	5 S 00	Su rises at long.	7 E 07	and sets at long. 175 W 44
At	4 S 00	Su rises at long.	6 E 50	and sets at long. 175 W 27
At	3 S 00	Su rises at long.	6 E 33	and sets at long. 175 W 10
At	2 S 00	Su rises at long.	6 E 16	and sets at long. 174 W 53
At	1 S 00	Su rises at long.	5 E 59	and sets at long. 174 W 36
At	0 N 00	Su rises at long.	5 E 42	and sets at long. 174 W 18
At	1 N 00	Su rises at long.	5 E 24	and sets at long. 174 W 01
At	2 N 00	Su rises at long.	5 E 07	and sets at long. 173 W 44
At	3 N 00	Su rises at long.	4 E 50	and sets at long. 173 W 27
At	4 N 00	Su rises at long.	4 E 33	and sets at long. 173 W 10
At	5 N 00	Su rises at long.	4 E 16	and sets at long. 172 W 53
At	6 N 00	Su rises at long.	3 E 59	and sets at long. 172 W 36
At	7 N 00	Su rises at long.	3 E 42	and sets at long. 172 W 18
At	8 N 00	Su rises at long.	3 E 24	and sets at long. 172 W 01
At	9 N 00	Su rises at long.	3 E 07	and sets at long. 171 W 44
At	10 N 00	Su rises at long.	2 E 49	and sets at long. 171 W 26
At	11 N 00	Su rises at long.	2 E 32	and sets at long. 171 W 08
At	12 N 00	Su rises at long.	2 E 14	and sets at long. 170 W 51
At	13 N 00	Su rises at long.	1 E 56	and sets at long. 170 W 33
At	14 N 00	Su rises at long.	1 E 38	and sets at long. 170 W 15
At	15 N 00	Su rises at long.	1 E 20	and sets at long. 169 W 56
At	16 N 00	Su rises at long.	1 E 01	and sets at long. 169 W 38
At	17 N 00	Su rises at long.	0 E 42	and sets at long. 169 W 19
At	18 N 00	Su rises at long.	0 E 24	and sets at long. 169 W 01
At	19 N 00	Su rises at long.	0 E 05	and sets at long. 168 W 42
At	20 N 00	Su rises at long.	0 W 15	and sets at long. 168 W 22
At	21 N 00	Su rises at long.	0 W 34	and sets at long. 168 W 03
At	22 N 00	Su rises at long.	0 W 54	and sets at long. 167 W 43
At	23 N 00	Su rises at long.	1 W 14	and sets at long. 167 W 23
At	24 N 00	Su rises at long.	1 W 35	and sets at long. 167 W 02

At 25 N 00 Su rises at long. 1 W 55 and sets at long. 166 W 42
At 26 N 00 Su rises at long. 2 W 16 and sets at long. 166 W 20

At 27 N 00 Su rises at long. 2 W 38 and sets at long. 165 W 59
At 28 N 00 Su rises at long. 3 W 00 and sets at long. 165 W 37
At 29 N 00 Su rises at long. 3 W 22 and sets at long. 165 W 15
At 30 N 00 Su rises at long. 3 W 45 and sets at long. 164 W 52

At 31 N 00 Su rises at long. 4 W 08 and sets at long. 164 W 29
At 32 N 00 Su rises at long. 4 W 32 and sets at long. 164 W 05
At 33 N 00 Su rises at long. 4 W 57 and sets at long. 163 W 40
At 34 N 00 Su rises at long. 5 W 21 and sets at long. 163 W 15

At 35 N 00 Su rises at long. 5 W 47 and sets at long. 162 W 50
At 36 N 00 Su rises at long. 6 W 13 and sets at long. 162 W 24
At 37 N 00 Su rises at long. 6 W 40 and sets at long. 161 W 57
At 38 N 00 Su rises at long. 7 W 08 and sets at long. 161 W 29

At 39 N 00 Su rises at long. 7 W 37 and sets at long. 161 W 00
At 40 N 00 Su rises at long. 8 W 06 and sets at long. 160 W 31
At 41 N 00 Su rises at long. 8 W 37 and sets at long. 160 W 00
At 42 N 00 Su rises at long. 9 W 08 and sets at long. 159 W 29

At 43 N 00 Su rises at long. 9 W 40 and sets at long. 158 W 56
At 44 N 00 Su rises at long. 10 W 14 and sets at long. 158 W 23
At 45 N 00 Su rises at long. 10 W 49 and sets at long. 157 W 48
At 46 N 00 Su rises at long. 11 W 25 and sets at long. 157 W 12

At 47 N 00 Su rises at long. 12 W 03 and sets at long. 156 W 34
At 48 N 00 Su rises at long. 12 W 42 and sets at long. 155 W 55
At 49 N 00 Su rises at long. 13 W 23 and sets at long. 155 W 14
At 50 N 00 Su rises at long. 14 W 06 and sets at long. 154 W 31
At 51 N 00 Su rises at long. 14 W 51 and sets at long. 153 W 46

At 52 N 00 Su rises at long. 15 W 38 and sets at long. 152 W 59
At 53 N 00 Su rises at long. 16 W 28 and sets at long. 152 W 09
At 54 N 00 Su rises at long. 17 W 20 and sets at long. 151 W 17
At 55 N 00 Su rises at long. 18 W 15 and sets at long. 150 W 22

At 56 N 00 Su rises at long. 19 W 13 and sets at long. 149 W 23
At 57 N 00 Su rises at long. 20 W 15 and sets at long. 148 W 21
At 58 N 00 Su rises at long. 21 W 21 and sets at long. 147 W 15
At 59 N 00 Su rises at long. 22 W 32 and sets at long. 146 W 05

At 60 N 00 Su rises at long. 23 W 48 and sets at long. 144 W 49
At 61 N 00 Su rises at long. 25 W 09 and sets at long. 143 W 28
At 62 N 00 Su rises at long. 26 W 37 and sets at long. 142 W 00
At 63 N 00 Su rises at long. 28 W 12 and sets at long. 140 W 25

At 64 N 00 Su rises at long. 29 W 57 and sets at long. 138 W 40
At 65 N 00 Su rises at long. 31 W 51 and sets at long. 136 W 46
At 66 N 00 Su rises at long. 33 W 58 and sets at long. 134 W 39
At 67 N 00 Su rises at long. 36 W 20 and sets at long. 132 W 17

At 68 N 00 Su rises at long. 39 W 00 and sets at long. 129 W 37
At 69 N 00 Su rises at long. 42 W 04 and sets at long. 126 W 33
At 70 N 00 Su rises at long. 45 W 38 and sets at long. 122 W 59

Diagram 9A

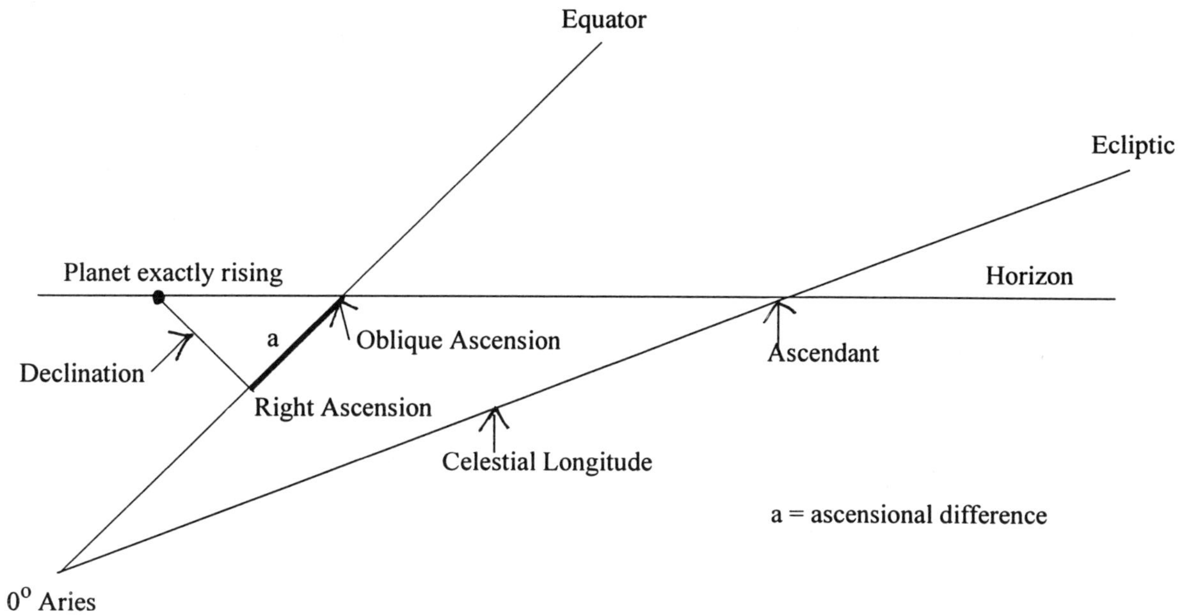

Diagram 9B

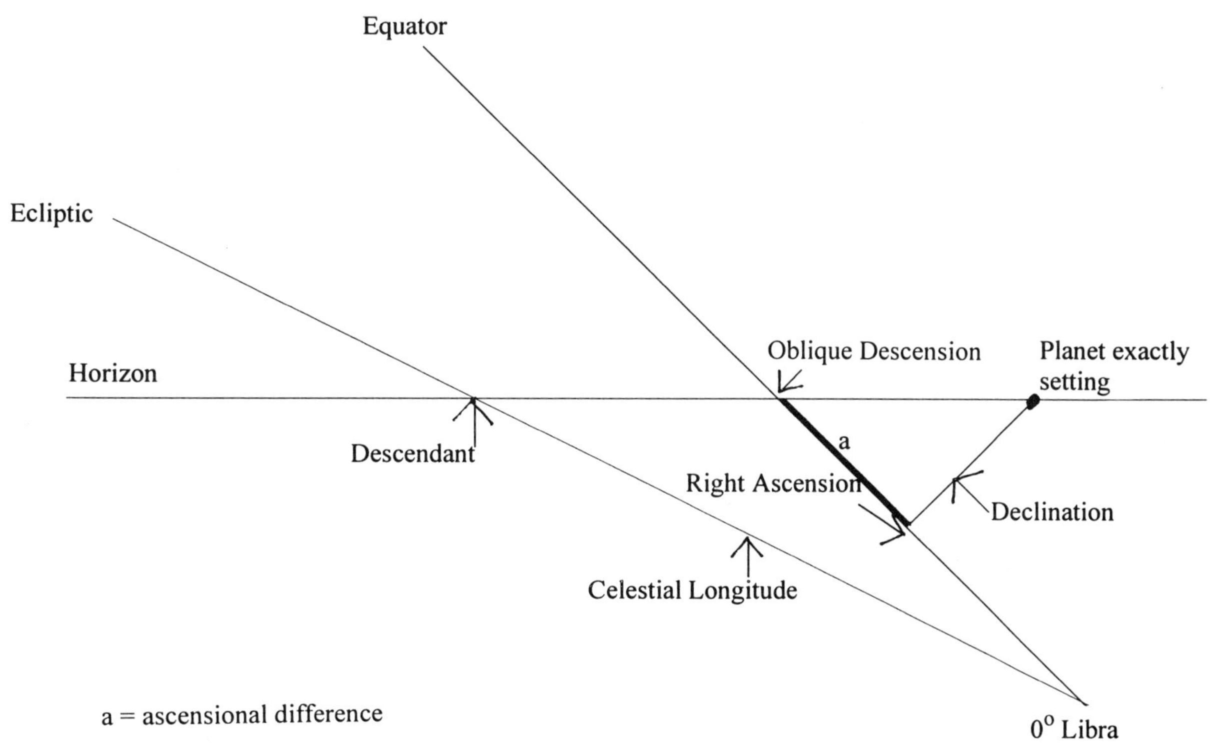

Diagram 10 illustrates how the Sun lines would be plotted on Simon's Astro*Carto*Graphy map. If you examine Simon's Astro*Carto*Graphy map at the end of this chapter and compare it with the listing of the Sun lines you will see how his map is drawn.

I have gone into a great amount of detail in explaining how an Astro*Carto*Graphy map is drawn and how it differs from relocation astrology. When I give lectures on Astro*Carto*Graphy the question most asked is: ''How are the lines drawn on a world map?'' So I decided to include a chapter in this book explaining how the planetary lines appear on a map of the world.

Much of the information in this chapter will be new and maybe difficult to understand, but it is not essential that you understand everything to be able to interpret an Astro*Carto*Graphy map. You may need to read this chapter a few times before acquiring a clear picture of what is used to calculate and draw the map.

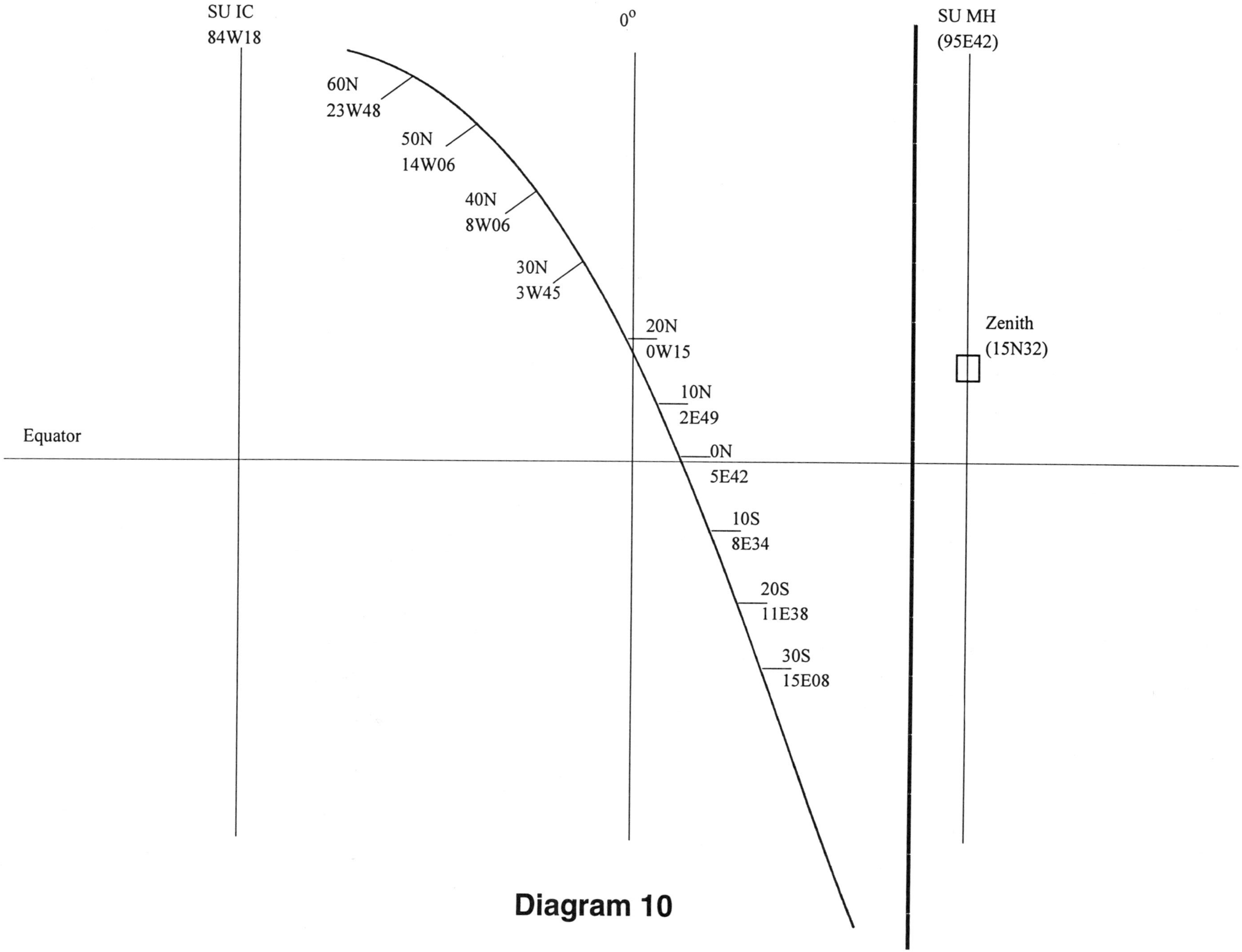
SU IC
84W18
0°
SU MH
(95E42)
60N
23W48
50N
14W06
40N
8W06
30N
3W45
20N
0W15
Zenith
(15N32)
10N
2E49
Equator
0N
5E42
10S
8E34
20S
11E38
30S
15E08

Diagram 10

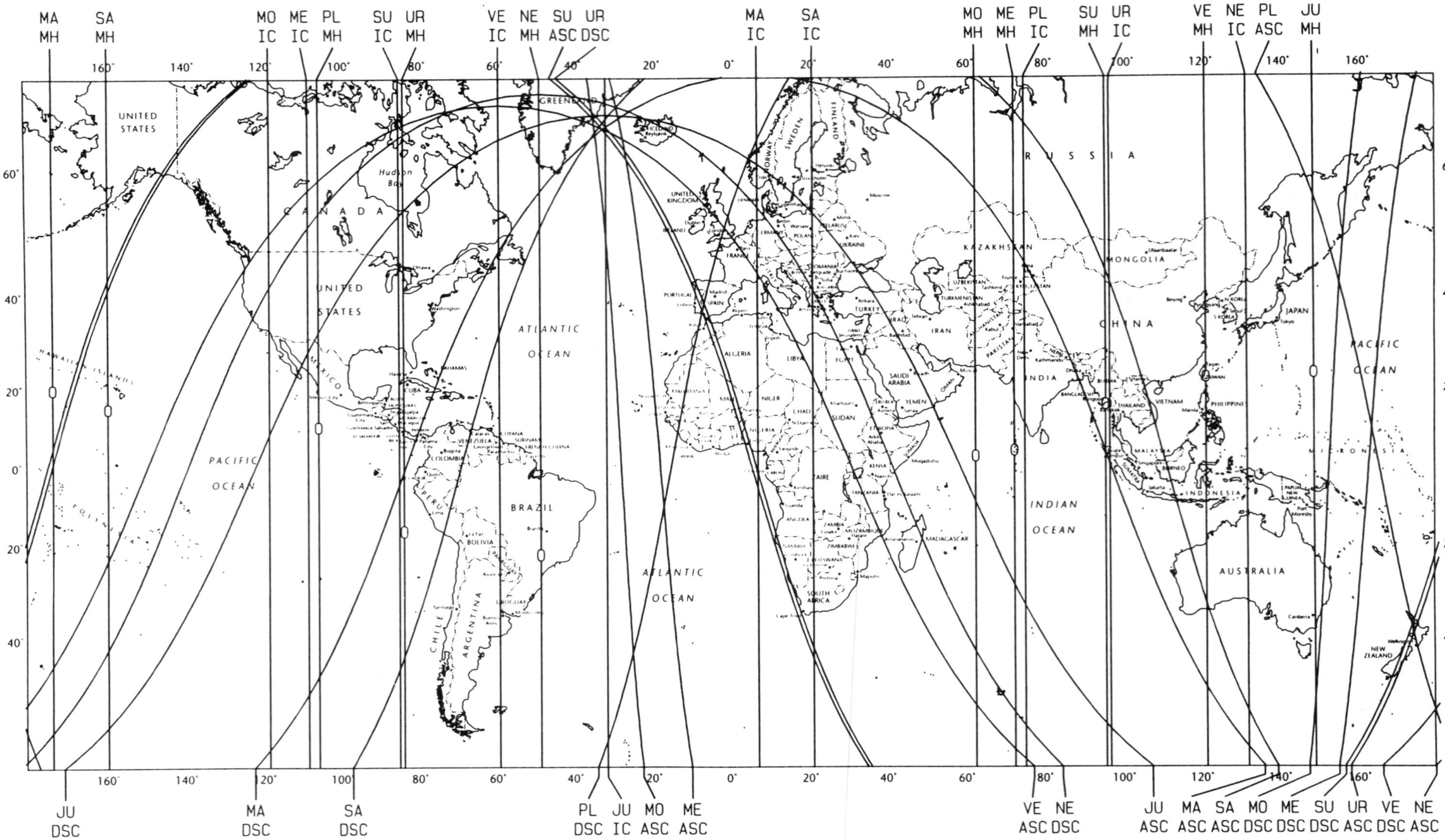

Simon

May 4, 1978 6:34 AM GDT

04/16/96

Simon
May 4, 1978 6:34 AM GDT

ASTRO*CARTO*GRAPHY LATITUDE CROSSINGS

NORTH LATITUDES

69N39 SA/IC * MA/ASC
68N24 SU/ASC * NE/DSC
68N18 UR/DSC * NE/DSC
68N17 VE/IC * NE/ASC
68N05 MO/ASC * NE/DSC
67N43 JU/IC * NE/DSC
67N23 ME/ASC * NE/DSC
67N23 SA/DSC * NE/DSC
67N23 ME/ASC * SA/DSC
67N11 NE/MH * VE/ASC
67N07 VE/DSC * NE/ASC
66N56 MO/ASC * UR/DSC
66N55 MA/DSC * NE/DSC
66N51 ME/MH * SA/ASC
66N34 JU/ASC * SA/DSC
66N33 JU/DSC * UR/DSC
66N32 ME/ASC * JU/ASC
66N32 MO/ASC * JU/DSC
66N31 SU/ASC * JU/DSC
66N31 MA/DSC * JU/ASC
66N21 VE/ASC * JU/DSC
66N16 ME/ASC * MA/DSC
66N15 MO/ASC * SU/ASC
66N15 JU/ASC * NE/DSC
66N08 JU/IC * SA/DSC
65N46 SA/DSC * UR/DSC
65N37 NE/MH * JU/DSC
65N36 PL/IC * SA/ASC
65N34 MO/ASC * VE/ASC
65N34 UR/MH * VE/DSC
65N25 SU/ASC * SA/DSC
65N24 JU/IC * UR/DSC
65N22 SU/IC * VE/DSC
65N21 MA/DSC * UR/DSC
65N20 JU/IC * MA/DSC
65N19 VE/ASC * SA/DSC
65N12 SU/ASC * VE/ASC
65N08 VE/ASC * MA/DSC
65N06 SU/ASC * MA/DSC
65N01 JU/IC * VE/ASC
64N44 VE/ASC * UR/DSC
64N43 JU/IC * SU/ASC
64N42 MO/ASC * SA/DSC
64N29 UR/MH * NE/ASC
64N28 MO/ASC * MA/DSC
64N10 SU/IC * NE/ASC
64N03 ME/ASC * VE/ASC
64N03 MA/DSC * SA/DSC
63N59 VE/IC * JU/DSC
63N52 MA/IC * PL/DSC
62N44 ME/ASC * UR/DSC
61N42 JU/ASC * PL/DSC
61N36 SU/ASC * ME/ASC
61N13 Anchorage
61N02 MA/IC * JU/ASC

59N20 Stockholm
59N05 PL/MH * VE/DSC
58N14 NE/DSC * PL/DSC
57N49 ME/IC * VE/DSC
57N43 MO/MH * MA/ASC
57N34 NE/MH * MA/DSC
56N34 JU/MH * PL/ASC
55N45 Moscow
55N41 Copenhagen
55N19 MA/IC * NE/DSC
54N57 UR/MH * JU/DSC
54N30 PL/MH * NE/ASC
54N29 SA/IC * JU/ASC
54N14 SU/IC * JU/DSC
52N30 ME/IC * NE/ASC
52N30 Berlin
52N21 VE/ASC * PL/DSC
52N18 NE/MH * SA/DSC

CENTRAL LATITUDES

51N30 London
51N22 MO/IC * VE/DSC
50N48 JU/IC * MO/ASC

49N22 ME/MH * MA/ASC
49N14 Vancouver
48N50 Paris
48N39 VE/IC * MA/DSC
47N36 Seattle
46N23 PL/IC * MA/ASC
45N32 Montreal
45N02 SU/MH * SA/ASC
44N59 Minneapolis
44N39 MA/IC * VE/ASC
43N39 Toronto
43N26 SA/MH * SU/DSC
42N54 UR/IC * SA/ASC
42N22 MO/DSC * PL/ASC
42N22 Boston
42N20 Detroit
42N12 MO/IC * NE/ASC
41N54 Rome
41N53 Chicago
41N35 SA/IC * NE/DSC
41N28 SA/MH * UR/ASC
40N43 New York

39N44 Denver
39N00 UR/DSC * PL/DSC
38N53 Washington
38N37 St. Louis
37N47 San Francisco
37N26 SU/ASC * PL/DSC
36N10 Las Vegas
35N40 Tokyo
34N04 Los Angeles
33N45 Atlanta
33N27 Phoenix
32N54 PL/MH * JU/DSC
32N49 VE/IC * SA/DSC
32N47 Dallas
30N03 Cairo

29N58 New Orleans
29N46 Houston
28N39 Delhi
28N36 ME/IC * JU/DSC
25N47 Miami
22N43 SA/IC * VE/ASC
22N34 Calcutta
22N16 Hong Kong
21N19 Honolulu

19N26 Mexico City
14N35 Manila
10N30 Caracas

9N48 ME/DSC * PL/ASC
8N36 MO/IC * JU/DSC
6N27 Lagos
5N09 MA/MH * SU/DSC
4N04 SU/MH * MA/ASC
1N08 MA/MH * UR/ASC
0N53 UR/IC * MA/ASC

0S53 UR/MH * MA/DSC
1S08 MA/IC * UR/DSC
4S04 SU/IC * MA/DSC
5S09 MA/IC * SU/ASC
8S36 MO/MH * JU/ASC
9S48 ME/ASC * PL/DSC

22S43 SA/MH * VE/DSC
22S54 Rio De Janeiro
26S12 Johannesburg

SOUTH LATITUDES

28S36 ME/MH * JU/ASC

32S49 VE/MH * SA/ASC
32S54 PL/IC * JU/ASC
33S55 Sydney
34S36 Buenos Aires
36S52 Auckland
37S26 SU/DSC * PL/ASC
37S50 Melbourne
39S00 UR/ASC * PL/ASC

41S28 SA/IC * UR/DSC
41S35 SA/MH * NE/ASC
42S12 MO/MH * NE/DSC
42S22 MO/ASC * PL/DSC
42S54 UR/MH * SA/DSC
43S26 SA/IC * SU/ASC
44S39 MA/MH * VE/DSC
45S02 SU/IC * SA/DSC
46S23 PL/MH * MA/DSC
48S39 VE/MH * MA/ASC
49S22 ME/IC * MA/DSC

50S48 JU/MH * MO/DSC
51S22 MO/MH * VE/ASC
52S18 NE/IC * SA/ASC
52S21 VE/DSC * PL/ASC
52S30 ME/MH * NE/DSC
54S14 SU/MH * JU/ASC
54S29 SA/MH * JU/DSC
54S30 PL/IC * NE/DSC
54S57 UR/IC * JU/ASC
55S19 MA/MH * NE/ASC
56S34 JU/IC * PL/DSC
57S34 NE/IC * MA/ASC
57S43 MO/IC * MA/DSC
57S49 ME/MH * VE/ASC
58S14 NE/ASC * PL/ASC
59S05 PL/IC * VE/ASC

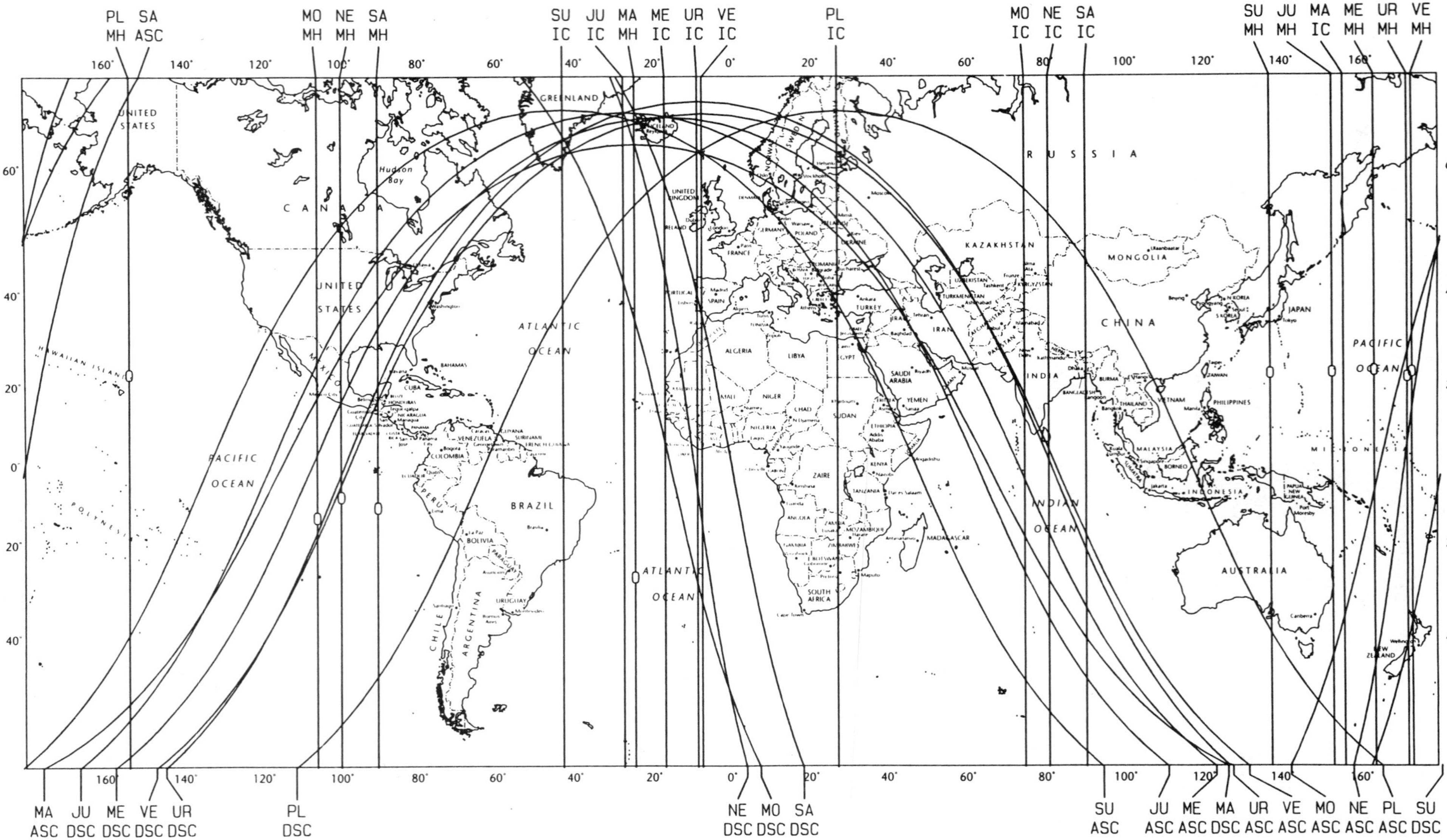

Wendy

June 11, 1954 3:50 AM CET

04/16/96

04/16/96

Wendy
June 11, 1954 3:50 AM CET

ASTRO*CARTO*GRAPHY LATITUDE CROSSINGS

NORTH LATITUDES

68N54 JU/IC * SA/DSC
68N25 PL/MH * SA/ASC
67N46 VE/IC * UR/ASC
67N34 ME/IC * UR/DSC
67N00 JU/IC * NE/DSC
66N59 MA/MH * UR/DSC
66N58 SA/DSC * UR/DSC
66N54 MO/DSC * SU/DSC
66N53 MA/MH * SA/DSC
66N41 UR/DSC * NE/DSC
66N41 JU/ASC * UR/DSC
66N41 JU/ASC * NE/DSC
66N39 JU/IC * UR/DSC
66N39 MA/MH * JU/ASC
66N39 JU/ASC * SA/DSC
66N36 SU/ASC * JU/DSC
66N34 UR/IC * VE/DSC
66N23 SU/ASC * UR/DSC
66N19 ME/IC * JU/ASC
66N18 VE/DSC * JU/ASC
66N17 ME/IC * VE/DSC
66N14 UR/ASC * PL/DSC
66N12 ME/DSC * VE/DSC
66N12 ME/ASC * JU/ASC
66N11 JU/IC * SU/ASC
66N06 SU/ASC * NE/DSC
66N04 ME/DSC * SA/DSC
66N01 MA/MH * ME/DSC
65N59 UR/IC * ME/ASC
65N58 SU/ASC * ME/DSC
65N55 VE/IC * ME/ASC
65N54 ME/DSC * NE/DSC
65N53 SU/IC * JU/DSC
65N51 MA/MH * SU/ASC
65N51 JU/IC * ME/DSC
65N45 VE/DSC * SA/DSC
65N44 VE/ASC * PL/DSC
65N43 SU/ASC * VE/DSC
65N40 MO/DSC * JU/DSC
65N40 SU/ASC * SA/DSC
65N36 MA/MH * VE/DSC
65N31 UR/IC * JU/ASC
65N25 VE/DSC * NE/DSC
65N21 VE/IC * JU/ASC
65N14 JU/IC * VE/DSC
64N54 ME/ASC * PL/DSC
64N39 ME/IC * SU/ASC
64N17 ME/DSC * UR/DSC
64N16 SU/IC * MO/DSC
64N14 JU/ASC * PL/DSC
63N56 MO/DSC * ME/DSC
63N51 SU/IC * ME/DSC
63N50 MO/DSC * UR/DSC
63N41 SU/IC * UR/DSC
63N22 MA/MH * NE/DSC
63N20 PL/IC * UR/ASC
63N18 MA/DSC * NE/DSC
63N16 JU/IC * MA/ASC
63N15 MA/DSC * SA/DSC
63N05 ME/IC * MA/DSC
62N45 UR/IC * SU/ASC
62N36 VE/IC * PL/DSC
62N34 MO/DSC * VE/DSC
62N30 SU/ASC * PL/DSC
62N24 VE/IC * SU/ASC
62N23 UR/IC * MA/DSC
62N22 MO/DSC * MA/ASC
62N21 MA/DSC * PL/DSC
62N17 PL/IC * VE/ASC
62N15 UR/IC * PL/DSC
62N14 VE/IC * MA/DSC
62N10 VE/DSC * MA/ASC
62N07 SU/ASC * MA/DSC
62N04 SU/IC * MA/ASC

CENTRAL LATITUDES

61N58 SU/IC * VE/DSC
61N13 Anchorage
60N13 MA/ASC * UR/DSC

59N20 Stockholm
59N20 ME/IC * PL/DSC
59N10 SA/DSC * PL/DSC
58N37 ME/IC * SA/DSC
58N34 PL/IC * ME/ASC
58N06 SA/MH * SU/DSC
57N37 MO/IC * PL/ASC
57N19 NE/DSC * PL/DSC
56N31 ME/DSC * MA/ASC
55N45 Moscow
55N41 Copenhagen
55N29 MA/MH * PL/DSC
54N11 NE/IC * PL/ASC
53N42 JU/IC * PL/DSC
53N29 PL/IC * JU/ASC
52N30 Berlin
52N27 NE/MH * SU/DSC
51N30 London
51N02 PL/IC * MA/DSC
50N51 MO/DSC * PL/DSC

49N14 Vancouver
48N50 Paris
48N48 MO/ASC * NE/ASC
47N36 Seattle
47N21 MO/MH * SU/DSC
47N01 SA/MH * JU/DSC
46N32 SA/IC * PL/ASC
45N32 Montreal
44N59 Minneapolis
44N48 ME/IC * NE/DSC
43N39 Toronto
42N58 JU/IC * MO/DSC
42N22 Boston
42N20 Detroit
41N54 Rome
41N53 Chicago
40N43 New York
40N29 UR/IC * SA/DSC

39N44 Denver
39N44 SA/MH * MA/ASC
38N54 PL/IC * SU/ASC
38N53 Washington
38N44 SU/IC * PL/DSC
38N37 St. Louis
38N05 VE/ASC * UR/ASC
37N47 San Francisco
36N28 VE/IC * SA/DSC
36N10 Las Vegas
35N42 NE/MH * JU/DSC
35N40 Tokyo
35N26 MA/MH * MO/DSC
34N04 Los Angeles
33N45 Atlanta
33N34 SA/MH * ME/DSC
33N27 Phoenix
33N21 MA/DSC * JU/ASC
32N47 Dallas
32N05 VE/MH * MO/ASC
30N03 Cairo

29N58 New Orleans
29N46 Houston
28N39 Delhi
28N08 UR/MH * MO/ASC
27N32 NE/MH * MA/ASC
25N47 Miami
25N40 MO/MH * JU/DSC
22N34 Calcutta
22N16 Hong Kong

SOUTH LATITUDES

21N19 Honolulu
20N27 SA/MH * UR/DSC

19N26 Mexico City
17N50 MO/IC * VE/ASC
17N29 MO/MH * MA/ASC
16N55 SA/MH * VE/DSC
16N30 NE/MH * ME/DSC
16N14 MO/IC * UR/ASC
14N35 Manila
14N23 VE/MH * NE/ASC
10N30 Caracas

6N27 Lagos
6N21 ME/IC * MO/DSC
5N40 UR/MH * NE/ASC
4N27 NE/IC * VE/ASC
3N13 MO/MH * ME/DSC
1N50 NE/IC * UR/ASC

1S50 NE/MH * UR/DSC
3S13 MO/IC * ME/ASC
4S27 NE/MH * VE/DSC
5S40 UR/IC * NE/DSC
6S21 ME/MH * MO/ASC

14S23 VE/IC * NE/DSC
16S14 MO/MH * UR/DSC
16S30 NE/IC * ME/ASC
16S55 SA/IC * VE/ASC
17S29 MO/IC * MA/DSC
17S50 MO/MH * VE/DSC

20S27 SA/IC * UR/ASC
22S54 Rio De Janeiro
25S40 MO/IC * JU/ASC
26S12 Johannesburg
27S32 NE/IC * MA/DSC
28S08 UR/IC * MO/DSC

32S05 VE/IC * MO/DSC
33S21 MA/ASC * JU/DSC
33S34 SA/IC * ME/ASC
33S55 Sydney
34S36 Buenos Aires
35S26 MA/IC * MO/ASC
35S42 NE/IC * JU/ASC
36S28 VE/MH * SA/ASC
36S52 Auckland
37S50 Melbourne
38S05 VE/DSC * UR/DSC
38S44 SU/MH * PL/ASC
38S54 PL/MH * SU/DSC
39S44 SA/IC * MA/DSC

40S29 UR/MH * SA/ASC
42S58 JU/MH * MO/ASC
44S48 ME/MH * NE/ASC
46S32 SA/MH * PL/DSC
47S01 SA/IC * JU/ASC
47S21 MO/IC * SU/ASC
48S48 MO/DSC * NE/DSC

50S51 MO/ASC * PL/ASC

04/16/96

Wendy
June 11, 1954 3:50 AM CET

ASTRO*CARTO*GRAPHY LATITUDE CROSSINGS

NORTH LATITUDES

CENTRAL LATITUDES

SOUTH LATITUDES

51S02 PL/MH * MA/ASC
52S27 NE/IC * SU/ASC
53S29 PL/MH * JU/DSC
53S42 JU/MH * PL/ASC
54S11 NE/MH * PL/DSC
55S29 MA/IC * PL/ASC
56S31 ME/ASC * MA/DSC
57S19 NE/ASC * PL/ASC
57S37 MO/MH * PL/DSC
58S06 SA/IC * SU/ASC
58S34 PL/MH * ME/DSC
58S37 ME/MH * SA/ASC
59S10 SA/ASC * PL/ASC
59S20 ME/MH * PL/ASC

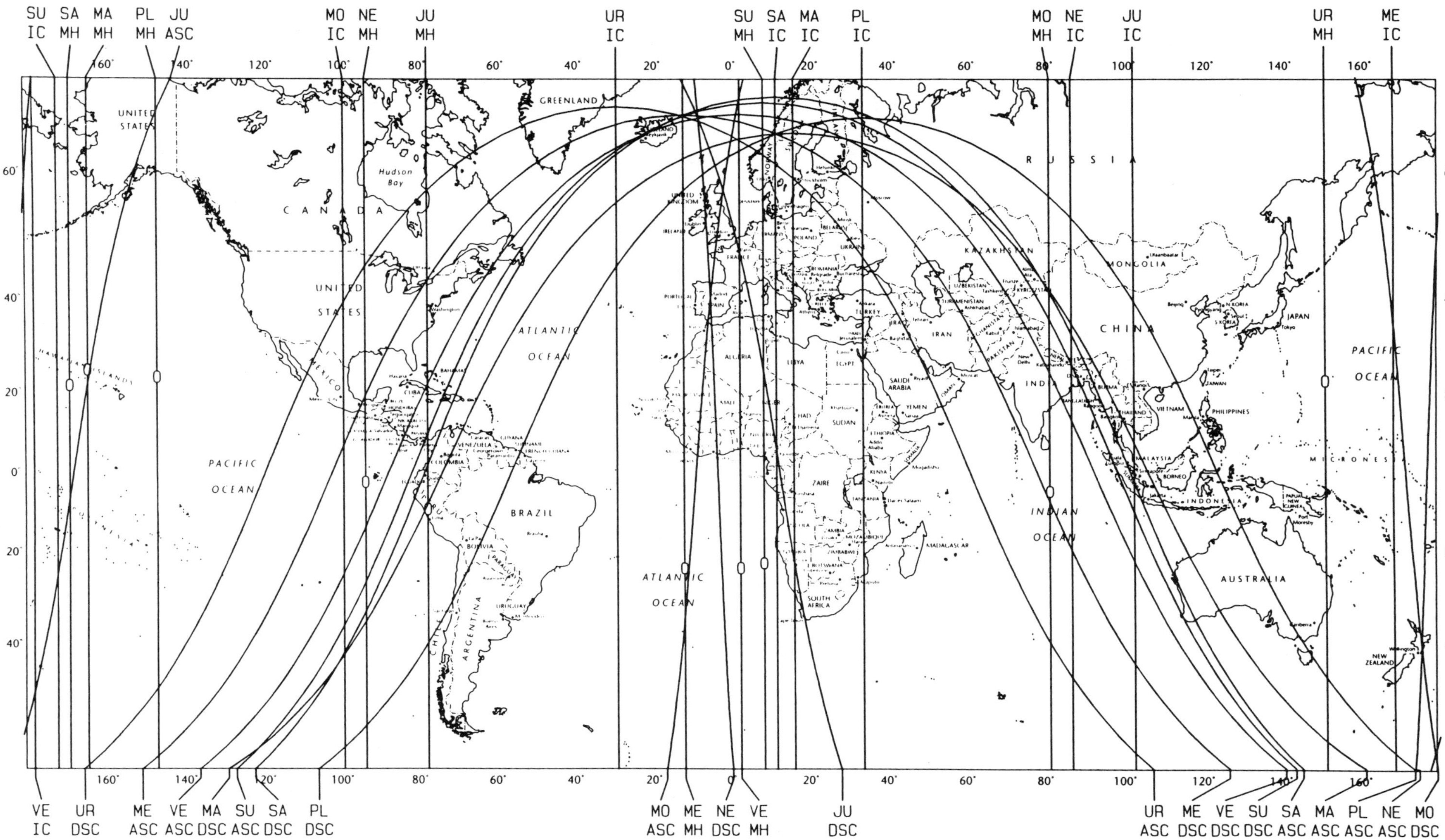

Nigel

January 9, 1946 11:32 AM GMT

04/16/96

04/16/96

Nigel
January 9, 1946 11:32 AM GMT

ASTRO*CARTO*GRAPHY LATITUDE CROSSINGS

NORTH LATITUDES

69N39 ME/MH * JU/DSC
68N21 SU/MH * SA/DSC
68N20 VE/MH * MO/ASC
68N20 MA/IC * SA/ASC
68N08 VE/MH * SA/DSC
68N08 MO/ASC * SA/DSC
67N49 SA/IC * SU/DSC
67N45 VE/MH * SU/ASC
67N44 MO/ASC * SU/ASC
67N40 MA/IC * SU/DSC
67N10 SA/DSC * NE/DSC
67N05 JU/DSC * SA/DSC
66N59 SU/ASC * NE/DSC
66N58 PL/MH * JU/ASC
66N56 SU/ASC * JU/DSC
66N51 PL/IC * SA/ASC
66N43 MO/ASC * VE/ASC
66N40 ME/DSC * SA/DSC
66N40 SU/ASC * ME/DSC
66N40 ME/ASC * UR/ASC
66N38 ME/MH * SA/DSC
66N38 ME/DSC * JU/DSC
66N37 ME/DSC * NE/DSC
66N36 SU/MH * VE/DSC
66N35 SA/DSC * UR/ASC
66N33 ME/MH * UR/ASC
66N32 SU/ASC * UR/ASC
66N32 ME/MH * SU/ASC
66N32 ME/DSC * VE/ASC
66N27 SA/IC * VE/DSC
66N22 VE/ASC * NE/DSC
66N22 VE/ASC * JU/DSC
66N20 JU/DSC * NE/DSC
66N20 SA/ASC * PL/ASC
66N19 SU/DSC * PL/ASC
66N15 VE/ASC * UR/ASC
66N07 MA/IC * VE/DSC
66N06 MO/ASC * ME/DSC
66N02 ME/MH * VE/ASC
66N02 UR/ASC * NE/DSC
66N00 VE/MH * ME/DSC
65N59 JU/DSC * UR/ASC
65N56 SU/ASC * SA/DSC
65N45 PL/IC * SU/DSC
65N44 VE/DSC * PL/DSC
65N43 UR/IC * ME/ASC
65N20 MA/IC * PL/DSC
65N18 SU/MH * ME/DSC
64N58 VE/ASC * SA/DSC
64N48 SA/IC * ME/DSC
64N45 ME/DSC * PL/DSC
64N45 ME/DSC * MA/DSC
64N45 MA/DSC * PL/DSC
64N44 SA/IC * MA/DSC
64N43 SA/IC * PL/DSC
64N36 SU/MH * MA/DSC
64N32 SU/ASC * VE/ASC
64N30 MO/ASC * UR/ASC
64N10 VE/MH * MA/DSC
64N10 VE/DSC * MA/ASC
64N10 SU/MH * PL/DSC
64N09 MA/DSC * UR/ASC
64N04 VE/MH * UR/ASC
63N57 MA/IC * ME/DSC
63N57 MO/ASC * MA/DSC
63N44 PL/IC * MA/ASC
63N27 UR/ASC * PL/DSC
63N17 PL/IC * VE/DSC
63N14 UR/IC * VE/ASC
63N13 MA/DSC * JU/DSC
62N54 VE/MH * PL/DSC
62N53 UR/IC * SU/ASC
62N52 MA/DSC * NE/DSC
62N27 UR/IC * SA/DSC

CENTRAL LATITUDES

62N26 SU/MH * UR/ASC
62N09 MO/ASC * PL/DSC
61N57 ME/MH * MA/DSC
61N21 SA/IC * UR/ASC
61N13 Anchorage
61N09 JU/DSC * PL/DSC
60N03 NE/DSC * PL/DSC

59N34 MA/IC * UR/ASC
59N20 Stockholm
58N19 PL/IC * ME/DSC
57N58 JU/MH * UR/DSC
57N57 MO/ASC * JU/DSC
57N56 ME/MH * PL/DSC
57N11 SU/DSC * MA/ASC
57N06 MO/MH * PL/ASC
56N20 UR/IC * MA/DSC
55N45 Moscow
55N41 Copenhagen
53N59 NE/IC * PL/ASC
52N30 Berlin
51N30 London

49N14 Vancouver
48N50 Paris
48N38 VE/MH * JU/DSC
47N51 PL/IC * UR/ASC
47N36 Seattle
46N20 UR/IC * PL/DSC
46N16 NE/MH * UR/DSC
45N32 Montreal
44N59 Minneapolis
43N52 JU/MH * ME/ASC
43N45 MA/ASC * SA/ASC
43N39 Toronto
42N22 Boston
42N20 Detroit
41N54 Rome
41N53 Chicago
41N51 MO/MH * MA/ASC
41N24 MO/MH * SA/ASC
40N43 New York
40N21 MO/ASC * NE/DSC

39N44 Denver
39N42 MO/IC * UR/DSC
39N33 JU/IC * PL/ASC
38N53 Washington
38N37 St. Louis
37N47 San Francisco
36N10 Las Vegas
36N02 MO/MH * SU/DSC
35N40 Tokyo
35N11 NE/IC * MA/ASC
34N51 ME/IC * MO/DSC
34N04 Los Angeles
33N45 Atlanta
33N27 Phoenix
32N58 NE/IC * SA/ASC
32N47 Dallas
30N03 Cairo

29N58 New Orleans
29N46 Houston
28N39 Delhi
26N25 NE/IC * SU/DSC
25N47 Miami
25N14 SU/MH * JU/DSC
24N31 MO/MH * VE/DSC
22N34 Calcutta
22N16 Hong Kong
22N05 JU/MH * VE/ASC
21N54 MA/MH * JU/ASC
21N19 Honolulu

SOUTH LATITUDES

19N26 Mexico City
19N15 NE/MH * ME/ASC
14N35 Manila
13N09 NE/IC * VE/DSC
10N30 Caracas
10N05 JU/MH * SU/ASC

7N29 JU/IC * MA/ASC
7N11 MO/IC * ME/ASC
6N27 Lagos
6N03 SA/IC * JU/DSC
2N22 JU/MH * SA/DSC

2S22 JU/IC * SA/ASC
6S03 SA/MH * JU/ASC
7S11 MO/MH * ME/DSC
7S29 JU/MH * MA/DSC

10S05 JU/IC * SU/DSC
13S09 NE/MH * VE/ASC
19S15 NE/IC * ME/DSC

21S54 MA/IC * JU/DSC
22S05 JU/IC * VE/DSC
22S54 Rio De Janeiro
24S31 MO/IC * VE/ASC
25S14 SU/IC * JU/ASC
26S12 Johannesburg
26S25 NE/MH * SU/ASC

32S58 NE/MH * SA/DSC
33S55 Sydney
34S36 Buenos Aires
34S51 ME/MH * MO/ASC
35S11 NE/MH * MA/DSC
36S02 MO/IC * SU/ASC
36S52 Auckland
37S50 Melbourne
39S33 JU/MH * PL/DSC
39S42 MO/MH * UR/ASC

40S21 MO/DSC * NE/ASC
41S24 MO/IC * SA/DSC
41S51 MO/IC * MA/DSC
43S45 MA/DSC * SA/DSC
43S52 JU/IC * ME/DSC
46S16 NE/IC * UR/ASC
46S20 UR/MH * PL/ASC
47S51 PL/MH * UR/DSC
48S38 VE/IC * JU/ASC

53S59 NE/MH * PL/DSC
56S20 UR/MH * MA/ASC
57S06 MO/IC * PL/DSC
57S11 SU/ASC * MA/DSC
57S56 ME/IC * PL/ASC
57S57 MO/DSC * JU/ASC
57S58 JU/IC * UR/ASC
58S19 PL/MH * ME/ASC
59S34 MA/MH * UR/DSC

Chapter 3/Interpretation of the Astro*Carto*Graphy Map

When you first see an Astro*Carto*Graphy map it looks complicated. There are lines all over the map, and this gives the impression that Astro*Carto*Graphy is difficult to understand.

As explained in Chapter 2, an Astro*Carto*Graphy map shows where the planets were angular around the world when a person was born. It's like transferring a horoscope onto a map of the world.

An Astro*Carto*Graphy map has forty lines. Sometimes a planetary line will go from one edge of the map to the other at the 180 degree longitude; for example, Uri Geller has SA DSC line passing through Australia and a small part of Russia. Then SA DSC leaves the right hand side edge of the map and continues on the other side passing through a small part of Alaska.

The forty lines on an Astro*Carto*Graphy map are:

Planets	**Angles**
SU (Sun)	MH (Midheaven)
MO (Moon)	IC (Imum Coeli)
ME (Mercury)	ASC (Ascendant)
VE (Venus)	DSC (Descendant)
MA (Mars)	
JU (Jupiter)	
SA (Saturn)	
UR (Uranus)	
NE (Neptune)	
PL (Pluto)	

10 Planets x 4 Angles = 40 lines

It is also possible to purchase Astro*Carto*Graphy maps that include the Moon's Nodes and Chiron lines. Many astrologers agree that the Moon's Nodes relate to connections with other people. Other astrologers suggest that the Nodes are linked to reincarnation. There is also the problem of deciding between the Mean Node and True Node. The difference between the Mean Node and the True Node can as much as one degree, forty-five minutes.

Chiron is used by many astrologers in natal horoscopes, and many books have also been written about how to interpret it. As Chiron became more popular, astrologers started to use it in relocation astrology, and this raised the question of Chiron lines and their effect in Astro*Carto*Graphy maps.

I asked Jim Lewis why he didn't include Chiron lines on Astro*Carto*Graphy maps, and he told me that further research needed to be done. He also said that if he included Chiron, then other astrologers would request asteroid lines (Ceres, Pallas, Juno and Vesta). Or the Uranian astrologers would ask for Cupido, Hades, Zeus, Kronus, Apollon, Admetos, Vulkanus and Poseidon. If he included extra lines, Astro*Carto*Graphy maps would soon become unreadable.

In *Essence and Application—A View from Chiron*, Zane Stein wrote about Chiron and relocation. He suggested that Chiron lines seem to pass through countries that a person always wanted to visit. He also found that life changing events take place in countries where a Chiron line is located.

As far as I know, there is only one book available in English that gives an interpretation for the Moon's Nodes and Chiron on an Astro*Carto*Graphy map: *What are Astrolocality Maps?* by Maritha Pottenger.

I have used Astro*Carto*Graphy for many years and always obtained accurate results with the ten planets. Before extra lines are included on an Astro*Carto*Graphy map more research needs to be done to find out what they mean. This will only be achieved by examining numerous maps and analysing what people experience on or near the lines of the Moon's Nodes and Chiron.

The Astro*Carto*Graphy techniques described in this book use the standard forty planetary lines. Astrologers new to Astro*Carto*Graphy will find this is enough to obtain accurate information. Extra lines should only be investigated after you are familiar with Astro*Carto*Graphy.

Natal Chart

Before beginning to interpret an Astro*Carto*Graphy map it is essential to examine the natal chart. When you move to a new location the planetary aspects in a natal chart do not change, so no matter where in the world you live, the natal chart will have an effect at that location.

What does change at a location are the positions of the Ascendant, Descendant, Midheaven and Imum Coeli. This means that different planets will be conjunct the angles, depending on how far away a person relocates.

Planets conjunct the angles in a natal chart have the strongest influence, so they will have more importance for the person concerned. If this person moves to a new location far away from the place of birth, then different planets will have a stronger influence.

If a person has a T-square involving Sun, Mars and Saturn in the natal chart, this aspect pattern will not be easy to deal with. It would be emphasised if one or more of the T-square planets were angular at the location where a person lives. If this person moved to where Jupiter was closely conjunct the Midheaven, then this planet would have a strong influence. The T-square (SU, MA and SA) would still have an influence at the location with JU MH, but the planetary energies would not be so dominant. It would be more of a minor influence, noticed occasionally.

Sometimes it can take up to a year for a person to readjust to the planetary energies at a new location. The person is used to the influences at the old location and needs to readjust to new experiences.

Other people experience the planetary influences at a location within days of arriving. This happens frequently when the location is a holiday destination. We all know people who return to the same holiday resort year after year. They obviously enjoyed the place so much on their first visit that they returned there every year for a holiday.

The planetary lines at a birthplace show the character and personality of a person. After moving to another location the birthplace planetary lines can still have an effect, although not so powerful as the new planetary energies.

Orbs for Angular Planets

On an Astro*Carto*Graphy map the planetary lines are used to interpret the influences at a location.

How close do you have to be for a planetary line to have an effect? Experience has shown that the planetary lines have an influence up to about 700 miles/1000 kilometres from a location. The closer the line is to a location the more powerful it will be. It's the same as orbs in aspects. If Mercury and Uranus are in an exact square in a natal chart, then that aspect will be a powerful influence for the person. When the orb is eight degrees, the aspect is much weaker and will probably not have much of an influence.

When a planetary line passes **exactly** through a location the effects will be very powerful. Locations where one or more lines are exact will usually have an immediate influence if the person concerned goes there.

The planetary lines can also have an effect at a location even when a person doesn't visit the place. This is known as remote activation. It is as if the planetary lines at a location come to a person no matter how far away he or she is.

A typical example of remote activation would be when a woman living in London meets an Australian man. They feel very attracted to each other and quickly fall in love. Then this woman sees on her Astro*Carto*Graphy map that she has VE DSC exact at Sydney, the place where the Australian man was born. She has never been to Australia, yet the VE DSC line has come to her by meeting this man.

What is the situation where there are no planetary lines at a location? This means that no planet is dominant at the location, so the natal chart needs to be examined in more detail. Sometimes, a location with no planetary lines can be a good place, especially if a person has a natal grand cross or T-square. Then, moving to a place with no lines is like getting away from it all.

I have an American client who has **seven** planetary lines within orb of London. She told me that her two visits to London were intense experiences, and feels a strong pull toward this place.

Usually, locations where many planetary lines cross each other within a narrow orb are near the North Pole. These locations are mostly in parts of the world that people never visit, either because the places are in remote areas or inhospitable climates.

It is possible that all the planetary lines can cross each other in holiday locations. I have seen Astro*Carto*Graphy maps where nine of the **planets** cross in northern Scandinavia or Iceland.

Karl Marx has **nine** planets passing through Finland. Even if he never visited Finland, these planetary lines would still have an effect. Experience has shown that locations where many planetary lines cross each other are very intense. With nine planets in Finland, Karl Marx's books would have a powerful impact there.

John Lennon has nine planets passing within orb of Iceland. It would be interesting to know how the Icelandic people regard his songs.

Orbs for Latitude Crossings

In Chapter 2 I mentioned that **paran** aspects are used in interpreting Astro*Carto*Graphy maps. Paran is from the Greek word *parantellonta*, which means "rising side by side" (page 155, *Planets in Locality* by Steve Cozzi).

In a natal chart, a planet that is conjunct an angle will have a very powerful influence. As explained in the previous chapter, it is important to use the **celestial latitude** and **declination** of a planet in a natal or relocation chart. Otherwise, the wrong information could be obtained as to which planets are actually conjunct the angles.

The advantage of using paran aspects is that celestial latitude and declination of planets are taken into consideration. This means that the two planets do not need to be on the angles of a chart for them to be in paran aspect. An important fact about paran aspects is that they depend on the **geographical latitude** of a location.

It is easy to understand why paran aspects have been ignored by astrologers because a great many calculations are involved. The NOVA computer program by Robert Hand calculates paran aspects (option 13—In Mundo Charts).

On an Astro*Carto*Graphy map, paran aspects are very easy to find because **no** calculations are involved. All the information for parans is on the world map, and all that is required is a ruler.

What is a paran aspect? It is when two planets are **simultaneously** conjunct the angles and have a combined influence.

In *Essays on Astrology*, Robert Hand writes that the seven types of paran aspects are:

	Paran Aspect	**Angle Planet A**	**Angle Planet B**
1.	Oblique Ascension conjunct Oblique Ascension	ASC	ASC
2.	Oblique Ascension opposition Oblique Descension	ASC DSC	DSC ASC
3.	Oblique Ascension square Right Ascension	ASC ASC MH IC	MH IC ASC ASC

4.	Oblique Descension conjunct Oblique Descension	DSC	DSC
5.	Oblique Descension square Right Ascension	DSC DSC MH IC	MH IC DSC DSC
6.	Right Ascension conjunct Right Ascension	MH IC	MH IC
7.	Right Ascension opposition Right Ascension	MH IC	IC MH

Paran aspect types 6 and 7 are not used in Astro*Carto*Graphy as **both** MH and IC lines are vertical on the world map.

An example of a paran aspect is shown in Simon's natal chart. Venus is conjunct the Ascendant at the same time that Pluto is conjunct the Descendant. On his Astro*Carto*Graphy map VE ASC and PL DSC lines cross exactly at Leicester. The paran printout for Simon shows that Venus rises at 303 degrees 42 minutes and Pluto sets at 304 degrees 17 minutes. An orb of one degree is used for paran aspects.

A paran aspect is located where two planetary lines cross each other on an Astro*Carto*Graphy map. In Astro*Carto*Graphy these aspects are called **latitude crossings**.

When we examine Simon's paran printout there are **four** paran aspects within an orb of one degree:

1.	VE ASC	303 42
	PL DSC	304 17
2.	NE MH	256 54
	SA DSC	257 11
3.	ME IC	197 37
	NE ASC	197 49
4.	MO ASC	273 44
	JU IC	273 57

The first paran aspect (VE/PL) is easy to see on Simon's Astro*Carto*Graphy map because both lines are exact at Leicester. The other three paran aspects make no sense at all when you examine his natal chart.

Paran aspects are also formed by the **rotation** of the earth during the course of a day. At the latitude for Leicester (52N39), these three paran aspects will be formed as the earth rotates.

The paran printout for Simon is calculated for geographical latitude 52N39. It shows where any two planets would **cross the angles together at this latitude**.

This will make more sense when you examine Simon's Astro*Carto*Graphy map. Look at his NE MH line and find where SA DSC crosses it. These two lines cross in the Atlantic Ocean, yet they have an effect at Leicester. As the earth rotates, this latitude crossing will pass through Leicester.

If you have a world globe, put a pencil at the point where NE MH and SA DSC cross, hold the pencil still and move the globe around slowly. As you do so you will see that Leicester passes under the pencil point. If you put a ruler at fifty-two and a half degrees latitude on Simon's Astro*Carto*Graphy map you will see which latitude crossings he has for Leicester.

The orb used for latitude crossings on an Astro*Carto*Graphy map is about seventy miles (100 km). This is the distance **north** or **south** of the geographical latitude where two planetary lines cross each other.

The influence of a latitude crossing is about a third of the power of a planetary line at a location. However, I have found that when a latitude crossing is **exact** its influence will be more important. An exact latitude crossing will have an influence about half that of a planetary line.

Computer Listing of Latitude Crossings

In the past, a ruler was needed to find the latitude crossings for a location on a full-sized Astro*Carto*Graphy map; otherwise it was not easy to see if the planets crossed at a certain latitude.

If you wanted to examine many locations, quite a lot of work was involved. It was not possible to see at a glance which latitudes had beneficial influences.

Jim Lewis decided that there must be an easier way to find the latitude crossings. Instead of looking at the Astro*Carto*Graphy map to find the latitude crossings, he thought it would be better to make a list of them.

The Astro*Carto*Graphy computer program was extended so that a list of all the latitude crossings could be printed out. This list starts at 70N00 and finishes at 60S00, which corresponds with an Astro*Carto*Graphy map.

A similar type of list is available in the Astro*Maps Hi-Res computer program from Matrix. This is the option Parans/Intersect Lat.

With this list it is now easy to find the crossings at **any** latitude, and an orb of one degree is used. This will be clear when we examine the computer listing of Simon's latitude crossings.

Leicester is at latitude 52N39, so the orb for his latitude crossings is from 51N39 to 53N39. Simon's latitude crossings are:

ME IC/NE ASC	52 N 30
VE ASC/PL DSC	52 N 21
NE MH/SA DSC	52 N 18

These latitude crossings are also paran aspects in Simon's natal chart. It is important to remember that the paran aspects are calculated for latitude 52N39 (Leicester's latitude).

Simon also has a MO ASC/JU IC paran in his natal chart (orb thirteen minutes). But this paran aspect is **not** a latitude crossing at Leicester (orb one degree, fifty-one minutes). Why is there a difference with this paran/latitude crossing?

Diagram 11 shows the difference between a paran and a crossing. The orb for a paran aspect is measured along the **geographical latitude**. But where the two planets in a paran aspect cross each other is **not** exactly the same

Diagram 11—Difference Between A Paran Aspect and A Latitude Crossing

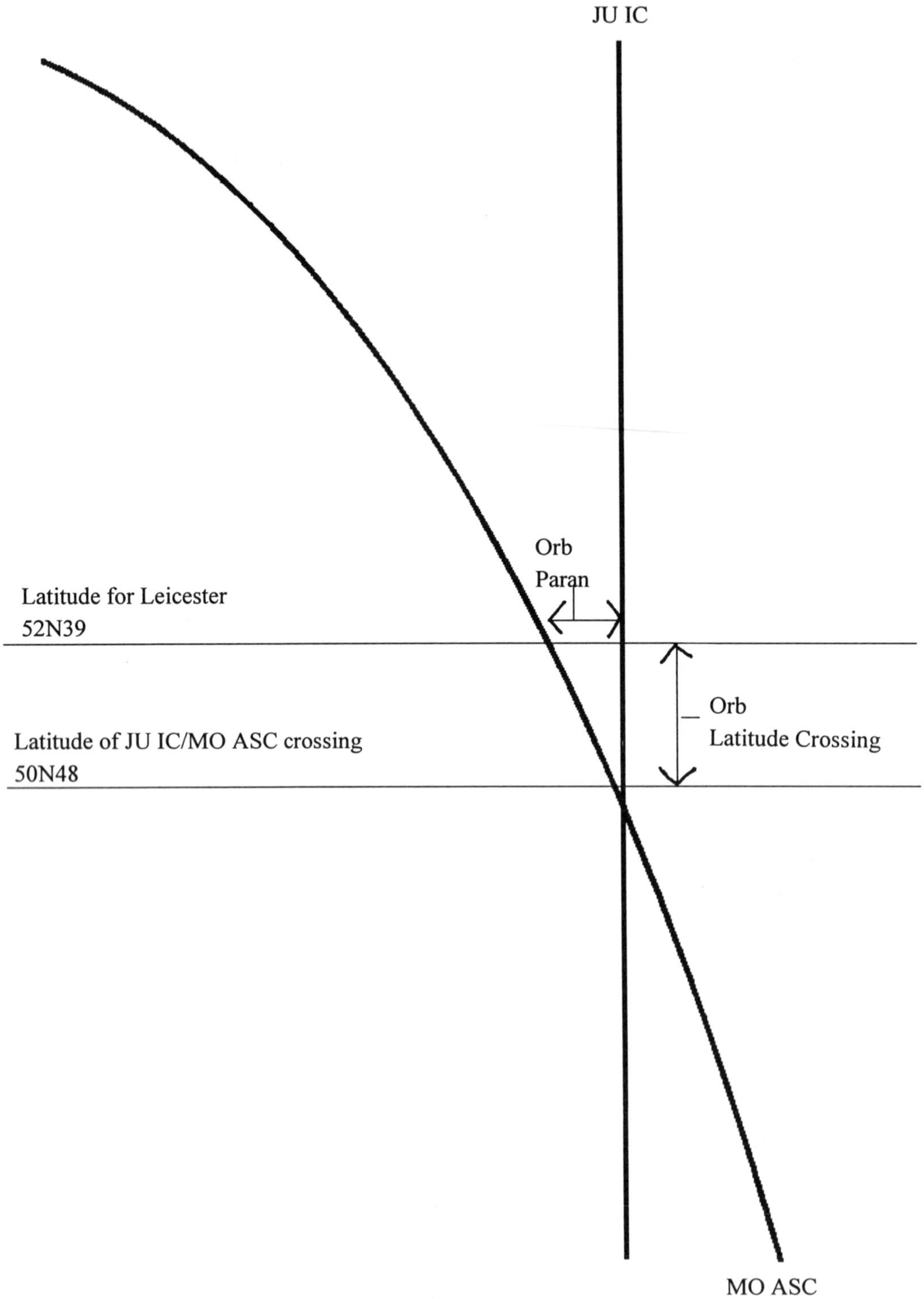

geographical latitude.

The orb for a latitude crossing is measured along the **geographical longitude**. Usually, the difference is so small that both the paran and latitude crossing remain in orb.

In Simon's chart we see the following:

	Orb-Paran	**Orb-Latitude Crossing**
ME IC/NE ASC	0 12	0 09
VE ASC/PL DSC	0 35	0 18
NE MH/SA DSC	0 17	0 21

In the previous chapter we saw that when a planet has a low declination the ASC and DSC lines will be less curved. In Simon's chart, the Moon has a declination of 3N03; hence his MO ASC and MO DSC lines will have very little curve. The exact latitude where Simon's JU IC and MO ASC lines cross will be further away from 52N39.

Astro*Carto*Graphy uses paran aspects, but in a slightly modified form. They are called latitude crossings because the orb is measured differently.

The Planets

Jim Lewis wrote a booklet about interpreting the planetary lines and latitude crossings on an Astro*Carto*Graphy map. It is available with the purchase of an official Astro*Carto*Graphy map.

There is also a brief interpretation of the planets in *The Astro*Carto*Graphy Book of Maps*, and in *Navigating by the Stars*, Edith Hathaway includes some interesting observations about the planets used in Astro*Carto*Graphy.

This volume is not intended to be a cookbook of interpretations for Astro*Carto*Graphy planets. Jim Lewis has already written an excellent booklet which gives accurate information for the planetary lines and latitude crossings.

By using keywords astrologers can form their **own** interpretations which are far better than relying on cookbook meanings. The **whole** horoscope needs to be taken into consideration when it is interpreted. Cookbook interpretations concentrate on only one factor of a chart. For example, an interpretation of Mercury trine Jupiter does not include the signs and houses of those planets. It's also possible that the rest of the horoscope shows contradictions to the general meaning of the Mercury trine Jupiter aspect.

Years ago, Terry Dwyer showed me how to use a "keyword grid" when interpretating a horoscope, and he also wrote a book (*How to Write an Astrological Synthesis*) including that information. I have adapted the keyword grid for use in interpreting Astro*Carto*Graphy maps.

The planets are used in interpretating an Astro*Carto*Graphy map, either in combination with an angle or with another planet (latitude crossings or two or more planets at a location).

First, write a list with words that represent the meaning of the planets. The more words you can list, the better, as they will give more combinations in the keyword grid.

Figure 1 shows a list of keywords for the ten planets that are used in Astro*Carto*Graphy. You can include your own keywords in the list if you wish.

On Simon's Astro*Carto*Graphy map he has VE ASC and PL DSC at Leicester. To find out what the combination of Venus and Pluto means, a keyword grid is formed from the list of words for these planets.

Figure 2 shows a keyword grid for Venus and Pluto. The keywords for Pluto are on the left side of the grid and those for Venus along the top. The squares are completed by writing in the two marginal keywords.

When this keyword grid is completed there will be sixty-four combinations for Venus and Pluto. Sometimes the combinations will be meaningless, and can be omitted. If similar meanings turn up in the **same** keyword grid, reduce them to one to avoid duplication.

Simon has ME IC/NE ASC and NE MH/SA DSC latitude crossings at Leicester. A keyword grid can also be formed for the combination of Mercury/Neptune and Neptune/Saturn.

The Angles

The same system of keywords is used in interpretating the ASC, DSC, MH and IC. Figure 3 shows a list of keywords that can be used for the four angles in a horoscope. Again, you can use your own keywords as the system is flexible.

In the last section a keyword grid was formed for the combination of two planets. This is used when the two planets are in a latitude crossing or two planetary lines are within orb of a location.

Now we will turn our attention to the planetary lines at a location. There are four types of planetary lines on an Astro*Carto*Graphy map, depending upon which angle is involved. This means that there is a combination of a planet and an angle.

A keyword grid can be formed for **any** combination of planet and angle. When a planet/angle keyword grid is completed you will have many meanings for this planetary line.

Simon has a PL DSC planetary line exactly at Leicester, so a keyword grid can be made for this combination of planet (PL) and angle (DSC). Figure 4 shows a completed keyword grid for PL DSC using the keywords from Figures 1 and 3.

The same can also be done for the VE ASC planetary line that Simon has at Leicester. At first you will find it easier to complete a keyword grid to see the different meanings.

As you get used to the system you will find it easier to combine the keywords in a different way. One method is to use a "keyword ladder." Have the two lists of keywords next to each other and work through them one at a time. As you do, write down what meaning is suggested by each combination.

Figure 5 shows a keyword ladder for PL DSC. One word is taken from each list, combined and then repeated with the next word on the list. When this is completed there will be forty different combinations.

Figure 1—Keywords for Planets

Sun
Leadership
Self-expression
Recognition
Creativity
Good health
Domination
Arrogance

Moon
Instincts
Nurture
Routine
Sensitive
Family
Mother
Habits

Mercury
Communication
Travel
Learning
Teaching
Adaptability
Analysis
Restlessness

Venus
Love
Relationships
Money
Friendship
Investments
Cooperaiton
Luxury
Laziness

Mars
Energy
Action
Initiatve
Courage
Self-assertion
Challenge
Aggresion
Impatience

Jupiter
Expansion
Good luck
Optimism
Business
Enthusiasm
Success
Prosperity
Extravagance

Saturn
Restriction
Discipline
Delay
Responsiblity
Hardship
Obstacles
Fear
Bad luck

Uranus
Individualtiy
Originality
Independence
Freedom
Sudden changes
Fluctuation
Shock
Unpredictability

Neptune
Confusion
Deception
Dishonesty
Misunderstanding
Creative imagination
Illusion
Idealism
Vagueness

Pluto
Transformation
Elimination
Rebirth
Crisis
Obsession
Intensification
Regeneration
Traumatic

Figure 2—Keyword Grid for Venus and Pluto

	Love	Relationships	Money	Friendship	Investments	Cooperation	Luxury	Laziness
Transformation	Transformation Love	Transformation Relationships	Transformation Money	Transformation Friendship	Transformation Investments	Transformation Cooperation	Transformation Luxury	Transformation Laziness
Elimination	Elimination Love	Elimination Relationships	Elimination Money	Elimination Friendship	Elimination Investments	Elimination Cooperation	Elimination Luxury	Elimination Laziness
Rebirth	Rebirth Love	Rebirth Relationships	Rebirth Money	Rebirth Friendship	Rebirth Investments	Rebirth Cooperation	Rebirth Luxury	Rebirth Laziness
Crisis	Crisis Love	Crisis Relationships	Crisis Money	Crisis Friendship	Crisis Investments	Crisis Cooperation	Crisis Luxury	Crisis Laziness
Obsession	Obsession Love	Obsession Relationships	Obsession Money	Obsession Friendship	Obsession Investments	Obsession Cooperation	Obsession Luxury	Obsession Laziness
Intensification	Intensification Love	Intensification Relationships	Intensification Money	Intensification Friendship	Intensification Investments	Intensification Cooperation	Intensification Luxury	Intensification Laziness
Regeneration	Regeneration Love	Regeneration Relationships	Regeneration Money	Regeneration Friendship	Regeneration Investments	Regeneration Cooperation	Regeneration Luxury	Regeneration Laziness
Traumatic	Traumatic Love	Traumatic Relationships	Traumatic Money	Traumatic Friendship	Traumatic Investments	Traumatic Cooperation	Traumatic Luxury	Traumatic Laziness

Figure 3—Keywords for Angles

Ascendant	**Descendant**
Self-image	Partnership
Personality	Marriage
Appearance	Cooperation
Characteritics	Confrontation
Self-expression	Business partner

Midheaven	**Immun Coeli**
Career	Home
Life direction	Family
Goals	Private life
Social status	Property
Reputation	Security

Figure 4—Keyword Grid for PL DSC

	Partnership	Marriage	Cooperation	Confrontation	Business Partner
Transformation	Transformation Partnership	Transformation Marriage	Transformation Cooperation	Transformation Confrontation	Transformation Business Partner
Elimination	Elimination Partnership	Elimination Marriage	Elimination Cooperation	Elimination Confrontation	Elimination Business Partner
Rebirth	Rebirth Partnership	Rebirth Marriage	Rebirth Cooperation	Rebirth Confrontation	Rebirth Business Partner
Crisis	Crisis Partnership	Crisis Marriage	Crisis Cooperation	Crisis Confrontation	Crisis Business Partner
Obsession	Obsession Partnership	Obsession Marriage	Obsession Cooperation	Obsession Confrontation	Obsession Business Partner
Intensification	Intensification Partnership	Intensification Marriage	Intensification Cooperation	Intensification Confrontation	Intensification Business Partner
Regeneration	Regeneration Partnership	Regeneration Marriage	Regeneration Cooperation	Regeneration Confrontation	Regeneration Business Partner
Traumatic	Traumatic Partnership	Traumatic Marriage	Traumatic Cooperation	Traumatic Confrontation	Traumatic Business Partner

Figure 5—Keyword Ladder for PL DSC

Pluto	**Descendant**
Transformation →	Partnership
Elimination	Marriage
Rebirth	Coperation
Crisis	Confrontation
Obsession	Business Partner
Intensification	
Regeneration	
Traumatic	

1) Transformation Partnership = Big change in a partnership

Pluto	**Descendant**
Transformation	Partnership
Elimination →	Marriage
Rebirth	Cooperation
Crisis	Confrontation
Obsession	Business Partner
Intensification	
Regeneration	
Traumatic	

2) Elimination Partnership = End of a partnership

Diagram 12—Latitude Crossings-Aspects

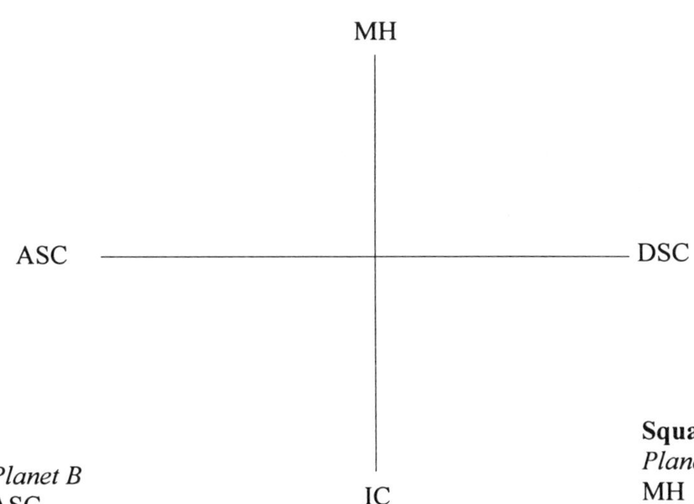

Conjunction

Planet A	*Planet B*
ASC	ASC
DSC	DSC

Opposition

Planet A	*Planet B*
ASC	DSC
DSC	ASC

Square

Planet A	*Planet B*
MH	ASC
ASC	MH
MH	DSC
DSC	MH
ASC	IC
IC	ASC
DSC	IC
IC	DSC

Sythesising the Information

Using keyword grids or ladders to interpret the planetary lines and latitude crossings on an Astro*Carto*Graphy map produces a great deal of information.

Now that the **analysis** of a location is complete, it is time to **synthesize** it. At first this may seem difficult because there is so much information to examine. Leicester on Simon's Astro*Carto*Graphy map will be used to show how the information is synthesized.

First, look for **repetition of themes** in the various keyword grids. When you see a certain theme repeated two or more times in the analysis, that theme is important.

Looking at Leicester on Simon's Astro*Carto*Graphy map he has the following planetary influences:

Planetary lines:	PL DSC, VE ASC
Planetary combinations:	PL/VE
Latitude crossings:	ME IC/NE ASC
	VE ASC/PL DSC
	NE MH/SA DSC

It is apparent that PL DSC and VE ASC have more importance in the synthesis. First, the PL DSC and VE ASC lines are **exact** at Leicester; hence their influence will be powerful. Next, there is a PL/VE planetary combination. Then, the VE ASC/PL DSC latitude crossing is **at** Leicester, so the crossing will have a stronger influence.

With the ME IC/NE ASC and NE MH/SA DSC latitude crossings, only the combination of the planets is considered. The angles are only examined to see if the latitude crossing is a conjunction, opposition or square.

Both ME IC/NE ASC and NE MH/SA DSC are square latitude crossings, so they are interpretated like a square aspect.

Diagram 12 shows the different kinds of latitude crossing aspects. When you compare this list with the different types of paran aspects mentioned earlier you will see how they are connected.

It is important to remember that the influence from latitude crossings is not as powerful as the planetary lines. This means that ME IC/NE ASC and NE MH/SA DSC for Simon will have less importance in the synthesis.

Sometimes a square latitude crossing will have a different aspect in the natal chart. In Simon's natal chart, ME and NE are in an **exact trine**, but ME IC/NE ASC is a **square** latitude crossing. SA and NE are also in **trine** aspect (six degrees) in his natal chart, but a latitude crossing square aspect.

Starting with VE ASC this suggests that he has a friendly personality, his appearance looks good and he likes to live a life of luxury. Simon gets along well with people and has many friends.

With PL DSC at Leicester, Simon doesn't find close personal relationships easy. When he is involved with a partner the relationship is experienced intensely. He tends to be obsessive with his partner, which can lead to confrontations and a great deal of trauma.

The VE/PL combination reiterates this theme about the difficulties in close personal relationships. It could also mean that his partnerships end in a traumatic way.

The background influences of the latitude crossings suggest that Simon has difficulty expressing himself clearly. This can lead to misunderstandings in partnerships, or that the other person is unsure about his feelings. Communication tends to be vague in the relationship, which can lead to problems.

Either Simon or the other person feels restricted in the partnership. When Simon starts to get involved in a close relationship it is quickly an intense and deeply loving partnership. But if one of them is dishonest or unfaithful there is soon a crisis in the partnership.

Simon doesn't understand why his close personal relationships are always intense and traumatic experiences. Partnerships are dominant for him at Leicester.

By the use of keywords we can get a clear picture of what Simon will experience in life living at Leicester. All you need do is combine keywords and use your imagination to work out the relevant meanings. When you can do this there is no need to refer to books to interpret locations on Astro*Carto*Graphy maps.

Chapter 4/Astro*Carto*Graphy in Action

In the last chapter I explained how an Astro*Carto*Graphy map is interpreted by using Simon's map to show how to analyze a location by using keywords. Then all the information was synthesized to show what Simon would experience in Leicester.

The Astro*Carto*Graphy map for Simon is fictitious! The reason I used it was that it showed impressively the importance of using celestial latitude and declination in a natal chart. His Astro*Carto*Graphy map also showed the difference between a paran and a latitude crossing.

It is not my intention to use this chapter to feature many examples to show how accurate Astro*Carto*Graphy is. This has already been done in *The Astro*Carto*Graphy Book of Maps* and *Navigating by the Stars*.

As this book explains how to use the various **techniques** used in Astro*Carto*Graphy, I do not want to repeat what has already been written. If you want to see how Astro*Carto*Graphy works in the lives of many well-known people, the two books mentioned are highly recommended.

Seven examples will be used in this chapter to explain the different Astro*Carto*Graphy techniques. Six are of well-known personalities, and the last example is one of my clients, whose name has been changed and birth data omitted. One example included is that of Prince Charles and Klosters. I use it in my lectures as it always convinces astrologers that Astro*Carto*Graphy is amazingly accurate.

Example 1—Prince Charles and Klosters

In March 1988, Prince Charles went to Klosters, Switzerland for a skiing holiday. It was a place he had visited many times before, which suggests he enjoyed it there.

During this holiday Prince Charles was warned that avalanches had been forecast for the area. Despite these warnings Prince Charles took a party of six onto dangerous ski slopes.

Early in the afternoon of March 10, 1988, while on dangerous ski slopes, an avalanche roared down on Prince Charles and his party. Prince Charles was lucky not to have been killed by the avalanche, but his close friend Major Hugh Lindsay lost his life. Another member of his party broke both her legs.

A small arrow on Prince Charles' Astro*Carto*Graphy map indicates the location of Klosters. I drew a horizontal line for the latitude of Klosters and circled the latitude crossings so that the planetary crossings are easier to see.

The nearest planetary line to Klosters is PL ASC, so this will have the strongest influence. Also within orb of Klosters are MO MH and VE IC planetary lines.

Prince Charles has seven planetary crossings on the line of latitude for Klosters.

Crossings: MO IC/UR DSC
MO IC/JU ASC
JU ASC/UR DSC
MA MH/PL DSC
MA MH/SA DSC
SA DSC/PL DSC
PL IC/MA DSC

The PL ASC line shows that Klosters is a potentially dangerous place for Prince Charles. This is a location that will eventually transform his personality, especially when he keeps returning for holidays. He could risk death or experience a total upheaval in life every time he visits Klosters, thereby changing his life direction and perception.

With MO MH line at Klosters he will be the center of attraction and become involved with people and their concerns. He would enjoy helping people there because of his deep sensitivity to public feelings and a desire to be useful. He likes being the center of attraction at Klosters, probably one of the reasons he keeps returning.

VE IC shows that Klosters is a place where Prince Charles is able to relax in a pleasurable environment, and have a peaceful holiday with his family.

The planetary lines show that there are two themes when Prince Charles visits Klosters. One is the great amount of danger to which he is exposed, and the other is the good social life he experiences there. The latitude crossings at Klosters reiterate these two themes from the planetary lines.

The two Mars/Pluto crossings make Prince Charles very competitive at Klosters, but this could lead to shocks and dramatic alterations of life direction.

The Saturn/Pluto crossing shows the possibility of death, mourning and unwanted burdens being experienced. With both these planets lying on the DSC, there is grave danger for him at Klosters.

The Mars/Saturn latitude crossings show that crowds, overexertion and high places (mountains) pose physical threats, and injury is likely.

All these crossings reiterate the potential danger shown by the close PL ASC planetary line.

Moon/Uranus explains Prince Charles' need for freedom and the desire for the unconventional. Strange and uncontrollable things happen here, and life at Klosters will be full of the unexpected, the exciting and emotional extremes.

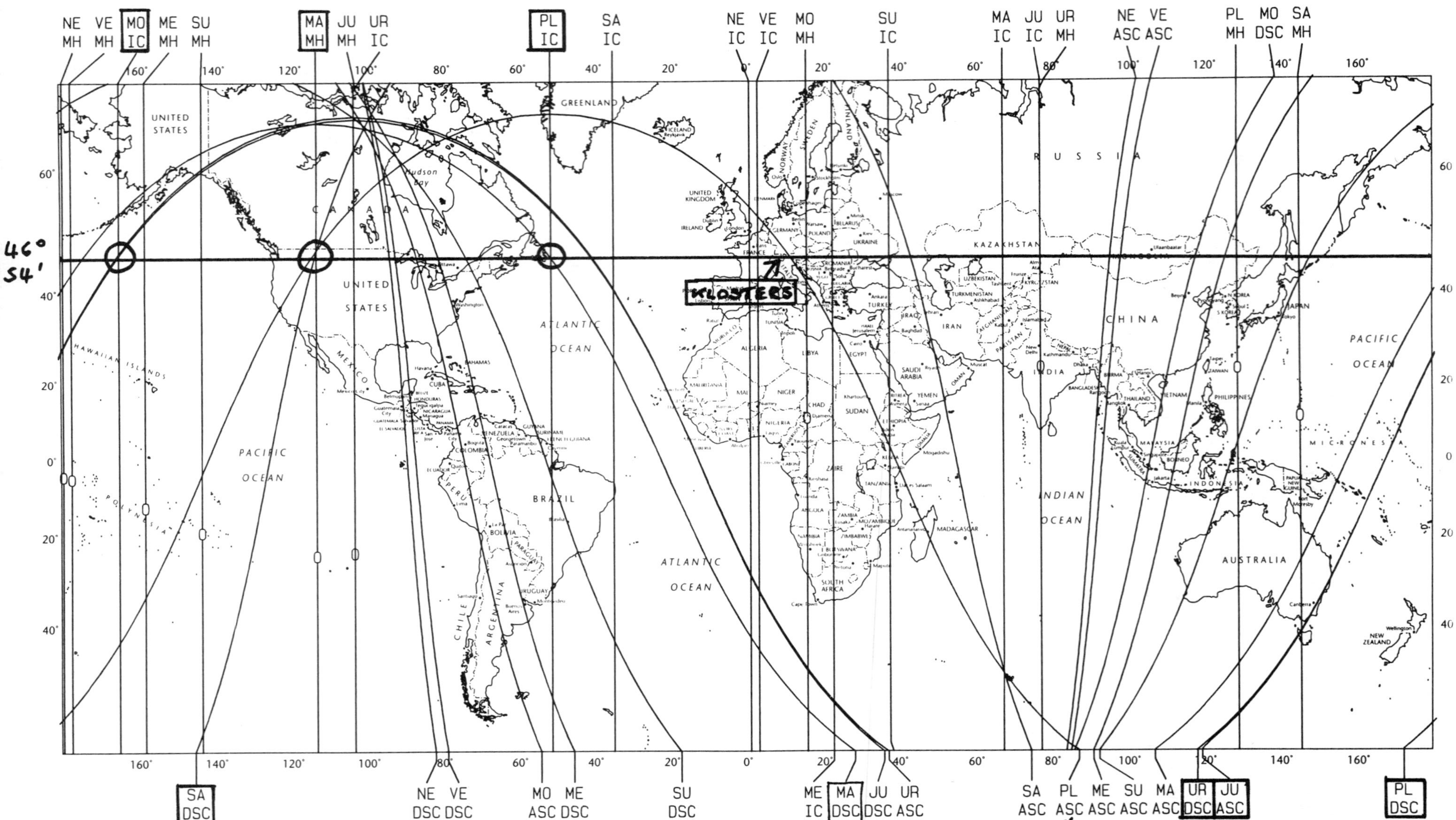

Prince Charles

November 14, 1948 9: 14 PM GMT

04/16/96

04/16/96

Prince Charles
November 14, 1948 9:14 PM GMT

ASTRO*CARTO*GRAPHY LATITUDE CROSSINGS

NORTH LATITUDES

69N34	JU/MH * ME/DSC
69N33	UR/IC * ME/DSC
69N00	MA/MH * SU/DSC
68N06	VE/MH * SU/ASC
68N06	SA/MH * ME/ASC
67N50	ME/DSC * NE/DSC
67N34	NE/MH * SU/ASC
67N28	JU/MH * MO/ASC
67N26	UR/IC * MO/ASC
67N12	ME/DSC * VE/DSC
67N01	ME/DSC * SA/DSC
66N55	VE/DSC * SA/DSC
66N37	UR/IC * JU/DSC
66N37	SU/DSC * JU/ASC
66N37	MO/ASC * JU/DSC
66N35	JU/MH * SU/DSC
66N35	UR/IC * SU/DSC
66N34	JU/DSC * NE/DSC
66N34	JU/DSC * SA/DSC
66N33	VE/DSC * JU/DSC
66N31	ME/DSC * JU/DSC
66N29	SA/DSC * NE/DSC
66N21	JU/MH * UR/DSC
66N21	SU/DSC * UR/ASC
66N21	MO/ASC * UR/ASC
66N19	MA/MH * JU/ASC
66N18	SA/DSC * UR/ASC
66N18	UR/ASC * NE/DSC
66N16	VE/DSC * UR/ASC
66N14	ME/DSC * UR/ASC
66N05	SA/IC * PL/ASC
66N02	MA/MH * UR/DSC
65N56	MA/ASC * UR/DSC
65N51	MO/ASC * SU/DSC
65N49	MA/ASC * JU/ASC
65N38	SU/DSC * SA/DSC
65N37	JU/MH * MA/DSC
65N37	UR/IC * MA/DSC
65N29	MO/ASC * SA/DSC
65N25	MA/DSC * SA/DSC
65N24	MO/ASC * MA/DSC
65N19	SU/DSC * MA/DSC
65N15	MA/DSC * NE/DSC
65N09	VE/DSC * MA/DSC
64N59	SU/DSC * NE/DSC
64N52	ME/DSC * MA/DSC
64N33	SU/DSC * VE/DSC
64N31	JU/DSC * PL/DSC
64N22	UR/ASC * PL/DSC
63N13	MO/ASC * NE/DSC
63N05	MA/DSC * PL/DSC
63N04	PL/MH * MO/DSC
63N03	UR/IC * SA/DSC
63N00	JU/MH * SA/DSC
62N42	SU/MH * MA/ASC
62N02	SU/DSC * ME/DSC
61N33	MO/ASC * VE/DSC
61N19	SU/DSC * PL/DSC
61N13	Anchorage
61N10	ME/DSC * PL/DSC
60N37	SU/MH * JU/ASC
60N17	SU/MH * UR/DSC
59N48	MO/ASC * PL/DSC
59N27	VE/DSC * PL/DSC
59N20	Stockholm
59N00	NE/DSC * PL/DSC
57N47	ME/MH * MA/ASC
57N29	SU/IC * SA/ASC
55N45	Moscow
55N41	Copenhagen
55N35	UR/IC * PL/DSC
55N33	JU/MH * PL/DSC
55N15	NE/IC * PL/ASC

CENTRAL LATITUDES

55N10	PL/IC * JU/DSC
54N51	PL/IC * UR/ASC
54N17	MO/IC * MA/ASC
53N52	VE/IC * PL/ASC
53N13	ME/MH * JU/ASC
52N50	ME/MH * UR/DSC
52N30	Berlin
52N26	PL/MH * ME/ASC
51N30	London
49N14	Vancouver
48N50	Paris
48N27	SA/DSC * PL/DSC
48N20	MA/MH * PL/DSC
48N11	MA/MH * SA/DSC
47N55	MO/IC * JU/ASC
47N36	Seattle
47N30	MO/IC * UR/DSC
46N59	PL/IC * MA/DSC
45N32	Montreal
44N59	Minneapolis
44N11	VE/MH * MA/ASC
43N39	Toronto
42N33	MO/MH * PL/ASC
42N22	Boston
42N20	Detroit
42N19	SA/MH * SU/ASC
41N54	Rome
41N53	Chicago
41N47	NE/MH * MA/ASC
41N13	SA/IC * JU/DSC
40N54	SA/IC * UR/ASC
40N43	New York
39N44	Denver
38N53	Washington
38N37	St. Louis
37N47	San Francisco
36N10	Las Vegas
35N40	Tokyo
34N04	Los Angeles
33N45	Atlanta
33N27	Phoenix
33N20	ME/IC * PL/ASC
32N47	Dallas
32N24	VE/MH * JU/ASC
32N00	VE/MH * UR/DSC
30N03	Cairo
29N58	New Orleans
29N46	Houston
28N45	NE/MH * JU/ASC
28N39	Delhi
28N22	NE/MH * UR/DSC
25N47	Miami
25N46	SA/IC * MA/DSC
22N34	Calcutta
22N16	Hong Kong
21N19	Honolulu
19N26	Mexico City
14N35	Manila
10N30	Caracas
8N30	JU/DSC * UR/ASC
6N27	Lagos
4N25	PL/MH * SU/ASC
3N27	SU/IC * PL/ASC
3S27	SU/MH * PL/DSC
4S25	PL/IC * SU/DSC
8S30	JU/ASC * UR/DSC
22S54	Rio De Janeiro
25S46	SA/MH * MA/ASC

SOUTH LATITUDES

26S12	Johannesburg
28S22	NE/IC * UR/ASC
28S45	NE/IC * JU/DSC
32S00	VE/IC * UR/ASC
32S24	VE/IC * JU/DSC
33S20	ME/MH * PL/DSC
33S55	Sydney
34S36	Buenos Aires
36S52	Auckland
37S50	Melbourne
40S54	SA/MH * UR/DSC
41S13	SA/MH * JU/ASC
41S47	NE/IC * MA/DSC
42S19	SA/IC * SU/DSC
42S33	MO/IC * PL/DSC
44S11	VE/IC * MA/DSC
46S59	PL/MH * MA/ASC
47S30	MO/MH * UR/ASC
47S55	MO/MH * JU/DSC
48S11	MA/IC * SA/ASC
48S20	MA/IC * PL/ASC
48S27	SA/ASC * PL/ASC
52S26	PL/IC * ME/DSC
52S50	ME/IC * UR/ASC
53S13	ME/IC * JU/DSC
53S52	VE/MH * PL/DSC
54S17	MO/MH * MA/DSC
54S51	PL/MH * UR/DSC
55S10	PL/MH * JU/ASC
55S15	NE/MH * PL/DSC
55S33	JU/IC * PL/ASC
55S35	UR/MH * PL/ASC
57S29	SU/MH * SA/DSC
57S47	ME/IC * MA/DSC
59S00	NE/ASC * PL/ASC
59S27	VE/ASC * PL/ASC
59S48	MO/DSC * PL/ASC

Moon/Jupiter indicates that this is one of the best places for Prince Charles. This is where he is loved and admired, and aspires to surround himself with the best in the social and material pleasures in life, yet in moderation.

The Jupiter/Uranus crossing makes Klosters one of the most exciting places for Prince Charles, where daring ideas work. This zone brings extraordinary good luck, and one of the possibilities is being rescued from difficult circumstances.

The obvious question is why did this tragic accident occur on his **tenth** visit to Klosters? The Astro*Carto*Graphy map shows lifelong and permanent influences at locations, so we need to use a different technique to answer this question.

In the next three chapters, the effect of temporary influences at locations on an Astro*Carto*Graphy map will be explained. It is important to consider **when** a person visits a location as it provides essential information.

To find the planetary crossings for Klosters I used a ruler for forty-seven degrees north latitude. For convenience I drew a line for this latitude and circled the crossings.

A square is drawn around the planets that are involved in the latitude crossings. An examination of the map shows clearly where these latitude crossings occur.

Does a latitude crossing listing provide the same information as using a ruler? To answer this question I will use the latitude crossings listing for Klosters. Using an orb of one degree the following information is obtained:

Klosters	46 N 54
PL IC/MA DSC	46 N 59
MO IC/UR DSC	47 N 30

Many latitude crossings are not picked up when an orb of one degree is used. This shows that the orb for latitude crossings needs to be increased. After examining the differences between the two systems I recommend that an orb up to three degrees be used with the latitude crossings listing.

The orbs for the latitude crossings work in the same way as orbs for the planetary lines. The closer an orb is for a latitude crossing the more powerful it will be. When the orb is approaching three degrees it will have a lesser influence. However, when there is a repetition of themes with the latitude crossings, then those near to three degrees orb will be important.

The exact latitude crossing where his JU ASC and UR DSC planetary lines cross is 8N30, which is a long way from the latitude of Klosters. However, in Chapter 5 we will see why **JU** and **UR** were significant for Prince Charles at Klosters.

When two planets are in exact conjunction or opposition with the **same** declination, then the ASC and/or DSC lines will always be together. Then the exact latitude crossing where the two planets cross each other can only be obtained by examining the latitude crossings listing.

Example 2—Norman Schwarzkopf and Kuwait

Until Iraq invaded Kuwait on August 2, 1990, Norman Schwarzkopf was an unknown person. After the invasion he was appointed the commander of Operation Desert Shield, in charge of the American and allied forces. The Gulf war began on January 17, 1991, when Desert Shield turned into Desert Storm. It was his skillful planning and clever tactics that brought the Gulf war to a quick end. It was the Iraq/Kuwait war where Norman Schwarzkopf achieved fame. When he returned to America he received numerous awards and was also knighted by Queen Elizabeth. Then publishers approached him for his autobiography, offering more than $5 million.

The Astro*Carto*Graphy map for Norman Schwarzkopf at Kuwait shows the following information:

Planetary lines:	SU MH, ME MH, SA IC
Crossings:	nil

Kuwait City is at the midpoint of the SU MH, ME MH and SA IC lines. Jim Lewis wrote in his Astro*Carto*Graphy interpretation booklet that SU MH is the most likely line for a person to become famous. Norman Schwarzkopf has his SU MH line passing through the Persian Gulf near to Kuwait, and be became a living legend as a result of the Gulf war.

The ME MH and SA IC lines near Kuwait show the careful and detailed planning of his operation. He also had a great deal of responsibility being in charge of 500,000 American and allied forces. The latitude crossings have no influence at Kuwait because the nearest ones have an orb of six to seven degrees.

Example 3—Bob Geldof and Ethiopia

In November 1984, Bob Geldof (singer of the group Boomtown Rats) brought together Britain's top rock stars to record a single. They called themselves Band Aid and produced a song called "Do They Know It's Christmas." The proceeds were designated to help starving refugees in Ethiopia where there was a famine and a civil war. The record was number one in the charts soon after it was released.

After the huge success of Band Aid, Bob Geldof decided to take his fund raising efforts a stage further. In July 1985, he arranged Live Aid where two concerts were held in Wembley Stadium London and JFK Stadium in Philadelphia. The Live Aid appeal raised more than £50 million.

On Bob Geldof's Astro*Carto*Graphy map, MO MH is nearest Ethiopia. The MO MH line is good for any kind of helping profession, and trying to help the starving people in Ethiopia is most appropriate.

He also has VE DSC near Ethiopia, which he used to raise money for the starving people there. With MA DSC near Ethiopia he was moved to do something about the terrible situation there. PL DSC is within orb of Ethiopia where death was very prominent.

Using Adis Abeba (capital of Ethiopia) for latitude crossings, Bob Geldof has the following:

NE MH/UR DSC	6 N 59
MA IC/MO DSC	7 N 15
Adis Abeba	9 N 02
SU IC/UR ASC	10 N 34

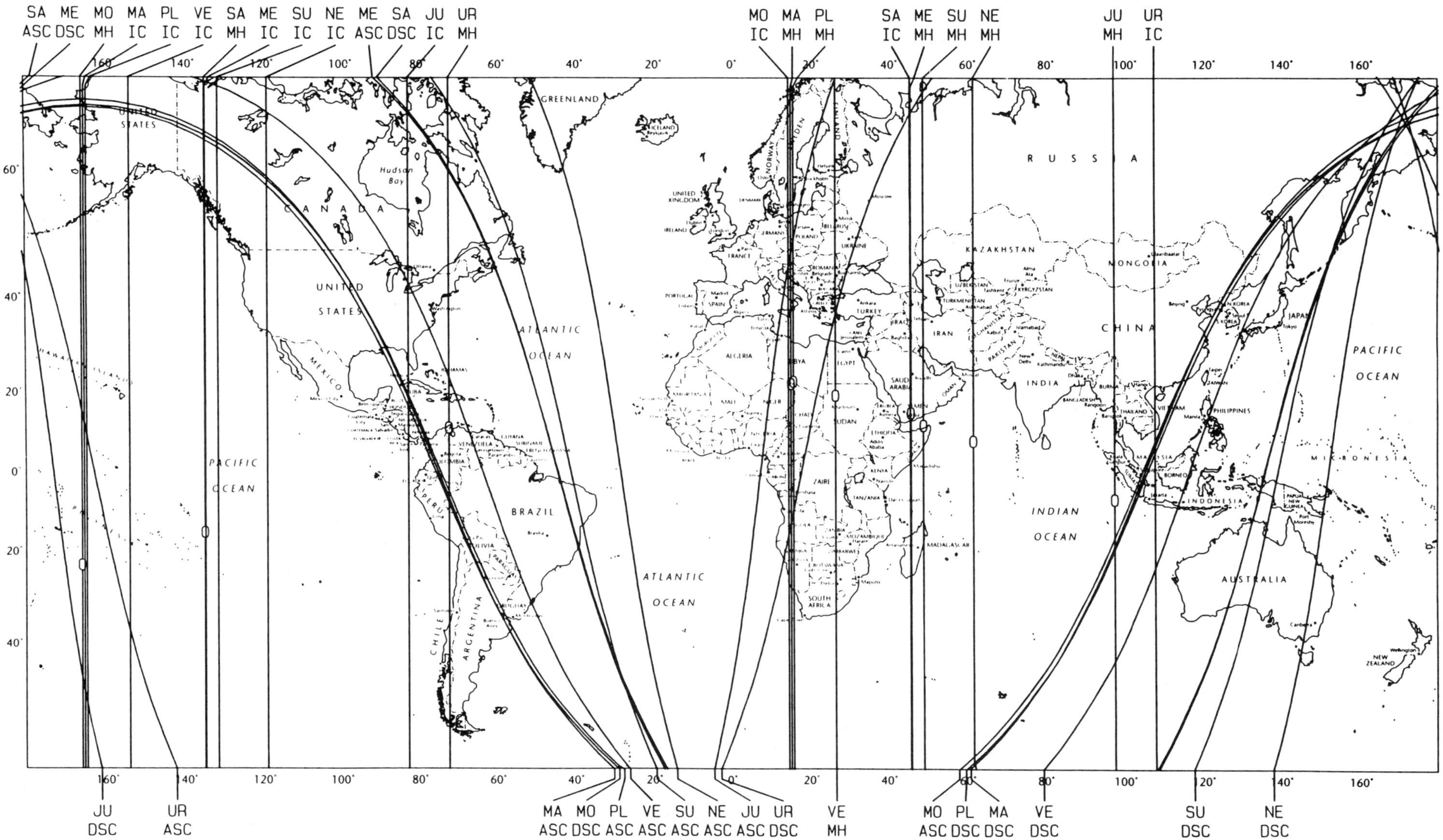

Norman Schwarzkopf

August 22, 1934 4:45 AM EDT

04/16/96

Norman Schwarzkopf
August 22, 1934 4:45 AM EDT

ASTRO*CARTO*GRAPHY LATITUDE CROSSINGS

NORTH LATITUDES

69N53 VE/MH * JU/ASC
69N48 SU/DSC * NE/DSC
69N41 SA/MH * VE/ASC
69N40 ME/IC * VE/ASC
69N15 SU/IC * VE/ASC
68N44 SU/MH * UR/DSC
68N16 ME/DSC * VE/DSC
67N56 MO/MH * MA/DSC
67N56 PL/IC * MA/ASC
67N53 VE/DSC * SA/ASC
67N47 UR/MH * SU/ASC
67N46 SU/DSC * JU/DSC
67N32 VE/IC * MA/ASC
67N25 MA/IC * PL/DSC
67N25 MO/MH * PL/DSC
67N14 MA/IC * MO/DSC
67N14 PL/IC * MO/DSC
67N05 SU/DSC * UR/ASC
67N03 VE/IC * PL/ASC
67N00 JU/IC * SA/DSC
66N54 JU/DSC * NE/DSC
66N50 NE/IC * VE/ASC
66N47 VE/IC * MO/DSC
66N46 VE/DSC * JU/DSC
66N43 ME/MH * UR/DSC
66N41 JU/IC * ME/ASC
66N41 VE/DSC * NE/DSC
66N39 SA/IC * UR/DSC
66N29 VE/DSC * UR/ASC
66N28 ME/DSC * MA/DSC
66N16 MA/DSC * SA/ASC
66N15 UR/ASC * NE/DSC
66N09 MA/DSC * JU/DSC
66N00 MA/DSC * UR/ASC
65N57 JU/DSC * SA/ASC
65N53 MA/DSC * NE/DSC
65N51 MO/ASC * PL/DSC
65N40 SA/ASC * UR/ASC
65N38 ME/DSC * JU/DSC
65N35 MO/ASC * JU/DSC
65N34 JU/DSC * PL/DSC
65N33 SU/DSC * VE/DSC
65N33 MO/ASC * ME/DSC
65N31 ME/DSC * PL/DSC
65N29 MO/ASC * UR/ASC
65N28 UR/ASC * PL/DSC
65N24 ME/DSC * UR/ASC
65N21 MO/ASC * SA/ASC
65N18 SA/ASC * PL/DSC
65N10 SU/DSC * MA/DSC
65N09 MO/ASC * NE/DSC
65N05 NE/DSC * PL/DSC
64N48 SA/MH * MA/ASC
64N47 ME/IC * MA/ASC
64N46 VE/DSC * MA/DSC
64N23 SA/MH * PL/ASC
64N21 ME/IC * PL/ASC
64N18 MO/ASC * SU/DSC
64N17 SA/ASC * NE/DSC
64N15 JU/DSC * UR/ASC
64N10 SU/DSC * PL/DSC
63N59 SU/IC * MA/ASC
63N54 SA/MH * MO/DSC
63N52 ME/IC * MO/DSC
63N34 SU/IC * PL/ASC
63N24 ME/DSC * NE/DSC
63N01 SU/IC * MO/DSC
62N58 MO/ASC * VE/DSC
62N27 VE/DSC * PL/DSC
61N13 Anchorage
60N41 UR/MH * SA/DSC
60N20 UR/MH * ME/ASC

59N29 NE/IC * MA/ASC

CENTRAL LATITUDES

59N20 Stockholm
59N08 NE/IC * PL/ASC
58N20 NE/IC * MO/DSC
55N45 Moscow
55N41 Copenhagen
52N30 Berlin
51N30 London
50N35 PL/MH * JU/ASC

49N14 Vancouver
48N50 Paris
48N22 MA/MH * JU/ASC
48N18 SU/DSC * SA/ASC
47N36 Seattle
46N01 MO/IC * JU/ASC
45N32 Montreal
45N02 SU/DSC * ME/DSC
44N59 Minneapolis
44N04 MA/ASC * PL/ASC
43N39 Toronto
43N09 JU/IC * VE/ASC
42N22 Boston
42N20 Detroit
41N54 Rome
41N53 Chicago
40N43 New York

39N44 Denver
38N53 Washington
38N37 St. Louis
37N47 San Francisco
37N08 VE/MH * UR/DSC
36N10 Las Vegas
35N51 MO/ASC * MA/DSC
35N40 Tokyo
34N04 Los Angeles
33N45 Atlanta
33N27 Phoenix
32N47 Dallas
30N03 Cairo

29N58 New Orleans
29N46 Houston
28N39 Delhi
25N47 Miami
23N38 UR/MH * VE/ASC
22N34 Calcutta
22N16 Hong Kong
21N54 ME/ASC * SA/DSC
21N19 Honolulu

19N26 Mexico City
19N23 JU/IC * PL/ASC
18N27 JU/IC * MA/ASC
16N32 JU/IC * MO/DSC
15N24 MO/MH * UR/ASC
14N35 Manila
12N32 MA/IC * UR/ASC
10N30 Caracas

9N29 PL/IC * UR/ASC
7N34 UR/IC * MO/ASC
6N27 Lagos
6N20 UR/IC * MA/DSC
4N39 UR/IC * PL/DSC

4S39 UR/MH * PL/ASC
6S20 UR/MH * MA/ASC
7S34 UR/MH * MO/DSC
9S29 PL/MH * UR/DSC

12S32 MA/MH * UR/DSC
15S24 MO/IC * UR/DSC
16S32 JU/MH * MO/ASC
18S27 JU/MH * MA/DSC

SOUTH LATITUDES

19S23 JU/MH * PL/DSC

21S54 ME/DSC * SA/ASC
22S54 Rio De Janeiro
23S38 UR/IC * VE/DSC
26S12 Johannesburg

33S55 Sydney
34S36 Buenos Aires
35S51 MO/DSC * MA/ASC
36S52 Auckland
37S08 VE/IC * UR/ASC
37S50 Melbourne

43S09 JU/MH * VE/DSC
44S04 MA/DSC * PL/DSC
45S02 SU/ASC * ME/ASC
46S01 MO/MH * JU/DSC
48S18 SU/ASC * SA/DSC
48S22 MA/IC * JU/DSC

50S35 PL/IC * JU/DSC
58S20 NE/MH * MO/ASC
59S08 NE/MH * PL/DSC
59S29 NE/MH * MA/DSC

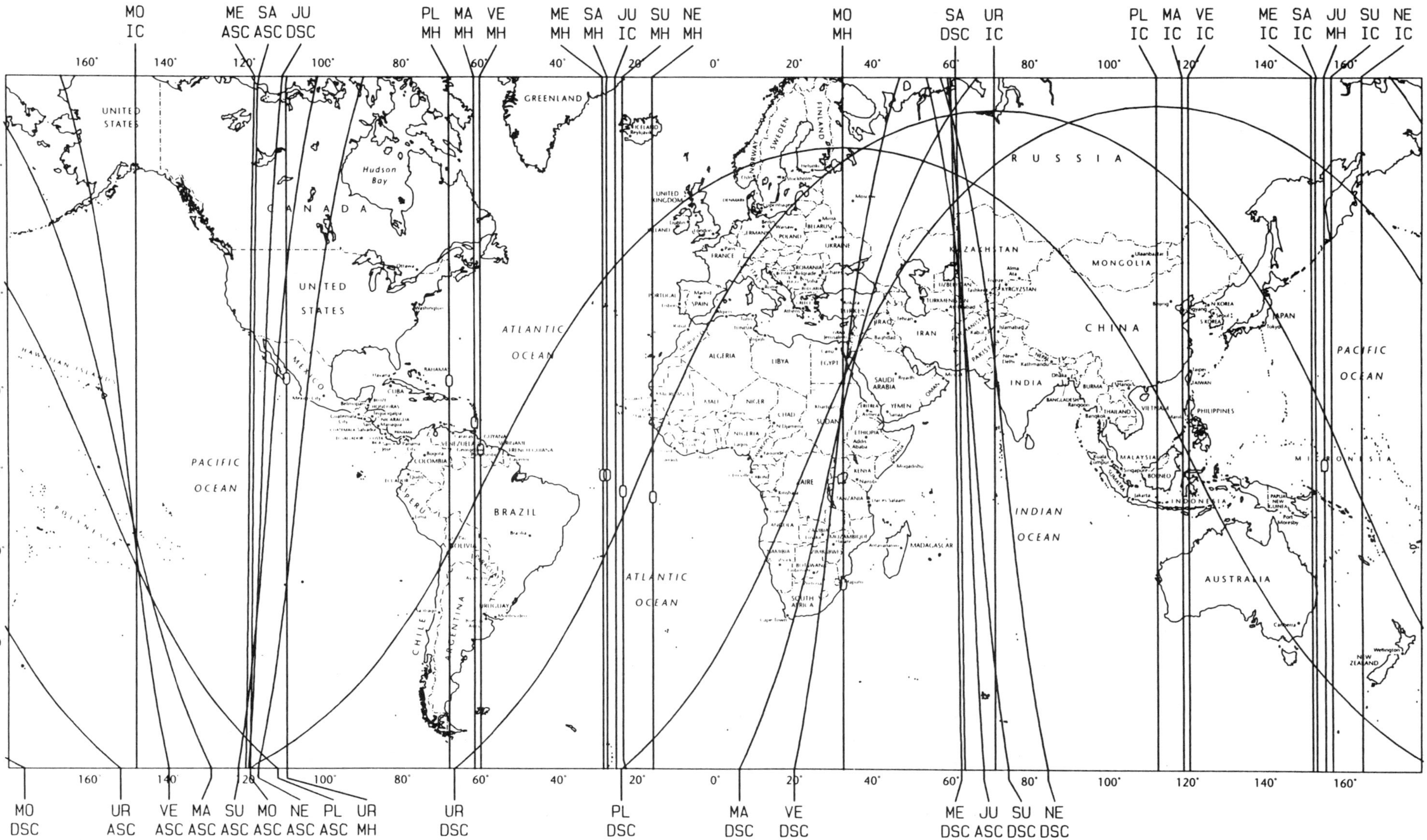

Bob Geldof

October 5, 1951 2:20 PM GDT

04/16/96

04/16/96

Bob Geldof
October 5, 1951 2:20 PM GDT

ASTRO*CARTO*GRAPHY LATITUDE CROSSINGS

NORTH LATITUDES

67N25	MA/DSC * SA/DSC
67N12	MA/IC * PL/ASC
67N07	VE/IC * PL/ASC
66N44	ME/DSC * MA/DSC
66N42	MA/DSC * JU/ASC
66N41	MA/DSC * NE/DSC
66N39	JU/ASC * NE/DSC
66N38	SA/DSC * UR/DSC
66N33	ME/DSC * NE/DSC
66N33	UR/DSC * NE/DSC
66N33	ME/DSC * UR/DSC
66N33	JU/ASC * UR/DSC
66N31	MA/DSC * UR/DSC
66N17	ME/DSC * JU/ASC
66N12	SU/DSC * UR/DSC
65N44	UR/ASC * PL/DSC
64N36	SU/DSC * MA/DSC
64N25	SA/DSC * NE/DSC
64N12	VE/DSC * UR/DSC
62N57	MO/DSC * UR/DSC
62N43	MO/DSC * VE/DSC
61N41	ME/IC * PL/ASC
61N29	MO/DSC * MA/DSC
61N26	MO/MH * UR/DSC
61N19	SA/IC * PL/ASC
61N13	Anchorage
60N57	UR/IC * PL/DSC
60N31	MO/DSC * SU/DSC
60N31	JU/MH * PL/ASC
60N28	PL/IC * UR/ASC
60N04	MO/DSC * ME/DSC
59N53	MO/DSC * JU/ASC
59N47	MO/DSC * SA/DSC
59N46	SU/IC * PL/ASC
59N20	Stockholm
59N12	MO/DSC * NE/DSC
58N43	MO/DSC * PL/DSC
58N05	NE/DSC * PL/DSC
57N38	MA/IC * UR/ASC
56N58	UR/IC * MO/DSC
56N50	VE/IC * UR/ASC
56N41	SA/DSC * PL/DSC
56N38	JU/ASC * PL/DSC
56N01	ME/DSC * PL/DSC
55N59	NE/IC * PL/ASC
55N45	Moscow
55N41	Copenhagen
55N34	SU/DSC * PL/DSC
55N32	JU/ASC * SA/DSC
52N30	NE/MH * MO/ASC
52N30	Berlin
51N39	SU/DSC * ME/DSC
51N30	London
49N14	Vancouver
48N50	Paris
47N41	SU/MH * MO/ASC
47N36	Seattle
46N18	JU/IC * MO/ASC
45N32	Montreal
44N59	Minneapolis
44N59	UR/MH * SU/ASC
44N33	SU/DSC * SA/DSC
44N29	SA/MH * MO/ASC
43N39	Toronto
43N34	ME/MH * MO/ASC
42N22	Boston
42N20	Detroit
41N54	Rome
41N53	Chicago
40N43	New York
39N44	Denver

CENTRAL LATITUDES

38N53	Washington
38N37	St. Louis
37N47	San Francisco
36N10	Las Vegas
35N40	Tokyo
35N22	SU/DSC * JU/ASC
34N04	Los Angeles
33N45	Atlanta
33N27	Phoenix
32N47	Dallas
30N03	Cairo
29N58	New Orleans
29N55	MA/DSC * PL/DSC
29N46	Houston
28N39	Delhi
26N34	UR/IC * NE/DSC
25N47	Miami
24N03	VE/DSC * PL/DSC
23N02	MO/MH * PL/DSC
22N34	Calcutta
22N16	Hong Kong
21N22	ME/IC * UR/ASC
21N19	Honolulu
19N49	MO/MH * VE/DSC
19N26	Mexico City
19N15	SA/IC * UR/ASC
19N15	PL/IC * MO/DSC
16N09	MO/MH * MA/DSC
14N37	JU/MH * UR/ASC
14N35	Manila
12N50	VE/DSC * MA/DSC
10N34	SU/IC * UR/ASC
10N30	Caracas
7N15	MA/IC * MO/DSC
6N59	NE/MH * UR/DSC
6N27	Lagos
4N13	VE/IC * MO/DSC
4S13	VE/MH * MO/ASC
6S59	NE/IC * UR/ASC
7S15	MA/MH * MO/ASC
10S34	SU/MH * UR/DSC
12S50	VE/ASC * MA/ASC
14S37	JU/IC * UR/DSC
16S09	MO/IC * MA/ASC
19S15	PL/MH * MO/ASC
19S15	SA/MH * UR/DSC
19S49	MO/IC * VE/ASC
21S22	ME/MH * UR/DSC
22S54	Rio De Janeiro
23S02	MO/IC * PL/ASC
24S03	VE/ASC * PL/ASC
26S12	Johannesburg
26S34	UR/MH * NE/ASC
29S55	MA/ASC * PL/ASC
33S55	Sydney
34S36	Buenos Aires
35S22	SU/ASC * JU/DSC
36S52	Auckland
37S50	Melbourne
43S34	ME/IC * MO/DSC
44S29	SA/IC * MO/DSC
44S33	SU/ASC * SA/ASC
44S59	UR/IC * SU/DSC
46S18	JU/MH * MO/DSC
47S41	SU/IC * MO/DSC
51S39	SU/ASC * ME/ASC

SOUTH LATITUDES

52S30	NE/IC * MO/DSC
55S32	JU/DSC * SA/ASC
55S34	SU/ASC * PL/ASC
55S59	NE/MH * PL/DSC
56S01	ME/ASC * PL/ASC
56S38	JU/DSC * PL/ASC
56S41	SA/ASC * PL/ASC
56S50	VE/MH * UR/DSC
56S58	UR/MH * MO/ASC
57S38	MA/MH * UR/DSC
58S05	NE/ASC * PL/ASC
58S43	MO/ASC * PL/ASC
59S12	MO/ASC * NE/ASC
59S46	SU/MH * PL/DSC
59S47	MO/ASC * SA/ASC
59S53	MO/ASC * JU/DSC

MA/MO reiterates that he would take action (MA) to help people (MO), and he would do something unusual to raise money (SU/UR). This he achieved through music, something that nobody had done before (NE/UR).

When Bob Geldof decided to do something to help the people in Ethiopia he had never even been to the country. It is probable that he was emotionally disturbed by TV news reports about the thousands of people dying of starvation in Ethiopia.

So the planetary influences in Ethiopia **came** to Bob Geldof and manifested as Band Aid and Live Aid. This is an excellent example of remote activation of the planetary influences at a location.

Example 4—Margaret Thatcher and the Falkland Islands

After Argentina invaded the Falkland Islands in April 1982, the British government had no choice but to respond. Margaret Thatcher was the British prime minister at the time of the Falklands war. Her Astro*Carto*Graphy map clearly shows why she was involved in a war over the Falkland Islands.

The two planetary lines nearest the Falkland Islands are SU ASC and MA ASC, both appropriate for getting involved in a battle. Her MA ASC line is also close to Buenos Aires, the capital of Argentina. This clearly shows the war with Argentina over the Falkland Islands.

Margaret Thatcher also has ME ASC at the Falkland Islands. Together with her SU ASC and MA ASC, it shows her anger over the invasion and determination to win the battle.

UR DSC and PL MH are just within orb of the Falkland Islands, so their influence will not be so strong. UR DSC could show that the war would only last a short time. After victory in the Falklands, Margaret Thatcher was recognised as a world leader. PL MH could show this change in her status after the Falklands war was won.

The capital of the Falkland Islands is Port Stanley. Using this location for the latitude crossings, Margaret Thatcher has the following influences at the Falkland Islands.

MO IC/SA DSC	53 S 29
SA MH/MO DSC	53 S 13
Port Stanley	51 S 42
SA IC/PL ASC	49 S 14

The two MO/SA crossings clearly show Margaret Thatcher's reputation as the "Iron Lady." The SA/PL crossing indicates the experiences with responsibility, duty and death which manifested during the war.

The Moon/Saturn combination for Margaret Thatcher is strongly emphasised at Britain. She has MO MH and SA ASC lines over Britain, and **two** MO/SA latitude crossings within orb of London.

SA MH/PL DSC	49 N 14
London	51 N 30
SA IC/MO ASC	53 N 13
MO MH/SA ASC	53 S 29

You will notice that both London and Port Stanley have the same planets involved in the latitude crossings. The exact latitude where the planets cross is also the same, the only difference being whether the latitude is north or south. This means that London (51N30) and Port Stanley (51S42) are both connected as they have the same latitude.

Latitude crossings north of the equator repeat themselves in the Southern Hemisphere. The only difference is the **angles** that are in a latitude crossing. The following list clarifies this:

49N14	SA **MH**/PL **DSC**	49S14	SA **IC**/PL **ASC**
51N30	London	51S42	Port Stanley
53N13	SA **IC**/MO **ASC**	53S13	SA **MH**/MO **DSC**
53N29	MO **MH**/SA **ASC**	53S29	MO **IC**/SA **DSC**

Also interesting is the Astro*Carto*Graphy map for the United Kingdom. On this map SU DSC and MA MH are at the Falkland Islands. With Margaret Thatcher also having SU ASC and MA ASC close to the Falkland Islands, the likelihood of conflict there was emphasised. The fact that Britain was involved in the Falklands war while Margaret Thatcher was prime minister is no real surprise.

Example 5—David Copperfield and Germany

David Copperfield has been a magicician for many years, and usually appears in Las Vegas, where he is one of the highest earning performers in America. At Las Vegas he has PL ASC, the location where he completely changed his life. It is a location where he tends to hide himself and his secrets (magic and personal). He has the following latitude crossings at Las Vegas:

MO ASC/UR DSC	34 N 20
JU DSC/PL DSC	34 N 21
MO MH/SA DSC	34 N 30
Las Vegas	36 N 10
NE IC/VE ASC	36 N 40
VE DSC/UR DSC	37 N 26

The JU/PL crossing shows that miracles can take place at Las Vegas. His magic shows there have transformed his career, and many of his illusions look like miracles.

MO/UR shows that life is never boring for him in Las Vegas. He is always working on new illusions which he can introduce into shows.

VE/NE is appropriate for artistic illusions and the entertainment industry, but his personal relationships would tend to be disappointing.

VE/UR indicates a wide circle of friends, but he would prefer to have his freedom in close personal relationships.

MO/SA suggests isolated self-study and planning his performances in detail. It could also be times when he feels lonely in life.

When all the planetary influences are synthesised it is clear why he is successful in his career. But close personal relationships are not easy for him in Las Vegas.

Prior to 1993, David Copperfield was unknown in Europe as he had never performed any shows there. That autumn he made his first European tour. A few months before the premiere performance in Munich he received a great deal of media publicity. Afterward, tickets for his shows sold

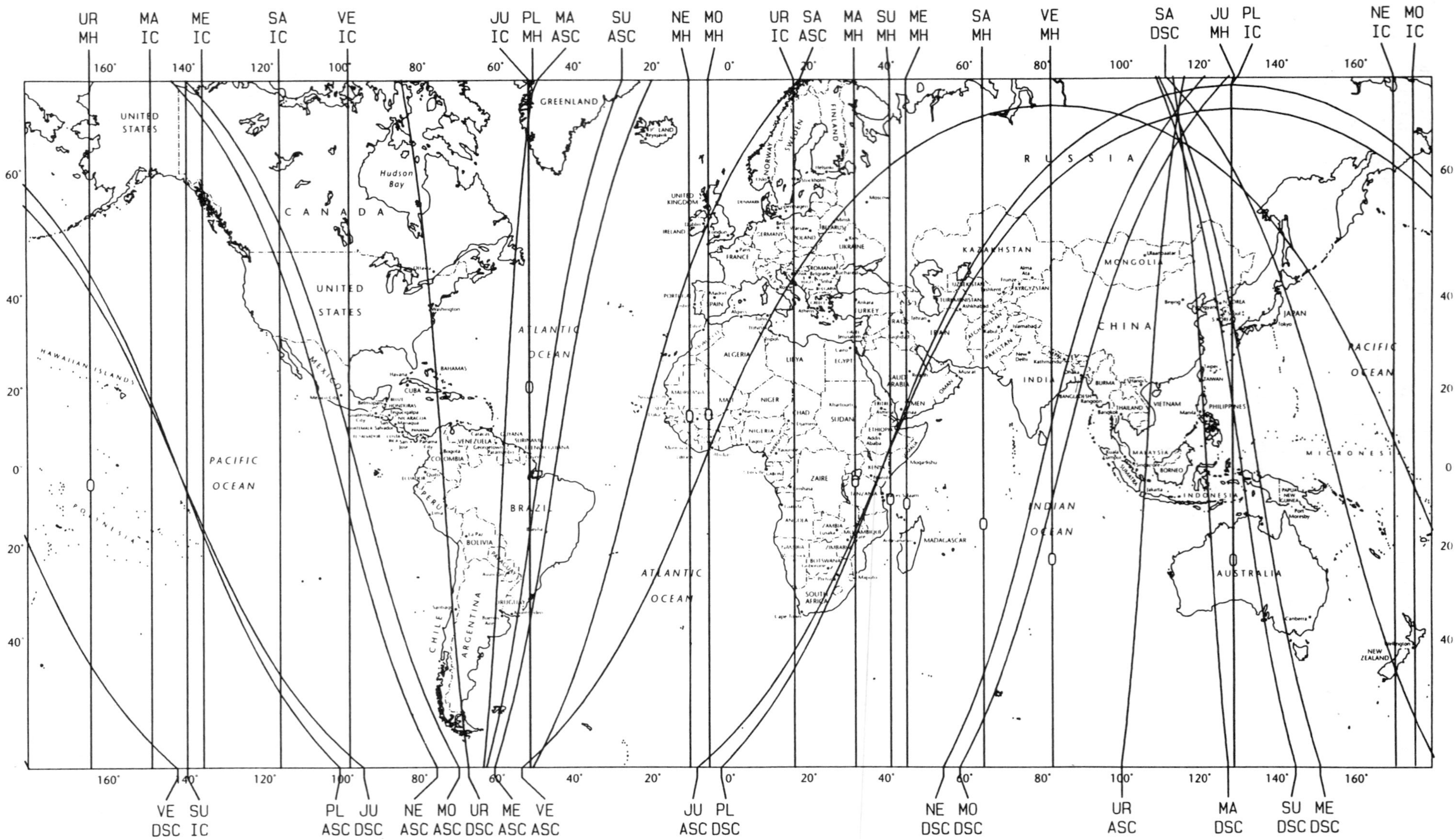

Margaret Thatcher

October 13, 1925 9:00 AM GMT

04/16/96

04/16/96

Margaret Thatcher
October 13, 1925 9:00 AM GMT

ASTRO*CARTO*GRAPHY LATITUDE CROSSINGS

NORTH LATITUDES

69N58 SU/IC * MO/ASC
69N24 UR/IC * SA/ASC
69N11 JU/MH * PL/DSC
69N09 MO/DSC * PL/DSC
69N09 MA/DSC * SA/DSC
68N54 NE/DSC * PL/DSC
68N42 UR/ASC * PL/DSC
68N39 SA/DSC * PL/DSC
68N34 MA/DSC * PL/DSC
68N34 SU/IC * NE/ASC
68N27 SA/DSC * UR/ASC
68N24 ME/IC * MO/ASC
68N22 ME/DSC * PL/DSC
68N16 SU/DSC * PL/DSC
67N59 SA/DSC * NE/DSC
67N09 UR/ASC * NE/DSC
67N00 MA/DSC * NE/DSC
66N59 PL/IC * JU/DSC
66N46 MO/DSC * JU/ASC
66N43 JU/ASC * SA/DSC
66N40 MA/DSC * UR/ASC
66N39 ME/DSC * NE/DSC
66N33 ME/IC * NE/ASC
66N32 MO/DSC * SA/DSC
66N31 VE/DSC * PL/DSC
66N29 SA/MH * VE/ASC
66N23 MA/DSC * JU/ASC
66N22 JU/ASC * UR/ASC
66N20 ME/DSC * JU/ASC
66N14 JU/ASC * NE/DSC
66N13 SU/DSC * JU/ASC
66N09 SU/DSC * NE/DSC
66N08 PL/MH * MA/ASC
66N07 ME/DSC * UR/ASC
66N04 JU/IC * MA/ASC
65N31 ME/DSC * MA/DSC
65N27 VE/DSC * JU/ASC
64N54 SU/DSC * UR/ASC
64N44 VE/DSC * NE/DSC
64N29 MO/DSC * ME/DSC
64N08 MO/DSC * MA/DSC
64N03 VE/DSC * UR/ASC
63N46 SU/DSC * VE/DSC
63N45 MO/DSC * VE/DSC
63N44 MO/DSC * SU/DSC
63N34 VE/DSC * MA/DSC
63N18 NE/IC * PL/ASC
63N09 ME/DSC * VE/DSC
62N35 MO/DSC * UR/ASC
62N27 ME/MH * VE/ASC
62N02 SU/DSC * MA/DSC
61N23 MO/IC * PL/ASC
61N13 Anchorage
61N12 VE/MH * PL/DSC
61N00 SU/MH * VE/ASC
60N40 NE/IC * JU/DSC
60N35 JU/MH * SA/DSC
60N34 PL/IC * SA/DSC

59N20 Stockholm
59N03 JU/MH * VE/DSC
59N03 PL/IC * VE/DSC
58N38 MO/IC * JU/DSC
58N27 VE/MH * JU/ASC
57N27 VE/DSC * SA/DSC
56N52 MA/MH * VE/ASC
55N45 Moscow
55N41 Copenhagen
53N29 MO/MH * SA/ASC
53N13 SA/IC * MO/ASC
52N30 Berlin
51N30 London

49N14 Vancouver

CENTRAL LATITUDES

49N14 SA/MH * PL/DSC
48N50 Paris
47N36 Seattle
46N05 SA/MH * JU/ASC
45N58 SA/IC * NE/ASC
45N33 NE/MH * SA/ASC
45N32 Montreal
45N02 UR/MH * PL/ASC
44N59 Minneapolis
44N17 UR/IC * VE/ASC
43N39 Toronto
42N22 Boston
42N20 Detroit
41N54 Rome
41N53 Chicago
41N51 UR/MH * JU/DSC
40N43 New York

39N44 Denver
38N53 Washington
38N37 St. Louis
37N47 San Francisco
36N10 Las Vegas
36N02 JU/MH * ME/DSC
35N57 PL/IC * ME/DSC
35N40 Tokyo
34N04 Los Angeles
33N45 Atlanta
33N27 Phoenix
32N47 Dallas
30N03 Cairo

29N58 New Orleans
29N46 Houston
28N39 Delhi
25N47 Miami
22N34 Calcutta
22N16 Hong Kong
21N19 Honolulu

19N26 Mexico City
17N14 MA/IC * PL/ASC
16N22 ME/MH * PL/DSC
16N10 JU/MH * SU/DSC
16N02 PL/IC * SU/DSC
15N28 MA/IC * JU/DSC
14N46 ME/MH * JU/ASC
14N35 Manila
11N37 VE/MH * NE/DSC
10N30 Caracas

8N07 VE/IC * MO/ASC
6N56 NE/IC * VE/DSC
6N27 Lagos
5N47 SU/MH * PL/DSC
5N14 SU/MH * JU/ASC
4N57 MO/MH * VE/ASC
0N26 JU/ASC * PL/DSC

0S26 JU/DSC * PL/ASC
4S57 MO/IC * VE/DSC
5S14 SU/IC * JU/DSC
5S47 SU/IC * PL/ASC
6S56 NE/MH * VE/ASC
8S07 VE/MH * MO/DSC

11S37 VE/IC * NE/ASC
14S46 ME/IC * JU/DSC
15S28 MA/MH * JU/ASC
16S02 PL/MH * SU/ASC
16S10 JU/IC * SU/ASC
16S22 ME/IC * PL/ASC
17S14 MA/MH * PL/DSC

22S54 Rio De Janeiro

SOUTH LATITUDES

26S12 Johannesburg

33S55 Sydney
34S36 Buenos Aires
35S57 PL/MH * ME/ASC
36S02 JU/IC * ME/ASC
36S52 Auckland
37S50 Melbourne

41S51 UR/IC * JU/ASC
44S17 UR/MH * VE/DSC
45S02 UR/IC * PL/DSC
45S33 NE/IC * SA/DSC
45S58 SA/MH * NE/DSC
46S05 SA/IC * JU/DSC
49S14 SA/IC * PL/ASC

53S13 SA/MH * MO/DSC
53S29 MO/IC * SA/DSC
56S52 MA/IC * VE/DSC
57S27 VE/ASC * SA/ASC
58S27 VE/IC * JU/DSC
58S38 MO/MH * JU/ASC
59S03 PL/MH * VE/ASC
59S03 JU/IC * VE/ASC

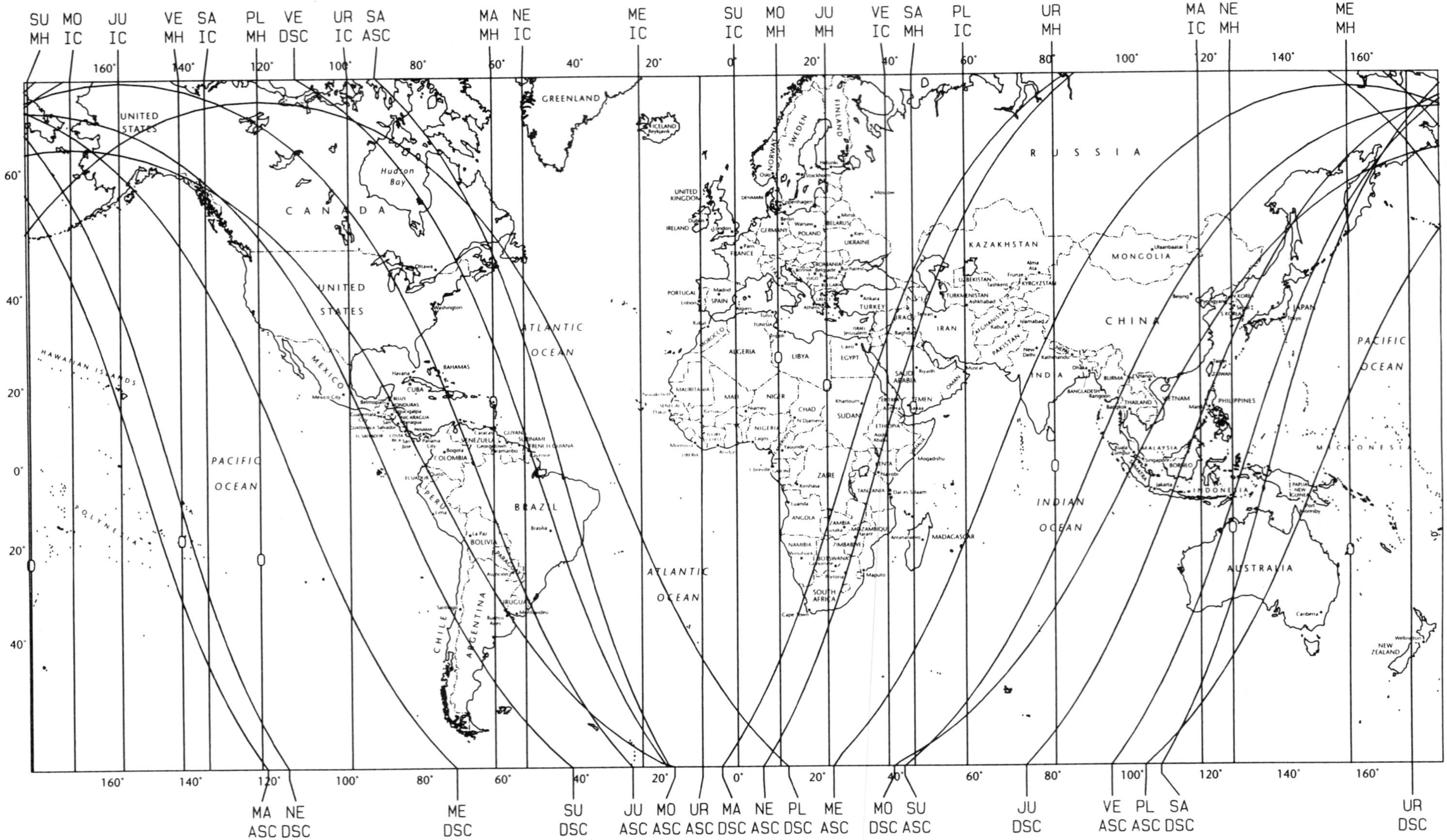

United Kingdom

January 1, 1801 12:00 AM GMT

09/18/98

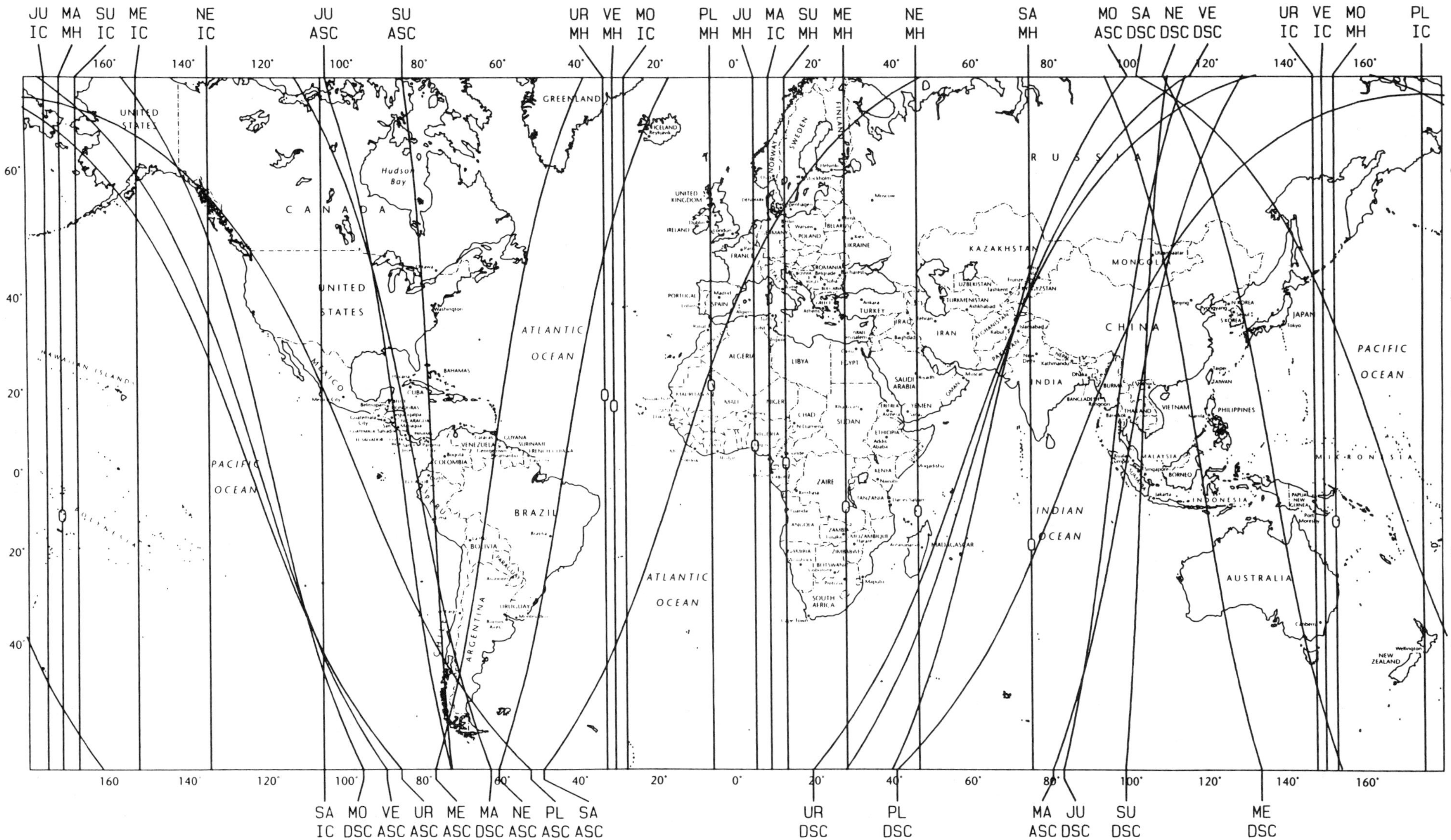

David Copperfield

September 16, 1956 7:02 AM EDT

04/16/96

04/16/96

David Copperfield
September 16, 1956 7:02 AM EDT

ASTRO*CARTO*GRAPHY LATITUDE CROSSINGS

NORTH LATITUDES

69N49 VE/DSC * JU/DSC
69N46 NE/MH * SA/ASC
69N37 MA/ASC * UR/DSC
69N34 JU/IC * VE/ASC
69N28 SU/DSC * NE/DSC
68N53 VE/DSC * NE/DSC
68N37 MA/MH * VE/ASC
68N35 SU/DSC * SA/DSC
68N34 VE/DSC * SA/DSC
68N33 SU/DSC * VE/DSC
68N23 PL/IC * UR/ASC
68N20 MO/ASC * ME/DSC
68N18 ME/IC * MO/DSC
68N17 UR/ASC * PL/ASC
68N00 SA/DSC * NE/DSC
67N55 JU/IC * PL/ASC
67N49 JU/DSC * SA/DSC
67N38 JU/DSC * NE/DSC
67N37 MA/MH * PL/ASC
67N21 SU/IC * VE/ASC
67N18 SA/DSC * UR/DSC
67N11 SU/IC * PL/ASC
67N05 VE/ASC * PL/ASC
66N59 UR/DSC * NE/DSC
66N46 MO/MH * PL/DSC
66N35 JU/DSC * UR/DSC
66N19 VE/IC * PL/DSC
65N54 UR/IC * PL/DSC
65N51 JU/IC * UR/ASC
65N48 SU/DSC * UR/DSC
65N28 SA/IC * MA/DSC
65N01 ME/DSC * VE/DSC
64N43 MA/ASC * SA/DSC
64N40 ME/MH * SA/ASC
64N40 MA/MH * UR/ASC
64N21 ME/IC * PL/ASC
63N09 ME/DSC * UR/DSC
63N06 SU/IC * UR/ASC
61N18 MA/ASC * NE/DSC
61N13 Anchorage
60N50 SA/DSC * PL/DSC
60N49 MO/DSC * PL/ASC
60N16 SU/DSC * JU/DSC

59N47 ME/IC * VE/ASC
59N20 Stockholm
58N35 MO/IC * NE/ASC
57N00 NE/IC * PL/ASC
56N46 NE/DSC * PL/DSC
55N55 ME/DSC * JU/DSC
55N45 Moscow
55N41 Copenhagen
55N25 SU/MH * SA/ASC
53N54 ME/IC * UR/ASC
53N49 SU/DSC * ME/DSC
53N24 VE/MH * NE/ASC
52N30 Berlin
51N30 London
51N21 MA/IC * SA/ASC
50N47 NE/IC * MO/DSC

49N14 Vancouver
48N50 Paris
48N32 MA/ASC * PL/DSC
48N01 UR/MH * NE/ASC
47N36 Seattle
46N47 JU/MH * SA/ASC
46N41 ME/DSC * PL/DSC
45N45 MA/DSC * JU/ASC
45N32 Montreal
45N12 ME/DSC * MA/ASC
44N59 Minneapolis
44N53 SA/MH * MO/ASC
43N44 UR/IC * SA/DSC

CENTRAL LATITUDES

43N39 Toronto
42N22 Boston
42N20 Detroit
42N13 SU/DSC * PL/DSC
41N54 Rome
41N53 SA/MH * VE/DSC
41N53 Chicago
41N15 SA/MH * UR/DSC
40N43 New York

39N57 VE/IC * SA/DSC
39N44 Denver
38N53 Washington
38N37 St. Louis
37N47 San Francisco
37N26 VE/DSC * UR/DSC
36N40 NE/IC * VE/ASC
36N10 Las Vegas
35N40 Tokyo
34N30 MO/MH * SA/DSC
34N21 JU/DSC * PL/DSC
34N20 MO/ASC * UR/DSC
34N04 Los Angeles
33N45 Atlanta
33N27 Phoenix
32N47 Dallas
32N08 MO/ASC * VE/DSC
30N03 Cairo

29N58 New Orleans
29N46 Houston
28N39 Delhi
27N45 NE/IC * UR/ASC
27N09 SU/DSC * MA/ASC
26N05 PL/MH * SA/ASC
25N47 Miami
22N34 Calcutta
22N16 Hong Kong
21N40 SA/IC * PL/ASC
21N19 Honolulu

19N26 Mexico City
14N35 Manila
10N30 Caracas

6N27 Lagos

21S40 SA/MH * PL/DSC
22S54 Rio De Janeiro
26S05 PL/IC * SA/DSC
26S12 Johannesburg
27S09 SU/ASC * MA/DSC
27S45 NE/MH * UR/DSC

32S08 MO/DSC * VE/ASC
33S55 Sydney
34S20 MO/DSC * UR/ASC
34S21 JU/ASC * PL/ASC
34S30 MO/IC * SA/ASC
34S36 Buenos Aires
36S40 NE/MH * VE/DSC
36S52 Auckland
37S26 VE/ASC * UR/ASC
37S50 Melbourne
39S57 VE/MH * SA/ASC

41S15 SA/IC * UR/ASC
41S53 SA/IC * VE/ASC
42S13 SU/ASC * PL/ASC
43S44 UR/MH * SA/ASC
44S53 SA/IC * MO/DSC
45S12 ME/ASC * MA/DSC
45S45 MA/ASC * JU/DSC
46S41 ME/ASC * PL/ASC
46S47 JU/IC * SA/DSC

SOUTH LATITUDES

48S01 UR/IC * NE/DSC
48S32 MA/DSC * PL/ASC

50S47 NE/MH * MO/ASC
51S21 MA/MH * SA/DSC
53S24 VE/IC * NE/DSC
53S49 SU/ASC * ME/ASC
53S54 ME/MH * UR/DSC
55S25 SU/IC * SA/DSC
55S55 ME/ASC * JU/ASC
56S46 NE/ASC * PL/ASC
57S00 NE/MH * PL/DSC
58S35 MO/MH * NE/DSC
59S47 ME/MH * VE/DSC

out fast in what became "Copperfield fever" in Germany. Copperfield agreed to put on extra shows in Germany, and no sooner had tickets gone on sale than they were sold out. David Copperfield was the sensation of the year when he visited Germany in 1993. In Munich, people were queuing over **six** hours to purchase tickets for his shows. Munich hadn't seen anything like this for years.

JU MH near Germany shows the spectacular success he obtained from the media publicity.

MA IC indicates the challenge and stamina he had to perform three shows a day when he was in Germany.

SA ASC is appropriate for the image he projected being dressed in black. He might have felt restricted that he couldn't speak German.

SU MH passes through Germany, which explains why he was an overnight success before he arrived in the country. This line is also good for promoting himself, and it was the publicity about his amazing illusions that made him a superstar.

The SU MH line passes through Berlin. In October 1993, he performed a special show in Berlin where he received a great amount of publicity for other reasons. In the audience was Claudia Schiffer (Germany's top model) and he invited her on stage to help him.

It was love at first sight for David Copperfield and Claudia Schiffer. After meeting each other in Berlin they were never out of the media limelight.

At first most people thought it was a big publicity stunt instead of true love. But it soon became clear they were deeply in love and couldn't be apart from each other. They are engaged and plan to marry.

Why did David Copperfield fall in love when he visited Germany? In the next chapter on Cyclo*Carto*Graphy the answer will be clear.

Example 6—Claudia Schiffer and Germany

Claudia Schiffer is famous for her modelling career, not only in Germany but also in other European countries. She is Germany's top model, having a high media profile. As a result she earns more than any other woman in Germany.

Amazingly, Claudia Schiffer has SU MH line passing through Berlin, the same place where David Copperfield has his SU MH line. They met in Berlin and, with both having SU MH exact at this location, their relationship will always be in the public spotlight.

Once again we see a SU MH line bringing a person fame and popularity. Claudia Schiffer's SU MH line has an influence throughout Germany where she achieved success at an early age. SU MH is good for any kind of promotion work and creativity, which Claudia Schiffer has certainly benefited from.

Claudia Schiffer also has MA MH passing through Germany and France. This shows that she works hard in her career, using the masculine energies in a very positive way. It is usually difficult for a woman to express the external energies of Mars lines, but Claudia Schiffer has no such problems. With her long blond hair and good figure she easily projects the sexual side of MA MH in society without much effort.

SA DSC is just within orb of Berlin, which would have some influence when she met David Copperfield. SA DSC suggests being cautious in close relationships, a late marriage or age difference with the partner. David Copperfield is fourteen years older than Claudia Schiffer, and despite what has been reported in the press they have not rushed into marriage.

At Berlin, Claudia Schiffer has the following latitude crossings:

SA MH/SU ASC	49 N 50
JU MH/MO DSC	50 N 15
Berlin	52 N 30
VE DSC/UR DSC	54 N 45
MO IC/SU DSC	54 N 51
SA ASC/PL DSC	54 N 59

The VE/UR crossing explains the "love at first sight" meeting with David Copperfield.

MO/JU indicates enjoying a good social life in Berlin. It was at a social event where she had the good fortune to be invited on stage by David Copperfield.

SU/MO shows that sexual and personal polarities are intensified, and it would be hard for her to relate to men through sexual or marital roles.

SU/SA suggests projecting seriousness, being drawn to serious ideas and studies, and seeing life basically as a long and tedious job, although one that merits careful execution.

SA/PL indicates seriousness, responsibility, duty and concern with death. There is also an unusual, unquestioning respect for authority.

The above interpretations for the latitude crossings are taken form the Astro*Carto*Graphy interpretation booklet. Little is known about Claudia Schiffer's visits to Berlin, only that she met David Copperfield there.

However, the latitude crossings for Berlin do give a good background picture about meeting her future husband.

Example 7—Matthew

I decided to include an example of one of my clients as his Astro*Carto*Graphy map is really amazing. The planetary influences at certain locations show very clearly the significant events that have taken place in his life.

I am most grateful that Matthew granted permission to include his Astro*Carto*Graphy in this book. For reasons of confidentiality, I have changed his name.

The most powerful influence in Matthew's life has been his SA conjunct UR in the tenth house. The SA and UR lines on his Astro*Carto*Graphy map seem to represent areas of high risk and sudden endings in his career.

Matthew has travelled all over the world. He is a master mariner, and in his career at sea he has visited many places where SA and UR influences are prominent.

When he was a junior officer working along the west coast of North America, he went ashore in Vancouver, got

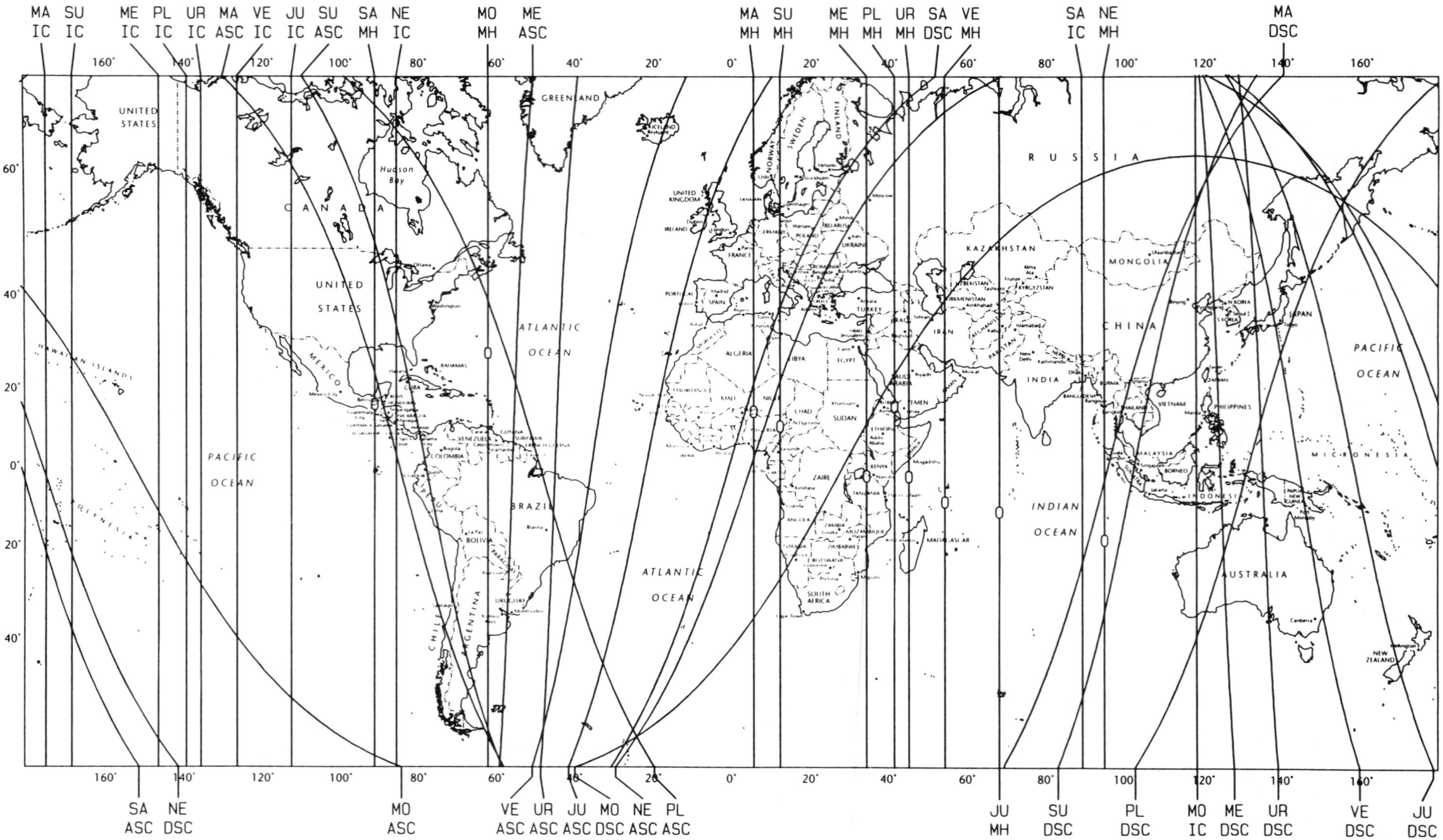

Claudia Schiffer

August 25, 1970 12:10 PM CET

04/16/96

04/16/96

Claudia Schiffer
August 25, 1970 12:10 PM CET

ASTRO*CARTO*GRAPHY LATITUDE CROSSINGS

NORTH LATITUDES

69N53 JU/MH * NE/ASC
69N15 SA/ASC * UR/DSC
68N40 VE/IC * MA/ASC
68N38 SU/DSC * SA/ASC
68N31 JU/DSC * NE/DSC
68N14 UR/DSC * NE/DSC
68N02 UR/MH * SA/DSC
68N01 SU/DSC * NE/DSC
67N39 JU/DSC * UR/DSC
67N39 MA/DSC * SA/ASC
67N37 SU/DSC * JU/DSC
67N35 SU/DSC * UR/DSC
67N24 SA/MH * PL/ASC
67N14 MA/DSC * NE/DSC
67N10 MA/MH * JU/ASC
66N42 VE/MH * NE/ASC
66N40 PL/MH * SA/DSC
66N22 MA/DSC * JU/DSC
65N39 MA/DSC * UR/DSC
64N53 NE/IC * PL/ASC
64N32 SU/DSC * VE/DSC
63N31 SA/ASC * NE/DSC
63N31 VE/DSC * MA/DSC
63N18 ME/MH * SA/DSC
63N08 UR/MH * NE/ASC
61N49 MO/ASC * ME/DSC
61N45 MO/ASC * SU/DSC
61N42 MO/ASC * MA/DSC
61N28 MO/ASC * VE/DSC
61N13 Anchorage
61N13 MO/ASC * UR/DSC
61N12 PL/MH * NE/ASC
60N48 JU/IC * MA/ASC
60N19 MO/ASC * JU/DSC

59N47 NE/MH * MO/DSC
59N20 Stockholm
59N17 SU/DSC * ME/DSC
59N04 ME/DSC * MA/DSC
58N35 SA/IC * MO/DSC
58N26 SU/DSC * MA/DSC
56N55 MO/ASC * SA/ASC
56N24 ME/MH * NE/ASC
55N59 NE/DSC * PL/DSC
55N58 MO/IC * MA/DSC
55N56 MO/ASC * PL/DSC
55N53 MO/ASC * NE/DSC
55N45 Moscow
55N41 Copenhagen
54N59 SA/ASC * PL/DSC
54N51 MO/IC * SU/DSC
54N45 VE/DSC * UR/DSC
52N30 Berlin
51N30 London
50N15 JU/MH * MO/DSC

49N50 SA/MH * SU/ASC
49N14 Vancouver
48N50 Paris
47N36 Seattle
45N32 Montreal
44N59 Minneapolis
44N20 JU/DSC * PL/DSC
43N39 Toronto
42N22 Boston
42N20 Detroit
41N54 Rome
41N53 Chicago
40N43 New York

39N44 Denver
39N31 VE/MH * MO/DSC
39N26 MO/MH * PL/ASC
38N53 Washington

CENTRAL LATITUDES

38N37 St. Louis
38N03 SU/MH * SA/DSC
37N47 San Francisco
36N10 Las Vegas
35N40 Tokyo
35N31 MA/IC * MO/ASC
34N21 NE/IC * SU/ASC
34N04 Los Angeles
33N45 Atlanta
33N27 Phoenix
32N47 Dallas
30N03 Cairo

29N58 New Orleans
29N46 Houston
28N39 Delhi
28N38 UR/MH * MO/DSC
27N41 VE/DSC * PL/DSC
26N55 SU/IC * MO/ASC
25N47 Miami
23N20 SA/MH * MA/ASC
23N14 PL/MH * MO/DSC
22N34 Calcutta
22N16 Hong Kong
21N46 SU/MH * NE/ASC
21N23 ME/ASC * PL/ASC
21N19 Honolulu
21N03 MA/MH * SA/DSC

19N26 Mexico City
14N35 Manila
11N34 UR/DSC * PL/DSC
11N30 ME/MH * MO/DSC
10N30 Caracas

6N27 Lagos
3N08 NE/IC * MA/ASC
2N28 MA/MH * NE/ASC

2S28 MA/IC * NE/DSC
3S08 NE/MH * MA/DSC

11S30 ME/IC * MO/ASC
11S34 UR/ASC * PL/ASC

21S03 MA/IC * SA/ASC
21S23 ME/DSC * PL/DSC
21S46 SU/IC * NE/DSC
22S54 Rio De Janeiro
23S14 PL/IC * MO/ASC
23S20 SA/IC * MA/DSC
26S12 Johannesburg
26S55 SU/MH * MO/DSC
27S41 VE/ASC * PL/ASC
28S38 UR/IC * MO/ASC

33S55 Sydney
34S21 NE/MH * SU/DSC
34S36 Buenos Aires
35S31 MA/MH * MO/DSC
36S52 Auckland
37S50 Melbourne
38S03 SU/IC * SA/ASC
39S26 MO/IC * PL/DSC
39S31 VE/IC * MO/ASC

44S20 JU/ASC * PL/ASC
49S50 SA/IC * SU/DSC

50S15 JU/IC * MO/ASC
54S45 VE/ASC * UR/ASC
54S51 MO/MH * SU/ASC
54S59 SA/DSC * PL/ASC
55S53 MO/DSC * NE/ASC
55S56 MO/DSC * PL/ASC

SOUTH LATITUDES

55S58 MO/MH * MA/ASC
55S59 NE/ASC * PL/ASC
56S24 ME/IC * NE/DSC
56S55 MO/DSC * SA/DSC
58S26 SU/ASC * MA/ASC
58S35 SA/MH * MO/ASC
59S04 ME/ASC * MA/ASC
59S17 SU/ASC * ME/ASC
59S47 NE/IC * MO/ASC

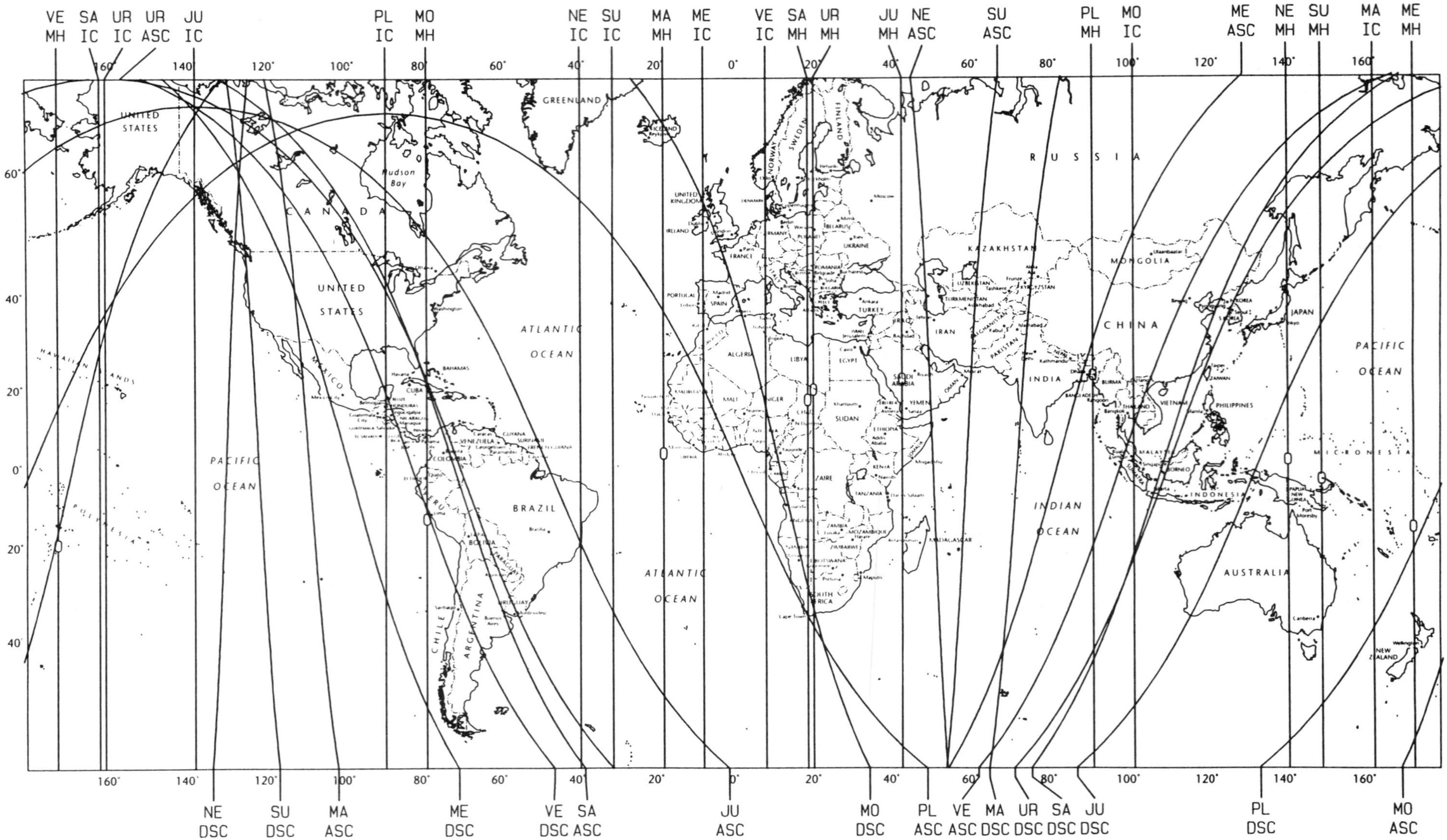

Matthew

04/16/96

ASTRO*CARTO*GRAPHY LATITUDE CROSSINGS

NORTH LATITUDES

69N54 MO/ASC * SA/ASC
69N40 VE/DSC * UR/ASC
69N37 VE/MH * UR/DSC
69N32 SU/DSC * SA/ASC
69N17 MA/IC * VE/ASC
69N12 ME/DSC * UR/ASC
68N31 JU/IC * UR/ASC
68N31 MA/IC * SA/DSC
68N26 SA/ASC * NE/DSC
68N09 ME/DSC * VE/DSC
68N07 MO/ASC * UR/ASC
67N37 ME/MH * UR/DSC
67N33 JU/IC * VE/DSC
67N32 VE/ASC * JU/DSC
67N32 ME/DSC * JU/DSC
67N32 MO/ASC * JU/ASC
67N24 JU/ASC * UR/ASC
67N17 MO/ASC * VE/DSC
67N17 SU/DSC * JU/ASC
67N04 JU/IC * ME/DSC
67N02 JU/ASC * NE/DSC
66N53 MO/ASC * ME/DSC
66N50 SU/DSC * UR/ASC
66N39 JU/IC * MO/ASC
66N29 MO/MH * PL/ASC
66N25 MA/ASC * SA/ASC
66N24 SU/MH * VE/ASC
66N17 MA/ASC * JU/ASC
66N13 JU/ASC * SA/ASC
66N04 UR/ASC * NE/DSC
65N52 UR/IC * JU/DSC
65N45 MA/MH * MO/DSC
65N38 SA/IC * JU/DSC
65N13 JU/ASC * PL/DSC
65N11 MA/IC * UR/DSC
64N47 SA/ASC * PL/DSC
64N25 SU/DSC * VE/DSC
64N10 MA/ASC * PL/DSC
64N04 SU/MH * SA/DSC
63N56 UR/ASC * PL/DSC
63N39 VE/DSC * NE/DSC
63N25 MA/ASC * UR/ASC
63N23 NE/MH * VE/ASC
63N17 VE/MH * JU/DSC
62N38 VE/DSC * PL/DSC
61N53 NE/DSC * PL/DSC
61N36 SU/DSC * PL/DSC
61N19 ME/DSC * PL/DSC
61N13 Anchorage
60N28 SU/DSC * ME/DSC
60N10 ME/DSC * NE/DSC

59N58 SU/MH * UR/DSC
59N20 Stockholm
59N16 NE/MH * SA/DSC
59N15 SU/DSC * NE/DSC
58N04 PL/IC * JU/ASC
57N50 VE/DSC * MA/ASC
57N10 JU/IC * PL/DSC
56N50 ME/MH * JU/DSC
56N38 NE/IC * PL/ASC
56N19 ME/IC * MO/DSC
55N45 Moscow
55N41 Copenhagen
54N26 NE/MH * UR/DSC
52N30 Berlin
51N30 London
51N23 MO/IC * ME/ASC
51N23 MO/MH * JU/ASC
51N11 SU/IC * PL/ASC

49N39 MA/IC * JU/DSC
49N14 Vancouver
48N50 Paris

CENTRAL LATITUDES

47N36 Seattle
45N32 Montreal
44N59 Minneapolis
43N39 Toronto
43N33 PL/IC * SA/ASC
42N22 Boston
42N20 Detroit
41N54 Rome
41N53 Chicago
41N47 PL/IC * UR/ASC
41N41 ME/DSC * MA/ASC
40N43 New York
40N21 MO/ASC * PL/DSC

39N44 Denver
38N53 Washington
38N37 St. Louis
37N47 San Francisco
37N42 MA/MH * PL/ASC
37N09 UR/IC * PL/DSC
36N10 Las Vegas
35N40 Tokyo
35N00 SA/IC * PL/DSC
34N10 SU/MH * JU/DSC
34N04 Los Angeles
33N45 Atlanta
33N38 UR/IC * MO/ASC
33N27 Phoenix
32N47 Dallas
30N50 PL/MH * ME/ASC
30N03 Cairo

29N58 New Orleans
29N46 Houston
28N48 SA/IC * MO/ASC
28N39 Delhi
28N07 SA/ASC * UR/ASC
25N47 Miami
22N34 Calcutta
22N16 Hong Kong
22N15 MO/MH * UR/ASC
21N19 Honolulu
21N17 MO/MH * SA/ASC
20N28 ME/IC * PL/ASC

19N26 Mexico City
19N14 PL/IC * VE/DSC
18N24 NE/MH * JU/DSC
15N37 VE/IC * MO/DSC
15N29 VE/MH * PL/DSC
14N35 Manila
10N30 Caracas
10N25 MO/IC * VE/ASC

6N27 Lagos

10S25 MO/MH * VE/DSC
15S29 VE/IC * PL/ASC
15S37 VE/MH * MO/ASC
18S24 NE/IC * JU/ASC
19S14 PL/MH * VE/ASC

20S28 ME/MH * PL/DSC
21S17 MO/IC * SA/DSC
22S15 MO/IC * UR/DSC
22S54 Rio De Janeiro
26S12 Johannesburg
28S07 SA/DSC * UR/DSC
28S48 SA/MH * MO/DSC

30S50 PL/IC * ME/DSC
33S38 UR/MH * MO/DSC
33S55 Sydney
34S10 SU/IC * JU/ASC
34S36 Buenos Aires

SOUTH LATITUDES

35S00 SA/MH * PL/ASC
36S52 Auckland
37S09 UR/MH * PL/ASC
37S42 MA/IC * PL/DSC
37S50 Melbourne

40S21 MO/DSC * PL/ASC
41S41 ME/ASC * MA/DSC
41S47 PL/MH * UR/DSC
43S33 PL/MH * SA/DSC
49S39 MA/MH * JU/ASC

51S11 SU/MH * PL/DSC
51S23 MO/IC * JU/DSC
51S23 MO/MH * ME/DSC
54S26 NE/IC * UR/ASC
56S19 ME/MH * MO/ASC
56S38 NE/MH * PL/DSC
56S50 ME/IC * JU/ASC
57S10 JU/MH * PL/ASC
57S50 VE/ASC * MA/DSC
58S04 PL/MH * JU/DSC
59S15 SU/ASC * NE/ASC
59S16 NE/IC * SA/ASC
59S58 SU/IC * UR/ASC

drunk, was arrested and spent the night in jail. Next day Matthew was charged with being under the influence of drugs, a crime with more serious consequences. He was faced with a large fine or jail term plus deportation. He was fined $200 (a lot of money in those days) and given a suspended sentence. At Vancouver Matthew has SU DSC and NE DSC, which show very clearly the events that happened. The NE connection to alcohol, drugs and being taken advantage of by others is very strong at Vancouver.

On the way back to New Zealand the captain learned about Matthew's ordeal in Vancouver. He informed the company, which made the decision to dismiss Matthew. The ship stopped at Rarotonga, a small mid-Pacific island in the Cook Islands, where Matthew was told he would be released at the end of the voyage. At Rarotonga Matthew has SA IC and UR IC.

A few years later Matthew was chief officer, and on a voyage experienced a considerable amount of trouble with the crew, some of whom were disobeying his orders and undermining his position. Matthew informed the captain about the dangerous situation that was developing, and asked if there were a possibility that he could get off the ship at Cape Town. The captain agreed and Matthew was taken ashore by boat. He was relieved to be in Cape Town, but his difficulties were far from over. The next day the authorities decided he was an illegal immigrant and wanted to arrest him immediately. They relented, giving him twenty-four hours to leave the country; he had no choice but to comply. At Cape Town Matthew has SA MH, UR MH, PL ASC and MO DSC. Once again trouble with authorities and sudden departure are shown by SA MH and UR MH. Things could have turned out much worse with PL ASC exact at Cape Town.

Three years later Matthew was promoted to captain in Lima, Peru, where he has MO MH and VE DSC. The zenith of MO MH is exactly at Lima, making this a powerful location for him. In Lima he is the center of attraction and people in high places recognize his abilities. After this high point, the people who favored him were removed and soon afterwards he was fired.

Matthew's last voyage began in Gdynia, Poland near where the captain was born (Kaliningrad). The SA MH and UR MH lines pass through Poland, with SA MH exact at Gdynia. Both lines are also near to Kaliningrad. It soon became apparent to Matthew that the captain was doing everything possible to undermine his job and lost no opportunity to make things as difficult as possible for him. Two months into the voyage the captain told the company about Matthew. The voyage ended in Barranquilla, Columbia, where Matthew was fired. At Barranquilla SA and UR turn up again, this time ASC lines. In Barranquilla Matthew experienced intense hardship. After learning his tickets were not in order, he had problems leaving the country. Eventually, he was allowed to leave and went to Miami, another location with SA and UR lines. Matthew was not allowed to stay in Miami because he didn't have a U.S. visa, and was forcibly detained at the airport for more than twelve hours while awaiting a flight to the United Kingdom. At that time the Miami airport did not have a separate lounge for transit passengers, so Matthew had a security guard with him all the time.

Matthew had no knowledge of Astro*Carto*Graphy while these events were taking place. The reoccurrence of events on different SA and UR lines is incredible.

Matthew has also experienced the effects of locations by remote activation. He has never been to Bangkok, where he has VE ASC and MO IC lines. These lines together suggest the possibility of a loving relationship, and a few years ago he met a woman from Bangkok and they were very much attracted to each other. He met her in England, so the planetary lines came to him through this woman.

Matthew has also been involved with two or three Italian women, all of whom came from the Turin/Genoa area; he met them in England. He visited Genoa briefly when he was on a tanker, but he didn't meet anyone then as he couldn't leave the ship. His VE IC line is exactly at Genoa, another example of remote activation.

Further north along his VE IC line Matthew met a woman in Emden, Germany. He wanted to stay with her, but when he returned she had married someone else.

Astro*Carto*Graphy is simple to use and provides accurate information about the planetary influences at various locations.

In the next three chapters, the same seven examples will be used to show the effects of temporary influences.

Chapter 5/Cyclo*Carto*Graphy

The seven examples in the last chapter showed that Astro*Carto*Graphy gives accurate information about how the planetary energies at locations are experienced. We also saw that it is not necessary to visit a location for the planetary influences to have an effect.

After a few years working with Astro*Carto*Graphy Jim Lewis realized that there were times when something seemed to be missing. His own research showed that sometimes the planetary influences at a location did not exactly correspond with the events and experiences that took place. Quite often he saw that major events occurred at locations with no planetary lines. Or residence under a certain line was not always a consistent experience, some years being different from others.

Feedback from clients showed that Astro*Carto*Graphy on its own was not a complete system, and sometimes planetary lines did not work out as expected. For example, a woman with a VE DSC line in Barbados read in the Astro*Carto*Graphy interpretation booklet that VE DSC is the best place to find a partner and get married. As this woman had not had much luck in partnerships she went to Barbados on holiday with great enthusiasm, knowing what was waiting for her. But during her two-week stay she found relationships frustrating and problematic. She did meet a man in Barbados, but he was already married, so nothing came of it, and she returned home extremely disappointed. She cannot understand why partnerships were so difficult in Barbados. Why didn't VE DSC work out the way it usually does? Obviously, this woman was not impressed with Astro*Carto*Graphy.

After receiving such letters from clients, Jim Lewis did further research to try to discover why the planetary lines were not always 100 percent accurate. He soon learned the reason, and it was such an important discovery that he decided to take Astro*Carto*Graphy one stage further. This was how Jim Lewis developed Cyclo*Carto*Graphy, which adds new information to the Astro*Carto*Graphy map.

Astro*Carto*Graphy Tells You Where, Cyclo*Carto*Graphy Tells You When

The Astro*Carto*Graphy map shows the planetary influences at locations around the world. These influences are lifelong and permanent, showing **where** in the world they will manifest.

Cyclo*Carto*Graphy shows the temporary influences taking place on the Astro*Carto*Graphy map. These temporary influences usually last for a year or two, and show times when the Astro*Carto*Graphy lines manifest more or less powerfully.

The Cyclo*Carto*Graphy planetary influences do not contradict the power of the Astro*Carto*Graphy lines. These influences are always at locations, whereas the moving Cyclo*Carto*Graphy lines show how the permanent influences are modified.

Cyclo*Carto*Graphy gives additional information about locations on the Astro*Carto*Graphy map and shows **when** these influences are most likely to manifest.

Returning to the woman with VE DSC at Barbados, at the time of her visit she had Cyclo*Carto*Graphy SA MH exact there. So during the two week period she was in Barbados the Cyclo*Carto*Graphy SA MH modified the influence of her VE DSC at that location. The temporary influence of SA MH would tend to suppress the potential of the permanent VE DSC line at Barbados. Hence, at that time the VE DSC line did not manifest as expected. Had she visited Barbados when a temporary JU line was there, then the permanent VE DSC would have been much more powerful and beneficial.

When an important move is planned it is essential to take Cyclo*Carto*Graphy into consideration. Having found a suitable location on the Astro*Carto*Graphy map to move to, the next question is **when** is the best time to do so.

Cyclo*Carto*Graphy is the most powerful technique to find out which temporary influences are in effect on the Astro*Carto*Graphy map. However, there are two other techniques which can be used: the Astro*Carto*Graphy solar return map and Astro*Carto*Graphy solar arc progressions. These will be explained in the next two chapters.

Progressions and Transits

Throughout life we all experience times when nothing seems to go right for us. This could be due to unemployment, a divorce or other experiences. They are all learning times for individual growth and development.

Astrologers use progressions and transits in the natal chart to find out when a person will be faced with challenges and opportunities in life. How these developments will manifest depends on the individual concerned. An astrologer can see from progressions and transits which area of life will be under the spotlight and how things are likely to work out. How this information is given to a client is most important. It is a subject that will be examined in more detail in Chapter 12.

Jim Lewis thought that if progressions and transits could be used in a natal chart, then surely they could also be used in Astro*Carto*Graphy. After finding out that this was the case he developed Cyclo*Carto*Graphy, the name indicating the planetary cycles.

Cyclo*Carto*Graphy uses a combination of progressions and transits which are then drawn on a transparent overlay the same size as an Astro*Carto*Graphy map. The overlay is then placed on top of the Astro*Carto*Graphy map, giving a worldwide view of the progressions and transits.

Some of the Cyclo*Carto*Graphy techniques have **no** equivalent methods in astrology. The only in-depth information written about Cyclo*Carto*Graphy is the interpretation booklet by Jim Lewis, which is available when a Cyclo*Carto*Graphy overlay is purchased.

First, we need to know some background information to the Cyclo*Carto*Graphy overlay. As mentioned earlier, Cyclo*Carto*Graphy uses a combination of progressions and transits, and these are as follows:

Progressions	**Transits**
Sun	Jupiter
Moon	Saturn
Mercury	Uranus
Venus	Neptune
Mars	Pluto

Notice that Cyclo*Carto*Graphy uses the same planets used in Astro*Carto*Graphy. The **new** positions of the planets are then drawn on an overlay, resulting in forty lines as on an Astro*Carto*Graphy map.

It's like using progressions and transits in a natal or relocated chart, but only looking for conjunctions to the four angles—that is, after allowing for the celestial latitude and declination of the progressed and transiting planets. Progressed Jupiter, Saturn, Uranus, Neptune and Pluto do not move much during a lifetime, so progressions for these planets are not considered in Cyclo*Carto*Graphy. On the other hand, the transiting Sun, Moon, Mercury, Venus and Mars move too fast to have a long term meaning. This combination of progressions and transits gives a good picture of the temporary influences on the Astro*Carto*Graphy map.

At this stage you might be asking if the progressed angles of a natal or relocated chart are used in Cyclo*Carto*Graphy. Yes, they are used, but not in the normal way. This will be explained later, but first it is important to understand how the progressed and transiting planets move on an Astro*Carto*Graphy map.

The normal direction the planets move across the Astro*Carto*Graphy map is **eastward.** If a planet is **retrograde,** then it moves **westward** across the Astro*Carto*Graphy map. The Sun and Moon always move in an **eastward** direction as they are never retrograde. The other planets can move both **eastward** and **westward.**

The Cyclo*Carto*Graphy planets move at varying rates of speed across an Astro*Carto*Graphy map. This is made more complex when progressions and transits change direction. Let's take a look at how the Cyclo*Carto*Graphy planets move across the Astro*Carto*Graphy map.

Sun: As the Sun only moves in an eastward direction its rate of motion is fairly constant. The progressed Sun moves about one degree per year, the equivalent of fifty-five miles.

Moon: This is one of the fastest moving lines. It moves about one degree per month, so in a year it moves approximately 660 miles.

Mercury: On average, Mercury moves about one and a half degrees per year, which is roughly seventy miles. However, twenty-five percent of the time it is retrograde and can slow down to virtually no movement. When Mercury is at full speed it moves about 110 miles a year (two degrees).

Venus: The rate of movement varies according to which direction it is moving. The average motion is about one and a quarter degrees a year (about seventy miles). Venus is retrograde less than ten percent of its time, and when it is, or when it is about to change direction, it moves very slowly.

Mars: This is the slowest moving line, on average moving about forty miles a year (three-quarters of a degree). As Mars spends a long period of time moving retrograde it can stay at a location for more than ten years. When it changes direction exactly at a location it will have an influence for more than twenty years!

Jupiter: This is the fastest moving line, its motion being about 1,700 miles per year (thirty degrees). Due to retrograde motion its effects at a location can vary. Sometimes Jupiter moves at a rapid speed, moving more than 2,000 miles in less than a year. Then it changes direction and moves about 500 miles westward. Jupiter can be at a location for one to two weeks when it is moving fast, or a few months when it is retrograde.

Saturn: In one year Saturn moves about 660 miles (about twelve degrees), but its motion is not constant. For about one-third of its time it is moving slowly, backwards and forwards. Then for the rest of the time its motion speeds up and it moves forward about 1,000 miles.

Uranus: On average Uranus moves about four and a half degrees a year, the equivalent of about 250 miles. Although Uranus moves about 440 miles in seven months, the remaining five months it moves about 220 miles in the opposite direction. So it can be said that Uranus moves in a constant eastward direction.

Neptune: In one year Neptune moves about two degrees, or just over 100 miles. Its movement is not constant due to retrograde motion. Over a two-year period Neptune can be at a location up to five times. Neptune tends to move 250 miles eastward and then westward for about 150 miles after changing direction.

Pluto: The speed at which Pluto moves is presently about two degrees, or about 100 miles. Its motion is about the same as that of Neptune, i.e. about 250 miles forward and then approximately 150 miles backward.

The Cyclo*Carto*Graphy overlay is valid for a period of approximately twelve months. After that time it is best to obtain a new overlay due to the movement of the planets.

At the present time it is not possible to do Cyclo*Carto*Graphy on a personal computer, mainly because it is virtually unknown among astrologers. The only way to learn about it is to purchase an overlay.

During the past few years more astrological software has become available that draws Astro*Carto*Graphy maps. But most astrologers who have such a program do not really know how to interpret them.

The best astrological software program available to draw Astro*Carto*Graphy maps is SolarMaps from Esoteric Technologies Pty Ltd. Although Cyclo*Carto*Graphy is **not** available on the SolarMaps program I am still able to obtain the necessary information for the temporary influences. (See appendix 3 for latest information about computer software for Astro*Carto*Graphy.)

The various Cyclo*Carto*Graphy techniques can be used without having to obtain an overlay, but more work is involved. As the information obtained from Cyclo*Carto*Graphy is vitally important, the extra work is worthwhile.

The techniques used in Cyclo*Carto*Graphy fall into three categories:

1. Anywhere aspects
2. Location aspects
3. Latitude aspects

When you look for these aspects it is best to write a list of the information obtained for each category. Afterwards, the information on these lists is synthesised so that a clear picture of the temporary influences can be acquired.

Anywhere Aspects

These are the most important temporary influences as they will have an effect **no matter where in the world you live**. I call these temporary influences "anywhere aspects" as they apply anywhere in the world.

When looking for the temporary influences **without** an overlay, it is essential to use a **right ascension horoscope**. This type of horoscope was mentioned briefly in Chapter 2, but it is only when working with Cyclo*Carto*Graphy that its importance will be understood.

Prince Charles' Astro*Carto*Graphy map will be used to explain how to find the Cyclo*Carto*Graphy temporary influences.

First, using right ascension, make a list in numerical order of the ten natal planets. This will make things easier in finding the Cyclo*Carto*Graphy aspects. The skiing tragedy at Klosters on March 10, 1988 that involved Prince Charles will be used as an example.

Next calculate the progressed positions in right ascension for March 10, 1988. Then, using the same date, calculate the right ascension positions of the transiting planets. Most computer software programs have a facility to calculate horoscopes in right ascension. When this has been done make a combined list of the progressions and transits in numerical order. You now have two lists.

If you have done this yourself the lists will look like:

A*C*G	**C*C*G**
ME 05 SCO 13	UR 00 CAP 50
SA 07 SAG 38	SA 01 CAP 51
NE 13 LIB 36	SU 02 CAP 31
VE 15 LIB 49	VE 03 SAG 09
SU 20 SCO 01	ME 09 CAP 43
MA 20 SAG 05	NE 10 CAP 42
PL 21 LEO 29	MO 14 LIB 31
MO 28 ARI 28	PL 14 SCO 55
JU 29 GEM 54	MA 22 CAP 52
UR 29 SAG 55	JU 28 ARI 33

The first aspects to look for are line superimpositions, i.e. a C*C*G planet exactly on top of, or very close to, an A*C*G planet. With an overlay this is easy to see by looking at both C*C*G and A*C*G MH and IC lines.

With the two lists we look for either a **conjunction** or **opposition** aspect between a C*C*G planet and A*C*G planet. An orb of one degree is used.

By doing so we find the following aspects:

1. C*C*G MO = A*C*G NE
2. C*C*G JU = A*C*G MO
3. C*C*G UR = A*C*G JU
4. C*C*G UR = A*C*G UR

The next aspects to look for are C*C*G lines crossing A*C*G lines **at the equator**. This is known as a mundane square aspect. Using the two lists look for a **square aspect** between a C*C*G planet and an A*C*G planet.

Prince Charles has no crossings at the equator. If an equator crossing had been found, then "EQU" is used instead of "=". If C*C*G MO was at 14 **CAP** 31, then the aspect would be C*C*G MO **EQU** A*C*G NE.

Next, look at the birthplace progressed angles to see if they make aspects to A*C*G planets. This is achieved as follows:

The birthplace is moving **eastward** across the map the same number of degrees as a person's age. Prince Charles was thirty-nine years old when the skiing tragedy occurred in Klosters. Moving London (0W10) thirty-nine degrees **eastward** brings the birthplace close to his A*C*G SU IC line (orb between one and one and a half degrees). About three years later the progressed birthplace would be conjunct his SA ASC line.

Progressing the birthplace is easy to do with the Astro*Carto*Graphy map. First, draw a line at London's latitude; then measure the distance for thirty-nine degrees along the top of the map. Using this distance, measure eastward along the line of latitude for London.

Alternatively, the same information can be obtained by using the SolarMaps or AstroMaps Hi-Res programs.

Having found this aspect it is written as follows:

5. prog. birthplace = A*C*G SU IC

Now look for C*C*G lines at the birthplace. This can be achieved by using AstroMaps Hi-Res—first, progressions and then transits for March 10, 1988. It's not possible to combine both progressions and transits with AstroMaps Hi-Res, so they have to be used separately.

Prince Charles has no C*C*G lines at his birthplace as MO IC is two degrees away from London. Therefore, we can move onto the next part.

The next technique will the most difficult to understand as

there is no astrological equivalent. Jim Lewis calls this technique the "Displaced Position" (local aspects to the progressed angles). The title on its own makes no sense.

With an overlay the **C*C*G SU MH** line is placed on top of the **A*C*G SU MH** line. The overlay is moved westward until both SU MH lines are on top of each other. Both SU IC lines will also be on top of each other. This is the displaced reading position. With the overlay in this position it is easy to see if a C*C*G line is exact, or nearly exact, on top of an A*C*G line.

What is happening astrologically when this technique is used? All the C*C*G planets are being moved westwards by the same distance. With C*C*G SU MH moving to A*C*G SU MH, this distance is how far the progressed Sun has moved in right ascension.

By calculating the difference between C*C*G SU and A*C*G SU it is then possible to use this technique. This value is **subtracted** from each of the **C*C*G** planets. For Prince Charles we have the following:

C*C*G SU	2 CAP 31
A*C*G SU	20 SCO 01
Difference	42 30

This value is then subtracted from the C*C*G planets (apart from SU).

C*C*G MO	14 LIB 31	-42 30	02 VIR 01
C*C*G ME	09 CAP 43	-42 30	27 SCO 13
C*C*G VE	03 SAG 09	-42 30	20 LIB 39
C*C*G MA	22 CAP 52	-42 30	10 SAG 22
C*C*G JU	28 ARI 33	-42 30	16 PIS 03
C*C*G SA	01 CAP 51	-42 30	19 SCO 21
C*C*G UR	00 CAP 50	-42 30	18 SCO 20
C*C*G NE	10 CAP 42	-42 30	28 SCO 12
C*C*G PL	14 SCO 55	-42 30	02 LIB 25

Next make a list of the A*C*G planets and displaced A*C*G planets.

A*C*G	**(42 30)**	**displaced C*C*G**
ME 5 SCO 13		MO 2 VIR 01
SA 7 VIR 38		PL 2 LIB 25
NE 13 LIB 36		MA 10 SAG 22
VE 15 LIB 49		JU 16 PIS 03
SU 20 SCO 01		UR 18 SCO 20
MA 20 SAG 05		SA 19 SCO 21
PL 21 LEO 29		VE 20 LIB 39
MO 28 ARI 28		ME 27 SCO 13
JU 29 SAG 54		NE 28 SCO 12
UR 29 SAG 55		

Now look for conjunction and opposition aspects between C*C*G and A*C*G planets.

Using one degree orb we find the following aspect:

6. Disp. C*C*G SA = A*C*G SU

The last technique in this section is to look for a displaced planet at the birthplace. This is the same as saying the progressed birthplace angles = C*C*G planets.

The difference between C*C*G SU and A*C*G SU is 42°30". Add this to 0W10 and it equals 42E20. Using AstroMaps Hi-Res, check if a C*C*G planet is at 51N20 and 42E20. Prince Charles has no C*C*G planets at these coordinates, so all anywhere aspects in effect for Prince Charles on March 10, 1988 have been found.

There are six techniques used in finding anywhere aspects, and with a little practice they can be found quickly.

Location Aspects

The techniques used in this category apply to specific locations. As we are only interested in the temporary influences that apply to Klosters the necessary information will be found very quickly.

The first location technique is to check for C*C*G planets at or very close to Klosters. This can be done quickly with the AstroMaps Hi-Res program.

Prince Charles has no C*C*G planets at Klosters, so we can move on.

Next, examine the progressed angles for Klosters for aspects to A*C*G planets. This is the same technique as the birthplace progressed angles, but with a different location.

Klosters is at 9E53. Add thirty-nine degrees (his age) and the new longitude is 48E53. Look at the Astro*Carto*Graphy map using 46N54 and 48E53. At these coordinates there are no A*C*G planets.

The last technique is to find out if progressed angles at Klosters make an aspect with the C*C*G planets.

Thus, 42°30" is added to 9E53, giving 52E23. Using 46N54 and 52E23 we find that C*C*G VE IC is at 50E48 (AstroMaps Hi-Res program). The orb is just over one and a half degrees. The aspect is applying to progressed Klosters, so it will start to have some effect a few months later.

7. prog. Klosters = C*C*G VE IC (applying)

That completes the techniques used to find location aspects, and of the three categories these are the easiest aspects to find.

Latitude Aspects

The first technique is only used when a location falls between the Tropic of Cancer (23N27) and Tropic of Capricorn (23S27). As Klosters is at 46N54, we cannot use this technique for Prince Charles.

When several zenith points of the C*C*G planets line up at a certain latitude between the tropics, the combined effect will be very powerful.

The zenith points are the declination of the planets. Make a list of the declinations of the C*C*G planets and put them in numerical order.

Examine the list and see if many planets are at the same declination. For example, if three planets are at 15N declination, then all locations at 15N geographical latitude will be powerfully influenced by these planets.

Obviously, the zenith points of the transiting planets will be the **same** for everyone at a certain date. It is the zenith points of the progressed planets that are different for each person. At least one of the progressed planets would need to be involved where several zenith points line up at a certain latitude.

This technique will be used when the temporary influences for Bob Geldof are examined. Ethiopia is located between the tropics, so this will be a good opportunity to use the zenith points.

The final technique is latitude crossings between C*C*G and A*C*G planets. There are two types of latitude crossings used to find the temporary influences.

a. Crossings between a C*C*G planet and an A*C*G planet

b. Crossings between two C*C*G planets

When an overlay is used these types of latitude crossings are easier to find. Latitude crossings can also be found by using the displaced C*C*G planets, which have no astrological equivalent.

Despite spending much time on this technique, I have not been able to find a way to obtain the same information. AstroMaps Hi-Res only calculates the C*C*G progressions **or** C*C*G transits, but not both combined.

When an overlay is not used, there is no easy way to find these types of latitude crossings. When displaced C*C*G planets are used things are too complicated to try to find latitude crossings.

There was only one way I could find to locate latitude crossings between C*C*G planets and A*C*G planets. First, look at the latitude crossings on an Astro*Carto*Graphy map. Find when **MH** and **IC** lines are involved in the latitude crossings. For Prince Charles at Klosters he has **MO IC**, **MA MH** and **PL IC** involved in the latitude crossings.

If a C*C*G planet is conjunct one of these A*C*G MH or IC lines, then there is a latitude crossing.

In the anywhere aspects Prince Charles has C*C*G JU = A*C*G MO (conjunction). This means that C*C*G JU IC line is exactly on top of the MO IC line. Therefore, there are two latitude crossings at Klosters.

8. C*C*G JU IC/A*C*G JU ASC
9. C*C*G JU IC/A*C*G UR DSC

Diagram 13 clarifies that these two latitude crossings have an influence at Klosters.

Apart from this type of C*C*G/A*C*G latitude aspect I suggest that the others be ignored when working without an overlay.

The other Cyclo*Carto*Graphy techniques provide plenty of information about the temporary influences, and the most important fact is that these techniques can be applied without having to use an overlay.

Still to be considered are the Astro*Carto*Graphy solar return map and Astro*Carto*Graphy solar arc progressions. These techniques more than make up for what is missed with latitude aspects.

If you take the time and effort to study these various techniques you will have no difficulty in using Cyclo*Carto*Graphy.

When you use each technique in a methodical way and write out a list of the aspects you will find Cyclo*Carto*Graphy easy to use.

Interpretation of Cyclo*Carto*Graphy Planets

As the same planets are used in Cyclo*Carto*Graphy as in Astro*Carto*Graphy no new keywords are required.

If there are no A*C*G lines at the birthplace, or place of residence, then C*C*G lines will be important when they are there. This temporary influence is like an A*C*G line being there, so the same keywords suggested in Chapter 3 can be used.

When there is one or more A*C*G line at the birthplace, or place of residence, then C*C*G lines will modify these influences. The C*C*G lines shows in which way the permanent influences will be modified.

When a C*C*G planet aspects an A*C*G planet, it will influence the natal planet powerfully. The C*C*G planet will bring out the potentials shown by the A*C*G planet. The most important factor is the A*C*G planet, whereas the C*C*G planet shows what modification will take place.

A **map** with the Cyclo*Carto*Graphy planets is included for each of the seven examples. Instead of these planets being drawn on an overlay they are on a world map to make it easier to understand how Cyclo*Carto*Graphy works.

Example 1—Prince Charles and Klosters

Having used his Astro*Carto*Graphy map to explain how Cyclo*Carto*Graphy works, we have all the information required.

The list of the temporary influences is:

1. C*C*G MO = A*C*G NE	anywhere aspects
2. C*C*G JU = A*C*G MO	
3. C*C*G UR = A*C*G JU	
4. C*C*G UR = A*C*G UR	
5. prog. birthplace = A*C*G SU IC	
6. Disp. C*C*G SA = A*C*G SU	
7. prog. Klosters = C*C*G VE IC	location aspects
8. C*C*G JU IC/A*C*G JU ASC	latitude aspects
9. C*C*G JU IC/ A*C*G UR DSC	

What do these temporary influences say about the skiing tragedy that occurred in Klosters? The anywhere aspects need to be considered first as they have a worldwide effect. At the time Prince Charles was in Klosters he was experiencing his Uranus half-return, the crisis of middle age. This is a time in life when a major review is made of what has been achieved. Shocks and surprises in life would also be possible at this time.

With C*C*G UR aspecting A*C*G JU, this shows that Prince Charles had a strong need for freedom and independence. With this influence he could experience sudden and unexpected events taking place in his life.

The C*C*G MO aspecting A*C*G NE indicates that it was a period of disillusionment and being misunderstood by others.

On the day of the tragedy C*C*G JU was exactly aspecting A*C*G MO. On this holiday he felt happy and enjoyed the social life.

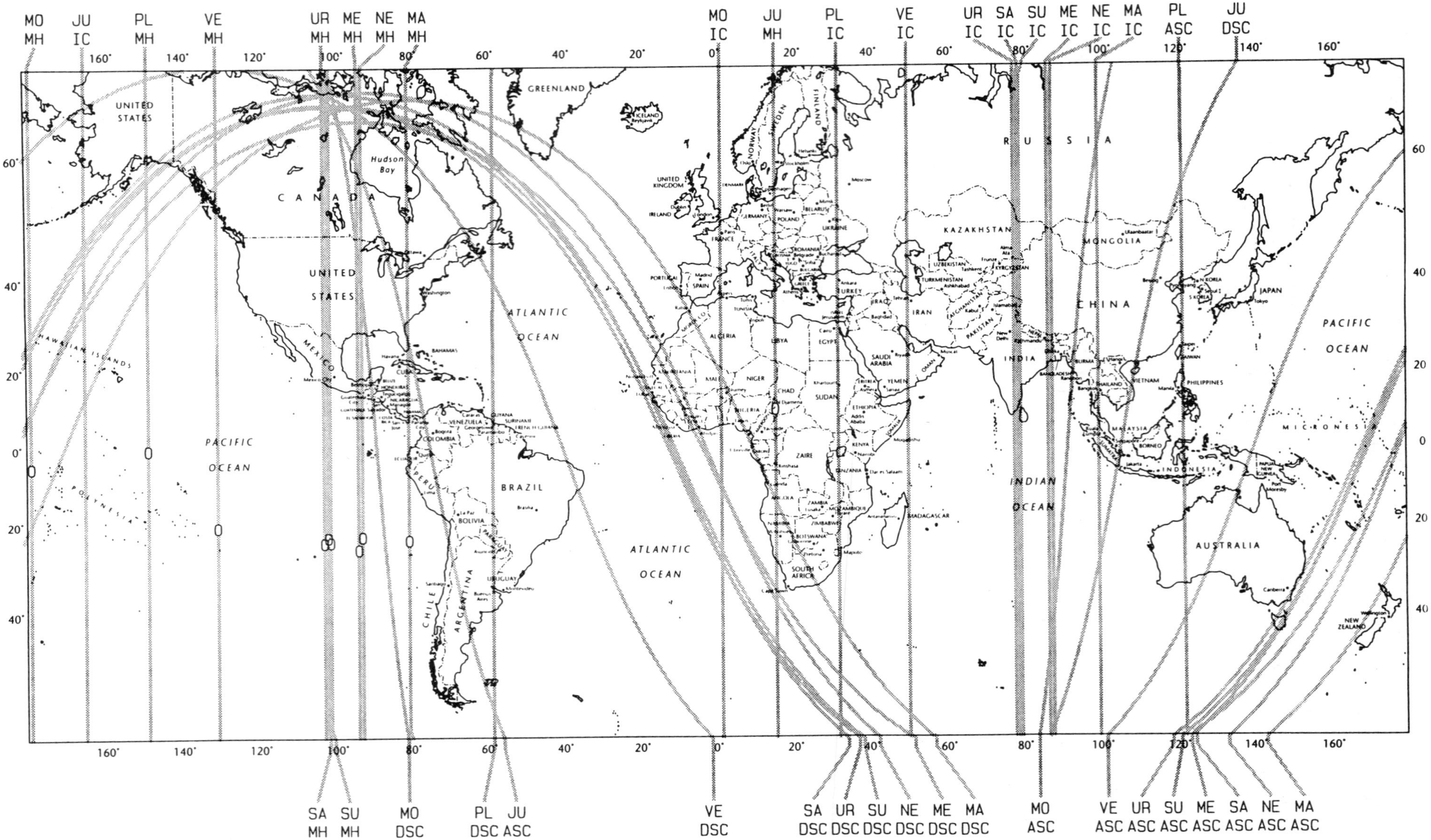

CYCLO*CARTO*GRAPHY

04/17/96

SU, MO, ME, VE AND MA ARE PROGRESSED
JU, SA, UR, NE AND PL ARE TRANSITING

Prince Charles

PROG	12	24	48	4:54 GMT
TRANSIT	3	10	88	0:00 GMT
NATAL	11	14	48	21:14 GMT

Diagram 13

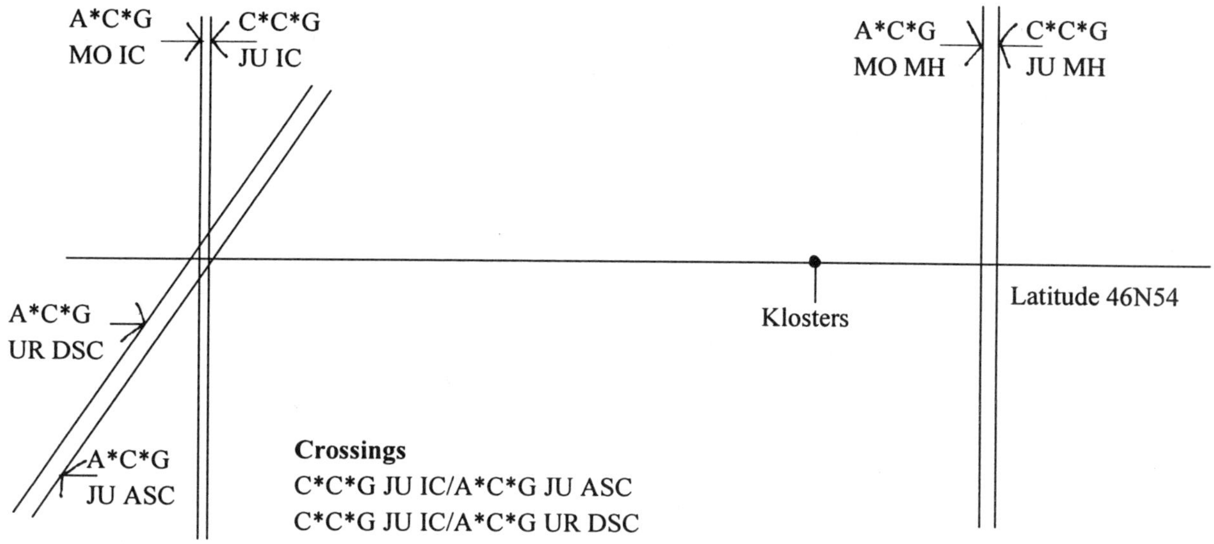

With the progressed birthplace on his A*C*G SU IC line, his family came under the spotlight.

Displaced C*C*G SA aspecting A*C*G SU suggests that Prince Charles was experiencing problems in his life. He was frustrated with the slow developments of what he wanted to achieve.

These temporary influences showed what Prince Charles was experiencing in life, we can only assume that they explain in some way his unhappy marriage to Diana.

Now we will examine the temporary influences that apply to Klosters. The progressed Klosters on his C*C*G VE IC is an applying aspect, so it would be more important in the next few years. Even after the tragedy in Klosters he still returns there for a holiday which is no surprise with this aspect becoming more powerful.

The two latitude crossings at Klosters show the extraordinary good luck and fortune he had at the time of the tragedy. Prince Charles himself said afterwards he was very lucky to be alive.

Example 2—Norman Schwarzkopf and Kuwait

The date the Gulf war started, January 17, 1991, will be used to examine the temporary influences for Norman Schwarzkopf.

Included are lists of A*C*G planets, C*C*G planets and displaced C*C*G planets in right ascension for ease in seeing how the aspects are found. The C*C*G planets drawn on a world map will be of assistance in finding the location aspects as not everyone has the AstroMaps Hi-Res program.

A*C*G	**C*C*G**
SU 00 VIR 43	MA 02 VIR 21
VE 08 LEO 14	UR 11 CAP 39
NE 13 VIR 23	JU 12 LEO 48
JU 19 LIB 04	ME 13 SCO 57
MO 26 CAP 13	VE 14 LIB 51
MA 26 CAN 50	NE 15 CAP 54
SA 27 AQU 19	MO 21 AQU 33
ME 27 LEO 27	PL 21 SCO 33
PL 27 CAN 29	SU 21 LIB 56
UR 29 ARI 24	SA 29 CAP 40

These lists show the following anywhere aspects:

1. C*C*G SA EQU A*C*G UR
2. C*C*G MA = A*C*G SU

Although the C*C*G MA = A*C*G SU has an orb just over one and a half degrees, it still has an effect. When Iraq invaded Kuwait this aspect had a narrower orb, so it was important in the buildup to the Gulf War. Most appropriate is that his C*C*G MA MH line passes through the Gulf showing where the war (MA) would take place.

A*C*G	**(51 13)**	**Displaced C*C*G**
SU 00 VIR 43		MO 00 CAP 20
VE 08 LEO 14		PL 00 LIB 20
NE 13 VIR 23		SA 08 SAG 27
JU 19 LIB 04		MA 11 CAN 08
MO 26 CAP 13		UR 20 SCO 26
MA 26 CAN 50		JU 21 GEM 35
SA 27 AQU 19		ME 22 VIR 44
ME 27 LEO 27		VE 23 LEO 38
PL 27 CAN 29		NE 24 SCO 41
UR 29 ARI 24		

Norman Schwarzkopf has **no** displaced aspects at the time of the Gulf War.

In his Cyclo*Carto*Graphy map he has an aspect that will have a worldwide affect.

3. C*C*G JU ASC = birthplace

Although Norman Schwarzkopf has no C*C*G planets at Kuwait he has the following passing through **Iraq**:

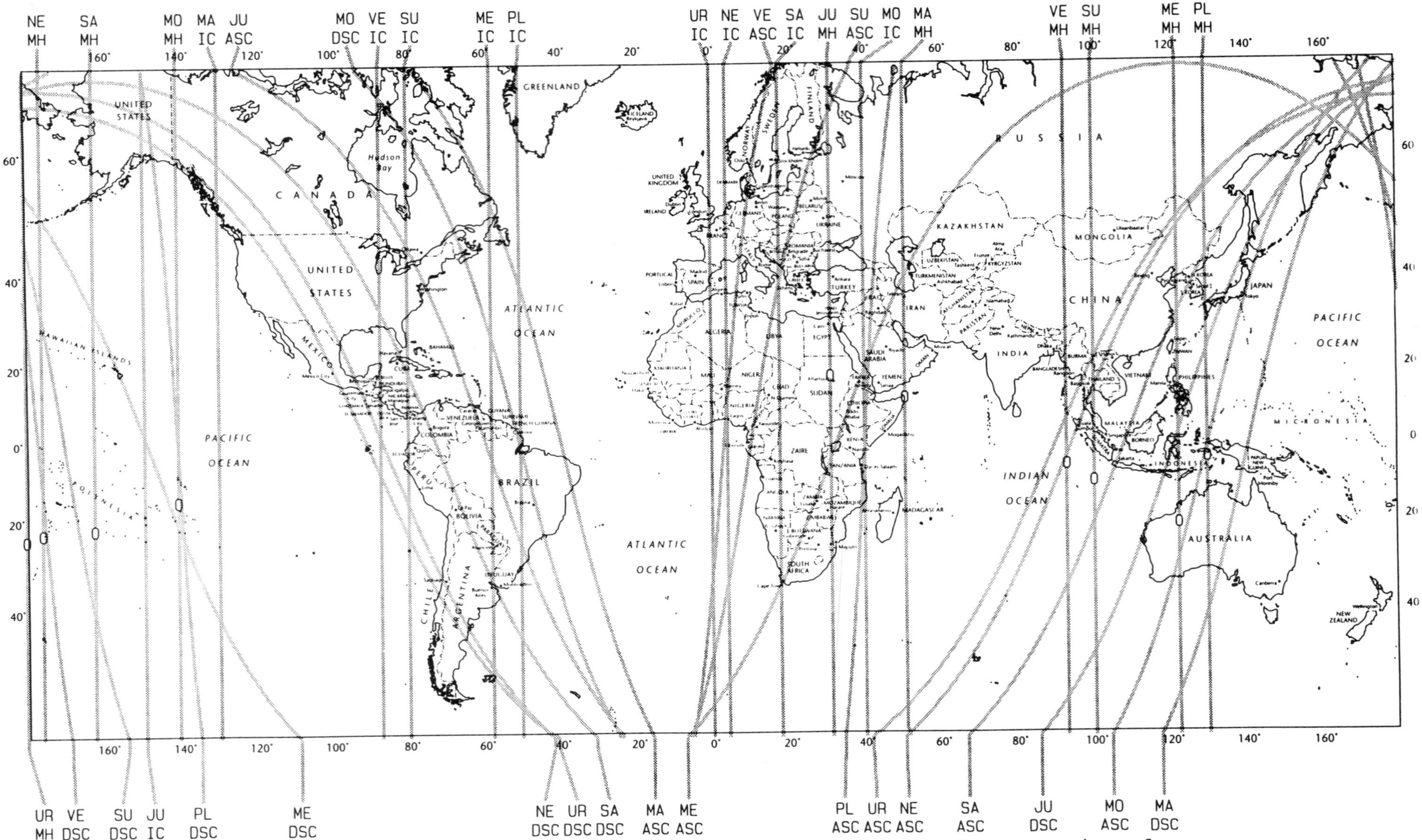

CYCLO*CARTO*GRAPHY

04/17/96

SU, MO, ME, VE AND MA ARE PROGRESSED
JU, SA, UR, NE AND PL ARE TRANSITING

Norman Schwarzkopf

PROG	10	17	34	18:31 GMT
TRANSIT	1	17	91	0:00 GMT
NATAL	8	22	34	8:45 GMT

4. C*C*G PL ASC in Iraq
5. C*C*G ME ASC in Iraq
6. C*C*G MO IC in Iraq

To sum up, the temporary influences show that it would be a successful time for Norman Schwarzkopf with C*C*G JU ASC at his birthplace. C*C*G MA MH passing through the Gulf and close to his A*C*G SU MH line explains how he became famous. He achieved fame (A*C*G SU MH) as a result of the Gulf war (C*C*G MA MH).

With C*C*G PL ASC, C*C*G ME ASC and C*C*G MO IC passing through Iraq, these lines would be dominant during the war. He was seen as a strong leader, powerful communicator and popular when dealing with Iraq.

The C*C*G SA crossing A*C*G UR at the equator manifested as the responsibility to liberate Kuwait from the control of Iraq. It also gave him the discipline and patience to plan everything before he went ahead with Desert Storm.

Example 3—Bob Geldof and Ethiopia

The song "Do They Know It's Christmas" was recorded on November 25, 1984, so this date will be used for the C*C*G planets.

A*C*G	C*C*G
MA 02 VIR 53	NE 00 CAP 08
VE 04 VIR 26	MO 04 PIS 36
ME 05 LIB 33	PL 06 SCO 33
SA 06 LIB 34	UR 11 SAG 44
MO 06 SAG 35	SU 12 SCO 00
JU 08 ARI 45	JU 14 CAP 29
SU 10 LIB 34	SA 18 SCO 50
UR 15 CAN 08	MA 21 VIR 44
NE 18 LIB 07	ME 26 SCO 39
PL 26 LEO 22	VE 28 VIR 23

These lists show the following anywhere aspects:

1. C*C*G MO = A*C*G VE
2. C*C*G JU = A*C*G UR
3. C*C*G ME EQU A*C*G PL

A*C*G	(31 26)	Displaced C*C*G
MA 02 VIR 53		MO 03 AQU 10
VE 04 VIR 26		PL 05 LIB 07
ME 05 LIB 33		UR 10 SCO 18
SA 06 LIB 34		JU 13 SAG 03
MO 06 SAG 35		SA 17 LIB 24
JU 08 ARI 45		MA 20 LEO 18
SU 10 LIB 34		ME 25 LIB 13
UR 15 CAN 08		VE 26 LEO 57
NE 18 LIB 07		NE 28 SCO 42
PL 26 LEO 22		

Bob Geldof has three displaced aspects:

4. Disp. C*C*G PL = A*C*G ME
5. Disp. C*C*G SA = A*C*G NE
6. Disp. C*C*G VE = A*C*G PL

As there are no temporary influences at his birthplace, the next step is to look at the location aspects.

7. C*C*G MO ASC near Ethiopia
8. C*C*G UR MH at Ethiopia (Adis Abeba)
9. prog. Adis Abeba = A*C*G UR IC (exact)
10. prog. Adis Abeba = C*C*G JU MH
11. C*C*G NE ASC near London
12. prog. London = A*C*G MO MH

With Ethiopia located between the tropics it is possible to use the zenith points technique. This is done by making a list of the declinations of the C*C*G planets.

VE	01 N 18
PL	02 N 45
MA	05 N 18
MO	11 S 11
SA	15 S 58
SU	16 S 11
ME	21 S 39
NE	22 S 20
UR	22 S 23
JU	22 S 59

There are no zenith points at Adis Abeba, but the technique is explained for completeness. Jim Lewis used the technique but I've not found it to give additional information. The other kind of latitude aspects show no crossings at Adis Abeba.

The temporary influences give a very clear picture why Bob Geldof became involved in helping the starving people in Ethiopia.

C*C*G UR MH is at Adis Abeba, showing the sudden and dramatic rise in his career and position in society. This is reiterated with progressed Adis Abeba exactly on his A*C*G UR IC line. At the time Band Aid recorded its single, Bob Geldof had an extremely powerful UR influence at Adis Abeba.

This powerful UR influence was activated by C*C*G JU = A*C*G UR. This is also shown with progressed Adis Abeba on his C*C*G JU MH line. The Band Aid project was an enormous success, and the fact that all these temporary influences are linked to Adis Abeba is quite incredible.

Although C*C*G MO ASC was near Ethiopia when the single was recorded, it was an important aspect. As C*C*G MO ASC moves eastward about fifty-five miles per month, it was passing through Ethiopia when Band Aid became a huge success. C*C*G MO ASC was still in Ethiopia when Bob Geldof arranged Live Aid. This shows his emotional appeal to the public, which helped him raise money.

C*C*G NE ASC was near London when the single was recorded. When the record was released this aspect was nearly exact at London. Bob Geldof used music to get his message across to other people.

Progressed London = A*C*G MO MH shows a connection between Band Aid and Ethiopia. Bob Geldof has A*C*G MO MH at Ethiopia, which was activated when progressed London was exactly conjunct this line.

He has two ME/PL temporary influences, C*C*G ME EQU A*C*G PL and disp. C*C*G PL = A*C*G ME. This helped him communicate in a powerful way to

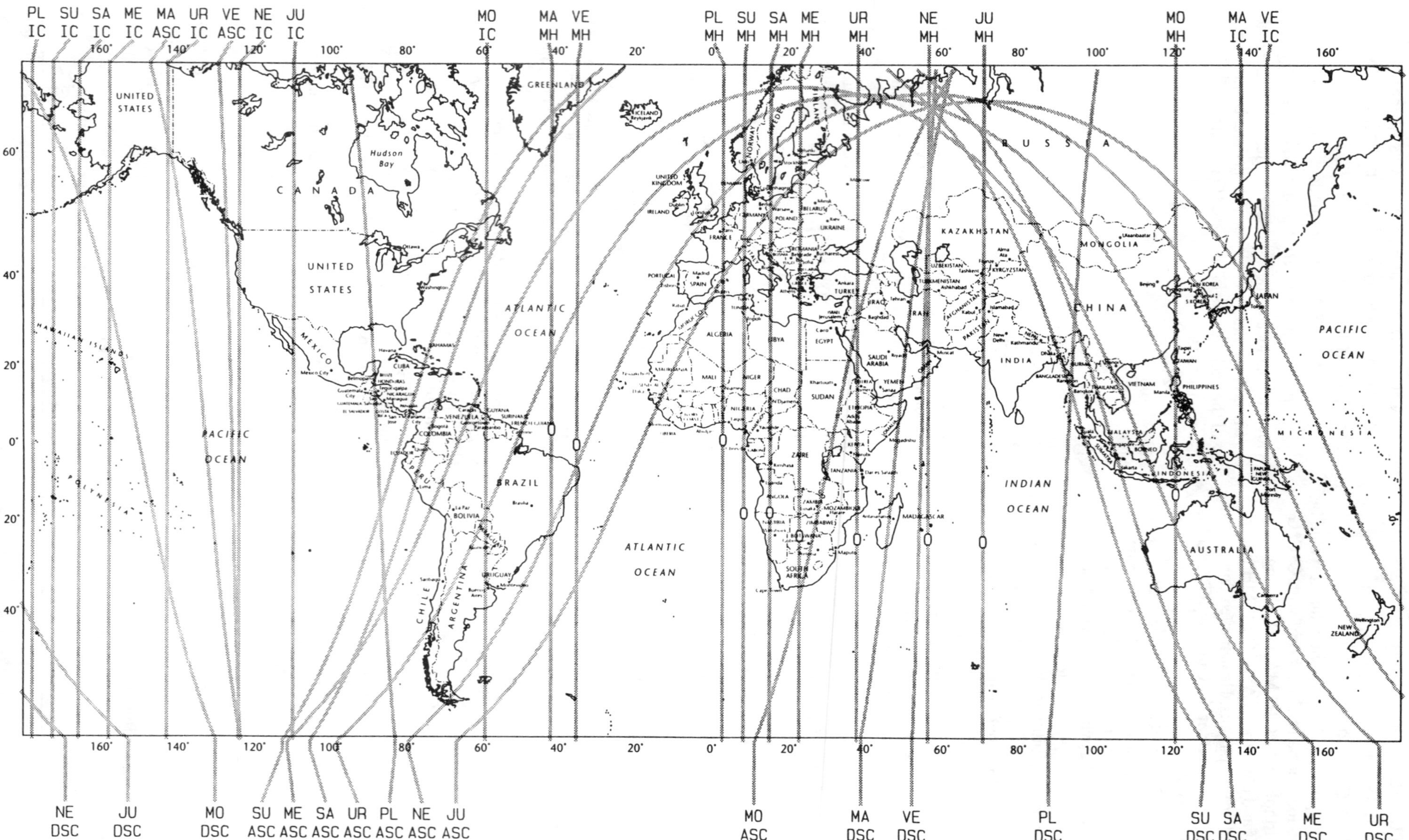

CYCLO*CARTO*GRAPHY

04/17/96

Bob Geldof

SU, MO, ME, VE AND MA ARE PROGRESSED
JU, SA, UR, NE AND PL ARE TRANSITING

PROG	11	7	51	16:45 GMT
TRANSIT	11	25	84	0:00 GMT
NATAL	10	5	51	13:20 GMT

explain what he wanted to do.

Bob Geldof also had C*C*G MO = A*C*G VE, a period when he increased his popularity. He was emotionally overcome with the generosity of other people.

Disp. C*C*G SA = A*C*G NE and disp. C*C*G VE = A*C*G PL could indicate the need to make sacrifices in his life. This he did by helping a country where many people were dying.

The location aspects for Bob Geldof accurately pinpoint Adis Abeba as significant at the time he formed Band Aid. This triggered the permanent influences that he has at Ethiopia.

Example 4—Margaret Thatcher and Falkland Islands

Argentina invaded the Falkland Islands on April 2, 1982, so this date is used for her C*C*G planets.

A*C*G	C*C*G
MO 01 VIR 37	PL 00 SCO 24
MA 09 LIB 02	UR 02 SAG 27
SA 11 SCO 58	VE 05 AQU 56
JU 15 CAP 47	JU 06 SCO 25
PL 15 CAN 48	MA 14 SCO 36
SU 18 LIB 00	SU 15 SAG 03
ME 22 LIB 13	SA 19 LIB 01
UR 23 PIS 25	ME 22 SAG 17
NE 26 LEO 40	MO 22 VIR 27
VE 29 SCO 34	NE 26 SAG 49

These lists show the following aspects:

1. C*C*G UR EQU A*C*G MO
2. C*C*G SA = A*C*G SU
3. C*C*G MO = A*C*G UR

C*C*G SA is moving **westward** (retrograde), so it is **applying** to the A*C*G SU line. This aspect was within one degree orb a few days after Argentina invaded the Falkland Islands.

A*C*G	(57 03)	Displaced C*C*G
MO 01 VIR 37		PL 03 VIR 21
MA 09 LIB 02		UR 05 LIB 24
SA 11 SCO 58		VE 08 SAG 53
JU 15 CAP 47		JU 09 VIR 22
PL 15 CAN 48		MA 17 VIR 33
SU 18 LIB 00		SA 21 LEO 58
ME 22 LIB 13		ME 25 LIB 14
UR 23 PIS 25		MO 25 CAN 24
NE 26 LEO 40		NE 29 LIB 46
VE 29 SCO 34		

Margaret Thatcher had **no** displaced aspects when Argentina invaded the Falkland Islands.

The anywhere aspects show that Margaret Thatcher was tested in a difficult situation (C*C*G SA = A*C*G SU). The Falklands War also gave her a great amount of political power on the world stage.

The two UR/MO aspects (C*C*G UR EQU A*C*G MO and C*C*G MO = A*C*G UR) suggest that she was under a lot of emotional trauma after what happened. But she responded quickly to free the people at the Falkland Islands from Argentinian rule.

There is only one location aspect on the Cyclo*Carto*Graphy map:

4. C*C*G MA ASC **exact** at London

Margaret Thatcher has A*C*G MA ASC at the Falkland Islands, and at the time Argentina invaded, this MA ASC line had progressed to London. As the British prime minister lives in London it was an important location for Margaret Thatcher.

It is incredible that Margaret Thatcher became involved in a war when C*C*G MA ASC was exact at London. The potential of war manifested as the MA ASC at the Falkland Islands had progressed all the way to London.

Normally, I do not consider C***C*G MA transits** as the progressed Mars is used. But for Margaret Thatcher it shows the exact date when the Falklands War started.

On April 2, 1982, transiting Mars was **exact** at Port Stanley in the Falkland Islands. Mars was moving westward, and when Britain declared war against Argentina, transiting Mars was exact at Buenos Aires. On April 2, 1982 Mars was at ten degrees Libra and moving retrograde. Margaret Thatcher has her natal Mars at 9 Libra 28, so transiting Mars was conjunct it. She has A*C*G MA ASC at the Falkland Islands, so transit Mars was also there.

Other astrologers have written about the fact that Margaret Thatcher has SU ASC and MA ASC at the Falkland Islands. The Astro*Carto*Graphy map for the United Kingdom has SU DSC and MA MH at the Falkland Islands. Thus, Margaret Thatcher activated the potential influences on the United Kingdom map.

Example 5—David Copperfield and Germany

On October 7, 1993, David Copperfield met Claudia Schiffer in Berlin. The C*C*G planets calculated for that date show the temporary influences for this meeting and his first European tour.

A*C*G	C*C*G
PL 5 VIR 14	ME 16 LIB 22
UR 8 LEO 07	MA 16 PIS 37
ME 9 LIB 11	MO 16 GEM 39
VE 10 LEO 24	NE 19 CAP 49
MO 13 AQU 17	UR 19 CAP 51
JU 16 VIR 14	JU 21 LIB 17
MA 19 PIS 57	VE 21 VIR 24
SU 24 VIR 04	PL 24 SCO 54
SA 26 SCO 07	SA 26 AQU 51
NE 27 LIB 25	SU 27 LIB 59

From these lists the following aspects are found:

1. C*C*G MA = A*C*G JU
2. C*C*G MO EQU A*C*G JU
3. C*C*G SU = A*C*G NE
4. C*C*G SA EQU A*C*G SA

A*C*G	(33 55)	Displaced C*C*G
PL 5 VIR 14		ME 12 VIR 27
UR 8 LEO 07		MA 12 AQU 42
ME 9 LIB 11		MO 12 TAU 44

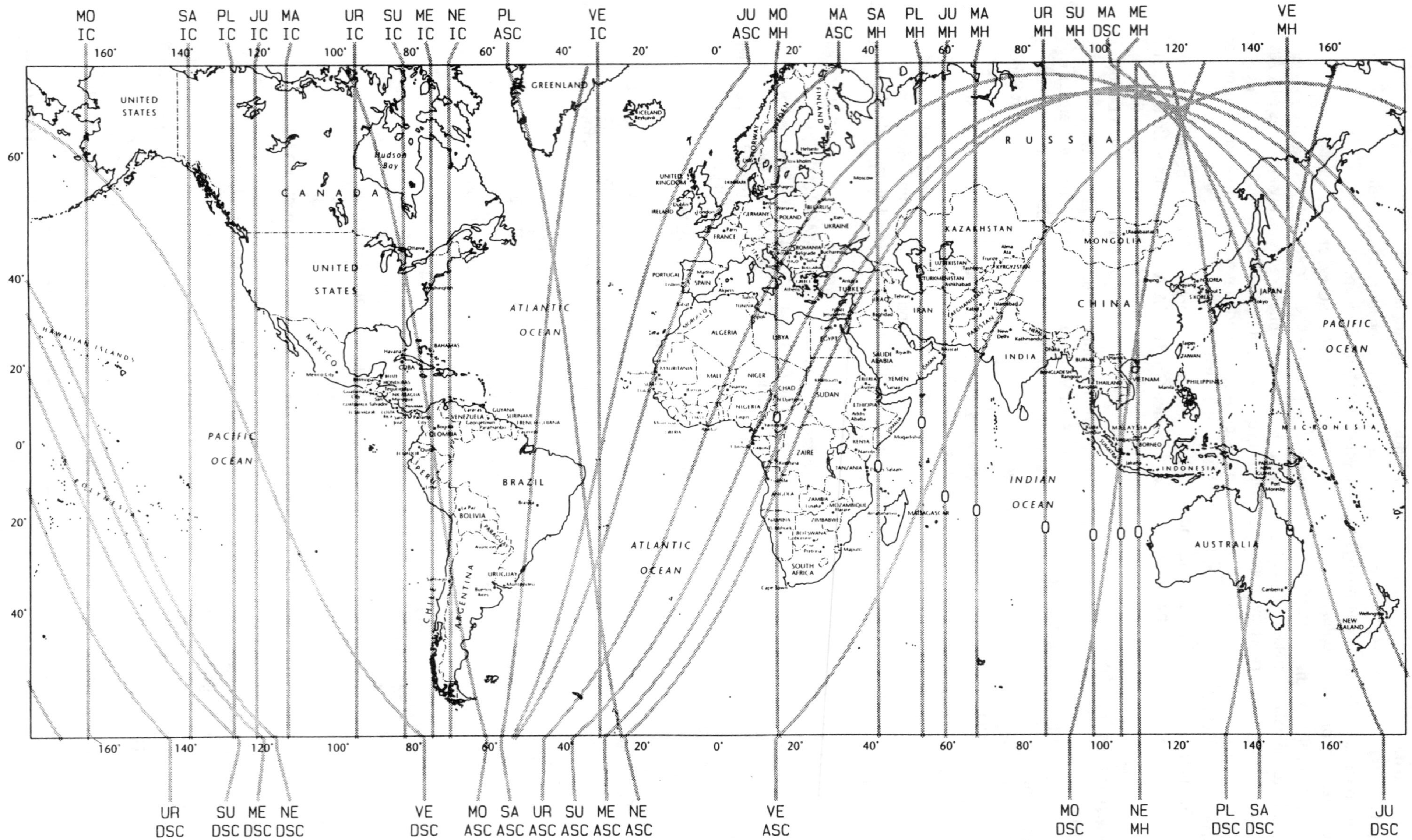

CYCLO*CARTO*GRAPHY

04/17/96

SU, MO, ME, VE AND MA ARE PROGRESSED
JU, SA, UR, NE AND PL ARE TRANSITING

Margaret Thatcher

PROG	12	8	25	20:16 GMT
TRANSIT	4	2	82	0:00 GMT
NATAL	10	13	25	9:00 GMT

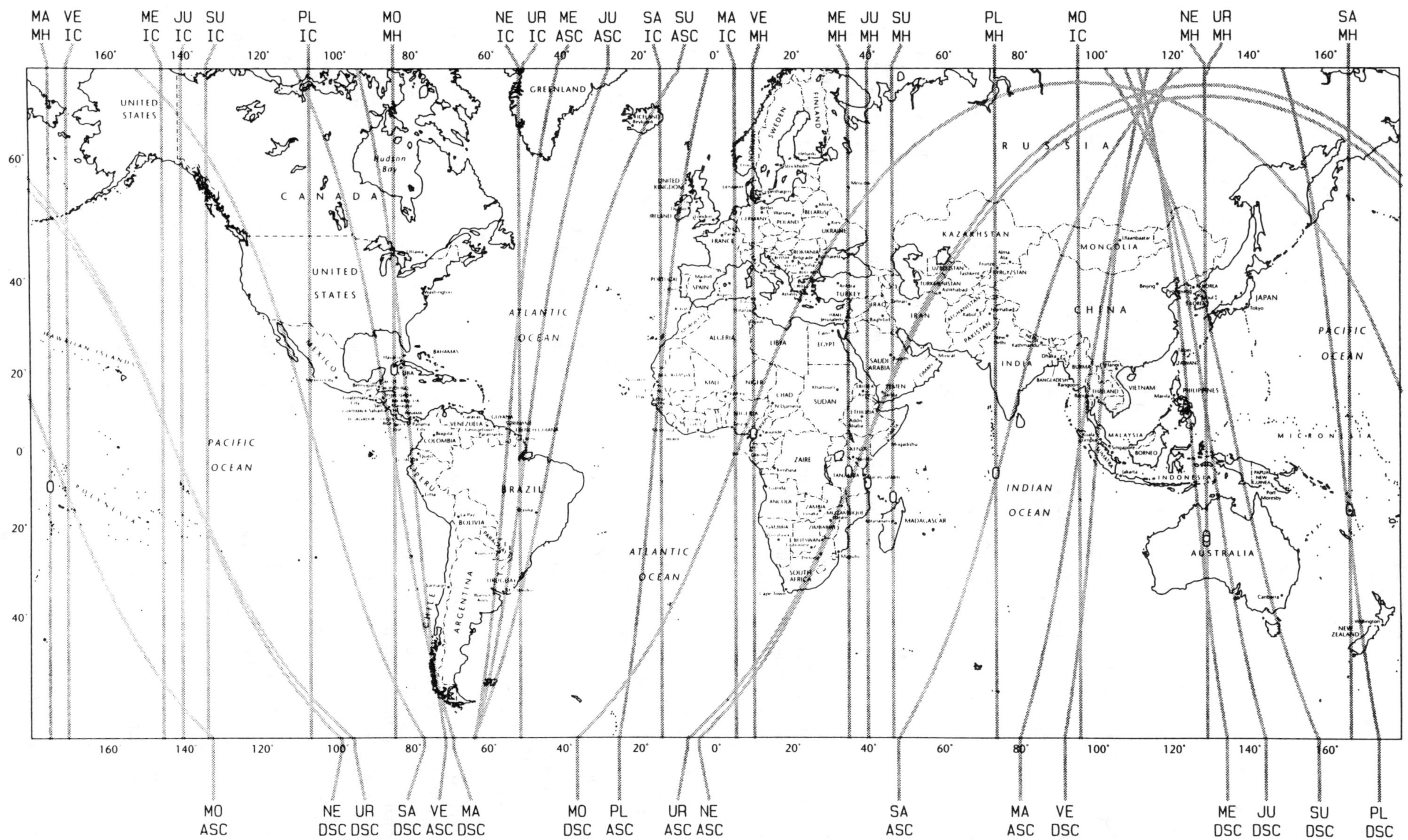

CYCLO*CARTO*GRAPHY

04/17/96

SU, MO, ME, VE AND MA ARE PROGRESSED
JU, SA, UR, NE AND PL ARE TRANSITING

David Copperfield

PROG	10	23	56	12:25 GMT
TRANSIT	10	7	93	0:00 GMT
NATAL	9	16	56	11:02 GMT

VE 10 LEO 24	NE 15 SAG 54
MO 13 AQU 17	UR 15 SAG 56
JU 16 VIR 14	JU 16 VIR 22
MA 19 PIS 57	VE 17 LEO 29
SU 24 VIR 04	PL 20 LIB 59
SA 26 SCO 07	SA 22 CAP 56
NE 27 LIB 25	

There are two displaced aspects:

5. Disp. C*C*G MA = A*C*G MO
6. Disp. C*C*G JU = A*C*G JU

The Cyclo*Carto*Graphy map shows the following location aspect:

7. C*C*G VE MH in Germany

There are no location aspects to explain why David Copperfield fell in love with Claudia Schiffer in Berlin. His C*C*G VE MH line is three degrees away from Berlin, out of orb.

During his tour in Germany he quickly won the hearts of many women with his charming personality. This is easy to understand with C*C*G VE MH passing through Germany.

I would have expected his C*C*G VE MH line to be nearer Berlin, but it seems it had an influence throughout Germany.

C*C*G MA = A*C*G JU indicates that David Copperfield had a great deal of stamina and energy during this European tour. Most days he was performing three shows, putting on extra performances when tickets sold out. This is also confirmed by disp. C*C*G MA = A*C*G MO.

He was feeling enthusiastic while in Germany, but had been reluctant to visit Europe because he didn't know how he would be received. His successful European tour is shown by disp. C*C*G JU = A*C*G JU.

A important aspect for David Copperfield is C*C*G SU = A*C*G NE. This was a period when he was unsure about his direction in life. Having met Claudia Schiffer, he was reluctant to rush into marriage even though he couldn't live without her. This is reiterated by C*C*G SA EQU A*C*G SA.

While David Copperfield was in Munich he had to cancel five shows due to food poisoning from a hamburger. Because it was the first time he had done that, and he was disappointed to let his fans down, he later returned to Munich to put on extra shows. The cancellation is explained by C*C*G SU = A*C*G NE.

Shortly after he completed his European tour David Copperfield experienced the C*C*G SU = A*C*G NE in another way. A magician had plans to publish a book about the secrets of many illusions. David Copperfield tried to stop its publication as some of his illusions were exposed in it. The magician claimed that David Copperfield allowed him to include some of his illusions, but he denies this and is now taking legal action.

David Copperfield felt the magician was taking advantage of him, and was disappointed that a fellow magician would damage his reputation.

Example 6—Claudia Schiffer and Germany

Her C*C*G planets are calculated for October 7, 1993, the same date that was used for David Copperfield.

A*C*G	C*C*G
PL 02 LIB 53	VE 05 SCO 23
SU 03 VIR 50	MA 11 VIR 10
UR 06 LIB 35	ME 15 VIR 42
VE 15 LIB 48	NE 19 CAP 49
MO 19 GEM 36	UR 19 CAP 51
SA 20 TAU 45	MO 20 ARI 26
ME 25 VIR 51	JU 21 LIB 17
NE 26 SCO 20	SU 24 VIR 44
MA 27 LEO 08	PL 24 SCO 54
JU 29 LIB 39	SA 26 AQU 51

The aspects found are:

1. C*C*G SU = A*C*G ME
2. C*C*G SA EQU A*C*G NE
3. C*C*G SA = A*C*G MA

Although C*C*G SU = A*C*G ME has an orb over one degree, it is an important aspect. It is an applying aspect and was within orb a few weeks later. C*C*G SA is moving westward and changed direction when it was exactly EQU A*C*G NE, so it had a powerful influence.

A*C*G	**(20 54)**	**Displaced C*C*G**
PL 02 LIB 53		JU 00 LIB 23
SU 03 VIR 50		PL 04 SCO 00
UR 06 LIB 35		SA 05 AQU 57
VE 15 LIB 48		VE 14 LIB 29
MO 19 GEM 36		MA 20 LEO 16
SA 20 TAU 45		ME 24 LEO 48
ME 25 VIR 51		NE 28 SAG 55
NE 26 SCO 20		UR 28 SAG 57
MA 27 LEO 08		MO 29 PIS 32
JU 29 LIB 39		

Claudia Schiffer had **no** displaced aspects at the time she met David Copperfield. At her birthplace there are two very important aspects:

4. prog. birthplace = A*C*G NE ASC
5. C*C*G SA IC at birthplace

The Cyclo*Carto*Graphy map shows the following aspects:

6. C*C*G VE ASC in Germany
7. C*C*G PL ASC in Germany

Like David Copperfield she has no location aspects at Berlin, and her C*C*G VE ASC is also about three degrees from Berlin. C*C*G VE ASC is passing through Germany for the next few years, which is appropriate for projecting her personality through modelling.

With C*C*G PL ASC also passing through Germany, the combined influence with C*C*G VE ASC shows her image as a sex object. It also indicated the possibility of a passionate loving relationship which would totally change her life.

The C*C*G SU = A*C*G ME was a good period for Claudia Schiffer to become involved in all kinds of

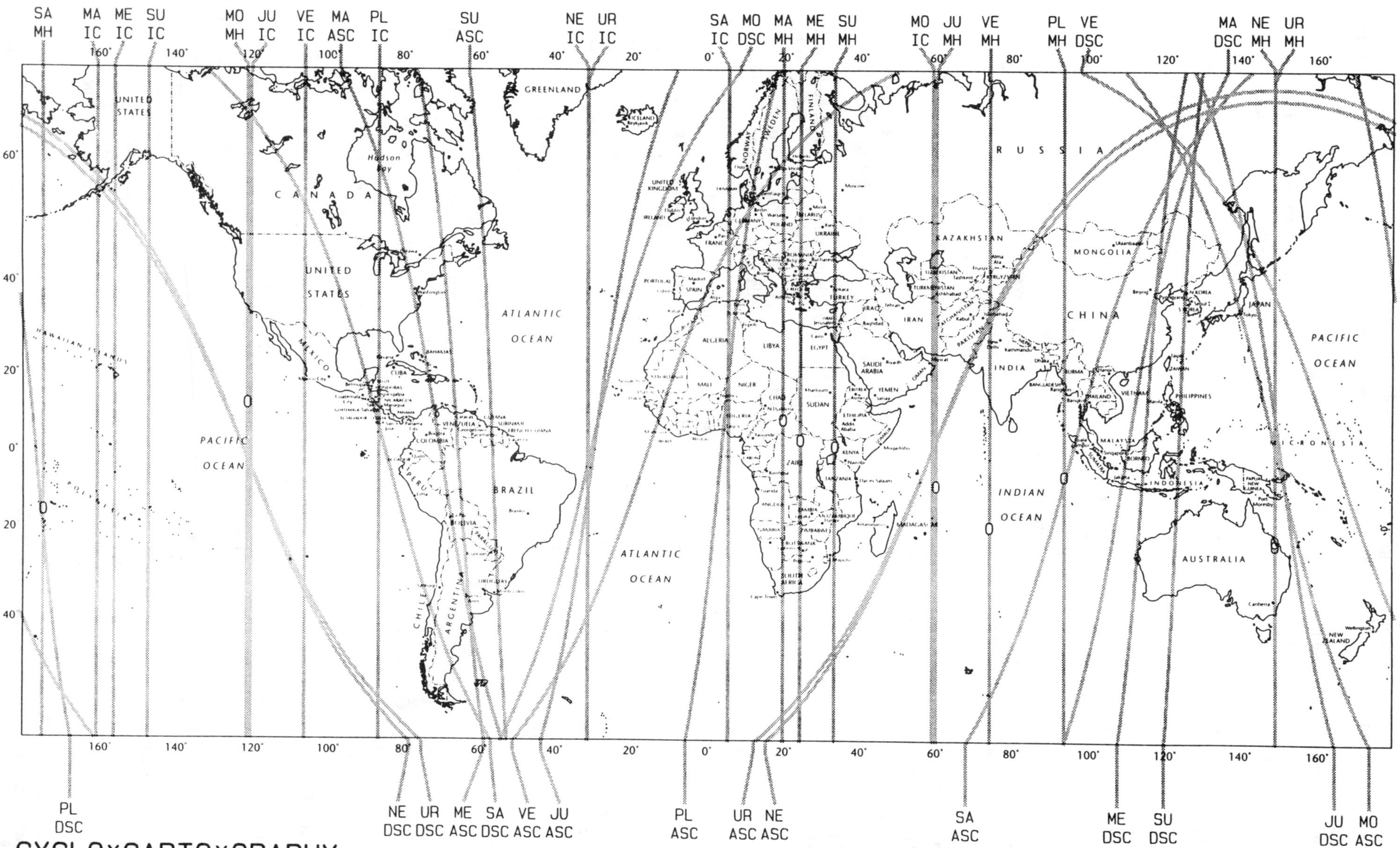

CYCLO*CARTO*GRAPHY

04/17/96

SU, MO, ME, VE AND MA ARE PROGRESSED
JU, SA, UR, NE AND PL ARE TRANSITING

Claudia Schiffer

PROG	9	17	70	14:01 GMT
TRANSIT	10	7	93	0:00 GMT
NATAL	8	25	70	11:10 GMT

communication, through advertising for companies, writing or publicity. She did write a book about her life which was released at the beginning of 1995.

The combined influence of SA and NE is emphasized for Claudia Schiffer. She has prog. birthplace = A*C*G NE ASC, C*C*G SA IC at her birthplace and C*C*G SA EQU A*C*G NE.

Prog. birthplace = A*C*G NE ASC could indicate that she was not sure of herself or other people, and could also show that her health may have been poor during that time. C*C*G SA IC at her birthplace suggests that family traditions and responsibilities were emphasised. It could also mean the need to make some sacrifices in life which she would have to do if she married David Copperfield.

C*C*G SA = A*C*G MA shows that Claudia Schiffer was not inclined to rush into marriage after meeting David Copperfield. The long distance between them was frustrating for her and could be a problem in the partnership.

Example 7—Matthew

The lists of A*C*G and C*C*G planets in right ascension are not included for Matthew. The time when he made his last voyage from Gdynia to Barranquilla will be examined.

The temporary influences that were in effect on Matthew's last voyage were:

1. C*C*G SA EQU A*C*G SA
2. Disp. C*C*G VE = A*C*G SA
3. Disp. C*C*G JU = A*C*G UR
4. Disp. C*C*G MO = A*C*G ME
5. C*C*G SA ASC at birthplace
6. C*C*G UR IC at birthplace

In the last chapter it was mentioned that Matthew has SA and UR lines at Gdynia, Barranquilla and Miami. With C*C*G SA ASC and C*C*G UR IC at his birthplace, this means that these temporary influences have a worldwide effect. They manifested at locations where he has A*C*G SA and A*C*G UR lines, and it is incredible that he made this voyage when C*C*G SA ASC and C*C*G UR IC were very powerful at his birthplace. This shows the importance of taking the temporary influences at the birthplace into consideration.

C*C*G SA EQU A*C*G SA indicates that Matthew experienced some problems and difficulties during this period. His problems started when he was in Gdynia where he has A*C*G SA MH exact.

Displaced C*C*G VE = A*C*G SA suggests that relationships of any kind would tend to be problematic and difficult. During the voyage the captain did everything to make things difficult for Matthew and undermine his job.

Although Matthew experienced intense hardship at Barranquilla and further problems in Miami, he was able to leave the A*C*G SA and A*C*G UR lines quite quickly. With displaced C*C*G JU = A*C*G UR, he had good fortune in what were extremely difficult situations.

Displaced C*C*G MO = A*C*G ME puts an emphasis on mental abilities and communication, so perhaps he was able to talk himself out of the difficult situations.

In summary, below is a checklist of the techniques used in Cyclo*Carto*Graphy.

Anywhere aspects:

1. C*C*G planet = A*C*G planet
2. C*C*G planet EQU A*C*G planet
3. C*C*G planet at birthplace
4. Disp. C*C*G planet = A*C*G planet
5. prog. birthplace = A*C*G planet
6. prog. birthplace = C*C*G planet

Location aspects:

1. C*C*G planet at location
2. prog. location = A*C*G planet
3. prog. location = C*C*G planet

Latitude aspects:

1. zenith points
2. latitude crossings
   a. C*C*G planet/A*C*G planet
   b. C*C*G planet/C*C*G planet

Anywhere aspects and location aspects usually give enough information about the temporary influences. The latitude aspects do not seem to be all that important, and I haven't found zenith points to provide additional information. The latitude crossings are impossible to work out unless a Cyclo*Carto*Graphy overlay is purchased. My recommendation is that you use only the anywhere aspects and location aspects as these can be found without having to purchase an overlay.

You may have noticed that the progressed birthplace and progressed location aspects have a different measurement according to whether an A*C*G or C*C*G planet is used. For an A*C*G planet the same number of degrees as a person's age is used, i.e. one degree equals one year. But for a C*C*G planet it is the solar arc in right ascension that is added to the birthplace and location. This is how these aspects are found when a Cyclo*Carto*Graphy overlay is used.

You would expect that the same measurement is used for both A*C*G and C*C*G planets in finding aspects to the progressed birthplace and progressed locations. As the solar arc in right ascension is used for C*C*G planets, I wondered if this should also be the same for A*C*G planets. To answer this question I have examined the above examples, with the exception of Matthew, to discover whether one degree a year or solar arc in right ascension works more effectively with A*C*G planets.

Prince Charles	**Age 39**	**Prog. SU 42 30**
a. London	A*C*G SU IC (orb 1-1½°)	A*C*G SA ASC (orb 1-1½°)
b. Klosters	nil	nil

It is difficult to say if progressed London = A*C*G SA ASC is more appropriate in 1988 (right ascension solar arc) or in 1991 (age 42). Perhaps the problems between Prince Charles and Princess Diana started in 1988, but it was only in 1991 that the general public became aware of them.

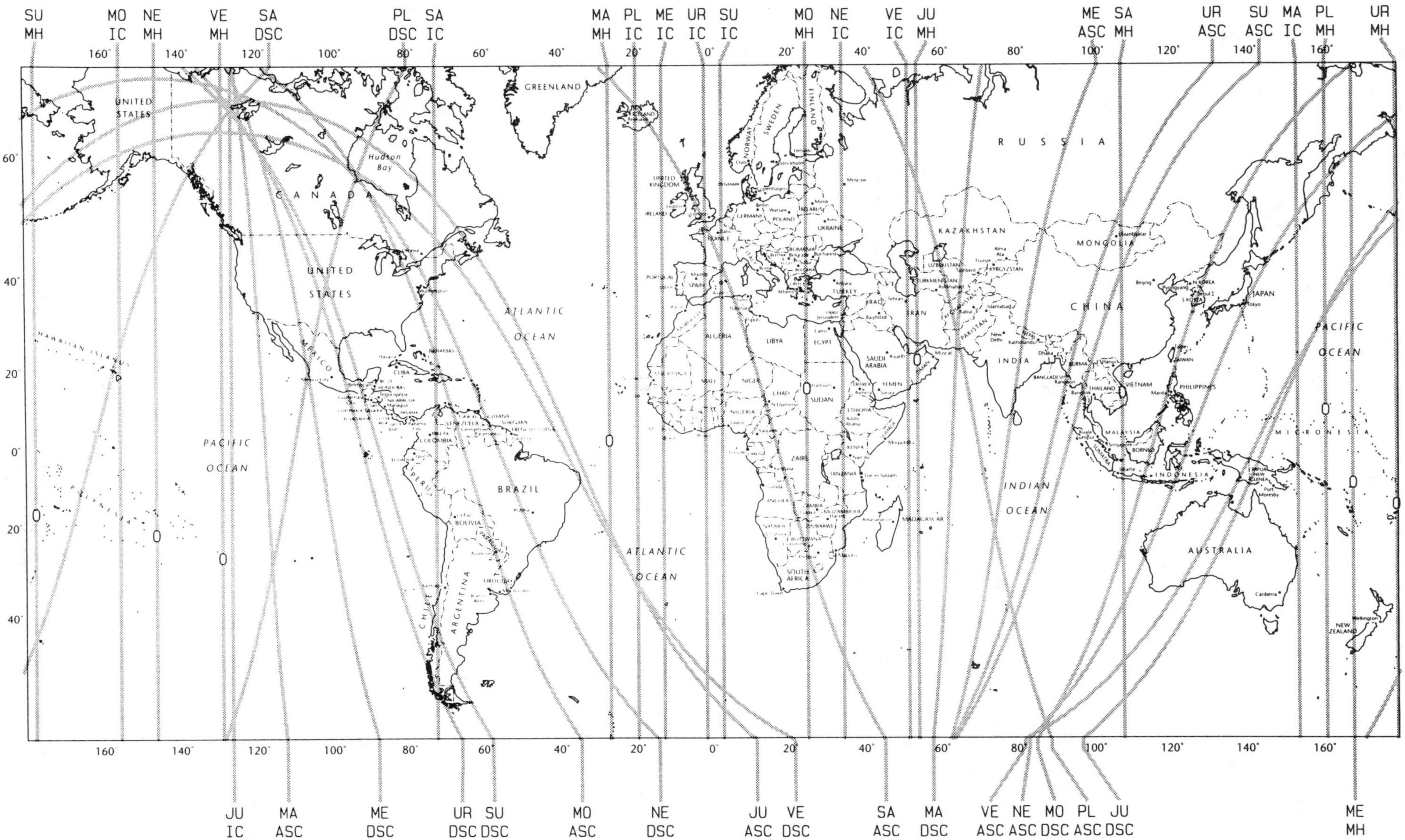

CYCLO*CARTO*GRAPHY

04/17/96

Matthew

SU, MO, ME, VE AND MA ARE PROGRESSED

JU, SA, UR, NE AND PL ARE TRANSITING

Norman Schwarzkopf	**Age 56**	**Prog. SU 51 13**
a. Kuwait	nil	A*C*G JU MH (orb 1½°)
b. Trenton	nil	nil

It is clear to see that the R.A. solar arc for Kuwait is more appropriate as it aspects JU MH. There is a difference of nearly five degrees between the two measurements, so this is a good example to test which works more effectively.

Bob Geldof	**Age 33**	**Prog. SU 31 26**
a. Adis Abeba	A*C*G UR IC (exact)	A*C*G UR IC (orb 1½°)
b. London	A*C*G MO MH (orb ½°)	A*C*G MO MH (orb 2°)
c. Dublin	nil	nil

Although the aspects are the same with both measurements, it is the age progressions that result in the closest orbs.

Margaret Thatcher	**Age 56**	**Prog. SU 57 03**
a. Falkland Islands	nil	nil
b. London	nil	nil

There are no aspects involving progressed Falkland Islands and London to A*C*G planets. Although Margaret Thatcher was the same age as Norman Schwarzkopf there is a difference of nearly six degrees with the right ascension solar arc.

David Copperfield	**Age 37**	**Prog. SU 33 55**
a. Berlin	nil	A*C*G NE MH (orb ¾°)
b. Metuchen	nil	nil

As C*C*G SU = A*C*G NE no decision can be made if prog. Berlin = A*C*G NE MH provides additional information.

Claudia Schiffer	**Age 23**	**Prog. SU 20 54**
a. Berlin	nil	A*C*G ME MH (orb 1½°)
b. Düsseldorf	A*C*G NE ASC (orb ½°)	A*C*G NE ASC (orb 1½°)

If Claudia Schiffer's book was published in Berlin, then this right ascension solar arc aspect would be appropriate. But the age progression aspect is closer to her birthplace, so no clear answer is shown in this example.

These examples do not show whether age progressions or right ascension solar arc should be used with A*C*G planets to find the progressed aspects involving locations. I suggest that you use both measurements to find out which one gives the most accurate results on a consistent basis.

The next two chapters explain other techniques which can be used to obtain information about the temporary influences.

Advanced Techniques

Chapter 6/Astro*Carto*Graphy Solar Return Map

Astrologers use various techniques when they examine how the natal chart is temporary influenced. These methods include secondary progressions, tertiary progressions, minor progressions, solar arc progressions, transits, solar returns and lunar returns. Most astrologers do not have time to apply all these techniques.

Solar Return Chart

A solar return chart is one that is calculated every year on a person's birthday. It can be called a birthday chart as it shows what a person will experience in life during the next twelve months.

The solar return chart is calculated for the date and time when the Sun returns to its exact position in the natal chart. This occurs once a year, either on the birthday, the day before or day after.

Astrologers cannot agree on which location should be used to calculate a solar return chart. Some say that the birthplace should always be used, no matter where a person lives. Others insist that the place of residence should be used for the solar return chart.

Then there is the theory that a person can change a solar return chart by being in a different location on the birthday. For example, a person goes on holiday in Spain and celebrates a birthday there. The coordinates of the holiday location are then used for the solar return.

What if a person is travelling on a long plane journey at the time of a solar return? This makes it difficult to calculate a solar return chart for the person's location at that time.

The problem of deciding which location to use for a solar return is avoided when Astro*Carto*Graphy is used. When a solar return is drawn on a map of the world the planetary influences can be found at any location.

Prince Charles' natal chart will be used to explain how a solar return is calculated.

Prince Charles has his natal Sun at 21 Scorpio 55:30.
Natal Sun = 21 55 30 SCO

Every year when the Sun returns to this position is when the solar return occurs. Prince Charles was thirty-nine years old when the skiing tragedy happened at Klosters.

Date of birth	November 14, 1948
	+39 years
	November 14, 1987

This means that his solar return for his thirty-ninth birthday took place around November 14, 1987. Either with a computer or ephemeris it is now possible to calculate the exact date and time for the solar return. I used the Astrocalc computer program to calculate the solar returns featured in this book.

It doesn't matter which location is used for the solar return, so it is recommended that you stay with the birthplace. Other locations can be examined after the solar return is drawn on a map of the world.

Using Prince Charles' birthplace for his thirty-ninth birthday solar return it is now possible to set up the chart.

This is done for 8:17 a.m. GMT, November 15, 1987, London. The following chart shows the solar return chart calculated for Prince Charles' thirty-ninth birthday. The solar return chart was valid for one year (until his fortieth birthday). From this information it is now possible to draw an Astro*Carto*Graphy map for the solar return.

With or Without Precession

The next problem that an astrologer faces when calculating a solar return is whether to allow for precession.

In astrology two types of zodiac can be used: tropical or sidereal. Each year when the Sun returns to zero degrees Aries, this is not the same place in relation to the fixed stars. The vernal equinox (zero degrees Aries) moves backward 50.25 seconds each year through the fixed stars. This movement is called precession.

The tropical zodiac is the one used by most astrologers, but very little thought is given about the effects of precession. The sidereal zodiac is used mainly in Hindu astrology.

There are two ways of overcoming the effects of precession:

1. Use the tropical zodiac and allow for precession
2. Use the precession-free sidereal zodiac

Using the tropical zodiac we need to consider the following:

As a person ages, the **natal planets** should be corrected for precession. As zero degrees Aries is moving backward, it means that the natal planets are moving forward in the tropical zodiac.

A few astrologers take precession into consideration when examining transits. The amount of precession is added to all the natal planets before looking at the transits.

Robert Hand gives a table for the precession correction values in *Planets in Transit.* At age seventy-two the amount of precession is one degree, which can make a difference in the timing of transits.

The effects of precession become more noticeable when a solar return is calculated. At age eighteen the difference between a solar return calculated with and without precession is six hours. The difference is twelve hours at age thirty-six, and an entire day at age seventy-two.

Prince Charles' thirty-ninth solar return chart is now calculated allowing for precession.

Natal Sun	22 25 21 SCO
+ precession (39 years)	+ 32 40
Precession-corrected Sun	22 58 01 SCO
Sun at 0 hour 16.11.1987	23 04 56
Sun at 0 hour 15.11.1987	22 04 29
Difference	1 00 27
Precession-corrected Sun	22 58 01
Sun at 0 hour 15.11.1987	22 04 29
Difference	0 53 32
1 00 27	= 24 hours
0 53 32	= X
X = 0 53 32 x 24 ÷ 1 00 27	9.15 pm

When the two types of solar return charts for Prince Charles are compared, the angles are in different positions.

With precession		**Without precession**
14 LIB 08	MH	27 VIR 47
14 ARI 08	IC	27 PIS 47
05 LEO 57	ASC	01 SAG 57
05 AQU 57	DSC	01 GEM 57

This will obviously make a tremendous difference when an Astro*Carto*Graphy map is drawn, but the key question is which type of solar return chart is used in Astro*Carto* Graphy? An Astro*Carto*Graphy solar return map **with** precession is the one that provides accurate information. This was Jim Lewis' conclusion as well.

If you examine Prince Charles' Astro*Carto*Graphy solar return map **without** precession (not shown here) you will see that VE ASC is nearest Klosters with SA ASC just within orb. These planetary lines do not explain why Prince Charles was involved in the skiing tragedy at Klosters. As will be seen, his Astro*Carto*Graphy solar return map **with** precession is significant.

Interpretation of the Astro*Carto*Graphy Solar Return Map

The solar return map is interpreted in the same way as the usual Astro*Carto*Graphy map, i.e. planetary lines and latitude crossings at a location.

It is important to remember that the Astro*Carto*Graphy solar return map shows the temporary influences for one year. The information obtained should be added to that found with Cyclo*Carto*Graphy to get a broader view of the temporary influences.

Using the Astro*Carto*Graphy solar return map, further information is found about the temporary influences at specific locations. It is also possible to examine the birthplace on a solar return map because it will have a worldwide effect.

In the seven examples that follow relevant locations or countries for the person concerned will be examined. If no planetary lines are within orb of a location, more importance will be placed on the temporary influences at the birthplace.

Quite a few astrologers use various techniques in a solar return chart to time events. I tried these techniques, but found they did not provide accurate information about when events would occur. For example, would it be possible to pinpoint March 10, 1988 as the date when Prince Charles experienced the skiing tragedy in Klosters? Despite trying many different techniques on the Astro*Carto*Graphy solar return map I haven't been able to find a way to time events.

Example 1—Prince Charles and Klosters

At Klosters Prince Charles has the following on his Astro*Carto*Graphy solar return map:

Lines:	MA IC (exact)
	JU MH
Crossings:	JU IC/UR ASC
	VE IC/MO ASC
	ME MH/NE ASC
	MA MH/UR ASC
	VE DSC/PL DSC

MA IC is **exact** at Klosters. This shows the possibility of accidents when he pushes himself to the limit, especially in any kind of sports. It could also show family arguments and disagreements.

JU MH indicates that Prince Charles enjoyed being in Klosters during this holiday. The combination of MA IC and JU MH shows that he was competitive and tended to take risks.

The good social life that Prince Charles always experiences at Klosters is reiterated by the VE/MO latitude crossing. On his Astro*Carto*Graphy map he has VE IC and MO MH within orb of Klosters.

ME/NE shows that Prince Charles did not listen to the advice of other people. He is independent, and wants to

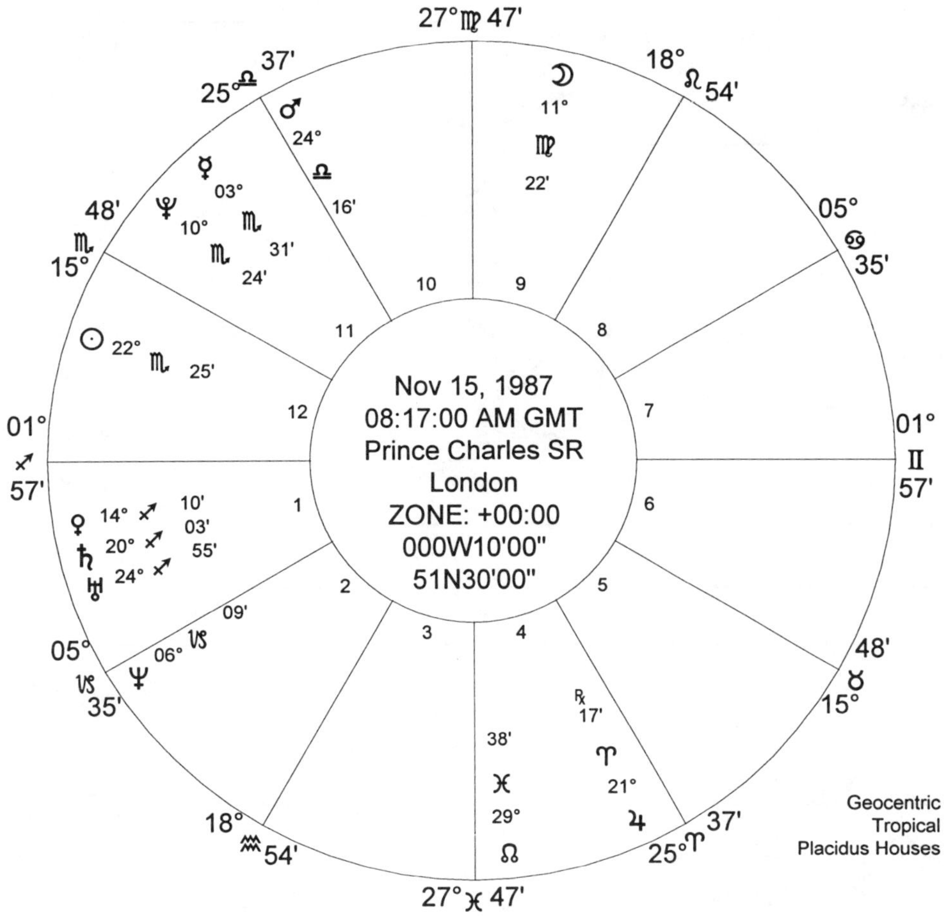
Nov 15, 1987
08:17:00 AM GMT
Prince Charles SR
London
ZONE: +00:00
000W10'00"
51N30'00"
Geocentric
Tropical
Placidus Houses

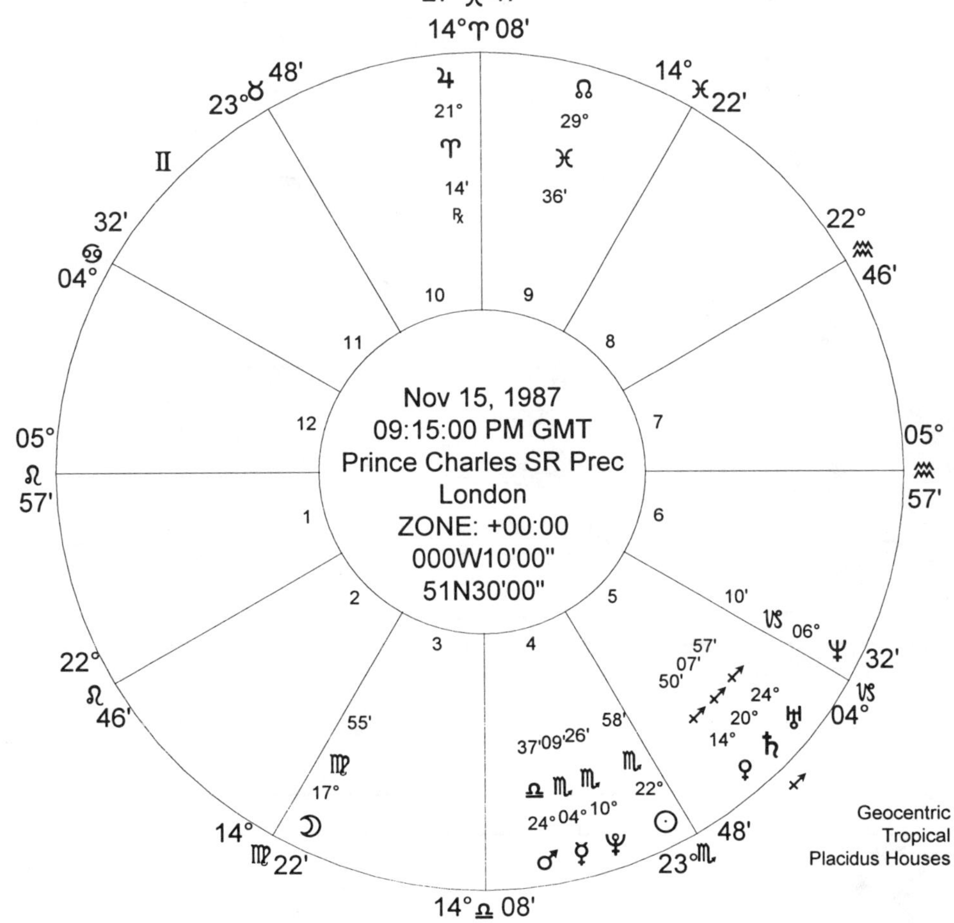
Nov 15, 1987
09:15:00 PM GMT
Prince Charles SR Prec
London
ZONE: +00:00
000W10'00"
51N30'00"
Geocentric
Tropical
Placidus Houses

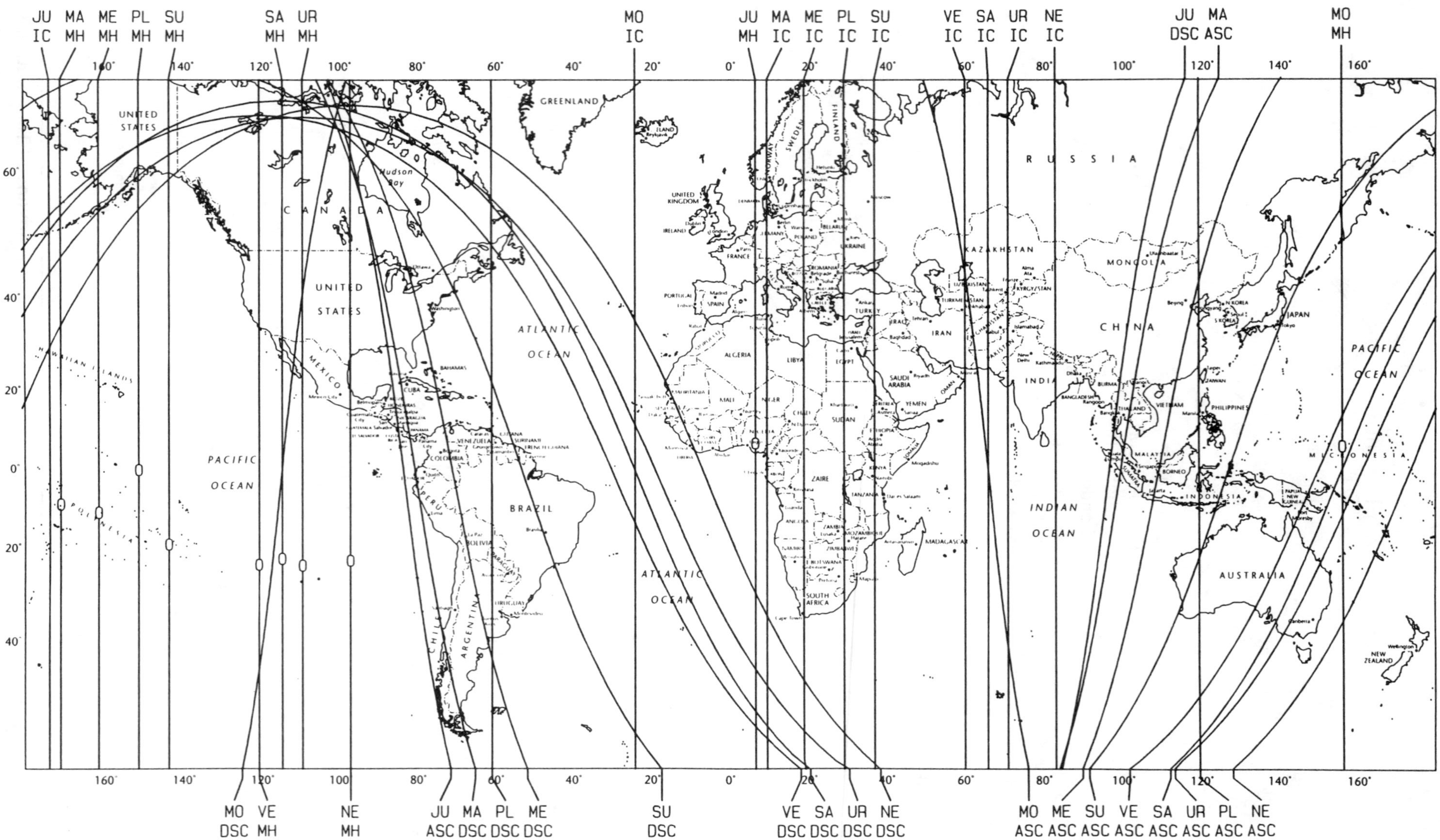

Prince Charles SR

November 15, 1987 9:15 PM GMT

04/16/96

ASTRO*CARTO*GRAPHY LATITUDE CROSSINGS

NORTH LATITUDES

69N19 MA/MH * SU/ASC
69N05 SA/MH * SU/DSC
68N47 JU/IC * SU/ASC
68N12 MO/DSC * ME/DSC
68N01 UR/MH * SU/DSC
68N01 UR/MH * SA/DSC
68N01 SU/DSC * SA/DSC
68N00 VE/MH * SA/ASC
67N41 MA/DSC * SA/DSC
67N40 ME/DSC * NE/ASC
67N39 MO/DSC * NE/ASC
67N38 JU/ASC * NE/ASC
67N36 SA/DSC * NE/ASC
67N35 MA/DSC * NE/ASC
67N30 JU/ASC * SA/DSC
67N28 MO/DSC * SA/DSC
67N23 SU/DSC * NE/ASC
67N16 ME/DSC * SA/DSC
67N13 UR/MH * NE/ASC
67N13 MO/DSC * JU/ASC
67N10 NE/MH * SA/DSC
66N43 SA/MH * NE/ASC
66N35 VE/DSC * NE/ASC
66N33 SA/MH * VE/DSC
66N31 NE/MH * ME/DSC
66N28 VE/DSC * UR/ASC
66N24 SA/MH * UR/ASC
66N22 SU/DSC * UR/DSC
66N17 MA/DSC * UR/DSC
66N16 MO/DSC * UR/DSC
66N16 UR/MH * VE/DSC
66N14 UR/ASC * NE/ASC
66N11 JU/ASC * UR/DSC
66N06 VE/MH * UR/ASC
66N05 MO/DSC * MA/DSC
66N01 NE/MH * UR/DSC
65N57 ME/DSC * UR/DSC
65N55 VE/MH * NE/ASC
65N45 MO/DSC * SU/DSC
65N33 MA/ASC * PL/ASC
65N29 SU/DSC * MA/DSC
65N29 MO/DSC * VE/DSC
65N22 SU/MH * SA/ASC
65N20 VE/DSC * MA/DSC
65N16 SU/DSC * VE/DSC
65N05 VE/DSC * JU/ASC
64N54 SU/MH * VE/ASC
64N51 NE/MH * VE/DSC
64N44 SU/DSC * JU/ASC
64N23 ME/DSC * VE/DSC
64N13 NE/MH * SU/DSC
63N26 PL/MH * SA/ASC
63N22 PL/MH * VE/ASC
63N07 VE/ASC * SA/ASC
63N03 NE/DSC * PL/DSC
62N42 NE/MH * JU/ASC
62N20 SU/MH * UR/ASC
61N51 SU/DSC * ME/DSC
61N24 NE/MH * MA/DSC
61N13 Anchorage
60N55 SA/DSC * UR/DSC
60N19 ME/MH * VE/ASC

59N49 PL/MH * UR/ASC
59N38 ME/MH * SA/ASC
59N22 SU/MH * NE/ASC
59N20 Stockholm
56N48 UR/DSC * PL/DSC
55N57 SA/DSC * PL/DSC
55N56 MA/MH * VE/ASC
55N55 MA/DSC * JU/ASC
55N45 Moscow
55N41 Copenhagen
55N11 PL/MH * NE/ASC

CENTRAL LATITUDES

55N11 MO/MH * SU/ASC
54N52 ME/MH * UR/ASC
54N11 JU/IC * VE/ASC
54N07 MA/MH * SA/ASC
52N30 Berlin
51N55 JU/IC * SA/ASC
51N30 London

49N43 VE/DSC * PL/DSC
49N14 Vancouver
48N50 Paris
47N45 MA/MH * UR/ASC
47N36 Seattle
46N53 ME/MH * NE/ASC
45N32 Montreal
45N03 VE/IC * MO/ASC
44N59 Minneapolis
44N55 JU/IC * UR/ASC
43N39 Toronto
43N00 ME/ASC * PL/ASC
42N22 Boston
42N20 Detroit
41N54 Rome
41N53 Chicago
41N05 UR/MH * MO/DSC
40N43 New York

39N44 Denver
38N53 Washington
38N37 St. Louis
37N47 San Francisco
36N10 Las Vegas
35N56 MO/IC * NE/DSC
35N40 Tokyo
34N34 MA/MH * NE/ASC
34N04 Los Angeles
33N45 Atlanta
33N27 Phoenix
32N47 Dallas
30N03 Cairo

29N58 New Orleans
29N46 Houston
29N40 JU/IC * NE/ASC
28N39 Delhi
25N47 Miami
22N34 Calcutta
22N16 Hong Kong
21N49 SU/DSC * PL/DSC
21N19 Honolulu

19N26 Mexico City
14N35 Manila
13N24 MO/MH * VE/ASC
11N38 MO/IC * UR/DSC
10N30 Caracas

6N27 Lagos
0N12 SA/IC * MO/ASC
0N03 MO/MH * SA/ASC

0S03 MO/IC * SA/DSC
0S12 SA/MH * MO/DSC

11S38 MO/MH * UR/ASC
13S24 MO/IC * VE/DSC

21S49 SU/ASC * PL/ASC
22S54 Rio De Janeiro
26S12 Johannesburg
29S40 JU/MH * NE/DSC

33S55 Sydney
34S34 MA/IC * NE/DSC
34S36 Buenos Aires

SOUTH LATITUDES

35S56 MO/MH * NE/ASC
36S52 Auckland
37S50 Melbourne

41S05 UR/IC * MO/ASC
43S00 ME/DSC * PL/DSC
44S55 JU/MH * UR/DSC
45S03 VE/MH * MO/DSC
46S53 ME/IC * NE/DSC
47S45 MA/IC * UR/DSC
49S43 VE/ASC * PL/ASC

51S55 JU/MH * SA/DSC
54S07 MA/IC * SA/DSC
54S11 JU/MH * VE/DSC
54S52 ME/IC * UR/DSC
55S11 MO/IC * SU/DSC
55S11 PL/IC * NE/DSC
55S55 MA/ASC * JU/DSC
55S56 MA/IC * VE/DSC
55S57 SA/ASC * PL/ASC
56S48 UR/ASC * PL/ASC
59S22 SU/IC * NE/DSC
59S38 ME/IC * SA/DSC
59S49 PL/IC * UR/DSC

express his own individuality (MA/UR). The need to prove himself to others can be seen from VE/PL.

Again, JU/UR turns up where he experienced good fortune in the tragedy.

The Astro*Carto*Graphy solar return map for his thirty-ninth birthday shows that it was a dangerous time for him to visit Klosters. The potential danger shown at Klosters on his Astro*Carto*Graphy map was triggered by the solar return MA IC exact there.

Example 2—Norman Schwarzkopf and Kuwait

On his Astro*Carto*Graphy map, SU MH is passing through the Gulf near Kuwait.

The Astro*Carto*Graphy solar return map shows that he had UR MH passing through the Gulf, exactly on top of his A*C*G SU MH line. This shows the sudden rise to fame he achieved as a result of the Gulf War.

The SU MH line in the Gulf turns up exactly at his birthplace on his solar return map.

During Norman Schwarzkopf's fifty-sixth birthday, the Gulf (A*C*G SU MH) and his birthplace (solar return SU MH) were connected.

At Kuwait Norman Schwarzkopf had the following latitude crossings on his solar return map:

Crossings: MA IC/VE DSC
ME IC/UR DSC
VE IC/MA ASC

The two VE/MA crossings confirm being in charge of the American and allied forces. With ME/UR he was able to use his intelligence and cleverness in bringing a quick end to the Gulf War.

Example 3—Bob Geldof and Ethiopia

Bob Geldof has UR IC passing through Ethiopia on his Astro*Carto*Graphy solar return map. Although MO DSC is more than eight degrees from Adis Adeba, it did have an influence in a small part of Ethiopia.

Cyclo*Carto*Graphy shows that UR had a strong influence for Bob Geldof in Ethiopia, especially at Adis Abeba. His C*C*G MO ASC was also moving through Ethiopia during the time that Band Aid and Live Aid were successful.

The Astro*Carto*Graphy solar return map shows a reiteration of the UR and MO temporary influences in Ethiopia, giving them greater importance.

Interestingly, the MO DSC line, which is just within orb of Ethiopia, is near London where the Band Aid single was recorded. Bob Geldof has MO MH near Ethiopia on his Astro*Carto*Graphy map, and the reiteration of MO for the temporary influences emphasizes the potential there.

At Adis Abeba Bob Geldof has one latitude crossing

Crossing: VE DSC/PL DSC

This shows the impossible fund raising efforts that Bob Geldof achieved in helping the starving people in Ethiopia.

Example 4—Margaret Thatcher and Falkland Islands

Cyclo*Carto*Graphy explains the powerful MA temporary influences when the Falklands War started.

Unbelievably, Margaret Thatcher had MA DSC near the Falkland Islands on her Astro*Carto*Graphy solar return map. This repetition of themes for the temporary influences shows that it was an important year for her.

The latitude crossings for Port Stanley on the solar return map are:

Crossings: MO MH/NE DSC
SU ASC/ME ASC
MO ASC/JU DSC
VE ASC/UR ASC
ME ASC/SA ASC
MO ASC/SU DSC
JU IC/NE DSC

When these crossings are interpreted they give background influences as to how Margaret Thatcher experienced MA DSC at the Falkland Islands. They show various types of influences which were effective before, during and after the Falklands war.

Example 5—David Copperfield and Germany

Using Cyclo*Carto*Graphy there are no location aspects to explain why David Copperfield met his future wife in Berlin.

On his Astro*Carto*Graphy solar return map David Copperfield has NE DSC and UR DSC very near to Berlin. With both planets on the DSC they would manifest externally through partnerships. Using keywords, there was a sudden (UR) and ideal (NE) partnership (DSC).

He also had solar return SU IC within orb of Berlin. This reiterates the exact SU MH line at Berlin on his Astro*Carto*Graphy map, thus putting more emphasis on this permanent influence.

With NE DSC, UR DSC and SU IC having an influence at Berlin, his partnership with Claudia Schiffer received plenty of publicity. There was uncertainty about the longevity of the relationship, as many people thought it was only for publicity.

At Berlin David Copperfield has the following latitude crossings:

Crossings: PL MH/UR ASC
VE ASC/SA DSC
PL MH/NE ASC

The influence from NE and UR at Berlin is highlighted by PL MH, which is involved in latitude crossings with both these planets. This could indicate that the new partnership with Claudia Schiffer would completely transform his life.

VE/SA indicates cautiousness in relationships. He would not rush into getting married, wanting to make sure that he didn't have unrealistic expectations (NE DSC). Independence in the partnership would also be important for him (UR DSC).

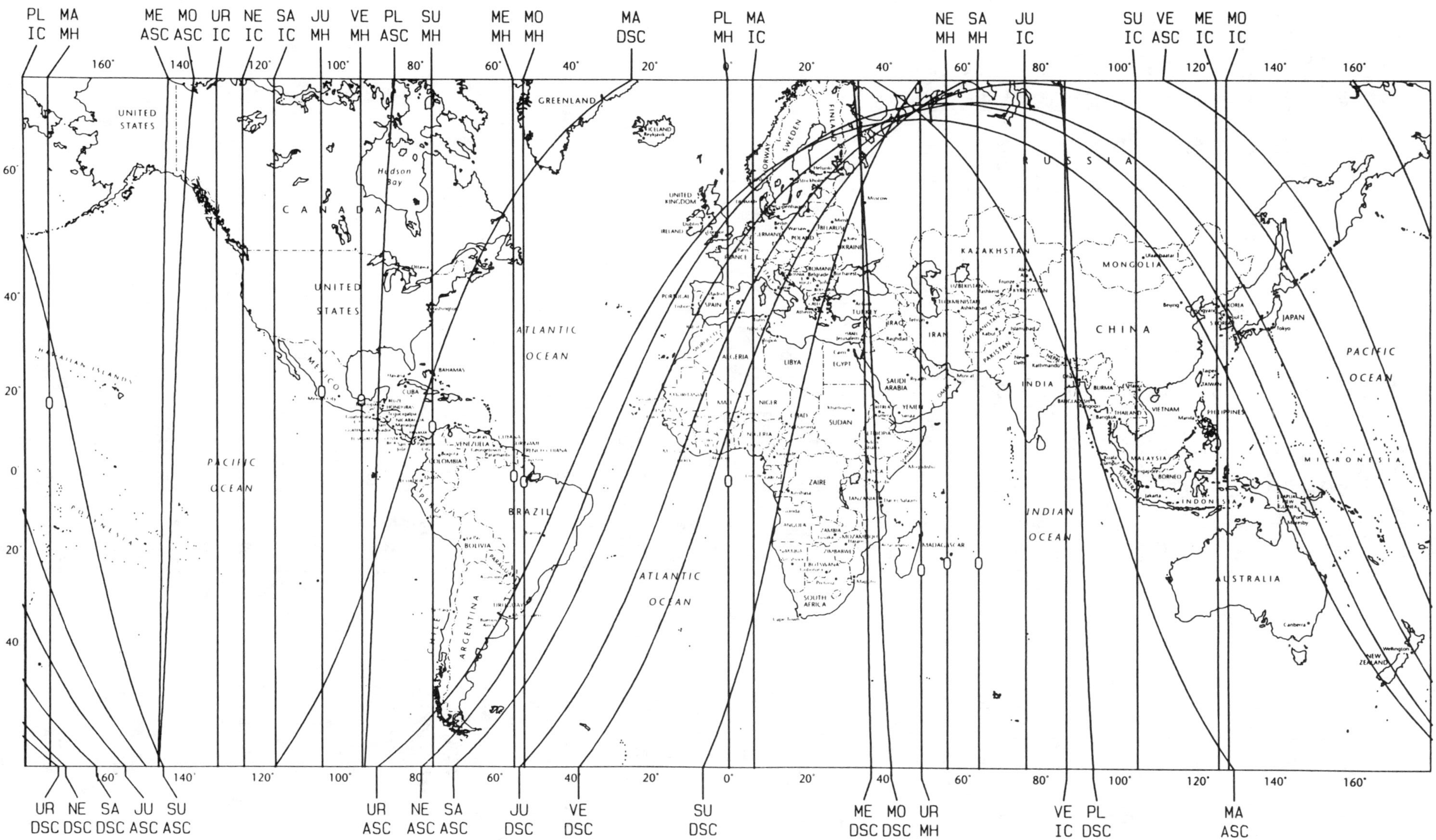

Norman Schwarzkopf SR

August 22, 1990 5:03 PM GMT

04/16/96

Norman Schwarzkopf SR
August 22, 1990 5:03 PM GMT

ASTRO*CARTO*GRAPHY LATITUDE CROSSINGS

NORTH LATITUDES

69N31 JU/ASC * PL/DSC
69N28 VE/IC * JU/ASC
69N20 SA/MH * JU/DSC
69N07 NE/MH * VE/DSC
68N33 NE/MH * JU/DSC
68N08 SU/DSC * MA/ASC
67N56 SA/ASC * NE/DSC
67N54 JU/DSC * NE/ASC
67N52 VE/DSC * NE/ASC
67N49 UR/MH * NE/ASC
67N48 NE/MH * SA/ASC
67N46 SA/MH * NE/DSC
67N43 MA/ASC * NE/ASC
67N35 UR/MH * VE/DSC
67N33 SU/DSC * NE/ASC
67N33 UR/MH * JU/DSC
67N32 JU/IC * SA/DSC
67N27 VE/DSC * JU/DSC
67N20 MA/ASC * JU/DSC
67N19 UR/MH * SA/ASC
67N18 VE/DSC * MA/ASC
67N17 SU/IC * JU/ASC
67N14 MA/ASC * SA/ASC
67N10 VE/DSC * SA/ASC
67N05 ME/IC * VE/ASC
67N03 JU/DSC * SA/ASC
66N58 UR/MH * MA/ASC
66N40 JU/IC * NE/DSC
66N35 SU/DSC * SA/ASC
66N27 SA/DSC * PL/DSC
66N24 VE/IC * SA/DSC
66N23 ME/DSC * NE/ASC
66N21 MA/ASC * UR/DSC
66N18 VE/DSC * UR/ASC
66N17 MO/IC * VE/ASC
66N16 MO/DSC * NE/ASC
66N14 JU/DSC * UR/ASC
66N13 NE/MH * UR/DSC
66N12 SU/DSC * UR/ASC
66N03 SA/ASC * UR/ASC
66N01 SU/DSC * JU/DSC
65N39 SA/MH * UR/DSC
65N34 ME/DSC * UR/ASC
65N31 MO/DSC * UR/ASC
64N56 ME/DSC * SA/ASC
64N53 NE/DSC * PL/DSC
64N52 VE/IC * NE/DSC
64N50 MO/DSC * SA/ASC
64N38 SU/DSC * VE/DSC
64N32 VE/IC * PL/DSC
64N21 NE/MH * MA/ASC
63N52 JU/IC * UR/DSC
63N32 ME/DSC * JU/DSC
63N25 MO/DSC * JU/DSC
63N05 UR/ASC * NE/ASC
62N03 SU/IC * SA/DSC
61N42 ME/DSC * VE/DSC
61N39 MO/DSC * VE/DSC
61N15 VE/IC * UR/DSC
61N13 Anchorage
61N05 UR/DSC * PL/DSC
60N49 MO/DSC * ME/DSC
60N37 ME/IC * JU/ASC

59N50 SA/MH * MA/ASC
59N20 Stockholm
59N16 MO/IC * JU/ASC
59N11 MA/IC * UR/ASC
59N03 MO/MH * MA/DSC
58N46 MO/DSC * SU/DSC
58N37 SU/IC * NE/DSC
58N31 SU/DSC * ME/DSC
58N04 MA/IC * NE/ASC
57N03 ME/MH * MA/DSC

CENTRAL LATITUDES

56N10 PL/MH * UR/ASC
55N45 Moscow
55N41 Copenhagen
54N10 PL/MH * NE/ASC
53N01 MA/IC * SA/ASC
52N34 SU/IC * UR/DSC
52N30 Berlin
51N30 London
50N04 PL/IC * SU/ASC
50N00 ME/IC * SA/DSC

49N14 Vancouver
48N50 Paris
48N05 JU/IC * MA/ASC
47N36 Seattle
47N33 MO/IC * SA/DSC
47N22 PL/MH * SA/ASC
45N32 Montreal
44N59 Minneapolis
43N39 Toronto
43N18 MA/IC * JU/DSC
42N22 Boston
42N20 Detroit
41N54 Rome
41N53 Chicago
41N10 ME/IC * NE/DSC
40N43 New York

39N44 Denver
38N53 Washington
38N37 St. Louis
37N47 San Francisco
37N35 MO/IC * NE/DSC
36N10 Las Vegas
35N40 Tokyo
34N04 Los Angeles
33N45 Atlanta
33N38 MA/MH * SU/ASC
33N27 Phoenix
33N08 PL/MH * JU/DSC
32N47 Dallas
30N03 Cairo

29N58 New Orleans
29N46 Houston
29N26 VE/IC * MA/ASC
29N01 ME/IC * UR/DSC
28N39 Delhi
28N08 MA/IC * VE/DSC
25N47 Miami
24N27 MO/IC * UR/DSC
23N50 SU/MH * MA/DSC
22N34 Calcutta
22N16 Hong Kong
21N38 MA/ASC * PL/DSC
21N19 Honolulu

19N26 Mexico City
14N35 Manila
11N11 PL/MH * VE/DSC
10N30 Caracas

6N27 Lagos

11S11 PL/IC * VE/ASC

21S38 MA/DSC * PL/ASC
22S54 Rio De Janeiro
23S50 SU/IC * MA/ASC
24S27 MO/MH * UR/ASC
26S12 Johannesburg
28S08 MA/MH * VE/ASC
29S01 ME/MH * UR/ASC
29S26 VE/MH * MA/DSC

SOUTH LATITUDES

33S08 PL/IC * JU/ASC
33S38 MA/IC * SU/DSC
33S55 Sydney
34S36 Buenos Aires
36S52 Auckland
37S35 MO/MH * NE/ASC
37S50 Melbourne

41S10 ME/MH * NE/ASC
43S18 MA/MH * JU/ASC
47S22 PL/IC * SA/DSC
47S33 MO/MH * SA/ASC
48S05 JU/MH * MA/DSC
50S00 ME/MH * SA/ASC

50S04 PL/MH * SU/DSC
52S34 SU/MH * UR/ASC
53S01 MA/MH * SA/DSC
54S10 PL/IC * NE/DSC
56S10 PL/IC * UR/DSC
57S03 ME/IC * MA/ASC
58S04 MA/MH * NE/DSC
58S31 SU/ASC * ME/ASC
58S37 SU/MH * NE/ASC
58S46 MO/ASC * SU/ASC
59S03 MO/IC * MA/ASC
59S11 MA/MH * UR/DSC
59S16 MO/MH * JU/DSC
59S50 SA/IC * MA/DSC

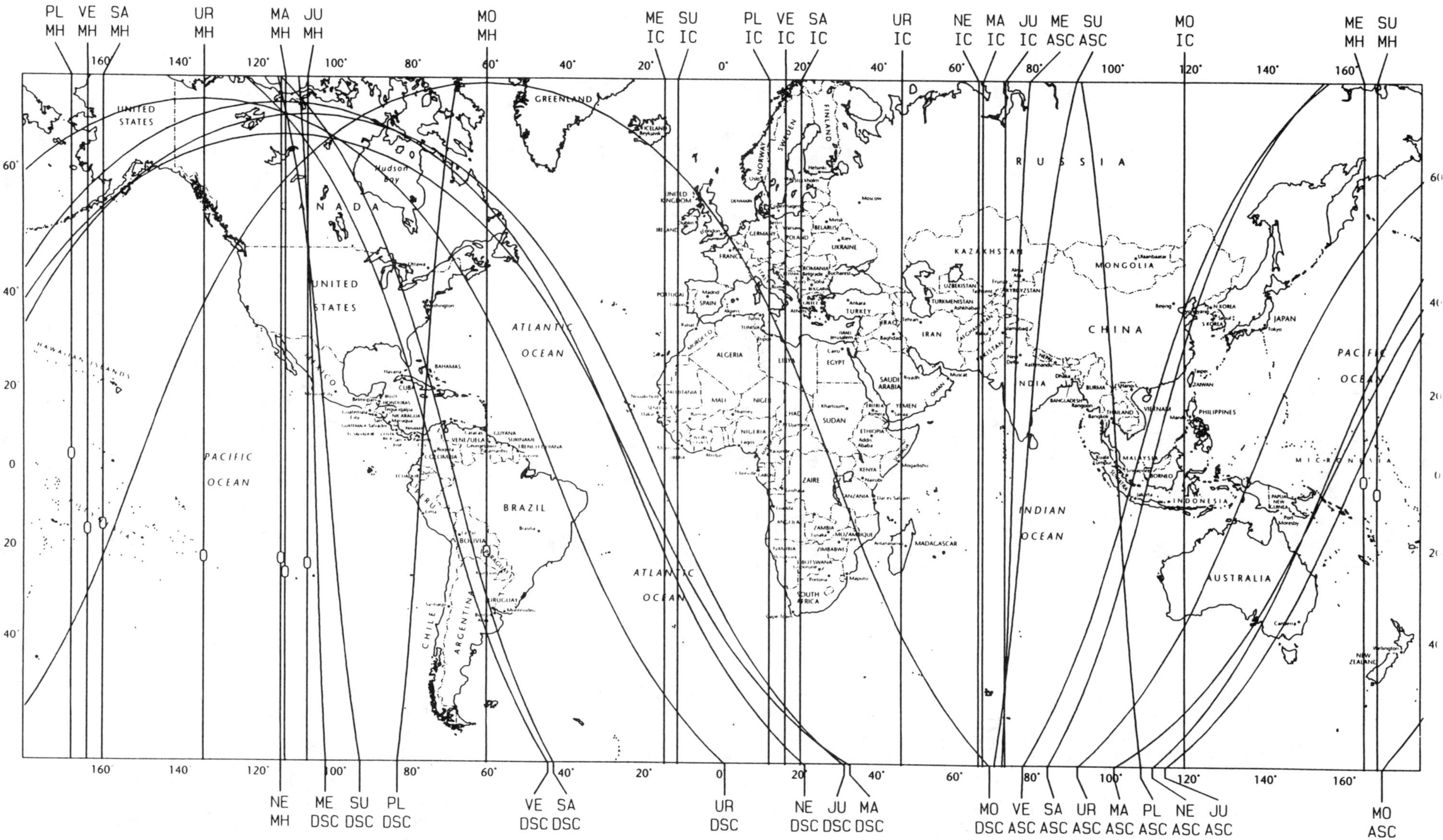

05/21/97

Bob Geldof SR

October 5, 1984 12:36 AM GMT

Bob Geldof SR
October 5, 1984 12:36 AM GMT

ASTRO*CARTO*GRAPHY LATITUDE CROSSINGS

NORTH LATITUDES

69N57	NE/MH * SA/DSC
69N35	MA/MH * SA/DSC
69N28	MO/ASC * PL/DSC
68N43	NE/MH * SU/DSC
68N08	ME/DSC * SA/DSC
67N43	SU/DSC * NE/DSC
67N43	MA/MH * NE/DSC
67N42	VE/DSC * NE/ASC
67N38	ME/DSC * NE/DSC
67N36	SA/DSC * NE/DSC
67N34	JU/MH * NE/DSC
67N32	UR/DSC * NE/ASC
67N16	JU/MH * SA/DSC
67N13	VE/ASC * SA/ASC
66N53	MA/MH * SU/DSC
66N46	NE/MH * VE/DSC
66N45	NE/MH * UR/DSC
66N44	VE/DSC * UR/DSC
66N36	MA/MH * UR/DSC
66N34	SU/DSC * UR/DSC
66N31	JU/DSC * SA/DSC
66N31	ME/DSC * JU/ASC
66N27	UR/MH * NE/ASC
66N27	JU/ASC * UR/DSC
66N26	SU/DSC * JU/ASC
66N25	MA/MH * JU/ASC
66N25	VE/DSC * JU/ASC
66N23	NE/MH * JU/ASC
66N20	MO/ASC * NE/DSC
66N17	MA/MH * VE/DSC
66N06	SU/DSC * VE/DSC
66N01	MO/ASC * JU/DSC
65N56	ME/DSC * UR/DSC
65N48	SA/MH * UR/ASC
65N39	JU/MH * UR/DSC
64N59	VE/MH * UR/ASC
64N50	JU/DSC * NE/DSC
64N35	SA/DSC * UR/DSC
64N25	NE/MH * MA/ASC
64N25	SU/DSC * MA/DSC
64N22	VE/DSC * MA/DSC
64N21	ME/DSC * MA/DSC
64N18	JU/MH * MA/DSC
64N15	MO/ASC * UR/DSC
64N05	UR/MH * JU/ASC
64N04	MO/ASC * SA/DSC
64N00	ME/DSC * VE/DSC
63N59	MO/ASC * MA/DSC
63N58	PL/MH * UR/ASC
63N56	MA/DSC * SA/DSC
63N44	MA/DSC * UR/DSC
63N16	JU/MH * VE/DSC
62N52	UR/MH * MA/ASC
62N25	MO/ASC * VE/DSC
61N57	ME/IC * MO/DSC
61N55	JU/DSC * PL/DSC
61N51	JU/MH * MO/ASC
61N26	MO/ASC * ME/DSC
61N13	Anchorage
61N04	NE/DSC * PL/DSC
60N28	MO/ASC * SU/DSC
60N22	SU/IC * MO/DSC
59N33	SA/MH * NE/ASC
59N20	Stockholm
59N00	MA/MH * MO/ASC
58N23	NE/MH * MO/ASC
57N56	MO/MH * JU/DSC
57N53	MA/DSC * PL/DSC
57N35	VE/MH * NE/ASC
57N31	MA/ASC * JU/ASC
55N47	MO/MH * NE/DSC
55N45	Moscow
55N41	Copenhagen

CENTRAL LATITUDES

55N06	PL/MH * NE/ASC
54N59	SA/MH * MA/ASC
54N25	SA/MH * JU/ASC
52N48	VE/MH * MA/ASC
52N45	SU/MH * UR/ASC
52N30	Berlin
52N13	MO/MH * MA/DSC
51N42	VE/MH * JU/ASC
51N30	London
51N24	JU/MH * ME/DSC
51N01	UR/DSC * PL/DSC
50N05	PL/MH * MA/ASC
49N53	ME/MH * UR/ASC
49N14	Vancouver
48N50	Paris
48N48	JU/MH * SU/DSC
48N17	PL/MH * JU/ASC
47N36	Seattle
47N23	SU/DSC * ME/DSC
45N32	Montreal
44N59	Minneapolis
43N39	Toronto
42N22	Boston
42N20	Detroit
41N54	Rome
41N53	Chicago
40N43	New York
39N44	Denver
38N53	Washington
38N37	St. Louis
38N33	UR/MH * MO/ASC
38N32	MO/IC * VE/ASC
38N11	PL/IC * MO/DSC
37N47	San Francisco
36N11	MO/MH * UR/DSC
36N10	Las Vegas
35N40	Tokyo
34N04	Los Angeles
33N45	Atlanta
33N27	Phoenix
32N47	Dallas
30N53	MO/IC * SA/ASC
30N45	VE/IC * MO/DSC
30N03	Cairo
29N58	New Orleans
29N46	Houston
28N39	Delhi
27N23	SU/MH * NE/ASC
25N47	Miami
24N32	SA/DSC * PL/DSC
22N34	Calcutta
22N18	SA/IC * MO/DSC
22N16	Hong Kong
22N02	SU/MH * MA/ASC
21N19	Honolulu
20N32	ME/MH * NE/ASC
19N26	Mexico City
15N42	ME/MH * MA/ASC
15N20	MA/DSC * NE/DSC
14N35	Manila
12N20	VE/DSC * PL/DSC
12N16	SU/MH * JU/ASC
10N30	Caracas
6N27	Lagos
4N34	ME/MH * JU/ASC
4S34	ME/IC * JU/DSC
12S16	SU/IC * JU/DSC
12S20	VE/ASC * PL/ASC

SOUTH LATITUDES

15S20	MA/ASC * NE/ASC
15S42	ME/IC * MA/DSC
20S32	ME/IC * NE/DSC
22S02	SU/IC * MA/DSC
22S18	SA/MH * MO/ASC
22S54	Rio De Janeiro
24S32	SA/ASC * PL/ASC
26S12	Johannesburg
27S23	SU/IC * NE/DSC
30S45	VE/MH * MO/ASC
30S53	MO/MH * SA/DSC
33S55	Sydney
34S36	Buenos Aires
36S11	MO/IC * UR/ASC
36S52	Auckland
37S50	Melbourne
38S11	PL/MH * MO/ASC
38S32	MO/MH * VE/DSC
38S33	UR/IC * MO/DSC
47S23	SU/ASC * ME/ASC
48S17	PL/IC * JU/DSC
48S48	JU/IC * SU/ASC
49S53	ME/IC * UR/DSC
50S05	PL/IC * MA/DSC
51S01	UR/ASC * PL/ASC
51S24	JU/IC * ME/ASC
51S42	VE/IC * JU/DSC
52S13	MO/IC * MA/ASC
52S45	SU/IC * UR/DSC
52S48	VE/IC * MA/DSC
54S25	SA/IC * JU/DSC
54S59	SA/IC * MA/DSC
55S06	PL/IC * NE/DSC
55S47	MO/IC * NE/ASC
57S31	MA/DSC * JU/DSC
57S35	VE/IC * NE/DSC
57S53	MA/ASC * PL/ASC
57S56	MO/IC * JU/ASC
58S23	NE/IC * MO/DSC
59S00	MA/IC * MO/DSC
59S33	SA/IC * NE/DSC

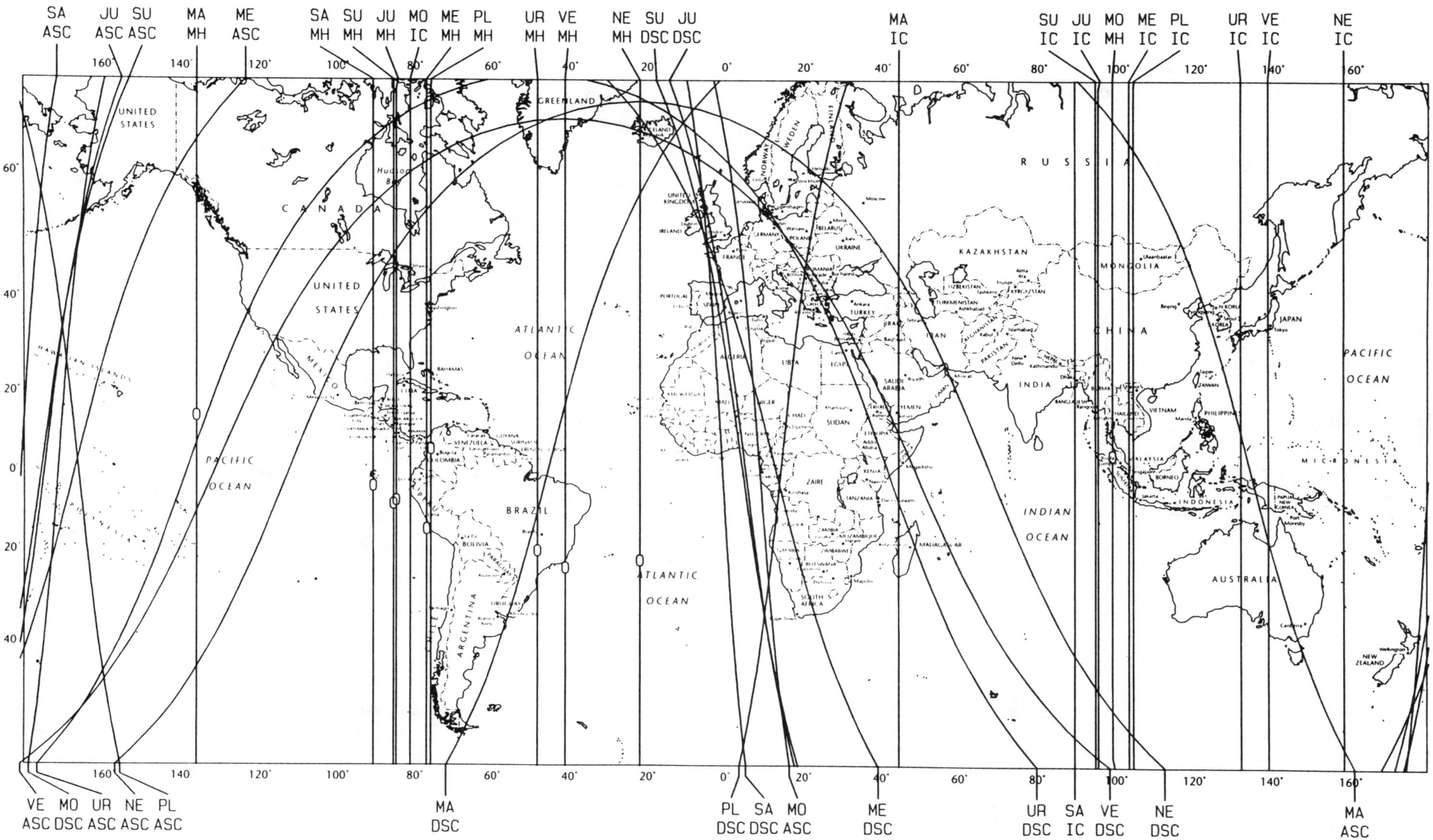

Margaret Thatcher SR

October 13, 1981 5:25 PM GMT

04/16/96

04/16/96

Margaret Thatcher SR
October 13, 1981 5:25 PM GMT

ASTRO*CARTO*GRAPHY LATITUDE CROSSINGS

NORTH LATITUDES

69N34 MO/ASC * MA/DSC
69N21 ME/DSC * UR/DSC
68N25 NE/MH * UR/DSC
68N17 SU/IC * MA/ASC
68N12 PL/MH * UR/ASC
68N02 UR/DSC * NE/DSC
68N01 ME/DSC * NE/ASC
67N59 ME/MH * UR/ASC
67N58 JU/IC * MA/ASC
67N55 SU/DSC * NE/DSC
67N45 JU/DSC * NE/DSC
67N30 SA/DSC * NE/DSC
67N26 MA/DSC * NE/DSC
67N07 MA/DSC * SA/DSC
67N07 MO/IC * UR/ASC
66N55 VE/MH * NE/ASC
66N50 SU/DSC * UR/DSC
66N38 MO/ASC * NE/DSC
66N29 NE/MH * ME/DSC
66N15 MO/MH * MA/ASC
66N15 VE/ASC * NE/ASC
66N13 JU/MH * UR/ASC
66N09 UR/MH * VE/ASC
66N02 MA/DSC * JU/DSC
66N00 SU/MH * UR/ASC
65N57 MA/DSC * UR/DSC
65N51 JU/DSC * UR/DSC
65N50 UR/MH * NE/ASC
65N06 NE/MH * VE/DSC
65N00 SA/DSC * UR/DSC
64N57 SU/DSC * MA/DSC
64N38 ME/DSC * VE/DSC
64N21 SA/MH * UR/ASC
64N03 MA/MH * ME/ASC
63N51 VE/DSC * MA/DSC
63N48 ME/IC * MA/ASC
63N10 SU/DSC * VE/DSC
63N04 PL/IC * MA/ASC
62N45 ME/DSC * MA/DSC
62N25 VE/DSC * JU/DSC
62N03 VE/DSC * SA/DSC
61N58 PL/MH * VE/ASC
61N39 ME/MH * VE/ASC
61N25 MO/ASC * UR/DSC
61N13 Anchorage
60N16 MO/IC * VE/ASC

59N30 NE/DSC * PL/DSC
59N20 Stockholm
59N03 MO/ASC * VE/DSC
58N53 JU/MH * VE/ASC
58N34 SU/MH * VE/ASC
57N51 JU/DSC * SA/DSC
57N30 NE/MH * MA/DSC
57N23 SA/ASC * PL/ASC
56N02 SA/MH * VE/ASC
55N51 PL/MH * NE/ASC
55N45 Moscow
55N41 Copenhagen
55N08 ME/MH * NE/ASC
52N30 Berlin
52N04 MO/IC * NE/ASC
52N04 SU/DSC * ME/DSC
51N55 MO/DSC * JU/ASC
51N49 VE/DSC * UR/DSC
51N30 London
50N41 ME/DSC * SA/DSC

49N33 MO/DSC * SU/ASC
49N14 Vancouver
48N56 JU/MH * NE/ASC
48N50 Paris
48N26 SU/DSC * SA/DSC
48N12 SU/MH * NE/ASC

CENTRAL LATITUDES

47N36 Seattle
46N56 VE/DSC * PL/DSC
46N11 ME/DSC * JU/DSC
45N34 UR/DSC * PL/DSC
45N32 Montreal
44N59 Minneapolis
44N44 MA/IC * NE/DSC
43N39 Toronto
42N24 SA/MH * NE/ASC
42N22 Boston
42N20 Detroit
41N54 Rome
41N53 Chicago
40N43 New York
40N17 SU/ASC * JU/ASC

39N44 Denver
38N53 Washington
38N37 St. Louis
37N47 San Francisco
36N10 Las Vegas
35N40 Tokyo
34N25 SU/ASC * PL/ASC
34N04 Los Angeles
33N59 JU/ASC * PL/ASC
33N45 Atlanta
33N27 Phoenix
32N47 Dallas
30N03 Cairo

29N58 New Orleans
29N46 Houston
28N39 Delhi
27N01 MO/DSC * PL/ASC
25N47 Miami
22N34 Calcutta
22N29 MO/ASC * ME/DSC
22N16 Hong Kong
21N19 Honolulu

19N26 Mexico City
17N04 VE/MH * MA/DSC
14N35 Manila
10N30 Caracas
10N06 MA/IC * VE/DSC
10N02 UR/IC * MA/ASC

7N13 MA/MH * UR/ASC
6N27 Lagos
3N04 ME/ASC * PL/ASC

3S04 ME/DSC * PL/DSC
7S13 MA/IC * UR/DSC

10S02 UR/MH * MA/DSC
10S06 MA/MH * VE/ASC
17S04 VE/IC * MA/ASC

22S29 MO/DSC * ME/ASC
22S54 Rio De Janeiro
26S12 Johannesburg
27S01 MO/ASC * PL/DSC

33S55 Sydney
33S59 JU/DSC * PL/DSC
34S25 SU/DSC * PL/DSC
34S36 Buenos Aires
36S52 Auckland
37S50 Melbourne

40S17 SU/DSC * JU/DSC
42S24 SA/IC * NE/DSC
44S44 MA/MH * NE/ASC
45S34 UR/ASC * PL/ASC
46S11 ME/ASC * JU/ASC

SOUTH LATITUDES

46S56 VE/ASC * PL/ASC
48S12 SU/IC * NE/DSC
48S26 SU/ASC * SA/ASC
48S56 JU/IC * NE/DSC
49S33 MO/ASC * SU/DSC

50S41 ME/ASC * SA/ASC
51S49 VE/ASC * UR/ASC
51S55 MO/ASC * JU/DSC
52S04 SU/ASC * ME/ASC
52S04 MO/MH * NE/DSC
55S08 ME/IC * NE/DSC
55S51 PL/IC * NE/DSC
56S02 SA/IC * VE/DSC
57S23 SA/DSC * PL/DSC
57S30 NE/IC * MA/ASC
57S51 JU/ASC * SA/ASC
58S34 SU/IC * VE/DSC
58S53 JU/IC * VE/DSC
59S03 MO/DSC * VE/ASC
59S30 NE/ASC * PL/ASC

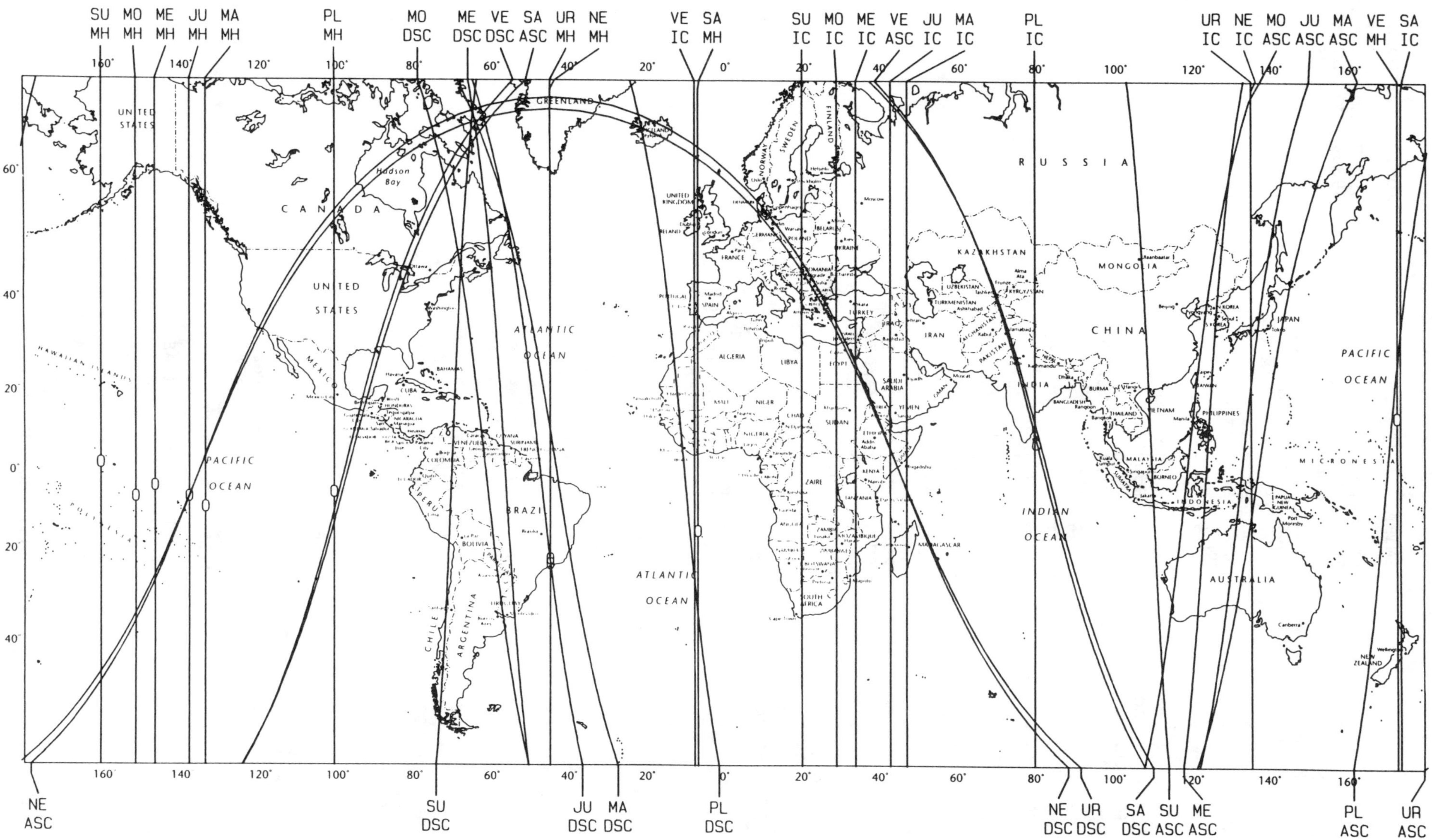

David Copperfield SR

September 16, 1993 10: 34 PM GMT

04/16/96

04/16/96

David Copperfield SR
September 16, 1993 10:34 PM GMT

ASTRO*CARTO*GRAPHY LATITUDE CROSSINGS

NORTH LATITUDES

69N12	SU/DSC * JU/DSC
68N49	UR/IC * MO/ASC
68N49	NE/IC * MO/ASC
68N30	JU/IC * VE/ASC
68N29	UR/MH * NE/DSC
68N05	SA/ASC * NE/ASC
68N04	JU/IC * SA/DSC
67N47	VE/DSC * NE/ASC
67N31	JU/DSC * NE/ASC
67N20	NE/MH * UR/ASC
67N17	SU/DSC * NE/ASC
67N16	ME/DSC * NE/ASC
67N13	MA/DSC * NE/ASC
67N10	SU/DSC * ME/DSC
66N55	SU/DSC * MA/DSC
66N52	NE/DSC * PL/DSC
66N48	VE/DSC * JU/DSC
66N40	ME/DSC * MA/DSC
66N38	SA/ASC * UR/ASC
66N34	MA/IC * VE/ASC
66N26	JU/DSC * UR/ASC
66N16	VE/DSC * UR/ASC
66N11	MA/DSC * UR/ASC
66N10	MA/IC * SA/DSC
66N08	ME/DSC * UR/ASC
66N03	SU/DSC * UR/ASC
65N58	VE/DSC * MA/DSC
65N49	JU/DSC * SA/ASC
65N45	ME/DSC * VE/DSC
65N33	UR/DSC * PL/DSC
65N27	MO/DSC * NE/ASC
65N08	SU/DSC * VE/DSC
65N03	MA/DSC * SA/ASC
64N33	ME/DSC * SA/ASC
64N20	MO/DSC * UR/ASC
63N43	VE/IC * NE/DSC
63N26	SA/MH * NE/DSC
63N25	SU/DSC * SA/ASC
62N24	VE/IC * UR/DSC
62N06	SA/MH * UR/DSC
61N13	Anchorage
59N54	MO/DSC * VE/DSC
59N47	MO/ASC * ME/ASC
59N20	Stockholm
58N33	MO/DSC * SA/ASC
56N31	MA/DSC * JU/DSC
55N45	Moscow
55N41	Copenhagen
55N11	PL/MH * NE/ASC
54N46	VE/ASC * SA/DSC
53N37	PL/MH * UR/ASC
52N30	Berlin
51N30	London
49N14	Vancouver
48N50	Paris
47N36	Seattle
47N19	SU/IC * NE/DSC
45N42	MO/DSC * SU/DSC
45N41	SU/IC * UR/DSC
45N32	Montreal
44N59	Minneapolis
43N39	Toronto
42N22	Boston
42N20	Detroit
41N54	Rome
41N53	Chicago
40N43	New York
39N44	Denver
38N53	Washington
38N37	St. Louis
37N47	San Francisco

CENTRAL LATITUDES

36N10	Las Vegas
35N40	MO/IC * NE/DSC
35N40	Tokyo
34N08	MO/IC * UR/DSC
34N04	Los Angeles
33N57	SA/IC * PL/ASC
33N45	Atlanta
33N27	Phoenix
32N47	Dallas
30N03	Cairo
29N58	New Orleans
29N46	Houston
28N39	Delhi
26N50	ME/IC * NE/DSC
26N13	VE/MH * PL/ASC
25N47	Miami
25N32	ME/IC * UR/DSC
22N34	Calcutta
22N16	UR/IC * JU/ASC
22N16	Hong Kong
22N15	NE/IC * JU/ASC
21N19	Honolulu
19N26	Mexico City
14N35	Manila
12N48	PL/IC * SA/DSC
10N37	NE/MH * MA/DSC
10N37	UR/MH * MA/DSC
10N30	Caracas
9N50	PL/IC * VE/ASC
6N27	Lagos
6N24	JU/IC * NE/DSC
6N03	JU/IC * UR/DSC
4N14	MA/MH * NE/ASC
4N00	MA/MH * UR/ASC
0N04	UR/ASC * NE/ASC
0S04	UR/DSC * NE/DSC
4S00	MA/IC * UR/DSC
4S14	MA/IC * NE/DSC
6S03	JU/MH * UR/ASC
6S24	JU/MH * NE/ASC
9S50	PL/MH * VE/DSC
10S37	UR/IC * MA/ASC
10S37	NE/IC * MA/ASC
12S48	PL/MH * SA/ASC
22S15	NE/MH * JU/DSC
22S16	UR/MH * JU/DSC
22S54	Rio De Janeiro
25S32	ME/MH * UR/ASC
26S12	Johannesburg
26S13	VE/IC * PL/DSC
26S50	ME/MH * NE/ASC
33S55	Sydney
33S57	SA/MH * PL/DSC
34S08	MO/MH * UR/ASC
34S36	Buenos Aires
35S40	MO/MH * NE/ASC
36S52	Auckland
37S50	Melbourne
45S41	SU/MH * UR/ASC
45S42	MO/ASC * SU/ASC
47S19	SU/MH * NE/ASC
53S37	PL/IC * UR/DSC
54S46	VE/DSC * SA/ASC
55S11	PL/IC * NE/DSC
56S31	MA/ASC * JU/ASC
58S33	MO/ASC * SA/DSC

SOUTH LATITUDES

59S47	MO/DSC * ME/DSC
59S54	MO/ASC * VE/ASC

Example 6—Claudia Schiffer and Germany

Claudia Schiffer has four planetary lines within orb of Berlin on her Astro*Carto*Graphy solar return map.

Lines:
UR IC
NE IC
JU ASC
VE MH

Like David Copperfield, she has UR and NE lines nearest Berlin, but this time on the IC angle. Home, family and property would be subject to unexpected changes and uncertainty.

Solar return VE MH reiterates the temporary influences from C*C*G VE ASC, so this would be an important planetary line. VE MH shows a successful time in her modelling career and having a high profile in society. It could also indicate that any partnership would be followed closely by the media.

With JU ASC she was self-confident and optimistic about the future. Other people saw her happiness in life, which was more noticeable after she met David Copperfield.

The latitude crossings for Berlin on her Astro*Carto*Graphy solar return map are:

Crossings:
JU MH/MO ASC
SU DSC/SA ASC
PL MH/UR ASC
PL MH/NE ASC
ME MH/PL ASC

Claudia Schiffer also had PL MH involved in latitude crossings with UR ASC and NE ASC (same as David Copperfield).

JU/MO reiterates her happiness in life and feeling optimistic about the future. She also has a JU/MO latitude crossing at Berlin on her Astro*Carto*Graphy map.

SU/SA is another crossing she has at Berlin on her Astro*Carto*Graphy map. This indicates that she is cautious in life, especially where important decisions are concerned.

ME/PL suggests that Claudia Schiffer can transform her life through any kind of communication. Having met David Copperfield, she needed to communicate in a foreign language. She also wrote a book about her life.

Example 8—Matthew

Matthew's Astro*Carto*Graphy solar return map for his last voyage from Gdynia to Barranquilla shows he had no solar return planetary lines within orb of Gdynia, Poland. Near his birthplace Matthew had MA IC and SA ASC, which would have a worldwide influence.

These solar return planetary lines show that it was not an easy year for Matthew. He was frustrated with the obstacles and difficulties that kept turning up in his life.

Within orb of Barranquilla, Columbia, he had a solar return SA IC line. This shows very clearly the problems he experienced there.

Solar return SA IC near Columbia shows where the difficulties, indicated by MA IC and SA ASC at his birthplace, could manifest.

With SA ASC and UR ASC passing through Barranquilla on his Astro*Carto*Graphy map the exact location can be pinpointed.

Most of the time the solar return map repeats some of the influences found with Cyclo*Carto*Graphy or Astro*Carto*Graphy.

When using different techniques to find the temporary influences, **repetition of themes** provides the most important information. When a specific theme keeps recurring, it will be a powerful influence.

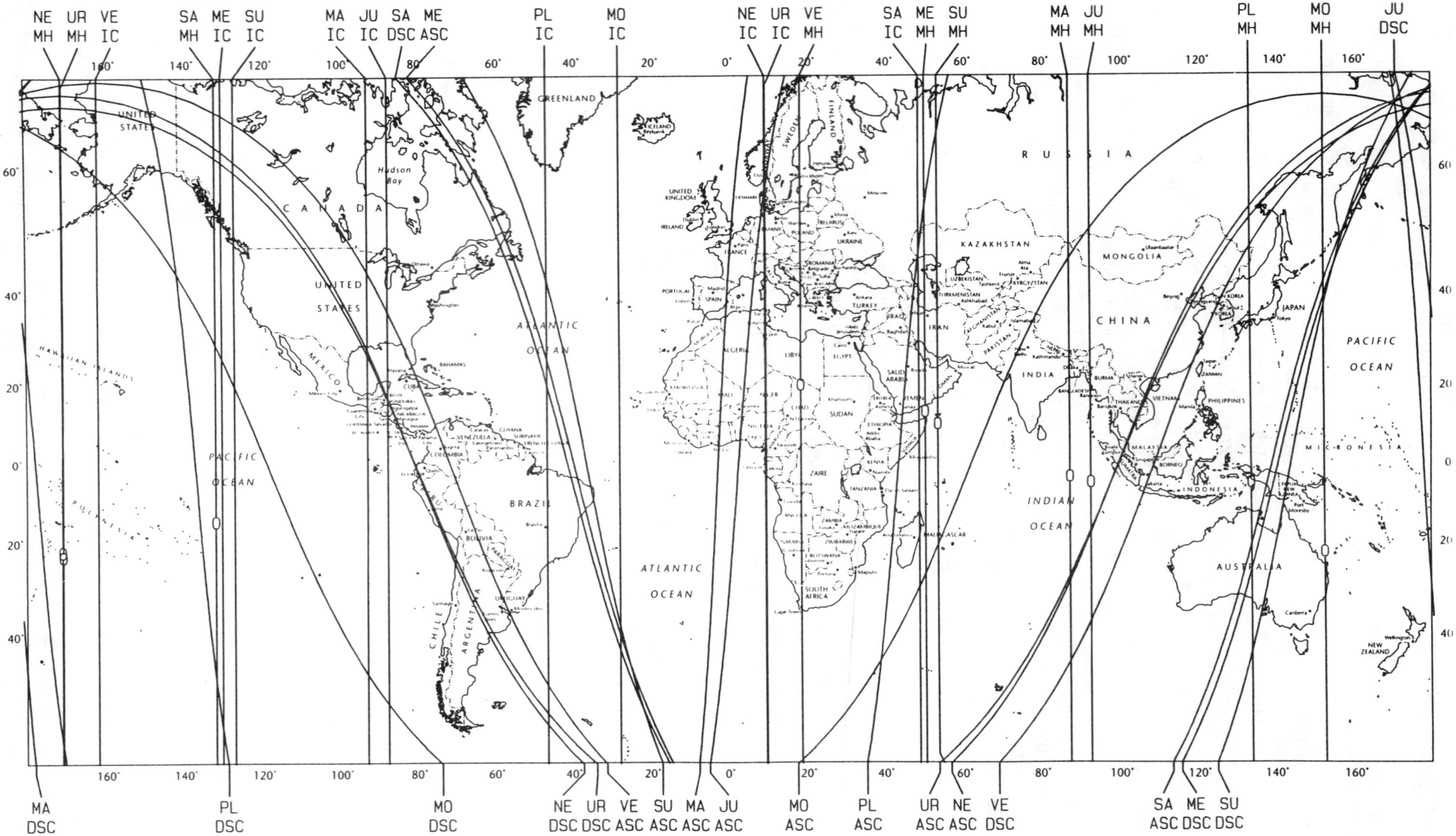

Claudia Schiffer SR

August 25, 1993 8:26 AM GMT

04/16/96

Claudia Schiffer SR
August 25, 1993 8:26 AM GMT

ASTRO*CARTO*GRAPHY LATITUDE CROSSINGS

NORTH LATITUDES

69N29 UR/MH * VE/DSC
69N28 NE/MH * VE/DSC
69N17 VE/ASC * PL/DSC
68N34 ME/DSC * VE/DSC
68N32 UR/MH * NE/DSC
68N29 VE/DSC * SA/ASC
68N17 VE/IC * NE/DSC
68N07 ME/DSC * NE/ASC
68N05 SA/ASC * NE/ASC
68N03 SU/DSC * JU/DSC
67N41 VE/DSC * NE/ASC
67N34 JU/DSC * NE/ASC
67N30 VE/DSC * JU/DSC
67N25 SU/DSC * NE/ASC
67N23 NE/MH * UR/ASC
67N22 MA/DSC * NE/ASC
67N13 SU/DSC * VE/DSC
67N12 VE/DSC * MA/DSC
67N11 SU/DSC * MA/DSC
67N10 MO/DSC * NE/ASC
67N08 VE/IC * UR/DSC
67N04 MO/DSC * VE/DSC
67N00 MO/DSC * SU/DSC
66N58 MO/DSC * MA/DSC
66N57 PL/MH * MO/ASC
66N56 NE/DSC * PL/DSC
66N52 SA/MH * VE/ASC
66N43 MO/DSC * JU/DSC
66N42 ME/DSC * UR/ASC
66N38 SA/ASC * UR/ASC
66N33 MO/DSC * UR/ASC
66N31 ME/IC * VE/ASC
66N29 MO/DSC * SA/ASC
66N26 JU/DSC * UR/ASC
66N24 MO/DSC * ME/DSC
66N13 MA/DSC * UR/ASC
65N57 SU/DSC * UR/ASC
65N47 JU/DSC * SA/ASC
65N44 SU/IC * VE/ASC
65N38 UR/DSC * PL/DSC
65N29 ME/DSC * JU/DSC
64N49 MA/DSC * SA/ASC
64N36 SU/MH * PL/ASC
64N25 ME/DSC * MA/DSC
63N38 VE/DSC * UR/ASC
63N36 MO/MH * NE/ASC
63N07 SA/MH * NE/DSC
63N00 NE/MH * MO/DSC
62N58 UR/MH * MO/DSC
62N34 ME/IC * NE/DSC
62N13 MO/MH * UR/ASC
61N48 SA/MH * UR/DSC
61N41 MO/MH * VE/DSC
61N21 SU/IC * NE/DSC
61N15 ME/IC * UR/DSC
61N13 Anchorage

59N59 SU/IC * UR/DSC
59N32 VE/IC * MO/DSC
59N20 Stockholm
57N49 UR/IC * JU/ASC
57N23 NE/IC * JU/ASC
55N45 Moscow
55N41 Copenhagen
55N09 ME/MH * PL/ASC
54N48 PL/MH * NE/ASC
53N09 PL/MH * UR/ASC
52N30 Berlin
52N23 SU/DSC * SA/ASC
51N39 JU/MH * MO/ASC
51N30 London

49N14 Vancouver
48N50 Paris

CENTRAL LATITUDES

48N37 PL/MH * VE/DSC
47N50 SA/IC * PL/ASC
47N36 Seattle
46N44 MA/MH * MO/ASC
45N32 Montreal
44N59 Minneapolis
44N32 SU/DSC * ME/DSC
44N30 MA/IC * VE/ASC
44N16 MO/DSC * PL/DSC
43N39 Toronto
42N51 PL/IC * SU/ASC
42N22 Boston
42N20 MO/MH * SA/ASC
42N20 Detroit
41N54 Rome
41N53 Chicago
40N43 New York

39N44 Denver
39N28 MO/MH * ME/DSC
38N53 Washington
38N37 St. Louis
37N47 San Francisco
37N43 MO/MH * SU/DSC
36N45 JU/IC * VE/ASC
36N10 Las Vegas
35N40 Tokyo
34N04 Los Angeles
33N45 Atlanta
33N27 Phoenix
32N47 Dallas
30N03 Cairo

29N58 New Orleans
29N46 Houston
29N34 SA/MH * MO/DSC
28N39 Delhi
28N19 MA/IC * NE/DSC
27N11 MA/IC * UR/DSC
26N34 ME/IC * MO/DSC
26N00 PL/IC * ME/ASC
25N47 Miami
22N34 Calcutta
22N16 Hong Kong
21N19 Honolulu
20N06 PL/IC * SA/DSC

19N55 SU/IC * MO/DSC
19N26 Mexico City
17N06 JU/IC * NE/DSC
16N28 JU/IC * UR/DSC
14N35 Manila
10N30 Caracas

6N27 Lagos
5N13 UR/DSC * NE/DSC

5S13 UR/ASC * NE/ASC

16S28 JU/MH * UR/ASC
17S06 JU/MH * NE/ASC
19S55 SU/MH * MO/ASC

20S06 PL/MH * SA/ASC
22S54 Rio De Janeiro
26S00 PL/MH * ME/DSC
26S12 Johannesburg
26S34 ME/MH * MO/ASC
27S11 MA/MH * UR/ASC
28S19 MA/MH * NE/ASC
29S34 SA/IC * MO/ASC

33S55 Sydney
34S36 Buenos Aires
36S45 JU/MH * VE/DSC

SOUTH LATITUDES

36S52 Auckland
37S43 MO/IC * SU/ASC
37S50 Melbourne
39S28 MO/IC * ME/ASC

42S20 MO/IC * SA/DSC
42S51 PL/MH * SU/DSC
44S16 MO/ASC * PL/ASC
44S30 MA/MH * VE/DSC
44S32 SU/ASC * ME/ASC
46S44 MA/IC * MO/DSC
47S50 SA/MH * PL/DSC
48S37 PL/IC * VE/ASC

51S39 JU/IC * MO/DSC
52S23 SU/ASC * SA/DSC
53S09 PL/IC * UR/DSC
54S48 PL/IC * NE/DSC
55S09 ME/IC * PL/DSC
57S23 NE/MH * JU/DSC
57S49 UR/MH * JU/DSC
59S32 VE/MH * MO/ASC
59S59 SU/MH * UR/ASC

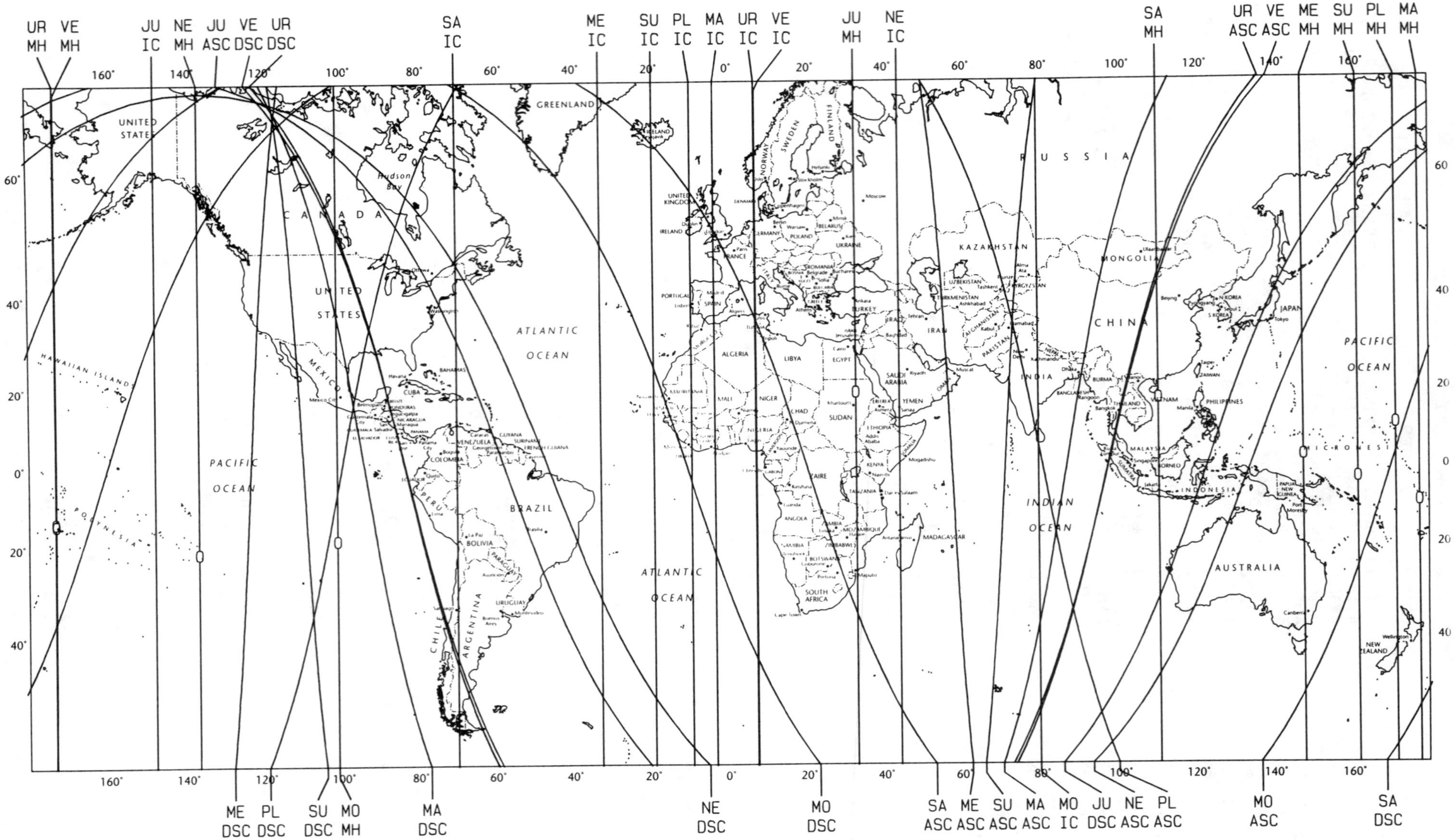

Matthew SR

04/16/96

ASTRO*CARTO*GRAPHY LATITUDE CROSSINGS

NORTH LATITUDES

69N55	MO/ASC * JU/ASC
69N53	ME/ASC * PL/ASC
69N44	SA/IC * MO/DSC
69N42	MO/IC * SU/ASC
69N37	MO/DSC * PL/DSC
69N17	SA/IC * PL/DSC
69N16	MO/ASC * NE/DSC
68N56	JU/IC * NE/ASC
68N51	VE/MH * JU/DSC
68N50	UR/MH * JU/DSC
68N46	ME/IC * SA/ASC
68N39	VE/DSC * JU/ASC
68N39	NE/MH * MO/ASC
68N34	VE/DSC * NE/DSC
68N30	JU/ASC * NE/DSC
68N22	UR/DSC * NE/DSC
68N17	MA/DSC * NE/DSC
68N12	SU/DSC * NE/DSC
68N12	JU/ASC * UR/DSC
68N04	ME/DSC * NE/DSC
68N03	MA/DSC * JU/ASC
67N57	SU/DSC * JU/ASC
67N47	ME/DSC * JU/ASC
67N43	SA/MH * MA/ASC
67N32	SU/DSC * MA/DSC
67N30	SU/DSC * UR/DSC
67N28	MA/DSC * UR/DSC
67N26	SA/DSC * NE/DSC
67N13	ME/DSC * UR/DSC
67N07	ME/DSC * MA/DSC
67N01	JU/ASC * SA/DSC
66N38	MA/MH * JU/DSC
66N36	SU/DSC * ME/DSC
66N35	ME/DSC * VE/DSC
66N34	SU/DSC * VE/DSC
66N21	SA/DSC * UR/DSC
66N10	MA/DSC * SA/DSC
66N03	VE/DSC * SA/DSC
65N48	SU/DSC * SA/DSC
65N34	ME/DSC * SA/DSC
65N23	VE/DSC * MA/DSC
65N06	JU/IC * MO/ASC
65N03	MO/MH * NE/DSC
64N47	PL/MH * JU/DSC
64N47	VE/MH * NE/ASC
64N44	UR/MH * NE/ASC
63N55	SU/IC * SA/ASC
62N57	MO/MH * JU/ASC
61N13	Anchorage
60N53	MA/MH * NE/ASC
60N34	SU/MH * JU/DSC
59N20	Stockholm
57N55	PL/IC * SA/ASC
57N41	PL/MH * NE/ASC
57N26	NE/DSC * PL/DSC
55N45	Moscow
55N41	Copenhagen
53N04	JU/ASC * PL/DSC
52N37	MO/MH * UR/DSC
52N30	Berlin
52N18	NE/MH * SA/DSC
52N15	MA/IC * SA/ASC
51N43	MO/MH * VE/DSC
51N30	London
50N16	SU/MH * NE/ASC
50N01	ME/MH * JU/DSC
49N59	ME/IC * MO/DSC
49N14	Vancouver
48N50	Paris
47N36	Seattle
46N43	SA/IC * NE/DSC
45N32	Montreal

CENTRAL LATITUDES

44N59	Minneapolis
44N31	SA/MH * UR/ASC
43N47	VE/MH * MO/ASC
43N39	Toronto
43N34	UR/MH * MO/ASC
43N05	SA/MH * VE/ASC
42N22	Boston
42N20	Detroit
41N54	Rome
41N53	Chicago
40N43	New York
39N44	Denver
38N53	Washington
38N37	St. Louis
37N47	San Francisco
36N58	MO/MH * MA/DSC
36N12	UR/IC * SA/ASC
36N10	Las Vegas
35N55	VE/IC * SA/ASC
35N40	Tokyo
34N50	JU/IC * SA/DSC
34N46	SU/ASC * PL/ASC
34N31	UR/DSC * PL/DSC
34N08	VE/DSC * PL/DSC
34N04	Los Angeles
33N45	Atlanta
33N27	Phoenix
32N47	Dallas
31N43	SA/IC * JU/ASC
31N32	ME/MH * NE/ASC
30N03	Cairo
29N58	New Orleans
29N46	Houston
28N39	Delhi
25N47	Miami
25N13	SU/IC * MO/DSC
22N34	Calcutta
22N16	Hong Kong
21N19	Honolulu
21N15	MA/MH * MO/ASC
19N26	Mexico City
16N41	MA/DSC * PL/DSC
14N35	Manila
14N04	VE/DSC * UR/DSC
10N30	Caracas
6N27	Lagos
5N39	MO/IC * PL/ASC
3N20	PL/MH * MO/ASC
3S20	PL/IC * MO/DSC
5S39	MO/MH * PL/DSC
14S04	VE/ASC * UR/ASC
16S41	MA/ASC * PL/ASC
21S15	MA/IC * MO/DSC
22S54	Rio De Janeiro
25S13	SU/MH * MO/ASC
26S12	Johannesburg
31S32	ME/IC * NE/DSC
31S43	SA/MH * JU/DSC
33S55	Sydney
34S08	VE/ASC * PL/ASC
34S31	UR/ASC * PL/ASC
34S36	Buenos Aires
34S46	SU/DSC * PL/DSC
34S50	JU/MH * SA/ASC
35S55	VE/MH * SA/DSC
36S12	UR/MH * SA/DSC
36S52	Auckland

SOUTH LATITUDES

36S58	MO/IC * MA/ASC
37S50	Melbourne
43S05	SA/IC * VE/DSC
43S34	UR/IC * MO/DSC
43S47	VE/IC * MO/DSC
44S31	SA/IC * UR/DSC
46S43	SA/MH * NE/ASC
49S59	ME/MH * MO/ASC
50S01	ME/IC * JU/ASC
50S16	SU/IC * NE/DSC
51S43	MO/IC * VE/ASC
52S15	MA/MH * SA/DSC
52S18	NE/IC * SA/ASC
52S37	MO/IC * UR/ASC
53S04	JU/DSC * PL/ASC
57S26	NE/ASC * PL/ASC
57S41	PL/IC * NE/DSC
57S55	PL/MH * SA/DSC

Chapter 7/Astro*Carto*Graphy Solar Arc Progressions

In the Cyclo*Carto*Graphy interpretation booklet Jim Lewis wrote that using solar arc progressions on the Astro*Carto*Graphy map can produce startlingly accurate results. This technique is easy to use and provides additional information about the temporary influences taking place on an Astro*Carto*Graphy map.

It is also possible to use arc directions for the other planets. In *Forecasting Backward and Forward*, Bruce F. Hammerslough gives examples of different planetary arc directions. He found that lunar arc directions time events more accurately than any other forecasting method.

I tried using lunar arc progressions on an Astro*Carto*Graphy map, but they didn't seem to provide any additional information. Other planetary arc progressions also proved to be disappointing, so I do not recommend that they be used with Astro*Carto*Graphy.

Solar Arc Progressions

In Chapter 5, it was explained that Cyclo*Carto*Graphy uses a combination of progressions and transits. The Cyclo*Carto*Graphy planets move at varying rates of speed, sometimes in the opposite direction. Only the progressed SU, MO, ME, VE and MA are used in Cyclo*Carto*Graphy. The other progressed planets (JU, SA, UR, NE and PL) are not considered because they move too slowly to have any meaningful effect.

With solar arc progressions, **all** the planets are used as they move at the same speed. The distance that the planets move is based on the progressed Sun, hence the name. As the Sun is never retrograde it means that solar arc progressions always move in the same direction.

The progressed Sun moves approximately one degree each year. The difference between the progressed Sun and the natal Sun is called the solar arc. The amount that the Sun moves forward is then added to the positions of the other planets in the natal chart. Once this has been done the solar arc progressed planets are examined to see if they make aspects with the natal planets. The aspects usually considered are conjunctions, oppositions and squares with an orb of one degree.

Simon's natal chart will be used to explain the use of solar arc progressions. If he had progressed Sun at 8 GEM 26, the solar arc is calculated as follows:

Progressed Sun	08 GEM 26
Natal Sun	13 TAU 24
Solar arc	= 25 02

You may find it easier to first convert these positions into absolute longitude, i.e. the number of degrees from zero Aries.

The solar arc is now added to the other planets in Simon's natal chart:

MO 08 ARI 17	+ 25 02	03 TAU 19
ME 17 ARI 58	+ 25 02	13 TAU 00
VE 08 GEM 24	+ 25 02	03 CAN 26
MA 09 LEO 25	+ 25 02	04 VIR 27
JU 03 CAN 38	+ 25 02	28 CAN 40
SA 23 LEO 43	+ 25 02	18 VIR 45
UR 14 SCO 26	+ 25 02	09 SAG 28
NE 17 SAG 49	+ 25 02	12 CAP 51
PL 14 LIB 34	+ 25 02	09 SCO 36

From these new positions it is possible to look for the solar arc progressed aspects (CJN, OPP and SQU).

1. solar arc progressed ME CJN natal SU
2. solar arc progressed VE CON natal JU
3. solar arc progressed SA SQU natal NE
4. solar arc progressed PL SQU natal MA

These aspects are then interpreted to describe what temporary influences he would experience at the time. The effects of solar arc progressed aspects have an influence for about two years.

The rate that the progressed Sun moves varies from 0°57" (Sun in Cancer) to just over 1°01" (Sun in Capricorn). Some astrologers use one degree equals one year for solar arc progressions to save having to do the necessary calculations, but this can lead to errors in determining at what age aspects will be made.

Bruce Hammerslough gives a good example to illustrate the difference between solar arc progressions and day-for-a-year directions (page 72, *Forecasting Backward and Forward*).

Person A, Born June 1

Person B, Born December 1

At age 50 the solar arc progressions are calculated:

Person A, solar arc = 47 43

Person B, solar arc = 50 53

This shows that person A will be nearly **fifty-three** years old when the solar arc is **fifty** degrees. A difference of three years means that the timing of aspects will be greatly affected.

Astro*Carto*Graphy Map On A Transparent Overlay

The easiest way to use Astro*Carto*Graphy solar arc progressions is by ordering a map drawn on a transparent overlay. When this map is drawn with a different color for the planetary lines it will be easy to find the solar arc progressed aspects.

This overlay is placed on top of the Astro*Carto*Graphy

map and moved **eastward** by the distance of the solar arc. If this were done for Simon the overlay would be moved twenty-five degrees (geographical longitude) eastward.

Instead of ordering a Astro*Carto*Graphy map drawn on an overlay, use tracing paper, marking the Astro*Carto*Graphy planetary lines on it. Then the tracing is put on top of the Astro*Carto*Graphy map and moved eastward according to the amount of solar arc.

Interpretation of Astro*Carto*Graphy Solar Arc Progressions

If you are using an overlay or tracing paper it is very easy to find the solar arc progressed aspects. You only need to look for the following:

1. solar arc lines on top of A*C*G lines
2. solar arc lines crossing A*C*G lines at the equator
3. solar arc lines at birthplace
4. solar arc lines at a location

Again, it is best when you make a list of the information found when looking for these solar arc progressed aspects.

Any of the above solar arc progressed aspects are then interpreted in the same manner as used in Cyclo*Carto*Graphy. Solar arc progressions type 1, 2 and 3 are anywhere aspects and type 4 location aspects.

If no solar arc progressed aspects are found, use the information from Cyclo*Carto*Graphy and the Astro*Carto*Graphy solar return map for the temporary influences.

Example 1—Prince Charles and Klosters

It is important to use the **right ascension** of planets to find the temporary influences, and it also needs to be used with Astro*Carto*Graphy solar arc progressions.

Make a list of the A*C*G planets and solar arc progressed planets in numerical order, as this will make it easier to look for aspects between solar arc progressed planets and A*C*G planets.

When I examined Prince Charles' displaced C*C*G planets, the difference between his **displaced** progressed Sun and natal Sun is 42°30". This value is the amount to use for solar arc progressions. The solar arc is now added to all the **A*C*G** planets, and results in:

A*C*G	**(42 30)**	**Solar arc A*C*G**
ME 5 SCO 13		MA 2 AQU 35
SA 07 SAG 38		PL 03 LIB 59
NE 13 LIB 36		MO 10 GEM 58
VE 15 LIB 49		JU 12 LEO 24
SU 20 SCO 01		UR 12 AQU 25
MA 20 SAG 05		ME 17 SAG 43
PL 21 LEO 29		SA 20 CAP 08
MO 28 ARI 28		NE 26 SCO 06
JU 29 GEM 54		VE 28 SCO 19
UR 29 SAG 55		

Now look for conjunction, opposition and square aspects between solar arc A*C*G and A*C*G planets using an orb of one degree.

Prince Charles has no solar arc progressed aspects, so we can examine his birthplace and Klosters. Without an Astro*Carto*Graphy map it is not as easy to find solar arc progressed aspects to the birthplace or other locations. There are two different ways that this can be achieved without using an Astro*Carto*Graphy map.

The first method is to use AstroMaps Hi-Res program from Matrix. By using the cursor keys it is possible to discover if solar arc progressed A*C*G lines are within orb of any location. The other way is to use the NOVA computer program to produce a paran printout (see Chapter 2).

The paran printout for Prince Charles shows the right ascension when the planets are exactly conjunct the four angles.

Using London (his birthplace) the procedure is:

1. Find R.A.M.C. (Right Ascension Medium Coeli). This is the amount for MC-upper, i.e. 12° 13" for Prince Charles.

2. **Subtract** solar arc from RAMC:

RAMC	12 13
Solar arc	42 30

This is not possible, so add 360 to RAMC.

RAMC	372 13
Solar arc	-42 30
Solar arc birthplace	329 43

The second part of the paran printout shows the right ascension of the angular planets in numerical order. Look at this list and to determine if a planet is within one degree orb of solar arc birthplace (329°43"). JU DSC (JU set) is at 326 58, but has an orb of 2°45". Therefore, there are no solar arc progressed aspects to his birthplace.

The same procedure is repeated for Klosters:

RAMC Klosters	22 16
Solar arc	42 30

This is not possible, so add 360 to RAMC:

RAMC Klosters	382 16
Solar arc	-42 30
Solar arc Klosters	339 46

The nearest planetary angle to solar arc Klosters is SA IC at 337°38", an orb of 2°08". Therefore, Prince Charles had no solar arc progressions within orb of Klosters.

Subtracting the solar arc from a location is the same as adding this amount to the A*C*G lines.

A*C*G SA IC	337 38
Solar arc	+42 30
Solar arc SA IC	380 08
	-360 00
Solar arc SA IC	20 08

RAMC Klosters 22 16 - 20 08 = 2 08

Prince Charles has no solar arc progressed aspects, either with A*C*G planets or at his birthplace and Klosters.

Example 2—Norman Schwarzkopf and Kuwait

A*C*G	**(51 13)**	**Solar arc A*C*G**
SU 00 VIR 43		NE 04 SCO 36
VE 08 LEO 14		JU 10 SAG 17

Prince Charles	Natal Chart	Paran Printout
Nov. 14, 1948	London	9:14:0 PM GMT
51N30'00"	0:48:50 ST	0W10'00"
Tropical		Geocentric

Pl	Rise d	m	h m s	Upper d	m	h m s	Set d	m	h m s	Lower d	m	h m s
☉	164	42	10:58:49	230	01	15:20:03	295	19	19:41:17	50	01	3:20:03
☽	284	00	18:56:02	28	28	1:53:53	132	56	8:51:44	208	28	13:53:53
☿	140	57	9:23:46	215	14	14:20:54	289	31	19:18:02	35	14	2:20:54
♀	111	49	7:27:16	195	49	13:03:15	279	49	18:39:14	15	49	1:03:15
♂	204	15	13:37:00	260	05	17:20:22	315	56	21:03:43	80	05	5:20:22
♃	212	49	14:11:16	269	54	17:59:35	326	58	21:47:54	89	54	5:59:35
♄	53	36	3:34:23	157	38	10:30:30	261	39	17:26:38	337	38	22:30:30
♅	326	32	21:46:06	89	55	5:59:42	213	19	14:13:17	269	55	17:59:42
♆	108	52	7:15:29	193	36	12:54:24	278	20	18:33:20	13	36	0:54:24
♇	19	07	1:16:29	141	29	9:25:56	263	50	17:35:22	321	29	21:25:56
☊	285	03	19:00:11	31	41	2:06:45	138	20	9:13:19	211	41	14:06:45
As	12	13	0:48:50	127	45	8:30:59	243	17	16:13:09	307	45	20:30:59
MC	275	35	18:22:21	12	13	0:48:50	108	50	7:15:19	192	13	12:48:50
Vx	204	55	13:39:41	262	13	17:28:53	319	31	21:18:06	82	13	5:28:53
EP	340	00	22:40:02	102	13	6:48:50	224	25	14:57:38	282	13	18:48:50
⊗	226	24	15:05:37	284	33	18:58:12	342	42	22:50:48	104	33	6:58:12

MC	upper	12	13	MC	set	108	50\	☊	lower	211	41	♀	set	279	49
As	rise	12	13	♆	rise	108	52/	♃	rise	212	49\	EP	lower	282	13
♆	lower	13	36	♀	rise	111	49	♅	set	213	19/	☽	rise	284	00\
♀	lower	15	49	As	upper	127	45	☿	upper	215	14	⊗	upper	284	33/\
♇	rise	19	07	☽	set	132	56	EP	set	224	25	☊	rise	285	03/
☽	upper	28	28	☊	set	138	20	⊗	rise	226	24	☿	set	289	31
☊	upper	31	41	☿	rise	140	57\	☉	upper	230	01	☉	set	295	19
☿	lower	35	14	♇	upper	141	29/	As	set	243	17	As	lower	307	45
☉	lower	50	01	♄	upper	157	38	♂	upper	260	05	♂	set	315	56
♄	rise	53	36	☉	rise	164	42	♄	set	261	39\	Vx	set	319	31
♂	lower	80	05	MC	lower	192	13	Vx	upper	262	13/	♇	lower	321	29
Vx	lower	82	13	♆	upper	193	36	♇	set	263	50	♄	rise	326	32\
♃	lower	89	54\	♀	upper	195	49	♃	upper	269	54\	♃	set	326	58/
♅	upper	89	55/	♂	rise	204	15\	♅	lower	269	55/	♄	lower	337	38
EP	upper	102	13	Vx	rise	204	55/	MC	rise	275	35	EP	rise	340	00
⊗	lower	104	33	☽	lower	208	28	♆	set	278	20	⊗	set	342	42

Prince Charles	Natal Chart	Paran Printout
Nov. 14,1948	Klosters	9:14:0 PM GMT
46N54’00"	1:29:02 ST	9E53’00"
Tropical		Geocentric

Pl	Rise			Upper			Set			Lower		
	d	m	h m s	d	m	h m s	d	m	h m s	d	m	h m s
☉	160	49	10:43:15	230	01	15:20:03	299	13	19:56:51	50	01	3:20:03
☽	286	13	19:04:51	28	28	1:53:53	130	44	8:42:54	208	28	13:53:53
☿	138	32	9:14:09	215	14	14:20:54	291	55	19:27:40	35	14	2:20:54
♀	110	55	7:23:39	195	49	13:03:15	280	43	18:42:51	15	49	1:03:15
♂	198	36	13:14:24	260	05	17:20:22	321	35	21:26:19	80	05	5:20:22
♃	207	25	13:49:38	269	54	17:59:35	332	23	22:09:31	89	54	5:59:35
♄	55	44	3:42:56	157	38	10:30:30	259	31	17:18:05	337	38	22:30:30
♅	332	02	22:08:06	89	55	5:59:42	207	49	13:51:17	269	55	17:59:42
♆	108	05	7:12:19	193	36	12:54:24	279	07	18:36:30	13	36	0:54:24
♇	24	25	1:37:41	141	29	9:25:56	258	33	17:14:11	321	29	21:25:56
☊	287	36	19:10:23	31	41	2:06:45	135	47	9:03:07	211	41	14:06:45
As	22	16	1:29:02	132	18	8:49:13	242	21	16:09:24	312	18	20:49:13
MC	282	09	18:48:36	22	16	1:29:02	122	22	8:09:29	202	16	13:29:02
Vx	205	55	13:43:42	268	20	17:53:18	330	44	22:02:55	88	20	5:53:18
EP	356	52	23:47:26	112	16	7:29:02	227	39	15:10:38	292	16	19:29:02
⊗	225	17	15:01:09	289	22	19:17:27	353	26	23:33:44	109	22	7:17:27

♆	lower	13	36	♀	rise	110	55	♅	set	207	49/\	MC	rise	282	09
♀	lower	15	49	EP	upper	112	16	☽	lower	208	28/	☽	rise	286	13
As	rise	22	16\	MC	set	122	22	☊	lower	211	41	☊	rise	287	36
MC	upper	22	16/	☽	set	130	44	☿	upper	215	14	⊗	upper	289	22
♇	rise	24	25	As	upper	132	18	⊗	rise	225	17	☿	set	291	55\
☽	upper	28	28	☊	set	135	47	EP	set	227	39	EP	lower	292	16/
☊	upper	31	41	☿	rise	138	32	☉	upper	230	01	☉	set	299	13
☿	lower	35	14	♇	upper	141	29	As	set	242	21	As	lower	312	18
☉	lower	50	01	♄	upper	157	38	♇	set	258	33\	♇	lower	321	29\
♄	rise	55	44	☉	rise	160	49	♄	set	259	31/\	♂	set	321	35/
♂	lower	80	05	♆	upper	193	36	♂	upper	260	05/	Vx	set	330	44
Vx	lower	88	20	♀	upper	195	49	Vx	upper	268	20	♅	rise	332	02\
♃	lower	89	54\	♂	rise	198	36	♃	upper	269	54\	♃	set	332	23/
♅	upper	89	55/	MC	lower	202	16	♅	lower	269	55/	♄	lower	337	38
♆	rise	108	05	Vx	rise	205	55	♆	set	279	07	⊗	set	353	26
⊗	lower	109	22	♃	rise	207	25\	♀	set	280	43	EP	rise	356	52

NE 13 VIR 23	MO 17 PIS 26
JU 19 LIB 04	MA 18 VIR 03
MO 26 CAP 13	SA 18 ARI 32
MA 26 CAN 50	ME 18 LIB 40
SA 27 AQU 19	PL 18 VIR 42
ME 27 LEO 27	UR 20 GEM 37
PL 27 CAN 29	VE 29 VIR 27
UR 29 ARI 24	

The solar arc progressed aspects are:

1. solar arc SA = A*C*G JU
2. solar arc ME = A*C*G JU

Using AstroMaps Hi-Res there are no solar arc progressions at his birthplace or at Kuwait and Iraq.

Norman Schwarzkopf has ME MH and SA IC near Kuwait on his Astro*Carto*Graphy map, but by solar arc progression these two lines have moved to his A*C*G JU MH line. His careful and detailed planning of Desert Storm is emphasized by the solar arc progressed aspect to JU MH. This shows that his operation was a big success during the Gulf War.

Example 3—Bob Geldof and Ethiopia

A*C*G	(31 26)	Solar arc A*C*G
MA 02 VIR 53		MA 04 LIB 19
VE 04 VIR 26		VE 05 LIB 52
ME 05 LIB 33		ME 06 SCO 59
SA 06 LIB 34		SA 08 SCO 00
MO 06 SAG 35		MO 08 CAP 01
JU 8 ARI 45		JU 10 TAU 11
SU 10 LIB 34		UR 16 LEO 34
UR 15 CAN 08		NE 19 SCO 33
NE 18 LIB 07		PL 27 VIR 48
PL 26 LEO 22		

Bob Geldof has the following solar arc progressed aspects:

1. solar arc VE = A*C*G ME
2. solar arc VE = A*C*G SA
3. solar arc MO EQU A*C*G JU

There are no solar arc progressed lines at his birthplace or in Ethiopia.

With solar arc MO applying EQU to A*C*G JU this influence was around throughout the time he was involved with Band Aid and Live Aid. He was feeling happy and enthusiastic about the success of his effort to raise money for the people in Ethiopia.

Solar arc VE = A*C*G ME is a separating aspect, but still had an influence when he formed Band Aid. This indicated the high media profile he had for his fund raising efforts. VE = A*C*G SA (conjunction) suggests taking on a great deal of responsibility to do something about the terrible situation in Ethiopia.

Example 4—Margaret Thatcher and Falkland Islands

A*C*G	(57 03)	Solar arc A*C*G
MO 01 VIR 37		MA 06 SAG 05
MA 09 LIB 02		SA 09 CAP 01
SA 11 SCO 58		JU 12 PIS 50
JU 15 CAP 47		PL 12 VIR 51
PL 15 CAN 48		ME 19 SAG 16
SU 18 LIB 00		UR 20 TAU 28
ME 22 LIB 13		NE 23 LIB 43
UR 23 PIS 25		VE 26 CAP 37
NE 26 LEO 40		MO 28 LIB 40
VE 29 SCO 34		

These lists show that Margaret Thatcher had one solar arc progressed aspect:

1. solar arc SA EQU A*C*G MA

She had solar arc progressed SA IC at Port Stanley in the Falkland Islands. This is confirmed by the paran printout for Port Stanley.

RAMC	98 34	
Solar arc	-57 03	
Solar arc Port Stanley	41 31	
A*C*G SA IC	41 58	(orb = 0 27)
Solar arc	57 03	
Solar arc SA IC	99 01	

This shows that solar arc progressed SA IC was moving away from Port Stanley, and would have an influence only for another six months.

In 1981, her solar arc progressed SA IC was within orb at Buenos Aires (58W27) and passing over the Falkland Islands. This suggests that difficulties were slowly increasing between Britain and Argentina. At the time Argentina invaded the Falkland Islands Margaret Thatcher had an exact solar arc progressed SA EQU A*C*G MA. This manifested like a square aspect. Although this was a difficult time for her, her great determination to regain the Falkland Islands was achieved in a short space of time.

Example 5—David Copperfield and Germany

A*C*G	(33 55)	Solar arc A*C*G
PL 05 VIR 14		SA 00 CAP 02
UR 08 LEO 07		NE 01 SAG 20
ME 09 LIB 11		PL 09 LIB 09
VE 10 LEO 24		UR 12 VIR 02
MO 13 AQU 17		ME 13 SCO 06
JU 16 VIR 14		VE 14 VIR 19
MA 19 PIS 57		MO 17 PIS 12
SU 24 VIR 04		JU 20 LIB 09
SA 26 SCO 07		MA 23 ARI 52
NE 27 LIB 25		

The solar arc progressed aspects are:

1. solar arc PL = A*C*G ME
2. solar arc ME EQU A*C*G MO
3. solar arc MO = A*C*G JU

There are no solar arc progressed aspects at his birthplace or in Germany.

ME/PL was exact at the time David Copperfield met Claudia Schiffer in Berlin. This aspect shows the intense publicity he received in the media. He was unknown in Germany at the beginning of 1993, but the publicity had a big influence on people. Afterward, tickets for his shows sold out quickly, even before he arrived in Germany. Near

Margaret Thatcher	Natal Chart	Paran Printout
Oct. 13, 1925	Port Stanley	9:0:0 AM GMT
51S42'00"	6:34:14 ST	57W51'00"
Tropical		Geocentric

Pl	Rise			Upper			Set			Lower		
	d	m	h m s	d	m	h m s	d	m	h m s	d	m	h m s
☉	98	14	6:32:55	198	00	13:12:00	297	46	19:51:05	18	00	1:12:00
☽	80	12	5::20:46	151	37	10:06:27	223	02	14:52:07	331	37	22:06:27
☿	100	58	6:43:53	202	13	13:28:53	303	28	20:13:54	22	13	1:28:53
♀	117	53	7:51:31	239	34	15:58:15	1	15	0:04:59	59	34	3:58:15
♂	95	14	6:20:58	189	02	12:36:08	282	49	18:51:18	9	02	0:36:08
♃	163	14	10:52:54	285	47	19:03:08	48	20	3:13:21	105	47	7:03:08
♄	113	36	7:34:24	221	58	14:47:53	330	21	22:01:22	41	58	2:47:53
♅	258	42	17:14:46	353	25	23:33:39	88	08	5:52:32	173	25	11:33:39
♆	74	45	4:59:01	146	40	9:46:40	218	35	14:34:18	326	40	21:46:40
♇	44	36	2:58:22	105	48	7:03:12	167	00	11:08:01	285	48	19:03:12
☊	60	15	4:00:59	122	44	8:10:55	185	13	12:20:50	302	44	20:10:55
As	98	34	6:34:14	198	42	13:14:49	298	51	19:55:23	18	42	1:14:49
MC	41	27	2:45:49	98	34	6:34:14	155	40	10:22:39	278	34	18:34:14
Vx	279	53	18:39:30	6	23	0:25:31	92	53	6:11:31	186	23	12:25:31
EP	93	52	6:15:29	188	34	12:34:14	283	15	18:52:59	8	34	0:34:14
⊗	76	42	5:06:47	151	30	10:06:01	226	19	15:05:15	331	30	22:06:01

♀	set	1	15	♅	set	88	08	MC	set	155	40	♅	rise	258	42
Vx	upper	6	23	Vx	set	92	53\	♃	rise	163	14	MC	lower	278	34
EP	lower	8	34\	EP	rise	93	52/	♇	set	167	00	Vx	rise	279	53
♂	lower	9	02/	♂	rise	95	14	♅	lower	173	25	♂	set	282	49\
☉	lower	18	00\	☉	rise	98	14\	☊	set	185	13	EP	set	283	15/
As	lower	18	42/	MC	upper	98	34/\	Vx	lower	186	23	♃	upper	285	47\
☿	lower	22	13	As	rise	98	34/	EP	upper	188	34\	♇	lower	285	48/
MC	rise	41	27\	☿	rise	100	58	♂	upper	189	02/	☉	set	297	46
♄	lower	41	58/	♃	lower	105	47\	☉	upper	198	00\	As	set	298	51
♇	rise	44	36	♇	upper	105	48/	As	upper	198	42/	☊	lower	302	44\
♃	set	48	20	♄	rise	113	36	☿	upper	202	13	☿	set	303	28/
♀	lower	59	34\	♀	rise	117	53	♆	set	218	35	♆	lower	326	40
☊	rise	60	15/	☊	upper	122	44	♄	upper	221	58	♄	set	330	21
♆	rise	74	45	♆	upper	146	40	☽	set	223	02	⊗	lower	331	30\
⊗	rise	76	42	⊗	upper	151	30\	⊗	set	226	19	☽	lower	331	37/
☽	rise	80	12	☽	upper	151	37/	♀	upper	239	34	♅	upper	353	25

the end of his European tour he met Claudia Schiffer, which made front page news in German newspapers.

With MO/JU David Copperfield was feeling happy about the huge success he achieved on his European tour. MO/ME suggests that his thoughts were influenced by his feelings and emotions. He was mentally stimulated by the response he received from the general public when he was in Europe.

Example 6—Claudia Schiffer and Germany

A*C*G	(20 54)	Solar arc A*C*G
PL 2 LIB 53		VE 6 SCO 42
SU 3 VIR 50		MO 10 CAN 30
UR 6 LIB 35		SA 11 GEM 39
VE 15 LIB 48		ME 16 LIB 45
MO 19 GEM 36		NE 17 SAG 14
SA 20 TAU 45		MA 18 VIR 02
ME 25 VIR 51		JU 20 SCO 33
NE 26 SCO 20		PL 23 LIB 47
MA 27 LEO 08		UR 27 LIB 29
JU 29 LIB 39		

The aspects found are:

1. solar arc ME = A*C*G VE
2. solar arc JU = A*C*G SA

Claudia Schiffer had no solar arc progressed aspects at her birthplace or in Berlin.

The ME/VE aspect had almost moved out of orb, so its influence would have lasted only for a few more weeks. Primarily, this indicates her high media profile as a model. JU/SA suggests she achieves success through perseverance and patience. She was also be making long term plans for her future security.

Example 7—Matthew

Matthew had only one solar arc progressed aspect.

1. solar arc VE = A*C*G JU

This aspect indicates that Matthew was enjoying a good social life and meeting many people. Also possible is that he was in a loving relationship during that time.

Astro*Carto*Graphy solar arc progressions give important additional information only when they are at **locations**. The solar arc progressed "anywhere aspects" tend to be minor influences, unless they reiterate themes shown by Cyclo*Carto*Graphy and the Astro*Carto*Graphy solar return map.

It is essential that the temporary influences be taken into consideration when examining the planetary influences on an Astro*Carto*Graphy map. Otherwise, you will miss vital information.

To sum up, the techniques used to find the temporary influences are:

1. Cyclo*Carto*Graphy (very important)
2. Astro*Carto*Graphy solar return map (important)
3. Astro*Carto*Graphy solar arc progressions (less important)

It is only when by trying these techniques that you will understand how important they are. The techniques concerning the temporary influences in Astro*Carto*Graphy are breaking new ground in astrology, so work your way through them to discover how easy they are to use.

Chapter 8/Astro*Carto*Graphy and Synastry

How can Astro*Carto*Graphy be used in synastry? Astrologers use various techniques to find out how two people will get along with each other. There are different types of relationships which can be examined astrologically: marriage partners, a loving relationship, friendship, work colleagues, business partners, family relationships, etc.

The usual way an astrologer approaches synastry is by comparing the natal charts. Suppose an astrologer is doing synastry for two people planning to marry, Person A and Person B.

Chart comparison is carried out as follows:

1. Person A's planets in Person B's houses
2. Person B's planets in Person A's houses
3. Person A's inter-aspects with Person B's

The information obtained will explain how Person A and Person B will get along with each other in the relationship.

An astrologer also can calculate **one** horoscope from both natal charts. This new chart then explains the relationship in more detail. There are two different techniques to calculate this new horoscope.

1. Composite chart
2. Relationship chart

Of the two charts, the composite seems to be the most popular among astrologers. In his book *Planets in Composite*, Robert Hand explains how this chart is calculated and interpretated.

What is the difference between a composite chart and a relationship chart? This is a question that astrologers have difficulty answering because both charts can be used to provide additional information about a relationship.

Other charts that can be used in synastry are:

1. Marriage chart
2. First meeting chart

It is important to remember that these types of charts cannot be used on their own without consideration of other synastry techniques. If the natal chart comparison and composite show that the relationship will be difficult and problematic, then a "good" marriage chart will not change the situation.

The Importance of Using Maps of Both Partners

When using synastry techniques astrologers are concerned about **how** two people will get along with each other in the relationship. Also important is **where** the relationship is taking place, but this question is not usually considered in synastry. A potentially beneficial and harmonious relationship could fall to pieces when the couple lives at a location with difficult planetary energies.

It could be that Person A is happy living at location X. Person B, however, feels unhappy in life at that location, but accepts that's the way things have to be. After a while there could be strains on the relationship, especially when Person B tries to make some changes in life.

The most important technique in finding a suitable location for two people is to examine their individual Astro*Carto*Graphy maps. Then you can see the permanent influences at a location for each person.

When Cyclo*Carto*Graphy, the Astro*Carto*Graphy solar return map and Astro*Carto*Graphy solar arc progressions are used, you will obtain a great deal of information for each person. This was done when the Astro*Carto*Graphy maps for David Copperfield and Claudia Schiffer were examined. They haven't yet decided where they will live when they are married. David Copperfield wants to live in an English speaking country, but Claudia Schiffer doesn't want to live far away from her family in Germany.

Using the Astro*Carto*Graphy maps of both partners to find a suitable location is nothing new as far as technique is concerned. The only difference is that two maps are being considered to find the planetary influences at locations. The interesting question is whether other synastry techniques can be used in Astro*Carto*Graphy.

Composite Chart

This chart is based on the midpoints between pairs of planets in two natal horoscopes. The composite Sun is the midpoint between the Sun of Person A and the Sun of Person B. This procedure is repeated to find the other composite planets. The charts of David Copperfield and Claudia Schiffer will be used as an example.

First, change the positions of the planets and angles into zodiacal longitude, i.e. the number of degrees from zero Aries.

Aries	0	Taurus	30	Gemini	60
Cancer	90	Leo	120	Virgo	150
Libra	180	Scorpio	210	Sagittarius	240
Capricorn	270	Aquarius	300	Pisces	330

For example, if a planet is at 20 SAG 42 it is converted into zodiacal longitude as follows:

00 SAG 00	240 00
20 SAG 42	+20 42
Zodiacial longitude	260 42

David Copperfield:

SU	23 VIR 33	+ 150	173 33
MO	12 AQU 11	+ 300	312 11

ME	11 LIB 44	+ 180	191 44
VE	08 LEO 22	+ 120	128 22
MA	16 PIS 40	+ 330	346 40
JU	14 VIR 38	+ 150	164 38
SA	27 SCO 56	+ 210	237 56
UR	05 LEO 36	+ 120	125 36
NE	28 LIB 51	+ 180	208 51
PL	29 LEO 08	+ 120	149 08
MH	26 GEM 47	+ 60	86 47
ASC	27 VIR 13	+ 150	177 13

Claudia Schiffer:

SU	01 VIR 50	+ 150	151 50
MO	20 GEM 48	+ 60	80 48
ME	26 VIR 57	+ 150	176 57
VE	17 LIB 49	+ 180	197 49
MA	24 LEO 26	+ 120	144 26
JU	01 SCO 26	+ 210	211 26
SA	22 TAU 32	+ 30	52 32
UR	06 LIB 53	+ 180	186 53
NE	28 SCO 11	+ 210	238 11
PL	26 VIR 17	+ 150	176 17
MH	25 LEO 19	+ 120	145 19
ASC	10 SCO 33	+ 210	220 33

	Copperfield	Schiffer			Composite
SU	173 33	+ 151 50	325 23 ÷ 2	162 41	12 VIR 41
MO	312 11	+ 80 48	392 59 ÷ 2	196 29	16 LIB 29*
ME	191 44	+ 176 57	368 41÷ 2	184 20	04 LIB 20
VE	128 22	+ 197 49	326 11÷ 2	163 05	13 VIR 05
MA	346 40	+ 144 26	491 06 ÷ 2	245 33	05 SAG 33*
JU	164 38	+ 211 26	376 04 ÷ 2	188 02	08 LIB 02
SA	237 56	+ 52 32	290 28 ÷ 2	145 14	25 LEO 14*
UR	125 36	+ 186 53	312 29 ÷ 2	156 14	06 VIR 14
NE	208 51	+ 238 11	447 02 ÷ 2	223 31	13 SCO 31
PL	149 08	+ 176 17	325 25 ÷ 2	162 42	12 VIR 42
MH	86 47	+ 145 19	232 06 ÷ 2	116 03	26 CAN 03
ASC	177 13	+ 220 33	397 46 ÷ 2	198 53	18 LIB 53 *

Calculating the composite planets doesn't always give the correct sign position. For example, the composite Moon is calculated at 16 **LIB** 29. What is actually required is the **nearer** midpoint between the two Moons, which is 16 **ARI** 29. The same happened with composite Mars and Saturn.

It's always best to look at the two natal charts to find the approximate position where a composite planet will be. In *Planets in Composite* Robert Hand writes that when a pair of planets are more than 150 degrees apart in the natal charts, that **both** midpoints should be used. In the composite chart for David Copperfield and Claudia Schiffer **two** midpoints each for Mars and Saturn should be used, according to the 150 degree rule.

David Copperfield MA	16 PIS 40
Claudia Schiffer MA	24 LEO 26
Orb	157 46
David Copperfield SA	27 SCO 56
Claudia Schiffer SA	22 TAU 32
Orb	174 36

Another difference is the position of the Ascendant in the composite chart. There are two ways to calculate the Ascendant for the composite chart:

1. Midpoint of both Ascendants
2. Ascendant calculated from composite Midheaven, using the **latitude** where the relationship is taking place.

For David Copperfield and Claudia Schiffer the Ascendant in their composite chart is calculated to be:

1. 18 LIB 53
2. 19 LIB 30 (using latitude for Berlin, 52N30)

Robert Hand suggests putting the composite Ascendant (midpoint of the two Ascendants) in the composite chart and using it as an important sensitive point. Most astrologers prefer to use either one or the other Ascendant for the composite chart.

As the composite is a midpoint chart from pairs of planets it is **not** possible to draw an Astro*Carto*Graphy map. The composite chart has no date and time of birth, both of which are needed to be able to draw an Astro*Carto*Graphy map.

The only possibility is to work out the midpoints of the MH and IC lines. For example, the midpoint of the two SU MH lines, both SU IC lines, the two MO MH lines, and so on. But this wouldn't be possible with the ASC and DSC lines due to the lines being curved.

For practical purposes it is clear that a composite chart cannot be drawn on an Astro*Carto*Graphy map.

Relationship Chart

This is also a midpoint type chart calculated from two natal horoscopes, but it is based on a completely different midpoint technique than used for the composite chart. The relationship chart is calculated by using the midpoint in **time** and **space** between two natal horoscopes. This new chart will have a date, time and place of birth, so it is possible to draw an Astro*Carto*Graphy map. The only book that explains in detail how to calculate a relationship chart is *Synastry* by Ronald Davison. The natal charts of David Copperfield and Claudia Schiffer will be used to explain how a relationship chart is calculated.

The relationship chart is calculated in four stages. Once this is done it is possible to set up a new horoscope.

1. Find midpoint between both dates of birth
2. Find midpoint between both times of birth
3. Find midpoint between both latitudes
4. Find midpoint between both longitudes

1. Midpoint—Dates of Birth

David Copperfield September 16, 1956
Claudia Schiffer August 25, 1970

a. Convert dates of birth into "Julian days" (*The American Ephemeris for the 20th Century* gives this information).

David Copperfield	
Julian days on September 16, 1956	20713
Claudia Schiffer	
Julian days on August 25, 1970	25804

b. Add both amounts of Julian days and divide by 2

20713 + 25804	46517*
46517 ÷ 2	23258.5

*When the total of the added Julian days is **odd**, the midpoint date of birth will be one of two days.

e.g. Julian days = 23258.5

23258	September 5, 1963
23259	September 6, 1963

2. Midpoint—Times of Birth

a. convert times of birth to GMT

David Copperfield	7.02 EDT	11.02 GMT
Claudia Schiffer	12.10 CET	11.10 GMT

*When the midpoint date of birth is between two dates, then give first date oldest person and second date the other person.

David Copperfield
11.02 GMT September **5**, 1963
Claudia Schiffer
11.10 GMT September **6**, 1963

b. Calculate the difference between both times.

Difference between David Copperfield and Claudia Schiffer is:

24 Hours 08 minutes

c. Divide difference by 2.

24.08 ÷ 2 = **12 04**

d. Add this amount to first date and time:

11 02	September 5, 1963
12 04	
23 06	September 5, 1963

This means that the midpoint in **time** between David Copperfield and Claudia Schiffer is:

11:06 PM GMT on September 5, 1963

Now we need to calculate the midpoint in **space** between David Copperfield and Claudia Schiffer.

3. Midpoint—Latitudes

David Copperfield	latitude at birth = 40N33
Claudia Schiffer	latitude at birth = 51N12

a. Add both latitudes:

40 33 + 51 12 = 91 45

b. Divide total by 2:

91 45 ÷ 2 = 45 52½

The half remainder is not so important, so midpoint latitude is:

45N52.

4. Midpoint—Longitudes

David Copperfield	longitude at birth = 74W22
Claudia Schiffer	longitude at birth = 6E47

When one longitude is **W** and the other is **E**, then the midpoint is calculated slightly differently. This also applies when latitudes are **N** and **S**.

a. Add both longitudes (ignore W and E).

74 22 + 6 47 = 81 09

b. Divide total by 2

81 09 ÷ 2 = 40 34½

c. Subtract this total from the **higher** longitude.

74W22 - 40 34½ = 33W47½

The half remainder is not so important, so midpoint longitude is

33W47.

The relationship chart can now be set up for David Copperfield and Claudia Schiffer using the following data:

Date September 5, 1963
Time 11:06 p.m. GMT
Place 45N52 - 33W47

With astrological computer software it is no problem to calculate a relationship chart. The detailed explanation of calculating a relationship chart is included because the angles are the most important part of a horoscope. So, the angles of a relationship chart are equally important.

The method of calculating the angles in a relationship chart is **not** correct! Ronald Davison explained that the midpoint in space is calculated by finding the halfway points between both latitudes and longitudes.

This way of calculating the midpoint in space is fairly accurate when both locations are a short distance from each other (no more than 100 miles).

When the two people are born on different continents, then the usual way of calculating the midpoint in space will not give the correct information.

The midpoint between two locations on earth is measured along a great circle. To calculate the true midpoint on earth between two latitudes and longitudes involves complex spherical trigonometry.

The difference between great circle midpoints and the Davison method can be very large. If two people are born on different continents, the two methods of calculating the midpoint in space will produce different angles in the relationship chart. Very often the two methods will give angles that differ by more than thirty degrees.

Using great circle midpoints for David Copperfield and Claudia Schiffer, the cordinates for the relationship chart are **53N31** and **38W31** (calculated by a computer program called Hermes). Win*Star 1.0 has an option to calculate both types of relationship charts, but the great circle midpoint chart is not calculated correctly. Using Win*Star 1.0, both types of relationship charts produce exactly the same angles which is obviously an error in the program.

The two types of relationship charts for David Copperfield and Claudia Schiffer have the following coordinates and angles.

Great circle	**midpoints**	**Mathematical**	**midpoints**
Latitude	53 N 31	Latitude	45 N 52
Longitude	38 W 31	Longitude	33 W 47
MH	20 CAP 38	MH	25 CAP 06
IC	20 CAN 38	IC	25 CAN 06
ASC	20 TAU 42	ASC	18 TAU 10
DSC	20 SCO 42	DSC	18 SCO 10

Using Placidus for both these charts, only SA is in a different house (mathematical midpoints = 10th - great circle midpoints = 11th). The difference of 4½° (MH) and 2½° (ASC) will be more important for the timing of transits and progressions to the angles of the relationship chart.

Another problem is that sometimes it is not possible to calculate a relationship chart with houses when the great circle midpoint is used. The true midpoint between some locations in America and Europe will be in the Arctic where most house systems break down. The problem of calculating a chart for locations above the arctic circle was explained in Chapter 2.

How important are the angles in a relationship chart? As the true midpoint between both birthplaces can often result in a different set of angles, this is obviously an important question. Astrologers who use the great circle midpoints for a relationship chart report that this method gives more accurate results.

Another question that needs to be examined is the **new location** used for a relationship chart. Regardless which system is used to calculate the midpoint in space, the new location could be in the middle of the ocean. The coordinates for the David Copperfield/Claudia Schiffer relationship chart are 45N52 and 33W47. The new location is in the Atlantic Ocean.

Will the angles of a relationship chart have an effect when the midpoint birthplace is in the ocean? More important is the location **where** the relationship is taking place, or other locations they have dealings with.

All you need to draw an Astro*Carto*Graphy map is time (GMT) and date of birth. Any location can be used for drawing an Astro*Carto*Graphy map, so there is no need to calculate the midpoint between the birthplaces for a relationship chart. Regardless which location is used the resulting map will be the same.

When a relationship chart is drawn on an Astro*Carto*Graphy map you can examine the planetary effects at **any** location. It is also possible to use Cyclo*Carto*Graphy, Astro*Carto*Graphy solar return and solar arc progressions with a Astro*Carto*Graphy relationship chart map.

Astrologers use composite charts to explain how the world sees two people together in a relationship. But a composite chart cannot explain how two people will get along with each other at a particular location. For this reason the relationship chart is more effective.

For a long time I have tried to discover the difference between a composite and relationship chart. Quite often both charts will show some similarities. When you look at the composite and relationship charts for David Copperfield and Claudia Schiffer both horoscopes show a close conjunction between Sun, Venus and Pluto. Other times, the same planets will be angular in both the composite and relationship chart.

In an article in *Meridian* magazine, Mona Riegger wrote that she has found the relationship chart to be more accurate in explaining the connection between two people. She explained that this chart shows what the two people will experience together and how the relationship develops. Mona Riegger's definition of the composite chart is that it shows **how** the relationship starts. Transits to the composite chart at the time two people first meet shows how things will develop between them.

Through my study, it is apparent that Mona Riegger's definition of a composite chart is accurate. Looking at composite charts this way is amazing since my impression had been that it showed how others saw the relationship.

Other Charts

When using Astro*Carto*Graphy in synastry, both individual maps need to be considered first as they are the most important.

An Astro*Carto*Graphy map for the relationship chart will show how both people will get along with each other at a certain location.

Astrologers can also set up a chart for the time two people marry. This chart shows how things will be for them as a married couple. Progressions and transits can also be applied to the marriage chart.

Perhaps an Astro*Carto*Graphy map of the marriage chart can be drawn to obtain additional information about planetary influences at locations. As an Astro*Carto*Graphy marriage chart map can only be drawn when an exact time is known, the time to use is when the couple is officially recognized as husband and wife. However, most couples are unable to remember the exact time they were married. In Example 2 an Astro*Carto*Graphy marriage chart map will be examined.

Another chart that can be used is one for when people meet each other for the first time. This is called the first meeting chart, and if the exact time is known it is possible to draw an Astro*Carto*Graphy map. It is rare when an exact time is known for a first meeting, so I doubt if this kind of map will be used much. In Chapter 11 a first meeting Astro*Carto*Graphy map will be examined.

For practical purposes both individual maps and an Astro*Carto*Graphy relationship chart map should be used in synastry. Only when an exact time is known for the marriage and first meeting can an Astro*Carto*Graphy map be drawn for these events.

Example 1—David Copperfield/ Claudia Schiffer

David Copperfield and Claudia Schiffer met on October 7, 1993 in Berlin when David Copperfield invited Claudia Schiffer on stage to help him with a prediction. The exact time they met is not known. The Astro*Carto*Graphy relationship chart map for David Copperfield and Claudia Schiffer shows the following planetary influences at Berlin.

Lines:	SU IC
	VE IC
	PL IC
	UR IC
Crossings:	ME DSC/VE DSC
	MA DSC/JU ASC

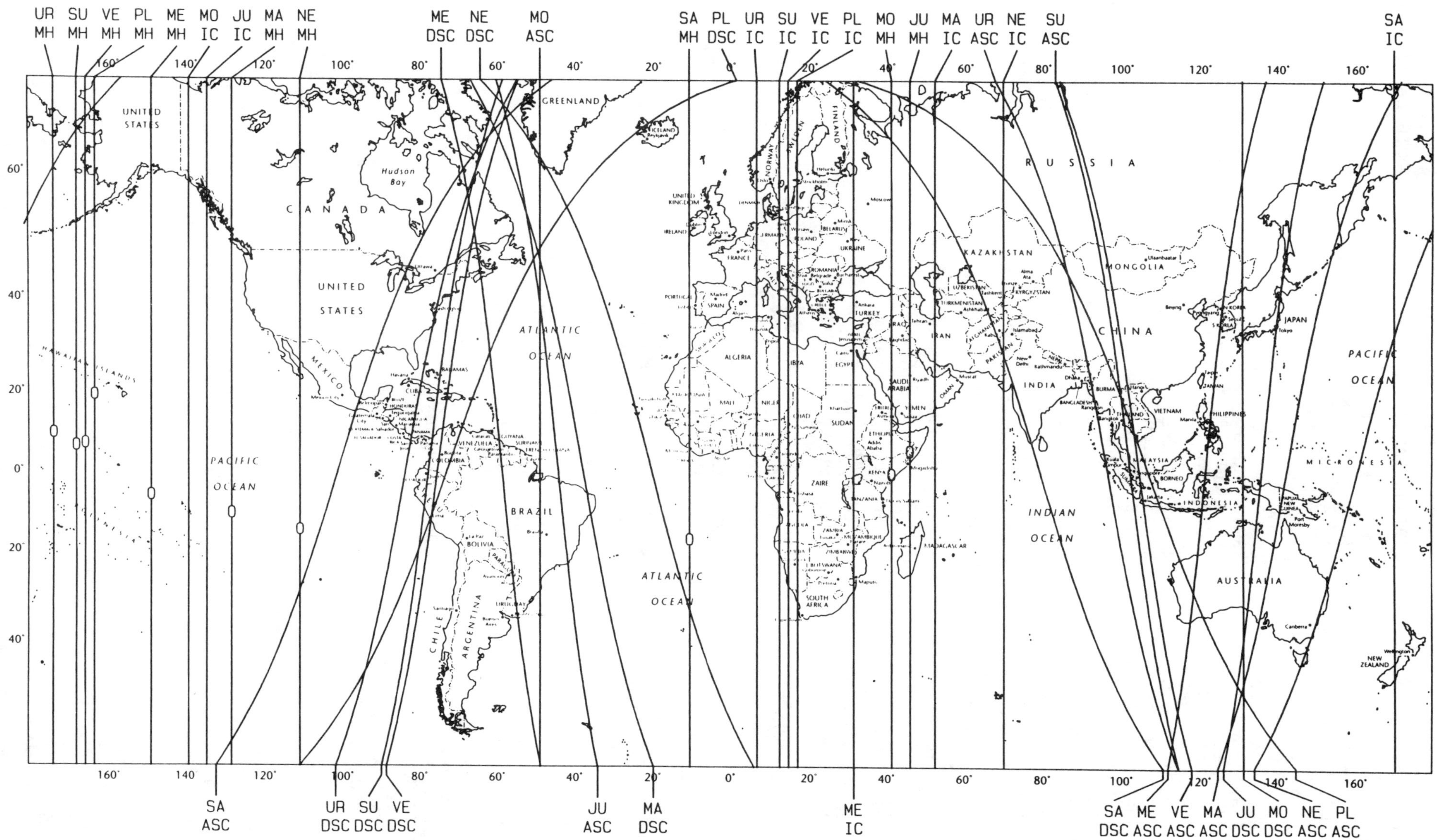

Copperfield/Schiffer

September 5, 1963 11:06 PM GMT

04/16/96

ASTRO*CARTO*GRAPHY LATITUDE CROSSINGS

NORTH LATITUDES

69N14 MO/ASC * SA/ASC
69N08 SU/DSC * JU/ASC
69N00 MO/MH * PL/ASC
68N47 SA/IC * MA/ASC
68N42 NE/IC * UR/ASC
68N27 SU/DSC * NE/DSC
68N23 SA/MH * PL/DSC
68N12 JU/MH * PL/ASC
68N04 ME/IC * SA/DSC
67N43 JU/ASC * NE/DSC
67N41 UR/DSC * NE/DSC
67N39 JU/ASC * UR/DSC
67N19 PL/MH * NE/ASC
67N16 VE/DSC * NE/DSC
66N52 SA/ASC * NE/DSC
66N50 VE/DSC * JU/ASC
66N47 MA/IC * PL/ASC
66N29 JU/ASC * SA/ASC
66N27 SU/DSC * MA/DSC
66N10 VE/DSC * SA/ASC
66N08 VE/MH * NE/ASC
65N37 MA/DSC * UR/DSC
65N17 MA/DSC * SA/ASC
64N50 SU/MH * NE/ASC
64N36 SA/ASC * UR/DSC
64N30 VE/DSC * MA/DSC
64N14 MO/MH * SA/DSC
63N40 SU/DSC * SA/ASC
62N13 MO/ASC * NE/DSC
61N35 JU/MH * SA/DSC
61N13 Anchorage
60N40 ME/DSC * SA/ASC
60N39 UR/MH * NE/ASC
60N26 SU/DSC * UR/DSC

59N54 NE/IC * PL/ASC
59N27 MO/DSC * ME/ASC
59N20 Stockholm
57N33 ME/DSC * UR/DSC
56N45 MA/IC * SA/DSC
56N39 SU/DSC * ME/DSC
55N45 Moscow
55N45 NE/DSC * PL/DSC
55N41 Copenhagen
54N55 MA/DSC * JU/ASC
52N30 Berlin
51N57 ME/DSC * VE/DSC
51N30 London

49N20 MO/ASC * PL/DSC
49N14 Vancouver
48N50 Paris
48N45 MA/DSC * PL/DSC
47N36 Seattle
47N32 JU/ASC * PL/DSC
47N27 MO/ASC * MA/DSC
45N32 Montreal
45N11 UR/ASC * PL/ASC
44N59 Minneapolis
43N39 Toronto
42N22 Boston
42N20 Detroit
41N54 Rome
41N53 Chicago
40N43 New York

39N44 Denver
39N23 MO/ASC * JU/ASC
38N53 Washington
38N37 St. Louis
37N47 San Francisco
36N10 Las Vegas
35N40 Tokyo
34N04 Los Angeles

CENTRAL LATITUDES

33N45 Atlanta
33N27 Phoenix
33N02 SA/IC * NE/ASC
32N47 Dallas
30N03 Cairo

29N58 New Orleans
29N46 Houston
29N10 ME/DSC * PL/DSC
28N52 NE/IC * SA/DSC
28N39 Delhi
25N47 Miami
22N34 Calcutta
22N16 Hong Kong
21N19 Honolulu

19N26 Mexico City
19N18 SU/ASC * PL/ASC
14N35 Manila
10N31 VE/ASC * PL/ASC
10N30 Caracas

6N27 Lagos

10S31 VE/DSC * PL/DSC
19S18 SU/DSC * PL/DSC

22S54 Rio De Janeiro
26S12 Johannesburg
28S52 NE/MH * SA/ASC
29S10 ME/ASC * PL/ASC

33S02 SA/MH * NE/DSC
33S55 Sydney
34S36 Buenos Aires
36S52 Auckland
37S50 Melbourne
39S23 MO/DSC * JU/DSC

45S11 UR/DSC * PL/DSC
47S27 MO/DSC * MA/ASC
47S32 JU/DSC * PL/ASC
48S45 MA/ASC * PL/ASC
49S20 MO/DSC * PL/ASC

51S57 ME/ASC * VE/ASC
54S55 MA/ASC * JU/DSC
55S45 NE/ASC * PL/ASC
56S39 SU/ASC * ME/ASC
56S45 MA/MH * SA/ASC
57S33 ME/ASC * UR/ASC
59S27 MO/ASC * ME/DSC
59S54 NE/MH * PL/DSC

SOUTH LATITUDES

The relationship SU IC line is exact at Berlin which is no surprise as both of them have SU MH there.

In calculating their relationship chart the midpoint date of birth was between two days. This means that SU **IC** will be at Berlin on the relationship chart map. If David Copperfield had been born one day earlier, the SU MH would be at Berlin on the relationship map. The relationship SU IC at Berlin emphasises the SU MH line they both have at this location.

David Copperfield and Claudia Schiffer fell in love at first sight when they met. This can be explained by relationship VE IC and UR IC being within orb of Berlin. PL IC is also within orb of Berlin and shows the passionate and deeply-loving relationship that developed between them.

Interestingly, both of them have an UR line at their birthplace on the relationship map. UR ASC is exact at Metuchen (Copperfield) and UR IC is very near to Düsseldorf (Schiffer). This is a clear indication that if they ever met each other it would be unexpected and in unusual circumstances.

The latitude crossings at Berlin will give additional information to the themes shown by the planetary lines.

ME/VE shows that they are able to communicate love and affection to each other. It also explains the media interest in their relationship. They were front page news in German newspapers.

MA/JU indicates that they will have an active life together. This has manifested in the great amount of travelling they have to do to be together.

The Astro*Carto*Graphy relationship chart map for David Copperfield and Claudia Schiffer gives accurate information about when they met each other in Berlin. It shows under what circumstances they met and how their relationship would develop.

The location where two people meet each other is the beginning of their relationship. By using an Astro*Carto*Graphy relationship chart map it is possible to obtain information about the start and development of a relationship.

Calculating the angles for a relationship chart is problematic because complicated trigonometry is used. Also, the midpoint between both birthplaces doesn't seem to mean anything, especially when it is in the ocean.

My recommendation is to ignore the calculated midpoint between birthplaces in a relationship chart. It is more logical to use the location where two people meet for the first time because this is where the "birth" of the relationship takes place.

It is also possible to find out which temporary influences are taking place on an Astro*Carto*Graphy relationship chart map.

First, a look at Cyclo*Carto*Graphy, which, as explained in Chapter 5, involves the use of the planet's right ascension positions. The positions on October 7, 1993 are:

A*C*G	**Age 30**	**C*C*G**
ME 03 LIB 06		SU 11 LIB 06
UR 08 VIR 11		MA 13 SCO 31
NE 11 SCO 27		NE 19 CAP 49
MO 12 ARI 45		UR 19 CAP 51
SU 14 VIR 00		VE 20 LIB 41
VE 16 VIR 19		MO 20 TAU 47
JU 17 ARI 27		JU 21 LIB 17
PL 18 VIR 41		PL 24 SCO 54
SA 20 AQU 55		ME 25 VIR 14
MA 23 LIB 51		SA 26 AQU 51

A*C*G	**(27 06)**	**Displaced C*C*G**
ME 03 LIB 06		MA 16 LIB 25
UR 08 VIR 11		NE 22 SAG 43
NE 11 SCO 27		UR 22 SAG 45
MO 12 ARI 45		VE 23 VIR 35
SU 14 VIR 00		MO 23 ARI 41
VE 16 VIR 19		JU 24 VIR 11
JU 17 ARI 27		PL 27 LIB 48
PL 18 VIR 41		ME 28 LEO 08
SA 20 AQU 55		SA 29 CAP 45
MA 23 LIB 51		

From these lists, the following aspects are found:

1. C*C*G MO EQU A*C*G SA
2. Disp. C*C*G MA = A*C*G JU
3. Disp. C*C*G MO = A*C*G MA

There are **no** location aspects at Berlin, so only the three anywhere aspects need to be considered for the temporary influences.

Disp. C*C*G MA = A*C*G JU puts emphasis on the A*C*G MA DSC/JU ASC latitude crossing at Berlin. This also indicated the travelling necessary to be together, and because it was an applying aspect, the influence increased in power during the next twelve months.

The MA/JU temporary influence also indicated engagement or marriage. A few months after they met, David Copperfield and Claudia Schiffer announced their engagement.

C*C*G MO EQU A*C*G SA shows they experienced periods when they could not be together. This was the result of David Copperfield living in Las Vegas and Claudia Schiffer being in Germany. Their professions made it difficult for them to spend much time together.

The C*C*G MO EQU A*C*G SA temporary influence was nearly exact at the time they met, but it didn't last long. A couple of months later this influence moved away, suggesting that the long distance between them was a major problem for the first few weeks.

Disp. C*C*G MO = A*C*G MA suggests that there could have been disagreements and quarrels regarding where to live after marriage; they were not easy to resolve.

The Astro*Carto*Graphy relationship chart solar return is calculated for the year they first met. The following planetary influences are found at Berlin:

Lines:	UR ASC
	NE ASC
	PL MH
crossings:	ME IC/NE DSC
	PL MH/UR ASC

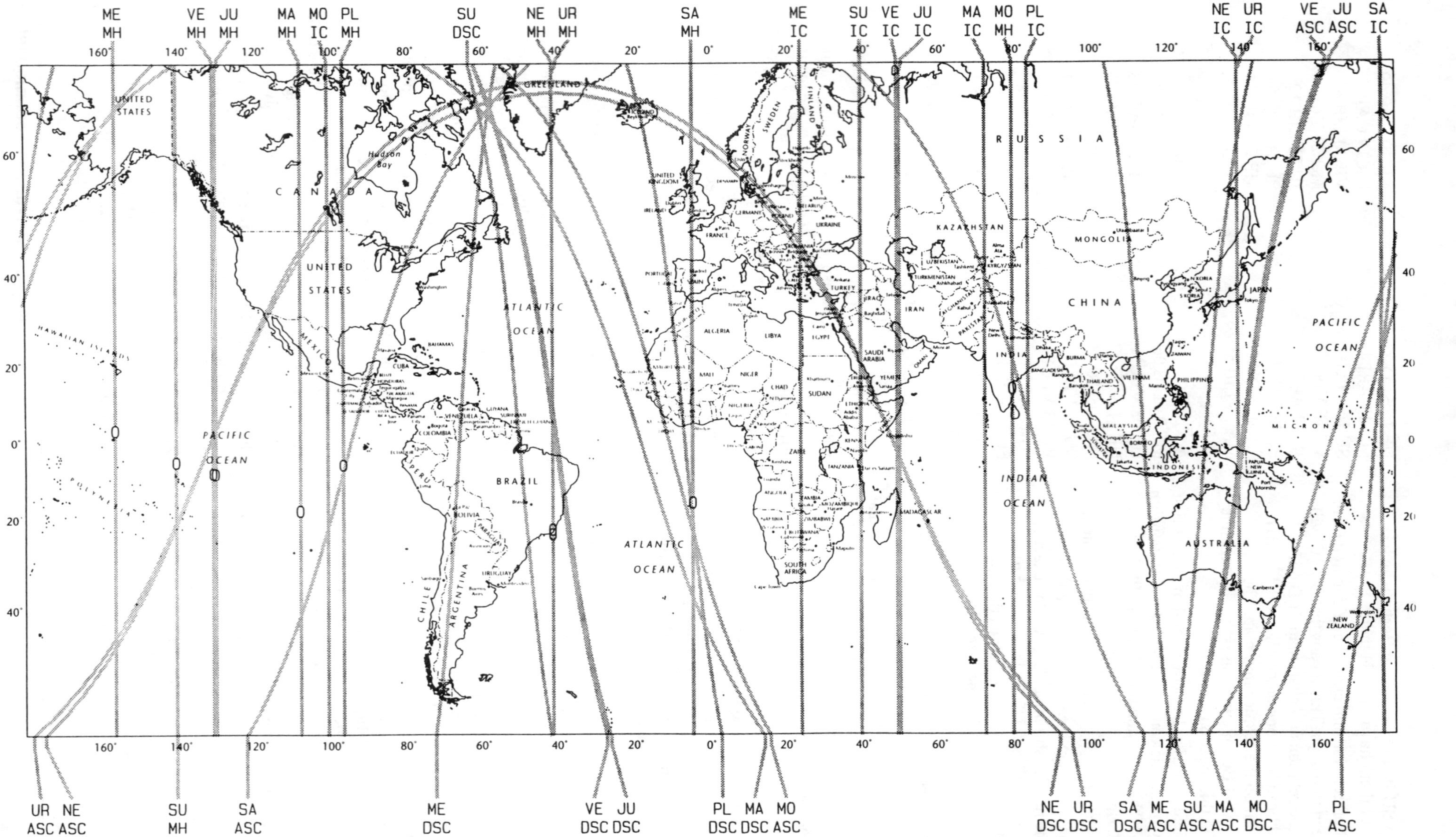

CYCLO*CARTO*GRAPHY

04/17/96

SU, MO, ME, VE AND MA ARE PROGRESSED
JU, SA, UR, NE AND PL ARE TRANSITING

Copperfield/Schiffer

PROG	10	6	63	1:15 GMT
TRANSIT	10	7	93	0:00 GMT
NATAL	9	5	63	23:06 GMT

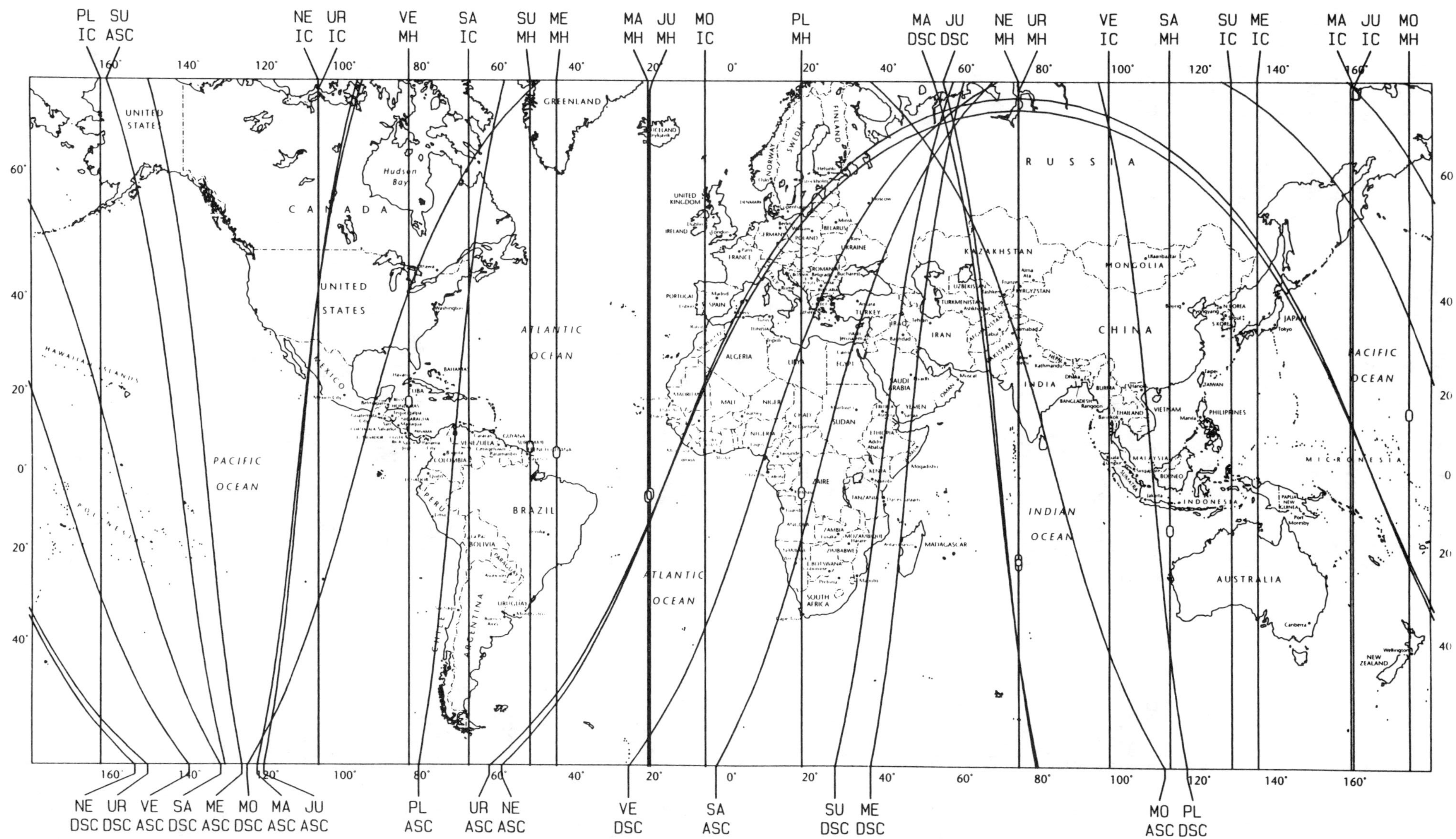

Copperfield/Schiffer SR

September 5, 1993 3:24 PM GMT

04/16/96

Copperfield/Schiffer SR
September 5, 1993 3:24 PM GMT

ASTRO*CARTO*GRAPHY LATITUDE CROSSINGS

NORTH LATITUDES

69N41	MA/IC * SA/DSC
69N31	SU/IC * VE/ASC
69N29	JU/IC * SA/DSC
69N28	VE/DSC * SA/ASC
69N13	SU/MH * MO/DSC
68N30	SU/DSC * JU/DSC
68N30	UR/MH * NE/DSC
68N05	SA/ASC * NE/ASC
67N57	ME/IC * VE/ASC
67N48	VE/DSC * NE/ASC
67N44	ME/DSC * NE/ASC
67N37	ME/DSC * VE/DSC
67N32	JU/DSC * NE/ASC
67N21	NE/MH * UR/ASC
67N20	SU/DSC * NE/ASC
67N18	MA/DSC * NE/ASC
67N09	VE/DSC * JU/DSC
67N05	SU/DSC * MA/DSC
66N54	NE/DSC * PL/DSC
66N48	VE/IC * NE/DSC
66N38	SA/ASC * UR/ASC
66N35	VE/DSC * MA/DSC
66N28	ME/DSC * UR/ASC
66N26	JU/DSC * UR/ASC
66N14	SU/DSC * VE/DSC
66N11	MA/DSC * UR/ASC
66N06	ME/DSC * JU/DSC
66N01	MO/ASC * NE/ASC
66N00	SU/DSC * UR/ASC
65N53	VE/DSC * UR/ASC
65N48	JU/DSC * SA/ASC
65N46	VE/IC * PL/DSC
65N35	VE/IC * UR/DSC
65N34	UR/DSC * PL/DSC
65N31	ME/DSC * SA/ASC
65N11	MO/DSC * PL/ASC
65N06	MO/ASC * UR/ASC
64N56	MA/DSC * SA/ASC
64N28	MO/ASC * VE/DSC
64N17	ME/DSC * MA/DSC
63N16	SA/MH * NE/DSC
62N59	MO/ASC * SU/DSC
62N36	MO/ASC * SA/ASC
62N12	MO/MH * SA/DSC
61N56	SA/MH * UR/DSC
61N36	SU/DSC * SA/ASC
61N13	Anchorage
61N08	SA/IC * MO/DSC
60N50	MO/ASC * ME/DSC
59N20	Stockholm
56N17	MO/ASC * MA/DSC
56N16	MA/IC * VE/ASC
55N55	SU/IC * NE/DSC
55N46	JU/IC * VE/ASC
55N45	Moscow
55N41	Copenhagen
54N59	PL/MH * NE/ASC
54N23	SU/IC * UR/DSC
53N40	MO/ASC * JU/DSC
53N23	PL/MH * UR/ASC
52N30	Berlin
51N30	London
50N48	ME/IC * NE/DSC
49N14	Vancouver
49N12	ME/IC * UR/DSC
48N50	Paris
47N36	Seattle
45N32	Montreal
44N59	Minneapolis
43N39	Toronto
42N22	Boston
42N20	Detroit

CENTRAL LATITUDES

42N13	UR/IC * JU/ASC
42N07	UR/IC * MA/ASC
41N58	NE/IC * JU/ASC
41N54	Rome
41N54	NE/IC * MA/ASC
41N53	Chicago
41N41	VE/MH * MO/DSC
41N26	SA/IC * PL/ASC
41N19	MA/ASC * JU/ASC
40N43	New York
39N44	Denver
38N53	Washington
38N37	St. Louis
37N47	San Francisco
36N48	MO/MH * VE/ASC
36N10	Las Vegas
35N40	Tokyo
34N04	Los Angeles
33N45	Atlanta
33N27	Phoenix
32N47	Dallas
31N21	NE/MH * MO/ASC
31N14	UR/MH * MO/ASC
30N46	PL/MH * VE/DSC
30N03	Cairo
29N58	New Orleans
29N46	Houston
28N39	Delhi
25N47	Miami
22N34	Calcutta
22N32	MO/IC * NE/ASC
22N16	Hong Kong
21N19	Honolulu
21N18	MO/IC * UR/ASC
19N26	Mexico City
16N25	PL/IC * SA/DSC
14N35	Manila
13N19	MA/IC * NE/DSC
12N41	MA/IC * UR/DSC
11N56	JU/IC * NE/DSC
11N23	JU/IC * UR/DSC
10N30	Caracas
6N27	Lagos
1N51	UR/DSC * NE/DSC
1S51	UR/ASC * NE/ASC
11S23	JU/MH * UR/ASC
11S56	JU/MH * NE/ASC
12S41	MA/MH * UR/ASC
13S19	MA/MH * NE/ASC
16S25	PL/MH * SA/ASC
21S18	MO/MH * UR/DSC
22S32	MO/MH * NE/DSC
22S54	Rio De Janeiro
26S12	Johannesburg
30S46	PL/IC * VE/ASC
31S14	UR/IC * MO/DSC
31S21	NE/IC * MO/DSC
33S55	Sydney
34S36	Buenos Aires
36S48	MO/IC * VE/DSC
36S52	Auckland
37S50	Melbourne
41S19	MA/DSC * JU/DSC
41S26	SA/MH * PL/DSC
41S41	VE/IC * MO/ASC
41S54	NE/MH * MA/DSC

SOUTH LATITUDES

41S58	NE/MH * JU/DSC
42S07	UR/MH * MA/DSC
42S13	UR/MH * JU/DSC
49S12	ME/MH * UR/ASC
50S48	ME/MH * NE/ASC
53S23	PL/IC * UR/DSC
53S40	MO/DSC * JU/ASC
54S23	SU/MH * UR/ASC
54S59	PL/IC * NE/DSC
55S46	JU/MH * VE/DSC
55S55	SU/MH * NE/ASC
56S16	MA/MH * VE/DSC
56S17	MO/DSC * MA/ASC

MO ASC/JU DSC
SU IC/UR DSC
PL MH/NE ASC

On their individual Astro*Carto*Graphy solar return maps, both David Copperfield and Claudia Schiffer had UR and NE at Berlin. David Copperfield had UR DSC and NE DSC, while Claudia Schiffer had UR IC and NE IC, so the relationship solar return UR ASC and NE ASC reiterates what they experienced individually.

Solar return PL MH is close to their relationship chart PL IC line, so this strengthened the influence from PL.

Both David Copperfield and Claudia Schiffer had PL MH/UR ASC and PL MH/NE ASC latitude crossings at Berlin on their individual solar returns. This is repeated on the relationship solar return map.

MO/JU confirms the feelings of happiness when they first met. They probably felt lucky to have met each other.

SU/UR reiterates the unusual and unexpected circumstances under which they met. Both have a UR line at their birthplace on the relationship chart map.

ME/NE indicates the uncertainty that the media had about David Copperfield and Claudia Schiffer staying together. Most thought it was a publicity stunt.

Finally, solar arc progressions are calculated for additional information about the temporary influences.

A*C*G	**(27 06)**	**Solar arc A*C*G**
ME 03 LIB 06		ME 00 SCO 12
UR 08 VIR 11		UR 05 LIB 17
NE 11 SCO 27		NE 08 SAG 33
MO 12 ARI 45		MO 09 TAU 51
SU 14 VIR 00		VE 13 LIB 25
VE 16 VIR 19		JU 14 TAU 33
JU 17 ARI 27		PL 15 LIB 47
PL 18 VIR 41		SA 18 PIS 01
SA 20 AQU 55		MA 20 SCO 57
MA 23 LIB 51		

1. solar arc A*C*G NE EQU A*C*G UR
2. solar arc A*C*G VE = A*C*G MO
3. solar arc A*C*G SA = A*C*G PL
4. solar arc A*C*G MA EQU A*C*G SA

NE/UR reiterated the UR and NE lines on the three solar return maps which show that these are important influences.

VE/MO is appropriate for a harmonious and loving relationship. This aspect would have an influence no matter where they met.

SA/PL indicated difficult circumstances in their relationship that were not be easy to overcome. They both needed a great deal of self-discipline to overcome the restrictions that result from these difficult circumstances.

MA/SA shows that they were cautious and let the relationship develop before making the commitment to marriage. It could also indicate some difficulties to overcome before marriage, such as where to live as married partners.

Example 2—Prince Charles/ Princess Diana

It was a sad day in Britain when John Major announced the official separation between Prince Charles and Princess Diana. The announcement was made at 3:30 p.m. on December 9, 1992 (live broadcast). They had been married at 11:17 a.m. on July 29, 1981 at Westminster. To examine the separation between Prince Charles and Princess Diana, the following will be used:

1. a. Astro*Carto*Graphy map for Prince Charles
   b. Temporary influences for December 9, 1992
2. a. Astro*Carto*Graphy map for Princess Diana
   b. Temporary influences for December 9, 1992
3. a. Astro*Carto*Graphy relationship chart map
   b. Temporary influences for December 9, 1992
4. a. Astro*Carto*Graphy marriage chart map
   b. Temporary influences for December 9, 1992

It is best to examine an Astro*Carto*Graphy marriage chart map with the other synastry techniques. Although the marriage chart map explains why Prince Charles and Princess Diana separated, it should not be used on its own in synastry.

London will be used with these Astro*Carto*Graphy techniques as that is the location where their relationship took place.

Prince Charles has three planetary lines within orb of London:

Lines: NE IC
VE IC
PL ASC

The line nearest London is NE IC. People who live on or near NE lines tend to not receive the recognition they deserve. Prince Charles is waiting to become king of England, but there are doubts that this will become reality.

VE IC is clear as he lives under luxurious conditions. He is surrounded by art and beauty in the home, and waited upon by servants.

As far as relationships are concerned, the combination of VE and NE indicates that Prince Charles is a romantic person. It could also indicate disappointments in relationships from unrealistic expectations.

PL ASC could show power struggles within a relationship. He wants to achieve success in life at all costs, even when he has to do so by being secretive.

At London Prince Charles had five latitude crossings.

Crossings: PL MH/ME ASC
ME MH/UR DSC
ME MH/JU ASC
VE IC/PL ASC
MO IC/MA ASC

VE/PL reiterated the VE IC and PL ASC planetary lines that he has within orb of London. This shows the secretive love affair he had with another woman.

PL/ME shows that Prince Charles is a convincing and influential speaker when he makes public appearances.

ME/UR and ME/JU shows his ability to be a good speaker and his interest in new ideas.

From these planetary influences at London it is clear that Prince Charles could experience disappointments in close personal relationships.

What temporary influences were taking place when Prince Charles separated from Princess Diana?

Cyclo*Carto*Graphy, an Astro*Carto*Graphy solar return map and Astro*Carto*Graphy solar arc progressions will be used.

Prince Charles' right ascension horoscope was featured in Chapter 5 when the Cyclo*Carto*Graphy temporary influences at Klosters were examined.

A*C*G	Age 44	C*C*G
ME 05 SCO 13		SU 07 CAP 47
SA 07 SAG 38		VE 09 SAG 21
NE 13 LIB 36		JU 10 LIB 34
VE 15 LIB 49		SA 17 AQU 00
SU 20 SCO 01		UR 17 CAP 50
MA 20 SAG 05		ME 18 CAP 07
PL 21 LEO 29		MO 18 SAG 27
MO 28 ARI 28		NE 18 CAP 56
JU 29 GEM 54		PL 24 SCO 56
UR 29 SAG 55		MA 26 CAP 50

Prince Charles had no aspects between C*C*G and A*C*G planets.

A*C*G	(47 46)	Displaced C*C*G
ME 05 SCO 13		UR 00 SAG 04
SA 07 SAG 38		ME 00 SAG 21
NE 13 LIB 36		MO0 0 SCO 41
VE 15 LIB 49		NE0 1 SAG 10
SU 20 SCO 01		PL 07 LIB 10
MA 20 SAG 05		MA 09 SAG 04
PL 21 LEO 29		VE 21 LIB 35
MO 28 ARI 28		JU 22 LEO 48
JU 29 GEM 54		SA 29 SAG 14
UR 29 SAG 55		

1. Disp. C*C*G SA = A*C*G JU
2. Disp. C*C*G SA = A*C*G UR

These are the only temporary influences that Prince Charles had at the time he separated from Princess Diana.

Disp. C*C*G SA is on top of his A*C*G JU/UR conjunction, a temporary influence that lasted until the end of 1992, so it was not a dominant aspect. This aspect suggests that it wasn't an easy decision for Prince Charles to make. It could also indicate that there were problems after the separation.

The Astro*Carto*Graphy solar return map for Prince Charles' forty-fourth birthday shows four planetary lines within orb of London.

Lines:

NE IC
UR IC
MO MH
MA MH

It is important to remember that the solar return map will have an influence for one year. Therefore, it shows the influences that Prince Charles experienced during and after the separation.

Solar return NE IC repeats the A*C*G NE IC line at London. After the separation there were doubts that he should be allowed to become king of England.

UR IC indicates that home and family life were subject to changes and disruptions. This was certainly the case when he separated from Princess Diana.

MO MH near London shows that he was deeply affected by public feelings. During the year that the solar return was in effect he experienced fluctuations in popularity.

MA MH shows the disagreements he had with Princess Diana about who should educate their children. It also indicates that he had to work hard to rebuild his damaged reputation.

Prince Charles had seven latitude crossings at London on his solar return map.

Crossings:

JU MH/ME ASC
SU MH/MO DSC
SU MH/UR ASC
MO IC/SU DSC
SU MH/NE ASC
MO DSC/UR ASC
MA IC/ME DSC

These latitude crossings reiterate the information obtained from the four planetary lines that were within orb of London.

A*C*G	(47 46)	Solar arc A*C*G
ME 05 SCO 13		NE 01 SAG 22
SA 07 SAG 38		VE 03 SAG 35
NE 13 LIB 36		MA 07 AQU 51
VE 15 LIB 49		PL 09 LIB 15
SU 20 SCO 01		MO 16 GEM 14
MA 20 SAG 05		JU 17 LEO 40
PL 21 LEO 29		UR 17 AQU 41
MO 28 ARI 28		ME 22 SAG 59
JU 29 GEM 54		SA 25 CAP 24
UR 29 SAG 55		

Prince Charles had no solar arc progressed aspects, so it was only his solar return chart that provided most of the information.

Next the separation will be examined by looking at Princess Diana's Astro*Carto*Graphy map and temporary influences. She has only one planetary line within orb of London:

ME DSC

ME DSC enabled her to promote herself as an important member of the Royal Family. It is also a good line for dealing with young people, and Princess Diana was a kindergarten teacher.

She has six latitude crossings at London:

Crossings:

PL MH/VE DSC
SU IC/PL DSC
SU ASC/ME ASC
NE IC/UR ASC
SA MH/VE ASC
MA IC/SU ASC

SU/ME supports the influence from her ME DSC line.

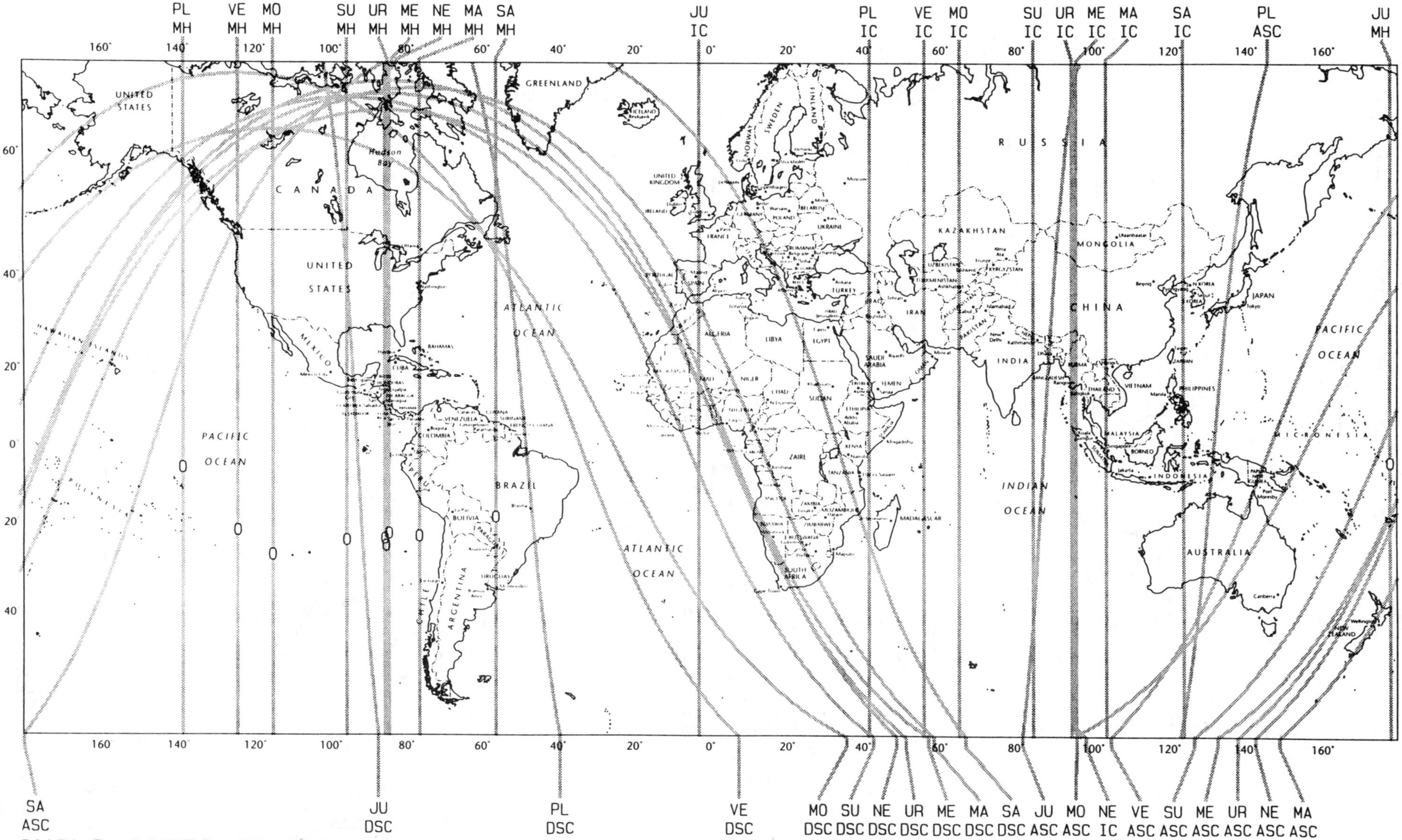

CYCLO*CARTO*GRAPHY

04/17/96

SU, MO, ME, VE AND MA ARE PROGRESSED
JU, SA, UR, NE AND PL ARE TRANSITING

Prince Charles

PROG	12	28	48	22:54 GMT
TRANSIT	12	9	92	0:00 GMT
NATAL	11	14	48	21:14 GMT

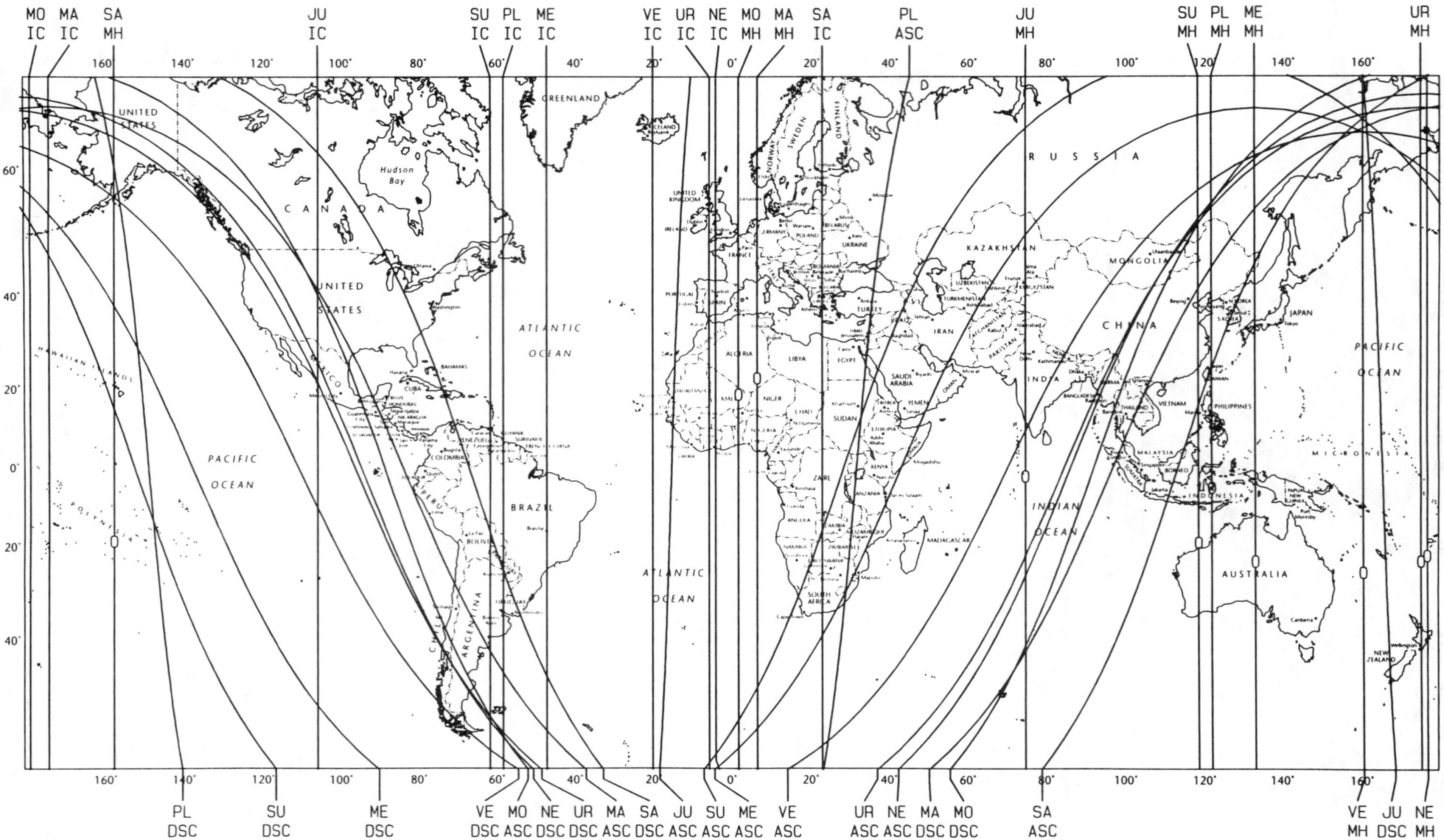

Prince Charles SR

November 15, 1992 3: 48 AM GMT

04/16/96

04/16/96

Prince Charles SR
November 15, 1992 3:48 AM GMT

ASTRO*CARTO*GRAPHY LATITUDE CROSSINGS

NORTH LATITUDES

69N55	NE/MH * SA/ASC
69N51	VE/MH * JU/DSC
69N51	SA/MH * MO/ASC
69N47	MO/DSC * JU/DSC
69N47	VE/MH * MO/DSC
69N39	UR/MH * SA/ASC
68N18	MO/DSC * SU/DSC
68N18	UR/MH * NE/ASC
68N12	MO/IC * NE/DSC
68N04	SA/ASC * NE/ASC
67N58	MA/IC * NE/DSC
67N34	JU/DSC * NE/ASC
67N31	VE/MH * NE/ASC
67N15	MO/IC * MA/DSC
67N04	SU/DSC * NE/ASC
67N03	MA/ASC * NE/DSC
67N01	NE/MH * UR/DSC
67N00	MA/DSC * UR/DSC
66N58	NE/MH * MA/DSC
66N51	UR/MH * MA/DSC
66N50	MO/IC * UR/DSC
66N42	MA/ASC * PL/DSC
66N37	MO/DSC * ME/DSC
66N36	PL/MH * ME/ASC
66N32	MA/IC * UR/DSC
66N29	SA/ASC * UR/ASC
66N27	SA/MH * MA/ASC
66N25	JU/DSC * UR/ASC
66N21	VE/MH * UR/ASC
66N20	NE/DSC * PL/DSC
66N20	SU/MH * ME/ASC
66N15	SU/DSC * UR/ASC
66N14	JU/DSC * SA/ASC
66N05	ME/DSC * NE/ASC
65N58	VE/MH * SA/ASC
65N55	SU/DSC * SA/ASC
65N54	SA/MH * NE/DSC
65N53	VE/MH * SU/DSC
65N36	ME/DSC * UR/ASC
65N34	SU/DSC * JU/DSC
65N16	SU/DSC * MA/DSC
65N08	MA/DSC * JU/DSC
65N07	JU/MH * SU/ASC
65N01	ME/DSC * SA/ASC
64N57	VE/MH * MA/DSC
64N44	ME/DSC * MA/DSC
64N32	VE/MH * ME/DSC
64N30	ME/DSC * VE/DSC
64N30	VE/DSC * JU/DSC
64N30	UR/DSC * PL/DSC
64N29	VE/ASC * MA/DSC
64N27	SU/DSC * VE/DSC
64N27	VE/ASC * SA/ASC
64N18	ME/DSC * JU/DSC
64N10	SA/MH * UR/DSC
63N47	UR/MH * VE/DSC
63N41	MA/DSC * SA/ASC
63N37	NE/MH * VE/DSC
63N25	VE/ASC * UR/ASC
62N48	MO/IC * VE/DSC
62N33	ME/MH * MO/DSC
62N30	SU/DSC * ME/DSC
62N23	JU/IC * SA/DSC
62N18	VE/ASC * NE/ASC
61N55	MA/IC * VE/DSC
61N51	ME/MH * VE/ASC
61N28	ME/MH * NE/ASC
61N16	MO/DSC * VE/ASC
61N13	Anchorage
60N43	SA/MH * PL/DSC
60N31	ME/MH * UR/ASC
60N30	UR/MH * ME/DSC
59N53	NE/MH * ME/DSC

CENTRAL LATITUDES

59N20	Stockholm
59N07	UR/MH * SU/DSC
58N46	PL/MH * VE/ASC
58N02	NE/MH * SU/DSC
57N30	SU/MH * VE/ASC
57N11	MO/IC * ME/DSC
56N49	SA/MH * VE/DSC
56N35	MO/DSC * NE/ASC
56N11	VE/DSC * PL/DSC
55N50	PL/MH * NE/ASC
55N45	Moscow
55N41	Copenhagen
55N39	PL/MH * MO/DSC
55N07	PL/MH * UR/ASC
54N51	ME/MH * MA/DSC
54N25	MA/IC * ME/DSC
53N32	MO/DSC * UR/ASC
53N29	SU/MH * NE/ASC
53N11	MO/IC * SU/DSC
52N53	SU/MH * UR/ASC
52N39	SU/MH * MO/DSC
52N30	Berlin
51N30	London
51N07	JU/MH * ME/ASC
49N14	Vancouver
48N50	Paris
48N00	MA/IC * SU/DSC
47N36	Seattle
46N12	ME/MH * SA/ASC
45N35	PL/MH * MA/DSC
45N32	Montreal
45N12	MO/ASC * MA/ASC
44N59	Minneapolis
44N05	UR/ASC * NE/ASC
43N39	Toronto
42N22	Boston
42N20	Detroit
41N54	Rome
41N53	Chicago
41N42	SU/MH * MA/DSC
41N41	JU/IC * MA/ASC
40N43	New York
40N42	JU/IC * MO/ASC
39N44	Denver
38N53	Washington
38N38	SA/MH * ME/DSC
38N37	St. Louis
37N47	San Francisco
36N10	Las Vegas
35N40	Tokyo
34N04	Los Angeles
33N45	Atlanta
33N27	Phoenix
32N47	Dallas
30N03	Cairo
29N58	New Orleans
29N46	Houston
29N29	ME/DSC * PL/DSC
28N39	Delhi
26N03	JU/IC * NE/DSC
25N47	Miami
24N54	PL/MH * SA/ASC
22N34	Calcutta
22N16	Hong Kong
21N31	JU/IC * UR/DSC
21N19	Honolulu
19N26	Mexico City
15N50	SU/MH * SA/ASC
15N23	SA/MH * SU/DSC
14N35	Manila
13N16	SU/ASC * PL/ASC

SOUTH LATITUDES

10N30	Caracas
9N53	JU/MH * VE/ASC
6N27	Lagos
9S53	JU/IC * VE/DSC
13S16	SU/DSC * PL/DSC
15S23	SA/IC * SU/ASC
15S50	SU/IC * SA/DSC
21S31	JU/MH * UR/ASC
22S54	Rio De Janeiro
24S54	PL/IC * SA/DSC
26S03	JU/MH * NE/ASC
26S12	Johannesburg
29S29	ME/ASC * PL/ASC
33S55	Sydney
34S36	Buenos Aires
36S52	Auckland
37S50	Melbourne
38S38	SA/IC * ME/ASC
40S42	JU/MH * MO/DSC
41S41	JU/MH * MA/DSC
41S42	SU/IC * MA/ASC
44S05	UR/DSC * NE/DSC
45S12	MO/DSC * MA/DSC
45S35	PL/IC * MA/ASC
46S12	ME/IC * SA/DSC
48S00	MA/MH * SU/ASC
51S07	JU/IC * ME/DSC
52S39	SU/IC * MO/ASC
52S53	SU/IC * UR/DSC
53S11	MO/MH * SU/ASC
53S29	SU/IC * NE/DSC
53S32	MO/ASC * UR/DSC
54S25	MA/MH * ME/ASC
54S51	ME/IC * MA/ASC
55S07	PL/IC * UR/DSC
55S39	PL/IC * MO/ASC
55S50	PL/IC * NE/DSC
56S11	VE/ASC * PL/ASC
56S35	MO/ASC * NE/DSC
56S49	SA/IC * VE/ASC
57S11	MO/MH * ME/ASC
57S30	SU/IC * VE/DSC
58S02	NE/IC * SU/ASC
58S46	PL/IC * VE/DSC
59S07	UR/IC * SU/ASC
59S53	NE/IC * ME/ASC

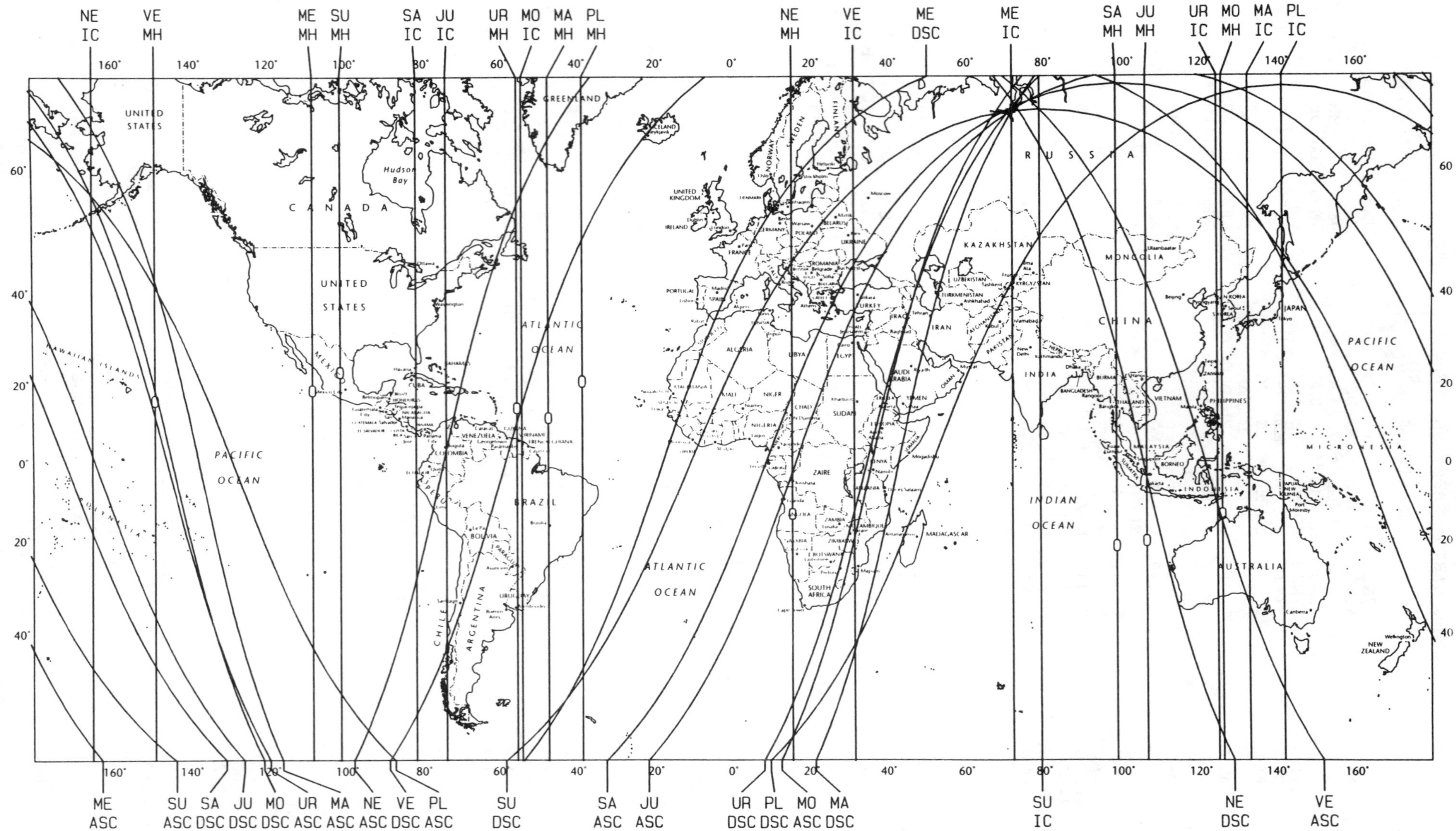

Princess Diana

July 1, 1961 7:45 PM GDT

04/16/96

04/16/96

Princess Diana
July 1, 1961 7:45 PM GDT

ASTRO*CARTO*GRAPHY LATITUDE CROSSINGS

NORTH LATITUDES

69N54 SU/IC * UR/DSC
69N31 UR/IC * JU/DSC
69N22 MO/MH * JU/DSC
69N21 ME/IC * VE/ASC
69N17 SA/MH * ME/ASC
69N17 ME/DSC * SA/DSC
69N08 MO/ASC * VE/ASC
69N06 JU/MH * SA/DSC
69N02 PL/MH * NE/ASC
68N47 ME/IC * MO/ASC
68N46 MA/IC * PL/DSC
68N40 JU/DSC * PL/DSC
68N26 VE/ASC * MA/DSC
68N25 VE/ASC * UR/DSC
68N23 MA/IC * JU/DSC
68N19 MO/ASC * NE/DSC
68N17 MA/DSC * UR/DSC
68N17 SU/IC * JU/ASC
68N16 MO/MH * PL/DSC
68N08 SU/IC * SA/ASC
68N08 UR/IC * PL/DSC
67N50 VE/ASC * JU/ASC
67N50 VE/ASC * SA/ASC
67N49 ME/IC * NE/DSC
67N49 JU/ASC * SA/ASC
67N45 JU/MH * ME/ASC
67N43 SA/DSC * PL/DSC
67N26 UR/DSC * NE/DSC
67N17 MA/DSC * NE/DSC
67N15 SA/ASC * NE/DSC
67N14 MA/DSC * SA/ASC
67N10 UR/IC * SA/DSC
67N06 ME/IC * UR/DSC
67N06 JU/ASC * NE/DSC
67N04 ME/IC * SA/ASC
67N03 SA/ASC * UR/DSC
66N57 SU/IC * VE/ASC
66N56 MO/MH * SA/DSC
66N55 SU/ASC * VE/ASC
66N51 MA/DSC * JU/ASC
66N49 SU/DSC * NE/DSC
66N47 ME/IC * JU/ASC
66N45 SU/DSC * MA/DSC
66N45 ME/IC * SU/DSC
66N44 SU/DSC * JU/ASC
66N43 SU/DSC * UR/DSC
66N40 ME/IC * MA/DSC
66N37 SU/DSC * SA/ASC
66N32 PL/IC * JU/DSC
66N29 MO/ASC * SU/DSC
66N26 JU/ASC * UR/DSC
66N11 ME/ASC * PL/DSC
66N01 MO/ASC * SA/ASC
66N00 VE/IC * ME/DSC
65N41 SA/MH * SU/ASC
65N26 MA/IC * SA/DSC
65N13 NE/IC * MA/ASC
64N49 JU/MH * PL/DSC
64N42 MO/ASC * JU/ASC
64N35 SU/ASC * PL/DSC
64N25 JU/MH * SU/ASC
63N49 MA/MH * NE/ASC
63N07 SU/IC * NE/DSC
62N42 PL/IC * SA/DSC
62N21 SA/MH * PL/DSC
61N34 UR/ASC * PL/ASC
61N13 Anchorage
60N55 UR/IC * ME/ASC
60N10 MO/MH * ME/ASC

59N34 VE/ASC * PL/DSC
59N20 Stockholm
58N44 UR/IC * SU/ASC
58N29 NE/MH * ME/DSC

CENTRAL LATITUDES

58N08 MO/MH * SU/ASC
57N38 VE/IC * SU/DSC
57N08 MO/IC * NE/ASC
56N56 MO/DSC * PL/ASC
56N47 NE/IC * PL/ASC
56N39 NE/IC * MO/DSC
55N57 NE/DSC * PL/DSC
55N45 Moscow
55N41 Copenhagen
55N27 UR/MH * NE/ASC
55N21 MA/IC * ME/ASC
54N20 MA/IC * SU/ASC
53N37 SA/MH * VE/ASC
52N30 Berlin
51N48 NE/IC * UR/ASC
51N30 London
51N13 SU/ASC * ME/ASC
50N28 SU/IC * PL/DSC
50N02 PL/MH * VE/DSC

49N14 Vancouver
48N50 Paris
47N36 Seattle
47N30 PL/IC * SU/ASC
46N08 NE/MH * SU/DSC
46N07 PL/IC * ME/ASC
45N46 VE/IC * SA/ASC
45N32 Montreal
44N59 Minneapolis
43N39 Toronto
42N46 VE/MH * MA/ASC
42N29 JU/MH * VE/ASC
42N28 ME/IC * PL/DSC
42N22 Boston
42N20 Detroit
41N54 Rome
41N53 Chicago
41N40 VE/MH * PL/ASC
40N43 New York
40N22 MA/ASC * PL/ASC

39N44 Denver
38N53 Washington
38N48 MO/ASC * UR/DSC
38N37 St. Louis
37N47 San Francisco
36N29 VE/IC * JU/ASC
36N10 Las Vegas
35N40 Tokyo
34N18 MA/MH * VE/DSC
34N04 Los Angeles
33N45 Atlanta
33N27 Phoenix
32N47 Dallas
30N03 Cairo

29N58 New Orleans
29N46 Houston
28N39 Delhi
28N10 SA/MH * NE/DSC
25N47 Miami
22N34 Calcutta
22N16 Hong Kong
21N19 Honolulu

19N26 Mexico City
18N36 VE/MH * MO/DSC
17N39 NE/MH * SA/ASC
15N04 MO/IC * VE/DSC
14N35 Manila
11N56 VE/MH * UR/ASC
10N44 UR/MH * VE/DSC
10N30 Caracas

6N27 Lagos

SOUTH LATITUDES

3N18 JU/IC * NE/ASC
2N06 NE/IC * JU/DSC

2S06 NE/MH * JU/ASC
3S18 JU/MH * NE/DSC

10S44 UR/IC * VE/ASC
11S56 VE/IC * UR/DSC
15S04 MO/MH * VE/ASC
17S39 NE/IC * SA/DSC
18S36 VE/IC * MO/ASC

22S54 Rio De Janeiro
26S12 Johannesburg
28S10 SA/IC * NE/ASC

33S55 Sydney
34S18 MA/IC * VE/ASC
34S36 Buenos Aires
36S29 VE/MH * JU/DSC
36S52 Auckland
37S50 Melbourne
38S48 MO/DSC * UR/ASC

40S22 MA/DSC * PL/DSC
41S40 VE/IC * PL/DSC
42S28 ME/MH * PL/ASC
42S29 JU/IC * VE/DSC
42S46 VE/IC * MA/DSC
45S46 VE/MH * SA/DSC
46S07 PL/MH * ME/DSC
46S08 NE/IC * SU/ASC
47S30 PL/MH * SU/DSC

50S02 PL/IC * VE/ASC
50S28 SU/MH * PL/ASC
51S13 SU/DSC * ME/DSC
51S48 NE/MH * UR/DSC
53S37 SA/IC * VE/DSC
54S20 MA/MH * SU/DSC
55S21 MA/MH * ME/DSC
55S27 UR/IC * NE/DSC
55S57 NE/ASC * PL/ASC
56S39 NE/MH * MO/ASC
56S47 NE/MH * PL/DSC
56S56 MO/ASC * PL/DSC
57S08 MO/MH * NE/DSC
57S38 VE/MH * SU/ASC
58S08 MO/IC * SU/DSC
58S29 NE/IC * ME/ASC
58S44 UR/MH * SU/DSC
59S34 VE/DSC * PL/ASC

Princess Diana was praised often in the press for her work and royal duties.

SU/PL shows the possibility of a total transformation taking place in her life. This new beginning in life manifested in the area of relationships (VE/PL).

SA/VE indicated relationships where there is a difference in age. It started as an unusual and ideal romance (NE/UR), but could lead to difficulties later through emotional inhibition (SA/VE).

MA/SU suggested that Princess Diana would be self-assertive in her life and take charge of situations.

A*C*G	**Age 31**	**C*C*G**
ME 3 CAN 22		JU 10 LIB 34
MA 4 VIR 04		SU 12 LEO 07
NE 6 SCO 51		SA 17 AQU 00
JU 7 AQU 36		UR 17 CAP 50
SU 10 CAN 31		NE 18 CAP 56
PL 12 LEO 47		MA 21 VIR 46
VE 22 TAU 48		MO 23 ARI 37
UR 25 LEO 55		PL 24 SCO 56
MO 27 AQU 13		VE 28 GEM 17
SA 29 CAP 56		ME 28 CAN 37

1. C*C*G JU EQU A*C*G SU
2. C*C*G SU = A*C*G PL
3. C*C*G PL EQU A*C*G UR

A*C*G	**(31 36)**	**Displaced C*C*G**
ME 3 CAN 22		JU 8 VIR 58
MA 4 VIR 04		SA 15 CAP 24
NE 6 SCO 51		UR 16 SAG 14
JU 7 AQU 36		NE 17 SAG 20
SU 10 CAN 31		MA 20 LEO 10
PL 12 LEO 47		MO 22 PIS 01
VE 22 TAU 48		PL 23 LIB 20
UR 25 LEO 55		VE 26 TAU 41
MO 27 AQU 13		ME 27 GEM 01
SA 29 CAP 56		

Princess Diana had no aspects between displaced C*C*G and A*C*G planets.

4. prog. birthplace = A*C*G VE IC
5. prog. birthplace = C*C*G UR ASC

C*C*G SU = A*C*G PL was an important temporary influence for Princess Diana as it shows great changes taking place in her life. It was a time to make a new beginning, which manifested in separating from Prince Charles. Other indications of separation in a partnership are the aspects the progressed birthplace makes. The progressed birthplace is conjunct A*C*G VE IC and C*C*G UR ASC which manifested as powerful influences.

Setting up a new life is shown by C*C*G PL EQU A*C*G UR, reiterating the theme from the C*C*G SU = A*C*G PL temporary influence.

C*C*G JU EQU A*C*G SU suggests conflicts and differences in opinion about the terms for a separation.

There are no planets within orb of London on her Astro*Carto*Graphy solar return map. Sometimes a solar return map doesn't have any planetary lines at a location, but the latitude crossings can still be used to obtain background information.

Crossings:	JU IC/SU ASC
	MA IC/SU DSC
	PL MH/MO DSC
	JU IC/VE ASC
	PL MH/UR ASC
	PL MH/VE DSC
	PL MH/NE ASC
	JU IC/UR DSC
	NE MH/MA ASC

PL MH was in latitude crossings with MO DSC, UR ASC, VE DSC and NE ASC. These combined influences with PL MH emphasized starting a new life as the result of a disappointing separation. This reiterates the information obtained from Cyclo*Carto*Graphy.

JU IC was with SU ASC, VE ASC and UR DSC, another indication of changes in partnerships.

MA/SU shows that she was determined to achieve what she wanted, but this could lead to disputes and disagreements. NE/MA could suggest that her health would not be good during this time.

A*C*G	**(31 36)**	**Solar arc A*C*G**
ME 3 CAN 22		SA 1 PIS 32
MA 4 VIR 04		ME 4 LEO 58
NE 6 SCO 51		MA 5 LIB 40
JU 7 AQU 36		NE 8 SAG 27
SU 10 CAN 31		JU 9 PIS 12
PL 12 LEO 47		PL 14 VIR 23
VE 22 TAU 48		VE 24 GEM 24
UR 25 LEO 55		UR 27 VIR 31
MO 27 AQU 13		MO 28 PIS 49
SA 29 CAP 56		

Princess Diana had no solar arc progressed aspects, so the most important information is obtained from Cyclo*Carto*Graphy.

The separation had a big impact on Princess Diana as shown by her Cyclo*Carto*Graphy temporary influences. For Prince Charles it was his solar return map which showed the changes that took place in his life. This shows the importance of using all the techniques to find the temporary influences. If only Cyclo*Carto*Graphy has been used for Prince Charles it wouldn't have shown the possibility of a separation.

Next, their relationship chart will be examined. The data for Prince Charles/Princess Diana relationship chart is:

Date	March 9, 1955
Time	8:00 p.m. GMT
Latitude	52N10
Longitude	0E10

An Astro*Carto*Graphy relationship chart map for Prince Charles and Princess Diana shows the following planetary influences at London:

Lines:	JU MH
Crossings:	MA MH/ME DSC
	ME IC/MA DSC
	VE IC/NE ASC
	SU MH/JU ASC
	NE IC/PL ASC

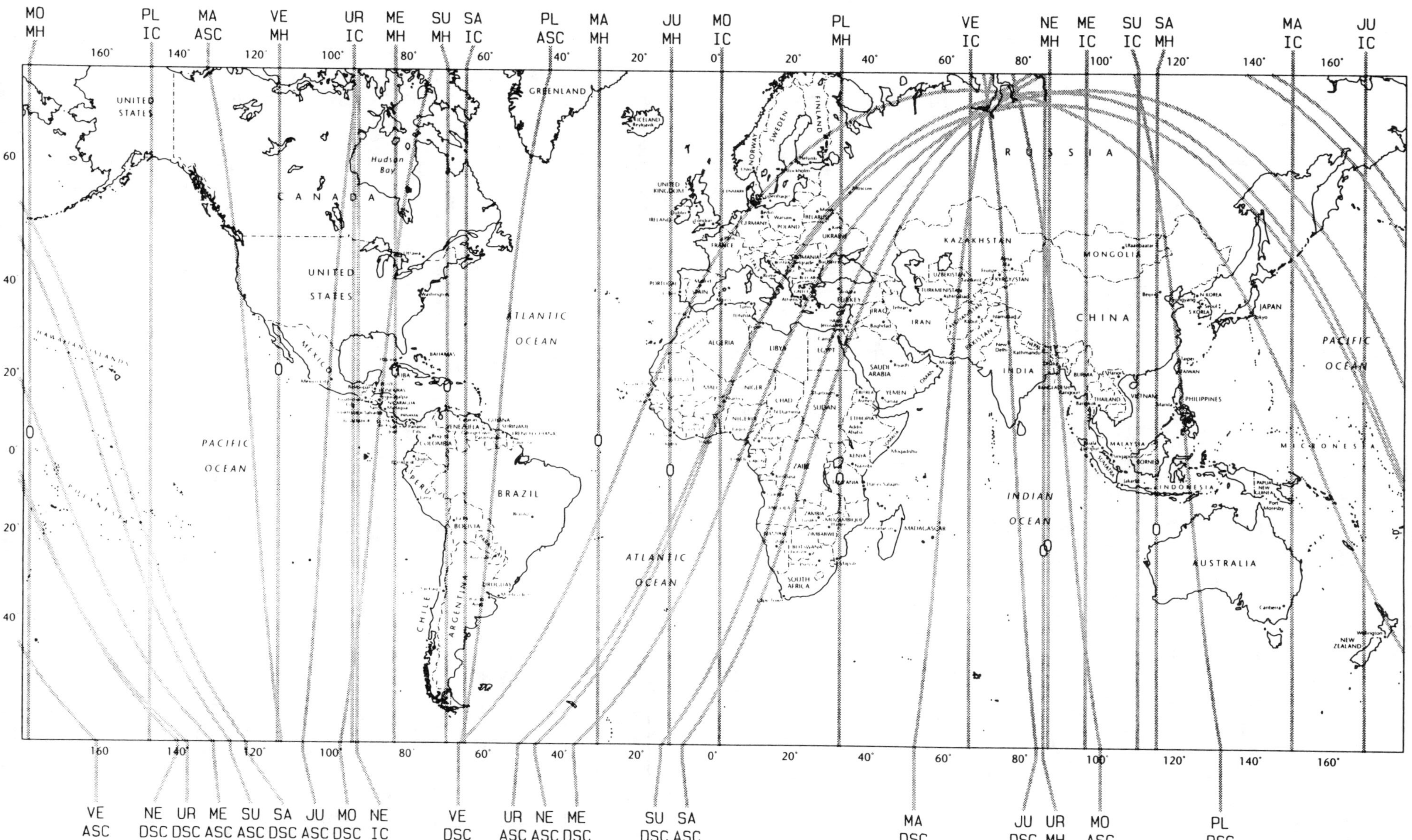

CYCLO*CARTO*GRAPHY

04/17/96

SU, MO, ME, VE AND MA ARE PROGRESSED
JU, SA, UR, NE AND PL ARE TRANSITING

Princess Diana

PROG	8	2	61	5:22 GMT
TRANSIT	12	9	92	0:00 GMT
NATAL	7	1	61	18:45 GMT

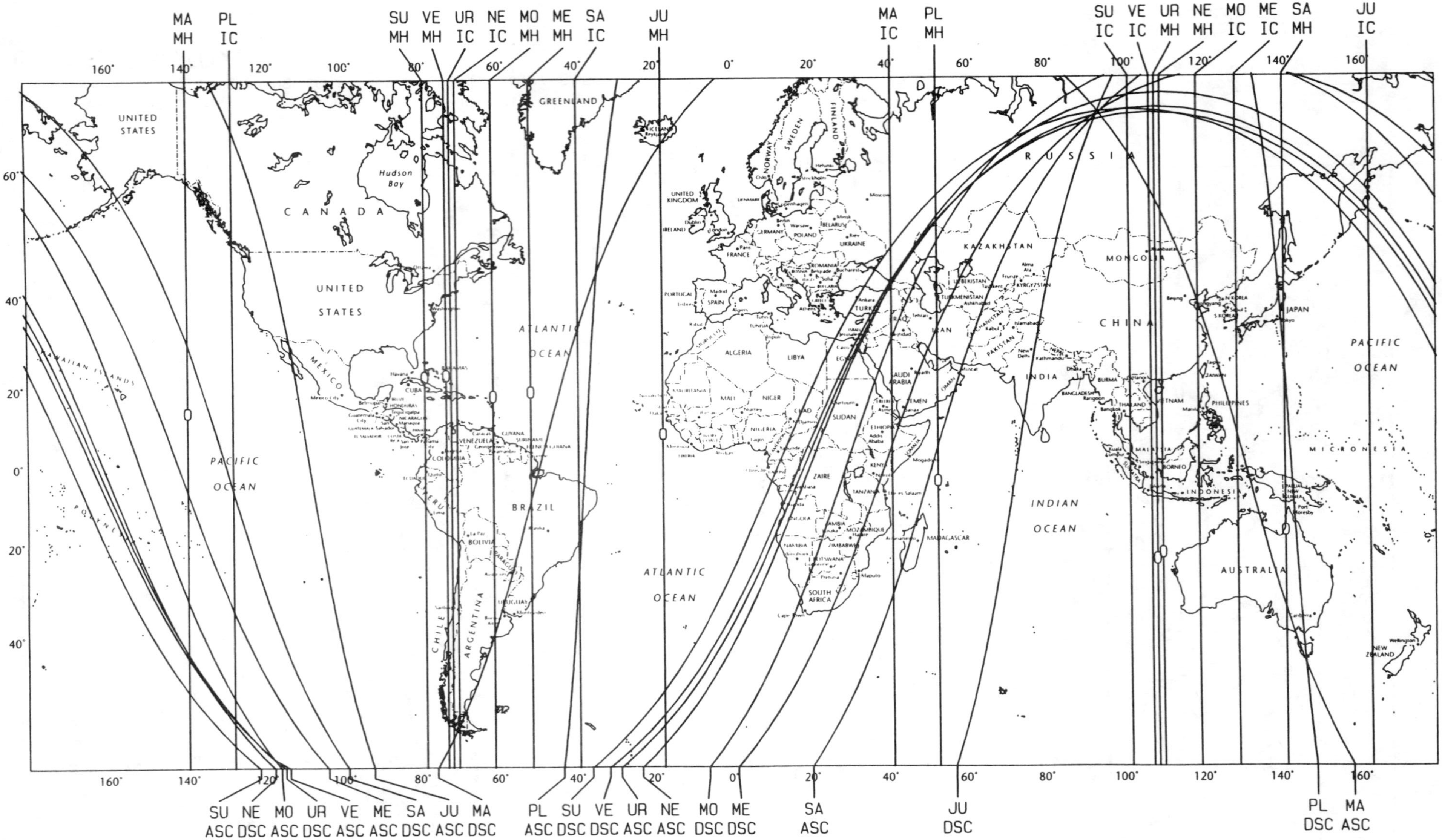

Princess Diana SR

July 1, 1992 5:15 PM GMT

04/16/96

04/16/96

Princess Diana SR
July 1, 1992 5:15 PM GMT

ASTRO*CARTO*GRAPHY LATITUDE CROSSINGS

NORTH LATITUDES

69N29 NE/MH * ME/DSC
69N19 UR/MH * ME/DSC
69N09 VE/IC * ME/DSC
69N05 SU/IC * SA/ASC
68N52 MO/DSC * MA/ASC
68N29 UR/MH * NE/ASC
68N28 VE/IC * NE/ASC
68N18 SU/IC * ME/DSC
68N17 SU/IC * NE/ASC
68N17 ME/DSC * NE/ASC
68N14 MO/IC * NE/DSC
68N05 SA/ASC * NE/ASC
67N50 JU/DSC * NE/ASC
67N32 MA/ASC * NE/ASC
67N22 ME/IC * NE/DSC
67N07 VE/IC * UR/ASC
67N07 NE/MH * UR/DSC
67N00 JU/IC * ME/ASC
66N59 SU/IC * UR/ASC
66N56 SU/DSC * UR/ASC
66N51 ME/DSC * SA/ASC
66N51 VE/IC * SU/ASC
66N48 UR/MH * SU/ASC
66N47 SU/DSC * SA/ASC
66N47 SU/DSC * ME/DSC
66N46 MA/ASC * JU/DSC
66N46 MO/IC * UR/DSC
66N45 SU/DSC * MA/ASC
66N45 SU/DSC * JU/DSC
66N44 NE/MH * SU/ASC
66N42 ME/DSC * MA/ASC
66N39 MA/ASC * SA/ASC
66N37 ME/DSC * JU/DSC
66N35 UR/MH * VE/ASC
66N34 NE/MH * VE/ASC
66N30 MA/ASC * UR/ASC
66N30 SU/IC * VE/DSC
66N29 NE/DSC * PL/DSC
66N26 SA/ASC * UR/ASC
66N25 JU/DSC * UR/ASC
66N25 SU/ASC * VE/ASC
66N23 JU/DSC * SA/ASC
66N20 ME/DSC * UR/ASC
66N08 VE/DSC * MA/ASC
66N08 MO/IC * VE/ASC
65N57 VE/DSC * JU/DSC
65N57 MO/IC * SU/ASC
65N53 VE/DSC * SA/ASC
65N45 ME/IC * UR/DSC
65N41 ME/DSC * VE/DSC
65N18 JU/IC * MO/ASC
65N16 SA/MH * NE/DSC
65N15 PL/IC * JU/ASC
65N00 ME/IC * VE/ASC
65N00 SU/DSC * NE/ASC
64N38 UR/DSC * PL/DSC
64N25 ME/IC * SU/ASC
63N59 MO/DSC * SU/DSC
63N44 VE/ASC * PL/DSC
63N39 JU/MH * MA/DSC
63N24 SA/MH * UR/DSC
62N42 SU/ASC * PL/DSC
62N32 MO/DSC * NE/ASC
62N28 SA/MH * VE/ASC
62N10 SU/IC * MA/ASC
61N13 Anchorage
61N10 SA/MH * SU/ASC

59N20 Stockholm
58N56 MO/DSC * UR/ASC
58N41 MO/DSC * VE/DSC
57N26 VE/IC * MA/ASC
57N00 PL/MH * SU/DSC
56N59 JU/IC * NE/DSC

CENTRAL LATITUDES

56N58 VE/DSC * UR/ASC
56N04 UR/MH * MA/ASC
55N45 Moscow
55N41 Copenhagen
54N22 NE/MH * MA/ASC
54N16 JU/IC * UR/DSC
53N55 PL/MH * NE/ASC
53N17 PL/MH * VE/DSC
53N07 PL/MH * UR/ASC
52N42 JU/IC * VE/ASC
52N30 Berlin
51N30 London

49N59 PL/MH * MO/DSC
49N14 Vancouver
49N02 MA/IC * SU/DSC
48N52 JU/IC * SU/ASC
48N50 Paris
47N42 VE/DSC * NE/ASC
47N36 Seattle
45N32 Montreal
44N59 Minneapolis
44N49 MA/DSC * PL/ASC
43N39 Toronto
43N06 MA/IC * VE/DSC
42N29 MA/IC * NE/ASC
42N22 Boston
42N21 MA/IC * UR/ASC
42N20 Detroit
41N54 Rome
41N53 Chicago
40N43 New York
40N35 UR/ASC * NE/ASC

39N44 Denver
38N53 Washington
38N37 St. Louis
38N30 MO/IC * MA/ASC
37N47 San Francisco
36N10 Las Vegas
35N40 Tokyo
34N04 Los Angeles
33N51 SA/IC * MA/DSC
33N45 Atlanta
33N45 PL/MH * ME/DSC
33N27 Phoenix
32N47 Dallas
31N37 MA/IC * MO/DSC
30N25 MA/MH * SA/DSC
30N03 Cairo

29N58 New Orleans
29N46 Houston
28N39 Delhi
25N47 Miami
25N33 SA/MH * PL/DSC
22N34 Calcutta
22N16 Hong Kong
21N19 Honolulu

19N26 Mexico City
14N35 Manila
10N30 Caracas

7N59 ME/IC * MA/ASC
6N27 Lagos
5N52 MA/IC * ME/DSC
5N22 PL/MH * SA/ASC

5S22 PL/IC * SA/DSC
5S52 MA/MH * ME/ASC
7S59 ME/MH * MA/DSC

22S54 Rio De Janeiro
25S33 SA/IC * PL/ASC

SOUTH LATITUDES

26S12 Johannesburg

30S25 MA/IC * SA/ASC
31S37 MA/MH * MO/ASC
33S45 PL/IC * ME/ASC
33S51 SA/MH * MA/ASC
33S55 Sydney
34S36 Buenos Aires
36S52 Auckland
37S50 Melbourne
38S30 MO/MH * MA/DSC

40S35 UR/DSC * NE/DSC
42S21 MA/MH * UR/DSC
42S29 MA/MH * NE/DSC
43S06 MA/MH * VE/ASC
44S49 MA/ASC * PL/DSC
47S42 VE/ASC * NE/DSC
48S52 JU/MH * SU/DSC
49S02 MA/MH * SU/ASC
49S59 PL/IC * MO/ASC

52S42 JU/MH * VE/DSC
53S07 PL/IC * UR/DSC
53S17 PL/IC * VE/ASC
53S55 PL/IC * NE/DSC
54S16 JU/MH * UR/ASC
54S22 NE/IC * MA/DSC
56S04 UR/IC * MA/DSC
56S58 VE/ASC * UR/DSC
56S59 JU/MH * NE/ASC
57S00 PL/IC * SU/ASC
57S26 VE/MH * MA/DSC
58S41 MO/ASC * VE/ASC
58S56 MO/ASC * UR/DSC

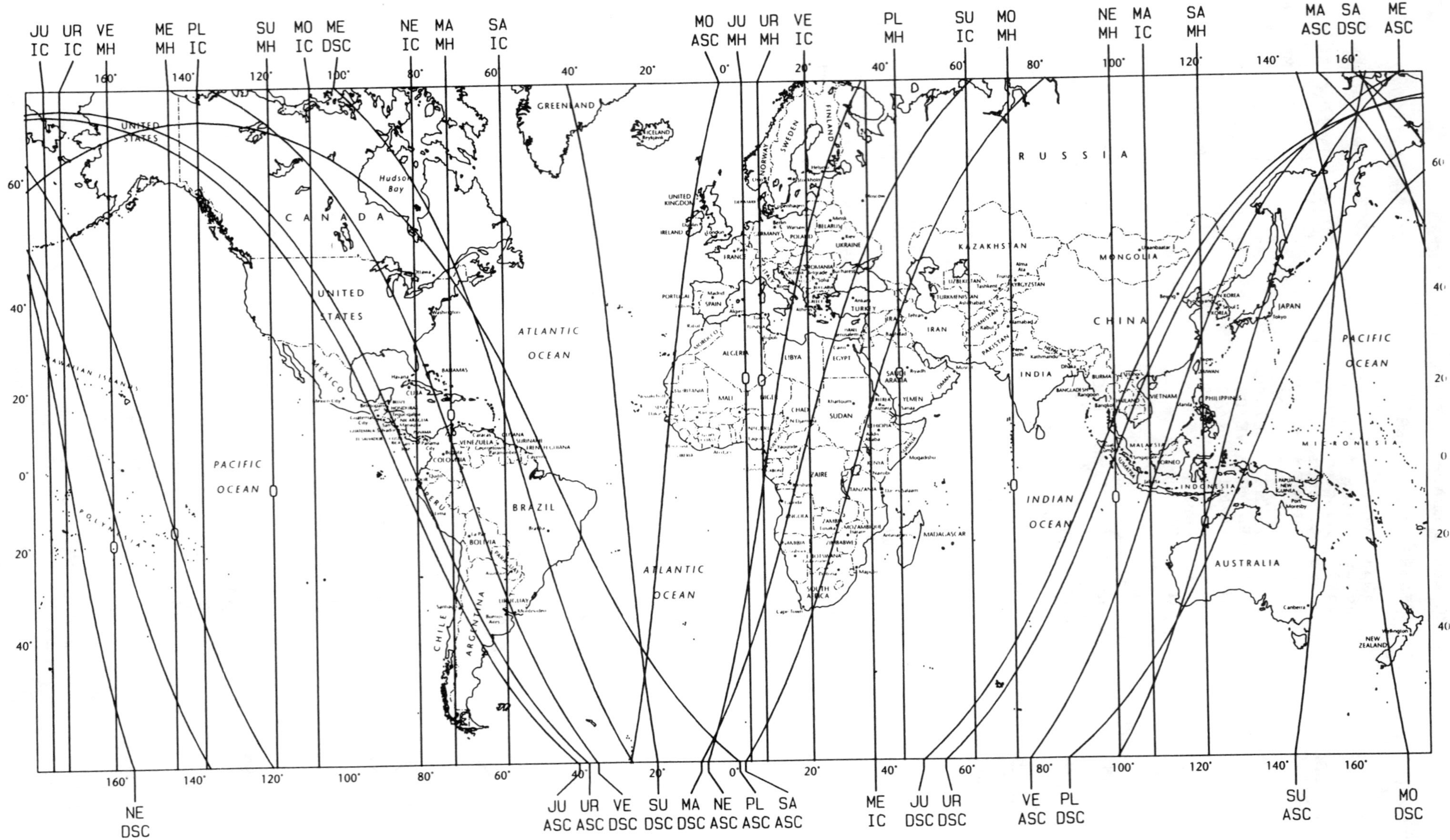

04/16/96

Charles/Diana

March 9, 1955 8:00 PM GMT

04/16/96

Charles/Diana
March 9, 1955 8:00 PM GMT

ASTRO*CARTO*GRAPHY LATITUDE CROSSINGS

NORTH LATITUDES

69N41 ME/ASC * VE/ASC
69N26 SU/IC * MA/DSC
69N16 SU/ASC * NE/DSC
69N13 SU/ASC * SA/DSC
69N09 SA/DSC * NE/DSC
68N30 VE/ASC * SA/DSC
68N17 VE/ASC * NE/DSC
68N04 JU/IC * UR/DSC
68N04 ME/ASC * SA/DSC
67N42 SU/ASC * VE/ASC
67N40 VE/MH * UR/ASC
67N33 ME/ASC * NE/DSC
67N31 UR/IC * JU/ASC
67N05 SA/DSC * UR/DSC
66N58 JU/DSC * SA/DSC
66N54 VE/ASC * MA/ASC
66N46 MO/MH * SA/ASC
66N44 ME/MH * PL/DSC
66N43 VE/MH * JU/ASC
66N43 UR/DSC * NE/DSC
66N39 JU/DSC * NE/DSC
66N29 SU/MH * VE/DSC
66N26 UR/ASC * PL/DSC
66N22 JU/DSC * UR/DSC
66N08 ME/ASC * JU/DSC
66N03 ME/ASC * UR/DSC
66N00 SU/ASC * JU/DSC
65N59 MA/ASC * JU/DSC
65N58 SU/ASC * MA/ASC
65N57 JU/ASC * PL/DSC
65N56 MA/ASC * UR/DSC
65N54 SU/ASC * UR/DSC
65N50 SU/MH * PL/ASC
65N49 ME/ASC * MA/ASC
65N35 ME/MH * UR/ASC
65N34 SU/ASC * ME/ASC
65N25 VE/DSC * PL/ASC
65N08 VE/MH * PL/DSC
64N12 VE/ASC * JU/DSC
64N03 ME/MH * JU/ASC
63N59 MO/IC * PL/ASC
63N49 MO/DSC * JU/DSC
63N48 PL/IC * UR/ASC
63N34 MO/DSC * VE/ASC
63N23 MO/DSC * UR/DSC
62N56 VE/ASC * UR/DSC
62N25 UR/IC * PL/DSC
61N55 MO/IC * VE/DSC
61N52 PL/IC * JU/ASC
61N13 Anchorage
61N06 JU/IC * PL/DSC
60N41 SU/IC * SA/ASC
60N23 SA/DSC * PL/DSC

59N41 NE/IC * ME/DSC
59N38 MO/DSC * ME/ASC
59N20 Stockholm
59N00 PL/MH * MA/DSC
58N48 JU/IC * SA/DSC
57N03 NE/DSC * PL/DSC
56N53 MA/ASC * PL/DSC
55N55 SU/MH * UR/ASC
55N47 MA/ASC * NE/DSC
55N45 Moscow
55N41 Copenhagen
54N51 UR/IC * SA/DSC
53N08 NE/IC * PL/ASC
52N30 Berlin
52N17 SU/MH * JU/ASC
51N30 London
51N12 VE/IC * NE/ASC
50N08 ME/IC * MA/DSC

49N29 MA/MH * ME/DSC

CENTRAL LATITUDES

49N14 Vancouver
48N50 Paris
48N21 SA/MH * JU/DSC
47N36 Seattle
47N01 MO/IC * UR/ASC
46N52 MO/DSC * SU/ASC
45N32 Montreal
45N29 MA/MH * PL/ASC
44N59 Minneapolis
44N51 SA/MH * UR/DSC
43N39 Toronto
43N29 MO/DSC * PL/DSC
42N25 JU/IC * MA/ASC
42N22 Boston
42N20 Detroit
41N54 Rome
41N53 Chicago
41N46 SU/ASC * PL/DSC
41N28 MO/IC * JU/ASC
40N43 New York

39N44 Denver
38N53 Washington
38N37 St. Louis
38N17 ME/DSC * PL/ASC
37N47 San Francisco
36N10 Las Vegas
35N59 PL/MH * SA/ASC
35N40 Tokyo
35N08 VE/MH * SA/DSC
34N04 Los Angeles
33N45 Atlanta
33N27 Phoenix
32N47 Dallas
32N45 UR/IC * MA/ASC
31N19 NE/IC * VE/DSC
31N15 SA/MH * VE/ASC
30N03 Cairo

29N58 New Orleans
29N57 MA/IC * JU/DSC
29N46 Houston
28N39 Delhi
27N45 JU/IC * NE/DSC
25N47 Miami
25N36 SA/IC * PL/ASC
22N38 MA/IC * UR/DSC
22N34 Calcutta
22N16 Hong Kong
21N19 Honolulu

19N26 Mexico City
15N35 SA/IC * ME/DSC
14N49 ME/IC * SA/ASC
14N35 Manila
11N29 NE/MH * JU/DSC
10N30 Caracas

9N29 VE/IC * MA/DSC
7N36 MA/MH * VE/DSC
6N27 Lagos
4N19 UR/IC * NE/DSC
1N43 NE/MH * UR/DSC

1S43 NE/IC * UR/ASC
4S19 UR/MH * NE/ASC
7S36 MA/IC * VE/ASC
9S29 VE/MH * MA/ASC

11S29 NE/IC * JU/ASC
14S49 ME/MH * SA/DSC
15S35 SA/MH * ME/ASC

22S38 MA/MH * UR/ASC
22S54 Rio De Janeiro

SOUTH LATITUDES

25S36 SA/MH * PL/DSC
26S12 Johannesburg
27S45 JU/MH * NE/ASC
29S57 MA/MH * JU/ASC

31S15 SA/IC * VE/DSC
31S19 NE/MH * VE/ASC
32S45 UR/MH * MA/DSC
33S55 Sydney
34S36 Buenos Aires
35S08 VE/IC * SA/ASC
35S59 PL/IC * SA/DSC
36S52 Auckland
37S50 Melbourne
38S17 ME/ASC * PL/DSC

41S28 MO/MH * JU/DSC
41S46 SU/DSC * PL/ASC
42S25 JU/MH * MA/DSC
43S29 MO/ASC * PL/ASC
44S51 SA/IC * UR/ASC
45S29 MA/IC * PL/DSC
46S52 MO/ASC * SU/DSC
47S01 MO/MH * UR/DSC
48S21 SA/IC * JU/ASC
49S29 MA/IC * ME/ASC

50S08 ME/MH * MA/ASC
51S12 VE/MH * NE/DSC
52S17 SU/IC * JU/DSC
53S08 NE/MH * PL/DSC
54S51 UR/MH * SA/ASC
55S47 MA/DSC * NE/ASC
55S55 SU/IC * UR/DSC
56S53 MA/DSC * PL/ASC
57S03 NE/ASC * PL/ASC
58S48 JU/MH * SA/ASC
59S00 PL/IC * MA/ASC
59S38 MO/ASC * ME/DSC
59S41 NE/MH * ME/ASC

The JU MH line shows the huge success their relationship created in society. The British public was happy and enthusiastic about Prince Charles marrying Princess Diana.

SU/JU also shows a successful and beneficial partnership, which was good publicity for the royal family.

The two ME/MA crossings indicated that they were involved in many arguments and quarrels. It wasn't until many years later that their disagreements became known to the public (NE/PL).

VE/NE suggests the ideal romance and partnership in the eyes of the general public. After they were married this influence manifested as disappointment when the reality of the love between them was not as they expected.

The relationship chart shows a grand trine featuring SU, JU, SA and UR which confirms the successful image of their marriage. But a T-square with ME, SA and PL shows that communication between each other was a difficult area.

A*C*G	**Age 37**	**C*C*G**
MO 00 LIB 17		MA 01 GEM 38
PL 01 VIR 07		MO 10 AQU 17
MA 05 TAU 26		JU 10 LIB 34
VE 07 AQU 56		SA 17 AQU 00
SA 19 SCO 20		ME 17 ARI 43
SU 19 PIS 24		UR 17 CAP 50
JU 21 CAN 40		NE 18 CAP 56
ME 23 AQU 36		VE 21 PIS 40
UR 25 CAN 48		SU 23 ARI 56
NE 26 LIB 29		PL 24 SCO 56

1. C*C*G MA EQU A*C*G PL

A*C*G	**(34 32)**	**Displaced C*C*G**
MO 00 LIB 17		MO 05 CAP 45
PL 01 VIR 07		JU 06 VIR 02
MA 05 TAU 26		SA 12 CAP 28
VE 07 AQU 56		ME 13 PIS 11
SA 19 SCO 20		UR 13 SAG 18
SU 19 PIS 24		NE 14 SAG 24
JU 21 CAN 40		VE 17 AQU 08
ME 23 AQU 36		PL 20 LIB 24
UR 25 CAN 48		MA 27 ARI 06
NE 26 LIB 29		

2. Disp. C*C*G MA EQU A*C*G NE
3. C*C*G UR IC = London
4. prog. London = A*C*G MA DSC
5. prog. London = A*C*G ME IC

There were many temporary influences indicating that Prince Charles and Princess Diana could no longer get along with each other. Prog. London activated A*C*G MA DSC and A*C*G ME IC, which shows there were many arguments and disagreements between them. Disp. C*C*G MA EQU A*C*G NE indicates the deception going on in their marriage. As this was a separating aspect it shows it was going on for some time. C*C*G MA EQU A*C*G PL explains the explosive situation between them. They argued over everything, making a life together impossible.

On the date that they officially separated C*C*G UR IC was about one degree from London. The decision to separate was probably made a few days before the official announcement.

I do not know the location where Prince Charles and Princess Diana first met each other. If it was in southeastern England, then C*C*G UR IC would have been exact at the place where the relationship was born.

The relationship chart solar return map shows the following planetary lines within orb of London.

Lines: MO ASC
UR MH
NE MH

With MO ASC near London their marriage and the subsequent difficulties were the center of attention. The media focused its attention on the scandals and illusions that surrounded the breakdown of the marriage between Prince Charles and Princess Diana (UR MH + NE MH). Interestingly, Prince Charles had the same three planets nearest to London on his solar return map. This suggests the media attention would focus more on Prince Charles.

The latitude crossings for London are:

Crossings: SU MH/UR DSC
SU MH/NE DSC
JU MH/MO DSC
PL MH/UR ASC

The SU MH is with UR DSC and NE DSC which reiterates the scandal, separation and disappointment connected with royalty (SU MH).

PL/UR can also indicate separation after going through a crisis in the relationship. JU/MO expressed negatively also shows marital conflicts.

A*C*G	**(34 32)**	**Solar arc A*C*G**
MO 0 LIB 17		UR 0 VIR 20
PL 1 VIR 07		NE 1 SAG 01
MA 5 TAU 26		MO 4 SCO 49
VE 7 AQU 56		PL 5 LIB 39
SA 19 SCO 20		MA 9 GEM 58
SU 19 PIS 24		VE 12 PIS 28
JU 21 CAN 40		SA 23 SAG 52
ME 23 AQU 36		JU 26 LEO 12
UR 25 CAN 48		ME 28 PIS 08
NE 26 LIB 29		

1. solar arc A*C*G UR = A*C*G PL
2. solar arc A*C*G NE EQU A*C*G PL
3. solar arc A*C*G MO = A*C*G MA

Solar arc A*C*G MO = A*C*G MA is another aspect that reiterates the continuous arguments between them. The other two solar arc aspects reiterate the information obtained from the relationship chart solar return map. With both aspecting A*C*G PL the marriage could no longer survive. The Astro*Carto*Graphy relationship chart map for Prince Charles and Princess Diana shows the circumstances that led to the separation.

On the Astro*Carto*Graphy marriage chart map there are five planetary lines within orb of London:

Lines: MO MH
MA MH

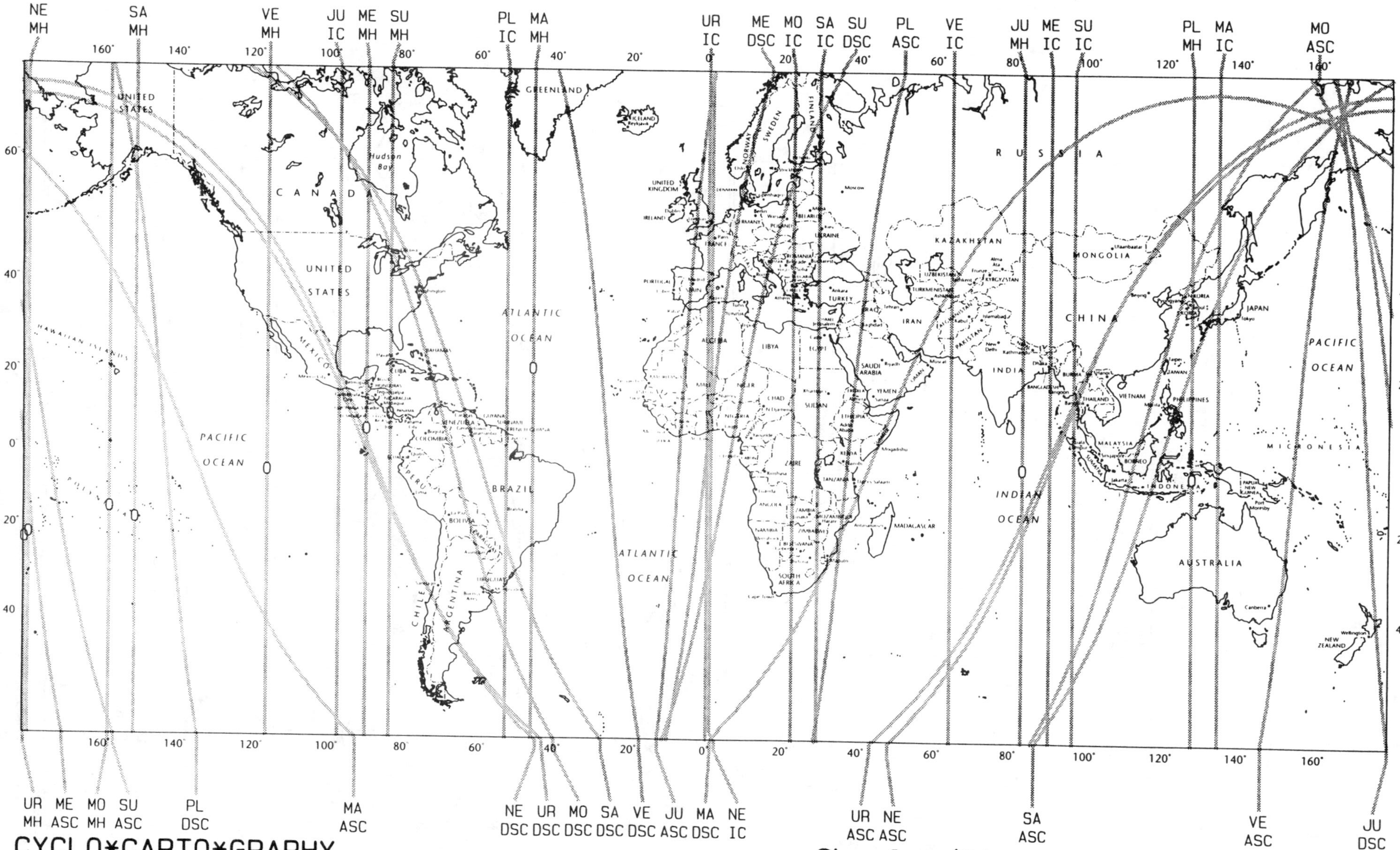

CYCLO*CARTO*GRAPHY

04/17/96

SU, MO, ME, VE AND MA ARE PROGRESSED
JU, SA, UR, NE AND PL ARE TRANSITING

Charles/Diana

PROG	4	16	55	14:09 GMT
TRANSIT	12	9	92	0:00 GMT
NATAL	3	9	55	20:00 GMT

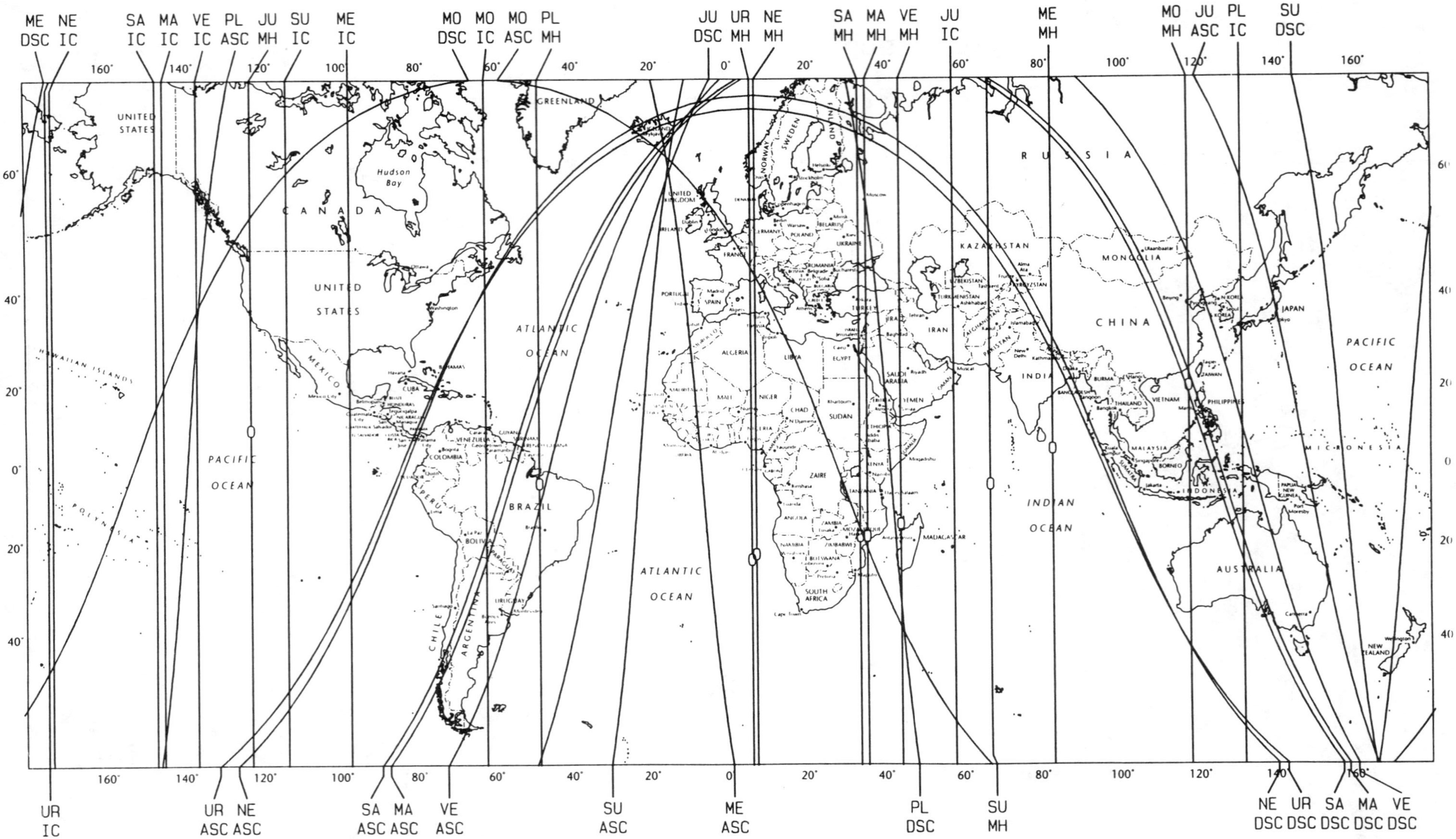

Charles/Diana SR

March 9, 1992 7:40 AM GMT

04/16/96

Charles/Diana SR
March 9, 1992 7:40 AM GMT

ASTRO*CARTO*GRAPHY LATITUDE CROSSINGS

NORTH LATITUDES

69N59 SU/MH * MA/DSC
69N30 PL/MH * MO/ASC
69N29 SU/MH * SA/DSC
68N33 UR/MH * NE/ASC
68N13 MA/ASC * NE/ASC
68N05 SA/ASC * NE/ASC
67N58 VE/ASC * NE/ASC
67N56 JU/DSC * NE/ASC
67N39 VE/ASC * JU/DSC
67N23 SU/ASC * NE/ASC
67N17 JU/DSC * SA/ASC
67N17 NE/MH * UR/DSC
66N56 VE/ASC * SA/ASC
66N38 ME/ASC * NE/ASC
66N37 MA/ASC * UR/ASC
66N35 JU/DSC * UR/ASC
66N31 NE/DSC * PL/DSC
66N27 VE/ASC * UR/ASC
66N26 MA/ASC * JU/DSC
66N24 SA/ASC * UR/ASC
66N18 SA/MH * NE/DSC
66N09 ME/IC * MO/DSC
66N08 SU/ASC * UR/ASC
65N57 MA/MH * NE/DSC
65N34 ME/ASC * UR/ASC
65N25 SU/ASC * SA/ASC
65N18 MO/ASC * NE/ASC
65N17 ME/MH * MA/DSC
64N54 VE/ASC * MA/ASC
64N48 UR/DSC * PL/DSC
64N42 SA/MH * UR/DSC
64N37 MO/ASC * UR/ASC
64N37 SU/ASC * MA/ASC
64N28 SU/ASC * VE/ASC
64N22 ME/MH * SA/DSC
64N18 MA/MH * UR/DSC
64N07 ME/ASC * SA/ASC
63N50 VE/MH * NE/DSC
63N29 ME/ASC * MA/ASC
63N23 SA/MH * PL/DSC
63N13 MO/ASC * SA/ASC
63N03 ME/ASC * VE/ASC
62N51 MO/ASC * MA/ASC
62N34 MO/ASC * VE/ASC
62N08 MO/ASC * ME/ASC
61N59 VE/MH * UR/DSC
61N47 MO/ASC * SU/ASC
61N25 MO/ASC * JU/DSC
61N13 Anchorage
60N48 SU/ASC * ME/ASC
60N11 ME/ASC * JU/DSC
60N08 SU/IC * MO/DSC

59N20 Stockholm
59N09 SU/ASC * JU/DSC
58N48 PL/IC * JU/ASC
58N11 JU/IC * NE/DSC
56N48 MA/MH * PL/DSC
55N51 JU/IC * UR/DSC
55N45 Moscow
55N41 Copenhagen
55N11 PL/MH * NE/ASC
54N16 PL/MH * UR/ASC
53N59 JU/MH * MO/DSC
52N30 Berlin
51N52 SU/MH * NE/DSC
51N30 London

49N14 Vancouver
49N03 SU/MH * UR/DSC
48N50 Paris
47N36 Seattle
47N35 MO/MH * VE/DSC
45N47 UR/MH * MO/ASC

CENTRAL LATITUDES

45N32 Montreal
44N59 Minneapolis
44N16 NE/MH * MO/ASC
43N39 Toronto
42N22 Boston
42N20 Detroit
42N03 MO/IC * NE/ASC
41N54 Rome
41N53 Chicago
41N44 MO/IC * UR/ASC
40N43 New York

39N44 Denver
38N53 Washington
38N37 St. Louis
38N27 MO/DSC * PL/ASC
37N47 San Francisco
37N30 VE/IC * MO/DSC
37N13 UR/ASC * NE/ASC
36N10 Las Vegas
35N40 Tokyo
34N04 Los Angeles
33N45 Atlanta
33N27 Phoenix
32N57 VE/IC * PL/ASC
32N47 Dallas
31N43 ME/MH * NE/DSC
30N03 Cairo

29N58 New Orleans
29N46 Houston
28N39 Delhi
28N00 ME/MH * UR/DSC
25N47 Miami
23N39 PL/MH * SA/ASC
23N10 MO/MH * MA/DSC
22N34 Calcutta
22N16 Hong Kong
21N19 Honolulu
20N05 MA/IC * MO/DSC

19N26 Mexico City
18N39 PL/MH * MA/ASC
17N54 MO/MH * SA/DSC
15N31 SA/IC * MO/DSC
14N35 Manila
10N30 Caracas

9N53 PL/IC * VE/DSC
6N27 Lagos

9S53 PL/MH * VE/ASC

15S31 SA/MH * MO/ASC
17S54 MO/IC * SA/ASC
18S39 PL/IC * MA/DSC

20S05 MA/MH * MO/ASC
22S54 Rio De Janeiro
23S10 MO/IC * MA/ASC
23S39 PL/IC * SA/DSC
26S12 Johannesburg
28S00 ME/IC * UR/ASC

31S43 ME/IC * NE/ASC
32S57 VE/MH * PL/DSC
33S55 Sydney
34S36 Buenos Aires
36S52 Auckland
37S13 UR/DSC * NE/DSC
37S30 VE/MH * MO/ASC
37S50 Melbourne
38S27 MO/ASC * PL/DSC

41S44 MO/MH * UR/DSC

SOUTH LATITUDES

42S03 MO/MH * NE/DSC
44S16 NE/IC * MO/DSC
45S47 UR/IC * MO/DSC
47S35 MO/IC * VE/ASC
49S03 SU/IC * UR/ASC

51S52 SU/IC * NE/ASC
53S59 JU/IC * MO/ASC
54S16 PL/IC * UR/DSC
55S11 PL/IC * NE/DSC
55S51 JU/MH * UR/ASC
56S48 MA/IC * PL/ASC
58S11 JU/MH * NE/ASC
58S48 PL/MH * JU/DSC
59S09 SU/DSC * JU/ASC

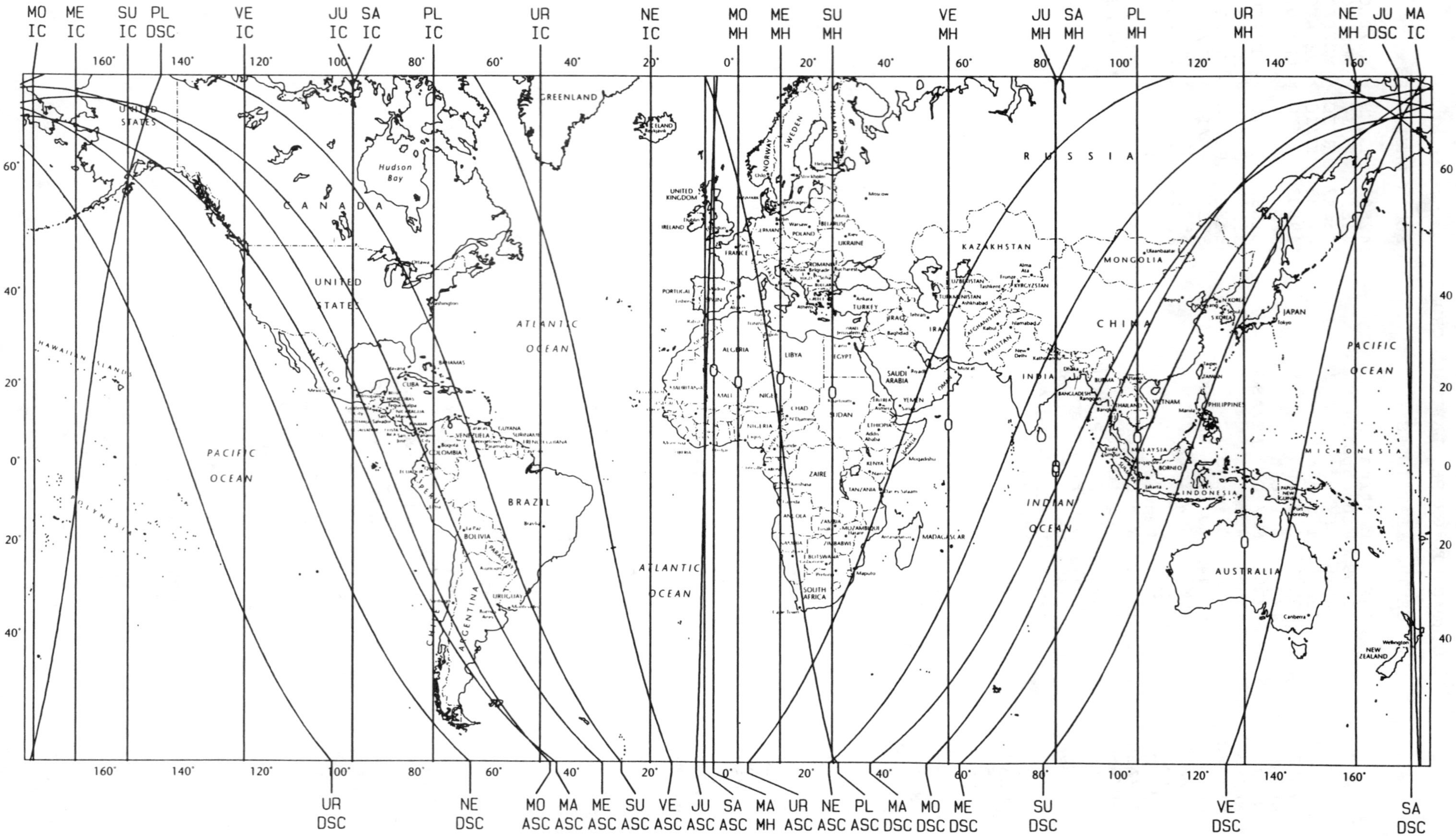

Charles/Diana Wedding

July 29, 1981 11:17 AM GDT

04/16/96

Charles/Diana Wedding
July 29, 1981 11:17 AM GDT

ASTRO*CARTO*GRAPHY LATITUDE CROSSINGS

NORTH LATITUDES

69N38 MO/IC * SU/DSC
69N13 MA/IC * VE/DSC
68N50 MO/DSC * SU/DSC
68N47 MA/IC * MO/DSC
68N45 MO/DSC * VE/DSC
68N44 PL/MH * UR/ASC
68N42 VE/IC * SU/ASC
68N40 MO/DSC * SA/DSC
68N37 NE/MH * UR/DSC
68N33 ME/IC * MO/ASC
68N32 MA/IC * SU/DSC
68N29 MO/DSC * JU/DSC
68N11 SU/DSC * VE/DSC
68N06 UR/DSC * NE/DSC
68N03 SU/DSC * SA/DSC
68N01 MO/DSC * NE/DSC
67N50 MA/MH * PL/ASC
67N50 MO/ASC * ME/ASC
67N50 MO/DSC * UR/DSC
67N47 VE/DSC * SA/DSC
67N45 JU/DSC * NE/DSC
67N40 MO/IC * ME/DSC
67N39 SU/DSC * NE/DSC
67N33 VE/DSC * NE/DSC
67N31 SA/DSC * NE/DSC
67N29 SU/IC * ME/ASC
67N24 JU/ASC * PL/ASC
67N21 SU/DSC * JU/DSC
67N18 NE/MH * MO/DSC
67N18 MA/IC * NE/DSC
67N12 ME/DSC * NE/DSC
67N06 MA/IC * ME/DSC
67N05 SU/IC * MO/ASC
66N56 ME/ASC * PL/DSC
66N49 ME/DSC * SA/DSC
66N41 SU/DSC * UR/DSC
66N36 ME/DSC * VE/DSC
66N29 MO/IC * NE/DSC
66N26 ME/DSC * JU/DSC
66N16 ME/DSC * UR/DSC
66N13 MO/ASC * PL/DSC
66N09 MA/DSC * SA/DSC
66N06 VE/DSC * MA/DSC
66N05 MA/DSC * JU/DSC
66N04 MA/DSC * UR/DSC
66N02 MO/IC * MA/ASC
65N58 ME/DSC * MA/DSC
65N55 JU/DSC * UR/DSC
65N53 VE/DSC * UR/DSC
65N50 MA/ASC * NE/DSC
65N49 VE/DSC * JU/DSC
65N48 SU/DSC * MA/DSC
65N32 UR/MH * NE/ASC
65N17 NE/MH * MA/DSC
65N11 ME/IC * MA/ASC
65N03 SA/DSC * UR/DSC
64N59 SU/DSC * ME/DSC
64N17 ME/IC * NE/DSC
64N12 MA/IC * UR/DSC
64N11 NE/MH * ME/DSC
64N02 MA/MH * JU/ASC
63N41 NE/MH * SU/DSC
62N52 SU/IC * MA/ASC
62N41 SA/MH * UR/ASC
62N39 JU/MH * UR/ASC
62N23 MA/ASC * PL/DSC
61N23 MO/IC * UR/DSC
61N13 Anchorage
61N10 VE/IC * ME/ASC

59N38 SU/IC * NE/DSC
59N28 NE/DSC * PL/DSC
59N20 Stockholm
58N45 SU/IC * PL/DSC

CENTRAL LATITUDES

58N40 UR/MH * MO/DSC
58N19 MO/MH * PL/ASC
58N19 UR/MH * MA/DSC
58N00 JU/IC * SU/ASC
57N56 SA/IC * SU/ASC
56N53 MO/DSC * MA/DSC
56N53 VE/IC * MO/ASC
55N45 Moscow
55N41 Copenhagen
54N43 PL/MH * NE/ASC
54N04 ME/IC * UR/DSC
52N40 UR/IC * VE/ASC
52N30 Berlin
51N30 London
51N15 NE/MH * VE/DSC

49N46 UR/MH * ME/DSC
49N14 Vancouver
48N50 Paris
48N30 VE/IC * MA/ASC
47N36 Seattle
45N32 Montreal
44N59 Minneapolis
44N48 UR/DSC * PL/DSC
43N39 Toronto
42N22 Boston
42N20 Detroit
41N54 Rome
41N53 Chicago
40N43 New York

39N44 Denver
39N43 JU/IC * ME/ASC
39N35 SA/IC * ME/ASC
38N53 Washington
38N37 St. Louis
37N47 San Francisco
37N45 UR/MH * SU/DSC
37N13 SU/IC * UR/DSC
36N10 Las Vegas
35N42 VE/MH * UR/ASC
35N40 Tokyo
35N24 PL/MH * MA/DSC
34N04 Los Angeles
33N45 Atlanta
33N27 Phoenix
32N47 Dallas
31N51 PL/IC * SU/ASC
31N03 SA/MH * NE/ASC
30N53 JU/MH * NE/ASC
30N27 VE/IC * NE/DSC
30N03 Cairo

29N58 New Orleans
29N46 Houston
28N39 Delhi
28N13 PL/MH * MO/DSC
25N47 Miami
22N34 Calcutta
22N16 Hong Kong
21N29 JU/IC * MO/ASC
21N19 Honolulu
21N16 SA/IC * MO/ASC

19N26 Mexico City
14N35 Manila
10N30 Caracas

8N36 ME/MH * PL/ASC
6N27 Lagos
5N24 JU/IC * MA/ASC
5N12 SA/IC * MA/ASC
4N40 JU/ASC * SA/ASC
2N45 PL/MH * ME/DSC

SOUTH LATITUDES

2S45 PL/IC * ME/ASC
4S40 JU/DSC * SA/DSC
5S12 SA/MH * MA/DSC
5S24 JU/MH * MA/DSC
8S36 ME/IC * PL/DSC

21S16 SA/MH * MO/DSC
21S29 JU/MH * MO/DSC
22S54 Rio De Janeiro
26S12 Johannesburg
28S13 PL/IC * MO/ASC

30S27 VE/MH * NE/ASC
30S53 JU/IC * NE/DSC
31S03 SA/IC * NE/DSC
31S51 PL/MH * SU/DSC
33S55 Sydney
34S36 Buenos Aires
35S24 PL/IC * MA/ASC
35S42 VE/IC * UR/DSC
36S52 Auckland
37S13 SU/MH * UR/ASC
37S45 UR/IC * SU/ASC
37S50 Melbourne
39S35 SA/MH * ME/DSC
39S43 JU/MH * ME/DSC

44S48 UR/ASC * PL/ASC
48S30 VE/MH * MA/DSC
49S46 UR/IC * ME/ASC

51S15 NE/IC * VE/ASC
52S40 UR/MH * VE/DSC
54S04 ME/MH * UR/ASC
54S43 PL/IC * NE/DSC
56S53 VE/MH * MO/DSC
56S53 MO/ASC * MA/ASC
57S56 SA/MH * SU/DSC
58S00 JU/MH * SU/DSC
58S19 UR/IC * MA/ASC
58S19 MO/IC * PL/DSC
58S40 UR/IC * MO/ASC
58S45 SU/MH * PL/ASC
59S28 NE/ASC * PL/ASC
59S38 SU/MH * NE/ASC

JU ASC
SA ASC
PL ASC

MO MH shows that their marriage was the center of public attention. The wedding ceremony was viewed live on TV by 700 million people around the world.

MA MH indicates they led an active life in fulfilling royal family duties. Both had a great deal of energy to get through their busy schedules. JU ASC explains the great amount of interest their marriage created. Everyone was happy for them at the time they married.

SA ASC suggests responsibility, duty and a limited private life in their marriage. They had to maintain royal family traditions throughout married life.

PL ASC shows that their lives were transformed when they married. They both experienced deep personal changes as a result of the marriage. There are four latitude crossings at London.

Crossings: UR MH/ME DSC
NE MH/VE DSC
UR IC/VE ASC
ME IC/UR DSC

The two ME/UR crossings indicate that both Prince Charles and Princess Diana need independence and individuality. Their marriage was a surprise to many people (UR/VE), but like a fairy-tale becoming reality (NE/VE).

It is not so easy in deciding how to interpret an Astro*Carto*Graphy marriage chart map. Does it only show the start of the marriage? Or is it possible to see how the marriage will develop?

A*C*G	Age 11	C*C*G
JU 05 LIB 44		MO 07 SAG 44
SA 05 LIB 49		JU 10 LIB 34
MA 08 CAN 08		MA 16 CAN 19
VE 8 VIR 11		SA 17 AQU 00
SU 8 LEO 35		UR 17 CAP 50
MO 14 CAN 28		NE 18 CAP 56
NE 21 SAG 50		SU 19 LEO 32
UR 23 SCO 48		ME 19 LEO 36
ME 25 CAN 20		VE 20 VIR 49
PL 26 LIB 25		PL 24 SCO 56

1. C*C*G MO EQU A*C*G VE
2. C*C*G VE EQU A*C*G NE
3. C*C*G PL = A*C*G UR

A*C*G	(10 57)	Displaced C*C*G
JU 05 LIB 44		MA 05 CAN 22
SA 05 LIB 49		SA 06 AQU 03
MA 08 CAN 08		UR 06 CAP 53
VE 08 VIR 11		NE 07 CAP 59
SU 08 LEO 35		ME 08 LEO 39
MO 14 CAN 28		VE 09 VIR 52
NE 21 SAG 50		PL 13 SCO 59
UR 23 SCO 48		MO 26 SCO 47
ME 25 CAN 20		JU 29 VIR 37
PL 26 LIB 25		

4. Disp. C*C*G NE = A*C*G MA
5. Disp. C*C*G ME = A*C*G SU

C*C*G PL = A*C*G UR has an orb just over one degree, but was still an important aspect for quite a while, during which time they experienced a crisis in their marriage. When things became impossible to resolve they decided it would be best to separate.

C*C*G VE EQU A*C*G NE is an applying aspect, so this shows the disappointment experienced by people after the announcement was made.

C*C*G MO EQU VE is an indicator of disharmony in the marriage, and manifests more like a square. Disp. C*C*G ME = A*C*G SU shows the high media coverage before, during and after the separation. Disp. C*C*G NE = A*C*G MA confirms the deception taking place within the marriage. As this was an applying aspect further information was revealed about their marriage being an illusion.

The marriage chart solar return map shows these planetary influences at London.

Lines: VE ASC
ME ASC
SA DSC

Crossings: PL MH/MO DSC
MA IC/ME DSC
JU IC/UR DSC
ME ASC/VE ASC
JU IC/NE DSC
PL MH/UR ASC
PL MH/NE ASC
MO IC/MA ASC

The VE ASC and ME ASC planetary lines cross each other near London, so they had a powerful influence. This indicates that communication and cooperation between them was important. With SA DSC also within orb of London the attention was focused on the difficulties in their marriage. They were already living separate lives without communication and harmony.

PL MH crosses with MO DSC, UR ASC and NE ASC, suggesting that changes could take place within their marriage. For Prince Charles and Princess Diana it was a traumatic time in what had already become a non-existent marriage. When the media started to reveal the scandals that had been going on for years, most people were shocked and disappointed.

MA/ME and MO/MA explain that there were many disagreements and arguments between Prince Charles and Princess Diana. JU/UR can also indicate conflicts in a marriage which lead to sudden changes.

A*C*G	(10 57)	Solar arc A*C*G
JU 05 LIB 44		NE 02 CAP 47
SA 05 LIB 49		UR 04 SAG 45
MA 08 CAN 08		ME 06 LEO 17
VE 08 VIR 11		PL 07 SCO 22
SU 08 LEO 35		JU 16 LIB 41
MO 14 CAN 28		SA 16 LIB 46
NE 21 SAG 50		MA 19 LEO 05
UR 23 SCO 48		VE 19 VIR 08
ME 25 CAN 20		MO 25 CAN 25

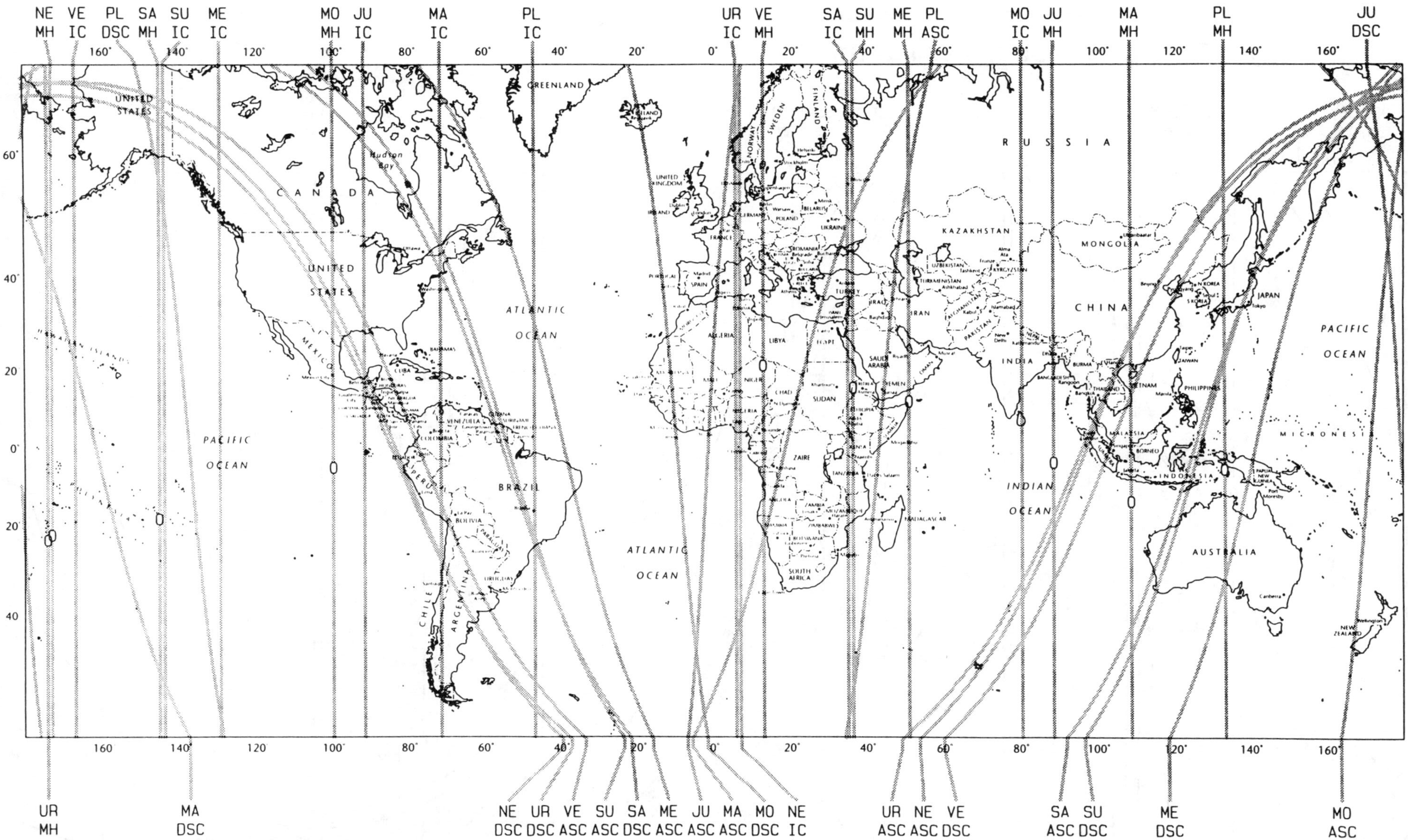

CYCLO*CARTO*GRAPHY

04/17/96

SU, MO, ME, VE AND MA ARE PROGRESSED
JU, SA, UR, NE AND PL ARE TRANSITING

Charles/Diana Wedding

PROG	8	8	82	19:04 GMT
TRANSIT	12	9	92	0:00 GMT
NATAL	7	29	82	10:17 GMT

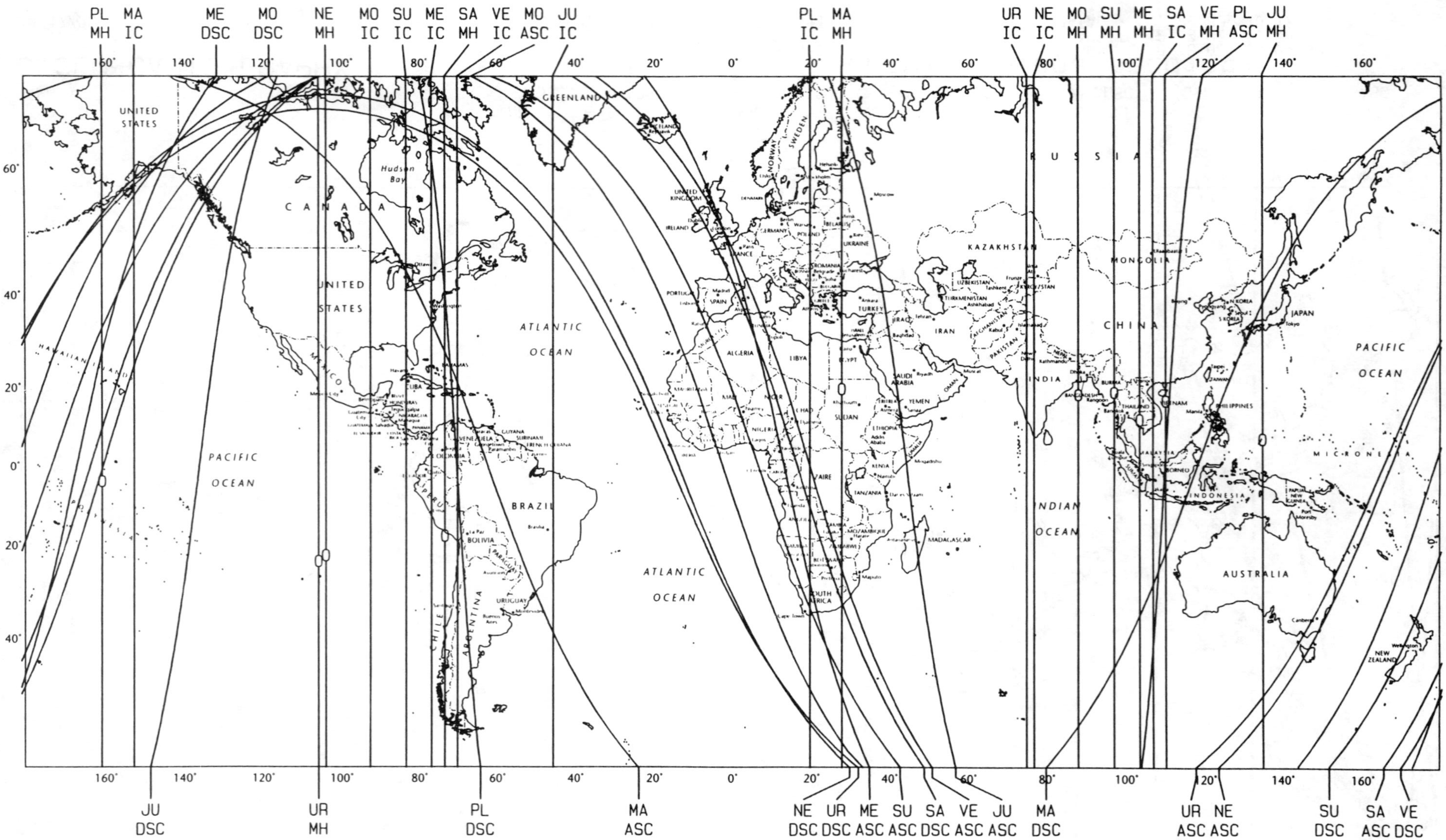

Charles/Diana Wedding SR

July 29, 1992 5:37 AM GMT

04/16/96

ASTRO*CARTO*GRAPHY LATITUDE CROSSINGS

NORTH LATITUDES

69N56 UR/MH * SU/DSC
69N48 SU/DSC * VE/DSC
69N36 SU/DSC * SA/ASC
69N13 ME/DSC * MA/ASC
68N23 UR/MH * NE/ASC
68N21 MO/DSC * MA/ASC
68N09 VE/DSC * NE/ASC
68N05 SA/ASC * NE/ASC
68N00 MO/IC * NE/DSC
67N49 SU/DSC * NE/ASC
67N44 JU/DSC * NE/ASC
67N33 SU/DSC * JU/DSC
67N24 MA/ASC * NE/ASC
67N06 SU/IC * NE/DSC
67N05 SU/DSC * MA/ASC
66N59 NE/MH * UR/DSC
66N54 JU/IC * SU/ASC
66N49 MA/ASC * JU/DSC
66N38 MA/ASC * SA/ASC
66N34 VE/DSC * UR/ASC
66N31 MA/ASC * UR/ASC
66N28 MA/MH * JU/ASC
66N27 MO/IC * UR/DSC
66N26 SA/ASC * UR/ASC
66N26 VE/DSC * MA/ASC
66N23 JU/DSC * UR/ASC
66N21 NE/DSC * PL/DSC
66N08 JU/DSC * SA/ASC
66N04 ME/IC * NE/DSC
65N50 SU/DSC * UR/ASC
65N22 SA/MH * NE/DSC
65N22 SU/IC * UR/DSC
65N14 VE/DSC * JU/DSC
64N37 VE/IC * NE/DSC
64N29 JU/IC * MO/ASC
64N22 UR/DSC * PL/DSC
64N11 ME/IC * UR/DSC
63N24 SA/MH * UR/DSC
63N20 MO/DSC * NE/ASC
63N10 ME/DSC * NE/ASC
62N58 MO/DSC * ME/DSC
62N32 VE/IC * UR/DSC
62N23 UR/MH * MA/ASC
61N35 ME/DSC * UR/ASC
61N33 NE/MH * MA/ASC
61N23 ME/IC * PL/DSC
61N13 Anchorage
60N27 MO/DSC * UR/ASC

59N20 Stockholm
59N03 MA/IC * NE/ASC
58N11 MA/IC * UR/ASC
56N46 MA/IC * MO/DSC
55N45 Moscow
55N41 Copenhagen
54N27 MO/IC * MA/ASC
54N09 PL/MH * NE/ASC
53N33 PL/MH * UR/ASC
53N19 JU/IC * NE/DSC
52N30 Berlin
51N50 ME/ASC * VE/ASC
51N30 London

49N55 JU/IC * UR/DSC
49N50 MA/IC * ME/DSC
49N14 Vancouver
48N50 Paris
48N45 PL/MH * MO/DSC
47N36 Seattle
46N39 MA/IC * SU/DSC
45N46 UR/ASC * NE/ASC
45N32 Montreal
45N02 SU/IC * MA/ASC
44N59 Minneapolis

CENTRAL LATITUDES

43N39 Toronto
42N22 Boston
42N20 Detroit
41N54 Rome
41N53 Chicago
41N03 SA/MH * PL/DSC
40N54 SU/DSC * ME/DSC
40N43 New York

39N44 Denver
38N53 Washington
38N37 St. Louis
38N28 JU/MH * MA/DSC
37N47 San Francisco
36N10 Las Vegas
35N40 Tokyo
35N00 ME/IC * MA/ASC
34N04 Los Angeles
33N45 Atlanta
33N27 Phoenix
33N27 PL/MH * SU/DSC
32N47 Dallas
32N11 MA/IC * SA/ASC
31N27 ME/ASC * SA/DSC
30N03 Cairo

29N58 New Orleans
29N46 Houston
28N39 Delhi
28N21 SA/MH * MA/ASC
27N54 PL/MH * ME/DSC
25N47 Miami
25N18 MA/ASC * PL/DSC
24N59 MA/IC * VE/DSC
22N34 Calcutta
22N16 Hong Kong
21N19 Honolulu
21N07 VE/IC * MA/ASC

19N26 Mexico City
14N35 Manila
10N30 Caracas

9N51 PL/MH * SA/ASC
6N27 Lagos
2N14 VE/MH * PL/ASC
0N28 PL/IC * VE/ASC

0S28 PL/MH * VE/DSC
2S14 VE/IC * PL/DSC
9S51 PL/IC * SA/DSC

21S07 VE/MH * MA/DSC
22S54 Rio De Janeiro
24S59 MA/MH * VE/ASC
25S18 MA/DSC * PL/ASC
26S12 Johannesburg
27S54 PL/IC * ME/ASC
28S21 SA/IC * MA/DSC

31S27 ME/DSC * SA/ASC
32S11 MA/MH * SA/DSC
33S27 PL/IC * SU/ASC
33S55 Sydney
34S36 Buenos Aires
35S00 ME/MH * MA/DSC
36S52 Auckland
37S50 Melbourne
38S28 JU/IC * MA/ASC

40S54 SU/ASC * ME/ASC
41S03 SA/IC * PL/ASC
45S02 SU/MH * MA/DSC
45S46 UR/DSC * NE/DSC
46S39 MA/MH * SU/ASC

SOUTH LATITUDES

48S45 PL/IC * MO/ASC
49S50 MA/MH * ME/ASC
49S55 JU/MH * UR/ASC

51S50 ME/DSC * VE/DSC
53S19 JU/MH * NE/ASC
53S33 PL/IC * UR/DSC
54S09 PL/IC * NE/DSC
54S27 MO/MH * MA/DSC
56S46 MA/MH * MO/ASC
58S11 MA/MH * UR/DSC
59S03 MA/MH * NE/DSC

PL 26 LIB 25

1. solar arc MO = A*C*G ME
2. solar arc MO EQU A*C*G PL

These two solar arc MO aspects activate the marriage chart A*C*G ME SQU A*C*G PL. This shows the emotional intensity in trying to communicate with each other, which led to emotional power struggles between them.

Now that the temporary influences have been examined the question about how to interpretate an Astro*Carto*Graphy marriage chart map can be explored.

The Astro*Carto*Graphy marriage chart map for Prince Charles and Princess Diana shows how the world saw their wedding. As both the wedding and married life took place in London, the Astro*Carto*Graphy marriage chart map also shows how things developed between them.

The combined influence of MO MH and MA MH shows the potential conflict that developed in their marriage. With PL ASC, SA ASC and JU ASC also having an influence there could be future problems in the outer image of their marriage. The temporary influences in effect at the time Prince Charles and Princess Diana separated indicated they would experience some problems within their marriage.

Although an Astro*Carto*Graphy marriage chart map with the temporary influences provides interesting information it shouldn't be used on its own in synastry. It is essential that the marriage chart map is considered with the other Astro*Carto*Graphy techniques.

On its own the marriage chart map temporary influences do not give strong indications for separation, but when the information from the various techniques is synthesised, it provides a clear picture.

Using Astro*Carto*Graphy in synastry is a new way of looking at how two people relate with each other. The location where a relationship is taking place is very important, but this fact is not usually considered by astrologers. The traditional methods of using composite and relationship charts do not solve the problem of how to calculate the angles because different methods are used in calculating the angles.

When Astro*Carto*Graphy is used, the relationship chart provides additional information without the problem of how to calculate the angles. It is also important to apply the temporary influences techniques to the Astro*Carto*Graphy relationship chart map.

Since I've started using the Astro*Carto*Graphy relationship chart map I do not need the composite chart. I've obtained accurate results with the Astro*Carto*Graphy relationship chart map, so I can highly recommend using it in synastry.

Chapter 9/Astro*Carto*Graphy and Mundane Astrology

One of the most interesting areas in which Astro*Carto*Graphy can be used is in the area of world events. Known as mundane astrology, it explains when events take place in the world. Various charts are used to examine why events happen or to predict when one will occur. Charts of nations and leaders are used with eclipses, lunations, ingresses, planetary cycles, etc. When examining a past event it is possible to see the relevant planetary influences to explain why it took place. It is more difficult to predict events that will take place in the future.

Transits to a chart of a nation and its leader give guidelines about how things will develop in that country, but more than that is needed to predict that an event will happen in a certain country. When eclipses, lunations and ingresses are examined it means that many charts need to be calculated, usually for the country's capital. Then they are examined with other horoscopes, resulting in a great deal of work. Many astrologers are put off by mundane astrology because too many charts are involved. This was my reaction until I attended Jim Lewis' seminar and started to look at mundane astrology from a different approach.

When an Astro*Carto*Graphy map is drawn for an eclipse, lunation or ingress it gives visual information about where the planets are angular. This obviously saves having to calculate many different charts. An Astro*Carto*Graphy map used in this way shows which areas of the world have significant planetary influences. From this information a select few charts can then be calculated. An event can happen anywhere in the world, so what better technique to use than Astro*Carto*Graphy.

The Astro*Carto*Graphy Source Book of Mundane Maps

This annual publication contains Astro*Carto*Graphy maps of ingresses, lunations and eclipses, and maps for the Mars, Jupiter and Saturn declination cycles. At the back of the book are maps for various events and, in older books, the president and vice president of the United States.

Many astrologers make a fatal mistake by examining one particular map in the book and trying to make predictions from it. Using the book in this way will inevitably lead mostly to erroneous predictions.

Once in a while the *Astro*Carto*Graphy Source Book of Mundane Maps* will explain why a major event occurred. In *The Astrology of the Macrocosm* Jim Lewis gives an example where the *Astro*Carto*Graphy Source Book of Mundane Maps* provided accurate information.

The gas refinery explosion in Mexico City (November 20, 1984) and the fatal gas leak from a chemical factory in Bhophal, India (December 3, 1984) are shown on three maps.

1. Eclipse Full Moon—November 8, 1984

Mexico City	PL MH, MA ASC
Bhophal	PL IC

2. Eclipse New Moon—November 22, 1984

Mexico City	MA MH, PL DSC
Bhophal	PL ASC, MA IC

3. Full Moon—December 8, 1984

Mexico City	MA IC, SA ASC
Bhophal	MA MH

This example shows that repetition of themes is important when using *The Astro*Carto*Graphy Source Book of Mundane Maps*. More interesting is how the book is used to make predictions, along with other charts.

Maps of Nations and Leaders

When an Astro*Carto*Graphy map is drawn of a nation or its leader it provides a worldwide view of the planetary influences experienced by that country. Often, the potential planetary energies on a nation's map will be triggered by those of a leader. This was the case with the United Kingdom and Margaret Thatcher.

The United Kingdom map has MA MH at the Falkland Islands. Margaret Thatcher has MA ASC at the Falkland Islands, thus activating the U.K. MA MH line.

Another example of a leader's map triggering that of a nation is Harry Truman and the United States. Truman's PL ASC line is near Hiroshima, which activates the U.S. Declaration of War PL MH there.

The use of the temporary influences techniques on Astro*Carto*Graphy maps of nations and leaders is startlingly accurate, but the main problem in using Astro*Carto*Graphy in this way is not knowing an exact time of birth. Astro*Carto*Graphy works best for modern nations where a reliable time is known.

The best reference source for charts of countries is *The Book of World Horoscopes* by Nicholas Campion. It contains some accurately timed horoscopes, but in the absence of this information many are calculated for 12:00 noon. Thus, seldom can an Astro*Carto*Graphy map be drawn for a nation; but a chart calculated for 12:00 noon can still be used in mundane astrology when examining world events and future trends.

The same problem is experienced in obtaining accurate birth data for the leader of a country. Unless an exact time of birth is known it is not possible to draw an Astro*Carto*Graphy map of a leader. It was only after Bill Clinton was elected president of the United States that his exact time of birth became available. Even now there

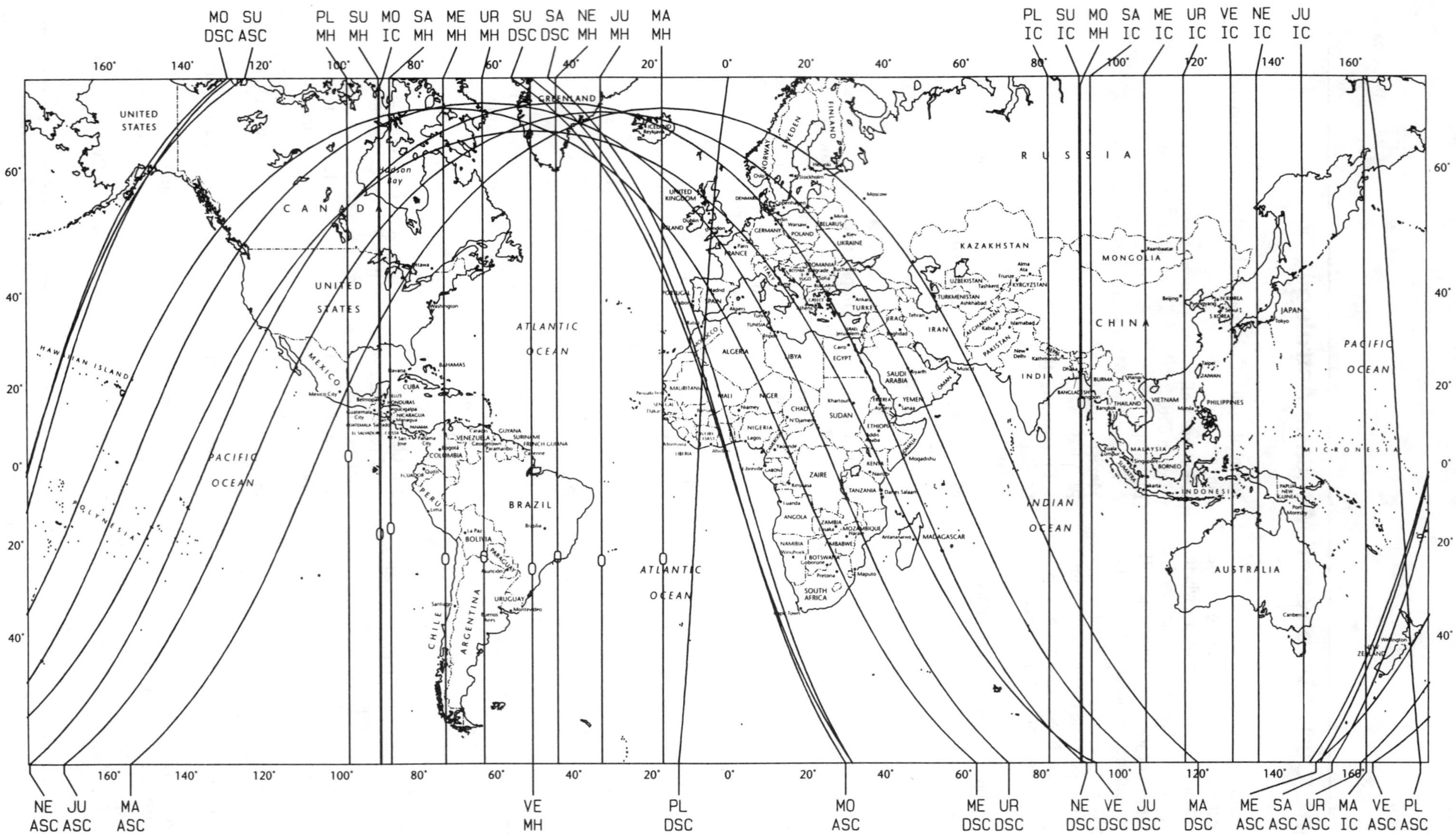

Eclipse Full Moon

November 8, 1984 5: 42 PM GMT

09/18/98

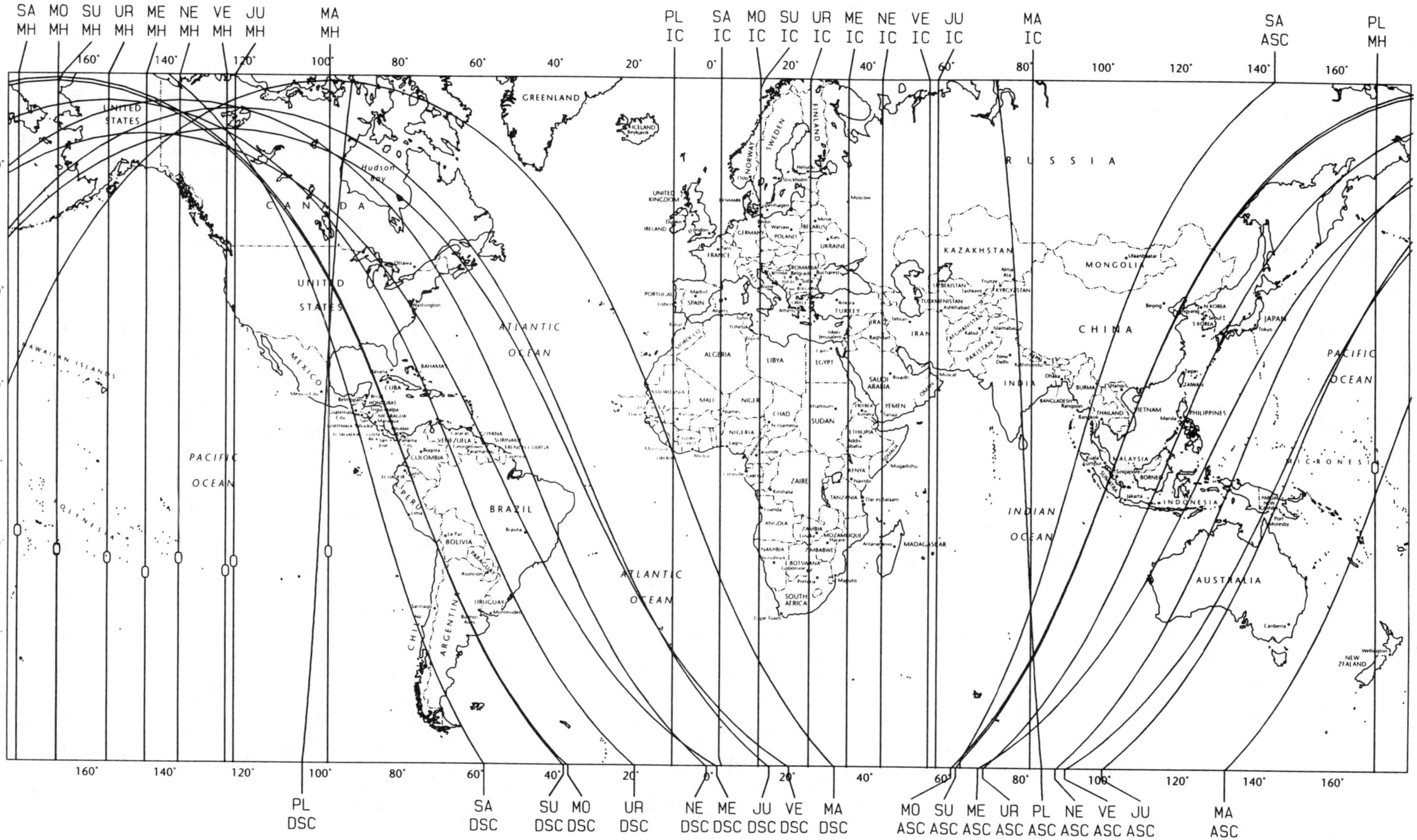

Eclipse New Moon

November 22, 1984 10:56 PM GMT

09/18/98

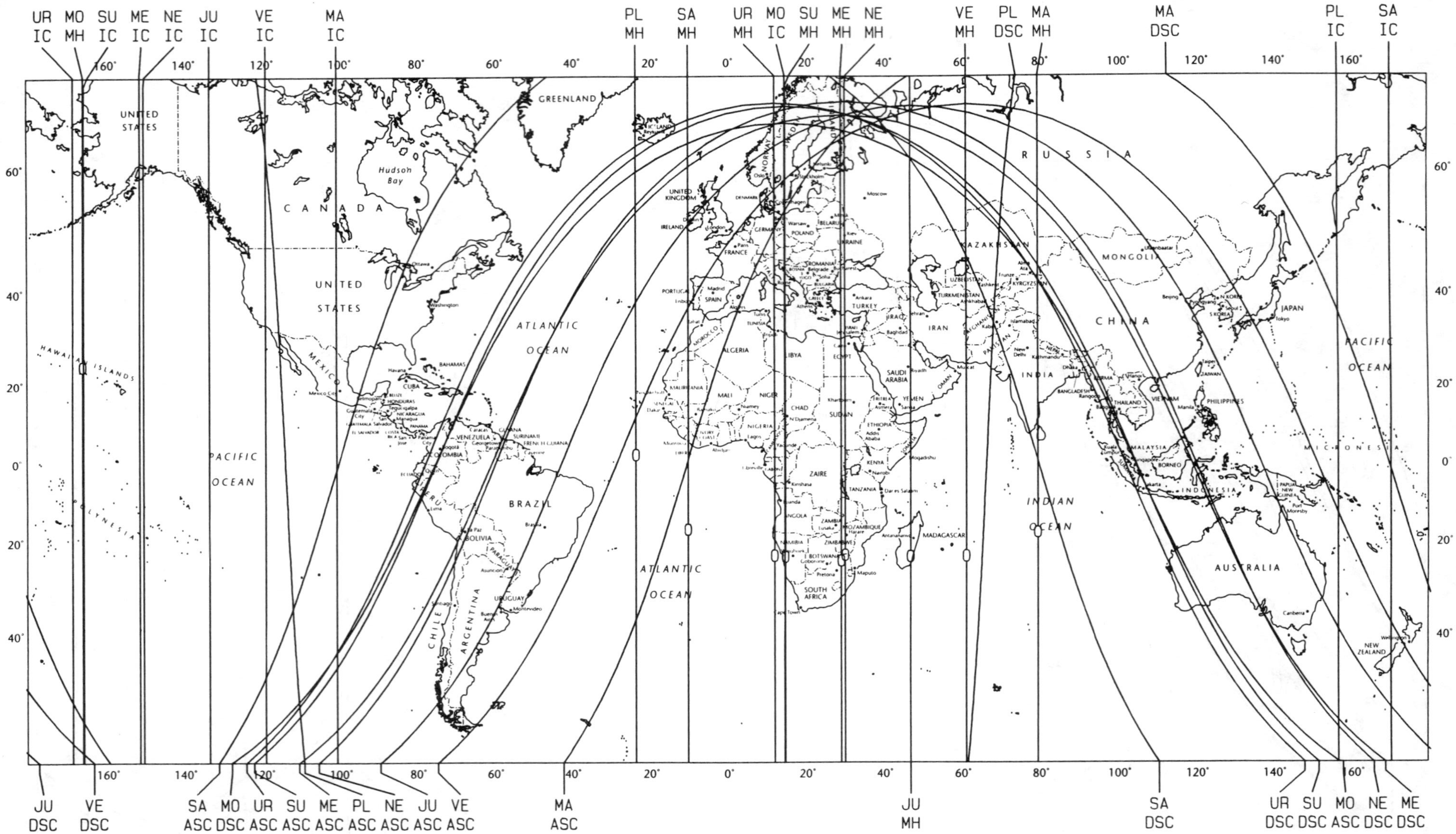

Full Moon

December 8, 1984 10:53 AM GMT

09/18/98

are some doubts about which birth time is the correct one for him.

When the leader's exact time of birth is unknown there is the possibility of using the inauguration chart. In the U.S. there is no problem in determining when a new president is inaugurated as it is a live event, but in other countries it is not so easy. Despite these limitations, there are times when it is possible to draw an Astro*Carto*Graphy map of a nation or leader.

Maps of Eclipses, Lunations and Ingresses

The use of eclipses, lunations and ingresses is explained in detail in Chapter 9 of *Mundane Astrology* by Baigent, Campion, Harvey.

These types of charts can be calculated for any location in the world and show the potential planetary influences for a period of time. Whether these potentials will manifest as events depends on what connections they make to the chart of the nation or leader concerned.

More importance is given to angular planets when examining charts of eclipses, lunations and ingresses. When these types of charts are drawn on an Astro*Carto*Graphy map the angular planets can be easily seen.

When examining an Astro*Carto*Graphy map of an eclipse, lunation or ingress it is essential that other charts be considered. This is especially true when looking for possible future events. My research has shown that eclipse maps seem to be the most effective, and lunations have also given relevant information to explain why certain events took place.

Disappointing were the ingress maps as they didn't seem to provide information on a consistent basis. Things are made more difficult as two types of zodiac can be used for ingresses.

*The Astro*Carto*Graphy Source Book of Mundane Maps* includes ingresses for both the tropical and sidereal zodiac, so there are eight different ingress charts that can be used.

1. Winter solstice
2. Vernal Equinox
3. Summer Solstice
4. Autumnal Equinox
5. Capsolar ingress
6. Arisolar ingress
7. Cansolar ingress
8. Libsolar ingress

The problem is knowing which ingress chart to use when looking ahead for the next twelve months. After an event has taken place it is easy to examine all eight ingress charts and find which one gives the best information.

Maps of Declination Cycle Charts

It was Joan Quigley who suggested that declination cycle charts be included in *The Astro*Carto*Graphy Source Book of Mundane Maps* because she finds them to be the most consistent and productive. This type of chart is calculated for when a planet moves from south to north declination. A declination cycle chart is used by examining the times when this planet makes transiting aspects. For example, any dates when Mars transits the other planets in its declination cycle chart should be examined in detail. If transiting Mars makes a conjunction to the Sun, then examine where the Sun lines are on the Mars declination cycle chart map. This will show the areas where a Mars type of event is likely to manifest.

My own research with Astro*Carto*Graphy declination cycle chart maps for Mars, Jupiter and Saturn has been disappointing. In the examination of many past events, the transiting cycle planet concerned didn't make the expected activation at the time the event took place.

Other Techniques

An interesting technique is using an Astro*Carto*Graphy map for the Mars/Pluto cycle, the exact time when Mars and Pluto form a conjunction or opposition with each other. I learned about this technique in an article by Charles Harvey in *New Astrologer* magazine. He wrote that he had used an Astro*Carto*Graphy map for the Mars/Pluto opposition at 5:14 p.m. on February 14, 1989, and as a result sent out warnings to some of the main fire departments and emergency services in the United Kingdom. He predicted that the February 12-14 would be a dangerous time with the possibility of violent storms. Also possible were oil and gas explosions, fires, accidents, assassinations and terrorist attacks.

On February 13, 1989, Britain experienced one of the strongest hurricanes ever recorded, which created a great amount of damage. As a result, the emergency services were overloaded and nearly broke down. This map showed that MA MH, PL IC and SU DSC had a powerful influence for Great Britain. After a natural storm had hit Britain a human one occurred a few hours later.

On February 14, 1989, the Ayatollah Khomeini sentenced Salmon Rushdie to death. Britain hadn't recovered from the destructive hurricane when it felt the effects of a different kind of violent energy. UR IC was exact at Tehran with SA IC and NE IC not too far away. This shows the sudden and revolutionary violence to traditional religion. A few days later the IRA destroyed army barracks in mainland Britain, the first terrorist attack in Britain in some time. The SU DSC line passes exactly through Market Drayton where the bomb attack occurred.

Charles Harvey has noticed that the effects of a MA/PL cycle chart start to have an effect a few days before the aspect is exact. It then continues to have an influence until the next cycle starts. Transits to the cycle chart show times when the potential influences are likely to be activated.

The MA/PL cycle is connected to the violent release of energy, sudden explosions and serious accidents. These kind of events are always taking place somewhere in the world, so an Astro*Carto*Graphy map of the MA/PL cycle has enormous potential.

Another example of using Astro*Carto*Graphy for a

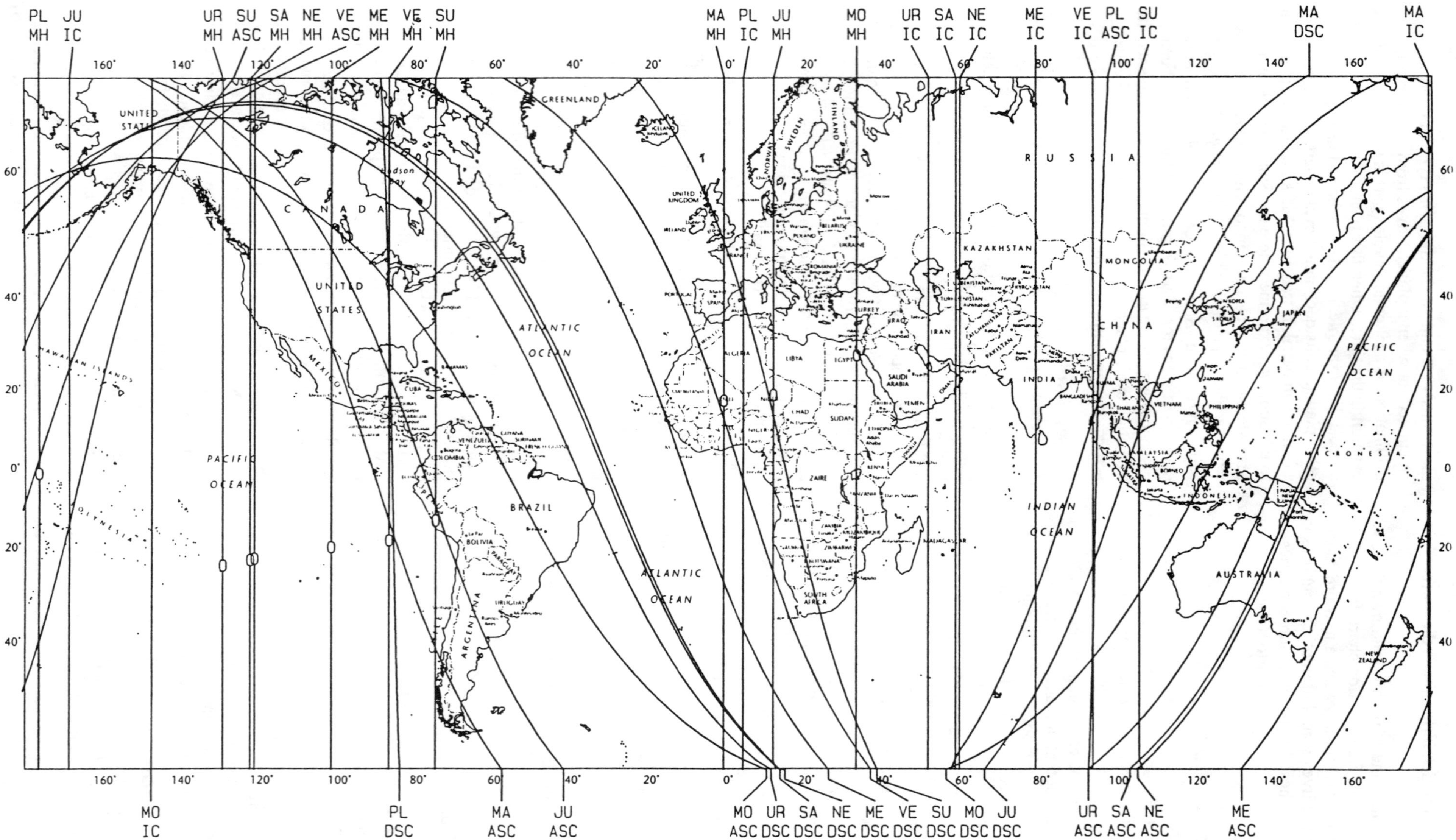

Mars Opp Pluto

February 14, 1989 5: 14 PM GMT

04/16/96

Mars Opp Pluto
February 14, 1989 5:14 PM GMT

ASTRO*CARTO*GRAPHY LATITUDE CROSSINGS

NORTH LATITUDES

69N54 MO/IC * JU/ASC
69N38 NE/MH * ME/ASC
69N30 SA/MH * ME/ASC
69N13 MO/IC * MA/ASC
69N01 NE/MH * VE/ASC
68N54 UR/MH * SU/ASC
68N46 SA/MH * VE/ASC
68N41 SU/MH * ME/DSC
68N29 UR/MH * ME/ASC
68N13 SU/ASC * ME/ASC
67N57 SA/MH * NE/ASC
67N54 VE/ASC * NE/ASC
67N46 UR/MH * NE/ASC
67N39 NE/MH * SA/DSC
67N37 SU/ASC * NE/ASC
67N36 VE/ASC * SA/ASC
67N31 UR/MH * SA/ASC
67N30 ME/ASC * JU/ASC
67N28 JU/ASC * NE/ASC
67N25 ME/ASC * NE/ASC
67N21 SU/ASC * SA/ASC
67N17 JU/ASC * SA/ASC
67N08 SU/ASC * JU/ASC
67N05 ME/ASC * SA/ASC
66N57 MA/ASC * NE/ASC
66N51 UR/MH * VE/ASC
66N49 MA/ASC * SA/ASC
66N45 ME/MH * NE/DSC
66N33 ME/ASC * MA/ASC
66N29 VE/ASC * JU/ASC
66N23 JU/ASC * UR/ASC
66N23 VE/ASC * UR/ASC
66N19 SU/ASC * UR/ASC
66N18 ME/MH * SA/DSC
66N15 SA/MH * UR/DSC
66N12 MA/ASC * UR/ASC
66N11 NE/MH * UR/DSC
66N09 UR/MH * JU/ASC
66N01 ME/ASC * UR/ASC
65N42 MO/IC * NE/ASC
65N35 MO/IC * SA/ASC
65N34 SU/ASC * MA/ASC
65N17 MO/IC * UR/ASC
65N04 VE/ASC * MA/ASC
64N28 SA/ASC * UR/ASC
64N19 NE/DSC * PL/DSC
64N15 UR/ASC * NE/ASC
63N56 VE/MH * NE/DSC
63N47 ME/MH * UR/DSC
63N43 SA/MH * JU/ASC
63N42 SA/DSC * PL/DSC
63N22 MO/IC * ME/ASC
63N18 VE/MH * SA/DSC
63N16 NE/MH * JU/ASC
62N48 UR/MH * MA/ASC
62N41 SU/ASC * VE/ASC
62N19 SA/ASC * NE/ASC
61N57 MO/DSC * ME/ASC
61N49 MO/ASC * VE/ASC
61N47 MO/ASC * SU/ASC
61N13 Anchorage
60N53 MO/DSC * NE/ASC
60N52 MO/DSC * SA/ASC
60N44 UR/MH * MO/ASC
60N29 JU/IC * UR/ASC
60N18 JU/IC * MO/DSC
60N07 MO/DSC * UR/ASC

59N58 UR/DSC * PL/DSC
59N56 MO/ASC * MA/ASC
59N50 SU/MH * NE/DSC
59N33 SA/MH * MO/ASC
59N32 VE/MH * UR/DSC
59N20 Stockholm

CENTRAL LATITUDES

59N20 NE/MH * MO/ASC
59N15 JU/IC * SA/ASC
59N08 JU/IC * NE/ASC
58N58 SU/MH * SA/DSC
58N43 PL/MH * MO/DSC
58N33 SA/MH * MA/ASC
57N44 NE/MH * MA/ASC
57N22 MA/IC * MO/DSC
57N19 MO/IC * VE/ASC
57N16 PL/MH * UR/ASC
55N45 Moscow
55N41 Copenhagen
55N37 MO/ASC * JU/ASC
55N02 PL/MH * SA/ASC
54N44 PL/MH * NE/ASC
54N33 MA/IC * UR/ASC
53N32 SU/MH * UR/DSC
53N28 MO/IC * SU/ASC
52N40 ME/MH * MO/ASC
52N30 Berlin
51N30 London
51N26 MA/IC * SA/ASC
50N58 MA/IC * NE/ASC
50N10 MA/MH * SU/DSC

49N14 Vancouver
49N00 ME/MH * JU/ASC
48N50 Paris
48N17 JU/IC * ME/ASC
47N36 Seattle
45N32 Montreal
44N59 Minneapolis
43N39 Toronto
43N07 MO/ASC * PL/DSC
42N56 VE/MH * MO/ASC
42N22 Boston
42N20 Detroit
41N54 Rome
41N53 Chicago
40N57 SU/IC * MA/DSC
40N43 New York

39N44 Denver
39N41 PL/IC * SU/DSC
38N53 Washington
38N37 St. Louis
37N47 San Francisco
37N11 PL/MH * ME/ASC
36N22 VE/MH * PL/DSC
36N10 Las Vegas
35N40 Tokyo
34N04 Los Angeles
33N45 Atlanta
33N27 Phoenix
32N47 Dallas
30N04 SU/MH * MO/ASC
30N03 Cairo

29N58 New Orleans
29N51 ME/MH * MA/ASC
29N46 Houston
28N39 Delhi
27N20 MA/IC * ME/ASC
25N47 Miami
25N02 JU/IC * VE/ASC
23N30 VE/MH * JU/ASC
22N38 JU/ASC * PL/DSC
22N34 Calcutta
22N16 Hong Kong
21N19 Honolulu

19N26 Mexico City
16N18 MA/DSC * PL/ASC
14N35 Manila
13N44 JU/MH * SU/DSC

SOUTH LATITUDES

12N56 VE/IC * MA/DSC
12N50 MA/MH * VE/DSC
10N30 Caracas

9N17 SU/IC * JU/DSC
6N27 Lagos
2N34 PL/MH * VE/ASC

2S34 PL/IC * VE/DSC
9S17 SU/MH * JU/ASC

12S50 MA/IC * VE/ASC
12S56 VE/MH * MA/ASC
13S44 JU/IC * SU/ASC
16S18 MA/ASC * PL/DSC

22S38 JU/DSC * PL/ASC
22S54 Rio De Janeiro
23S30 VE/IC * JU/DSC
25S02 JU/MH * VE/DSC
26S12 Johannesburg
27S20 MA/MH * ME/DSC
29S51 ME/IC * MA/DSC

30S04 SU/IC * MO/DSC
33S55 Sydney
34S36 Buenos Aires
36S22 VE/IC * PL/ASC
36S52 Auckland
37S11 PL/IC * ME/DSC
37S50 Melbourne
39S41 PL/MH * SU/ASC

40S57 SU/MH * MA/ASC
42S56 VE/IC * MO/DSC
43S07 MO/DSC * PL/ASC
48S17 JU/MH * ME/DSC
49S00 ME/IC * JU/DSC

50S10 MA/IC * SU/ASC
50S58 MA/MH * NE/DSC
51S26 MA/MH * SA/DSC
52S40 ME/IC * MO/DSC
53S28 MO/MH * SU/DSC
53S32 SU/IC * UR/ASC
54S33 MA/MH * UR/DSC
54S44 PL/IC * NE/DSC
55S02 PL/IC * SA/DSC
55S37 MO/DSC * JU/DSC
57S16 PL/IC * UR/DSC
57S19 MO/MH * VE/DSC
57S22 MA/MH * MO/ASC
57S44 NE/IC * MA/DSC
58S33 SA/IC * MA/DSC
58S43 PL/IC * MO/ASC
58S58 SU/IC * SA/ASC
59S08 JU/MH * NE/DSC
59S15 JU/MH * SA/DSC
59S20 NE/IC * MO/DSC
59S32 VE/IC * UR/ASC
59S33 SA/IC * MO/DSC
59S50 SU/IC * NE/ASC
59S56 MO/DSC * MA/DSC
59S58 UR/ASC * PL/ASC

MA/PL cycle chart is explained in *Working with Astrology* by Michael Harding and Charles Harvey.

A Mars/Pluto conjunction took place at 12:24 p.m. on December 11, 1987. An Astro*Carto*Graphy map drawn for this chart shows the Piper Alpha oil rig explosion in the North Sea. Michael Harding and Charles Harvey explained that UR MH and MO DSC pass exactly over the Piper Alpha oil field. The Mars/Pluto conjunction chart shows in which way the UR MH energy would manifest. UR is exactly semisquare the Mars/Pluto conjunction (orb two minutes). This indicates a destructive and sudden explosion. On the day the Piper Alpha oil rig exploded transiting Mars activated this destructive potential. Transiting Mars was exactly square UR conjunct MH and sesquiquadrate MA/PL.

In *The Astrology of the Macrocosm* Jim Lewis wrote a chapter about how to use Astro*Carto*Graphy in mundane astrology. On page 105 he explained the addition of other data to Astro*Carto*Graphy maps, a difficult concept to understand. On page 7 of *The Astro*Carto*Graphy Source Book of Mundane Maps* there is also an explanation about the addition of other data to the maps.

The technique involves "mundane square" aspects to the MH, IC, ASC and DSC. These four additional degree axes can be activated by transits and pinpoint when an event is likely to take place.

This technique is most effective when there is a repetition of themes, and mundane square aspects successfully explain why events happen on particular days.

This is how to find the four additional degree axes for the example described in the *Astro*Carto*Graphy Source Book of Mundane Maps*: The latitude for San Francisco is 37N47. In *Raphael's Tables of Houses for Northern Latitudes* the positions are given for Athens (37N58), which is nearest San Francisco. The positions for Athens are then used to find the four additional degree axes for San Francisco.

Step 1: Find mundane square to MH.

Look at tenth cusp for 6 VIR.

When 6 VIR is the Midheaven, then approx **25 SCO** is on the Ascendant.

Step 2: Find mundane square to ASC.

Look at Ascendant column for 6 VIR.

When 6 VIR is the Ascendant, then approx **3 GEM** is on the Midheaven.

Step 3: Find mundane square to IC.

Look at the tenth cusp for 6 PIS (IC is 06 VIR).

When 6 VIR is on the IC, then approximately **28 GEM** is on the Ascendant.

Step 4: Find mundane square to DSC.

Look at the Ascendant column for 6 PIS (DSC is 6 VIR).

When 6 VIR is the Descendant, then approx **16 SAG** is on the Midheaven.

The four additional degree axes for Saturn at San Francisco are:

a. 1st degree axis for Saturn = 25 SCO/TAU

b. 2nd degree axis for Saturn = 03 GEM/SAG

c. 3rd degree axis for Saturn = 28 GEM/SAG

d. 4th degree axis for Saturn = 16 SAG/GEM

To calculate the four additional degree axes accurately the right ascension of the planets is used, but a tables of houses will give reasonably accurate information.

My experience with this technique has shown that the transiting **Sun** is the planet to use for timing an event. When the Sun is conjunct one of these additional axes an event is likely to take place.

This Astro*Carto*Graphy technique is only used in mundane astrology. If you are unsure how mundane squares work in timing an event the following example will clarify the technique.

Example 1—Fall of Berlin Wall

The opening of the Berlin Wall on November 9, 1989 was an unbelievable event that many people thought they would never see. Charles Harvey provided the exact time when East Germany was founded (source: *The Fate of the German Nation* by Reinhold Ebertin): 1:45 p.m., October 7, 1949, East Berlin. More difficult was the exact time of birth for the Berlin Wall because various times had been quoted. *The Berlin Wall* by Norman Gelb quotes 1:11 a.m. (pages 144-145, the time that the East German news agency, ADN, sent a communication to Reuters. This was the official announcement that the East German authorities were taking action to safeguard their security. Another book, *Die Mauer* by Jürgen Petschull, cites the same time: Berlin Wall, 1.11 a.m., August 13, 1961, East Berlin (page 47).

Although Erich Honecker was no longer the East German leader on November 9, 1989, his Astro*Carto*Graphy map shows important information. After eighteen years, he resigned his position on October 18, 1989. Honecker was responsible for building the Berlin Wall, and as his exact time of birth is known an Astro*Carto*Graphy map can be drawn: Erich Honecker, 6:00 p.m., August 25, 1912, Neunkirchen. Egon Krenz succeeded him as the East German leader, but only his date of birth is known (Astrologie Heute Nr. 22). It is doubtful whether his chart would have much effect as there was nothing he could do to stop the changes taking place.

East Germany Astro*Carto*Graphy Map

The UR DSC line that passes through Berlin explains a divided city. UR DSC also passes through Bonn (capital of West Germany), confirming that East Germany was its own separate country.

Close latitude crossings (within one degree orb) at Berlin are:

Crossings: SA DSC/PL DSC
UR IC/SU DSC

SA/PL had more of an influence after the Berlin Wall was built, as there was grave danger for people tying to escape to West Berlin. SU/UR emphasizes the influence of UR DSC. Before and after the Berlin Wall was built people

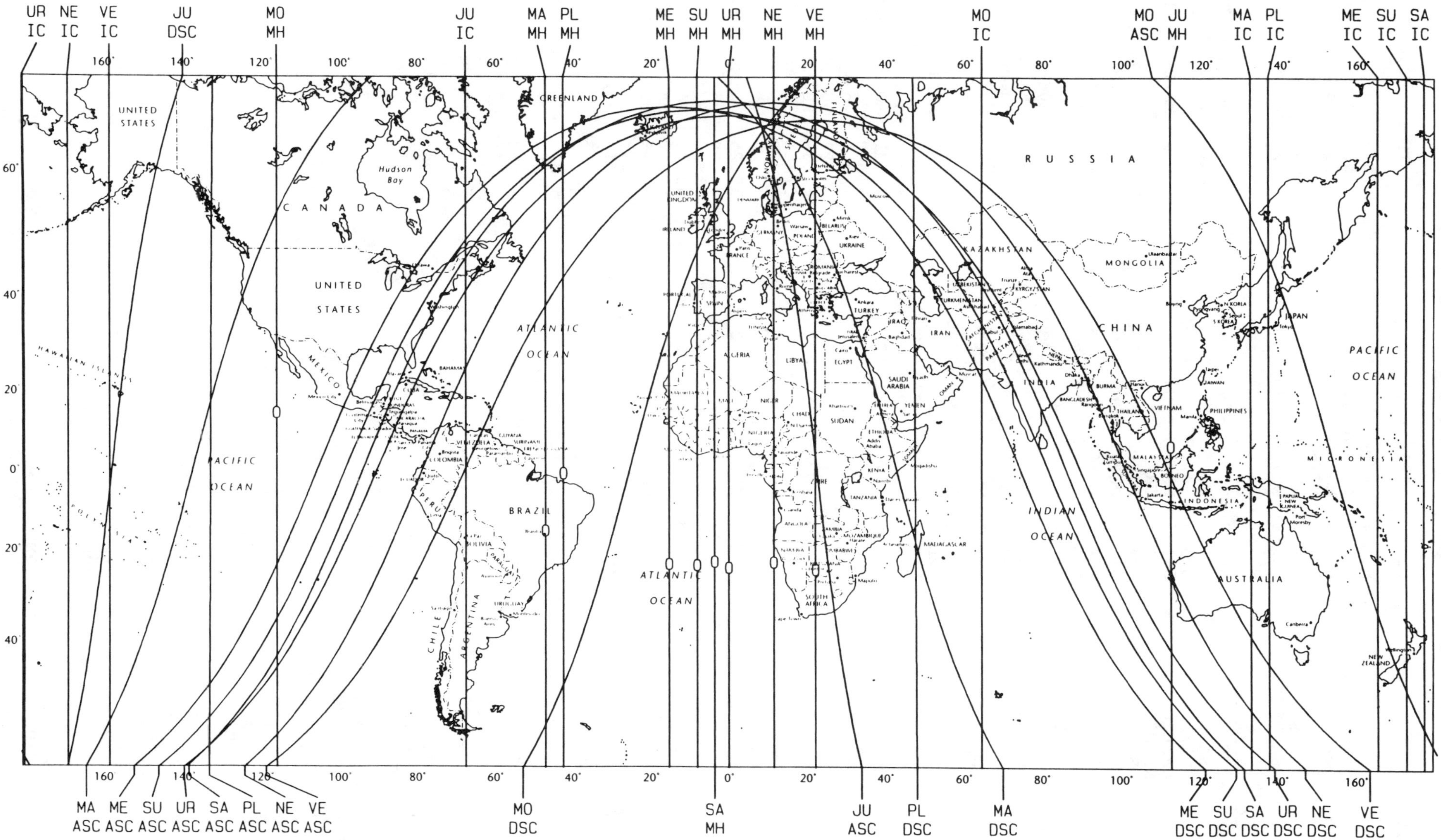

Mars Cjn Pluto

December 11, 1987 12:24 PM GMT

04/16/96

04/16/96

Mars Cjn Pluto
December 11, 1987 12:24 PM GMT

ASTRO*CARTO*GRAPHY LATITUDE CROSSINGS

NORTH LATITUDES

69N05	UR/MH * MA/DSC
68N24	JU/MH * MO/ASC
67N50	UR/MH * SA/DSC
67N49	SU/MH * SA/ASC
67N43	MA/DSC * SA/DSC
67N41	MO/DSC * NE/DSC
67N38	JU/ASC * NE/ASC
67N36	SA/DSC * NE/ASC
67N32	JU/ASC * SA/DSC
67N32	MA/DSC * NE/ASC
67N27	ME/MH * SA/ASC
67N21	VE/MH * NE/DSC
67N19	ME/ASC * SA/ASC
67N18	UR/MH * NE/ASC
67N10	NE/MH * SA/DSC
67N10	SU/MH * ME/DSC
67N01	MO/DSC * SA/DSC
66N59	SA/MH * NE/ASC
66N59	SU/DSC * ME/DSC
66N58	SU/DSC * NE/ASC
66N57	SA/MH * SU/DSC
66N56	ME/DSC * NE/ASC
66N54	SA/MH * ME/DSC
66N52	ME/MH * SU/ASC
66N49	UR/MH * SU/DSC
66N35	UR/MH * ME/DSC
66N28	SU/MH * NE/ASC
66N26	ME/DSC * UR/DSC
66N24	SA/MH * UR/ASC
66N20	SU/DSC * MA/DSC
66N19	NE/MH * MO/DSC
66N18	MA/DSC * UR/DSC
66N17	SU/DSC * UR/DSC
66N15	JU/ASC * UR/DSC
66N14	SU/MH * UR/ASC
66N12	SU/DSC * JU/ASC
66N03	MO/DSC * UR/DSC
66N02	UR/ASC * NE/ASC
66N01	NE/MH * UR/DSC
66N00	VE/ASC * SA/DSC
65N52	MO/DSC * SU/DSC
65N47	VE/ASC * UR/DSC
65N46	NE/MH * SU/DSC
65N42	VE/MH * SA/DSC
65N41	ME/MH * UR/ASC
65N40	SU/DSC * VE/ASC
65N37	NE/MH * VE/ASC
65N32	ME/DSC * MA/DSC
65N30	MO/DSC * VE/ASC
65N29	ME/DSC * JU/ASC
65N25	ME/DSC * VE/ASC
65N22	VE/ASC * JU/ASC
65N22	VE/ASC * MA/DSC
65N20	ME/MH * NE/ASC
65N19	MA/DSC * JU/ASC
65N17	MO/DSC * ME/DSC
65N01	MO/DSC * MA/DSC
64N55	NE/MH * ME/DSC
64N53	VE/DSC * NE/DSC
64N53	MO/DSC * JU/ASC
64N52	PL/MH * ME/ASC
64N49	SU/ASC * SA/ASC
64N47	VE/MH * UR/DSC
64N19	UR/MH * VE/ASC
63N51	MA/MH * ME/ASC
63N50	VE/DSC * PL/DSC
63N50	VE/MH * SU/DSC
63N41	SA/MH * VE/ASC
63N30	NE/MH * MA/DSC
63N12	NE/DSC * PL/DSC
62N52	PL/MH * SU/ASC
62N42	SU/MH * VE/ASC
62N28	PL/MH * SA/ASC
62N14	VE/MH * ME/DSC

CENTRAL LATITUDES

61N27	MA/MH * SU/ASC
61N13	Anchorage
60N48	MA/MH * SA/ASC
60N43	ME/MH * VE/ASC
59N49	NE/MH * JU/ASC
59N28	PL/MH * UR/ASC
59N20	Stockholm
58N57	MO/IC * VE/DSC
58N29	UR/MH * MO/DSC
57N47	SA/DSC * UR/DSC
57N26	MA/MH * UR/ASC
57N15	UR/DSC * PL/DSC
57N11	SA/DSC * PL/DSC
55N48	JU/IC * ME/ASC
55N45	Moscow
55N43	MO/IC * NE/DSC
55N41	Copenhagen
55N13	PL/MH * NE/ASC
54N43	SA/MH * MO/DSC
54N28	MO/MH * MA/ASC
54N10	MA/IC * MO/ASC
53N45	VE/MH * MA/DSC
53N10	SU/DSC * PL/DSC
52N30	Berlin
51N56	MA/MH * NE/ASC
51N30	London
50N25	JU/IC * SU/ASC
49N14	Vancouver
48N50	Paris
48N29	SU/MH * MO/DSC
47N42	ME/DSC * PL/DSC
47N39	PL/IC * MO/ASC
47N36	Seattle
47N27	JU/IC * SA/ASC
45N32	Montreal
44N59	Minneapolis
44N38	MO/IC * UR/DSC
44N00	PL/MH * VE/ASC
43N39	Toronto
42N31	MO/IC * SA/DSC
42N22	Boston
42N20	Detroit
41N54	Rome
41N53	Chicago
41N31	JU/IC * UR/ASC
40N43	New York
39N44	Denver
38N53	Washington
38N39	MA/MH * VE/ASC
38N37	St. Louis
37N47	San Francisco
36N10	Las Vegas
35N40	Tokyo
35N06	MO/IC * SU/DSC
34N04	Los Angeles
33N59	ME/MH * MO/DSC
33N45	Atlanta
33N27	Phoenix
32N47	Dallas
30N03	Cairo
29N58	New Orleans
29N46	Houston
28N39	Delhi
25N47	Miami
25N22	JU/IC * NE/ASC
23N03	MO/IC * ME/DSC
22N34	Calcutta
22N16	Hong Kong
21N19	Honolulu
19N26	Mexico City

SOUTH LATITUDES

17N22	MA/ASC * PL/ASC
14N35	Manila
10N30	Caracas
6N27	Lagos
4N20	VE/MH * JU/ASC
1N06	JU/IC * VE/ASC
1S06	JU/MH * VE/DSC
4S20	VE/IC * JU/DSC
17S22	MA/DSC * PL/DSC
22S54	Rio De Janeiro
23S03	MO/MH * ME/ASC
25S22	JU/MH * NE/DSC
26S12	Johannesburg
33S55	Sydney
33S59	ME/IC * MO/ASC
34S36	Buenos Aires
35S06	MO/MH * SU/ASC
36S52	Auckland
37S50	Melbourne
38S39	MA/IC * VE/DSC
41S31	JU/MH * UR/DSC
42S31	MO/MH * SA/ASC
44S00	PL/IC * VE/DSC
44S38	MO/MH * UR/ASC
47S27	JU/MH * SA/DSC
47S39	PL/MH * MO/DSC
47S42	ME/ASC * PL/ASC
48S29	SU/IC * MO/ASC
50S25	JU/MH * SU/DSC
51S56	MA/IC * NE/DSC
53S10	SU/ASC * PL/ASC
53S45	VE/IC * MA/ASC
54S10	MA/MH * MO/DSC
54S28	MO/IC * MA/DSC
54S43	SA/IC * MO/ASC
55S13	PL/IC * NE/DSC
55S43	MO/MH * NE/ASC
55S48	JU/MH * ME/DSC
57S11	SA/ASC * PL/ASC
57S15	UR/ASC * PL/ASC
57S26	MA/IC * UR/DSC
57S47	SA/ASC * UR/ASC
58S29	UR/IC * MO/ASC
58S57	MO/MH * VE/ASC
59S28	PL/IC * UR/DSC
59S49	NE/IC * JU/DSC

Lunar Eclipse
September 16, 1978
San Francisco, California
37N47, 122W25

1

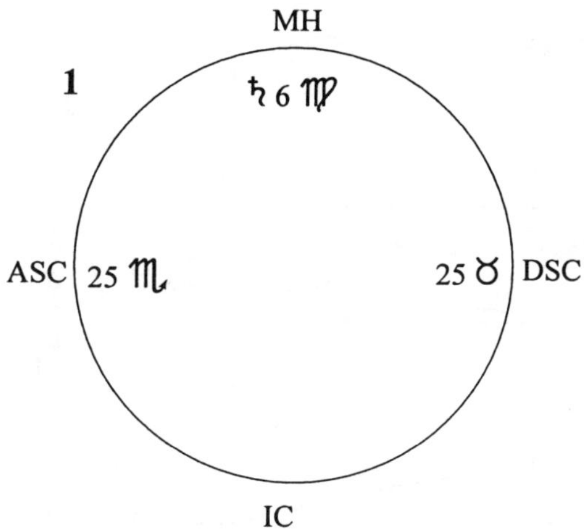

2

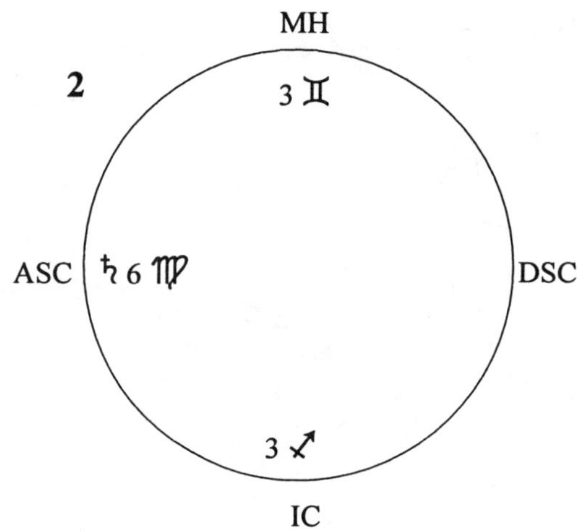

3

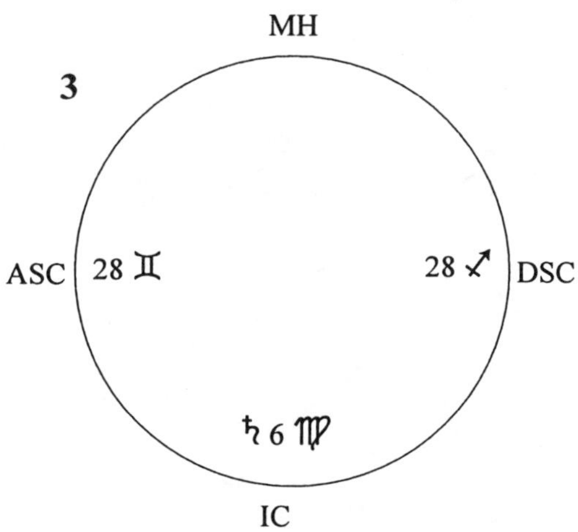

4

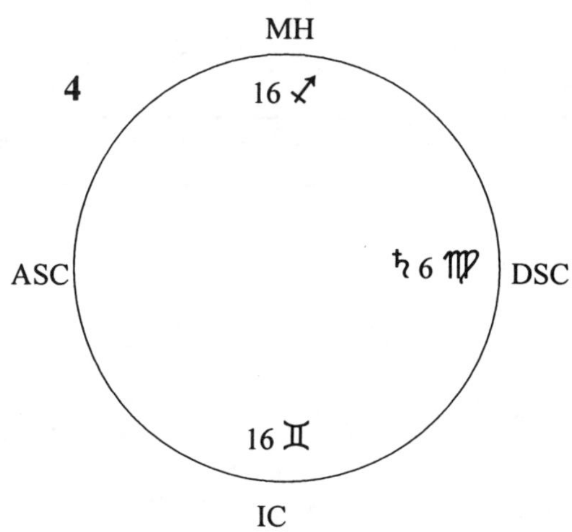

1st degree axis = 25 ♏/25 ♉
2nd degree axis = 3 ♊/3 ♐
3rd degree axis = 28 ♊/28 ♐
4th degree axis = 16 ♐/16 ♊

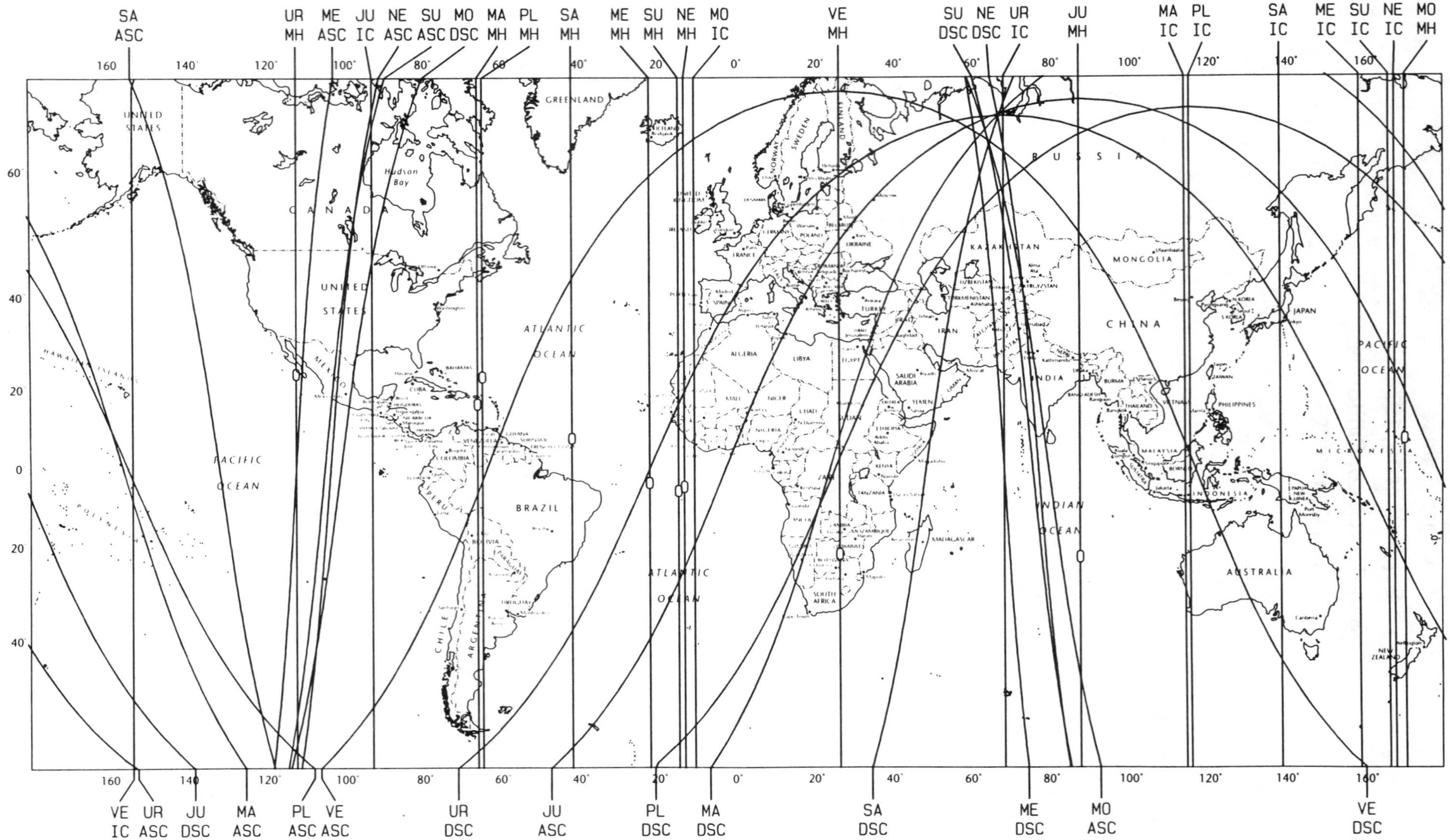

East Germany

October 7, 1949 1:45 PM CET

04/16/96

East Germany
October 7, 1949 1:45 PM CET

ASTRO*CARTO*GRAPHY LATITUDE CROSSINGS

NORTH LATITUDES

69N36 VE/IC * SA/ASC
67N38 ME/IC * MA/ASC
67N36 UR/IC * SA/DSC
67N04 MA/IC * PL/DSC
66N58 JU/IC * NE/ASC
66N55 UR/IC * MA/DSC
66N49 JU/DSC * PL/DSC
66N46 UR/IC * JU/ASC
66N41 MA/DSC * JU/ASC
66N38 JU/ASC * SA/DSC
66N34 JU/ASC * NE/DSC
66N31 MA/DSC * SA/DSC
66N24 SA/DSC * UR/DSC
66N24 MA/DSC * UR/DSC
66N24 UR/DSC * NE/DSC
66N23 JU/ASC * UR/DSC
66N22 MA/DSC * NE/DSC
66N19 SU/DSC * UR/DSC
66N18 MO/ASC * UR/DSC
66N12 SA/DSC * NE/DSC
66N12 ME/DSC * UR/DSC
66N00 SU/DSC * JU/ASC
65N57 MO/ASC * JU/ASC
65N56 VE/DSC * UR/DSC
65N37 MA/IC * JU/DSC
65N33 ME/DSC * JU/ASC
65N21 PL/IC * JU/DSC
65N19 SA/IC * PL/ASC
65N15 VE/DSC * JU/ASC
65N14 JU/IC * SU/ASC
65N09 JU/MH * UR/ASC
65N00 SU/DSC * MA/DSC
64N59 MO/ASC * MA/DSC
64N50 UR/ASC * PL/DSC
64N49 ME/DSC * VE/DSC
64N39 SU/IC * MA/ASC
64N32 MO/ASC * SU/DSC
64N24 JU/MH * PL/DSC
64N24 VE/DSC * MA/DSC
64N02 MO/IC * VE/ASC
63N54 NE/IC * MA/ASC
63N52 SU/DSC * VE/DSC
63N49 MO/ASC * VE/DSC
63N47 ME/DSC * MA/DSC
63N41 VE/DSC * SA/DSC
63N27 MO/ASC * SA/DSC
63N17 SU/DSC * SA/DSC
63N14 NE/MH * VE/ASC
63N08 UR/IC * NE/DSC
62N43 SU/MH * VE/ASC
62N32 UR/IC * VE/DSC
62N28 MO/MH * MA/ASC
62N27 VE/DSC * NE/DSC
61N13 Anchorage
60N21 VE/DSC * PL/DSC
60N05 ME/IC * PL/ASC

59N43 ME/DSC * SA/DSC
59N43 VE/MH * UR/DSC
59N39 ME/MH * VE/ASC
59N20 Stockholm
58N44 NE/DSC * PL/DSC
58N02 UR/IC * PL/DSC
57N50 MA/IC * UR/ASC
57N49 MO/ASC * PL/DSC
57N23 SU/DSC * PL/DSC
57N15 SA/IC * JU/DSC
57N12 PL/IC * UR/ASC
56N53 UR/IC * MO/ASC
56N39 SU/IC * PL/ASC
55N48 NE/IC * PL/ASC
55N45 Moscow
55N41 Copenhagen
55N19 ME/DSC * PL/DSC

CENTRAL LATITUDES

54N29 SU/ASC * NE/ASC
54N13 MO/MH * PL/ASC
53N01 UR/IC * SU/DSC
52N30 Berlin
52N28 SA/DSC * PL/DSC
51N30 London
51N20 JU/MH * VE/DSC
50N10 VE/MH * JU/ASC

49N14 Vancouver
48N50 Paris
47N36 Seattle
47N01 JU/IC * MO/DSC
45N32 Montreal
44N59 Minneapolis
43N50 SA/MH * VE/ASC
43N39 Toronto
42N22 Boston
42N20 Detroit
41N54 Rome
41N53 Chicago
41N10 MO/ASC * NE/DSC
40N43 New York

39N44 Denver
39N16 ME/IC * JU/DSC
38N53 Washington
38N37 St. Louis
37N47 San Francisco
37N37 SA/IC * UR/ASC
36N10 Las Vegas
35N40 Tokyo
34N04 Los Angeles
33N45 Atlanta
33N27 Phoenix
32N47 Dallas
30N03 Cairo

29N58 New Orleans
29N46 Houston
28N39 Delhi
27N13 SU/IC * JU/DSC
25N47 Miami
24N21 MO/IC * UR/DSC
24N17 NE/IC * JU/DSC
22N34 Calcutta
22N16 Hong Kong
21N19 Honolulu

19N26 Mexico City
19N21 NE/MH * UR/DSC
18N52 MO/MH * JU/DSC
16N17 SU/MH * UR/DSC
14N35 Manila
10N30 Caracas
10N14 MA/ASC * PL/ASC

8N05 VE/MH * MA/DSC
6N27 Lagos
6N10 MA/IC * VE/DSC
2N48 PL/IC * VE/DSC
2N35 VE/MH * PL/DSC
0N11 UR/IC * ME/DSC
0N01 ME/MH * UR/DSC

0S01 ME/IC * UR/ASC
0S11 UR/MH * ME/ASC
2S35 VE/IC * PL/ASC
2S48 PL/MH * VE/ASC
6S10 MA/MH * VE/ASC
8S05 VE/IC * MA/ASC

10S14 MA/DSC * PL/DSC
16S17 SU/IC * UR/ASC
18S52 MO/IC * JU/ASC

SOUTH LATITUDES

19S21 NE/IC * UR/ASC

22S54 Rio De Janeiro
24S17 NE/MH * JU/ASC
24S21 MO/MH * UR/ASC
26S12 Johannesburg
27S13 SU/MH * JU/ASC

33S55 Sydney
34S36 Buenos Aires
36S52 Auckland
37S37 SA/MH * UR/DSC
37S50 Melbourne
39S16 ME/MH * JU/ASC

41S10 MO/DSC * NE/ASC
43S50 SA/IC * VE/DSC
47S01 JU/MH * MO/ASC

50S10 VE/IC * JU/DSC
51S20 JU/IC * VE/ASC
52S28 SA/ASC * PL/ASC
53S01 UR/MH * SU/ASC
54S13 MO/IC * PL/DSC
54S29 SU/DSC * NE/DSC
55S19 ME/ASC * PL/ASC
55S48 NE/MH * PL/DSC
56S39 SU/MH * PL/DSC
56S53 UR/MH * MO/DSC
57S12 PL/MH * UR/DSC
57S15 SA/MH * JU/ASC
57S23 SU/ASC * PL/ASC
57S49 MO/DSC * PL/ASC
57S50 MA/MH * UR/DSC
58S02 UR/MH * PL/ASC
58S44 NE/ASC * PL/ASC
59S39 ME/IC * VE/DSC
59S43 VE/IC * UR/ASC
59S43 ME/ASC * SA/ASC

had a strong desire for freedom in restrictive East Germany.

JU ASC is at Moscow, capital of the former USSR, with which East Germany had a good relationship. MA IC and PL IC are at Beijing, China, and East Germany was one of the few countries that supported the Chinese government's brutal action in Tiananmen Square. When the East German people demonstrated in the streets the government threatened to use "Tiananmen Square" action to bring things back to order.

The temporary influences of Cyclo*Carto*Graphy show the events leading up to the opening of the Berlin Wall:

A*C*G	Age 40	C*C*G
UR 05 CAN 28		MO 02 LIB 55
ME 05 LIB 29		UR 03 CAP 06
SU 12 LIB 48		SA 10 CAP 53
NE 14 LIB 18		NE 11 CAP 06
SA 15 VIR 48		JU 11 CAN 33
MO 16 ARI 53		VE 12 CAP 19
MA 21 LEO 29		MA 13 VIR 28
PL 22 LEO 48		PL 17 SCO 07
VE 23 SCO 54		ME 18 SCO 28
JU 24 CAP 50		SU 21 SCO 33

1. C*C*G VE EQU A*C*G SU
2. C*C*G SU EQU A*C*G MA
3. C*C*G UR = Berlin + West/East German border
4. C*C*G SA ASC = (a) East German/Czechoslovakian border
5. C*C*G JU DSC = (b) Prague
6. C*C*G NE ASC = (c) Czechoslovakian/West German border
7. C*C*G PL MH = (a) Warsaw
8. C*C*G ME MH = (b) Budapest
9. C*C*G VE ASC = Hungarian/Austrian border

A*C*G	(38 45)	Displaced C*C*G
UR 05 CAN 28		SA 02 SAG 08
ME 05 LIB 29		NE 02 SAG 21
SU 12 LIB 48		JU 02 GEM 48
NE 14 LIB 18		VE 03 SAG 34
SA 15 SCO 48		MA 04 LEO 43
MO 16 ARI 53		PL 08 LIB 22
MA 21 LEO 29		ME 09 LIB 43
PL 22 LEO 48		MO 24 LEO 10
VE 23 SCO 54		UR 24 SCO 21
JU 24 CAP 50		

10. Disp. C*C*G UR = A*C*G VE

A*C*G	(38 45)	Solar arc A*C*G
UR 05 CAN 28		MA 0 LIB 14
ME 05 LIB 29		PL 01 LIB 33
SU 12 LIB 48		VE 02 CAP 39
NE 14 LIB 18		JU 03 PIS 35
SA 15 SCO 48		UR 14 AQU 13
MO 16 ARI 53		ME 14 SCO 14
MA 21 LEO 29		NE 23 SCO 03
PL 22 LEO 48		SA 24 SAG 33
VE 23 SCO 54		MO 25 TAU 38
JU 24 CAP 50		

11. solar arc A*C*G NE EQU A*C*G PL
12. solar arc A*C*G NE = A*C*G VE

C*C*G UR ASC passes through the West German/East German border and is near Berlin. This activated the A*C*G UR DSC at Berlin which resulted in demonstrations for reform and more freedom. It also shows the opening of the borders with West Germany and the sweeping changes forced upon the East German government.

C*C*G SA ASC, C*C*G NE ASC and C*C*G JU DSC pass through the East German/Czechoslovakian border and Prague. East Germans had found an escape route to West Germany via Czechoslovakia and Prague, first by occupying the West German embassy in Prague and then travelling through Czechoslovakia to West Germany. C*C*G SA ASC and C*C*G NE ASC show the dissolving of barriers and restrictions which thousands of East Germans took advantage of (C*C*G JU DSC).

C*C*G PL MH and C*C*G ME MH passed through Budapest and Warsaw; Hungary and Poland also offered an escape route for East Germans. In Hungary leaflets were given to East German holiday-makers explaining how they could escape to West Germany via Austria. The opening of the border between Hungary and Austria made East Germans realize they could flee to West Germany.

C*C*G VE ASC passes through the Hungarian/Austrian border where thousands of East Germans experienced a happy ending to the years of a restricted lifestyle.

C*C*G SU EQU A*C*G MA is appropriate for the battle the government had in trying to stop the sweeping changes taking place. There were also disagreements and disputes within the government.

The solar return map for the year the Berlin Wall was opened had PL DSC within orb of East Germany. This PL DSC line passed through Eastern Europe where it had a strong effect in other Communist countries.

Gorbachev was in East Germany to celebrate the country's fortieth anniversary, and he urged the East German government to introduce reforms in what was a changing world. At Moscow the solar return map had JU ASC, NE DSC, SA DSC, MO DSC and UR DSC. These planets were in square aspect to SU CJN MA, thus forming a powerful T-square. These planetary influences are most appropriate for what Gorbachev told Honecker during the fortieth anniversary celebrations.

Berlin Wall Astro*Carto*Graphy Map

At first glance it seems inappropriate as MO IC and VE ASC are close to Berlin, but the Berlin Wall was a work of art on the West Berlin side, where artists had the opportunity to create their own paintings. In *The Berlin Wall Book*, published by Thames and Hudson, there are more than 100 color photos of the Wall.

PL IC is also within orb of Berlin, where people risked their lives to get to the other side. The Berlin Wall created upheaval in many families.

The Cyclo*Carto*Graphy temporary influences for the Berlin Wall map were also most impressive for November

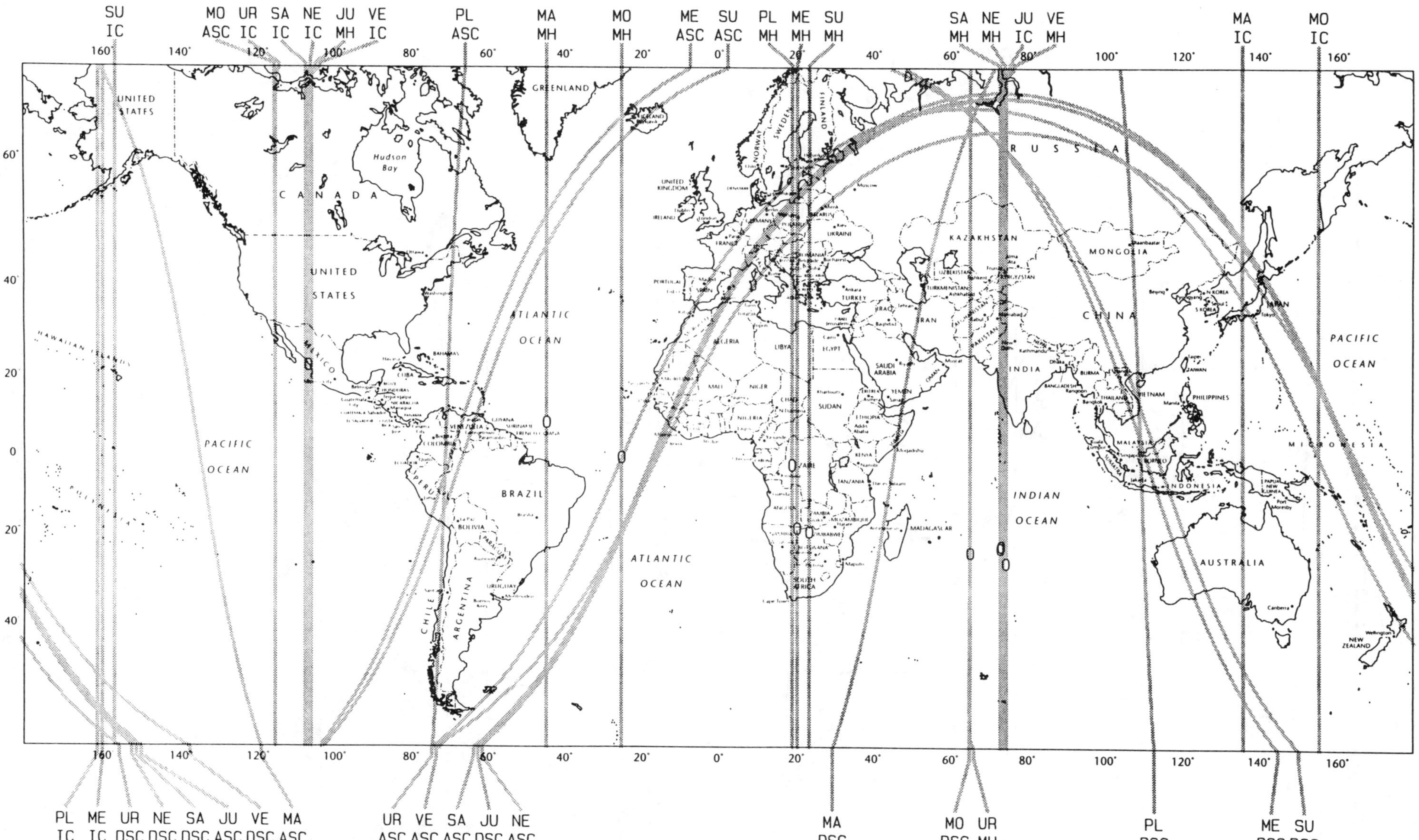

CYCLO*CARTO*GRAPHY

04/17/96

East Germany

SU, MO, ME, VE AND MA ARE PROGRESSED
JU, SA, UR, NE AND PL ARE TRANSITING

PROG	11	16	49	14:57 GMT
TRANSIT	11	9	89	0:00 GMT
NATAL	10	7	49	12:45 GMT

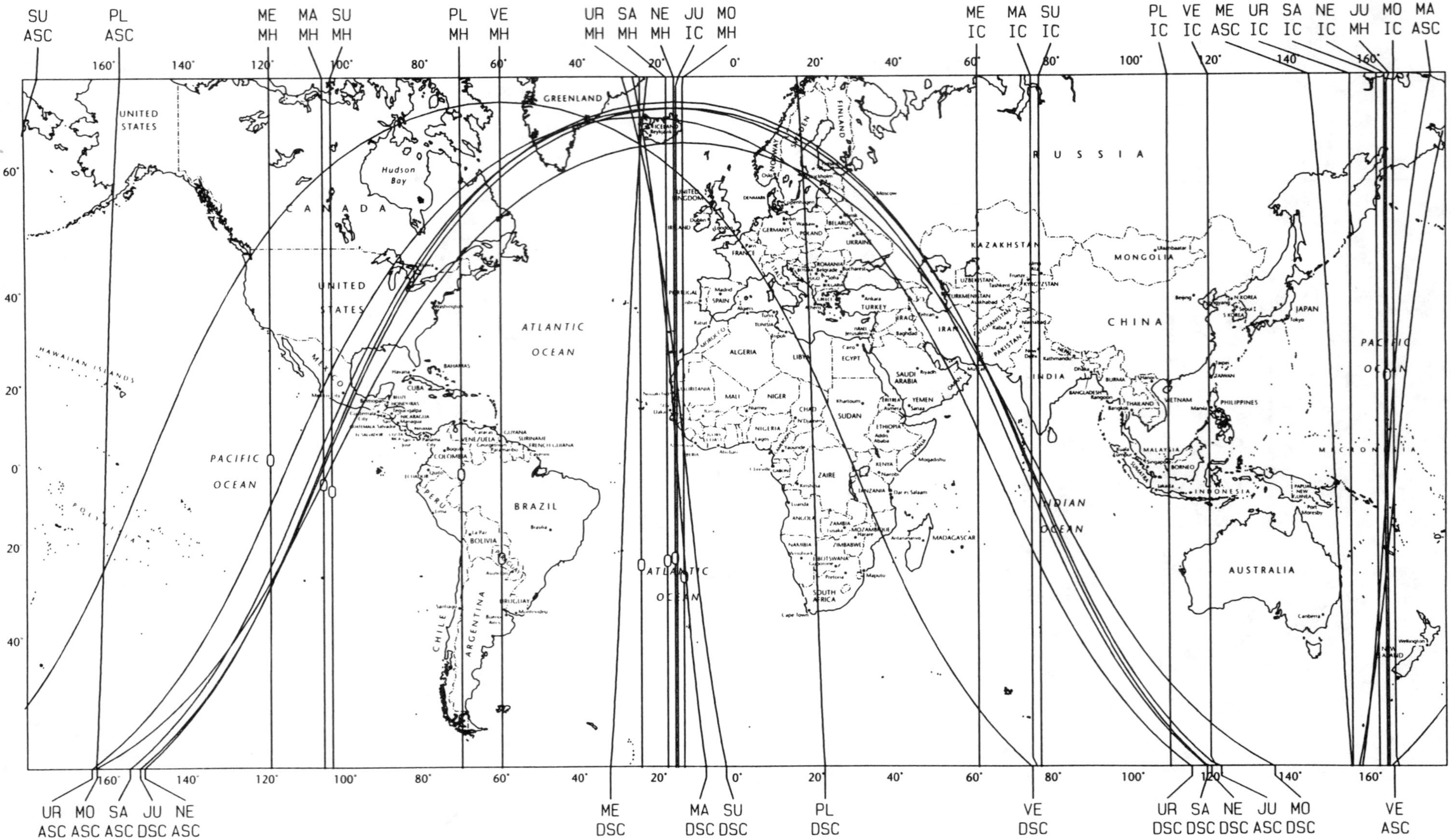

East Germany SR

October 7, 1989 6:36 PM GMT

05/21/97

East Germany SR
October 7, 1989 6:36 PM GMT

ASTRO*CARTO*GRAPHY LATITUDE CROSSINGS

NORTH LATITUDES

67N48 JU/IC * NE/DSC
67N47 SA/MH * NE/ASC
67N47 MO/MH * NE/DSC
67N38 ME/DSC * NE/ASC
67N34 UR/MH * NE/ASC
67N33 MA/DSC * NE/ASC
67N31 PL/MH * VE/ASC
67N25 SU/DSC * NE/ASC
67N13 NE/MH * JU/DSC
67N13 JU/DSC * SA/DSC
67N13 NE/MH * SA/DSC
67N12 MO/MH * JU/ASC
67N12 JU/IC * SA/DSC
67N12 SA/MH * JU/DSC
67N10 MO/MH * SA/DSC
67N07 ME/DSC * SA/ASC
67N05 UR/MH * SA/ASC
67N04 MA/DSC * SA/ASC
67N00 ME/DSC * JU/DSC
66N59 SU/DSC * SA/ASC
66N57 UR/MH * JU/DSC
66N56 MA/DSC * JU/DSC
66N49 SU/DSC * JU/DSC
66N40 UR/MH * MA/DSC
66N18 MA/DSC * UR/DSC
66N18 ME/DSC * UR/DSC
66N17 SU/DSC * UR/ASC
66N15 VE/DSC * NE/ASC
66N09 SA/MH * UR/DSC
66N04 VE/DSC * SA/ASC
66N03 NE/MH * UR/DSC
66N02 JU/IC * UR/DSC
65N56 JU/DSC * UR/ASC
65N55 MO/MH * UR/DSC
65N54 VE/DSC * UR/ASC
65N53 VE/DSC * JU/DSC
65N18 SA/ASC * UR/ASC
65N10 ME/DSC * MA/DSC
64N46 UR/ASC * NE/ASC
64N11 NE/DSC * PL/DSC
63N39 JU/ASC * PL/DSC
63N32 JU/IC * MO/ASC
63N32 NE/MH * MO/ASC
63N29 SA/MH * MO/ASC
63N24 SU/DSC * VE/DSC
63N24 UR/MH * VE/DSC
63N24 UR/MH * SU/DSC
63N19 ME/DSC * VE/DSC
63N14 MO/ASC * MA/DSC
63N11 MO/ASC * VE/DSC
63N09 MO/ASC * ME/DSC
63N08 MO/ASC * SU/DSC
63N08 UR/MH * MO/ASC
63N00 SU/DSC * ME/DSC
63N00 SA/DSC * PL/DSC
62N58 VE/DSC * MA/DSC
61N55 UR/MH * ME/DSC
61N39 VE/MH * UR/ASC
61N34 SA/ASC * NE/ASC
61N13 Anchorage
61N09 SA/MH * VE/DSC
60N55 SU/MH * VE/ASC
60N39 MO/DSC * UR/DSC
60N25 VE/MH * SA/ASC
60N23 NE/MH * VE/DSC
60N20 VE/MH * NE/ASC
60N09 JU/IC * VE/DSC
60N01 MA/MH * VE/ASC

59N57 MO/DSC * PL/DSC
59N34 UR/DSC * PL/DSC
59N23 MO/MH * VE/DSC
59N23 VE/MH * JU/DSC
59N20 Stockholm

CENTRAL LATITUDES

57N46 PL/MH * UR/ASC
55N45 Moscow
55N41 Copenhagen
55N21 PL/MH * SA/ASC
54N53 PL/MH * NE/ASC
54N08 VE/MH * MO/ASC
53N44 PL/MH * JU/DSC
52N30 Berlin
52N00 ME/MH * VE/ASC
51N30 SU/DSC * MA/DSC
51N30 London

49N14 Vancouver
48N50 Paris
47N45 PL/MH * MO/ASC
47N36 Seattle
45N32 Montreal
44N59 Minneapolis
43N39 Toronto
42N22 Boston
42N20 Detroit
41N54 Rome
41N53 Chicago
40N52 MO/DSC * SA/DSC
40N43 New York

39N49 SA/MH * SU/DSC
39N44 Denver
38N53 Washington
38N37 St. Louis
37N47 San Francisco
36N58 JU/ASC * NE/DSC
36N10 Las Vegas
35N40 Tokyo
34N04 Los Angeles
33N45 Atlanta
33N27 Phoenix
33N03 SA/MH * MA/DSC
32N47 Dallas
30N03 Cairo

29N58 New Orleans
29N46 Houston
28N48 ME/IC * JU/ASC
28N39 Delhi
28N31 ME/IC * NE/DSC
27N37 ME/IC * MO/DSC
26N28 NE/MH * SU/DSC
25N47 Miami
25N03 VE/DSC * PL/DSC
24N26 SU/MH * UR/ASC
24N14 ME/IC * SA/DSC
23N29 MO/DSC * NE/DSC
22N34 Calcutta
22N16 Hong Kong
21N59 JU/IC * SU/DSC
21N33 MO/IC * MA/ASC
21N19 Honolulu
20N59 MO/DSC * JU/ASC
20N17 MA/MH * UR/ASC

19N26 Mexico City
14N35 Manila
11N13 SU/MH * SA/ASC
10N30 Caracas

9N42 NE/MH * MA/DSC
9N24 ME/IC * UR/DSC
6N57 SU/MH * NE/ASC
6N27 Lagos
6N11 MA/MH * SA/ASC
5N48 MO/MH * SU/DSC
5N29 SU/MH * JU/DSC
3N11 MA/IC * MO/DSC
2N05 JU/IC * MA/DSC

SOUTH LATITUDES

1N41 MA/MH * NE/ASC
1N10 SU/MH * MO/ASC
0N21 MA/MH * JU/DSC

0S21 MA/IC * JU/ASC
1S10 SU/IC * MO/DSC
1S41 MA/IC * NE/DSC
2S05 JU/MH * MA/ASC
3S11 MA/MH * MO/ASC
5S29 SU/IC * JU/ASC
5S48 MO/IC * SU/ASC
6S11 MA/IC * SA/DSC
6S57 SU/IC * NE/DSC
9S24 ME/MH * UR/ASC
9S42 NE/IC * MA/ASC

11S13 SU/IC * SA/DSC

20S17 MA/IC * UR/DSC
20S59 MO/ASC * JU/DSC
21S33 MO/MH * MA/DSC
21S59 JU/MH * SU/ASC
22S54 Rio De Janeiro
23S29 MO/ASC * NE/ASC
24S14 ME/MH * SA/ASC
24S26 SU/IC * UR/DSC
25S03 VE/ASC * PL/ASC
26S12 Johannesburg
26S28 NE/IC * SU/ASC
27S37 ME/MH * MO/ASC
28S31 ME/MH * NE/ASC
28S48 ME/MH * JU/DSC

33S03 SA/IC * MA/ASC
33S55 Sydney
34S36 Buenos Aires
36S52 Auckland
36S58 JU/DSC * NE/ASC
37S50 Melbourne
39S49 SA/IC * SU/ASC

40S52 MO/ASC * SA/ASC
47S45 PL/IC * MO/DSC

51S30 SU/ASC * MA/ASC
52S00 ME/IC * VE/DSC
53S44 PL/IC * JU/ASC
54S08 VE/IC * MO/DSC
54S53 PL/IC * NE/DSC
55S21 PL/IC * SA/DSC
57S46 PL/IC * UR/DSC
59S23 VE/IC * JU/ASC
59S23 MO/IC * VE/ASC
59S34 UR/ASC * PL/ASC
59S57 MO/ASC * PL/ASC

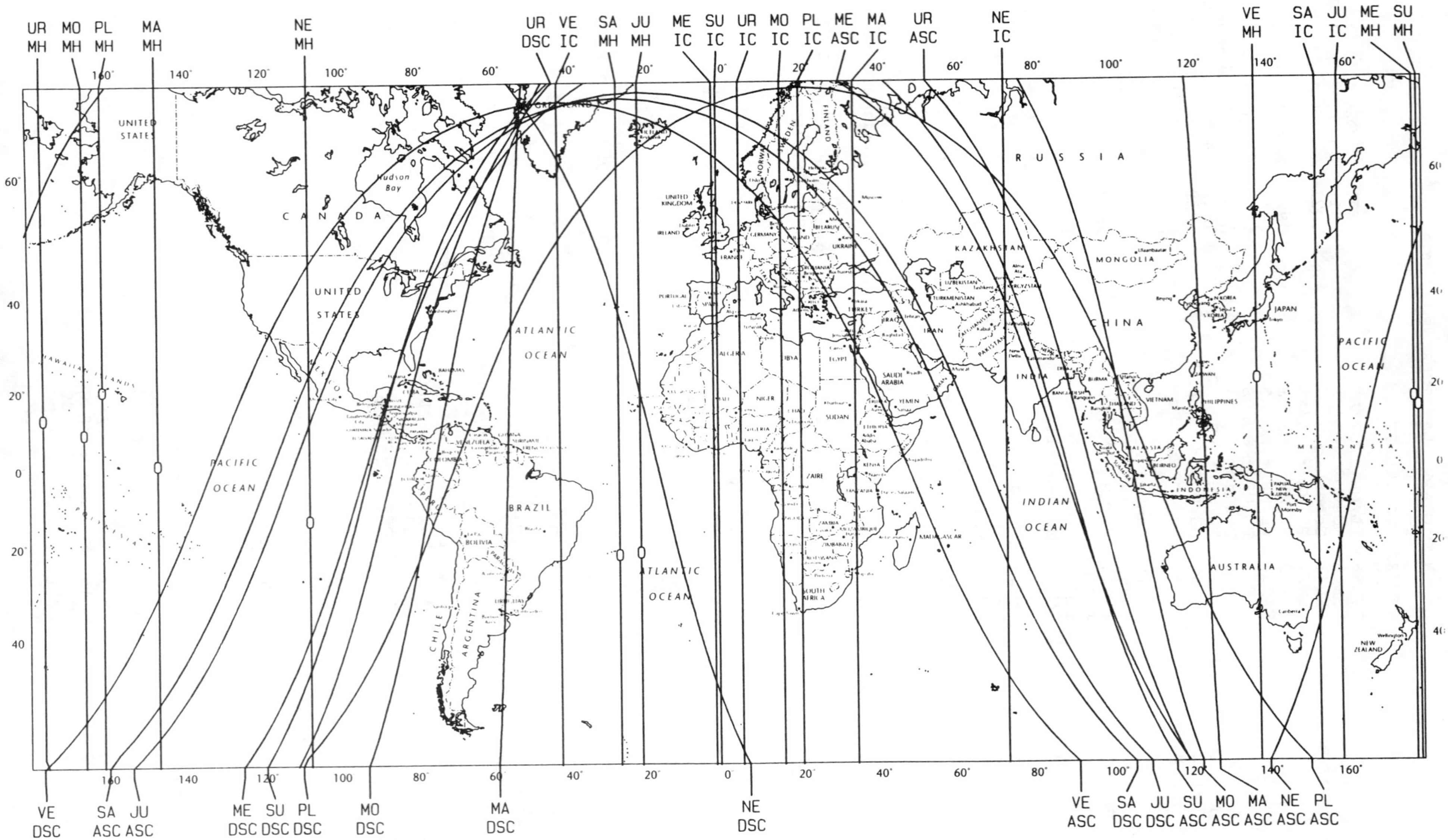

Berlin Wall

August 13, 1961 1:11 AM CET

04/16/96

04/16/96

Berlin Wall
August 13, 1961 1:11 AM CET

ASTRO*CARTO*GRAPHY LATITUDE CROSSINGS

NORTH LATITUDES

69N28 MO/IC * PL/DSC
69N25 PL/MH * NE/ASC
69N09 MA/IC * ME/ASC
69N07 SA/MH * JU/ASC
68N58 MA/IC * PL/ASC
68N52 ME/ASC * PL/ASC
68N52 MO/DSC * SU/DSC
68N49 UR/IC * PL/DSC
68N34 JU/MH * SA/DSC
68N28 MA/DSC * NE/DSC
68N21 ME/DSC * VE/ASC
68N19 VE/IC * ME/DSC
68N19 VE/ASC * JU/ASC
68N17 VE/ASC * SA/ASC
68N14 VE/DSC * UR/DSC
68N12 MO/DSC * VE/DSC
68N10 SU/DSC * VE/DSC
68N09 SU/IC * PL/DSC
68N07 VE/DSC * NE/DSC
68N06 ME/IC * JU/DSC
68N04 VE/DSC * MA/DSC
68N04 JU/ASC * SA/ASC
68N03 JU/DSC * PL/DSC
68N00 ME/IC * PL/DSC
67N58 SU/IC * JU/DSC
67N56 VE/IC * SA/ASC
67N54 SU/DSC * NE/DSC
67N53 VE/IC * JU/ASC
67N45 ME/DSC * SA/ASC
67N38 SA/MH * VE/ASC
67N33 MO/DSC * NE/DSC
67N31 ME/DSC * JU/ASC
67N28 UR/DSC * NE/DSC
67N21 SU/DSC * MA/DSC
67N21 SA/DSC * PL/DSC
67N06 SA/ASC * NE/DSC
67N06 MO/DSC * UR/DSC
67N05 UR/IC * JU/DSC
67N00 MO/MH * NE/ASC
67N00 JU/MH * VE/ASC
66N51 JU/ASC * NE/DSC
66N51 MO/DSC * SA/ASC
66N48 SA/ASC * UR/DSC
66N45 ME/IC * SA/DSC
66N38 MA/DSC * SA/ASC
66N34 SU/IC * SA/DSC
66N25 ME/DSC * NE/DSC
66N25 SU/ASC * PL/ASC
66N18 MO/DSC * JU/ASC
66N14 MA/DSC * UR/DSC
66N13 SU/DSC * SA/ASC
66N10 MA/DSC * JU/ASC
66N07 JU/ASC * UR/DSC
65N44 VE/ASC * PL/DSC
65N38 MO/DSC * MA/DSC
65N23 UR/IC * SA/DSC
65N05 SU/DSC * JU/ASC
64N37 MO/IC * JU/DSC
64N35 ME/DSC * MA/DSC
63N43 MO/DSC * ME/DSC
63N28 JU/MH * PL/DSC
63N04 PL/IC * JU/DSC
62N41 ME/IC * VE/ASC
62N17 SU/IC * VE/ASC
62N13 VE/IC * NE/DSC
62N10 MO/IC * SA/DSC
61N17 SA/MH * PL/DSC
61N13 Anchorage
60N08 PL/IC * SA/DSC
60N08 UR/ASC * PL/ASC

59N55 UR/IC * VE/ASC
59N24 ME/DSC * UR/DSC
59N20 Stockholm

CENTRAL LATITUDES

58N12 UR/MH * NE/ASC
58N12 NE/IC * PL/ASC
56N26 NE/IC * UR/ASC
56N02 MA/IC * JU/DSC
55N56 NE/DSC * PL/DSC
55N45 Moscow
55N41 Copenhagen
53N31 MO/IC * VE/ASC
52N30 Berlin
51N30 London
51N10 VE/IC * PL/DSC
51N00 MA/IC * SA/DSC

49N52 SU/MH * NE/ASC
49N26 PL/IC * VE/ASC
49N14 Vancouver
48N50 Paris
47N44 ME/MH * NE/ASC
47N36 Seattle
47N12 NE/MH * VE/DSC
45N32 Montreal
45N22 NE/IC * SU/ASC
44N59 Minneapolis
43N39 Toronto
42N22 Boston
42N20 Detroit
41N54 Rome
41N53 Chicago
40N43 New York

39N44 Denver
39N07 NE/IC * ME/ASC
38N53 Washington
38N37 St. Louis
37N49 SA/MH * NE/DSC
37N47 San Francisco
36N10 Las Vegas
35N40 Tokyo
35N02 MA/DSC * PL/DSC
34N04 Los Angeles
33N45 Atlanta
33N27 Phoenix
32N47 Dallas
30N36 MA/IC * VE/ASC
30N03 Cairo

29N58 New Orleans
29N46 Houston
28N39 Delhi
25N47 SU/DSC * ME/DSC
25N47 Miami
24N09 NE/MH * SA/ASC
22N43 MO/ASC * PL/ASC
22N34 Calcutta
22N16 Hong Kong
21N19 Honolulu

19N40 JU/MH * NE/DSC
19N26 Mexico City
14N35 Manila
12N00 NE/MH * JU/ASC
10N30 Caracas

6N27 Lagos

12S00 NE/IC * JU/DSC
19S40 JU/IC * NE/ASC

22S43 MO/DSC * PL/DSC
22S54 Rio De Janeiro
24S09 NE/IC * SA/DSC
25S47 SU/ASC * ME/ASC
26S12 Johannesburg

30S36 MA/MH * VE/DSC

SOUTH LATITUDES

33S55 Sydney
34S36 Buenos Aires
35S02 MA/ASC * PL/ASC
36S52 Auckland
37S49 SA/IC * NE/ASC
37S50 Melbourne
39S07 NE/MH * ME/DSC

45S22 NE/MH * SU/DSC
47S12 NE/IC * VE/ASC
47S44 ME/IC * NE/DSC
49S26 PL/MH * VE/DSC
49S52 SU/IC * NE/DSC

51S00 MA/MH * SA/ASC
51S10 VE/MH * PL/ASC
53S31 MO/MH * VE/DSC
55S56 NE/ASC * PL/ASC
56S02 MA/MH * JU/ASC
56S26 NE/MH * UR/DSC
58S12 NE/MH * PL/DSC
58S12 UR/IC * NE/DSC
59S24 ME/ASC * UR/ASC
59S55 UR/MH * VE/DSC

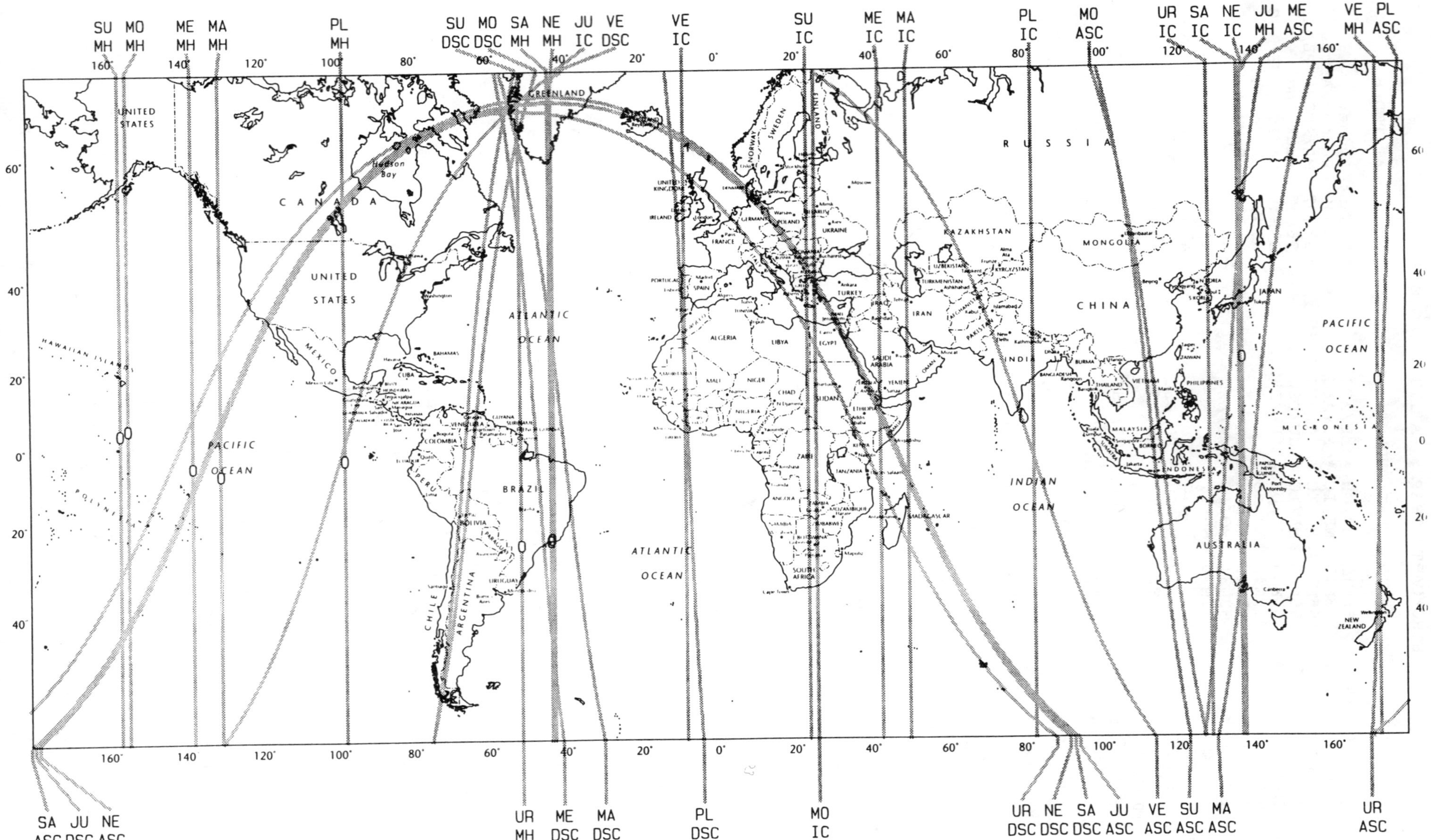

CYCLO*CARTO*GRAPHY

04/17/96

SU, MO, ME, VE AND MA ARE PROGRESSED
JU, SA, UR, NE AND PL ARE TRANSITING

Berlin Wall

PROG	9	10	61	5:59 GMT
TRANSIT	11	9	89	0:00 GMT
NATAL	8	13	61	0:11 GMT

9, 1989.

A*C*G	Age 28	C*C*G
JU 02 AQU 16		UR 03 CAP 06t
NE 06 SCO 53		ME 07 LIB 15
MO 09 VIR 02		SA 10 CAP 53
VE 11 CAN 31		NE 11 CAP 06
PL 13 VIR 53		JU 11 CAN 33
ME 21 LEO 17		MA 14 LIB 32
SU 22 LEO 25		VE 16 LEO 24
SA 26 CAP 47		PL 17 SCO 07
MA 27 VIR 56		SU 18 VIR 19
UR 28 LEO 16		MO 20 VIR 25

1. C*C*G SA = A*C*G VE
2. C*C*G NE = A*C*G VE
3. C*C*G JU = A*C*G VE
4. C*C*G NE DSC = Berlin
5. C*C*G SA DSC = Berlin
6. C*C*G JU ASC = Berlin

A*C*G	(25 54)	Displaced C*C*G
JU 02 AQU 16		UR 07 SAG 12
NE 06 SCO 53		ME 11 VIR 21
MO 09 VIR 02		SA 14 SAG 59
VE 11 CAN 31		NE 15 SAG 12
PL 13 VIR 53		JU 15 GEM 39
ME 21 LEO 17		MA 18 VIR 38
SU 22 LEO 25		VE 20 CAN 30
SA 26 CAP 47		PL 21 LIB 13
MA 27 VIR 56		MO 24 LEO 31
UR 28 LEO 16		

A*C*G	(25 54)	Solar arc A*C*G
JU 2 AQU 16		NE 2 SAG 47
NE 6 SCO 53		MO 4 LIB 56
MO 9 VIR 02		VE 7 LEO 25
VE 11 CAN 31		PL 9 LIB 47
PL 13 VIR 53		ME 17 VIR 11
ME 21 LEO 17		SA 22 AQU 41
SU 22 LEO 25		MA 23 LIB 50
SA 26 CAP 47		UR 24 VIR 10
MA 27 VIR 56		JU 28 AQU 10
UR 28 LEO 16		

7. solar arc A*C*G SA = A*C*G SU
8. solar arc A*C*G JU = A*C*G UR

C*C*G NE DSC, C*C*G SA DSC and C*C*G JU ASC were exactly at Berlin, which shows the opening of the Wall. A week later as many as two million East Germans visited West Berlin in one day (solar arc JU EQU A*C*G UR).

C*C*G NE DSC, C*C*G SA DSC and C*C*G JU ASC were all on top of A*C*G VE ASC. Large queues had formed at the border between East and West Berlin, so parts of the Wall were removed to create extra crossings. These planetary influences show that the world's largest art gallery was dismantled.

Solar arc SA CJN A*C*G SU indicated that difficult negotiations were ahead for the two German countries.

The solar return map had SU ASC and PL IC within orb of Berlin. These planetary influences are relevant for the end of the Berlin Wall and a new beginning for the city. PL IC passes through Eastern Europe, reiterating the information from the East German solar return map.

Erich Honecker's Astro*Carto*Graphy Map

An Astro*Carto*Graphy map for a leader represents the country itself. It's as if it can be used as an auxiliary map for the country for as long as the leader remains in power. This means that Erich Honecker's map can be used as an additional map for East Germany. His Cyclo*Carto*Graphy planets are calculated for September 11, 1989, when the border between Hungary and Austria was opened.

He has SA IC exact at Berlin which is a clear indication of the responsibility he had to build a wall. During the eighteen years that he was leader, the Berlin Wall prevented East Germans from travelling to the west. Also exact at Berlin is UR ASC, which manifested in the strong urge for freedom that the East German people had.

Within orb of Berlin are JU MH and NE DSC which also had an effect. NE DSC passes through West Germany, which East Germans dreamed of going to. JU MH is near the border between Hungary and Austria, which turned out to be the easiest way for East Germans to move to West Germany.

A*C*G	Age 77	C*C*G
SA 02 GEM 05		UR 01 CAP 27
UR 02 AQU 37		MO 05 SAG 53
SU 04 VIR 05		ME 06 SAG 17
JU 04 SAG 35		SA 07 CAP 55
MO 12 AQU 35		JU 07 CAN 58
VE 17 VIR 41		NE 10 CAP 25
MA 25 VIR 36		MA 13 SCO 48
NE 26 CAN 46		PL 15 SCO 12
ME 27 LEO 09		SU 15 SCO 32
PL 29 GEM 50		VE 19 SAG 59

1. C*C*G MO MH = Hungarian/Austrian border
2. C*C*G ME MH = Hungarian/Austrian border
3. Prog. Berlin = C*C*G VE DSC
4. Prog. Berlin = C*C*G PL DSC

A*C*G	(71 27)	Displaced C*C*G
SA 02 GEM 05		MA 02 VIR 21
UR 02 AQU 37		PL 03 VIR 45
SU 04 VIR 05		VE 08 LIB 32
JU 04 SAG 35		UR 20 LIB 00
MO 12 AQU 35		MO 24 VIR 26
VE 17 VIR 41		ME 24 VIR 50
MA 25 VIR 36		SA 26 LIB 28
NE 26 CAN 46		JU 26 ARI 31
ME 27 LEO 09		NE 28 LIB 58
PL 29 GEM 50		

5. Disp. C*C*G PL = A*C*G SU
6. Disp. C*C*G MO = A*C*G MA (Oct)
7. Disp. C*C*G ME = A*C*G MA

A*C*G	(71 27)	Solar arc A*C*G
SA 02 GEM 05		MA 07 SAG 03

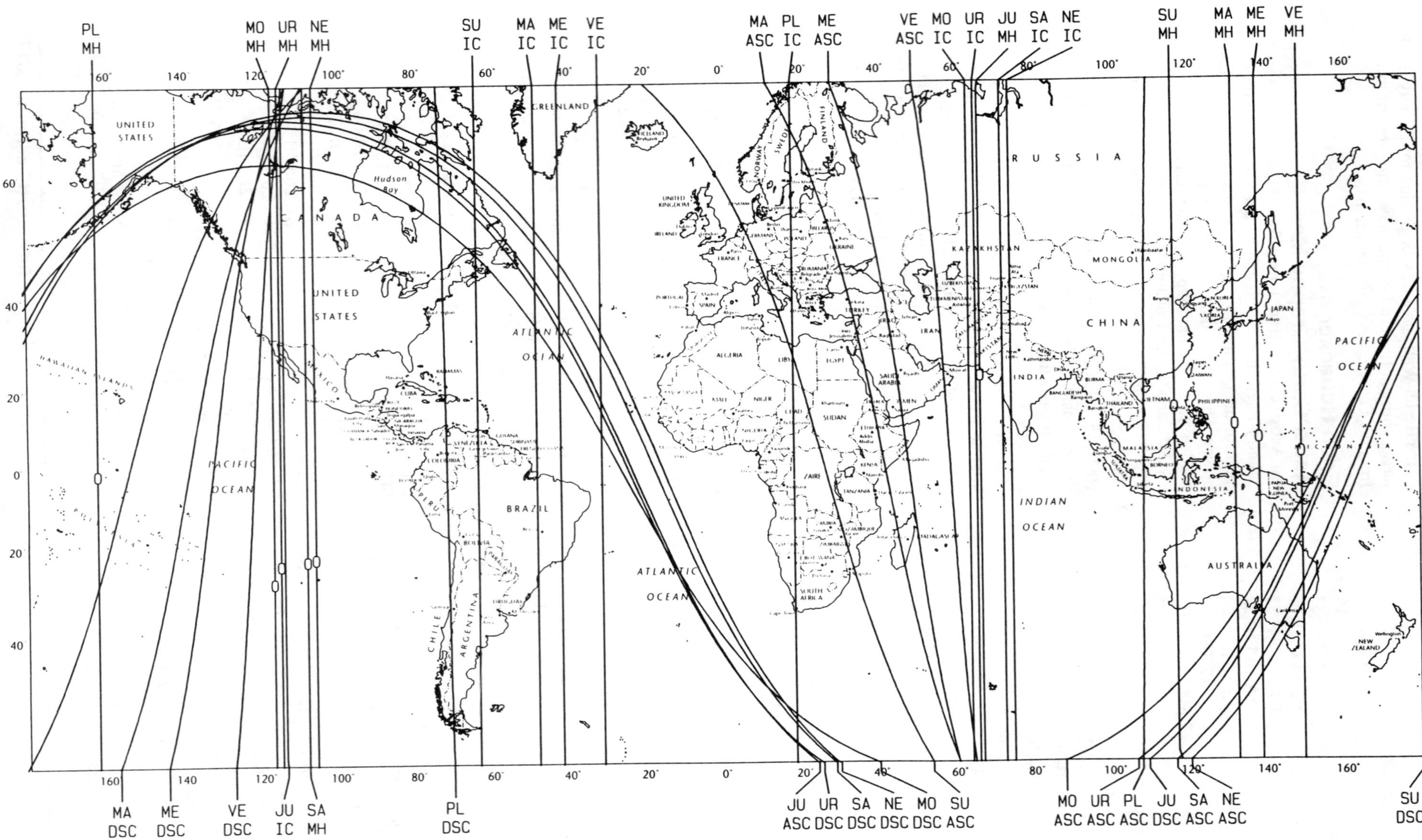

Berlin Wall SR

August 13, 1989 4:17 AM GMT

04/16/96

04/16/96

Berlin Wall SR
August 13, 1989 4:17 AM GMT

ASTRO*CARTO*GRAPHY LATITUDE CROSSINGS

NORTH LATITUDES

69N56 SA/MH * SU/DSC
69N26 SU/DSC * MA/DSC
69N04 JU/IC * ME/DSC
68N00 UR/MH * ME/DSC
67N54 SU/DSC * VE/DSC
67N50 SA/MH * NE/ASC
67N49 JU/IC * SU/DSC
67N44 MA/DSC * NE/ASC
67N40 VE/DSC * NE/ASC
67N39 JU/IC * NE/ASC
67N38 SU/DSC * NE/ASC
67N36 UR/MH * NE/ASC
67N35 ME/DSC * NE/ASC
67N30 MO/MH * NE/ASC
67N26 JU/IC * VE/DSC
67N23 UR/MH * SU/DSC
67N20 NE/MH * SA/DSC
67N16 MA/DSC * SA/ASC
67N14 JU/IC * SA/ASC
67N14 VE/DSC * SA/ASC
67N12 UR/MH * SA/ASC
67N11 SU/DSC * SA/ASC
67N10 ME/DSC * SA/ASC
67N07 MO/MH * SA/ASC
67N00 SU/DSC * ME/DSC
66N54 VE/DSC * JU/DSC
66N54 MA/DSC * JU/ASC
66N54 UR/MH * JU/DSC
66N54 ME/DSC * JU/DSC
66N54 SU/DSC * JU/DSC
66N53 MO/MH * JU/DSC
66N48 SA/MH * JU/ASC
66N42 NE/MH * JU/ASC
66N39 JU/DSC * SA/ASC
66N34 MO/MH * SU/DSC
66N28 JU/IC * MA/DSC
66N18 VE/DSC * UR/DSC
66N18 MA/DSC * UR/DSC
66N17 JU/IC * UR/DSC
66N17 ME/DSC * UR/ASC
66N17 MO/MH * UR/ASC
66N17 SU/DSC * UR/ASC
66N14 JU/DSC * NE/ASC
66N09 SA/MH * UR/DSC
66N03 VE/DSC * MA/DSC
66N02 NE/MH * UR/DSC
65N49 MO/MH * ME/DSC
65N37 UR/MH * MA/DSC
65N33 PL/IC * MA/ASC
65N01 SA/ASC * UR/ASC
64N47 UR/ASC * NE/ASC
64N29 UR/MH * VE/DSC
64N13 NE/DSC * PL/DSC
63N58 MO/MH * MA/DSC
63N38 SA/ASC * NE/ASC
63N03 SA/DSC * PL/DSC
62N10 MO/DSC * VE/DSC
62N10 MO/ASC * MA/DSC
62N10 UR/MH * MO/DSC
62N10 MO/ASC * ME/DSC
62N09 JU/IC * MO/DSC
61N58 MO/ASC * SU/DSC
61N55 SA/MH * MO/DSC
61N45 NE/MH * MO/DSC
61N17 SU/IC * NE/DSC
61N13 Anchorage
60N38 JU/ASC * PL/DSC

59N50 ME/DSC * MA/DSC
59N48 SU/IC * SA/DSC
59N30 UR/DSC * PL/DSC
59N20 Stockholm
57N28 PL/MH * JU/DSC
57N12 PL/MH * UR/ASC

CENTRAL LATITUDES

57N01 MO/MH * VE/DSC
56N27 SU/IC * JU/ASC
55N45 Moscow
55N41 Copenhagen
55N06 SU/IC * UR/DSC
54N42 PL/MH * SA/ASC
53N55 PL/MH * NE/ASC
53N49 MO/DSC * PL/DSC
53N09 PL/MH * MO/ASC
52N54 MA/IC * NE/DSC
52N30 Berlin
51N32 MO/ASC * NE/ASC
51N30 London
50N29 MA/IC * SA/DSC
50N00 JU/DSC * UR/ASC

49N14 Vancouver
48N54 MO/ASC * SA/ASC
48N53 SU/IC * MO/DSC
48N50 Paris
47N36 Seattle
47N31 ME/IC * NE/DSC
45N32 Montreal
44N59 Minneapolis
44N31 ME/IC * SA/DSC
44N17 MA/IC * JU/ASC
43N39 Toronto
42N22 Boston
42N20 Detroit
42N19 MA/IC * UR/DSC
41N54 Rome
41N53 Chicago
40N43 New York

39N44 Denver
38N53 Washington
38N37 St. Louis
37N47 San Francisco
36N31 ME/IC * JU/ASC
36N10 Las Vegas
35N40 Tokyo
35N06 MA/IC * MO/DSC
34N15 ME/IC * UR/DSC
34N04 Los Angeles
33N45 VE/IC * NE/DSC
33N45 Atlanta
33N27 Phoenix
32N47 Dallas
30N03 Cairo

29N58 New Orleans
29N46 Houston
29N26 VE/IC * SA/DSC
28N39 Delhi
28N06 PL/IC * SU/ASC
26N59 ME/IC * MO/DSC
25N47 Miami
25N16 MO/ASC * JU/DSC
22N34 Calcutta
22N16 Hong Kong
21N19 Honolulu

19N26 Mexico City
19N16 MO/ASC * UR/ASC
17N39 VE/IC * JU/ASC
15N04 VE/IC * UR/DSC
14N35 Manila
10N30 Caracas

9N21 VE/IC * MO/DSC
6N27 Lagos

9S21 VE/MH * MO/ASC

15S04 VE/MH * UR/ASC

SOUTH LATITUDES

17S39 VE/MH * JU/DSC
19S16 MO/DSC * UR/DSC

22S54 Rio De Janeiro
25S16 MO/DSC * JU/ASC
26S12 Johannesburg
26S59 ME/MH * MO/ASC
28S06 PL/MH * SU/DSC
29S26 VE/MH * SA/ASC

33S45 VE/MH * NE/ASC
33S55 Sydney
34S15 ME/MH * UR/ASC
34S36 Buenos Aires
35S06 MA/MH * MO/ASC
36S31 ME/MH * JU/DSC
36S52 Auckland
37S50 Melbourne

42S19 MA/MH * UR/ASC
44S17 MA/MH * JU/DSC
44S31 ME/MH * SA/ASC
47S31 ME/MH * NE/ASC
48S53 SU/MH * MO/ASC
48S54 MO/DSC * SA/DSC
50S00 JU/ASC * UR/DSC

50S29 MA/MH * SA/ASC
51S32 MO/DSC * NE/DSC
52S54 MA/MH * NE/ASC
53S09 PL/IC * MO/DSC
53S49 MO/ASC * PL/ASC
53S55 PL/IC * NE/DSC
54S42 PL/IC * SA/DSC
55S06 SU/MH * UR/ASC
56S27 SU/MH * JU/DSC
57S01 MO/IC * VE/ASC
57S12 PL/IC * UR/DSC
57S28 PL/IC * JU/ASC
59S30 UR/ASC * PL/ASC
59S48 SU/MH * SA/ASC
59S50 ME/ASC * MA/ASC

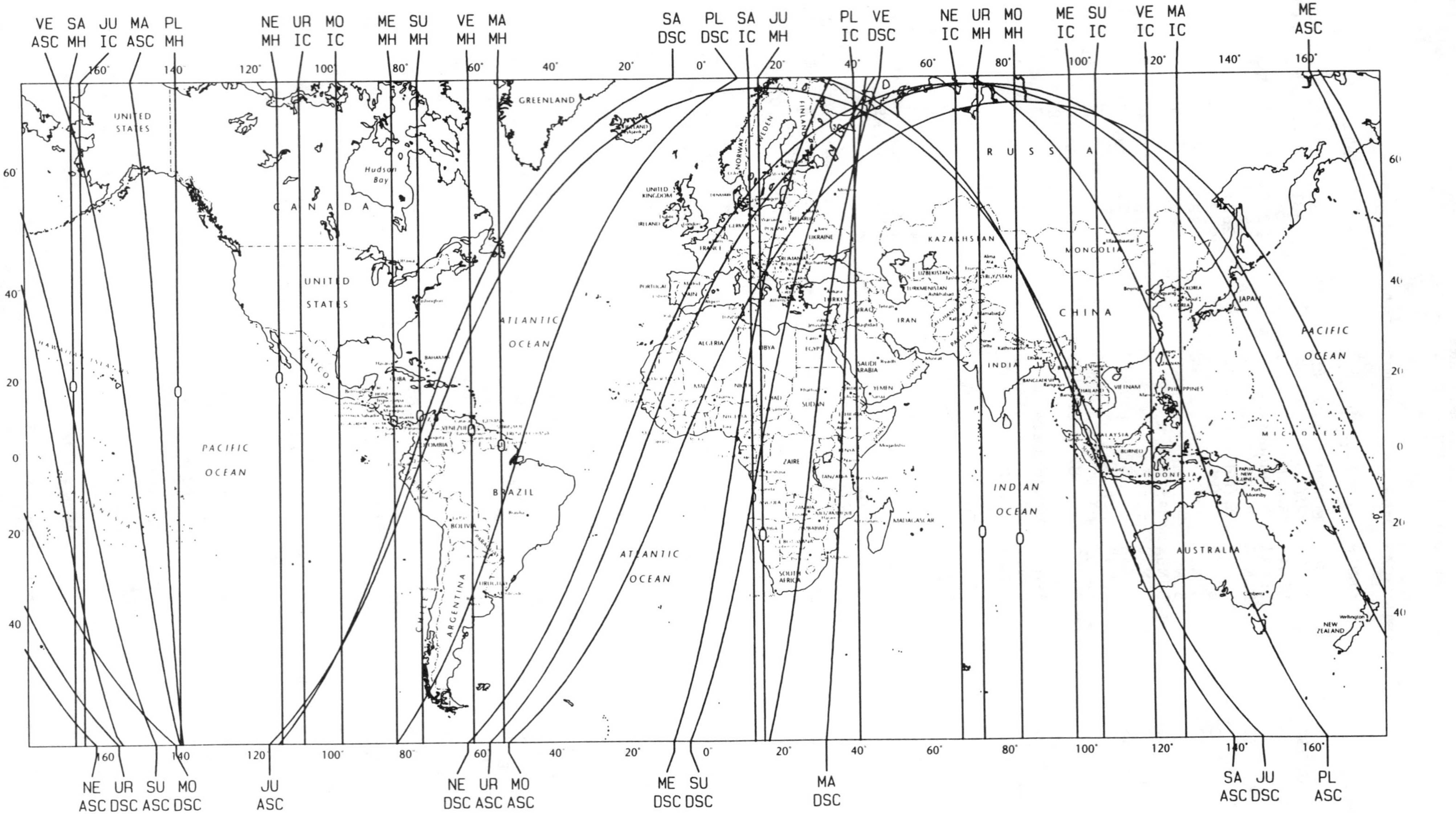

Erich Honecker

August 25, 1912 6:00 PM CET

04/16/96

04/16/96

Erich Honecker
August 25, 1912 6:00 PM CET

ASTRO*CARTO*GRAPHY LATITUDE CROSSINGS

NORTH LATITUDES

69N48 ME/DSC * SA/ASC
69N35 UR/MH * PL/ASC
69N18 UR/DSC * PL/ASC
69N18 UR/ASC * NE/ASC
69N15 UR/MH * NE/ASC
69N13 NE/IC * UR/ASC
69N10 NE/ASC * PL/ASC
69N06 SA/IC * JU/ASC
69N01 MO/MH * UR/DSC
68N53 PL/IC * SA/ASC
68N45 JU/ASC * PL/DSC
68N36 MO/MH * NE/ASC
68N22 MA/DSC * SA/ASC
68N21 SU/DSC * SA/ASC
68N18 SU/DSC * MA/DSC
68N16 ME/DSC * JU/DSC
67N54 VE/DSC * SA/ASC
67N53 SA/ASC * NE/DSC
67N52 VE/DSC * NE/DSC
67N30 MA/DSC * NE/DSC
67N27 ME/IC * UR/DSC
67N27 MO/ASC * PL/ASC
67N19 SA/ASC * UR/ASC
67N17 SU/DSC * NE/DSC
67N09 UR/MH * MO/ASC
67N08 PL/IC * JU/DSC
67N06 JU/DSC * NE/DSC
67N05 MO/DSC * NE/ASC
67N05 PL/IC * NE/DSC
67N00 SU/DSC * JU/DSC
66N55 MO/MH * PL/ASC
66N48 ME/IC * MO/DSC
66N45 MA/DSC * JU/DSC
66N40 NE/IC * MO/ASC
66N40 VE/DSC * UR/ASC
66N32 JU/DSC * UR/ASC
66N27 VE/DSC * JU/DSC
66N25 PL/IC * SU/DSC
66N23 ME/IC * NE/ASC
66N17 MA/DSC * UR/ASC
66N07 SU/IC * UR/DSC
66N05 SA/MH * VE/ASC
65N58 SU/IC * MO/DSC
65N49 PL/IC * UR/ASC
65N41 MO/DSC * UR/DSC
65N32 SU/DSC * UR/ASC
64N56 MO/ASC * SA/ASC
64N38 SU/IC * NE/ASC
64N18 MO/ASC * JU/DSC
63N45 ME/DSC * NE/DSC
63N07 VE/IC * MO/DSC
62N28 VE/DSC * MA/DSC
62N15 JU/IC * VE/ASC
61N53 VE/IC * UR/DSC
61N13 Anchorage
60N48 MO/ASC * MA/DSC
60N32 MO/ASC * VE/DSC
60N32 PL/IC * MO/ASC
60N30 PL/IC * VE/DSC
60N26 MA/IC * MO/DSC
60N10 ME/DSC * UR/ASC

59N55 ME/IC * PL/ASC
59N24 NE/IC * SA/ASC
59N20 Stockholm
59N07 VE/IC * NE/ASC
58N25 JU/MH * NE/DSC
58N07 NE/IC * JU/DSC
57N55 MA/IC * UR/DSC
56N57 PL/IC * MA/DSC
56N54 SA/IC * NE/DSC
55N51 MO/ASC * SU/DSC
55N45 Moscow
55N41 Copenhagen

CENTRAL LATITUDES

55N12 UR/MH * SA/ASC
54N30 JU/MH * UR/ASC
54N15 SU/IC * PL/ASC
54N13 UR/MH * JU/DSC
53N56 MA/IC * NE/ASC
52N30 Berlin
52N29 SA/IC * UR/ASC
51N30 London

49N24 MA/MH * SA/DSC
49N14 Vancouver
48N50 Paris
47N36 Seattle
45N40 JU/DSC * SA/ASC
45N32 Montreal
44N59 Minneapolis
44N28 MO/MH * JU/DSC
44N18 MO/MH * SA/ASC
44N07 MO/ASC * ME/DSC
43N39 Toronto
43N14 MA/MH * JU/ASC
42N22 Boston
42N20 Detroit
42N04 JU/MH * MO/ASC
41N54 Rome
41N53 Chicago
40N43 New York

39N44 Denver
38N53 Washington
38N48 SA/IC * MO/ASC
38N37 St. Louis
38N12 JU/MH * ME/DSC
38N10 VE/MH * SA/DSC
37N47 San Francisco
36N10 Las Vegas
35N40 Tokyo
34N04 Los Angeles
33N58 VE/IC * PL/ASC
33N45 Atlanta
33N27 Phoenix
32N47 Dallas
30N42 VE/MH * JU/ASC
30N03 Cairo

29N58 New Orleans
29N46 Houston
28N39 Delhi
27N38 SA/IC * ME/DSC
25N47 Miami
22N34 Calcutta
22N16 Hong Kong
21N19 Honolulu

19N26 Mexico City
18N44 ME/IC * JU/DSC
14N35 Manila
14N08 ME/IC * SA/ASC
13N18 MA/IC * PL/ASC
10N30 Caracas
10N24 SA/MH * SU/ASC

6N27 Lagos
5N49 SU/MH * SA/DSC
2N38 JU/MH * SU/DSC
1N19 SU/IC * JU/DSC

1S19 SU/MH * JU/ASC
2S38 JU/IC * SU/ASC
5S49 SU/IC * SA/ASC

10S24 SA/IC * SU/DSC
13S18 MA/MH * PL/DSC
14S08 ME/MH * SA/DSC
18S44 ME/MH * JU/ASC

SOUTH LATITUDES

22S54 Rio De Janeiro
26S12 Johannesburg
27S38 SA/MH * ME/ASC

30S42 VE/IC * JU/DSC
33S55 Sydney
33S58 VE/MH * PL/DSC
34S36 Buenos Aires
36S52 Auckland
37S50 Melbourne
38S10 VE/IC * SA/ASC
38S12 JU/IC * ME/ASC
38S48 SA/MH * MO/DSC

42S04 JU/IC * MO/DSC
43S14 MA/IC * JU/DSC
44S07 MO/DSC * ME/ASC
44S18 MO/IC * SA/DSC
44S28 MO/IC * JU/ASC
45S40 JU/ASC * SA/DSC
49S24 MA/IC * SA/ASC

52S29 SA/MH * UR/DSC
53S56 MA/MH * NE/DSC
54S13 UR/IC * JU/ASC
54S15 SU/MH * PL/DSC
54S30 JU/IC * UR/DSC
55S12 UR/IC * SA/DSC
55S51 MO/DSC * SU/ASC
56S54 SA/MH * NE/ASC
56S57 PL/MH * MA/ASC
57S55 MA/MH * UR/ASC
58S07 NE/MH * JU/ASC
58S25 JU/IC * NE/ASC
59S07 VE/MH * NE/DSC
59S24 NE/MH * SA/DSC
59S55 ME/MH * PL/DSC

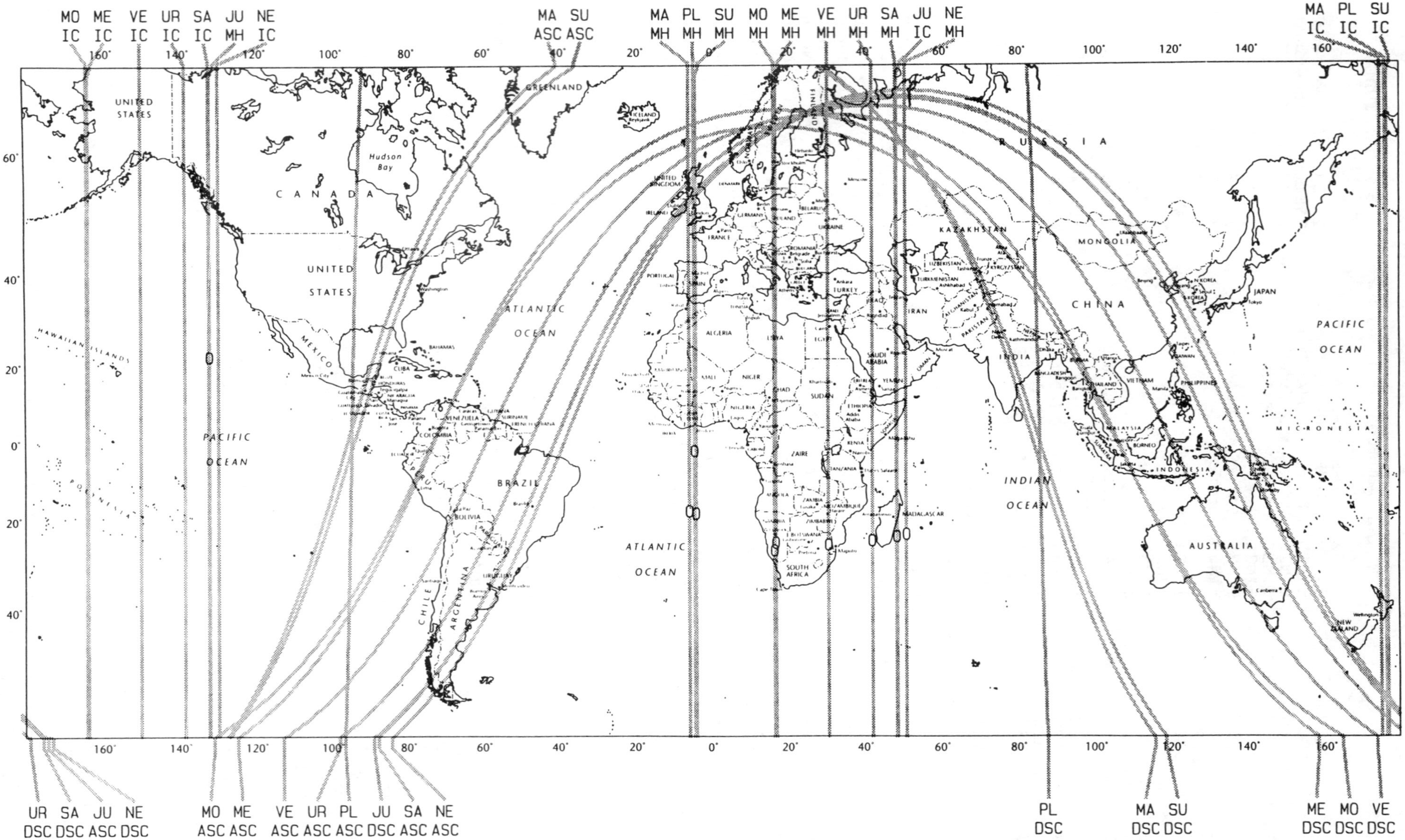

CYCLO*CARTO*GRAPHY

04/17/96

SU, MO, ME, VE AND MA ARE PROGRESSED
JU, SA, UR, NE AND PL ARE TRANSITING

Erich Honecker

PROG	11	10	12	18:09 GMT
TRANSIT	9	11	89	0:00 GMT
NATAL	8	25	12	17:00 GMT

UR 02 AQU 37	NE 08 LIB 13
SU 04 VIR 05	ME 08 SCO 36
JU 04 SAG 35	PL 11 VIR 17
MO 12 AQU 35	SA 13 LEO 42
VE 17 VIR 41	UR 14 ARI 04
MA 25 VIR 36	JU 16 AQU 02
NE 26 CAN 46	MO 24 ARI 02
ME 27 LEO 09	VE 29 SCO 08
PL 29 GEM 50	

C*C*G MO MH and C*C*G ME MH were at the Hungarian/Austrian border when thousands of East Germans fled to the West. These temporary influences activated the A*C*G JU MH line at the border. East Germans couldn't believe how easy it was to cross and start a new life in the west.

Prog. Berlin = C*C*G VE DSC and C*C*G PL DSC occurred at a time when East Germany was facing economic ruin. The mass exodus of East Germans made the economic situation even worse.

Disp. C*C*G PL = A*C*G SU showed that East Germany would experience some kind of transformation. After Honecker resigned as leader on October 18, 1989, the changes that were taking place were irreversible.

Disp. C*C*G MO = A*C*G MA was exact at the beginning of October 1989 when the East German government warned of a "Tiananmen Square" crackdown. This aspect is appropriate for the street demonstrations calling for change. In restrictive East Germany people were not allowed to voice their opinions, but Gorbachev's visit gave them hope.

There were no planetary influences within orb of East Germany on Honecker's solar return map, however the Cyclo*Carto*Graphy temporary influences provided a clear picture of what was happening in East Germany.

The Astro*Carto*Graphy maps for East Germany, the Berlin Wall and Erich Honecker give accurate information about the end of Communism in East Germany. When T*he Astro*Carto*Graphy Source Book of Mundane Maps* is used, there is additional information about the fall of the Berlin Wall.

The winter solstice map of December 21, 1988 has SA DSC, NE DSC and ME DSC at Berlin. C*C*G SA and C*C*G NE were significant for East Germany and the Berlin Wall maps, so the winter solstice map has a repetition of themes.

The SA/NE cycle is connected with the development of socialism/communism (*Mundane Astrology*, page 182). When the previous conjunctions between SA and NE are examined the connection to Communism is clear.

1846	Formation of Marxism
1882	Formation of Socialist Parties in Europe
1882	Marx died
1917	Russian revolution
1953	Stalin died
1953	Revolution in East Germany

At the previous SA/NE conjunction in 1953, there was unrest in East Germany. In 1989, the SA/NE conjunction took place three times, which coincided with the revolution in Eastern Europe. The SA/NE conjunction shown on the winter solstice map was important for East Germany, and it needed only an eclipse or lunation with SA and NE at Berlin to trigger it.

Although the eclipse new Moon on August 31, 1989 does not show a repetition of themes, it activates the Berlin Wall chart. The SU CJN MO at 07 VIR 48 is conjunct the Berlin Wall IC.

When examining an eclipse map it is also important to look at the position of the Sun and Moon. The *Book of World Horoscopes* gives an integrated planetary index in zodiacal order, so it is easy to see which national horoscopes are activated. Looking at the lunation maps there is one that is dominant for Berlin: a new Moon which took place on October 29, 1989. The map shows **seven** planetary lines at Berlin:

Lines:	SA MH
	JU IC
	NE MH
	SU DSC
	MO DSC
	ME DSC
	MA DSC

Once again, SA and NE were powerful at Berlin, activating the winter solstice map. With various maps showing a repetition of themes I decided to use this new Moon chart to calculate the mundane square aspects to SA and NE.

The appropriate midpoint for SA/NE is nine and a half degrees Capricorn. Using a tables of houses for Birmingham (nearest latitude to Berlin) the four additional degree axes are:

1. SA/NE (MH)	25 ARI/25 LIB
2. SA/NE (ASC)	17 SCO/17 TAU
3. SA/NE (DSC)	05 PIS/05 VIR
4. SA/NE (IC)	07 LIB/07 ARI

At a press conference at 6:57 p.m. on November 9, 1989, it was announced that East Germans could travel to the West without any restrictions. The Sun was at 17 SCO 19 conjunct the second SA/NE degree axis.

The mundane squares technique pinpointed the exact date that the Berlin Wall was opened. After the new Moon occurred on October 29, 1989, the first sensitive point that the Sun activated was 17 SCO.

This event happened four days before the final conjunction of Saturn and Neptune, completing a cycle that had started thirty-six years earlier when an uprising by East German workers was crushed by the Soviets.

Example 2—Challenger Explosion

On January 28, 1986, the American space shuttle lifted off from Cape Canaveral in what seemed to be a routine flight. Shortly after lift-off the Challenger suddenly exploded into a fireball which destroyed it and killed all seven crew members.

Does Astro*Carto*Graphy explain why this tragic event occurred? An appropriate chart to examine is that of NASA (National Aeronautics and Space Administration).

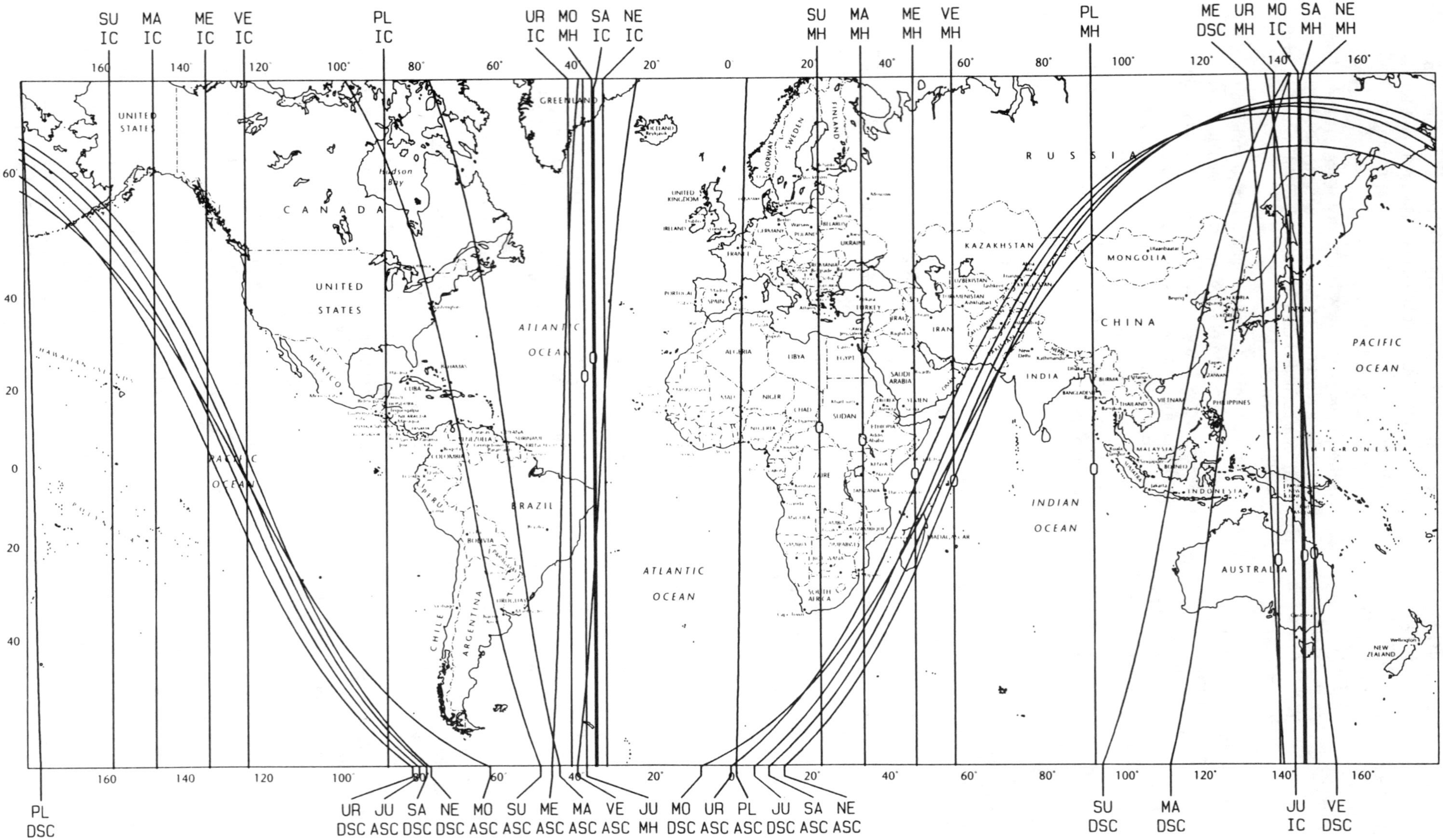

Erich Honecker SR

August 26, 1989 10:29 AM GMT

04/16/96

Erich Honecker SR
August 26, 1989 10:29 AM GMT

ASTRO*CARTO*GRAPHY LATITUDE CROSSINGS

NORTH LATITUDES

69N34 JU/IC * MA/DSC
67N49 SA/MH * NE/ASC
67N48 MO/IC * NE/ASC
67N45 JU/IC * NE/ASC
67N43 UR/MH * SU/DSC
67N41 MA/DSC * NE/ASC
67N35 UR/MH * NE/ASC
67N34 SU/DSC * NE/ASC
67N31 VE/DSC * NE/ASC
67N17 MO/IC * SA/ASC
67N16 NE/MH * SA/DSC
67N16 JU/IC * SA/ASC
67N13 MA/DSC * SA/ASC
67N09 UR/MH * SA/ASC
67N07 SU/DSC * SA/ASC
67N07 VE/DSC * SA/ASC
67N06 ME/DSC * NE/ASC
67N06 SU/DSC * VE/DSC
66N57 MO/IC * JU/ASC
66N57 MA/DSC * JU/DSC
66N56 SA/MH * JU/ASC
66N55 UR/MH * JU/DSC
66N54 VE/DSC * JU/DSC
66N53 SU/DSC * JU/DSC
66N53 NE/MH * JU/ASC
66N47 ME/DSC * SA/ASC
66N39 ME/DSC * JU/DSC
66N18 VE/DSC * UR/ASC
66N17 MA/DSC * UR/DSC
66N17 SU/DSC * UR/ASC
66N15 JU/IC * UR/DSC
66N11 ME/DSC * UR/ASC
66N10 MO/IC * UR/DSC
66N09 SA/MH * UR/DSC
66N02 NE/MH * UR/DSC
66N02 JU/DSC * SA/ASC
65N16 JU/DSC * NE/ASC
65N13 UR/MH * MA/DSC
65N05 VE/DSC * MA/DSC
65N02 SA/ASC * UR/ASC
64N51 UR/MH * VE/DSC
64N50 UR/ASC * NE/ASC
64N23 JU/DSC * UR/ASC
64N11 NE/DSC * PL/DSC
63N59 SA/ASC * NE/ASC
62N57 SU/DSC * ME/DSC
62N54 SA/DSC * PL/DSC
62N45 SA/MH * MO/ASC
62N44 JU/IC * MO/DSC
62N43 NE/MH * MO/ASC
62N39 MO/DSC * VE/DSC
62N37 UR/MH * MO/DSC
62N31 MO/DSC * MA/DSC
62N16 MO/DSC * ME/DSC
62N12 MO/DSC * SU/DSC
62N07 PL/IC * SU/ASC
61N39 JU/ASC * PL/DSC
61N13 Anchorage

59N27 UR/DSC * PL/DSC
59N20 Stockholm
57N32 MO/ASC * PL/DSC
57N24 PL/MH * UR/ASC
56N32 ME/DSC * MA/DSC
56N15 PL/MH * JU/DSC
55N45 Moscow
55N41 Copenhagen
55N02 PL/MH * SA/ASC
54N56 SU/IC * NE/DSC
54N09 PL/MH * NE/ASC
52N30 Berlin
52N29 SU/IC * SA/DSC
51N30 London
50N27 UR/IC * ME/ASC

CENTRAL LATITUDES

49N52 SU/IC * JU/ASC
49N37 PL/MH * MO/DSC
49N17 MO/ASC * UR/DSC
49N14 Vancouver
48N50 Paris
47N36 Seattle
46N58 JU/IC * VE/DSC
46N15 SU/IC * MO/ASC
45N32 Montreal
45N29 MA/IC * NE/DSC
45N25 SU/IC * UR/DSC
44N59 Minneapolis
43N39 Toronto
42N22 Boston
42N20 Detroit
41N54 Rome
41N53 MA/IC * SA/DSC
41N53 Chicago
40N43 New York

39N44 Denver
38N53 Washington
38N37 St. Louis
37N51 MA/IC * JU/ASC
37N47 San Francisco
36N10 Las Vegas
35N40 Tokyo
35N29 MA/IC * MO/ASC
34N04 Los Angeles
33N45 Atlanta
33N27 Phoenix
32N47 Dallas
31N16 MA/IC * UR/DSC
30N03 Cairo

29N58 New Orleans
29N46 Houston
28N39 Delhi
25N47 Miami
25N20 ME/IC * NE/DSC
24N54 MO/DSC * NE/ASC
24N05 MO/IC * VE/DSC
23N11 MO/ASC * JU/ASC
22N34 Calcutta
22N16 Hong Kong
21N19 Honolulu
20N09 NE/IC * VE/ASC

19N51 ME/IC * SA/DSC
19N26 Mexico City
17N10 SA/MH * VE/DSC
16N58 VE/MH * UR/ASC
15N27 ME/IC * MO/ASC
14N35 Manila
13N40 ME/IC * JU/ASC
10N30 Caracas

9N00 VE/MH * JU/DSC
6N27 Lagos
5N08 MO/DSC * SA/ASC
4N47 ME/IC * UR/DSC
3N14 VE/IC * NE/DSC
3N07 VE/MH * MO/DSC
2N39 VE/MH * SA/ASC

2S39 VE/IC * SA/DSC
3S07 VE/IC * MO/ASC
3S14 VE/MH * NE/ASC
4S47 ME/MH * UR/ASC
5S08 MO/ASC * SA/DSC
9S00 VE/IC * JU/ASC

13S40 ME/MH * JU/DSC
15S27 ME/MH * MO/DSC

SOUTH LATITUDES

16S58 VE/IC * UR/DSC
17S10 SA/IC * VE/ASC
19S51 ME/MH * SA/ASC

20S09 NE/MH * VE/DSC
22S54 Rio De Janeiro
23S11 MO/DSC * JU/DSC
24S05 MO/MH * VE/ASC
24S54 MO/ASC * NE/DSC
25S20 ME/MH * NE/ASC
26S12 Johannesburg

31S16 MA/MH * UR/ASC
33S55 Sydney
34S36 Buenos Aires
35S29 MA/MH * MO/DSC
36S52 Auckland
37S50 Melbourne
37S51 MA/MH * JU/DSC

41S53 MA/MH * SA/ASC
45S25 SU/MH * UR/ASC
45S29 MA/MH * NE/ASC
46S15 SU/MH * MO/DSC
46S58 JU/MH * VE/ASC
49S17 MO/DSC * UR/ASC
49S37 PL/IC * MO/ASC
49S52 SU/MH * JU/DSC

50S27 UR/MH * ME/DSC
52S29 SU/MH * SA/ASC
54S09 PL/IC * NE/DSC
54S56 SU/MH * NE/ASC
55S02 PL/IC * SA/DSC
56S15 PL/IC * JU/ASC
56S32 ME/ASC * MA/ASC
57S24 PL/IC * UR/DSC
57S32 MO/DSC * PL/ASC
59S27 UR/ASC * PL/ASC

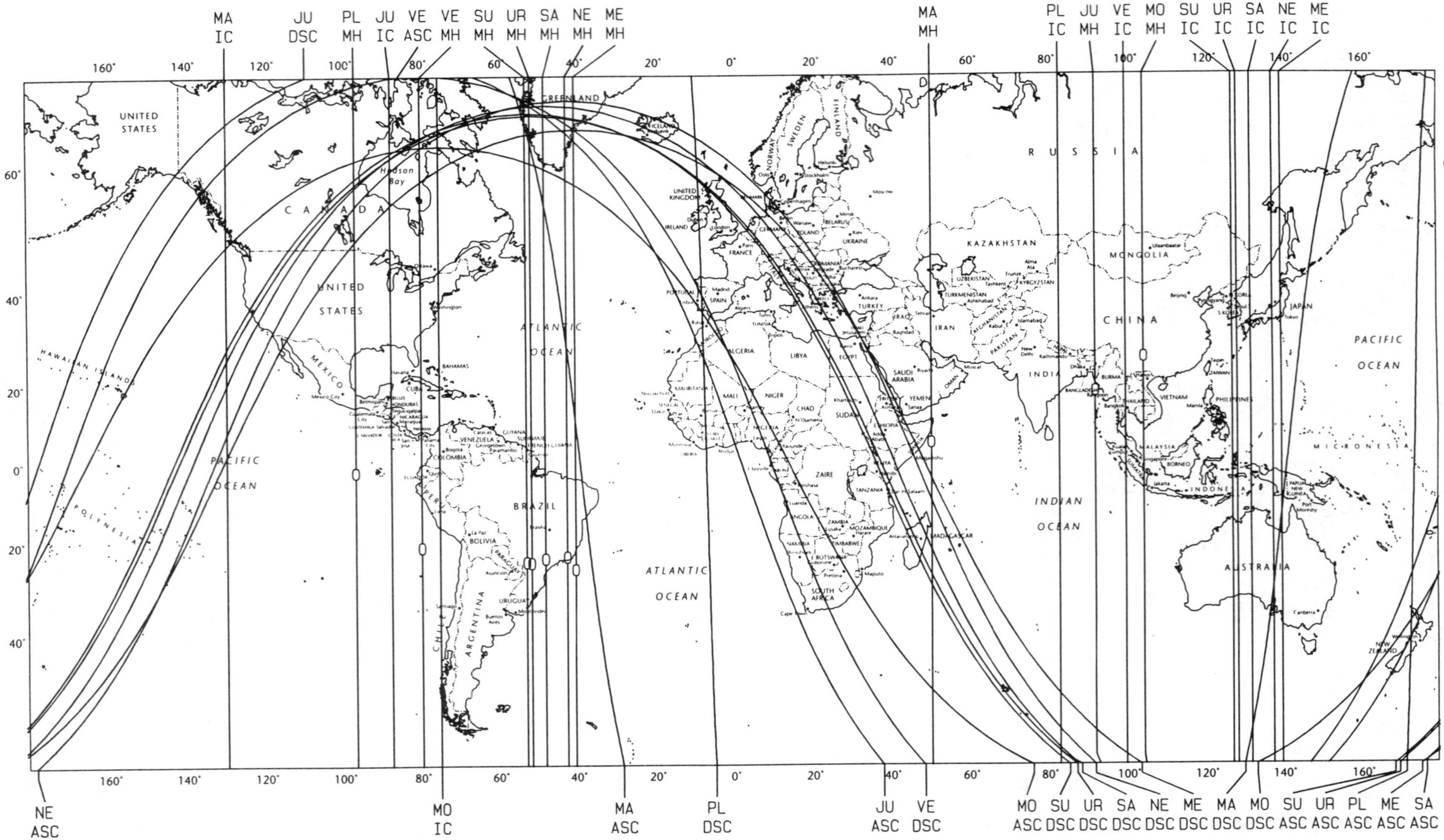

Winter Solstice

December 21, 1988 3:29 PM GMT

09/18/98

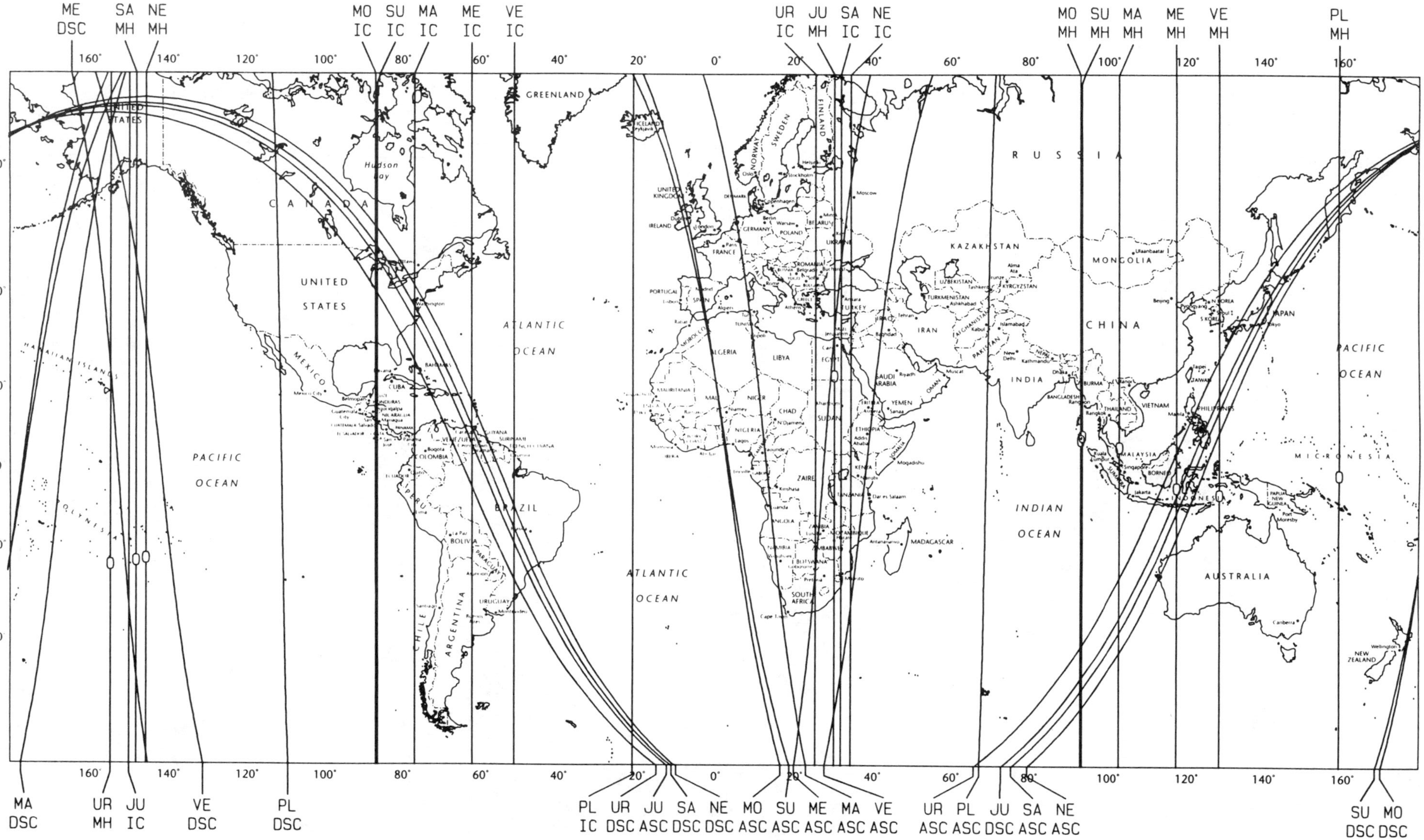

Eclipse New Moon

August 31, 1989 5:44 AM GMT

09/18/98

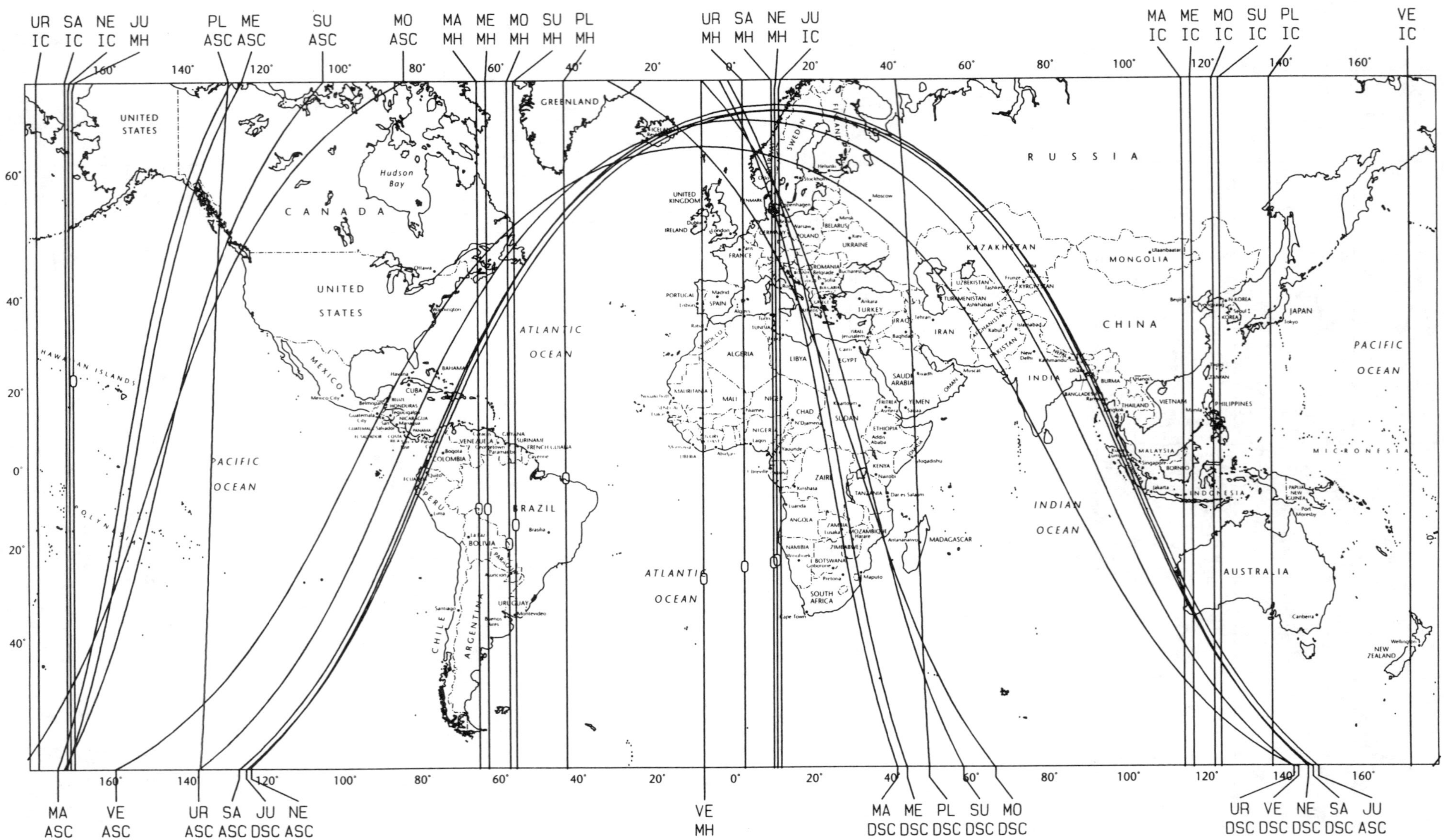

New Moon

October 29, 1989 3: 27 PM GMT

09/18/98

New Moon
October 29, 1989
Berlin, Germany
52N30, 13E22

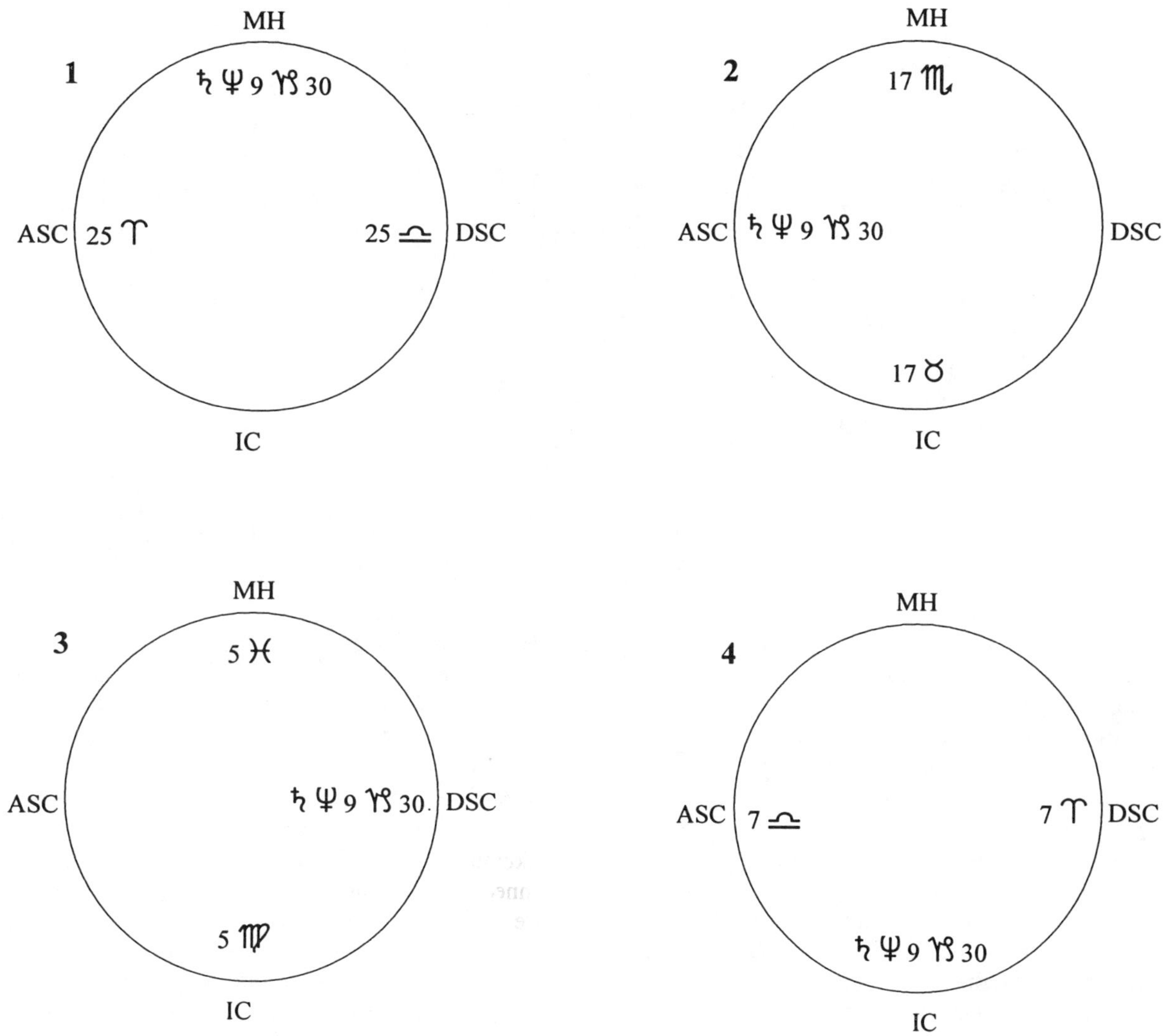

1st degree axis = 25 ♈/25 ♎
2nd degree axis = 17 ♏/17 ♉
3rd degree axis = 5 ♓/5 ♍
4th degree axis = 7 ♎/7 ♈

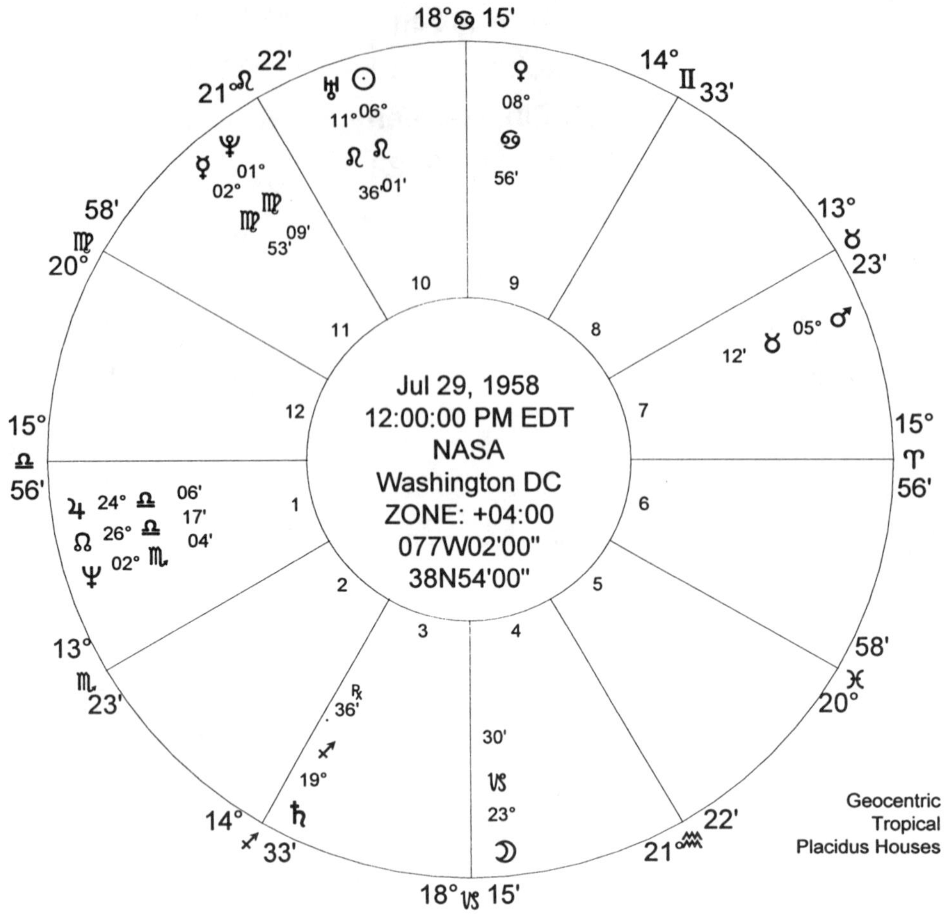

Unfortunately, an exact time is not known for when the National Aeronautics and Space Act of 1958 was signed by President Eisenhower. Although no Astro*Carto*Graphy map can be drawn for NASA it is still worthwhile to calculate a mid-day chart (July 29, 1958, White House, Washington).

In the *Book of World Horoscopes* by Nicholas Campion there is an appendix that deals with the era of manned flight and space travel. The first manned space flight was launched at 9:07 a.m. (Moscow time) on April 12, 1961 from the Balkonur Cosmodrome in Kazakhstan. Campion is of the opinion that this chart is more important than the one for the first U.S. manned space flight, May 5, 1961, 9:24 a.m. EST.

Another chart that can be used is for the First Artificial Space Satellite (Sputnik), which was launched on October 4, 1957. The Americans calculated the launch time and site as 9:00 p.m., north of the Caspian Sea.

I decided to draw an Astro*Carto*Graphy map for the first manned space flight as it provided accurate information for the Challenger explosion. JU DSC passes exactly through the location where the first manned space flight was launched. SA DSC is also within orb of Balkonur showing that a great deal of organization and planning would be used in the Soviet space program.

The same planetary influences are repeated in the American space program where JU DSC and SA ASC are near to Cape Canaveral.

It is very interesting that the two countries most involved with manned space flights both had a JU line at their launch sites. JU is connected with long distance travel which is the case after every lift-off from Balkonur and Cape Canaveral.

Although the Challenger was an American space shuttle it is very important to examine the temporary influences at Balkonur. This location was the birthplace of the first manned space flight, so temporary influences there will have a worldwide effect.

This point is emphasized because it's easy to concentrate on examining the temporary influences at Cape Canaveral and forget about the birthplace. The temporary influences for January 28, 1986 are amazing.

A*C*G	Age 24	C*C*G
SA 01 AQU 24		MA 03 LEO 10
ME 03 ARI 53		NE 04 CAP 57
JU 6 AQU 43		SA 6 SAG 14
NE 08 SCO 37		PL 10 SCO 19
PL 12 VIR 44		MO 10 AQU 38
MO 15 PIS 03		VE 11 ARI 32
VE 15 ARI 49		SU 13 TAU 49
MA 20 CAN 15		ME 19 TAU 49
SU 20 ARI 28		UR 20 SAG 10
UR 24 LEO 24		JU 27 AQU 06

1. C*C*G MA DSC = Cape Canaveral
2. Prog. birthplace = A*C*G NE IC (1½ degrees)
3. Prog. birthplace = A*C*G UR ASC (exact)

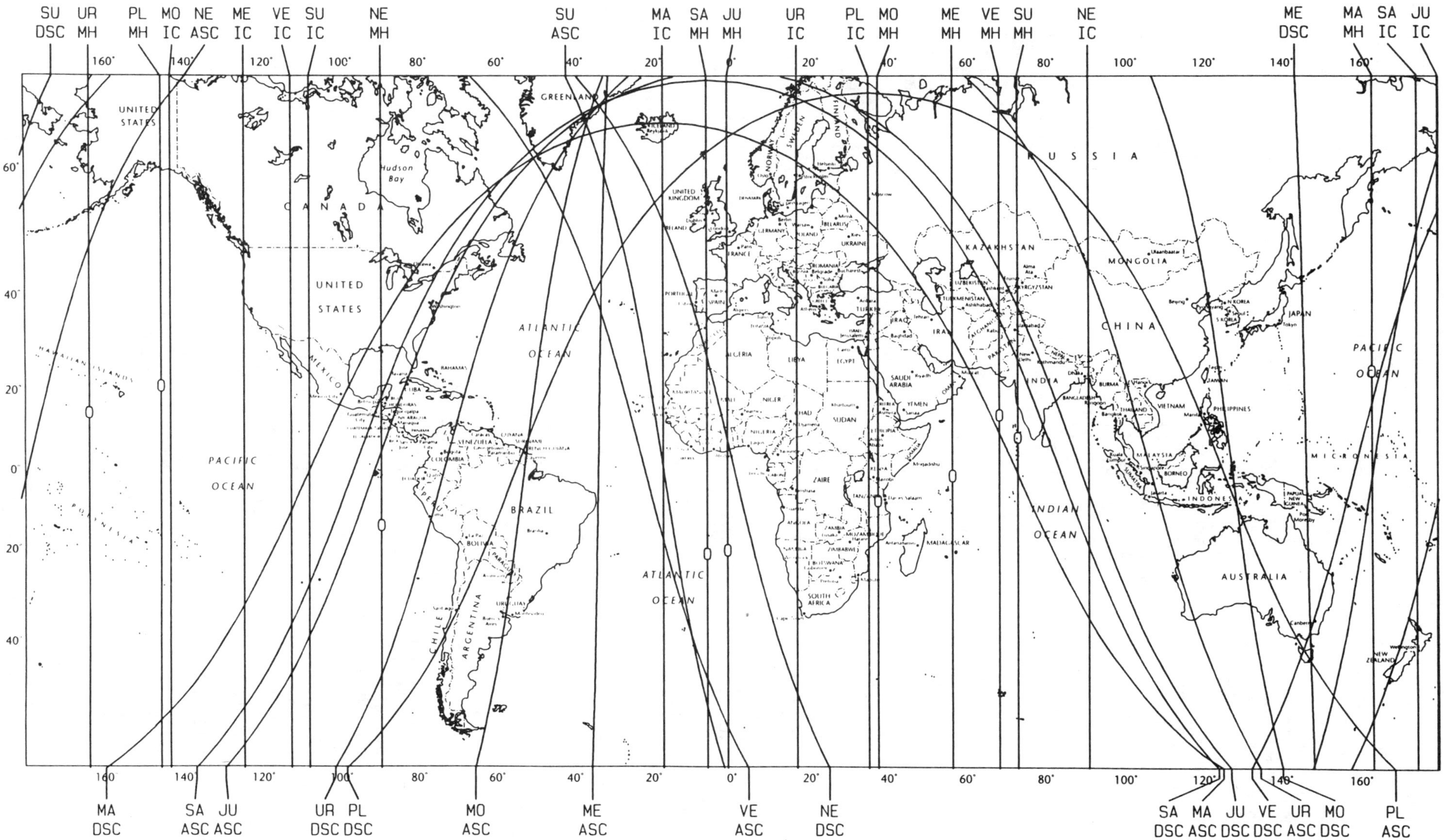

First Manned Space Flight

April 12, 1961 9:07 AM Time Zone -2:00

04/17/96

First Manned Space Flight
April 12, 1961 9:07 AM Time Zone -2:00

ASTRO*CARTO*GRAPHY LATITUDE CROSSINGS

NORTH LATITUDES

69N46 MA/IC * JU/ASC
69N39 UR/IC * JU/DSC
69N36 JU/MH * SA/DSC
69N19 MA/IC * SA/ASC
68N32 MO/MH * PL/ASC
68N26 MO/IC * NE/ASC
68N19 JU/DSC * PL/DSC
68N11 ME/ASC * UR/DSC
68N05 UR/IC * SA/DSC
67N57 MO/ASC * NE/DSC
67N57 JU/ASC * SA/ASC
67N50 UR/MH * VE/DSC
67N45 SA/DSC * PL/DSC
67N45 ME/ASC * SA/ASC
67N41 ME/ASC * JU/ASC
67N31 UR/IC * PL/DSC
67N26 SA/ASC * UR/DSC
67N24 SA/ASC * NE/DSC
67N23 UR/DSC * NE/DSC
67N19 JU/ASC * NE/DSC
67N17 PL/MH * NE/ASC
67N09 JU/ASC * UR/DSC
67N09 ME/MH * PL/ASC
67N03 MO/ASC * SA/ASC
66N49 VE/MH * UR/ASC
66N45 MO/ASC * JU/ASC
66N33 SU/ASC * SA/ASC
66N25 ME/ASC * NE/DSC
66N23 PL/IC * JU/DSC
66N17 SU/ASC * JU/ASC
66N10 MO/ASC * UR/DSC
65N45 SU/ASC * UR/DSC
65N44 MO/MH * JU/DSC
65N28 MO/ASC * SU/ASC
65N14 SA/MH * MA/ASC
65N10 MA/DSC * NE/DSC
64N58 ME/ASC * MA/DSC
64N52 VE/MH * PL/ASC
64N44 JU/MH * MA/ASC
64N32 MA/ASC * PL/DSC
64N31 SU/ASC * MA/DSC
64N26 SU/MH * UR/ASC
64N16 MO/ASC * MA/DSC
64N05 JU/MH * PL/DSC
63N50 MA/DSC * UR/DSC
63N45 PL/IC * SA/DSC
63N35 SU/MH * PL/ASC
62N54 MO/MH * SA/DSC
62N38 UR/ASC * PL/ASC
62N28 VE/ASC * SA/ASC
62N22 SA/MH * PL/DSC
61N53 MA/DSC * JU/ASC
61N45 VE/ASC * MA/DSC
61N40 VE/ASC * JU/ASC
61N33 JU/IC * SU/DSC
61N19 UR/IC * MA/ASC
61N13 Anchorage

59N35 SU/ASC * ME/ASC
59N23 MA/DSC * SA/ASC
59N20 Stockholm
59N00 VE/ASC * UR/DSC
57N09 MA/IC * PL/DSC
56N54 ME/MH * JU/DSC
55N56 NE/DSC * PL/DSC
55N45 Moscow
55N41 Copenhagen
54N59 NE/IC * PL/ASC
54N33 JU/IC * VE/DSC
54N16 MO/ASC * VE/ASC
53N22 PL/IC * MA/ASC
53N19 MA/IC * NE/DSC
52N30 Berlin
51N51 MO/MH * MA/ASC

CENTRAL LATITUDES

51N30 London
51N22 SA/IC * SU/DSC
51N19 ME/MH * SA/DSC

49N35 SU/ASC * PL/DSC
49N14 Vancouver
49N13 UR/MH * NE/ASC
48N50 Paris
47N36 Seattle
46N37 SA/IC * VE/DSC
45N32 NE/IC * UR/ASC
45N32 Montreal
45N16 VE/MH * JU/DSC
44N59 Minneapolis
43N41 ME/ASC * PL/DSC
43N39 Toronto
42N22 Boston
42N20 Detroit
41N54 Rome
41N53 Chicago
41N17 VE/ASC * PL/DSC
40N43 New York

39N44 Denver
38N53 Washington
38N37 St. Louis
38N22 SU/MH * JU/DSC
38N05 SU/DSC * VE/DSC
37N47 San Francisco
37N26 ME/ASC * VE/ASC
36N10 Las Vegas
35N59 VE/MH * SA/DSC
35N40 Tokyo
34N51 NE/MH * MA/DSC
34N04 Los Angeles
33N45 Atlanta
33N27 Phoenix
32N47 Dallas
31N55 ME/MH * MA/ASC
30N03 Cairo

29N58 New Orleans
29N46 Houston
28N39 Delhi
28N06 SA/MH * NE/DSC
27N08 SU/MH * SA/DSC
25N47 Miami
22N34 Calcutta
22N16 Hong Kong
21N19 Honolulu

19N26 Mexico City
18N43 NE/MH * SA/ASC
16N54 MA/MH * VE/DSC
14N35 Manila
10N30 Caracas

9N41 VE/MH * MA/ASC
8N36 MO/ASC * PL/DSC
8N01 JU/MH * NE/DSC
6N27 Lagos
5N21 NE/MH * JU/ASC
1N31 MA/IC * SU/ASC
0N30 SU/IC * MA/DSC

0S30 SU/MH * MA/ASC
1S31 MA/MH * SU/DSC
5S21 NE/IC * JU/DSC
8S01 JU/IC * NE/ASC
8S36 MO/DSC * PL/ASC
9S41 VE/IC * MA/DSC

16S54 MA/IC * VE/ASC
18S43 NE/IC * SA/DSC

SOUTH LATITUDES

22S54 Rio De Janeiro
26S12 Johannesburg
27S08 SU/IC * SA/ASC
28S06 SA/IC * NE/ASC

31S55 ME/IC * MA/DSC
33S55 Sydney
34S36 Buenos Aires
34S51 NE/IC * MA/ASC
35S59 VE/IC * SA/ASC
36S52 Auckland
37S26 ME/DSC * VE/DSC
37S50 Melbourne
38S05 SU/ASC * VE/ASC
38S22 SU/IC * JU/ASC

41S17 VE/DSC * PL/ASC
43S41 ME/DSC * PL/ASC
45S16 VE/IC * JU/ASC
45S32 NE/MH * UR/DSC
46S37 SA/MH * VE/ASC
49S13 UR/IC * NE/DSC
49S35 SU/DSC * PL/ASC

51S19 ME/IC * SA/ASC
51S22 SA/MH * SU/ASC
51S51 MO/IC * MA/DSC
53S19 MA/MH * NE/ASC
53S22 PL/MH * MA/DSC
54S16 MO/DSC * VE/DSC
54S33 JU/MH * VE/ASC
54S59 NE/MH * PL/DSC
55S56 NE/ASC * PL/ASC
56S54 ME/IC * JU/ASC
57S09 MA/MH * PL/ASC
59S00 VE/DSC * UR/ASC
59S23 MA/ASC * SA/DSC
59S35 SU/DSC * ME/DSC

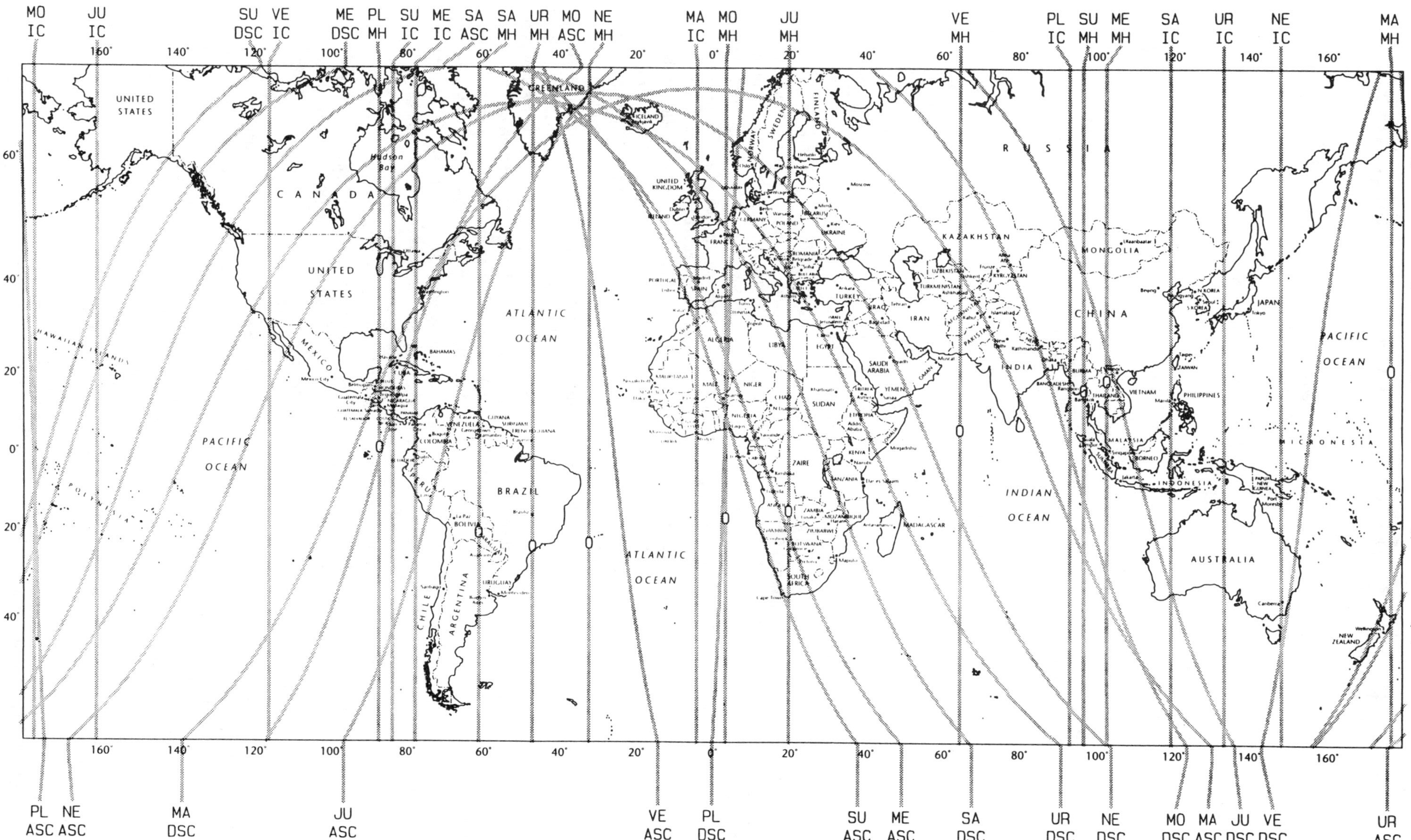

CYCLO*CARTO*GRAPHY

04/18/96

SU, MO, ME, VE AND MA ARE PROGRESSED
JU, SA, UR, NE AND PL ARE TRANSITING

First Manned Space Flight

PROG	5 7 61	2:15 GMT
TRANSIT	1 28 86	0:00 GMT
NATAL	4 12 61	7:07 GMT

Diagram 14—Full Moon, January 26, 1986, NASA

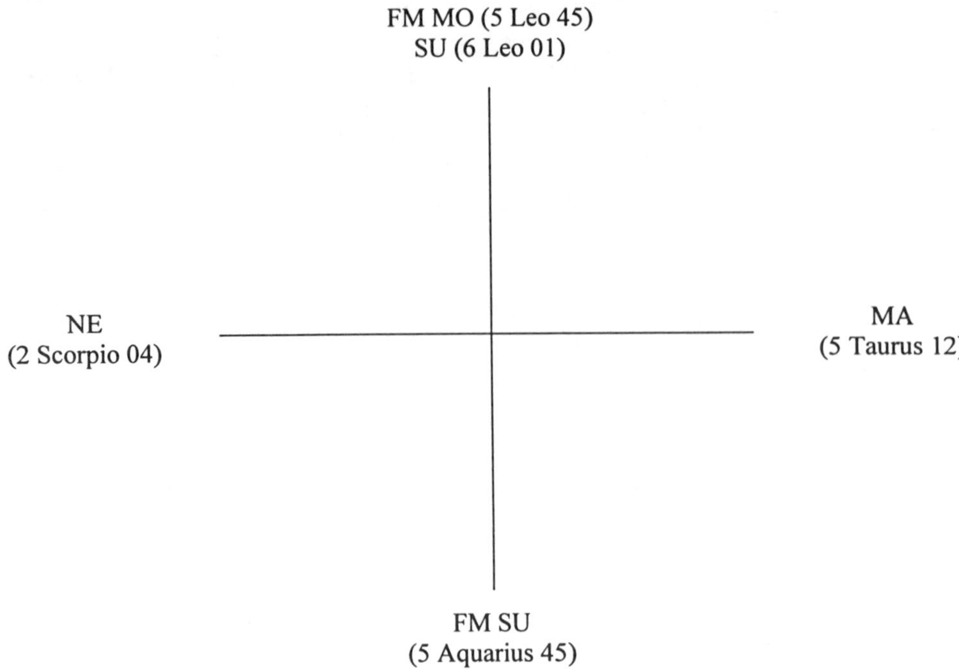

A*C*G	**(23 21)**	**Displaced C*C*G**
SA 01 AQU 24		JU 03 AQU 45
ME 03 ARI 53		MA 09 CAN 49
JU 06 AQU 43		NE 11 SAG 36
NE 08 SCO 37		SA 12 SCO 53
PL 12 VIR 44		PL 16 LIB 58
MO 15 PIS 03		MO 17 CAP 17
VE 25 ARI 49		VE 18 PIS 11
MA 20 CAN 15		ME 26 ARI 28
SU 20 ARI 28		UR 26 AQU 49
UR 24 LEO 24		

A*C*G	**(23 21)**	**Solar arc A*C*G**
SA 01 AQU 24		JU 00 PIS 04
ME 03 ARI 53		NE 01 SAG 58
JU 06 AQU 43		PL 06 LIB 05
NE 08 SCO 37		MO 08 ARI 24
PL 12 VIR 44		VE 09 TAU 10
MO 15 PIS 03		MA 13 LEO 36
VE 25 ARI 49		UR 17 VIR 45
MA 20 CAN 15		SA 24 AQU 45
SU 20 ARI 28		ME 27 ARI 14
UR 24 LEO 24		

4. solar arc A*C*G VE EQU A*C*G NE
5. solar arc SA = A*C*G UR
6. solar arc MA ASC = birthplace

On the day that the Challenger exploded there were significant temporary influences in effect at Balkonur.

a. Prog. birthplace = A*C*G UR ASC
b. Solar arc MA ASC = birthplace
c. Prog. birthplace = A*C*G NE IC (orb 1½ degrees)

These temporary influences had a powerful effect for a period of about two years, during which time both American and Soviet manned space flights needed to be conducted with extreme caution. The possibility of a sudden (UR) accident (MA) involving a manned space flight is emphasized by these temporary influences.

Another important temporary influence involving Balkonur is the applying prog. birthplace = A*C*G NE IC (orb one and a half degrees). This could indicate that the safety controls for a manned space flight were not up to the usual high quality. A few months after the Challenger explosion a 256-page report of the disaster was published. The report said that a faulty ring on one of the booster rockets was the reason for the explosion. NASA knew about this problem but didn't do anything to solve it. The prog. birthplace = A*C*G NE IC also had an effect after the explosion. As a result of the disaster, the American space program was grounded with a great deal of uncertainty surrounding its future.

C*C*G MA DSC at Cape Canaveral shows that an American manned space flight would be involved if an unexpected accident occurred.

The other temporary influences give background information to the dangerous situation developing with the American space program. Solar arc SA = A*C*G UR indicates problems involving an American manned space flight. After the Challenger explosion this aspect manifested as the suspension of the space shuttle program. Solar arc VE EQU A*C*G NE is a separating aspect, so this had more influence leading up to the disaster, and could show that NASA was trying to cut down on costs without careful consideration of the consequences. Solar return SU ASC has an influence at Cape Canaveral which does not seem to be all that important. I can only assume that this puts emphasis on NASA, which is responsible for the American space program.

The *Astro*Carto*Graphy Source Book of Mundane Maps* can be used to find out if this potential danger was activated. The important thing to remember when examining these maps is repetition of themes.

The eclipse map for November 12, 1985 shows that MA MH is near Cape Canaveral. Eclipse Mars (10 LIB 00) and Sun/Moon (20 SCO 08) do not activate any of the charts relevant for space flights, although MA MH at Cape Canaveral does show a repetition of themes with C*C*G MA DSC of the first manned space flight.

The first lunation with MA at Cape Canaveral took place on January 26, 1986, two days before the Challenger explosion. Full Moon Mars is at 25 SCO 32 which doesn't activate any of the mentioned charts. The full Moon is at SU (05 AQU 45) + MO (05 LEO 45) which activates both NASA (Diagram 14) and the first manned space flight chart (Diagram 15).

NASA's MA/NE/SU T-square was triggered by this full Moon which activated the repetition of themes obtained from other maps.

The ASC/DSC axis (both Balkonur and Cape Canaveral) and JU was also activated by this full Moon. This is a clear indication that the American space program was in the news.

The Astro*Carto*Graphy map for the Mars/Pluto conjunction that took place at 18:42 GMT on December 25, 1985 shows that MA DSC was near Cape Canaveral. This conjunction occurred at 06 SCO 46 which activated NASA's and the first manned space flight charts. The full Moon on January 26, 1986 at 05 AQU/LEO 45 also triggered the MA/PL conjunction (see Diagram 16). As the full Moon MA didn't activate any of the charts mentioned, mundane squares were not used. In this event it was the full Moon SU/MO that activated everything.

Example 3—Gulf War

On August 2, 1990, Iraq invaded Kuwait and took control of this oil-rich country. The United Nations gave Iraq until January 15, 1991 to withdraw from Kuwait, which Saddam Hussein failed to do. As a result, the Gulf War started on January 16, 1991, and after six weeks Kuwait was liberated.

British astrologer Charles Harvey predicted that the Iraq/Kuwait situation would come to a head around the eclipse on January 15, 1991. He predicted that war was likely on that date.

It was not until November 28, 1990 that the UN security council announced the deadline for Iraq to withdraw its troops from Kuwait. Charles Harvey made his prediction on August 6, 1990, when he had no knowledge about what action the UN would decide to take.

After Iraq invaded Kuwait I used Astro*Carto*Graphy to find out if information could be obtained about future developments. The *Astro*Carto*Graphy Source Book of Mundane Maps* and the following three charts were used:

1. Kuwait Independence, 12:00 p.m., June 19, 1961, Kuwait
2. Iraq-Kingdom, 6:00 a.m. August 23, 1921, Baghdad
3. Saddam Hussein, April 28, 1937, Tikrit (24N36, 43E42), birthtime unknown

In the *Book of World Horoscopes* Nicholas Campion writes that accurate information has not been obtained for the time Kuwait was granted Independence. Thus, an Astro*Carto*Graphy map for Kuwait cannot be drawn.

The Iraq-Kingdom chart is set for the beginning of the proclamation ceremony that celebrated the forming of a new state. An Astro*Carto*Graphy map for the Iraq-Kingdom chart showed nothing significant for the invasion of Kuwait. When an Astro*Carto*Graphy map for a country doesn't work, the time used for the chart is incorrect.

Only the charts can be used for Kuwait Independence, Iraq-Kingdom and Saddam Hussein. The Astro*Carto*Graphy map for the solar eclipse on July 22, 1990 shows that MA MH passes through the Gulf near to Kuwait. This on its own means nothing, so the next thing to do is examine what aspects Mars makes to the three charts.

The eclipse Mars is located at 6 TAU 19. In *Mundane Astrology* (page 328) it says that 6/7 TAU/SCO and 9 LEO/AQU are associated with oil, so the eclipse Mars activated an axis connected to oil.

The Kuwait Independence chart has a T-square consisting of NE (08 SCO 48), JU (06 AQU 13) and VE (12 TAU 08). These planets are placed in the second, fifth and eighth houses, so it is clear that Kuwait receives a great deal of money from oil. The solar eclipse Mars also activates this T-square, and the fact that Mars is near Kuwait on this map is highly significant.

The Iraq-Kingdom chart has MA (12 LEO 38) and NE (14 LEO 06), thus falling on the missing leg of the Kuwait T-square.

Progressing the Iraq-Kingdom chart for the time of the solar eclipse on July 22, 1990 gives:

prog. SU 07 SCO 04
prog. MO 07 SCO 45
prog. ME 07 SCO 59

All three progressed planets are conjunct the degree associated with oil, and conjunct Kuwait's natal Neptune, thus activating the Kuwait T-square. Thus the eclipse Mars activates both Kuwait and Iraq charts.

Saddam Hussein's chart for April 28, 1937 has SU (07 TAU) and UR (09 TAU), again hitting the Kuwait T-square. Thus the solar eclipse Mars activated all three charts when Iraq invaded Kuwait.

Mundane Astrology says that the Sun indicates the day of the event (page 189). At 2:00 a.m. on August 2, 1990, the Sun was at 09 LEO 26, the degree of oil, and hitting the Kuwait T-square.

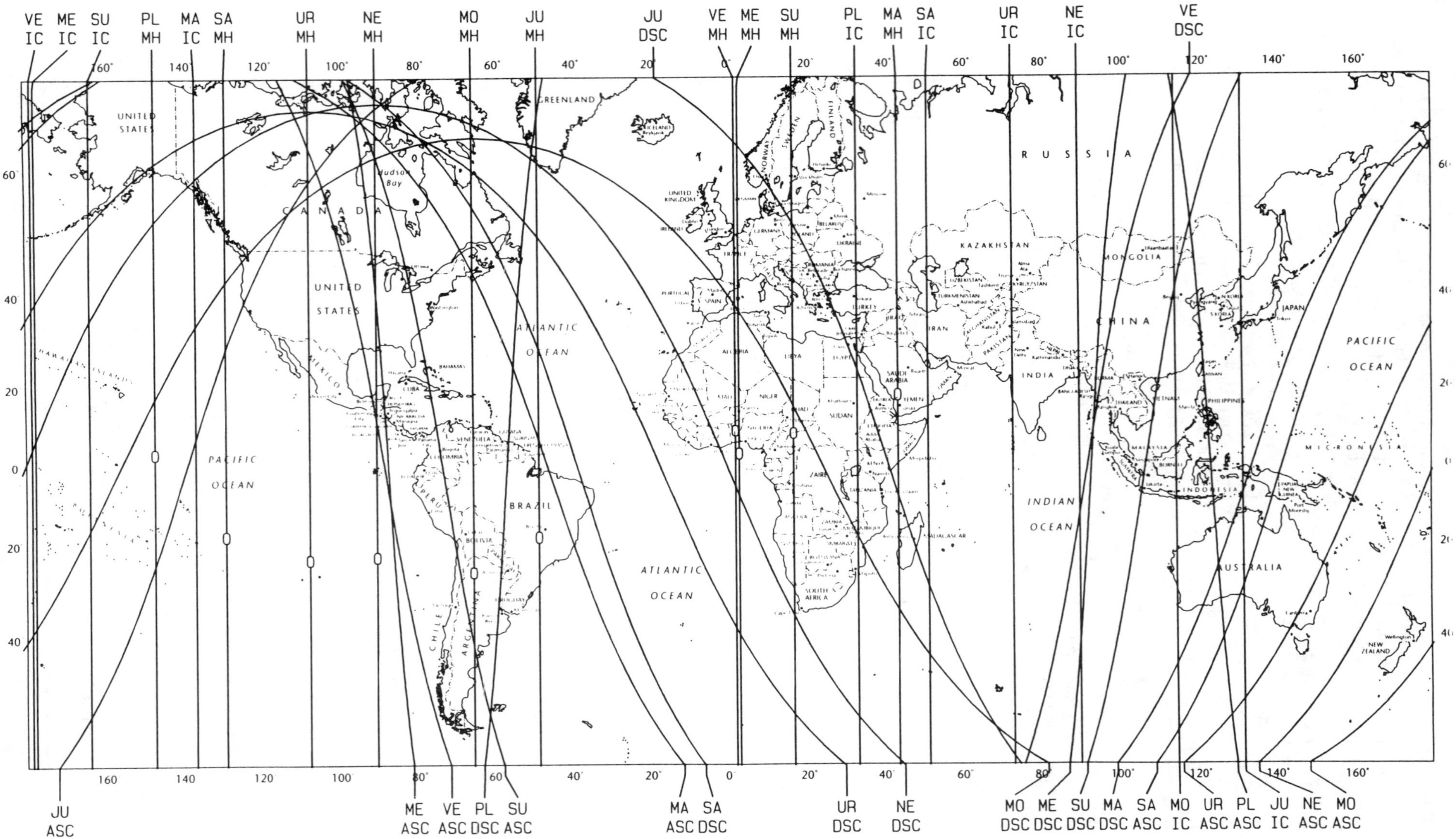

First Manned Space Flight SR

April 12, 1985 10:51 AM GMT

04/16/96

ASTRO*CARTO*GRAPHY LATITUDE CROSSINGS

NORTH LATITUDES

69N57 ME/ASC * SA/DSC
69N47 SU/IC * MA/DSC
69N42 JU/IC * SU/DSC
69N25 UR/MH * MA/ASC
69N22 SU/IC * SA/ASC
68N23 NE/MH * SA/DSC
68N21 SU/ASC * ME/ASC
67N50 JU/ASC * SA/DSC
67N45 MO/IC * PL/ASC
67N45 JU/ASC * NE/DSC
67N44 SA/DSC * NE/DSC
67N40 SU/ASC * NE/ASC
67N39 ME/ASC * NE/ASC
67N31 MA/ASC * NE/ASC
67N16 NE/MH * JU/ASC
67N05 UR/DSC * NE/ASC
67N03 VE/ASC * UR/ASC
67N01 VE/DSC * PL/ASC
66N49 UR/MH * NE/ASC
66N45 MO/IC * VE/DSC
66N42 MA/ASC * UR/DSC
66N39 ME/ASC * UR/DSC
66N33 ME/IC * MA/DSC
66N32 SU/ASC * UR/DSC
66N27 ME/ASC * MA/ASC
66N26 VE/ASC * NE/ASC
66N25 JU/ASC * UR/DSC
66N20 JU/MH * PL/DSC
66N14 VE/IC * MA/DSC
66N09 NE/MH * UR/DSC
66N06 SU/ASC * JU/ASC
66N00 MA/ASC * JU/ASC
65N47 SU/ASC * MA/ASC
65N46 MO/MH * NE/DSC
65N36 SA/MH * UR/ASC
65N35 ME/ASC * JU/ASC
64N56 ME/IC * SA/ASC
64N37 NE/MH * MA/ASC
64N29 VE/IC * SA/ASC
64N21 SA/DSC * UR/DSC
64N12 MA/IC * UR/ASC
63N58 MO/DSC * NE/DSC
63N48 MO/ASC * UR/DSC
63N43 VE/MH * JU/DSC
63N43 MO/ASC * SA/DSC
63N21 MO/DSC * PL/DSC
63N15 UR/MH * VE/ASC
63N14 ME/MH * JU/DSC
63N13 JU/MH * MO/DSC
62N44 MO/ASC * MA/ASC
62N24 SA/MH * NE/ASC
62N06 NE/MH * SU/ASC
62N01 MO/ASC * SU/ASC
62N01 NE/MH * MO/ASC
61N54 NE/DSC * PL/DSC
61N22 JU/MH * NE/DSC
61N18 MO/ASC * ME/ASC
61N18 PL/MH * UR/ASC
61N13 Anchorage
60N28 MO/MH * UR/DSC
60N26 VE/ASC * JU/ASC

59N25 MA/IC * NE/ASC
59N20 Stockholm
58N55 UR/MH * JU/ASC
58N25 MO/ASC * VE/ASC
57N04 UR/MH * MO/ASC
55N45 Moscow
55N41 Copenhagen
55N31 MO/MH * SA/DSC
54N50 MO/ASC * JU/ASC
53N26 UR/DSC * PL/DSC
53N21 PL/MH * NE/ASC
52N58 SU/IC * UR/ASC

CENTRAL LATITUDES

52N57 SU/MH * JU/DSC
52N56 SU/DSC * PL/ASC
52N30 Berlin
51N30 London
50N57 JU/MH * UR/DSC

49N14 Vancouver
48N50 Paris
47N36 Seattle
45N32 Montreal
44N59 Minneapolis
44N59 MO/MH * MA/ASC
43N39 Toronto
43N04 SA/MH * MO/ASC
42N22 Boston
42N20 Detroit
41N54 Rome
41N53 Chicago
41N13 MO/IC * SU/DSC
40N43 New York
40N20 NE/MH * ME/ASC
40N07 SA/DSC * PL/DSC

39N44 Denver
39N11 ME/IC * UR/ASC
38N53 Washington
38N37 St. Louis
38N33 VE/MH * MO/DSC
37N50 VE/IC * UR/ASC
37N47 San Francisco
37N22 ME/MH * MO/DSC
36N10 Las Vegas
35N40 Tokyo
35N23 SU/IC * NE/ASC
34N11 MA/IC * MO/ASC
34N04 Los Angeles
33N45 Atlanta
33N27 Phoenix
32N47 Dallas
30N03 Cairo

29N58 New Orleans
29N46 Houston
29N16 JU/MH * SA/DSC
28N58 SA/MH * JU/ASC
28N39 Delhi
25N47 Miami
24N58 MA/ASC * PL/DSC
24N14 PL/IC * JU/DSC
22N34 Calcutta
22N16 Hong Kong
21N19 Honolulu

19N26 Mexico City
17N24 PL/MH * MO/ASC
15N36 SU/MH * MO/DSC
14N35 Manila
11N56 NE/MH * VE/ASC
10N30 Caracas

9N27 ME/DSC * VE/DSC
7N22 MA/IC * JU/ASC
7N16 ME/IC * NE/ASC
7N02 JU/MH * MA/ASC
6N27 Lagos
4N53 VE/IC * NE/ASC

4S53 VE/MH * NE/DSC
7S02 JU/IC * MA/DSC
7S16 ME/MH * NE/DSC
7S22 MA/MH * JU/DSC
9S27 ME/ASC * VE/ASC

11S56 NE/IC * VE/DSC
15S36 SU/IC * MO/ASC

SOUTH LATITUDES

17S24 PL/IC * MO/DSC

22S54 Rio De Janeiro
24S14 PL/MH * JU/ASC
24S58 MA/DSC * PL/ASC
26S12 Johannesburg
28S58 SA/IC * JU/DSC
29S16 JU/IC * SA/ASC

33S55 Sydney
34S11 MA/MH * MO/DSC
34S36 Buenos Aires
35S23 SU/MH * NE/DSC
36S52 Auckland
37S22 ME/IC * MO/ASC
37S50 VE/MH * UR/DSC
37S50 Melbourne
38S33 VE/IC * MO/ASC
39S11 ME/MH * UR/DSC

40S07 SA/ASC * PL/ASC
40S20 NE/IC * ME/DSC
41S13 MO/MH * SU/ASC
43S04 SA/IC * MO/DSC
44S59 MO/IC * MA/DSC

50S57 JU/IC * UR/ASC
52S56 SU/ASC * PL/DSC
52S57 SU/IC * JU/ASC
52S58 SU/MH * UR/DSC
53S21 PL/IC * NE/DSC
53S26 UR/ASC * PL/ASC
54S50 MO/DSC * JU/DSC
55S31 MO/IC * SA/ASC
57S04 UR/IC * MO/DSC
58S25 MO/DSC * VE/DSC
58S55 UR/IC * JU/DSC
59S25 MA/MH * NE/DSC

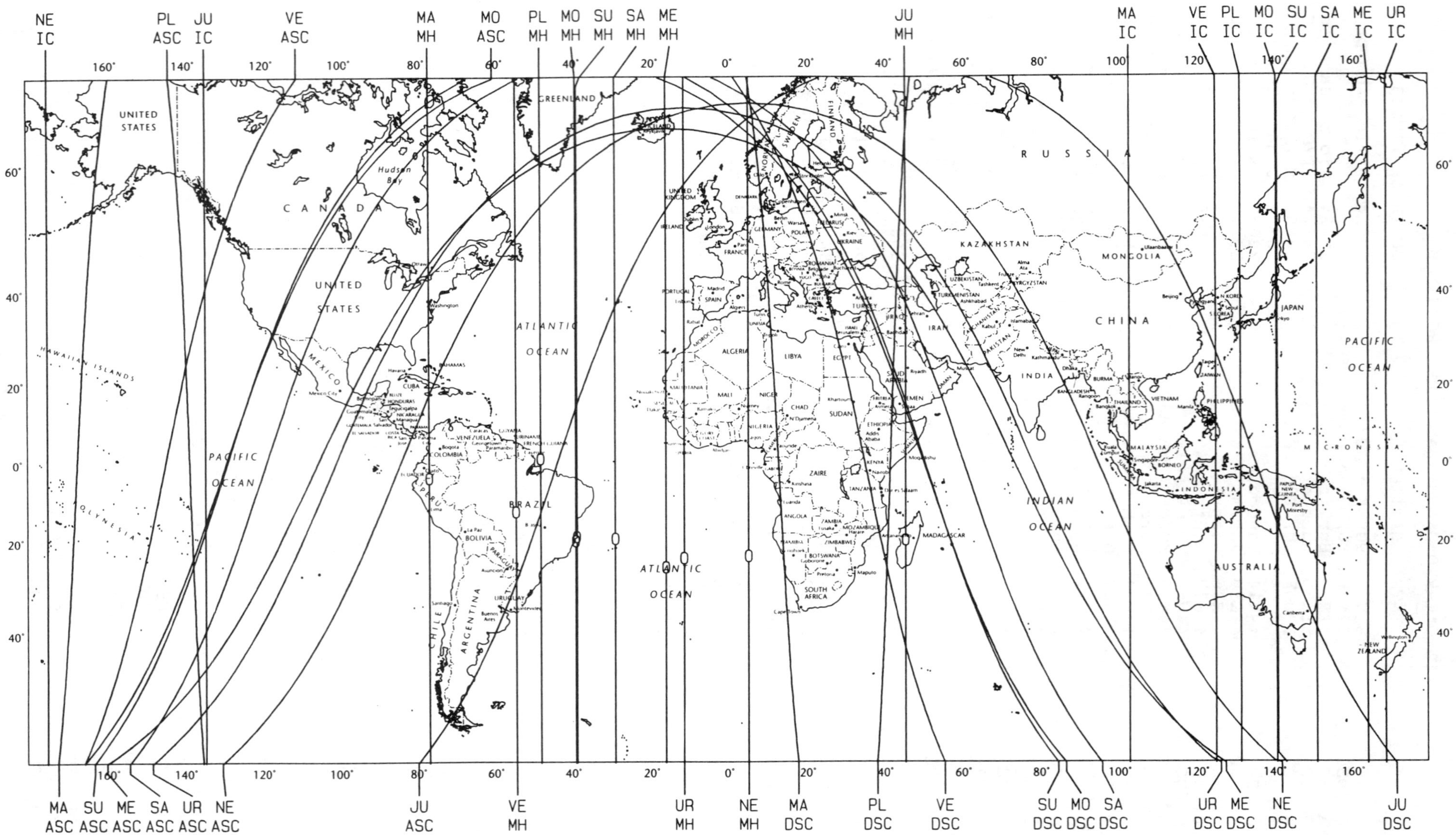

Eclipse New Moon

November 12, 1985 2:19 PM GMT

09/18/98

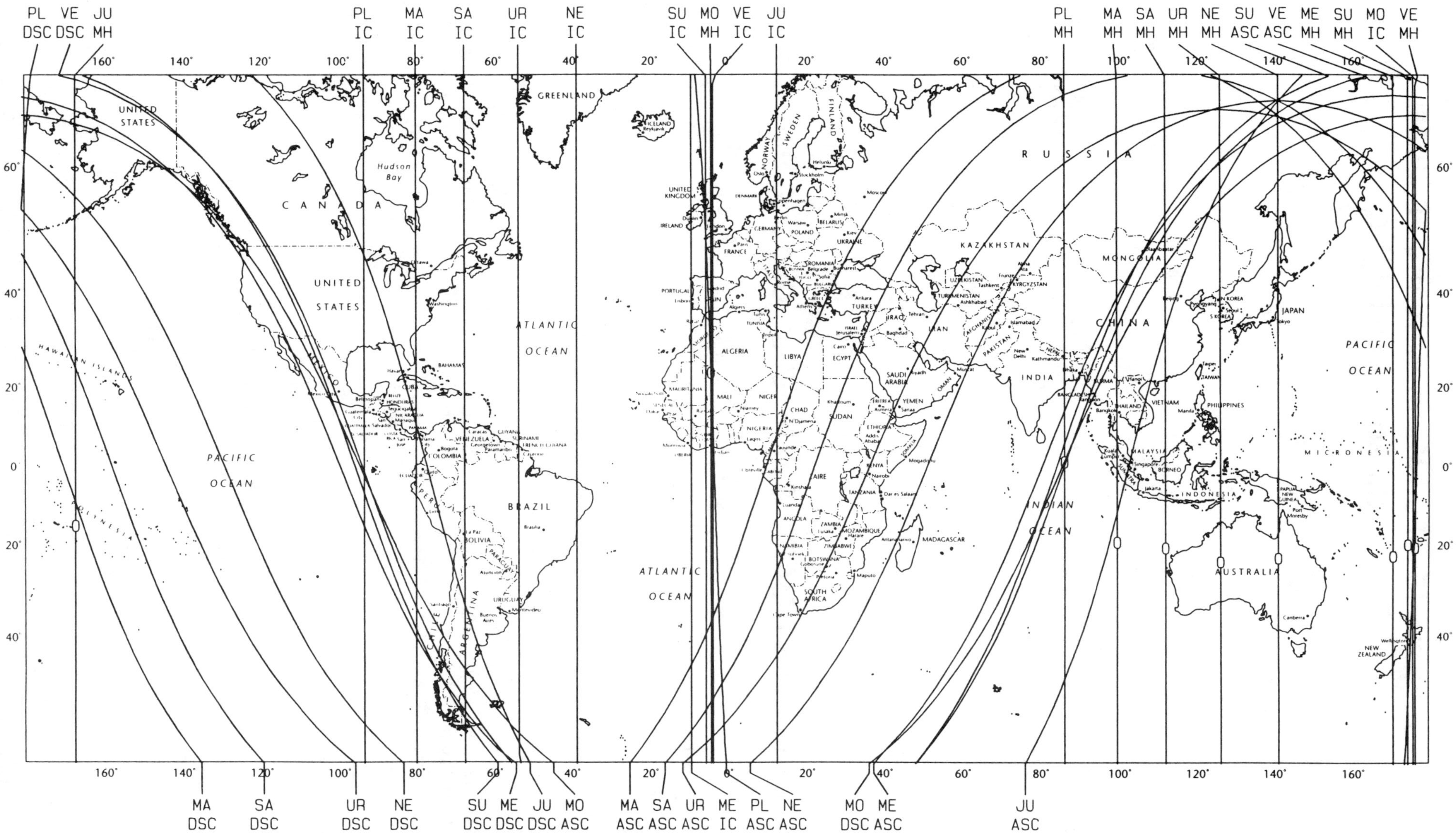

Full Moon

January 26, 1986 12:31 AM GMT

09/18/98

Diagram 15—Full Moon, January 26, 1986, First Manned Space Flight

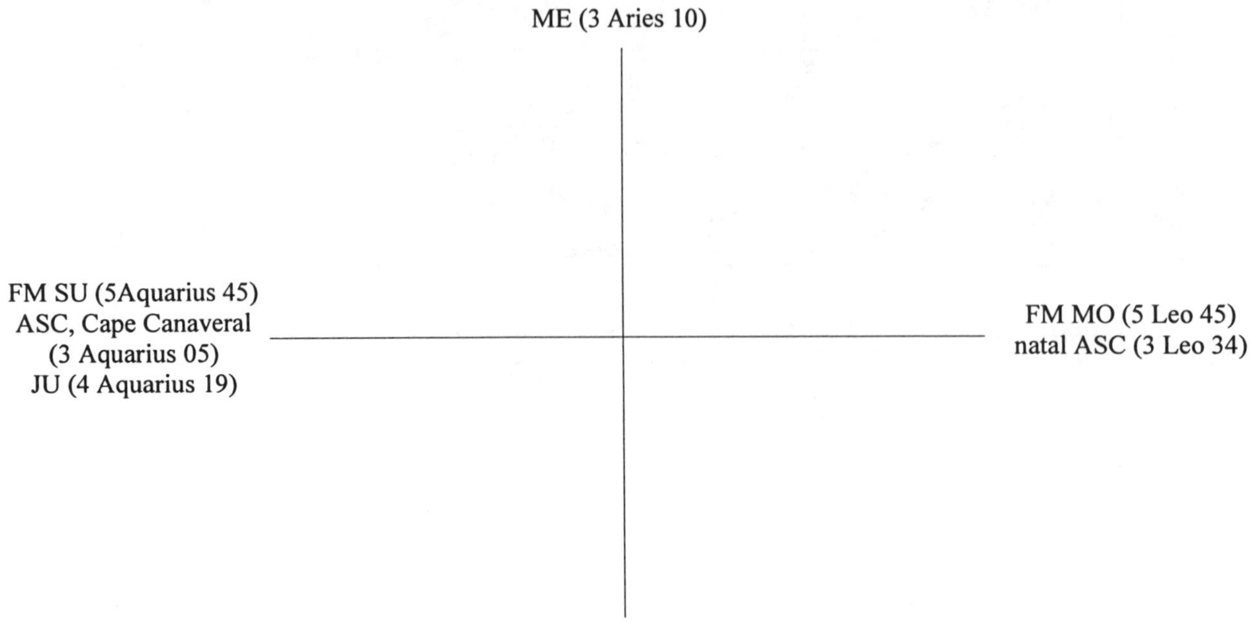

Diagram 16—MA/PL Conjunction Cycle (6 Scorpio 46)

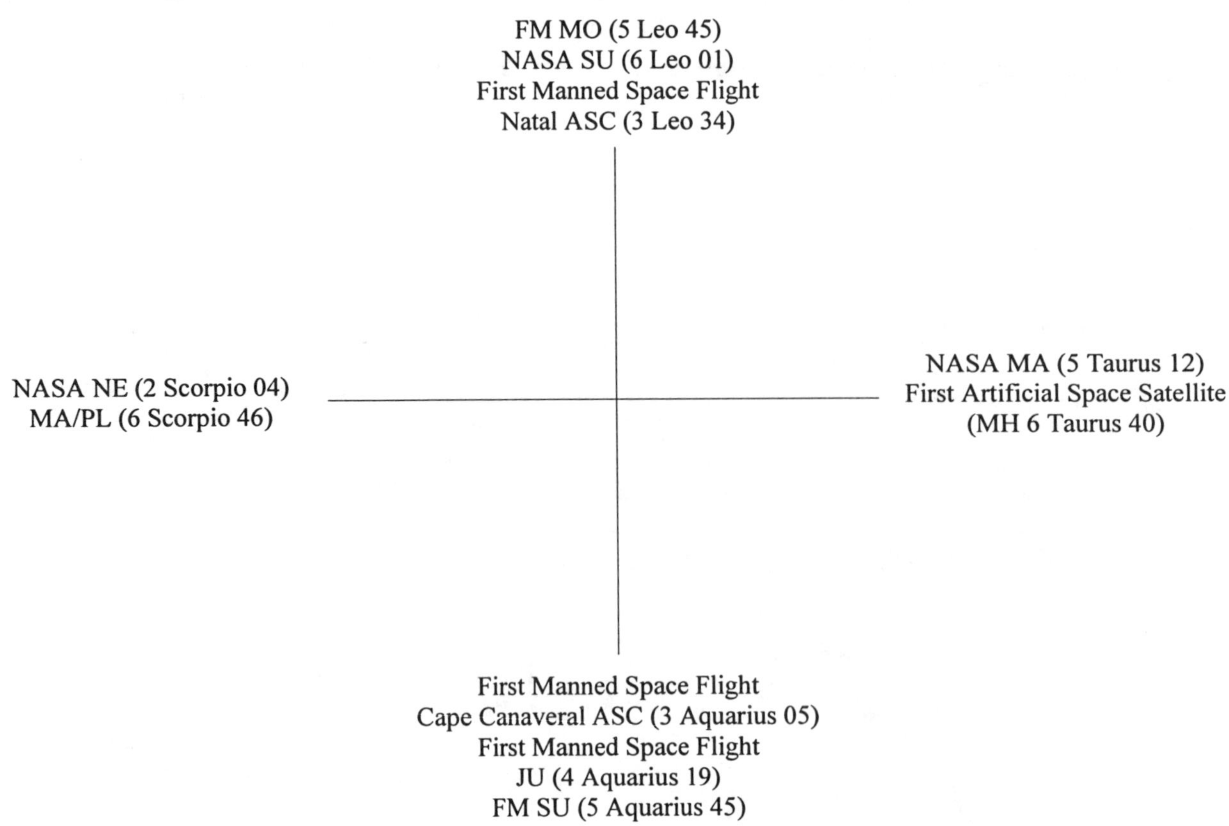

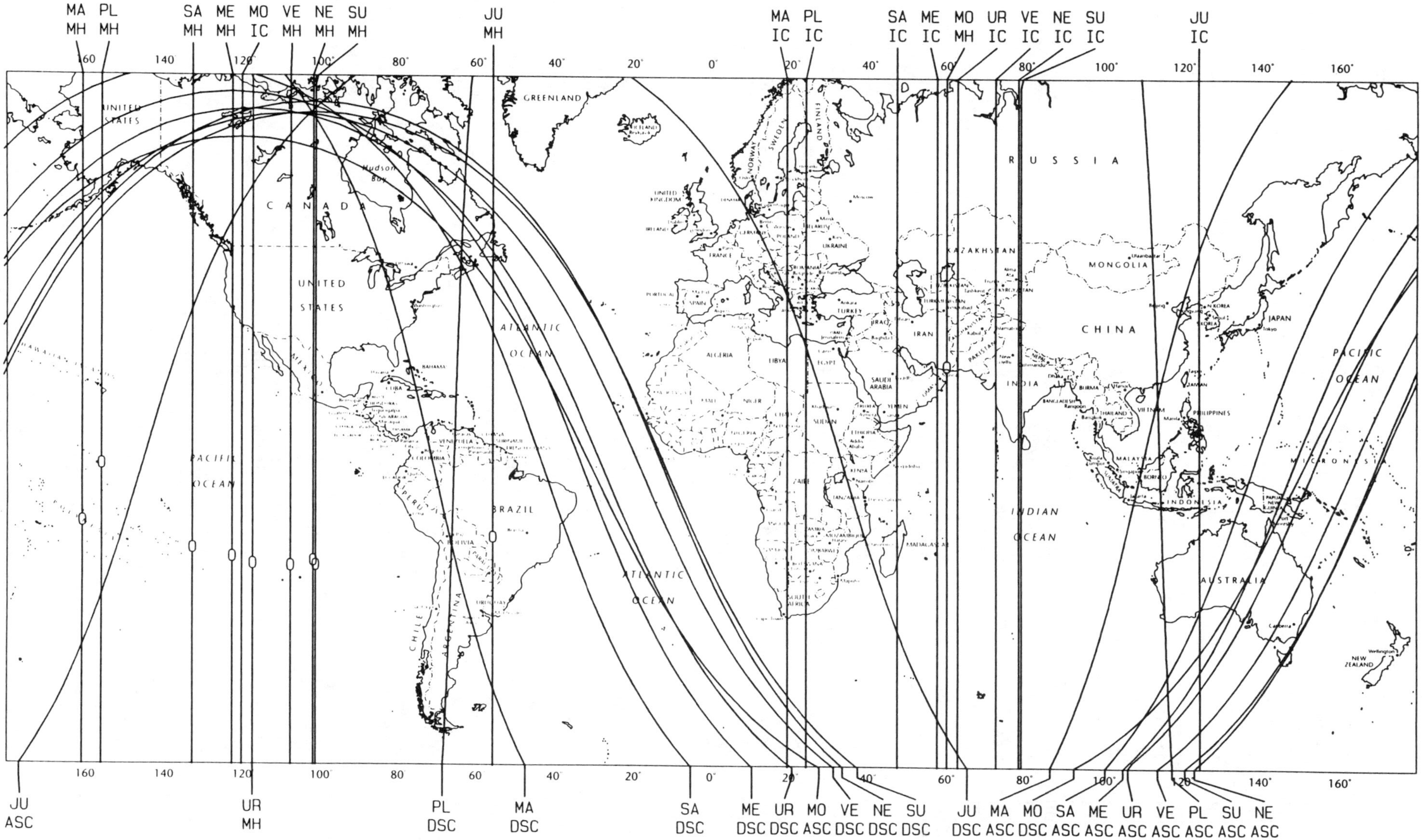

04/16/96

Mars Cjn Pluto

December 25, 1985 6:42 PM GMT

Mars Cjn Pluto
December 25, 1985 6:42 PM GMT

ASTRO*CARTO*GRAPHY LATITUDE CROSSINGS

NORTH LATITUDES

69N21	VE/MH * MA/DSC
69N11	PL/MH * SA/ASC
68N55	VE/MH * SA/DSC
68N41	MA/DSC * SA/DSC
68N35	MO/IC * ME/DSC
68N31	UR/MH * ME/DSC
68N23	MA/MH * SA/ASC
68N17	SA/MH * ME/ASC
67N56	VE/MH * ME/DSC
67N52	NE/MH * SA/DSC
67N43	SU/MH * SA/DSC
67N40	SU/MH * NE/DSC
67N40	SA/DSC * NE/DSC
67N39	JU/ASC * NE/DSC
67N39	MA/DSC * NE/ASC
67N38	ME/DSC * NE/ASC
67N33	VE/MH * NE/ASC
67N32	ME/DSC * MA/DSC
67N29	JU/ASC * SA/DSC
67N16	NE/MH * ME/DSC
67N11	SU/MH * ME/DSC
67N10	ME/DSC * JU/ASC
67N06	SU/MH * JU/ASC
66N57	UR/DSC * NE/ASC
66N55	MO/IC * UR/ASC
66N55	UR/MH * NE/ASC
66N52	NE/MH * JU/ASC
66N52	ME/MH * UR/ASC
66N40	MA/DSC * JU/ASC
66N39	VE/MH * UR/DSC
66N37	NE/MH * SU/ASC
66N37	SU/ASC * MA/DSC
66N37	SU/ASC * JU/ASC
66N36	MO/IC * NE/ASC
66N34	SU/DSC * ME/DSC
66N32	SU/ASC * UR/DSC
66N32	SU/DSC * SA/DSC
66N29	SU/ASC * VE/DSC
66N29	VE/MH * SU/ASC
66N28	VE/DSC * UR/DSC
66N25	VE/DSC * JU/ASC
66N23	VE/DSC * MA/DSC
66N23	NE/MH * VE/DSC
66N21	SU/MH * VE/DSC
66N20	NE/MH * MA/DSC
66N19	JU/ASC * UR/DSC
66N18	ME/MH * NE/ASC
66N11	SA/MH * UR/ASC
66N11	UR/MH * VE/ASC
66N11	NE/MH * UR/DSC
66N09	MA/DSC * UR/DSC
66N06	SU/MH * UR/DSC
66N06	ME/DSC * SA/DSC
65N59	MO/IC * VE/ASC
65N56	SU/MH * MA/DSC
65N52	VE/DSC * SA/DSC
65N48	ME/DSC * VE/DSC
65N46	ME/MH * VE/ASC
65N46	UR/MH * SU/ASC
65N26	MO/IC * SU/ASC
65N06	ME/MH * SU/ASC
64N53	PL/MH * ME/ASC
64N27	SA/MH * NE/ASC
64N22	SA/MH * VE/ASC
64N20	VE/MH * JU/ASC
64N10	SA/DSC * UR/DSC
64N06	VE/ASC * NE/ASC
63N59	ME/MH * MO/DSC
63N59	UR/MH * MO/ASC
63N38	MO/DSC * SU/ASC
63N35	MO/ASC * JU/ASC
63N29	MA/MH * ME/ASC
63N28	VE/MH * MO/ASC
63N28	SA/MH * MO/DSC

CENTRAL LATITUDES

63N08	SA/MH * SU/ASC
62N49	NE/MH * MO/ASC
62N44	MO/DSC * NE/ASC
62N44	SU/MH * MO/ASC
62N21	MO/DSC * VE/ASC
62N03	NE/DSC * PL/DSC
61N51	MO/ASC * MA/DSC
61N29	PL/MH * UR/ASC
61N13	Anchorage
61N06	SU/DSC * PL/DSC
61N04	ME/DSC * UR/DSC
59N39	MA/MH * UR/ASC
59N23	JU/MH * NE/DSC
59N20	Stockholm
59N02	PL/MH * MO/DSC
58N33	VE/DSC * PL/DSC
58N23	JU/MH * SU/DSC
58N09	UR/MH * JU/ASC
57N18	MA/MH * MO/DSC
56N56	PL/MH * VE/ASC
55N45	Moscow
55N41	Copenhagen
55N39	MO/IC * JU/ASC
55N06	PL/MH * NE/ASC
54N50	JU/MH * VE/DSC
54N11	UR/DSC * PL/DSC
54N08	MA/MH * VE/ASC
53N17	PL/MH * SU/ASC
53N11	ME/MH * JU/ASC
52N47	ME/DSC * PL/DSC
52N45	MO/ASC * SA/DSC
52N30	Berlin
51N33	MA/MH * NE/ASC
51N30	London
49N34	MA/MH * SU/ASC
49N14	Vancouver
48N50	Paris
48N44	MO/ASC * PL/DSC
48N33	JU/IC * MA/ASC
48N17	JU/MH * UR/DSC
47N36	Seattle
46N15	SA/DSC * PL/DSC
45N32	Montreal
45N18	JU/MH * ME/DSC
44N59	Minneapolis
43N39	Toronto
42N22	Boston
42N20	Detroit
41N54	Rome
41N53	Chicago
41N41	JU/MH * MO/ASC
40N52	MA/IC * JU/DSC
40N43	New York
39N44	Denver
38N53	Washington
38N37	St. Louis
37N47	San Francisco
37N38	SA/MH * JU/ASC
36N30	MO/DSC * UR/ASC
36N10	Las Vegas
35N40	Tokyo
34N04	Los Angeles
33N45	Atlanta
33N27	Phoenix
33N12	JU/MH * SA/DSC
32N47	Dallas
30N11	PL/IC * JU/DSC
30N03	Cairo
29N58	New Orleans
29N46	Houston
28N39	Delhi

SOUTH LATITUDES

27N25	SU/DSC * NE/DSC
25N47	Miami
23N10	MO/ASC * ME/DSC
22N34	Calcutta
22N16	Hong Kong
21N19	Honolulu
19N26	Mexico City
18N36	MA/ASC * PL/ASC
14N35	Manila
10N30	Caracas
6N27	Lagos
18S36	MA/DSC * PL/DSC
22S54	Rio De Janeiro
23S10	MO/DSC * ME/ASC
26S12	Johannesburg
27S25	SU/ASC * NE/ASC
30S11	PL/MH * JU/ASC
33S12	JU/IC * SA/ASC
33S55	Sydney
34S36	Buenos Aires
36S30	MO/ASC * UR/DSC
36S52	Auckland
37S38	SA/IC * JU/DSC
37S50	Melbourne
40S52	MA/MH * JU/ASC
41S41	JU/IC * MO/DSC
45S18	JU/IC * ME/ASC
46S15	SA/ASC * PL/ASC
48S17	JU/IC * UR/ASC
48S33	JU/MH * MA/DSC
48S44	MO/DSC * PL/ASC
49S34	MA/IC * SU/DSC
51S33	MA/IC * NE/DSC
52S45	MO/DSC * SA/ASC
52S47	ME/ASC * PL/ASC
53S11	ME/IC * JU/DSC
53S17	PL/IC * SU/DSC
54S08	MA/IC * VE/DSC
54S11	UR/ASC * PL/ASC
54S50	JU/IC * VE/ASC
55S06	PL/IC * NE/DSC
55S39	MO/MH * JU/DSC
56S56	PL/IC * VE/DSC
57S18	MA/IC * MO/ASC
58S09	UR/IC * JU/DSC
58S23	JU/IC * SU/ASC
58S33	VE/ASC * PL/ASC
59S02	PL/IC * MO/ASC
59S23	JU/IC * NE/ASC
59S39	MA/IC * UR/DSC

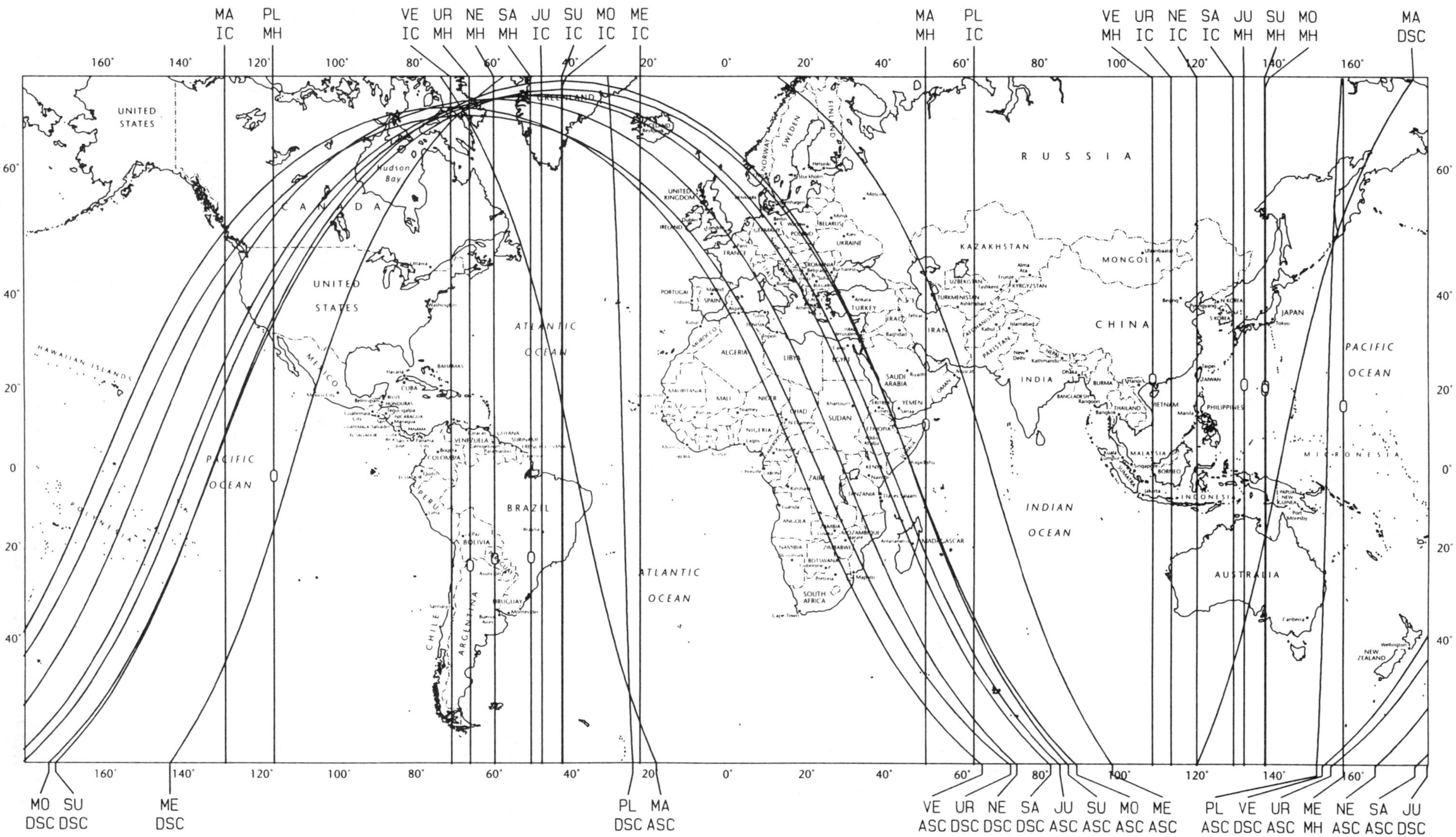

Eclipse New Moon

July 22, 1990 2:54 AM GMT

09/18/98

Diagram 17

Iraq Kingdom
♂ 7 12 ♌ 38 ♆ 14 ♌ 06
Kuwait Invasion
☉ 9 ♌ 26
Lunar Eclipse 1/30/91
☽ 9 ♌ 50
Oil 9 ♌

Kuwait
♆ 8 ♏ 48 (2nd)
Oil 6/7 ♏
Iraq Kingdom
(Progressed 7/22/90)
☉ 7 ♏ 04
☽ 7 ♏ 45
☿ 7 ♏ 59

Solar Eclipse 7/22/90
♂ 6 ♉ 19
Oil 6/7 ♉
Saddam Hussein
☉ 7 ♉
♅ 9 ♉
Kuwait
♀ 12 ♉ 08 (8th)

Oil 9 ♒
Lunar Eclipse 1/30/91
☉ 9 ♒ 50
Kuwait
♃ 6 ♒ 13 (5th)

The date of the invasion is also confirmed by using mundane squares to Mars on the solar eclipse Astro*Carto*Graphy map.

The four additional degree axes for solar eclipse Mars are:

1. 12 LEO/12 AQU
2. 24 CAP/24 CAN
3. 19 CAP/19 CAN
4. 9½ LEO/9½ AQU

These are calculated for the Iraq/Kuwait border, about 30N, so a tables of houses for Cairo (30N02) is used.

On the day that Iraq invaded Kuwait the Sun was at 9/10 LEO, thus hitting the Mars fourth degree axis. This is a clear indication of conflict in the Gulf arising from a dispute over oil. The lunar eclipse on August 6, 1990 shows **seven** angular planets in the Middle East.

Lines:

MA IC
PL MH
JU DSC
SA ASC
VE DSC
NE ASC
UR ASC

The eclipse is made more powerful as it features a grand cross with close orbs.

SU	13 LEO 51
MO	13 AQU 51
MA	16 TAU 06
PL	15 SCO 01

This grand cross triggers the Iraq-Kingdom MA (12 LEO 38) and NE (14 LEO 06) and Kuwait VE (12 TAU 08), so this eclipse shows intense MA and PL energies affecting both Iraq and Kuwait. The solar eclipse on January 15, 1991 had MA DSC exact at Kuwait, with MA making a SQU aspect to the Iraq Kingdom SU (29 LEO 26); war could break out sometime after January 15, 1991. The solar eclipse also shows the importance of examining repetition of themes.

1. Solar eclipse July 22, 1990

MA MH near to Kuwait

2. Solar eclipse January 15, 1991

Eclipse New Moon
July 22, 1990
Iraq/Kuwait Border

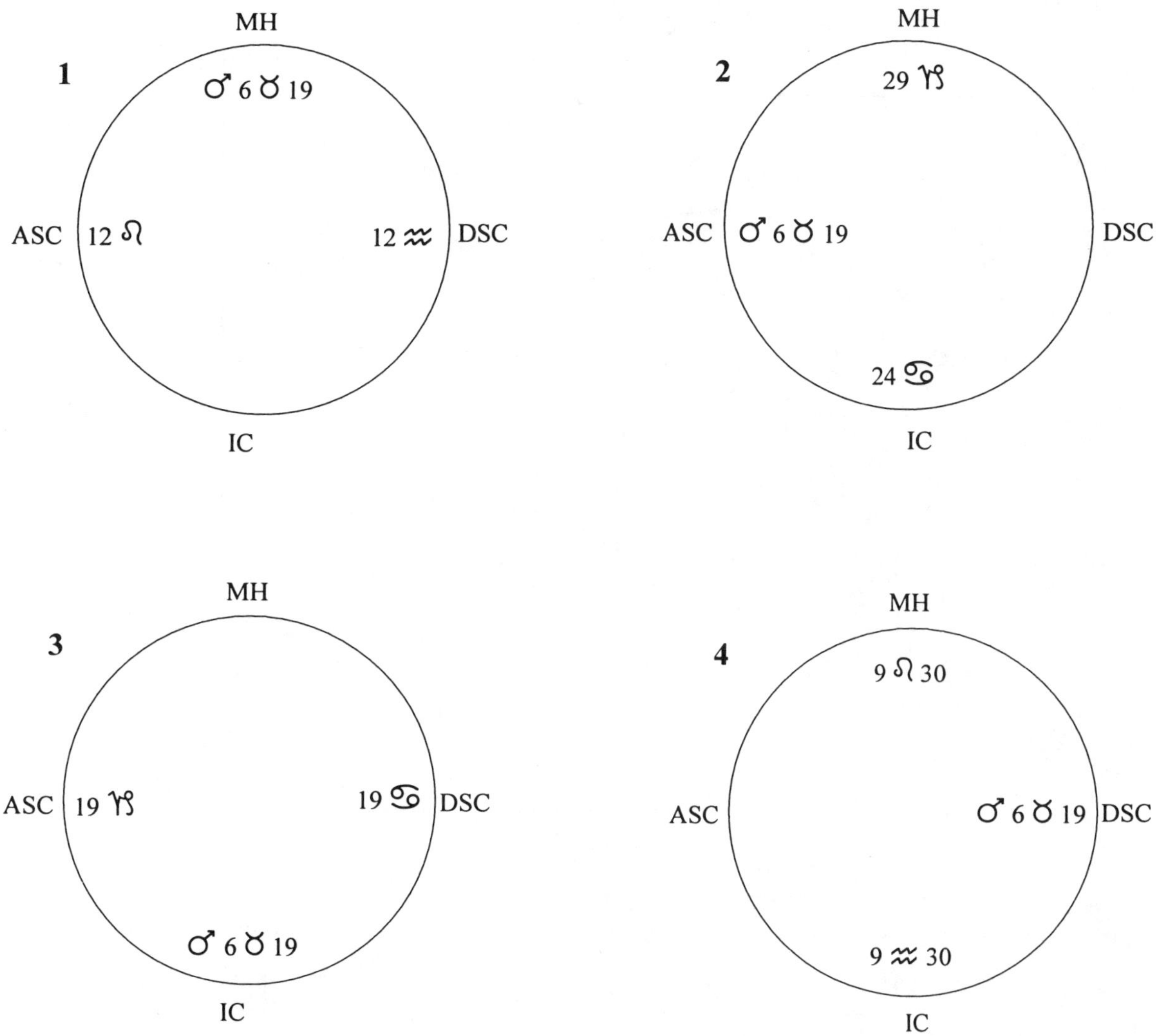

Notes:
1st degree axis = 12 ♌/12 ♒
2nd degree axis = 24 ♑/24 ♋
3rd degree axis = 19 ♑/19 ♋
4th degree axis = 9 ♌ 30/9 ♒ 30

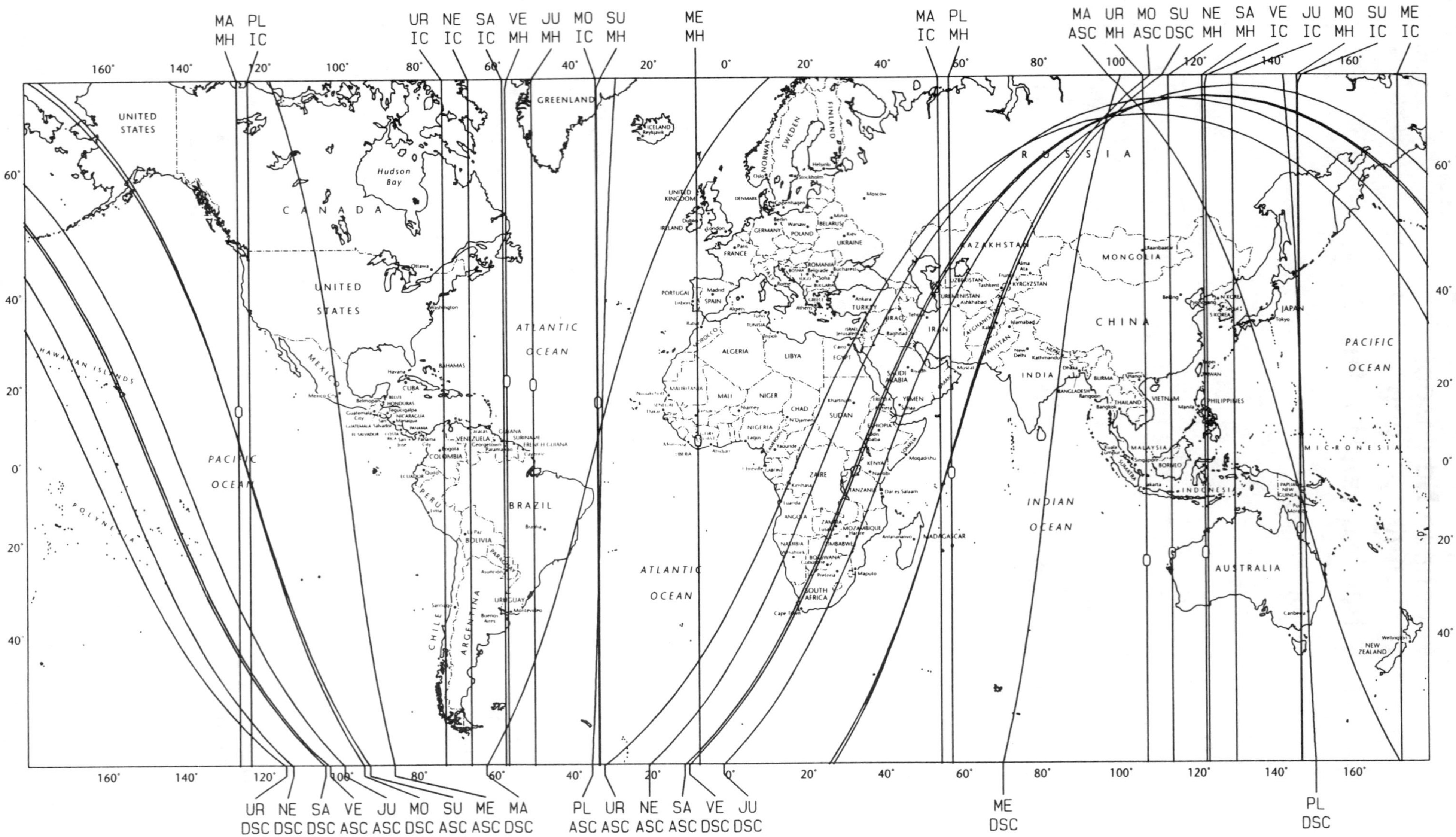

Eclipse Full Moon

August 6, 1990 2: 19 PM GMT

09/18/98

MA DSC exact at Kuwait

I thought war was likely to start at the time of the lunar eclipse on January 30, 1991, because the Sun/Moon position was at 09 AQU/LEO 50. This is conjunct a degree associated with oil, thus indirectly activating the solar eclipse Mars from July 22, 1990. The lunar eclipse Sun/Moon also triggered the invasion Sun (09 LEO 26).

At the time of this lunar eclipse Iraq started to release oil from Kuwait into the Gulf. Then Iraq set fire to oil fields, and at the beginning of March it was reported that 950 oil wells were burning or badly damaged. Diagram 17 shows how the degrees associated with oil were activated in many different ways.

An Astro*Carto*Graphy map drawn for the MA/PL opposition on August 4, 1990 shows that UR MH is at Baghdad. This indicated an unexpected and shocking MA/PL type of event connected to Iraq. NE MH passes through the Gulf near Kuwait where MA/PL energy was activated during the war when oil fields were set on fire.

This example shows that Astro*Carto*Graphy can be used most effectively in mundane astrology even when maps for countries and leaders are not possible.

Example 4—Lockerbie

On December 21, 1988, a Pan Am jumbo jet exploded in mid-air and crashed onto the Scottish town of Lockerbie. All 259 passengers on board and eleven people on the ground were killed, making this Britain's worst air disaster.

The day after this event I was scheduled to travel to West Berlin on the plane that blew up over Lockerbie. Before going to Heathrow Airport, I checked the *Astro*Carto*Graphy Source Book of Mundane Maps* to find out why this disaster occurred. On the winter solstice Astro*Carto*Graphy map for December 21, 1988, PL DSC, SU DSC and UR DSC were close to Lockerbie.

From this map I was certain that a bomb (PL) caused the explosion. The next question was who was responsible for the bomb. I concluded that Libya was involved as the winter solstice SU DSC and UR DSC both pass through this country, but at the time it was unknown why the plane had exploded.

In November 1991, two Libyan intelligence officers were accused of planting the bomb which blew up the plane. Libya refused to hand over the two men, which resulted in UN sanctions.

When the Lockerbie explosion was examined in more detail, it was apparent that this was not an isolated event for the United Kingdom. Fewer than six months earlier there was the Piper Alpha oil rig explosion, and nine days before the Lockerbie explosion Britain experienced its worse railway disaster in more than twenty years (Clapham Junction). Then, two weeks after Lockerbie, a planed crashed next to a motorway near East Midlands Airport.

The previous Mars/Pluto conjunction took place at 12:24 p.m. GMT on December 11, 1987. In *Working with Astrology*, Michael Harding and Charles Harvey explained that UR MH and MO DSC crossed exactly over the Piper Alpha oilfield. Both planets are also near Lockerbie, so these lines had an effect at more than one location.

The MA/PL conjunction (11 SCO 24) made a semisquare aspect with UR (26 SAG 26), a violent and destructive combination. At Lockerbie was SA MH, where security at airports became an important issue.

Why did the United Kingdom feel the full force from this MA/PL cycle chart? This becomes clear when you examine the United Kingdom's horoscope. In this chart Mars is at 11 TAU 45, which is activated by the MA/PL cycle chart. The MA/PL conjunction made a close opposition aspect to the United Kingdom's Mars. During the two years that this MA/PL cycle was in effect the United Kingdom was in the news when this violent potential manifested.

The 1987 winter solstice map had MA and PL near Ireland. Three months later the vernal equinox map showed that PL DSC, UR DSC and SA DSC were powerful for the United Kingdom. Although no significant event took place at the time of these ingresses the dangerous energy was building.

The summer solstice map saw seven planets having a powerful influence for the United Kingdom, and two weeks later the Piper Alpha oil rig explosion occurred. After Piper Alpha the destructive energy for the United Kingdom intensified.

The eclipse full Moon on August 27, 1988 had PL ASC near Lockerbie. A few months later the 1988 winter solstice map showed PL DSC, SU DSC and UR DSC close to Lockerbie. The repetition of themes within the space of a few months was significant for Lockerbie. With PL so prominent it is clear that a bomb was the reason why the Pan Am plane exploded.

In *Hymns to the Ancient Gods* Michael Harding explained the connection between the world's first powered flight and the 1988 winter solstice. Both Astro*Carto*Graphy maps have SU DSC and US DSC lines almost on top of each other near Lockerbie. This is an unbelievable example of planetary synchronization.

Michael Harding also wrote that an Astro*Carto*Graphy map of the world's first powered flight picks out locations where major disasters occur. With this fact in mind the temporary influences for this map at the time of the Lockerbie explosion will be examined.

A*C*G	**Age 85**	**C*C*G**
NE 05 CAN 12		UR 01 CAP 15
MA 06 AQU 44		SA 04 CAP 46
VE 07 SCO 19		MA 09 ARI 44
SA 09 AQU 05		ME 09 PIS 50
MO 09 SAG 17		NE 10 CAP 19
ME 09 CAP 41		PL 16 SCO 22
JU 17 PIS 25		MO 21 CAP 03
PL 19 GEM 23		SU 21 PIS 26
SU 24 SAG 03		VE 22 AQU 49
UR 25 SAG 21		JU 25 TAU 29

1. C*C*G PL DSC = Lockerbie
2. C*C*G SA = A*C*G NE

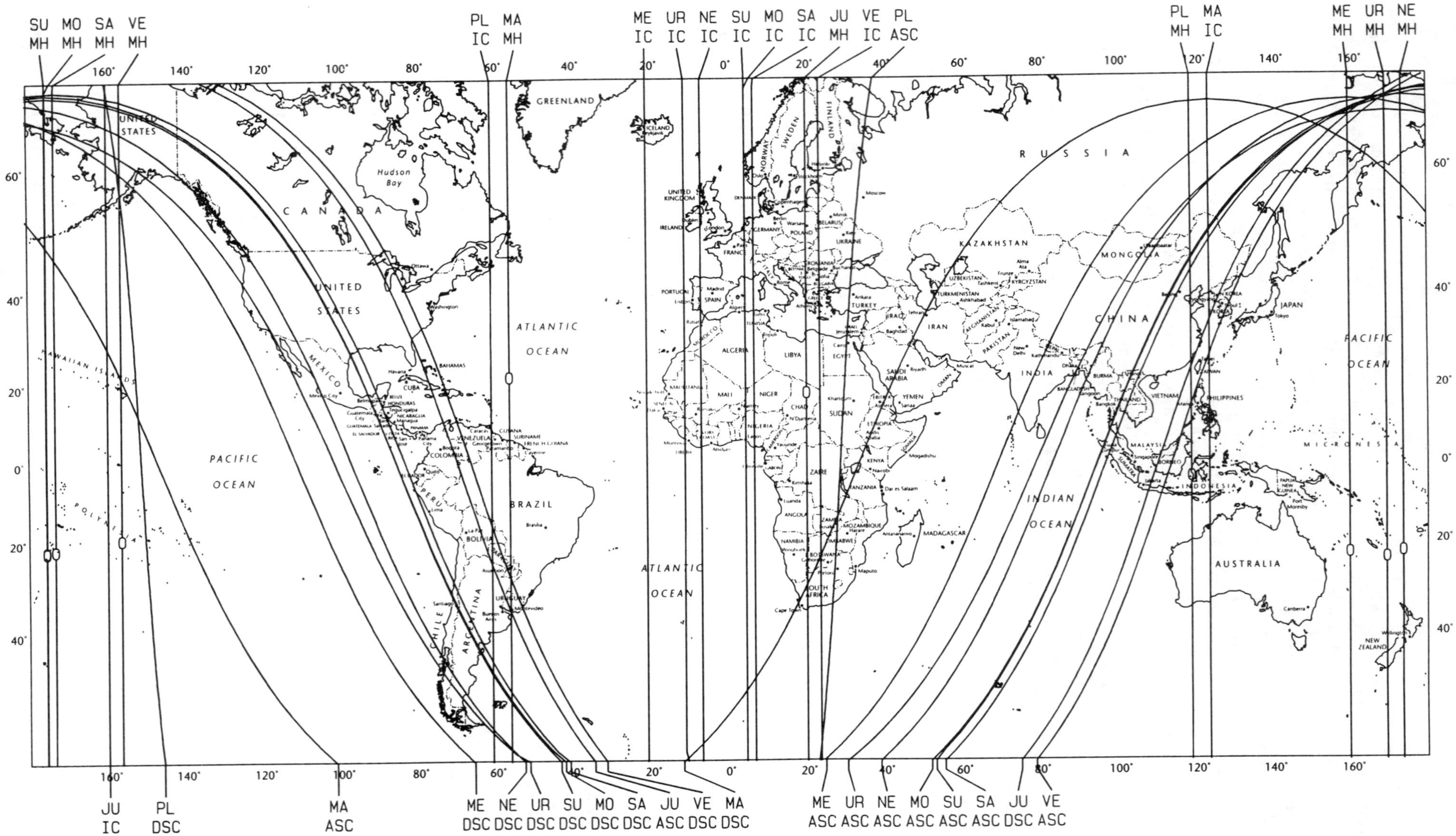

Eclipse New Moon

January 15, 1991 11:49 PM GMT

09/18/98

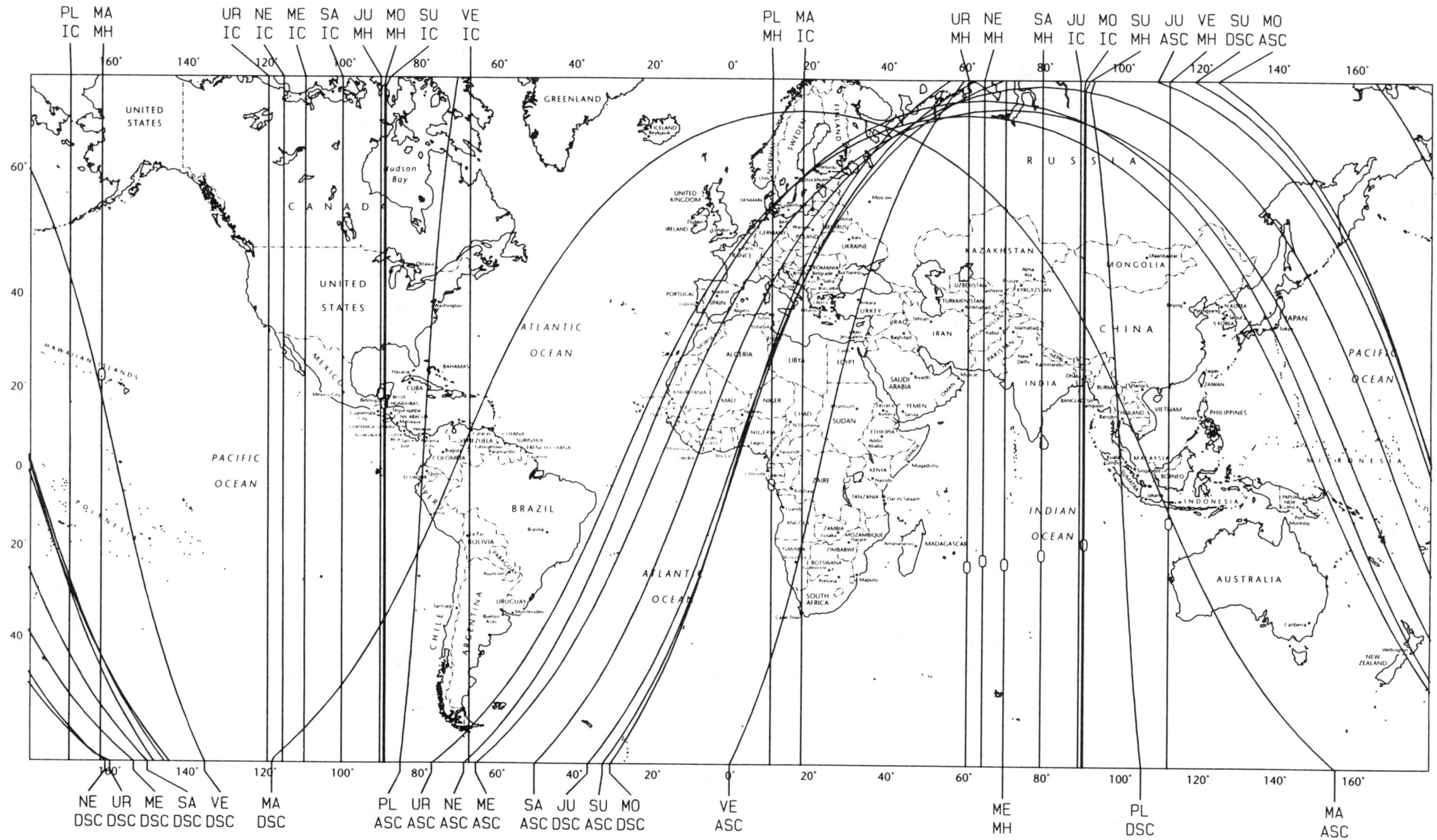

Eclipse Full Moon

January 30, 1991 6:09 AM GMT

09/18/98

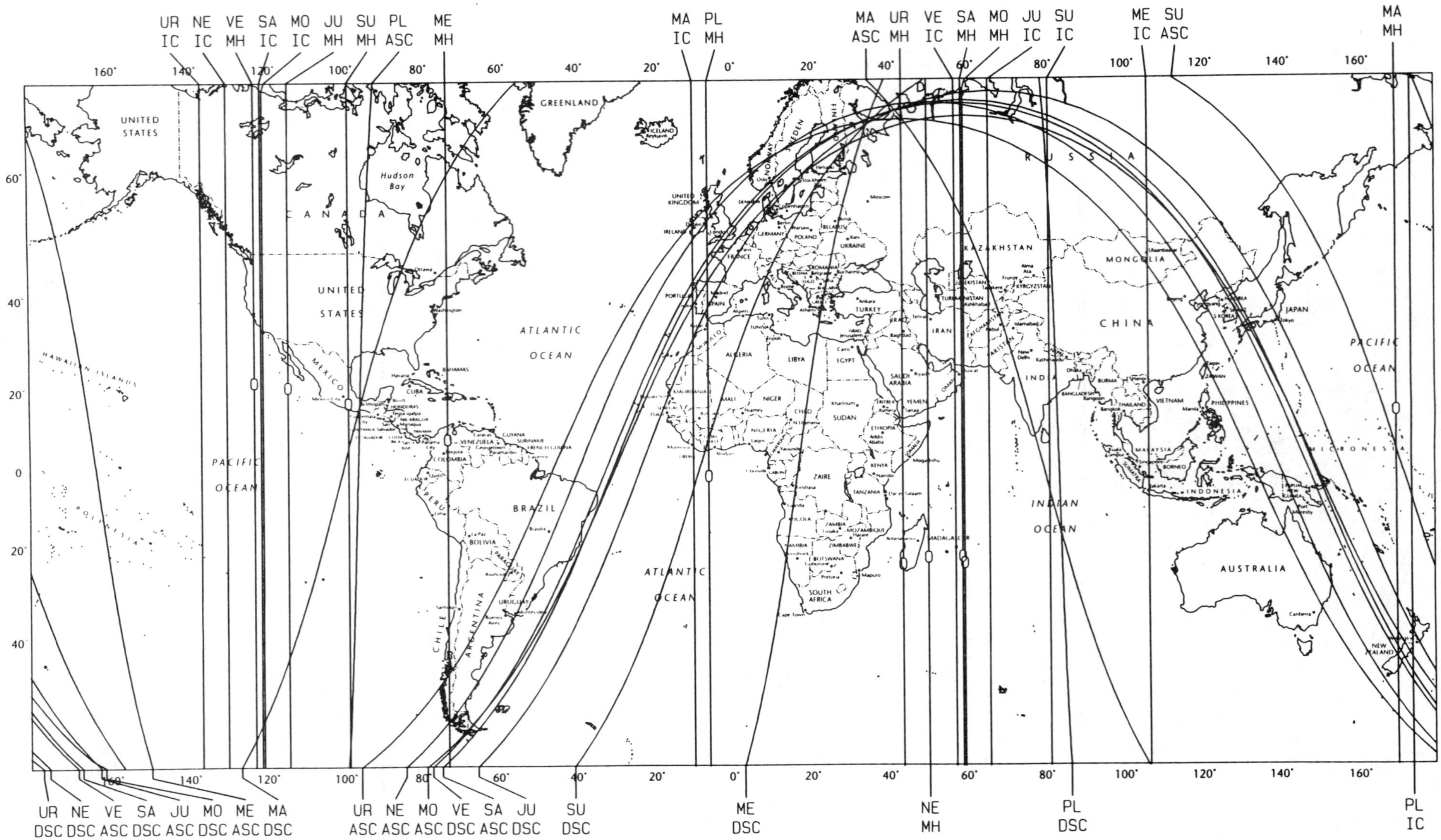

Mars OPP Pluto

August 4, 1990 6:39 PM GMT

05/21/97

Mars OPP Pluto
August 4, 1990 6:39 PM GMT

ASTRO*CARTO*GRAPHY LATITUDE CROSSINGS

NORTH LATITUDES

68N59 ME/DSC * MA/ASC
68N52 MO/MH * JU/DSC
68N51 SA/MH * JU/DSC
68N46 VE/IC * JU/DSC
68N40 UR/MH * SU/DSC
68N25 JU/ASC * PL/DSC
68N17 SU/IC * JU/ASC
68N15 NE/MH * JU/DSC
68N10 MO/MH * SA/DSC
68N10 VE/IC * SA/ASC
68N01 JU/IC * SA/DSC
68N00 SA/ASC * NE/DSC
67N59 JU/DSC * NE/ASC
67N56 NE/MH * SA/ASC
67N55 SA/MH * VE/ASC
67N55 VE/DSC * NE/DSC
67N55 MO/MH * VE/ASC
67N52 UR/MH * NE/ASC
67N51 SU/DSC * MA/ASC
67N51 VE/IC * NE/DSC
67N47 NE/MH * VE/DSC
67N46 SA/MH * NE/DSC
67N44 MA/ASC * NE/ASC
67N44 MO/MH * NE/DSC
67N42 SU/DSC * NE/ASC
67N42 JU/IC * VE/ASC
67N27 UR/MH * JU/DSC
67N27 UR/MH * SA/ASC
67N25 JU/DSC * SA/ASC
67N24 ME/DSC * NE/ASC
67N21 UR/MH * VE/DSC
67N17 MA/ASC * SA/ASC
67N14 MA/ASC * JU/DSC
67N14 JU/IC * NE/DSC
67N13 VE/DSC * MA/ASC
67N12 VE/DSC * JU/DSC
66N45 SA/DSC * PL/DSC
66N40 SA/MH * MO/ASC
66N39 VE/IC * MO/ASC
66N38 SU/DSC * SA/ASC
66N38 SU/DSC * VE/DSC
66N34 SU/IC * SA/DSC
66N33 VE/DSC * SA/ASC
66N32 JU/IC * MO/DSC
66N31 UR/MH * MA/ASC
66N24 NE/MH * MO/ASC
66N23 ME/DSC * VE/DSC
66N23 MA/ASC * UR/DSC
66N22 ME/DSC * SA/ASC
66N18 MO/ASC * UR/DSC
66N14 NE/MH * UR/DSC
66N13 VE/ASC * PL/DSC
66N13 JU/DSC * UR/ASC
66N11 SU/DSC * UR/ASC
66N08 ME/DSC * UR/ASC
66N02 SU/IC * VE/ASC
66N02 SA/ASC * UR/ASC
66N00 VE/DSC * UR/ASC
65N58 MO/ASC * MA/ASC
65N52 SU/DSC * JU/DSC
65N51 UR/MH * MO/ASC
65N50 ME/DSC * JU/DSC
65N47 SU/DSC * ME/DSC
65N46 VE/IC * UR/DSC
65N42 MO/DSC * NE/DSC
65N36 SA/MH * UR/DSC
65N33 MO/MH * UR/DSC
65N13 MO/DSC * PL/DSC
65N04 SU/IC * MO/DSC
64N56 NE/DSC * PL/DSC
64N43 SU/IC * NE/DSC
64N42 JU/IC * UR/DSC
64N23 MO/ASC * ME/DSC
63N37 MO/ASC * SU/DSC

CENTRAL LATITUDES

63N07 ME/IC * JU/ASC
62N54 UR/ASC * NE/ASC
62N35 NE/MH * MA/ASC
61N13 Anchorage
61N11 UR/DSC * PL/DSC
61N03 SU/IC * UR/DSC
60N18 ME/MH * MA/DSC

59N42 MO/ASC * JU/DSC
59N20 Stockholm
59N13 ME/IC * SA/DSC
58N01 ME/IC * VE/ASC
57N35 ME/IC * MO/DSC
57N12 SU/IC * PL/DSC
56N20 VE/IC * MA/ASC
55N52 PL/MH * UR/ASC
55N45 Moscow
55N41 Copenhagen
54N16 SA/MH * MA/ASC
53N52 PL/MH * NE/ASC
53N47 ME/IC * NE/DSC
53N39 MO/MH * MA/ASC
53N33 MA/IC * UR/ASC
52N30 Berlin
52N12 MO/DSC * VE/ASC
51N30 London
50N49 MA/IC * NE/ASC

49N14 Vancouver
48N50 Paris
47N49 PL/MH * VE/DSC
47N36 Seattle
46N21 PL/MH * SA/ASC
45N59 ME/IC * UR/DSC
45N32 Montreal
44N59 Minneapolis
43N41 PL/MH * MO/ASC
43N39 Toronto
43N35 MA/IC * VE/DSC
43N11 JU/IC * MA/ASC
42N22 Boston
42N20 Detroit
41N54 Rome
41N53 Chicago
41N45 MA/IC * SA/ASC
40N43 New York

39N44 Denver
39N01 MA/IC * MO/ASC
38N53 Washington
38N37 St. Louis
38N35 PL/MH * JU/DSC
37N47 San Francisco
36N10 Las Vegas
35N40 Tokyo
34N04 Los Angeles
33N45 Atlanta
33N27 Phoenix
32N47 Dallas
32N09 MA/IC * JU/DSC
30N03 Cairo

29N58 New Orleans
29N46 Houston
28N39 Delhi
25N47 Miami
22N34 Calcutta
22N16 Hong Kong
21N19 Honolulu

19N26 Mexico City
16N23 MA/DSC * PL/ASC
15N48 MO/DSC * SA/DSC
14N35 Manila
10N30 Caracas

SOUTH LATITUDES

7N48 PL/MH * SU/DSC
6N27 Lagos
5N37 SU/MH * MA/DSC
4N41 MA/MH * SU/ASC

4S41 MA/IC * SU/DSC
5S37 SU/IC * MA/ASC
7S48 PL/IC * SU/ASC

15S48 MO/ASC * SA/ASC
16S23 MA/ASC * PL/DSC

22S54 Rio De Janeiro
26S12 Johannesburg

32S09 MA/MH * JU/ASC
33S55 Sydney
34S36 Buenos Aires
36S52 Auckland
37S50 Melbourne
38S35 PL/IC * JU/ASC
39S01 MA/MH * MO/DSC

41S45 MA/MH * SA/DSC
43S11 JU/MH * MA/DSC
43S35 MA/MH * VE/ASC
43S41 PL/IC * MO/DSC
45S59 ME/MH * UR/ASC
46S21 PL/IC * SA/DSC
47S49 PL/IC * VE/ASC

50S49 MA/MH * NE/DSC
52S12 MO/ASC * VE/DSC
53S33 MA/MH * UR/DSC
53S39 MO/IC * MA/DSC
53S47 ME/MH * NE/ASC
53S52 PL/IC * NE/DSC
54S16 SA/IC * MA/DSC
55S52 PL/IC * UR/DSC
56S20 VE/MH * MA/DSC
57S12 SU/MH * PL/ASC
57S35 ME/MH * MO/ASC
58S01 ME/MH * VE/DSC
59S13 ME/MH * SA/ASC
59S42 MO/DSC * JU/ASC

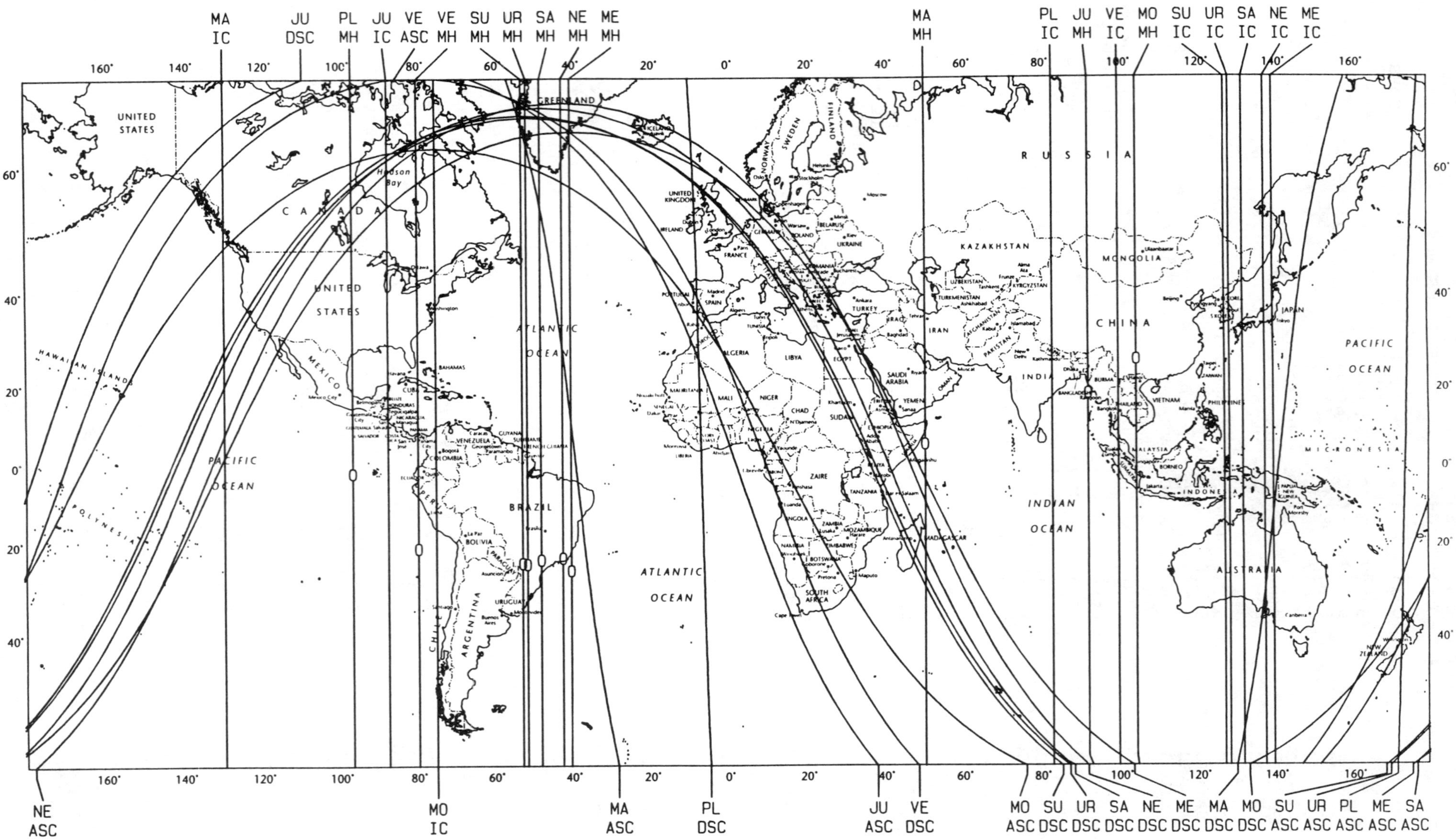

Winter Solstice

December 21, 1988 3:29 PM GMT

09/18/98

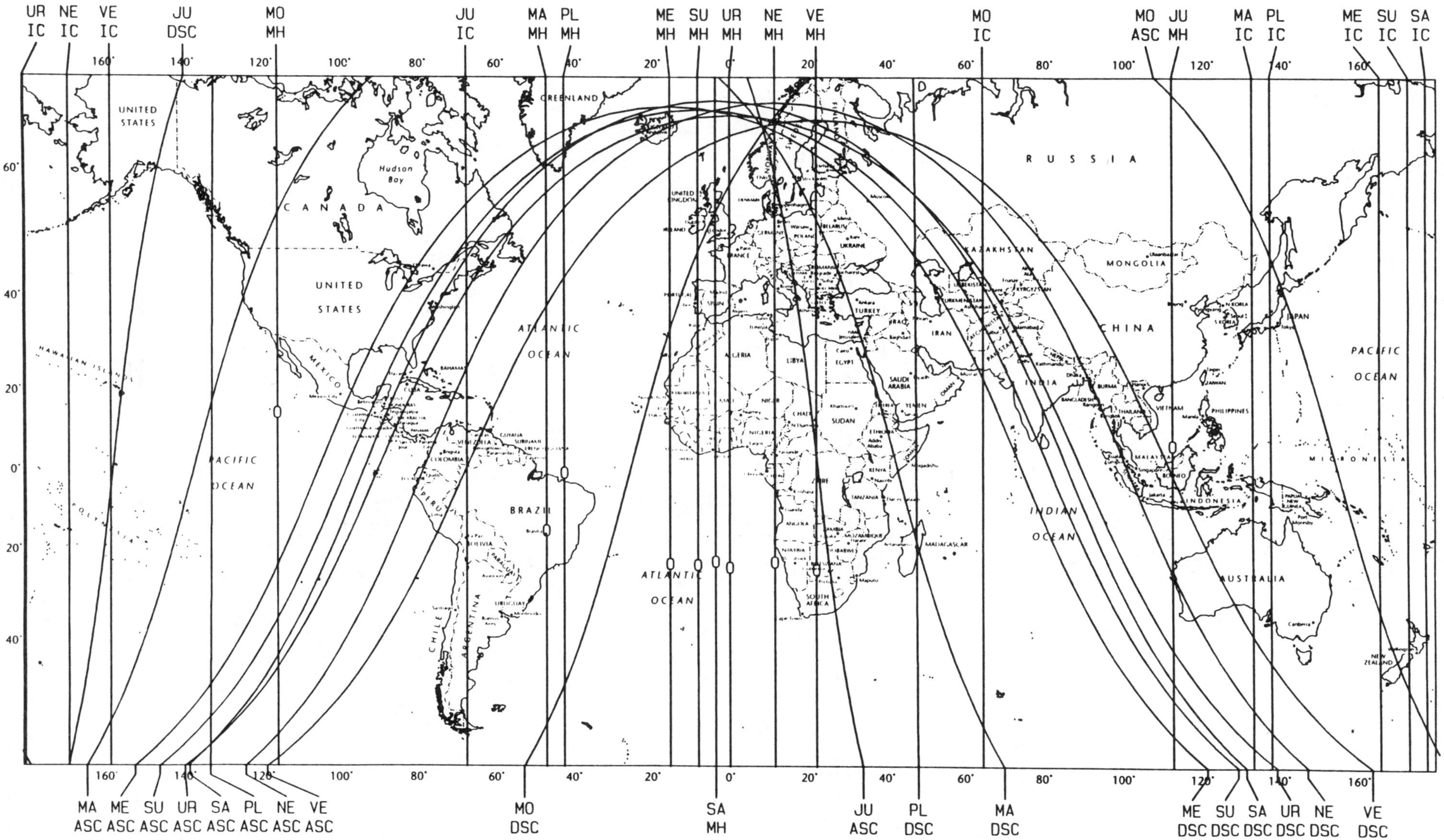

Mars Cjn Pluto

December 11, 1987 12:24 PM GMT

04/16/96

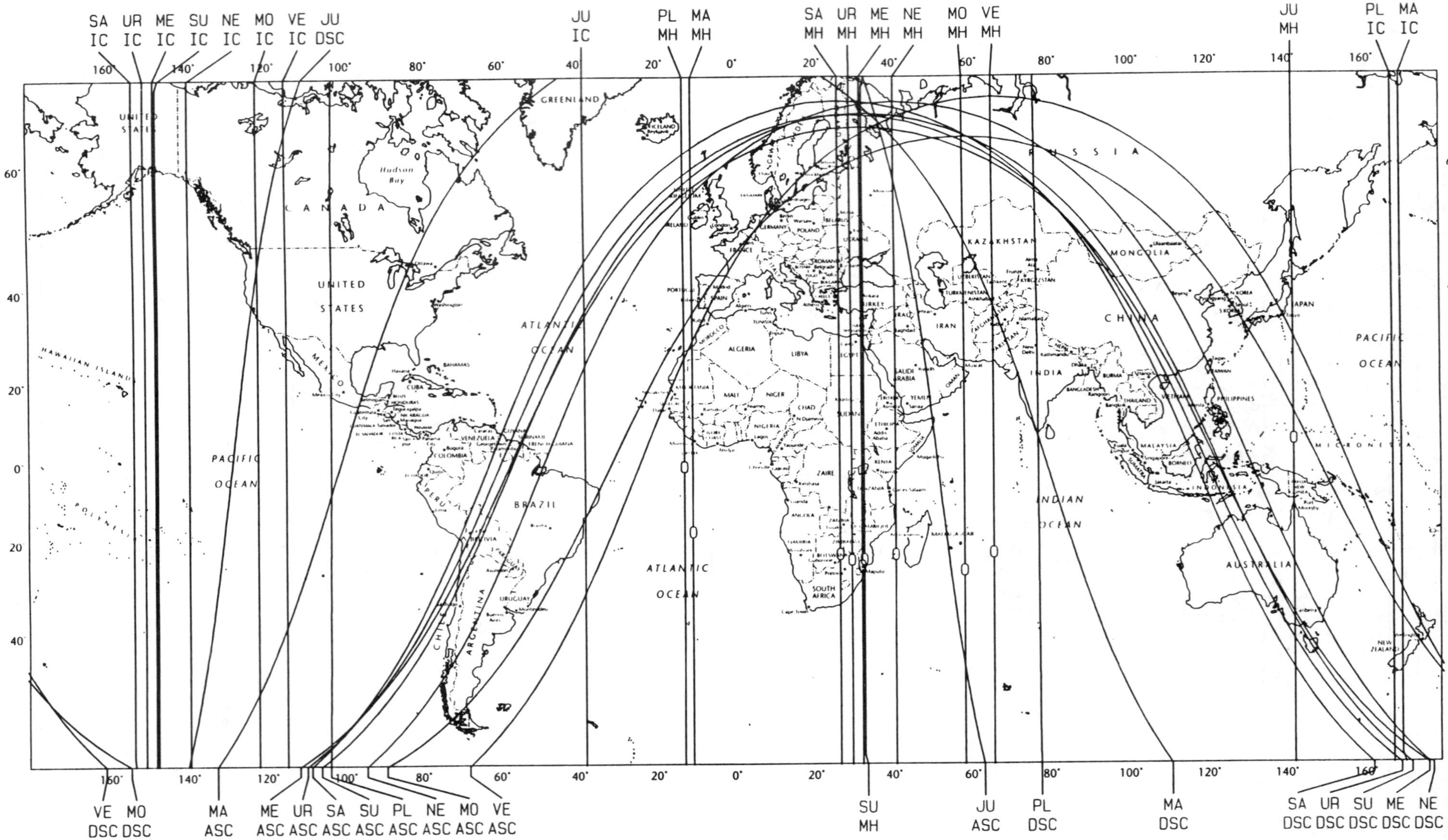

Winter Solstice

December 22, 1987 9:46 AM GMT

09/18/98

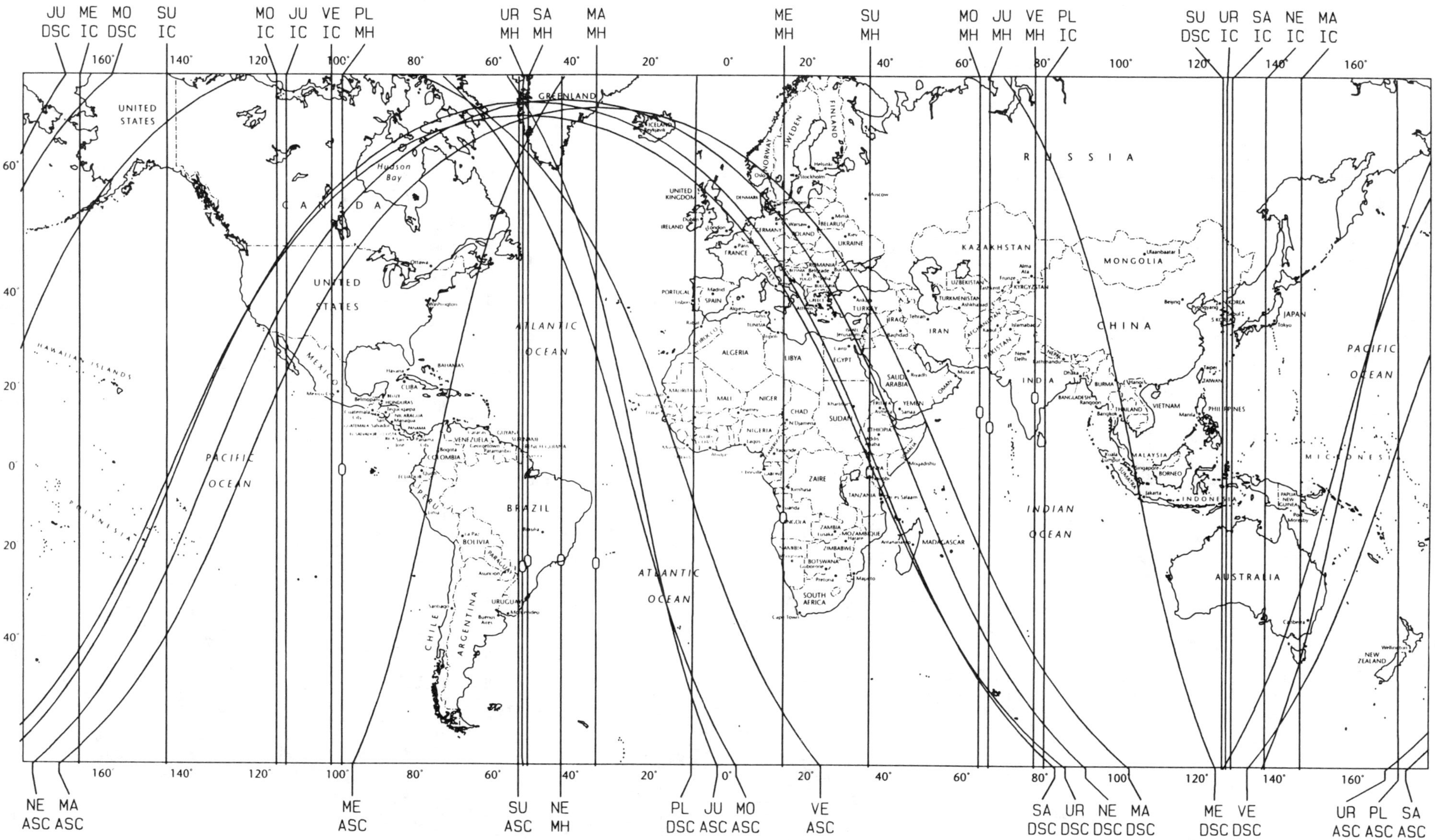

Vernal Equinox

March 20, 1988 9: 39 AM GMT

09/18/98

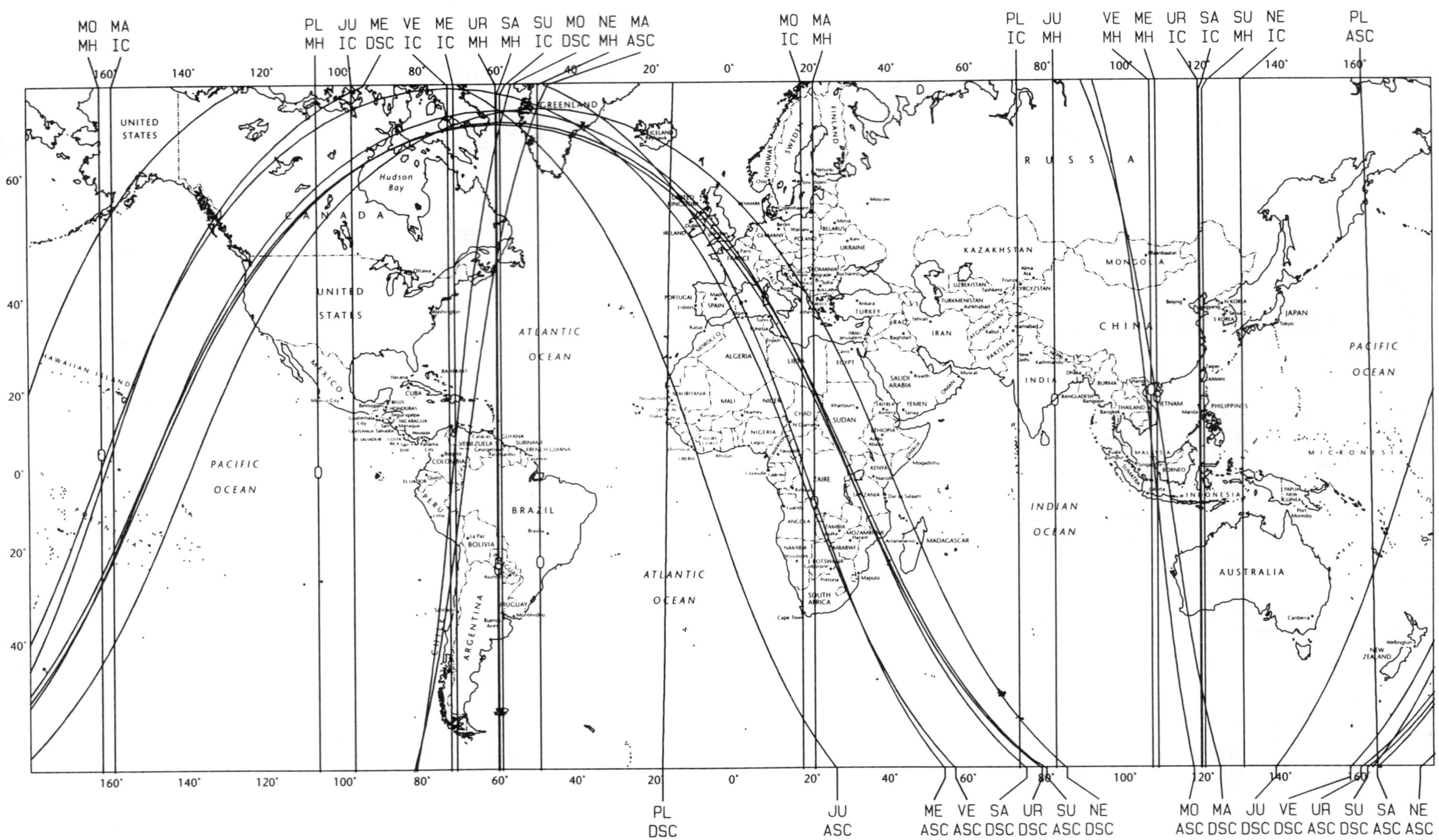

Summer Solstice

June 21, 1988 3:57 AM GMT

09/18/98

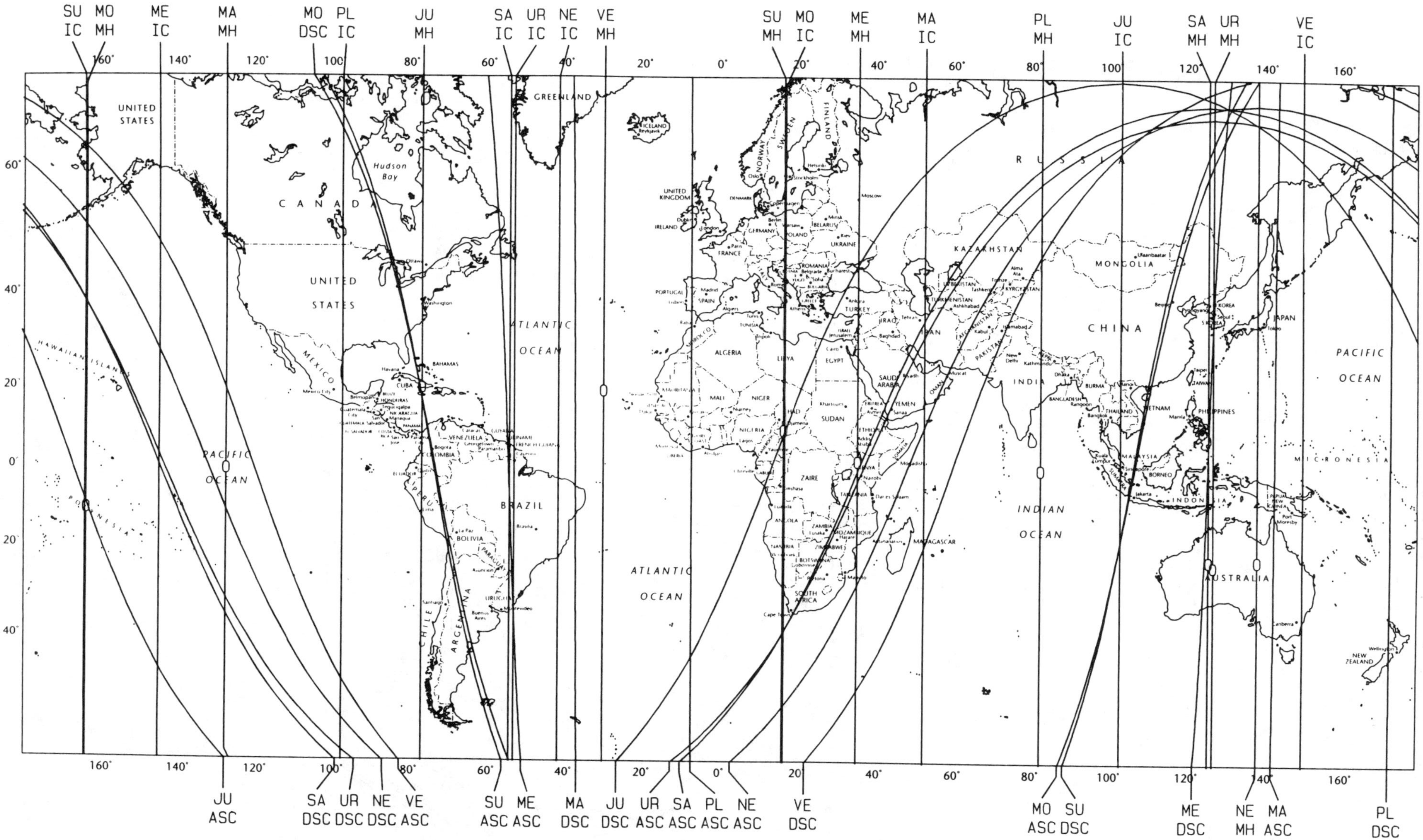

Eclipse Full Moon

August 27, 1988 10:55 AM GMT

09/18/98

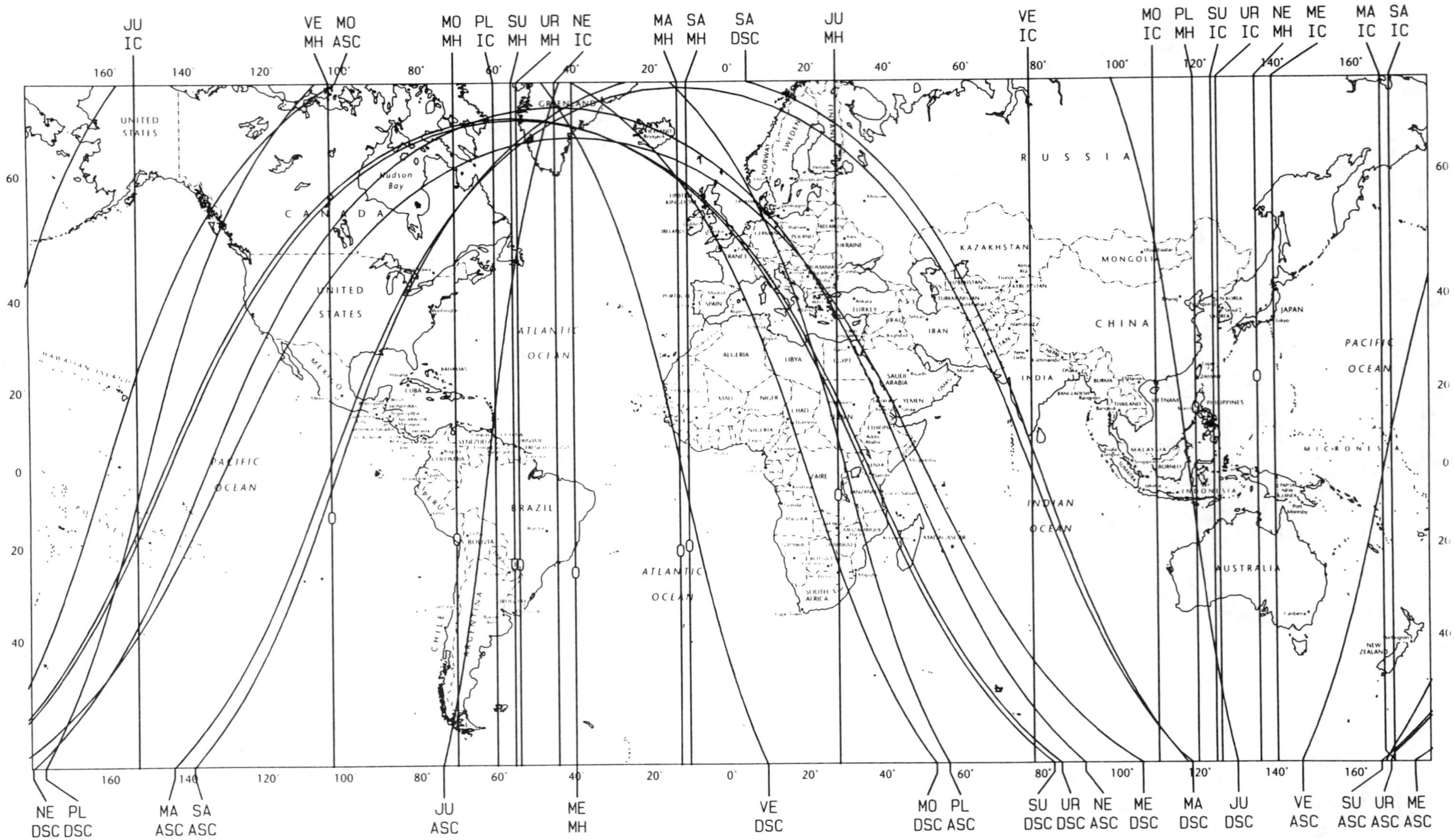

World's First Flight

December 17, 1903 10:35 AM EST

04/16/96

World's First Flight
December 17, 1903 10:35 AM EST

ASTRO*CARTO*GRAPHY LATITUDE CROSSINGS

NORTH LATITUDES

69N59 ME/MH * MO/DSC
69N41 VE/MH * MO/ASC
69N32 MA/ASC * PL/ASC
69N31 SA/MH * MA/DSC
69N24 MA/MH * PL/ASC
68N56 MO/DSC * SA/ASC
68N46 NE/IC * JU/ASC
68N30 SA/MH * PL/ASC
68N28 VE/DSC * JU/ASC
68N22 MO/DSC * MA/ASC
68N19 NE/IC * VE/DSC
68N04 ME/MH * SA/ASC
67N43 JU/ASC * NE/DSC
67N43 VE/DSC * NE/ASC
67N42 SA/ASC * NE/ASC
67N40 ME/MH * NE/ASC
67N36 MA/ASC * NE/ASC
67N27 VE/DSC * SA/ASC
67N25 UR/MH * NE/DSC
67N20 SU/MH * NE/DSC
67N16 ME/MH * MA/ASC
67N06 NE/IC * SA/ASC
66N56 PL/IC * NE/DSC
66N53 VE/DSC * MA/ASC
66N50 MO/DSC * NE/ASC
66N40 UR/MH * SU/DSC
66N38 JU/ASC * SA/ASC
66N36 PL/IC * SU/ASC
66N28 SU/MH * UR/ASC
66N26 SU/DSC * SA/ASC
66N24 SU/DSC * JU/ASC
66N23 SU/ASC * NE/DSC
66N22 NE/IC * MA/ASC
66N21 PL/IC * UR/ASC
66N19 SA/ASC * UR/DSC
66N17 SU/DSC * MA/ASC
66N16 JU/ASC * UR/DSC
66N16 NE/IC * SU/DSC
66N12 MA/ASC * UR/DSC
66N10 MO/ASC * PL/DSC
66N09 NE/IC * UR/DSC
65N57 MO/MH * SU/ASC
65N56 SU/DSC * VE/DSC
65N55 JU/MH * SA/DSC
65N52 ME/MH * SU/DSC
65N50 VE/DSC * UR/DSC
65N48 ME/MH * UR/DSC
65N46 UR/ASC * NE/DSC
65N46 MA/ASC * JU/ASC
65N37 MO/MH * UR/ASC
65N31 ME/MH * VE/DSC
65N31 MO/MH * NE/DSC
64N48 SU/DSC * UR/DSC
64N35 ME/DSC * VE/DSC
64N32 NE/IC * ME/ASC
64N23 ME/ASC * JU/ASC
64N20 MA/MH * NE/ASC
64N09 SU/DSC * ME/DSC
64N07 ME/DSC * UR/DSC
64N06 UR/MH * SA/ASC
64N01 ME/ASC * MA/ASC
63N53 UR/MH * ME/ASC
63N48 ME/ASC * SA/ASC
63N48 JU/MH * MA/DSC
63N45 SU/MH * ME/ASC
63N45 SA/MH * NE/ASC
63N36 SU/MH * SA/ASC
63N34 UR/MH * MA/ASC
63N09 PL/IC * ME/ASC
63N05 SU/MH * MA/ASC
62N56 MO/DSC * ME/DSC
61N56 MA/MH * ME/DSC
61N32 PL/IC * SA/ASC
61N24 SA/MH * ME/DSC

CENTRAL LATITUDES

61N13 Anchorage
61N10 MO/MH * ME/ASC
61N10 PL/IC * MA/ASC
60N03 MO/DSC * UR/DSC

59N52 MA/MH * UR/DSC
59N43 MA/MH * MO/DSC
59N36 MA/MH * SU/DSC
59N25 MO/DSC * SU/DSC
59N20 Stockholm
58N55 SA/MH * UR/DSC
58N36 SA/MH * SU/DSC
58N01 SA/MH * MO/DSC
57N49 VE/MH * SU/ASC
56N55 VE/MH * UR/ASC
56N47 NE/ASC * PL/ASC
56N47 MA/ASC * SA/ASC
55N45 Moscow
55N41 Copenhagen
55N16 MO/MH * MA/ASC
55N07 MO/MH * SA/ASC
53N59 ME/DSC * PL/ASC
52N30 Berlin
52N22 VE/MH * NE/DSC
51N30 London

49N56 MO/IC * JU/DSC
49N19 UR/MH * JU/ASC
49N14 Vancouver
48N50 Paris
47N36 Seattle
46N28 ME/DSC * NE/ASC
45N32 Montreal
44N59 Minneapolis
44N20 VE/MH * ME/ASC
44N14 SU/MH * JU/ASC
43N39 Toronto
42N22 Boston
42N20 Detroit
41N54 Rome
41N53 Chicago
40N43 New York

39N44 Denver
38N53 Washington
38N37 St. Louis
38N35 JU/MH * ME/DSC
37N47 San Francisco
36N45 JU/MH * NE/ASC
36N10 Las Vegas
35N40 Tokyo
34N04 Los Angeles
33N45 Atlanta
33N27 Phoenix
32N47 Dallas
30N03 Cairo

29N58 New Orleans
29N46 Houston
29N25 UR/DSC * PL/ASC
28N39 Delhi
25N47 Miami
24N26 SU/DSC * PL/ASC
24N12 JU/IC * MO/ASC
22N34 Calcutta
22N16 Hong Kong
21N19 Honolulu

19N26 Mexico City
17N36 JU/MH * UR/DSC
16N15 PL/IC * JU/ASC
15N01 JU/MH * SU/DSC
14N35 Manila
10N30 Caracas

SOUTH LATITUDES

8N18 SA/IC * VE/ASC
7N45 JU/MH * PL/ASC
6N27 Lagos
5N01 VE/IC * SA/DSC
2N45 MA/MH * VE/DSC
1N34 VE/MH * MA/ASC

1S34 VE/IC * MA/DSC
2S45 MA/IC * VE/ASC
5S01 VE/MH * SA/ASC
7S45 JU/IC * PL/DSC
8S18 SA/MH * VE/DSC

15S01 JU/IC * SU/ASC
16S15 PL/MH * JU/DSC
17S36 JU/IC * UR/ASC

22S54 Rio De Janeiro
24S12 JU/MH * MO/DSC
24S26 SU/ASC * PL/DSC
26S12 Johannesburg
29S25 UR/ASC * PL/DSC

33S55 Sydney
34S36 Buenos Aires
36S45 JU/IC * NE/DSC
36S52 Auckland
37S50 Melbourne
38S35 JU/IC * ME/ASC

44S14 SU/IC * JU/DSC
44S20 VE/IC * ME/DSC
46S28 ME/ASC * NE/DSC
49S19 UR/IC * JU/DSC
49S56 MO/MH * JU/ASC

52S22 VE/IC * NE/ASC
53S59 ME/ASC * PL/DSC
55S07 MO/IC * SA/DSC
55S16 MO/IC * MA/DSC
56S47 MA/DSC * SA/DSC
56S47 NE/DSC * PL/DSC
56S55 VE/IC * UR/DSC
57S49 VE/IC * SU/DSC
58S01 SA/IC * MO/ASC
58S36 SA/IC * SU/ASC
58S55 SA/IC * UR/ASC
59S25 MO/ASC * SU/ASC
59S36 MA/IC * SU/ASC
59S43 MA/IC * MO/ASC
59S52 MA/IC * UR/ASC

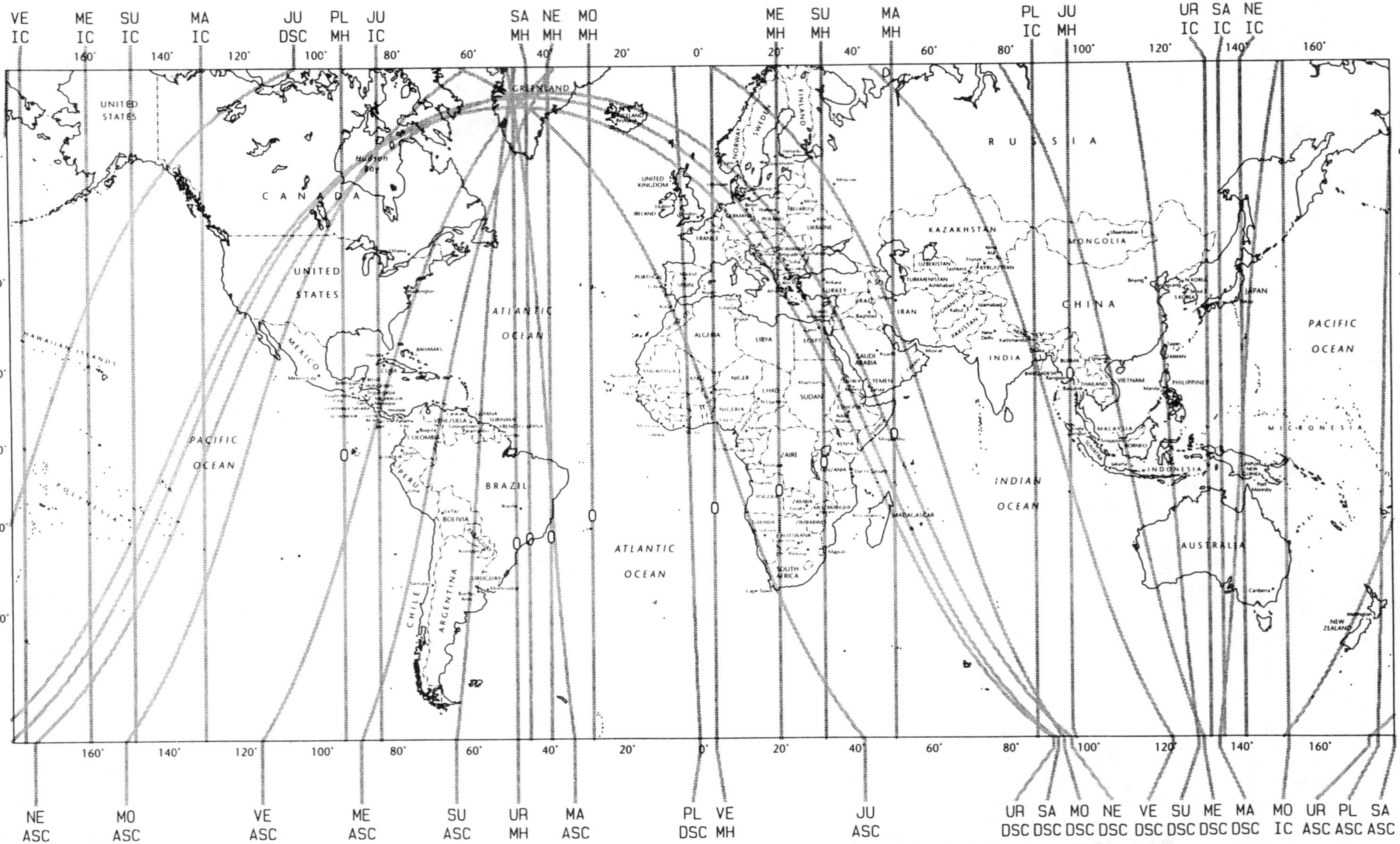

CYCLO*CARTO*GRAPHY

04/17/96

SU, MO, ME, VE AND MA ARE PROGRESSED
JU, SA, UR, NE AND PL ARE TRANSITING

World's First Flight

PROG	3	11	4	15:57 GMT
TRANSIT	12	21	88	0:00 GMT
NATAL	12	17	3	15:35 GMT

3. C*C*G MA EQU A*C*G ME
4. C*C*G ME EQU A*C*G MO
5. C*C*G NE = A*C*G ME

A*C*G	(87 23)	Displaced C*C*G
NE 05 CAN 12		UR 03 LIB 52
MA 06 AQU 44		SA 07 LIB 23
VE 07 SCO 19		MA 12 CAP 21
SA 09 AQU 05		ME 12 SAG 27
MO 09 SAG 17		NE 12 LIB 56
ME 09 CAP 41		PL 18 LEO 59
JU 17 PIS 25		MO 23 LIB 40
PL 19 GEM 23		VE 25 SCO 26
SU 24 SAG 03		JU 28 AQU 06
UR 25 SAG 21		

A*C*G	(87 23)	Solar arc A*C*G
NE 05 CAN 12		NE 02 LIB 35
MA 06 AQU 44		MA 04 TAU 07
VE 07 SCO 19		VE 04 AQU 42
SA 09 AQU 05		SA 06 TAU 28
MO 09 SAG 17		MO 06 PIS 40
ME 09 CAP 41		ME 07 ARI 04
JU 17 PIS 25		JU 14 GEM 48
PL 19 GEM 23		PL 16 VIR 46
SU 24 SAG 03		UR 22 PIS 44
UR 25 SAG 21		

6. Solar arc A*C*G SA EQU A*C*G MA
7. Solar arc A*C*G SA = A*C*G VE
8. Solar arc A*C*G PL = A*C*G JU

Amazingly, the world's first powered flight C*C*G PL DSC was within one degree orb of Lockerbie. This is the same planetary line that had a powerful influence at Lockerbie on the 1988 winter solstice Astro*Carto*Graphy map.

This shows that the world's first powered flight Astro*Carto*Graphy map had a strong connection to the dangerous influences effecting the United Kingdom. If these destructive energies erupted, the airline industry would most likely experience the most powerful impact.

Solar arc SA EQU A*C*G MA indicates that the airline industry experienced some problems. There was increased concern about the safety of Boeing jets, resulting in checks being carried out. Airline companies were also facing a difficult time due to many countries slipping into recession. This was reiterated by solar arc SA = A*C*G VE.

It was later revealed that Pan-Am had received warnings about bombs in cassette recorders. The American Embassy in Helsinki had received a telephone call informing them that a Pan Am flight from Frankfurt to the U.S. might be blown up. Despite this warning the British government decided there was no need to increase airport security.

The other temporary influences to the world's first powered flight Astro*Carto*Graphy map clearly explain these circumstances. C*C*G NE = A*C*G ME shows the faulty judgement made about the warning to blow up a Pan Am plane.

C*C*G MA EQU A*C*G ME is relevant for the communication received about the aggressive action that would be taken. Once the true facts were revealed the British government was criticized for doing nothing about the warnings (C*C*G ME EQU A*C*G MO).

C*C*G SA = NE shows that security at airports was not at a high standard. The bomb inside a cassette recorder was easily put on a plane, something airport security should prevent.

Solar arc PL = A*C*G JU indicated that a major airline could be ruined by international terrorism. After the Lockerbie disaster Pan American Airways faced financial ruin from claims for compensation. In December 1991, Pan Am stopped flying after a financial rescue plan failed to secure the airline.

Solar return UR MH and MA ASC near Helsinki showed that the warning received by the American Embassy should have been taken seriously by the British government. They were informed about this warning, but dismissed it as of "little credibility."

Solar return VE MH and JU IC at Lockerbie do not seem appropriate for the explosion that destroyed a plane. This shows that the solar return planetary influences should never be used on their own because incorrect information can be obtained.

The C*C*G PL DSC temporary influence is more significant for Lockerbie as it reiterates information obtained from other maps. Solar return VE MH and JU IC, when considered with C*C*G PL DSC, shows a different picture: danger (PL) for a large group of people (VE/JU).

More important is the repetition of themes as they have the most powerful influence. Other planetary influences will give background information to the theme that is emphasized in many different ways.

Example 5—Reunification of Germany

On October 3, 1990, the two German countries were reunited after more than forty years of division. After the East/West German borders were opened on November 9, 1989, it was inevitable that this would lead to German reunification. But the speed at which this was achieved surprised most people.

Using the *Astro*Carto*Graphy Source Book of Mundane Maps* it is not easy to find the relevant planetary influences that indicated the two German countries would be reunited.

The eclipse map for July 22, 1990 has JU ASC, SU ASC, MO ASC, SA DSC and NE DSC within orb of Germany. Eclipse JU ASC is near Berlin with SU ASC and MO ASC not too far away. But SA DSC is close to Bonn, suggesting that the West German government still had some obstacles to overcome before reunification was possible.

The full Moon map for September 5, 1990 shows that JU ASC is within orb of Bonn. One week later the four allied powers signed the treaty which allowed Germany to be reunited. The full Moon aspected Chancellor Kohl's Mars, which could explain the active role he had to negotiate a successful German reunification.

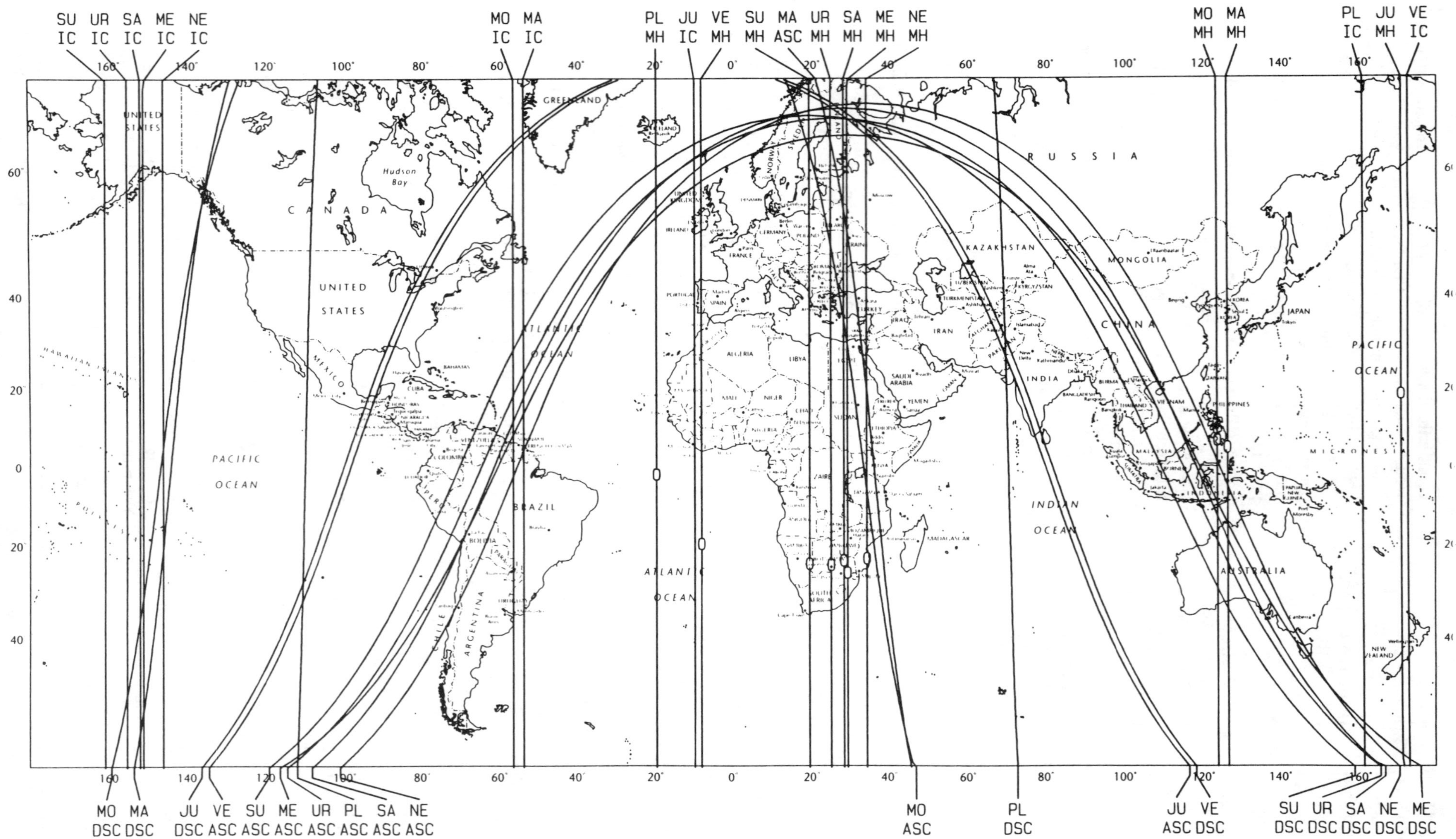

04/16/96

World's First Flight SR

December 17, 1988 10:35 AM GMT

04/16/96

World's First Flight SR
December 17, 1988 10:35 AM GMT

ASTRO*CARTO*GRAPHY LATITUDE CROSSINGS

NORTH LATITUDES

69N56 MO/ASC * VE/DSC
69N36 MO/ASC * JU/ASC
69N02 SU/MH * VE/DSC
68N41 SU/MH * JU/ASC
68N29 VE/DSC * MA/ASC
68N01 MA/ASC * JU/ASC
67N55 UR/MH * VE/DSC
67N43 ME/MH * NE/ASC
67N41 SA/MH * NE/ASC
67N36 VE/DSC * NE/ASC
67N32 UR/MH * NE/ASC
67N31 JU/ASC * NE/ASC
67N27 UR/MH * JU/ASC
67N25 MA/ASC * NE/ASC
67N21 VE/DSC * SA/ASC
67N21 ME/MH * SA/DSC
67N20 JU/ASC * SA/ASC
67N19 UR/MH * SA/ASC
67N16 MA/ASC * SA/ASC
67N14 NE/MH * SA/DSC
67N07 SA/MH * VE/DSC
67N06 SU/MH * NE/ASC
67N06 SU/MH * SA/ASC
67N06 SA/ASC * NE/ASC
66N55 MO/ASC * SA/ASC
66N52 ME/MH * VE/DSC
66N49 MO/ASC * NE/ASC
66N36 MO/ASC * SU/ASC
66N35 SA/MH * JU/ASC
66N34 SU/DSC * MA/ASC
66N33 SU/ASC * NE/ASC
66N32 UR/MH * SU/DSC
66N24 SU/ASC * SA/ASC
66N22 SA/MH * SU/DSC
66N20 MA/ASC * UR/ASC
66N19 SU/DSC * JU/ASC
66N19 ME/MH * SU/DSC
66N19 ME/MH * JU/ASC
66N19 SA/MH * UR/DSC
66N17 ME/MH * UR/DSC
66N17 JU/ASC * UR/DSC
66N16 SU/DSC * UR/DSC
66N14 SU/MH * UR/ASC
66N13 VE/DSC * UR/DSC
66N08 MO/ASC * UR/ASC
66N08 SU/DSC * VE/DSC
66N04 NE/MH * UR/DSC
65N55 NE/MH * SU/DSC
65N19 NE/MH * VE/DSC
65N08 UR/ASC * NE/ASC
64N45 MA/IC * JU/DSC
64N38 NE/MH * JU/ASC
64N35 SA/MH * ME/ASC
64N32 UR/MH * ME/ASC
64N31 ME/ASC * MA/ASC
64N30 NE/MH * ME/DSC
64N29 ME/DSC * JU/ASC
64N24 ME/DSC * VE/DSC
64N15 SU/MH * ME/ASC
64N12 MO/ASC * ME/ASC
64N06 MA/IC * VE/ASC
64N00 VE/MH * SU/ASC
64N00 UR/MH * MA/ASC
63N54 NE/DSC * PL/DSC
63N43 MO/IC * JU/DSC
63N37 JU/IC * SU/ASC
62N58 MO/IC * VE/ASC
62N52 SU/MH * MO/ASC
62N34 VE/MH * SA/ASC
62N31 SU/DSC * ME/DSC
62N22 VE/MH * UR/ASC
62N01 JU/IC * SA/ASC
61N52 JU/IC * UR/ASC
61N30 SA/DSC * PL/DSC

CENTRAL LATITUDES

61N13 Anchorage
61N04 VE/MH * NE/ASC
60N53 PL/MH * SU/ASC
60N36 SA/ASC * UR/ASC
60N21 JU/IC * NE/ASC

59N20 Stockholm
59N07 VE/MH * ME/ASC
58N59 UR/DSC * PL/DSC
58N32 ME/DSC * PL/DSC
58N29 JU/IC * ME/ASC
58N21 PL/MH * UR/ASC
58N04 PL/MH * SA/ASC
56N37 SU/DSC * PL/DSC
55N45 Moscow
55N42 MO/DSC * MA/DSC
55N41 Copenhagen
55N41 ME/DSC * UR/DSC
55N31 SA/MH * MA/ASC
55N17 PL/MH * NE/ASC
54N08 PL/MH * ME/ASC
52N30 Berlin
52N02 ME/MH * MA/ASC
51N30 London

49N44 UR/MH * MO/ASC
49N14 Vancouver
48N50 Paris
48N27 ME/ASC * NE/ASC
47N36 Seattle
45N32 Montreal
44N59 Minneapolis
43N39 Toronto
42N22 Boston
42N20 Detroit
41N54 Rome
41N53 Chicago
40N43 New York

39N44 Denver
38N53 Washington
38N37 St. Louis
37N47 San Francisco
36N10 Las Vegas
36N09 SA/MH * MO/ASC
35N40 Tokyo
34N04 Los Angeles
33N45 Atlanta
33N40 MA/IC * SU/ASC
33N27 Phoenix
32N47 Dallas
32N03 VE/DSC * PL/DSC
30N56 ME/MH * MO/ASC
30N03 Cairo

29N58 New Orleans
29N46 Houston
29N24 MO/IC * SU/ASC
28N39 Delhi
27N54 JU/ASC * PL/DSC
25N47 Miami
23N44 MA/IC * UR/ASC
22N34 Calcutta
22N16 Hong Kong
21N19 Honolulu

19N43 NE/MH * MA/ASC
19N26 Mexico City
18N36 MO/IC * UR/ASC
18N09 MA/IC * SA/ASC
15N47 ME/DSC * SA/DSC
14N35 Manila
14N14 MA/IC * ME/ASC
12N20 MO/IC * SA/ASC
10N30 Caracas

SOUTH LATITUDES

8N57 MO/IC * ME/ASC
6N27 Lagos
5N35 NE/IC * MO/DSC
4N43 MA/IC * NE/ASC
1N43 MO/MH * NE/DSC

1S43 MO/IC * NE/ASC
4S43 MA/MH * NE/DSC
5S35 NE/MH * MO/ASC
8S57 MO/MH * ME/DSC

12S20 MO/MH * SA/DSC
14S14 MA/MH * ME/DSC
15S47 ME/ASC * SA/ASC
18S09 MA/MH * SA/DSC
18S36 MO/MH * UR/DSC
19S43 NE/IC * MA/DSC

22S54 Rio De Janeiro
23S44 MA/MH * UR/DSC
26S12 Johannesburg
27S54 JU/DSC * PL/ASC
29S24 MO/MH * SU/DSC

30S56 ME/IC * MO/DSC
32S03 VE/ASC * PL/ASC
33S40 MA/MH * SU/DSC
33S55 Sydney
34S36 Buenos Aires
36S09 SA/IC * MO/DSC
36S52 Auckland
37S50 Melbourne

48S27 ME/DSC * NE/DSC
49S44 UR/IC * MO/DSC

52S02 ME/IC * MA/DSC
54S08 PL/IC * ME/DSC
55S17 PL/IC * NE/DSC
55S31 SA/IC * MA/DSC
55S41 ME/ASC * UR/ASC
55S42 MO/ASC * MA/ASC
56S37 SU/ASC * PL/ASC
58S04 PL/IC * SA/DSC
58S21 PL/IC * UR/DSC
58S29 JU/MH * ME/DSC
58S32 ME/ASC * PL/ASC
58S59 UR/ASC * PL/ASC
59S07 VE/IC * ME/DSC

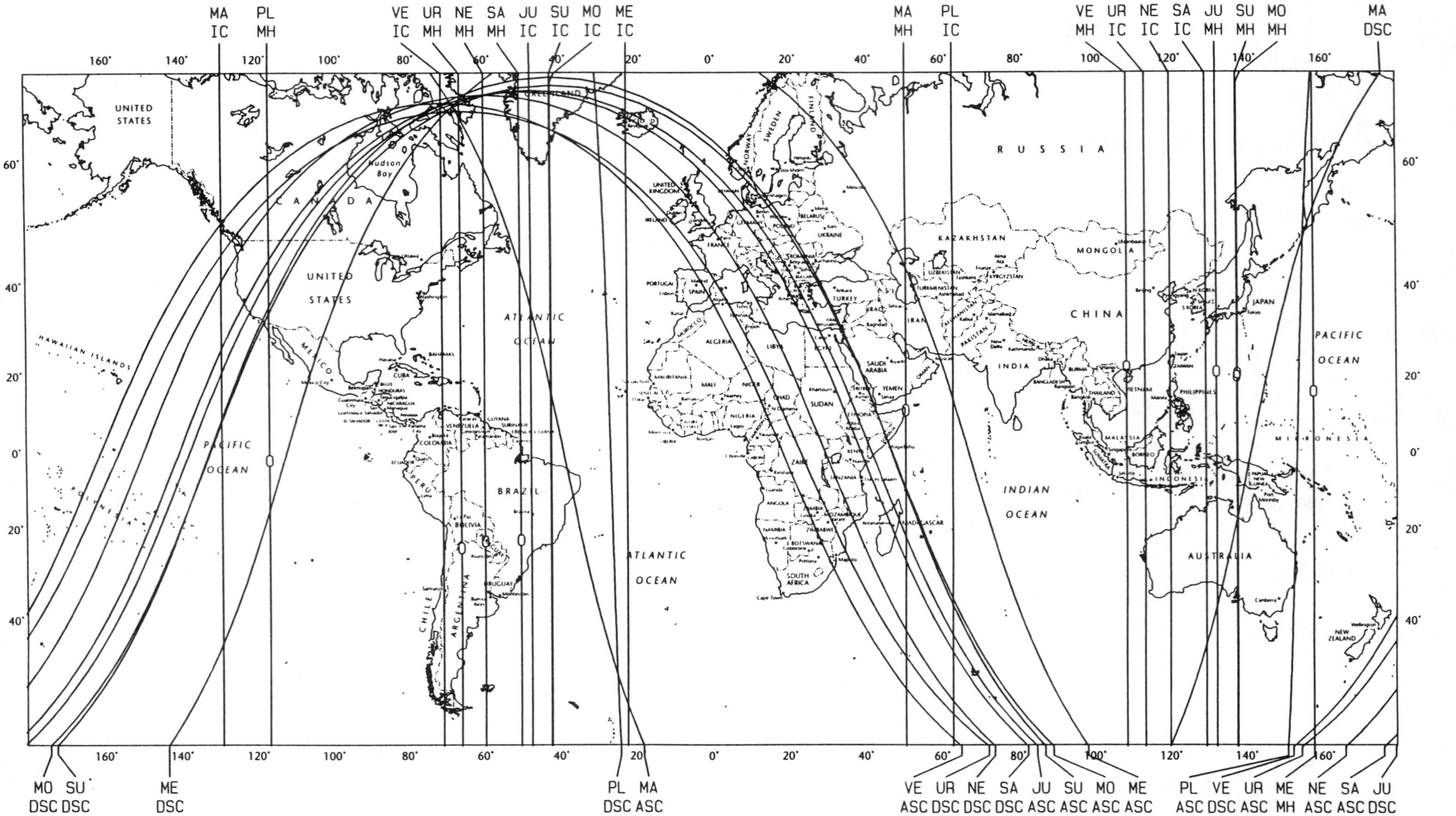

Eclipse New Moon

July 22, 1990 2:54 AM GMT

09/18/98

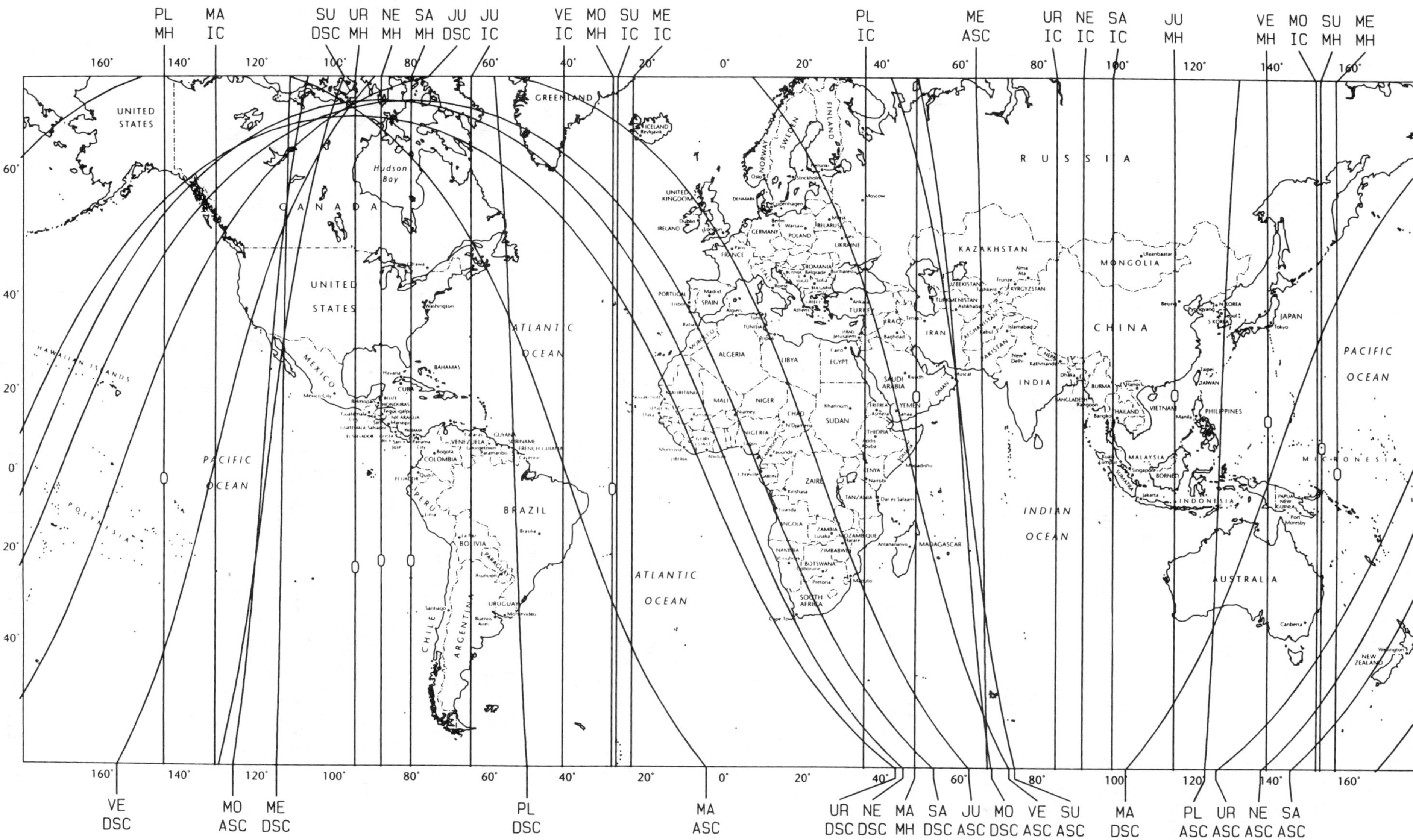

Full Moon

September 5, 1990 1: 45 AM GMT

09/18/98

The new Moon map for September 19, 1990 also has JU ASC within orb of Bonn, but this time it is much closer. Neither JU nor the new Moon activates any of the relevant charts, so this map doesn't seem to be significant to indicate German reunification. JU does show a repetition of themes from the eclipse and full Moon maps.

It is unlikely that German reunification could have been predicted just by using the *Astro*Carto*Graphy Source Book of Mundane Maps*. Therefore, this book should be used with extreme care when making predictions.

It is interesting to examine the Astro*Carto*Graphy maps for the two German countries and Chancellor Helmut Kohl. The Astro*Carto*Graphy map for East Germany gave accurate information for the opening of the Berlin Wall.

As a time of birth is known for Helmut Kohl it would be no problem to draw his Astro*Carto*Graphy map. He would be Chancellor of the reunited Germany, so his Astro*Carto*Graphy map should provide important information.

In the *Book of World Horoscopes* (first edition), Nicholas Campion writes that West Germany was proclaimed at 00:00 hours on May 23, 1949. Later, he wrote that the West Germany proclamation chart should be corrected to 00:00 on May 24, 1949. This was the legal beginning of West Germany's constitutional existence.

In the *Book of World Horoscopes* (second edition) Nicholas Campion includes a chart for when the constitution was signed at 4:00 p.m. on May 23, 1949. This represented a ceremonial beginning for West Germany.

Nicholas Campion includes another chart for West Germany calculated for 12:00 noon on May 5, 1955. This was when West Germany gained full independence and is regarded as the true beginning of the country's existence.

Which national horoscope for West Germany should be used to draw an Astro*Carto*Graphy map? Campion suggests that the gaining of full independence must be regarded as a national horoscope for West Germany together with that for May 23, 1949.

The West Germany proclamation map provides the most accurate information. Both the West Germany independence and the signing of the constitution maps were not all that impressive, so these charts don't seem to be very important. For these reasons the proclamation map is used to examine German reunification.

First, the Astro*Carto*Graphy map for East Germany will be examined. As reunification took place eleven months after the Berlin Wall was opened, it will be interesting to see how things changed during that time.

A*C*G	**Age 39**	**C*C*G**
UR 05 CAN 28		UR 06 CAP 16
ME 05 LIB 29		JU 11 LEO 08
SU 12 LIB 48		NE 12 CAP 46
NE 14 LIB 18		VE 13 CAP 20
SA 15 VIR 48		MA 13 VIR 55
MO 16 ARI 53		MO 14 LIB 34
MA 21 LEO 29		PL 18 SCO 01
PL 22 LEO 48		ME 19 SCO 54
VE 23 SCO 54		SA 20 CAP 21
JU 24 CAP 50		SU 22 SCO 29

1. C*C*G UR ASC = Berlin
2. C*C*G UR = A*C*G UR
3. C*C*G UR EQU A*C*G ME
4. C*C*G NE EQU A*C*G SU
5. C*C*G VE EQU A*C*G SU
6. C*C*G VE EQU A*C*G NE
7. C*C*G MO = A*C*G NE
8. C*C*G SU EQU A*C*G PL
9. Prog. Moscow = C*C*G VE MH

A*C*G	**(39 41)**	**Displaced C*C*G**
UR 05 CAN 28		JU 01 CAN 27
ME 05 LIB 29		NE 03 SAG 05
SU 12 LIB 48		VE 03 SAG 39
NE 14 LIB 18		MA 04 LEO 14
SA 15 SCO 48		MO 04 VIR 53
MO 16 ARI 53		PL 08 LIB 20
MA 21 LEO 29		ME 10 LIB 13
PL 22 LEO 48		SA 10 SAG 40
VE 23 SCO 54		UR 26 SCO 35
JU 24 CAP 50		

A*C*G	**(39 41)**	**Solar arc A*C*G**
UR 05 CAN 28		MA 01 LIB 10
ME 05 LIB 29		PL 02 LIB 29
SU 12 LIB 48		VE 03 CAP 35
NE 14 LIB 18		JU 04 PIS 31
SA 15 SCO 48		UR 15 AQU 09
MO 16 ARI 53		ME 15 SCO 10
MA 21 LEO 29		NE 23 AQU 59
PL 22 LEO 48		SA 25 SAG 29
VE 23 SCO 54		MO 26 TAU 34
JU 24 CAP 50		

10. Solar arc A*C*G UR EQU A*C*G SA
11. Solar arc A*C*G ME = A*C*G SA
12. Solar arc A*C*G NE EQU A*C*G VE

On the day that the two German countries were reunited C*C*G UR ASC was at Berlin. During the time that C*C*G UR ASC was passing through East Germany many changes took place which a few years earlier didn't seem possible. C*C*G UR = A*C*G UR shows that East Germany had the opportunity to introduce reforms. When the government refused to implement any Gorbachev-type reforms the East German people started to protest.

Generally, the East German people had no freedom of speech, so the fact that they started to voice their opinions was incredible. This is shown by C*C*G UR EQU A*C*G ME which also indicates the quick pace of events that resulted in German reunification.

The end of East Germany as a country is confirmed by C*C*G SU EQU A*C*G PL. The solar return map for the opening of the Berlin Wall was still in effect when German reunification took place. Solar return PL DSC within orb of East Germany is most appropriate for the total transformation that occurred during those twelve

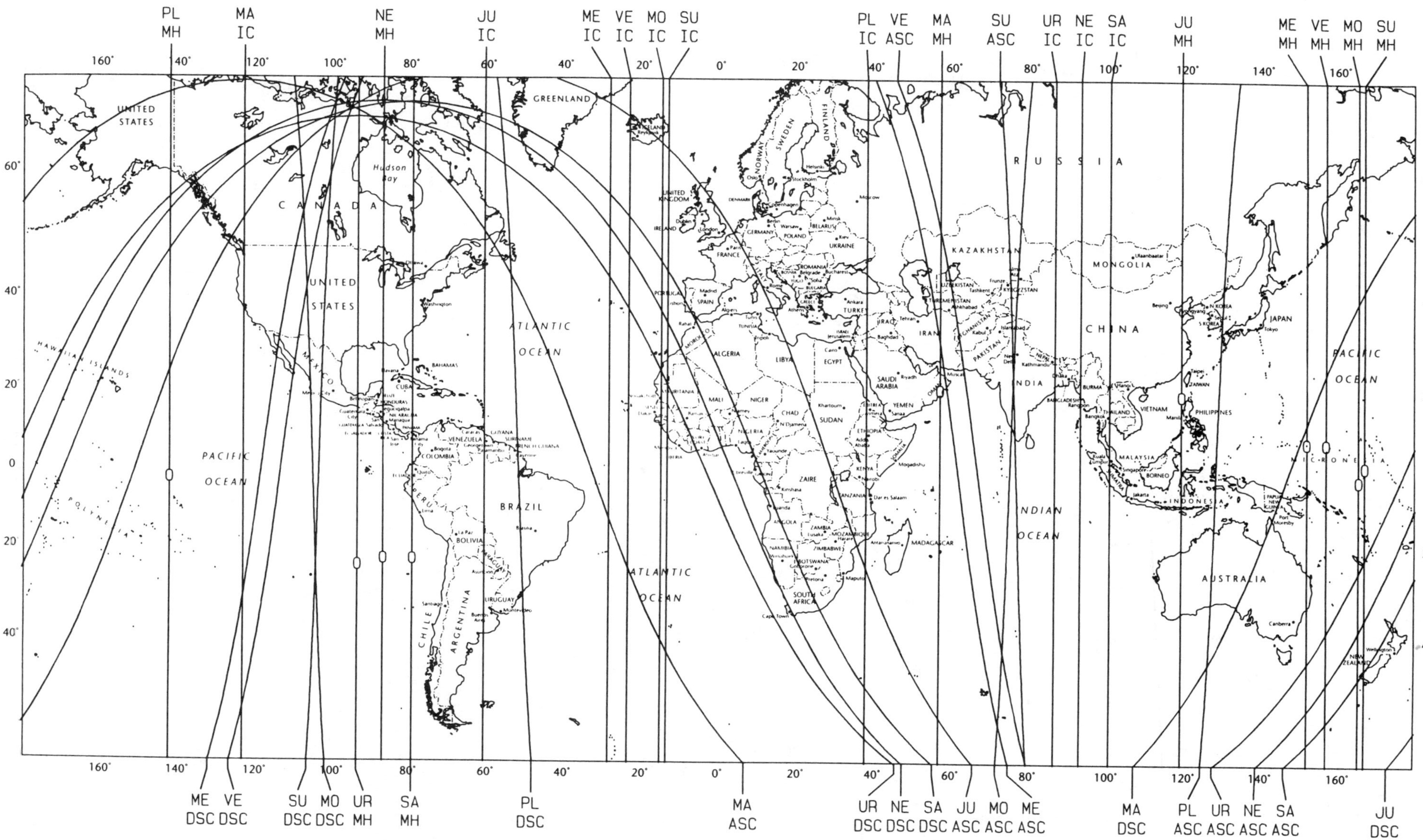

New Moon

September 19, 1990 12: 46 AM GMT

09/18/98

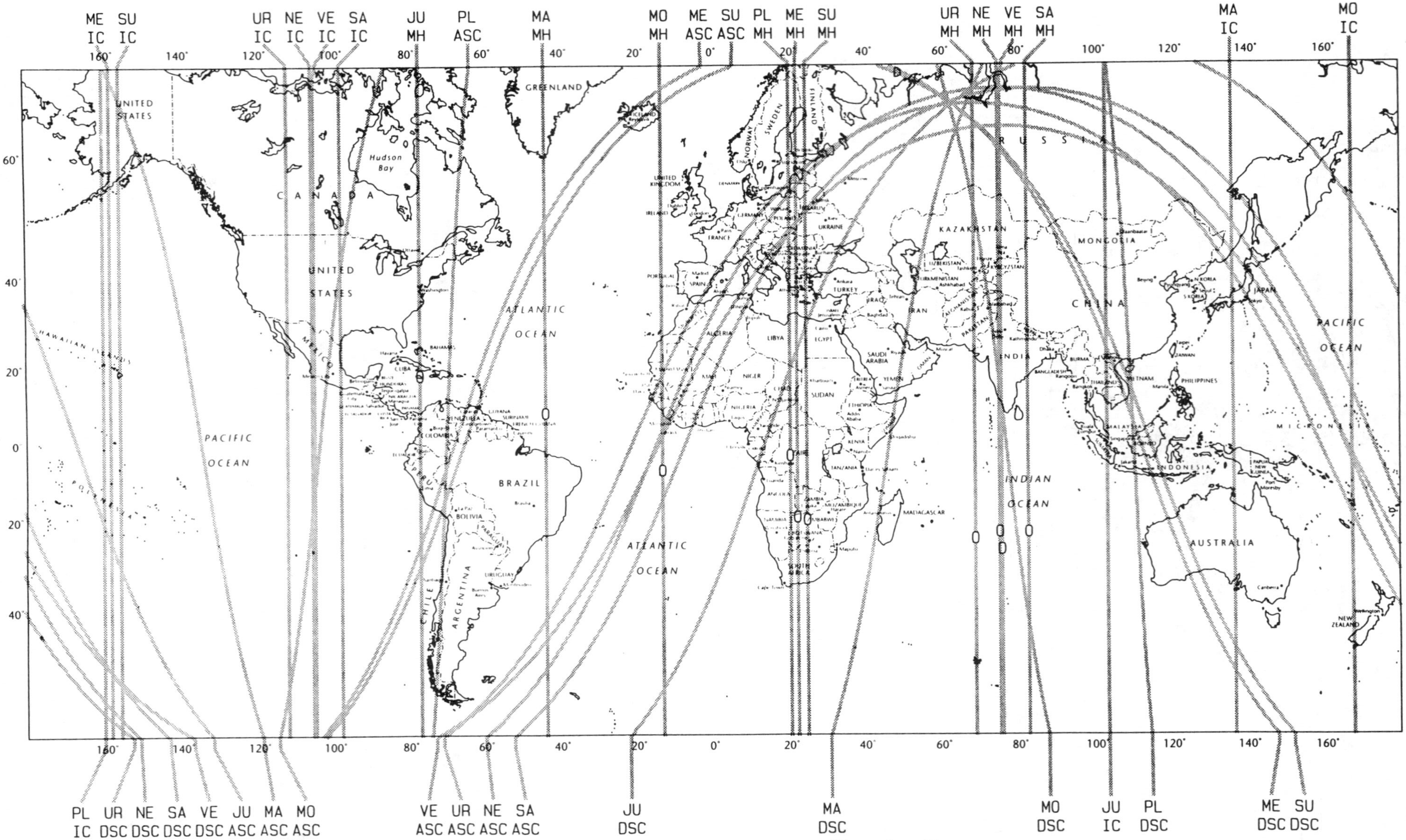

CYCLO*CARTO*GRAPHY

04/17/96

SU, MO, ME, VE AND MA ARE PROGRESSED
JU, SA, UR, NE AND PL ARE TRANSITING

East Germany

PROG	11	17	49	12: 30 GMT
TRANSIT	10	3	90	0: 00 GMT
NATAL	10	7	49	12: 45 GMT

months.

Quite a few temporary influences indicated that East Germans expected too much from the reunification with West Germany. Many East Germans had the impression that West Germany was a country where everything seemed to be fantastic. East Germans thought that after reunification they would enjoy the same lifestyle as in the West.

It was not long after reunification that East Germans realized how difficult it would be to adapt to a different way of life. Unemployment soon became a big problem in that part of reunited Germany, something most had never before experienced.

C*C*G NE is conjunct C*C*G VE and both are EQU A*C*G SU. This is reiterated by C*C*G VE EQU A*C*G NE and solar arc NE EQU A*C*G VE. These temporary influences are clear indications that East Germany would not immediately benefit from reunification. When the Deutschmark became the official currency of East Germany on July 1, 1990, it was as if people had changed their old money into gold.

The economic reality of reunification turned out to be a big disappointment for most East Germans (C*C*G MO = A*C*G NE).

Solar arc UR EQU A*C*G SA and solar arc ME = A*C*G SA show that it wouldn't be easy for them after reunification. It would take some time before they started to see the positive effects of reunification.

Prog. Moscow = A*C*G VE MH indicated that the East German government would experience no difficulties with the Soviet Union about German reunification. This activated the A*C*G JU ASC at Moscow and is an appropriate indication for peace.

On the solar return map JU ASC and NE DSC are exact at Moscow with SA DSC and MO DSC also within orb. East Germans saw Gorbachev as the great messiah who could rescue them from a restrictive lifestyle. But this would only be achieved after experiencing many problems.

An Astro*Carto*Graphy map for the proclamation of West Germany provides most interesting information. Astro*Carto*Graphy maps for West Germany Independence and the signing of the constitution were also examined, but they weren't as effective.

With MA IC near Berlin the West German government would continually experience confrontations with East Germany. There was a great deal of tension between the two countries, especially where the divided city of Berlin was concerned.

UR DSC is the nearest line to Bonn, showing that West Germany would have an unusual look to it. Although Berlin was in East Germany, the west part of this city belonged to West Germany.

VE IC and JU ASC at Moscow indicates that West Germany and the Soviet Union would have a harmonious partnership. It was more with the East German government itself that the challenging and troublesome relationship developed.

A*C*G	**Age 41**	**C*C*G**
SU 00 GEM 20		VE 05 LEO 28
SA 02 VIR 28		UR 06 CAP 16
JU 04 AQU 30		JU 11 LEO 08
VE 10 GEM 40		NE 12 CAP 46
NE 12 LIB 16		SU 12 CAN 59
MA 15 TAU 15		MA 15 GEM 55
ME 16 GEM 02		PL 18 SCO 01
MO 16 ARI 09		SA 20 CAP 21
PL 19 LEO 14		ME 20 GEM 32
UR 28 GEM 56		MO 22 LIB 27

1. C*C*G UR ASC = Bonn
2. C*C*G NE ASC = Berlin
3. C*C*G VE SQU A*C*G JU
4. C*C*G NE EQU A*C*G NE
5. C*C*G SU EQU A*C*G NE
6. C*C*G MA = A*C*G ME
7. Prog. Berlin = A*C*G PL DSC
8. Prog. Bonn = C*C*G ME IC
9. Prog. Moscow = A*C*G MA ASC
10. Prog. Moscow = C*C*G SA MH

A*C*G	**(42 39)**	**Displaced C*C*G**
SU 00 GEM 20		NE 00 SAG 07
SA 02 VIR 28		MA 03 TAU 16
JU 04 AQU 30		PL 05 LIB 22
VE 10 GEM 40		SA 07 SAG 42
NE 12 LIB 16		ME 07 TAU 53
MA 15 TAU 15		MO 09 VIR 48
ME 16 GEM 02		VE 22 GEM 49
MO 16 ARI 09		UR 23 SCO 37
PL 19 LEO 14		JU 28 GEM 29
UR 28 GEM 56		

11. disp. C*C*G NE = A*C*G SU
12. disp. C*C*G JU = A*C*G UR

A*C*G	**(42 39)**	**Solar arc A*C*G**
SU 00 GEM 20		PL 01 LIB 53
SA 02 VIR 28		UR 11 LEO 35
JU 04 AQU 30		SA 15 LIB 07
VE 10 GEM 40		JU 17 PIS 09
NE 12 LIB 16		VE 23 CAN 19
MA 15 TAU 15		NE 24 SCO 55
ME 16 GEM 02		MA 27 GEM 54
MO 16 ARI 09		ME 28 CAN 41
PL 19 LEO 14		MO 28 TAU 48
UR 28 GEM 56		

13. solar arc ME DSC = Moscow

The temporary influences for the date of German reunification are impressive, especially the location aspects for Berlin, Moscow and Bonn.

Berlin:

C*C*G NE ASC = Berlin

Prog. Berlin = A*C*G PL DSC

C*C*G NE ASC modifies the A*C*G MA IC line at Berlin. The West German government set out to achieve reunification in the shortest possible time, but in doing so they didn't obtain a clear picture about the real cost of

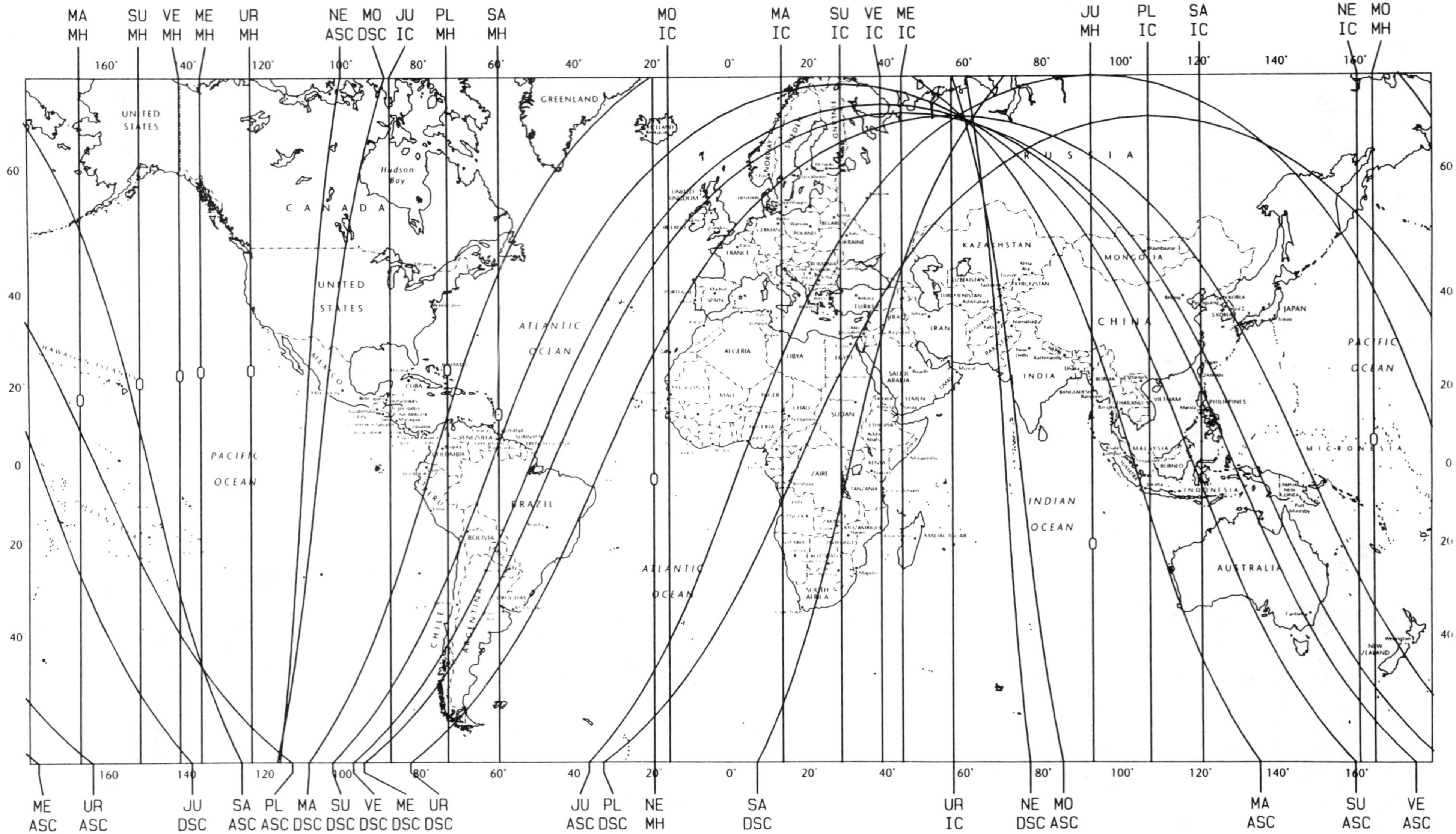

West Germany Proclamation

May 24, 1949 12:00 AM Time Zone -2:00

04/16/96

04/16/96

West Germany Proclamation
May 24, 1949 12:00 AM Time Zone -2:00

ASTRO*CARTO*GRAPHY LATITUDE CROSSINGS

NORTH LATITUDES

69N23 PL/IC * JU/DSC
69N02 VE/IC * SU/ASC
68N41 MA/IC * SU/DSC
68N37 ME/IC * SU/ASC
67N45 JU/ASC * SA/DSC
67N36 SA/IC * JU/DSC
67N24 ME/IC * VE/ASC
67N14 UR/IC * MA/ASC
67N10 SU/IC * VE/DSC
67N05 JU/ASC * NE/DSC
66N46 UR/IC * SU/ASC
66N38 MO/ASC * JU/ASC
66N33 VE/IC * ME/DSC
66N27 MA/ASC * JU/ASC
66N26 UR/IC * VE/ASC
66N23 SU/ASC * JU/ASC
66N21 VE/ASC * UR/ASC
66N20 JU/ASC * UR/ASC
66N20 SU/ASC * UR/ASC
66N19 MA/ASC * UR/ASC
66N19 ME/ASC * UR/DSC
66N19 MO/ASC * UR/ASC
66N15 VE/ASC * JU/ASC
66N15 UR/ASC * NE/DSC
66N15 SA/DSC * UR/ASC
66N08 SU/ASC * MA/ASC
66N07 SA/DSC * NE/DSC
66N06 UR/IC * ME/ASC
66N04 MO/ASC * SU/ASC
66N03 ME/ASC * JU/ASC
66N00 MO/ASC * MA/ASC
65N59 MO/ASC * VE/ASC
65N58 VE/ASC * MA/ASC
65N54 UR/IC * JU/ASC
65N51 SU/ASC * VE/ASC
65N50 SU/IC * ME/DSC
65N48 ME/IC * UR/DSC
65N46 MO/ASC * ME/ASC
65N46 VE/ASC * SA/DSC
65N43 SU/ASC * SA/DSC
65N42 ME/ASC * MA/ASC
65N38 ME/ASC * SA/DSC
65N38 VE/ASC * NE/DSC
65N33 MA/ASC * SA/DSC
65N32 SA/IC * PL/ASC
65N32 SU/ASC * ME/ASC
65N31 ME/ASC * NE/DSC
65N30 SU/ASC * NE/DSC
65N23 JU/MH * PL/DSC
65N22 MA/IC * VE/DSC
65N14 VE/IC * UR/DSC
65N11 MO/ASC * SA/DSC
65N09 ME/ASC * VE/ASC
64N58 MA/ASC * NE/DSC
64N03 UR/ASC * PL/DSC
63N29 SU/IC * UR/DSC
63N19 MA/IC * ME/DSC
62N57 JU/DSC * PL/ASC
62N49 ME/ASC * PL/DSC
62N33 VE/ASC * PL/DSC
62N20 UR/IC * SA/DSC
62N17 MO/IC * SU/DSC
61N56 SU/ASC * PL/DSC
61N42 JU/MH * UR/ASC
61N14 ME/IC * JU/ASC
61N13 Anchorage
60N57 MA/ASC * PL/DSC
60N34 NE/MH * SU/DSC

59N20 Stockholm
58N53 NE/DSC * PL/DSC
58N48 MA/IC * UR/DSC
58N43 MO/ASC * PL/DSC
58N20 VE/IC * JU/ASC

CENTRAL LATITUDES

56N55 JU/MH * ME/ASC
56N02 MO/ASC * NE/DSC
55N45 Moscow
55N41 Copenhagen
55N34 PL/IC * UR/ASC
55N16 UR/IC * PL/DSC
54N55 JU/MH * VE/ASC
54N30 MO/IC * VE/DSC
53N37 NE/IC * PL/ASC
52N30 Berlin
51N45 MA/MH * SA/ASC
51N41 NE/MH * VE/DSC
51N30 London
50N56 MO/MH * PL/ASC
50N08 SU/IC * JU/ASC

49N18 MO/IC * ME/DSC
49N14 Vancouver
49N09 JU/MH * SU/ASC
48N50 Paris
47N36 Seattle
46N15 PL/IC * ME/ASC
46N07 NE/IC * JU/DSC
45N53 SA/DSC * PL/DSC
45N40 NE/MH * ME/DSC
45N32 Montreal
45N31 ME/IC * PL/DSC
45N29 SA/IC * UR/ASC
45N07 ME/IC * SA/DSC
44N59 Minneapolis
44N19 SA/MH * MA/DSC
43N39 Toronto
42N22 Boston
42N20 Detroit
41N54 Rome
41N53 Chicago
41N24 PL/IC * VE/ASC
40N51 MO/MH * JU/DSC
40N43 New York

39N44 Denver
39N31 VE/IC * PL/DSC
38N53 Washington
38N37 St. Louis
37N47 San Francisco
36N10 Las Vegas
35N40 Tokyo
34N04 MO/IC * UR/DSC
34N04 Los Angeles
33N45 Atlanta
33N27 Phoenix
32N47 Dallas
31N33 JU/MH * MA/ASC
31N21 VE/IC * SA/DSC
30N03 Cairo

29N58 New Orleans
29N46 Houston
28N39 Delhi
28N30 SA/IC * ME/ASC
27N46 NE/MH * UR/DSC
27N07 MA/IC * JU/ASC
27N05 PL/IC * SU/ASC
25N47 Miami
23N31 SU/IC * PL/DSC
22N34 Calcutta
22N16 Hong Kong
21N19 Honolulu

19N26 Mexico City
18N57 SA/IC * VE/ASC
14N35 Manila
12N54 PL/MH * MA/DSC
10N30 Caracas

SOUTH LATITUDES

9N03 SU/MH * SA/ASC
8N55 MA/MH * PL/ASC
6N27 Lagos
5N38 SA/MH * SU/DSC

5S38 SA/IC * SU/ASC
8S55 MA/IC * PL/DSC
9S03 SU/IC * SA/DSC

12S54 PL/IC * MA/ASC
18S57 SA/MH * VE/DSC

22S54 Rio De Janeiro
23S31 SU/MH * PL/ASC
26S12 Johannesburg
27S05 PL/MH * SU/DSC
27S07 MA/MH * JU/DSC
27S46 NE/IC * UR/ASC
28S30 SA/MH * ME/DSC

31S21 VE/MH * SA/ASC
31S33 JU/IC * MA/DSC
33S55 Sydney
34S04 MO/MH * UR/ASC
34S36 Buenos Aires
36S52 Auckland
37S50 Melbourne
39S31 VE/MH * PL/ASC

40S51 MO/IC * JU/ASC
41S24 PL/MH * VE/DSC
44S19 SA/IC * MA/ASC
45S07 ME/MH * SA/ASC
45S29 SA/MH * UR/DSC
45S31 ME/MH * PL/ASC
45S40 NE/IC * ME/ASC
45S53 SA/ASC * PL/ASC
46S07 NE/MH * JU/ASC
46S15 PL/MH * ME/DSC
49S09 JU/IC * SU/DSC
49S18 MO/MH * ME/ASC

50S08 SU/MH * JU/DSC
50S56 MO/IC * PL/DSC
51S41 NE/IC * VE/ASC
51S45 MA/IC * SA/DSC
53S37 NE/MH * PL/DSC
54S30 MO/MH * VE/ASC
54S55 JU/IC * VE/DSC
55S16 UR/MH * PL/ASC
55S34 PL/MH * UR/DSC
56S02 MO/DSC * NE/ASC
56S55 JU/IC * ME/DSC
58S20 VE/MH * JU/DSC
58S43 MO/DSC * PL/ASC
58S48 MA/MH * UR/ASC
58S53 NE/ASC * PL/ASC

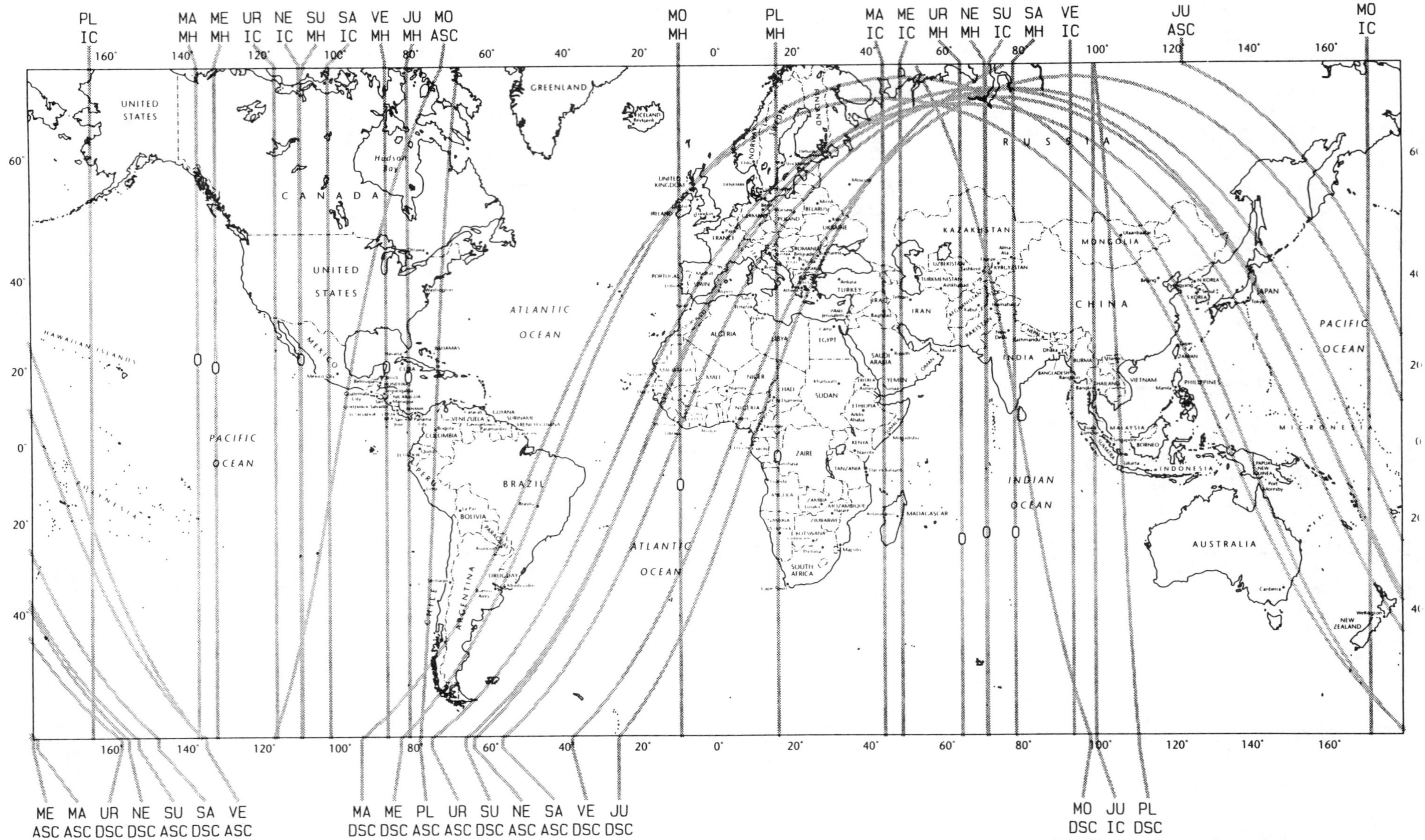

CYCLO*CARTO*GRAPHY

04/17/96

SU, MO, ME, VE AND MA ARE PROGRESSED
JU, SA, UR, NE AND PL ARE TRANSITING

West Germany Proclamation

PROG	7	4	49	6:45 GMT
TRANSIT	10	3	90	0:00 GMT
NATAL	5	23	49	22:00 GMT

rebuilding the East. Moving the capital from Bonn to Berlin also turned out to be more of a problem than first thought. These temporary influences at Berlin are reiterated by the following anywhere aspects:

C*C*G SU EQU A*C*G NE

C*C*G NE EQU A*C*G NE

disp. C*C*G NE = A*C*G SU

Prog. Berlin = A*C*G PL DSC showed the transformation that took place in a short space of time. This aspect had an effect when the Berlin Wall was opened, leading to German reunification.

Solar return PL IC was also near Berlin, which emphasized the deep changes taking place at the East German capital (East Berlin). It seemed like a miracle that German reunification was achieved less than a year after the Berlin Wall started to be dismantled.

Bonn:

C*C*G UR ASC = Bonn

Prog. Bonn = C*C*G ME IC

It is amazing that C*C*G UR ASC was exact at Bonn when German reunification occurred. Both countries had C*C*G UR ASC at their capital cities when the changes were put into effect.

East Germany	C*C*G UR ASC = East Berlin
West Germany	C*C*G UR ASC = Bonn

Prog. Bonn = C*C*G ME IC indicates the negotiations the West German government had about reunification. The fact that an agreement was obtained in a short space of time is shown by the combination of ME and UR. A quick settlement for reunification is reiterated by this anywhere aspect:

C*C*G MA = A*C*G ME

Solar return ME MH is near Bonn, which emphasizes the importance of negotiations with other countries. The four allied powers signed a treaty on September 12, 1990, which handed back full sovereignty allowing the two German countries to be reunited.

Moscow:

Prog. Moscow = A*C*G MA ASC

Prog. Moscow = C*C*G SA MH

Solar arc ME ASC = Moscow

Prog. Moscow = A*C*G MA ASC and C*C*G SA MH indicates that the negotiations for reunification were not easy. Many problems had to be overcome before a settlement was possible, but the determination and great effort from both governments meant that a solution was found enabling German reunification to go ahead.

Disp. C*C*G JU = A*C*G UR had a worldwide effect which showed that despite the many difficulties a successful settlement for reunification could be reached.

Solar arc ME ASC = Moscow is an appropriate location aspect. It was at Moscow that the four allied powers signed the treaty which allowed German reunification to take place.

On his Astro*Carto*Graphy map, Helmut Kohl has **six** planetary lines that have a powerful effect in Germany:

Lines: SA MH
VE ASC
SU ASC
PL IC
UR ASC
ME ASC

SA MH and VE ASC pass through West Germany. When Kohl became Chancellor in 1982, his first priority was to tackle the economic problems and reduce unemployment. This is aptly described by SA MH and VE ASC.

During the time that Kohl was chancellor, the country became an economic powerhouse. West Germans enjoyed a high standard of living which East Germans could only dream about. The German Mark was also a stable and powerful currency in the world financial markets.

Interestingly, Kohl's PL IC has a powerful influence in East Germany, Poland, Hungary and Czechoslovakia, all of which experienced the downfall of Communism in 1989. As East Germany was the Soviet occupation zone Chancellor Kohl had to deal with Moscow to obtain an agreement for German reunification. At Moscow Kohl has JU ASC which indicates that he had easy success and much luck there.

A*C*G	**Age 60**	**C*C*G**
NE 03 VIR 28		UR 06 CAP 16
JU 10 GEM 39		SU 09 GEM 53
UR 10 ARI 46		JU 11 LEO 08
SU 11 ARI 43		VE 11 CAN 21
SA 12 CAP 33		NE 12 CAP 46
ME 13 ARI 43		PL 18 SCO 01
MA 15 PIS 03		SA 20 CAP 21
PL 18 CAN 53		ME 22 TAU 26
VE 24 ARI 48		MA 27 ARI 59
MO 28 TAU 19		MO 29 LEO 43

1. C*C*G SU = A*C*G JU
2. C*C*G VE EQU A*C*G UR
3. C*C*G VE EQU A*C*G SU
4. C*C*G NE = A*C*G SA
5. C*C*G NE EQU A*C*G ME
6. Prog. Moscow = A*C*G UR MH
7. Prog. Moscow = A*C*G SU MH
8. Prog. Bonn = A*C*G SA DSC (orb 1¾°—app)
9. Prog. Berlin = A*C*G PL ASC
10. C*C*G VE IC = Bonn/birthplace
11. C*C*G MA ASC = border East/West Germany
12. C*C*G NE MH = Germany
13. C*C*G JU IC = Moscow

A*C*G	**(58 10)**	**Displaced C*C*G**
NE 03 VIR 28		MO 01 CAN 33
JU 10 GEM 39		UR 08 SCO 06
UR 10 ARI 46		JU 12 GEM 58
SU 11 ARI 43		VE 13 TAU 11
SA 12 CAP 33		NE 14 SCO 36
ME 13 ARI 43		PL 19 VIR 51
MA 15 PIS 03		SA 22 SCO 11

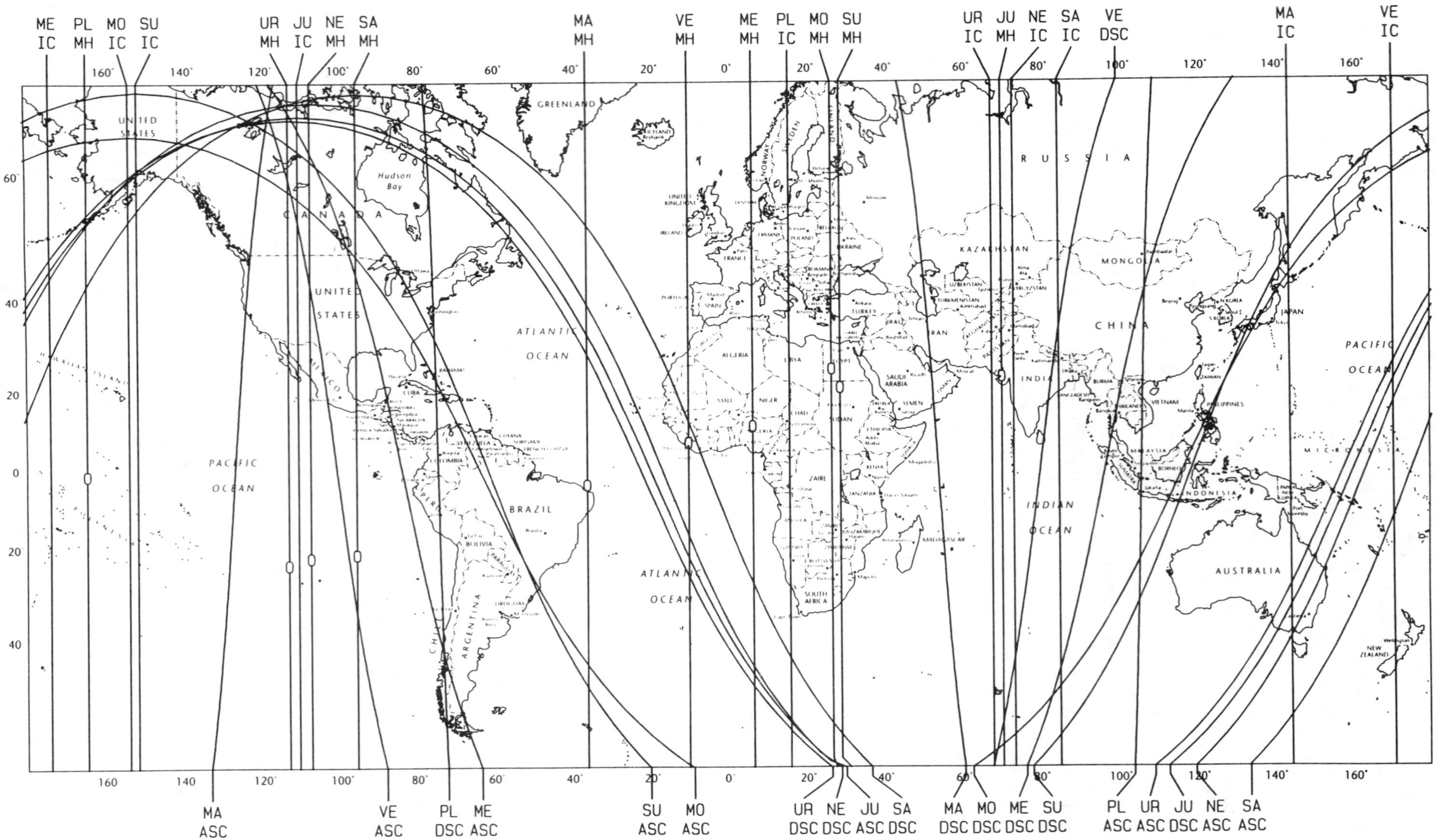

04/16/96

West Germany Proclamation SR

May 24, 1990 10:01 AM GMT

04/16/96

West Germany Proclamation SR
May 24, 1990 10:01 AM GMT

ASTRO*CARTO*GRAPHY LATITUDE CROSSINGS

NORTH LATITUDES

69N31 ME/ASC * MA/ASC
69N14 MO/IC * SU/DSC
68N45 PL/MH * SU/DSC
68N37 NE/MH * SA/ASC
68N22 JU/IC * SA/ASC
68N09 JU/IC * NE/ASC
68N07 UR/MH * SA/ASC
68N05 UR/MH * NE/ASC
68N05 SA/ASC * NE/ASC
68N03 SA/DSC * PL/DSC
68N03 ME/ASC * NE/ASC
68N01 ME/ASC * SA/ASC
67N52 MA/ASC * NE/ASC
67N49 VE/ASC * NE/ASC
67N46 SA/MH * NE/DSC
67N44 ME/IC * SU/DSC
67N32 MA/ASC * SA/ASC
67N28 VE/ASC * SA/ASC
67N22 UR/MH * ME/ASC
67N09 VE/ASC * MA/ASC
67N08 SU/ASC * NE/ASC
66N52 ME/ASC * JU/DSC
66N52 UR/MH * JU/DSC
66N51 NE/MH * JU/ASC
66N44 VE/ASC * JU/DSC
66N42 MA/ASC * JU/DSC
66N38 SU/ASC * SA/ASC
66N32 ME/ASC * UR/DSC
66N31 JU/IC * UR/DSC
66N29 VE/ASC * UR/ASC
66N27 MA/ASC * UR/ASC
66N27 SU/ASC * JU/DSC
66N27 NE/MH * UR/DSC
66N18 SU/ASC * UR/ASC
66N11 SA/MH * JU/ASC
66N08 JU/DSC * SA/ASC
65N57 SA/ASC * UR/ASC
65N36 SU/ASC * MA/ASC
65N36 JU/IC * ME/ASC
65N34 SA/MH * UR/DSC
65N20 JU/DSC * UR/ASC
65N13 NE/DSC * PL/DSC
65N00 SU/ASC * VE/ASC
64N52 SU/IC * MO/ASC
64N29 PL/MH * MO/DSC
64N19 MO/ASC * NE/ASC
64N09 MO/ASC * UR/ASC
64N06 MO/ASC * JU/DSC
64N04 UR/MH * SU/ASC
64N03 VE/IC * SU/DSC
63N39 MO/ASC * SA/ASC
63N28 ME/IC * MO/DSC
63N13 JU/IC * SU/ASC
63N02 NE/MH * ME/ASC
62N58 JU/ASC * PL/DSC
62N03 NE/MH * SU/ASC
61N50 UR/DSC * PL/DSC
61N32 UR/ASC * NE/ASC
61N21 SU/ASC * ME/ASC
61N13 Anchorage
60N53 SU/IC * UR/ASC
60N45 SU/IC * NE/ASC
60N34 MO/ASC * MA/ASC
60N23 SU/IC * JU/DSC
60N07 MO/IC * UR/ASC

59N48 MO/IC * NE/ASC
59N42 UR/MH * VE/ASC
59N41 VE/IC * MO/DSC
59N32 MO/IC * JU/DSC
59N20 Stockholm
58N10 UR/MH * MO/ASC
57N49 MO/ASC * VE/ASC
57N08 JU/IC * MO/ASC

CENTRAL LATITUDES

57N02 JU/DSC * NE/ASC
55N52 SA/MH * SU/ASC
55N45 Moscow
55N44 NE/MH * MO/ASC
55N41 Copenhagen
55N32 SU/IC * SA/ASC
54N59 PL/MH * UR/ASC
53N59 MO/IC * SA/ASC
53N51 PL/MH * JU/DSC
53N46 JU/IC * VE/ASC
53N28 PL/MH * NE/ASC
53N14 MA/MH * SA/DSC
52N30 Berlin
51N30 London
50N05 MO/ASC * ME/ASC

49N14 Vancouver
48N50 Paris
48N23 SA/MH * MO/ASC
48N16 MA/IC * SU/DSC
48N06 ME/IC * UR/ASC
47N36 Seattle
46N10 ME/IC * JU/DSC
45N32 Montreal
45N12 SA/MH * ME/ASC
44N59 Minneapolis
44N43 ME/IC * NE/ASC
44N21 MA/IC * MO/DSC
43N39 Toronto
43N17 NE/MH * VE/ASC
43N09 PL/MH * SA/ASC
42N22 Boston
42N20 Detroit
42N09 ME/DSC * PL/ASC
41N54 Rome
41N53 Chicago
40N43 New York

39N44 Denver
39N29 MA/MH * NE/DSC
38N53 Washington
38N37 St. Louis
37N47 San Francisco
36N10 Las Vegas
35N40 Tokyo
34N20 SA/IC * VE/DSC
34N04 Los Angeles
33N45 Atlanta
33N27 Phoenix
33N10 MA/MH * JU/ASC
32N47 Dallas
32N02 SU/ASC * PL/DSC
30N03 Cairo

29N58 New Orleans
29N46 Houston
28N39 Delhi
28N35 MA/MH * UR/DSC
27N54 ME/IC * SA/ASC
26N11 VE/IC * UR/ASC
25N47 Miami
22N42 MO/ASC * PL/DSC
22N34 Calcutta
22N16 Hong Kong
22N15 MO/DSC * SU/DSC
21N44 VE/IC * JU/DSC
21N19 Honolulu

19N26 Mexico City
16N21 VE/IC * NE/ASC
14N35 Manila
12N31 VE/MH * SA/DSC
10N30 Caracas

6N27 Lagos

SOUTH LATITUDES

12S31 VE/IC * SA/ASC
16S21 VE/MH * NE/DSC

21S44 VE/MH * JU/ASC
22S15 MO/ASC * SU/ASC
22S42 MO/DSC * PL/ASC
22S54 Rio De Janeiro
26S11 VE/MH * UR/DSC
26S12 Johannesburg
27S54 ME/MH * SA/DSC
28S35 MA/IC * UR/ASC

32S02 SU/DSC * PL/ASC
33S10 MA/IC * JU/DSC
33S55 Sydney
34S20 SA/MH * VE/ASC
34S36 Buenos Aires
36S52 Auckland
37S50 Melbourne
39S29 MA/IC * NE/ASC

42S09 ME/ASC * PL/DSC
43S09 PL/IC * SA/DSC
43S17 NE/IC * VE/DSC
44S21 MA/MH * MO/ASC
44S43 ME/MH * NE/DSC
45S12 SA/IC * ME/DSC
46S10 ME/MH * JU/ASC
48S06 ME/MH * UR/DSC
48S16 MA/MH * SU/ASC
48S23 SA/IC * MO/DSC

50S05 MO/DSC * ME/DSC
53S14 MA/IC * SA/ASC
53S28 PL/IC * NE/DSC
53S46 JU/MH * VE/DSC
53S51 PL/IC * JU/ASC
53S59 MO/MH * SA/DSC
54S59 PL/IC * UR/DSC
55S32 SU/MH * SA/DSC
55S44 NE/IC * MO/DSC
55S52 SA/IC * SU/DSC
57S02 JU/ASC * NE/DSC
57S08 JU/MH * MO/DSC
57S49 MO/DSC * VE/DSC
58S10 UR/IC * MO/DSC
59S32 MO/MH * JU/ASC
59S41 VE/MH * MO/ASC
59S42 UR/IC * VE/DSC
59S48 MO/MH * NE/DSC

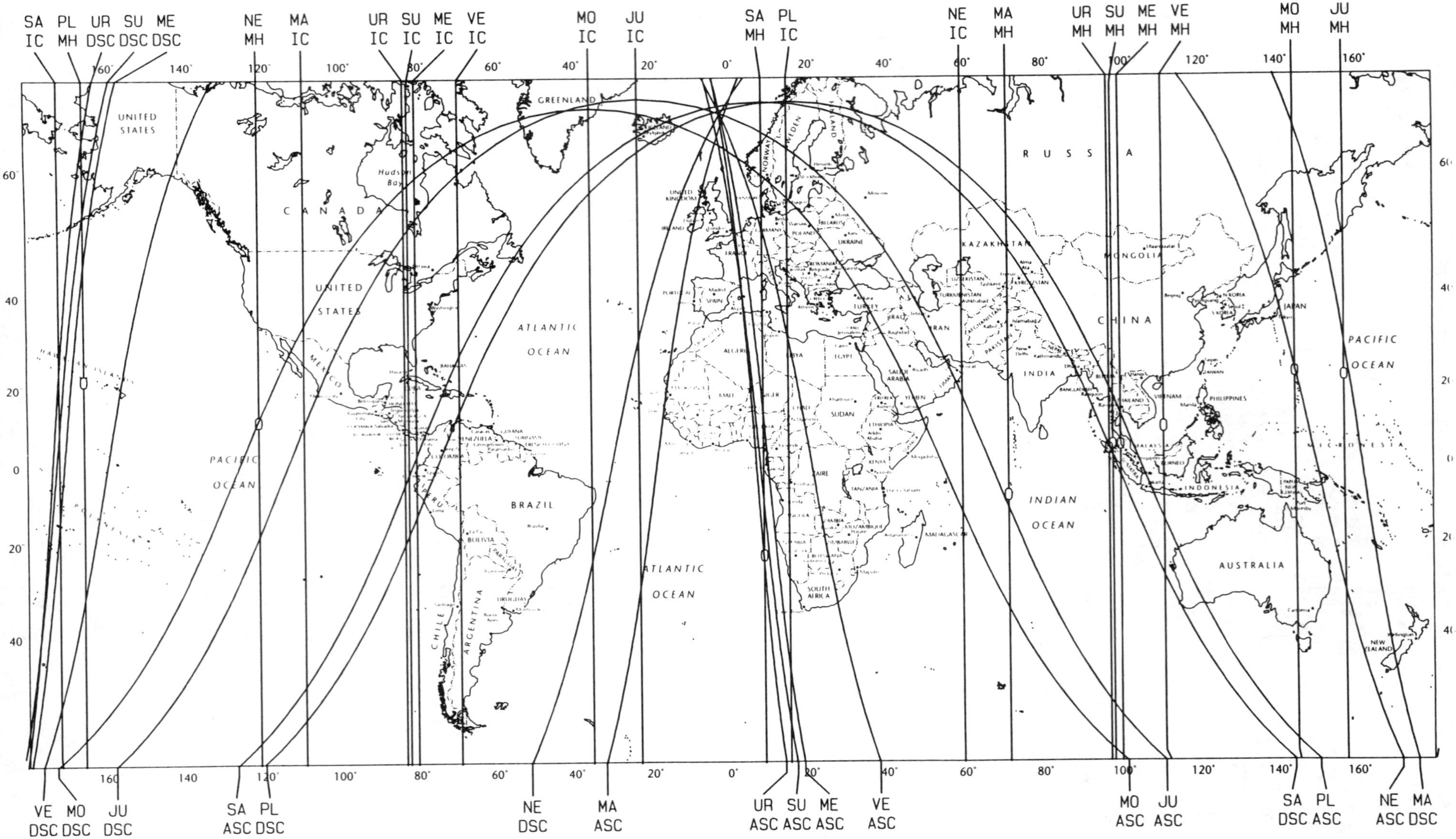

Helmut Kohl

April 3, 1930 6:30 AM CET

04/16/96

04/16/96

Helmut Kohl
April 3, 1930 6:30 AM CET

ASTRO*CARTO*GRAPHY LATITUDE CROSSINGS

NORTH LATITUDES

69N38	SU/ASC * VE/ASC
69N35	ME/ASC * UR/ASC
68N22	MA/ASC * NE/DSC
67N47	SA/DSC * PL/DSC
67N44	SA/MH * PL/DSC
67N40	PL/IC * SA/DSC
67N39	MO/IC * JU/DSC
67N34	MA/ASC * SA/ASC
67N32	SA/ASC * NE/DSC
67N25	ME/ASC * SA/ASC
67N24	SA/ASC * UR/ASC
67N23	VE/ASC * SA/ASC
67N16	SU/ASC * SA/ASC
67N09	JU/ASC * SA/ASC
67N08	VE/ASC * UR/ASC
67N08	MA/ASC * PL/DSC
67N05	MO/DSC * JU/DSC
67N02	NE/DSC * PL/DSC
66N58	SU/ASC * JU/ASC
66N57	ME/ASC * PL/DSC
66N55	VE/ASC * PL/DSC
66N54	UR/ASC * PL/DSC
66N50	JU/ASC * PL/DSC
66N44	JU/ASC * UR/ASC
66N43	SU/ASC * PL/DSC
66N43	JU/IC * MO/ASC
66N41	VE/ASC * JU/ASC
66N40	ME/ASC * JU/ASC
66N40	JU/ASC * NE/DSC
66N38	ME/ASC * NE/DSC
66N37	VE/ASC * NE/DSC
66N33	ME/ASC * VE/ASC
66N31	MA/ASC * JU/ASC
66N25	UR/ASC * NE/DSC
66N02	VE/ASC * MA/ASC
65N53	MO/ASC * SA/ASC
65N49	ME/ASC * MA/ASC
65N34	SU/ASC * NE/DSC
65N24	MA/ASC * UR/ASC
65N24	MO/ASC * PL/DSC
64N57	MO/MH * MA/DSC
64N41	SA/MH * JU/ASC
64N19	JU/IC * SA/ASC
64N12	SU/ASC * MA/ASC
64N10	PL/MH * UR/DSC
64N02	MO/ASC * NE/DSC
63N31	MO/ASC * MA/ASC
63N17	MO/ASC * SU/ASC
63N14	VE/IC * MO/DSC
62N55	PL/IC * JU/ASC
62N54	MO/ASC * UR/ASC
62N38	MO/ASC * ME/ASC
62N36	JU/IC * PL/DSC
61N50	MO/ASC * VE/ASC
61N13	Anchorage
60N20	MO/IC * SA/ASC
60N15	NE/IC * PL/ASC
60N02	VE/IC * JU/DSC
59N35	SA/MH * MO/ASC
59N26	ME/IC * MO/DSC
59N20	Stockholm
58N32	SU/IC * MO/DSC
58N04	UR/IC * MO/DSC
57N21	MO/IC * PL/DSC
57N05	NE/IC * SA/DSC
56N30	PL/IC * MO/ASC
55N45	Moscow
55N41	Copenhagen
54N46	PL/MH * SU/DSC
53N50	MA/MH * PL/ASC
53N38	ME/IC * JU/DSC
52N30	Berlin
52N07	SU/IC * JU/DSC

CENTRAL LATITUDES

51N42	SA/MH * VE/ASC
51N30	London
51N19	UR/IC * JU/DSC
49N14	Vancouver
48N50	Paris
48N31	MA/MH * SA/DSC
47N36	Seattle
46N01	PL/MH * ME/DSC
45N32	Montreal
44N59	Minneapolis
43N39	Toronto
42N22	Boston
42N20	Detroit
41N54	Rome
41N53	Chicago
40N43	New York
40N39	SU/ASC * UR/ASC
39N44	Denver
38N53	Washington
38N37	St. Louis
37N47	San Francisco
36N10	Las Vegas
35N40	Tokyo
34N24	MA/IC * MO/DSC
34N04	Los Angeles
33N45	Atlanta
33N27	Phoenix
32N47	Dallas
31N35	PL/IC * VE/ASC
31N08	JU/IC * NE/DSC
30N03	Cairo
29N58	New Orleans
29N46	Houston
29N44	JU/MH * MA/DSC
28N39	Delhi
27N26	VE/IC * SA/ASC
25N47	Miami
24N40	SA/IC * UR/DSC
23N30	MO/MH * NE/ASC
22N34	Calcutta
22N16	Hong Kong
21N19	Honolulu
19N26	Mexico City
19N09	UR/MH * PL/ASC
17N16	NE/IC * JU/ASC
17N02	SU/MH * PL/ASC
14N35	Manila
14N11	VE/IC * PL/DSC
12N53	SA/MH * ME/ASC
12N31	ME/MH * PL/ASC
12N04	NE/MH * MO/DSC
10N50	MA/IC * JU/DSC
10N30	Caracas
9N24	SA/IC * SU/DSC
6N27	Lagos
4N24	UR/MH * SA/DSC
2N48	ME/IC * SA/ASC
2N03	SU/MH * SA/DSC
2S03	SU/IC * SA/ASC
2S48	ME/MH * SA/DSC
4S24	UR/IC * SA/ASC
9S24	SA/MH * SU/ASC
10S50	MA/MH * JU/ASC
12S04	NE/IC * MO/ASC
12S31	ME/IC * PL/DSC
12S53	SA/IC * ME/DSC
14S11	VE/MH * PL/ASC
17S02	SU/IC * PL/DSC

SOUTH LATITUDES

17S16	NE/MH * JU/DSC
19S09	UR/IC * PL/DSC
22S54	Rio De Janeiro
23S30	MO/IC * NE/DSC
24S40	SA/MH * UR/ASC
26S12	Johannesburg
27S26	VE/MH * SA/DSC
29S44	JU/IC * MA/ASC
31S08	JU/MH * NE/ASC
31S35	PL/MH * VE/DSC
33S55	Sydney
34S24	MA/MH * MO/ASC
34S36	Buenos Aires
36S52	Auckland
37S50	Melbourne
40S39	SU/DSC * UR/DSC
46S01	PL/IC * ME/ASC
48S31	MA/IC * SA/ASC
51S19	UR/MH * JU/ASC
51S42	SA/IC * VE/DSC
52S07	SU/MH * JU/ASC
53S38	ME/MH * JU/ASC
53S50	MA/IC * PL/DSC
54S46	PL/IC * SU/ASC
56S30	PL/MH * MO/DSC
57S05	NE/MH * SA/ASC
57S21	MO/MH * PL/ASC
58S04	UR/MH * MO/ASC
58S32	SU/MH * MO/ASC
59S26	ME/MH * MO/ASC
59S35	SA/IC * MO/DSC

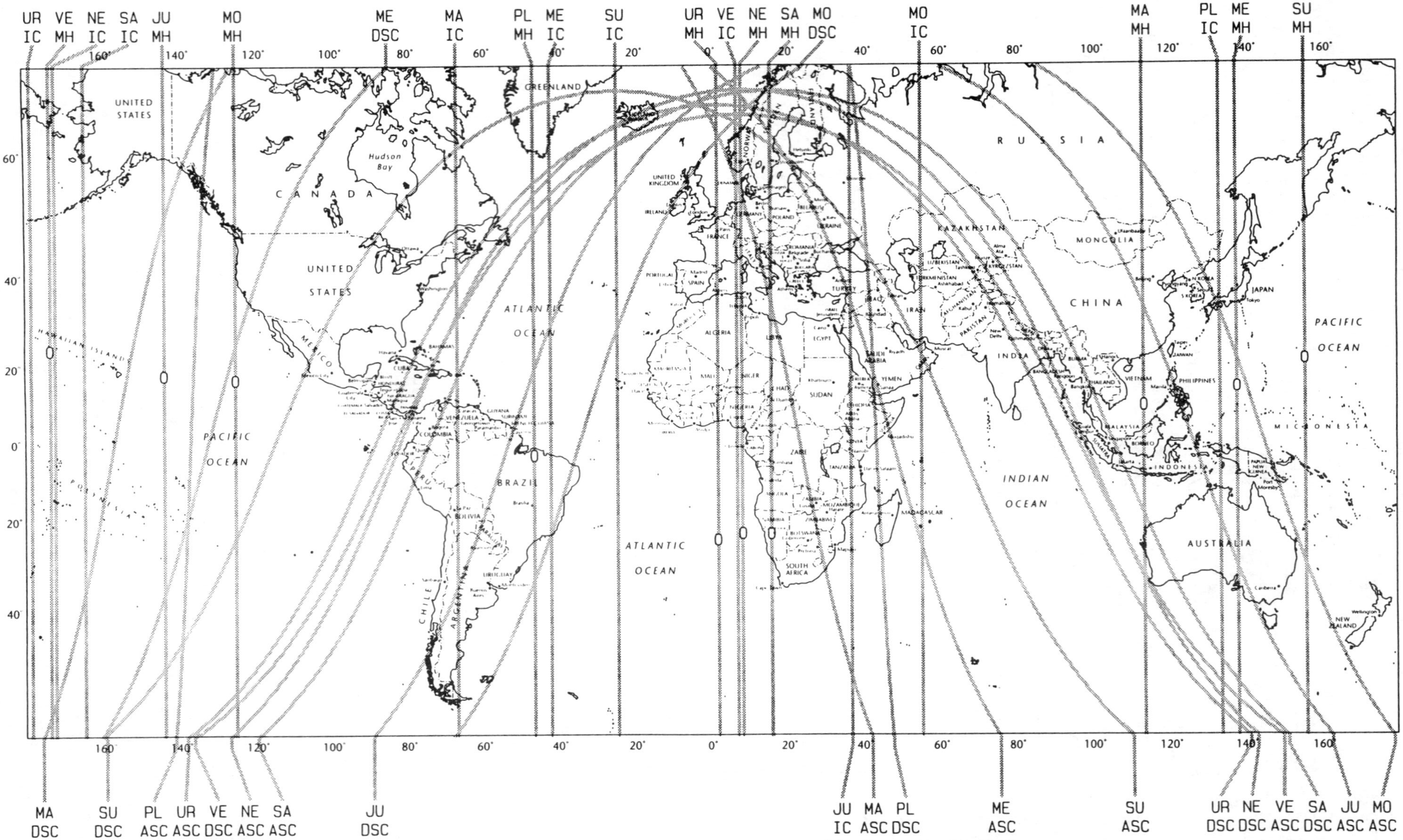

CYCLO*CARTO*GRAPHY

04/17/96

SU, MO, ME, VE AND MA ARE PROGRESSED
JU, SA, UR, NE AND PL ARE TRANSITING

Helmut Kohl

PROG	6	2	30	17:34 GMT
TRANSIT	10	3	90	0:00 GMT
NATAL	4	3	30	5:30 GMT

PL 18 CAN 53		ME 24 PIS 16
VE 24 ARI 48		MA 29 TAU 49
MO 28 TAU 19		
A*C*G	**(58 10)**	**Solar arc A*C*G**
NE 03 VIR 28		NE 01 SCO 38
JU 10 GEM 39		JU 08 LEO 49
UR 10 ARI 46		UR 08 GEM 56
SU 11 ARI 43		SA 10 PIS 43
SA 12 CAP 33		ME 11 GEM 53
ME 13 ARI 43		MA 13 TAU 13
MA 15 PIS 03		PL 17 VIR 03
PL 18 CAN 53		VE 22 GEM 58
VE 24 ARI 48		MO 26 CAN 29
MO 28 TAU 19		

14. solar arc SA EQU A*C*G JU
15. solar arc SA ASC = Bonn

The year 1990 was especially lucky for Helmut Kohl. He had C*C*G SU = A*C*G JU which would have a worldwide effect. This good luck for Chancellor Kohl was very powerful at Moscow where he had C*C*G JU IC. Solar return JU MH is also nearest to Moscow. To have an A*C*G, C*C*G and solar return JU line at a location is the best possible combination to achieve success so easily. Prog. Moscow = A*C*G UR MH and A*C*G SU MH shows that Kohl was able to achieve his goal quickly and unexpectedly. A quick agreement was also shown by disp. C*C*G VE EQU A*C*G UR and A*C*G SU.

Prog. Berlin = A*C*G PL DSC shows the new beginning for the reunited Germany. A few years earlier such an event didn't seem possible, so it was as if a miracle had taken place.

C*C*G MA ASC on the East/West German border indicates that Kohl was determined to achieve German reunification. When the East German government opened the borders to West Germany this gave Kohl the challenge to reunite the two German countries.

Other temporary influences show that German reunification was not an immediate success. Although prog. Bonn = A*C*G SA DSC had an orb of one and three-quarters degrees, it was an applying aspect that began to have an effect within a few months. Solar arc SA ASC = Bonn reiterates the difficulties the government had after German reunification.

C*C*G VE IC was at Bonn, thus activating the A*C*G SA MH and A*C*G VE ASC. This puts emphasis on the financial consequences in rebuilding East Germany. But with C*C*G NE MH passing through Germany near C*C*G VE IC, the real cost of reunification was much higher than originally expected.

As C*C*G NE MH started to pass through East Germany the people there became disappointed in Kohl. They expected too much from German reunification and were quickly dissatisfied when the reality became clear.

C*C*G NE EQU A*C*G ME indicates that Kohl underestimated the true cost of German reunification. It meant that a new tax had to be introduced to help finance the rebuilding of East Germany.

C*C*G NE = A*C*G SA reiterates the disappointment that reunification was not as easy as first imagined and, as the temporary influence aspect is near Bonn, it had a powerful effect there.

This example shows the effectiveness of using Astro*Carto*Graphy for countries and leaders.

Making Predictions with Astro*Carto*Graphy

There is no doubt that Astro*Carto*Graphy can be used effectively in mundane astrology. The question is how can Astro*Carto*Graphy be used to learn about possible future events? At first this seems difficult as many different Astro*Carto*Graphy techniques can be used. The main problem is knowing which one will provide the most reliable information.

Three books are needed when looking at world events with Astro*Carto*Graphy:

a. *Book of World Horoscopes*
b. *Mundane Astrology*
c. *Astro*Carto*Graphy Source Book of Mundane Maps*

There are two ways to go about using Astro*Carto*Graphy in trying to learn where future events are likely to occur.

The first is to draw an Astro*Carto*Graphy map of a country and/or its leader if an accurate time of birth is available. These maps provide the most reliable information about future trends.

When the time of birth is unknown for a country and the leader, use a "midday" chart. Although no Astro*Carto*Graphy map can be drawn, it is still possible to apply some of the techniques to find the temporary influences. Other relevant Astro*Carto*Graphy maps can then be considered if they are significant for the country concerned.

If the temporary influences for these maps indicate that some kind of event is possible, examine the *Astro*Carto*Graphy Source Book of Mundane Maps*. When you see a repetition of themes activating the type of event shown on the maps of the country or leader, it is likely to manifest.

The second way is to start with the *Astro*Carto*Graphy Source Book of Mundane Maps*. If there is a repetition of themes in one part of the world on many different maps it could provide important information. To find out if the potential will manifest in an event, examine charts of the countries concerned where the repetition of themes occurs. If a country's chart is activated by these planetary energies there is a strong possibility that an event will happen.

Below are some general guidelines in using Astro*Carto*Graphy to examine future trends in the world. It is best to have a methodical approach when trying to find out where events are likely to manifest.

1. Astro*Carto*Graphy map of country (or natal chart)
2. Astro*Carto*Graphy map of country's leader (or natal chart)

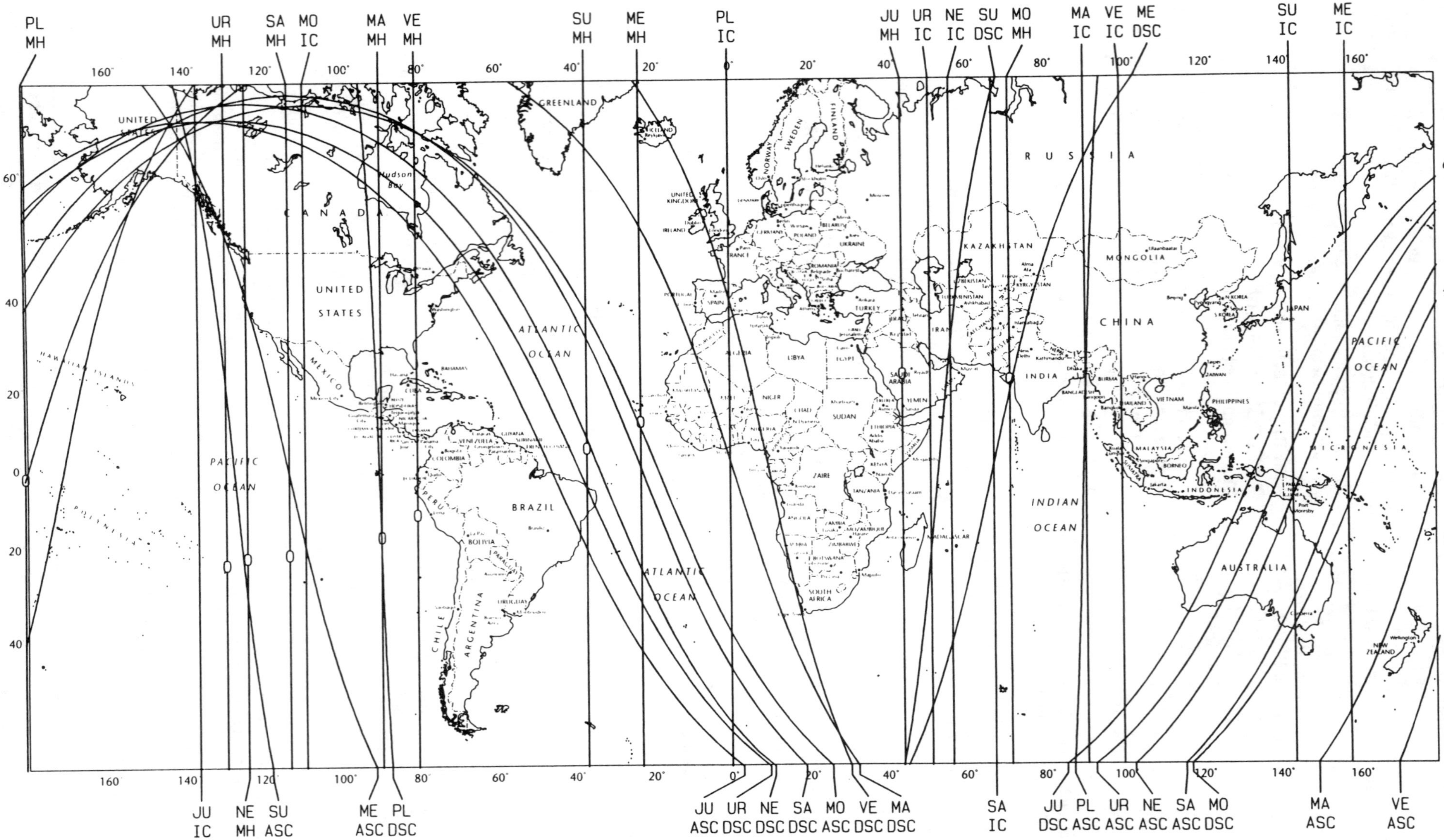

Helmut Kohl SR

April 3, 1990 2:30 PM GMT

04/16/96

04/16/96

Helmut Kohl SR
April 3, 1990 2:30 PM GMT

ASTRO*CARTO*GRAPHY LATITUDE CROSSINGS

NORTH LATITUDES

69N51 NE/MH * MA/ASC
69N24 ME/MH * VE/DSC
68N56 MO/IC * SA/DSC
68N43 SA/IC * SU/DSC
68N38 NE/MH * SA/ASC
68N32 UR/MH * MA/ASC
68N16 VE/IC * ME/DSC
68N13 UR/MH * SA/ASC
68N07 UR/MH * NE/ASC
68N04 SA/ASC * NE/ASC
68N04 MA/ASC * NE/ASC
68N02 MA/ASC * SA/ASC
67N59 SA/MH * MO/DSC
67N56 MO/DSC * NE/DSC
67N56 SU/ASC * VE/ASC
67N55 SA/DSC * PL/DSC
67N51 SA/MH * NE/DSC
67N44 JU/IC * NE/ASC
67N31 MO/IC * NE/DSC
67N24 SU/ASC * NE/ASC
67N22 MO/ASC * PL/DSC
67N21 NE/MH * MO/DSC
67N20 JU/IC * SA/ASC
67N16 MA/MH * SA/DSC
67N14 VE/ASC * NE/ASC
66N58 ME/ASC * NE/ASC
66N52 MA/MH * MO/ASC
66N45 SU/ASC * SA/ASC
66N45 UR/MH * MO/DSC
66N36 MO/DSC * UR/ASC
66N31 NE/MH * UR/DSC
66N31 MA/ASC * JU/ASC
66N30 JU/ASC * UR/ASC
66N29 SU/ASC * JU/DSC
66N29 MA/ASC * UR/ASC
66N27 JU/IC * UR/ASC
66N26 MO/DSC * JU/ASC
66N26 JU/DSC * SA/ASC
66N24 ME/ASC * VE/ASC
66N23 ME/ASC * JU/DSC
66N22 VE/ASC * JU/DSC
66N21 UR/MH * JU/ASC
66N16 SU/ASC * UR/ASC
66N12 ME/ASC * SA/ASC
66N11 JU/IC * MA/ASC
66N03 ME/ASC * UR/ASC
66N02 VE/ASC * SA/ASC
66N01 NE/MH * JU/ASC
65N56 VE/ASC * UR/ASC
65N53 SA/ASC * UR/ASC
65N48 JU/DSC * NE/ASC
65N46 SA/MH * UR/DSC
65N36 JU/IC * MO/DSC
65N29 VE/MH * SA/DSC
65N22 VE/MH * MO/ASC
65N15 MO/IC * UR/DSC
65N13 NE/DSC * PL/DSC
65N08 SU/ASC * MA/ASC
65N05 MO/DSC * SU/ASC
65N01 MO/DSC * MA/ASC
64N43 SA/MH * JU/ASC
64N42 MO/DSC * ME/ASC
64N39 MO/ASC * SA/DSC
64N30 ME/ASC * MA/ASC
64N12 MA/MH * NE/DSC
64N10 ME/DSC * PL/ASC
63N57 MO/IC * JU/ASC
63N27 SU/MH * MA/DSC
62N49 MO/DSC * VE/ASC
62N04 SU/ASC * ME/ASC
61N57 UR/DSC * PL/DSC
61N43 MA/IC * ME/DSC
61N15 UR/ASC * NE/ASC
61N13 Anchorage

CENTRAL LATITUDES

61N12 VE/MH * NE/DSC
61N05 JU/IC * ME/ASC
60N44 MA/MH * UR/DSC

59N32 JU/IC * SU/ASC
59N20 Stockholm
59N13 JU/ASC * PL/DSC
58N50 PL/MH * JU/DSC
57N42 MA/MH * JU/ASC
56N48 VE/MH * UR/DSC
55N48 VE/ASC * MA/ASC
55N45 Moscow
55N41 Copenhagen
55N24 PL/MH * UR/ASC
54N08 PL/MH * NE/ASC
52N51 ME/MH * MA/DSC
52N30 Berlin
52N22 VE/MH * JU/ASC
51N46 UR/MH * ME/ASC
51N30 London

49N14 Vancouver
48N50 Paris
47N36 Seattle
45N32 Montreal
45N17 PL/MH * SA/ASC
44N59 Minneapolis
43N39 Toronto
42N22 Boston
42N20 Detroit
41N55 ME/IC * JU/DSC
41N54 Rome
41N53 Chicago
40N43 New York
40N29 NE/MH * ME/ASC

39N48 PL/IC * VE/DSC
39N44 Denver
38N53 Washington
38N37 St. Louis
38N27 PL/MH * MO/DSC
37N47 San Francisco
37N36 SU/MH * MO/ASC
36N10 Las Vegas
35N40 Tokyo
34N04 Los Angeles
33N45 Atlanta
33N27 Phoenix
32N47 Dallas
32N18 ME/IC * UR/ASC
32N04 SU/MH * SA/DSC
30N39 NE/IC * SU/DSC
30N03 Cairo

29N58 New Orleans
29N46 Houston
28N39 Delhi
25N47 Miami
24N43 ME/IC * NE/ASC
22N34 Calcutta
22N16 Hong Kong
21N19 Honolulu
21N11 UR/MH * SU/ASC
20N38 MA/IC * PL/ASC
20N09 SU/IC * JU/DSC

19N26 Mexico City
19N12 MO/MH * ME/DSC
14N35 Manila
10N32 ME/MH * MO/ASC
10N30 Caracas

7N55 SU/MH * NE/DSC
6N27 Lagos
4N48 SU/IC * UR/ASC

SOUTH LATITUDES

2N04 PL/IC * MA/DSC
0N40 SA/IC * ME/DSC
0N23 ME/MH * SA/DSC

0S23 ME/IC * SA/ASC
0S40 SA/MH * ME/ASC
2S04 PL/MH * MA/ASC
4S48 SU/MH * UR/DSC
7S55 SU/IC * NE/ASC

10S32 ME/IC * MO/DSC
19S12 MO/IC * ME/ASC

20S09 SU/MH * JU/ASC
20S38 MA/MH * PL/DSC
21S11 UR/IC * SU/DSC
22S54 Rio De Janeiro
24S43 ME/MH * NE/DSC
26S12 Johannesburg

30S39 NE/MH * SU/ASC
32S04 SU/IC * SA/ASC
32S18 ME/MH * UR/DSC
33S55 Sydney
34S36 Buenos Aires
36S52 Auckland
37S36 SU/IC * MO/DSC
37S50 Melbourne
38S27 PL/IC * MO/ASC
39S48 PL/MH * VE/ASC

40S29 NE/IC * ME/DSC
41S55 ME/MH * JU/ASC
45S17 PL/IC * SA/DSC

51S46 UR/IC * ME/DSC
52S22 VE/IC * JU/DSC
52S51 ME/IC * MA/ASC
54S08 PL/IC * NE/DSC
55S24 PL/IC * UR/DSC
55S48 VE/DSC * MA/DSC
56S48 VE/IC * UR/ASC
57S42 MA/IC * JU/DSC
58S50 PL/IC * JU/ASC
59S13 JU/DSC * PL/ASC
59S32 JU/MH * SU/DSC

3. Astro*Carto*Graphy map of special event for a country (or natal chart)

By examining the temporary influences for a certain time period it is possible to learn what a country is likely to experience. The leader's Astro*Carto*Graphy map represents the country concerned, so it should be considered whenever possible. The other Astro*Carto*Graphy maps used in mundane astrology are drawn for a certain type of cycle. As accurate times are known for these charts there is no problem in drawing an Astro*Carto*Graphy map.

When examining these Astro*Carto*Graphy maps, **repetition of themes** is most important in trying to locate where certain planetary energies are likely to manifest.

4. Astro*Carto*Graphy map for the Mars/Pluto cycle (conjunction and opposition)

a. Locations with MA, PL and UR lines

b. Position of the MA/PL conjunction or opposition

a. This map shows where violent and destructive energies are most likely to manifest, so look for locations which have MA, PL and UR lines. It's possible that these planetary lines will be combined with others at a location, so they also need to be taken into consideration.

b. The position where the MA/PL conjunction or opposition takes place is a powerful point. Examine this in charts of countries and leaders to find out if significant activations are made. The *Book of World Horoscopes* has an integrated planetary index in zodiacal order which helps check many charts quickly. Use an orb of two degrees when using the zodiacal position of the MA/PL cycle.

5. Astro*Carto*Graphy map for an eclipse (solar and lunar eclipse)

a. Locations with MA, PL and UR lines

b. Position of the SU/MO conjunction or opposition

a. Most significant events take place where MA, PL and UR lines are angular. These are usually major tragedies, accidents or events that have a big impact around the world.

This is not always the case, so other planetary lines should be considered when they seem to be important. For the fall of the Berlin Wall it was SA/NE lines that were very significant.

If many planetary lines (five or more) appear bunched at one particular location, this can also be important. Much depends on which planets are involved and whether other maps show a repetition of themes.

b. The zodiacal position of the SU/MO conjunction is examined in the charts of countries and leaders. If any kind of activation is made it could be important for the country concerned.

6. Astro*Carto*Graphy maps for lunations

a. Repetition of themes with locations on MA/PL and eclipse maps.

b. Five or more planetary lines at a location.

c. Position of the SU/MO conjunction or opposition.

a. Lunation maps tend to show when an event will manifest, provided they have repetition of themes with other maps.

b. When five or more planetary lines are at a certain part of the world, it can be significant, especially when there is a repetition of themes at one particular location.

c. Examine the zodiacal position of the SU/MO conjunction in the charts of countries and leaders. If one of these charts is activated it could be important.

7. Astro*Carto*Graphy map for ingresses

a. Repetition of themes at locations on other maps.

b. Position of planetary lines at a location.

a. I tend to put less importance on the ingress maps unless they show a strong repetition of themes. I use ingress maps after first examining the other type of maps.

b. If a planetary line at a location activates that country's chart, it could have an important influence.

These guidelines do not provide a guaranteed way to use Astro*Carto*Graphy in making predictions. What they do offer is the opportunity to locate areas where significant events could manifest.

Sometimes an event will happen that wasn't seen in advance from the various Astro*Carto*Graphy maps. Then it is only possible to examine the event in hindsight to find out why it happened.

The 1999 Eclipse

For many years astrologers have been discussing what impact the eclipses in July and August 1999 will have. This interest has been created by the famous Nostradamus prediction for July 1999.

In the year 1999 and seven months
From the sky will come the great king of terror...
Before and afterwards war reigns happily

The eclipses are powerful as both charts have a grand cross (see diagram 18 for the planets involved).

Nostradamus predicts a war will break out in 1999, so Astro*Carto*Graphy maps for these two eclipses should provide significant information.

Mars lines on both eclipse Astro*Carto*Graphy maps are important as this planet is connected with wars. Mars is also extremely powerful as it is involved in the grand crosses. In both eclipse Astro*Carto*Graphy maps Mars has a potent influence for three countries:

1. United States
2. Russia
3. China

United States

a. Eclipse July 28, 1999, Washington = MA IC, UR DSC, SA MH

b. Eclipse August 11, 1999, Washington = MA IC, SA MH

Russia

a. Eclipse July 28, 1999, Moscow = VE MH, MA ASC, SA DSC

b. Eclipse August 11, 1999, Moscow = MA ASC,

Diagram 18—Eclipse Full Moon, July 28, 1999

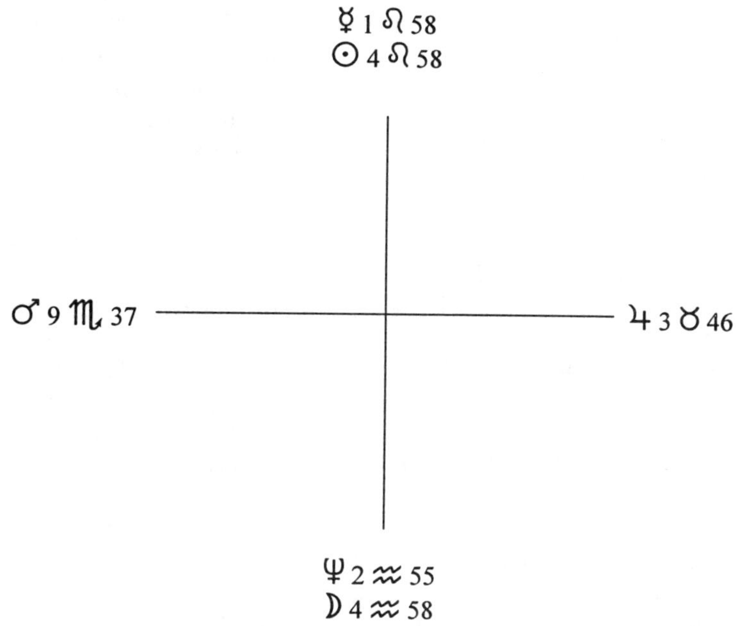

Eclipse New Moon, August 11, 1999

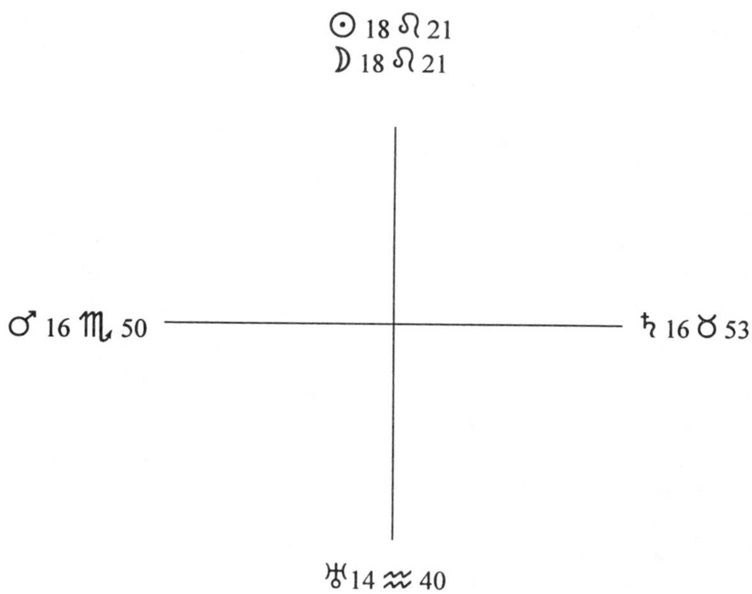

SA DSC China

a. Eclipse July 28, 1999, Beijing = SU DSC, MO ASC, NE ASC

MA MH passes through China

b. Eclipse August 11, 1999, Beijing = UR ASC, PL MH, MO DSC, SU DSC, VE DSC.

MA MH passes through China

The full Moon eclipse on July 28, 1999 (04 LEO/AQU 58) is conjunct China's ASC/DSC axis (05 AQU 57) and MO (03 AQU 10). This is significant for China as the eclipse SU DSC and MO ASC are exactly at Beijing with NE ASC.

The new Moon eclipse on August 11, 1999 (18 LEO 21) is conjunct Communist China's PL (17 LEO 39). It is also activating the USSR revolution T-square (SA OPP UR SQU SU CJN ME). Despite the breakup of the USSR in 1991, this chart is still important to consider.

The exact UR ASC line (14 AQU 40) on this eclipse map is SQU Communist China's MA (14 LEO 53), activating MA CJN PL (seventh house) and emphasizing relations with other countries.

During 1999, tr. PL passes through the Nuclear Axis (7-11 GEM/SAG), which could be potentially dangerous for mankind. At the same time, tr. PL will be opposite the United States UR.

The two eclipses indicate there will be some kind of conflict between China and the United States, possibly linked to nuclear energy. Russia will try to mediate, but talks are likely to fail and Russia will be drawn into the conflict (eclipse new Moon MA ASC exact at Moscow).

The dispute could start as a result of deceptive actions by the Chinese government which anger the United States. At the time of the eclipse on August 11, 1999, there is a strong indication of Chinese military action.

The key date will be on or near August 11, 1999 when the eclipse activates Communist China's MA CJN PL. The eclipse map is also significant for China as five planetary lines are powerful at Beijing.

Amazingly, the lunar eclipse on January 31, 1999 shows a repetition of themes for Washington and Beijing which could be when the dispute starts.

Washington	MA DSC
	SA ASC
Beijing	MA ASC (exact)
	NE IC
	ME IC
	MO MH
	SU IC
	SA DSC
	UR IC

The lunar eclipse (11 AQU/LEO 20) axis is conjunct UR (12 AQU 41) and ME (08 AQU 50), all of which are at Beijing. This eclipse will have a powerful impact on China as seven planetary lines are near Beijing.

With lunar eclipse VE DSC at Moscow it reiterates the peacemaker role of Russia in trying to resolve the dispute between China and the United States.

The solar eclipse on February 16, 1999 has PL ASC at Washington, which could intensify the dispute with China. SA ASC and MA DSC are at Moscow, an indication that Russia will experience difficulties in bringing about a peaceful settlement. JU MH at Beijing could show some kind of success for China.

There are four eclipses in 1999, three of which have a very powerful influence for China and the United States. If Nostradamus' prediction is correct, the war will be between China and America. With PL passing through the nuclear axis there is a strong possibility that atomic energy could be featured in some way.

Eclipses featuring a grand cross usually manifest as significant world events. The eclipse on August 6, 1990 had a grand cross (SU, MO, MA, PL) and seven planetary lines in the Middle East. This eclipse occurred shortly after Iraq invaded Kuwait.

In 1999, there are **two** eclipses with a grand cross, so a major event can be expected somewhere in the world. With five planetary lines at Beijing on the August 11, 1999 eclipse Astro*Carto*Graphy map, this could manifest as a significant event for China.

An Astro*Carto*Graphy map drawn for the Mars/Pluto opposition on June 2, 1998 has MA ASC and SU ASC within orb of Washington. This is another indication that America will become involved in a war.

The summer solstice takes place at 7:49 p.m. GMT on June 21, 1999. An Astro*Carto*Graphy map for this ingress shows the following information.

Washington	NE IC
	MA ASC
	JU DSC
Moscow	PL MH
	ME DSC
	NE ASC
Beijing	PL DSC
	UR MH
	VE IC

This map has an effect a few weeks before the two grand cross eclipses occur. On its own, the Cancer ingress map has little effect. But when it is considered with the two grand cross eclipse maps it has a powerful impact.

Before investigating the Nostradamus prediction with Astro*Carto*Graphy, I thought a war in the Middle East was likely. This was based on the fact that both the eclipses on August 6, 1990 and August 11, 1999 occur in the middle of the fixed signs. But Astro*Carto*Graphy does not emphasize the Middle East in 1999. What is incredible is that Nostradamus made his predictions more than 400 years ago. He rarely gave precise dates for his predictions, so "July 1999" is significant.

The use of Astro*Carto*Graphy in mundane astrology offers tremendous possibilities in locating where events are likely to occur in the world. It is certainly a branch of astrology where Astro*Carto*Graphy can be used effectively.

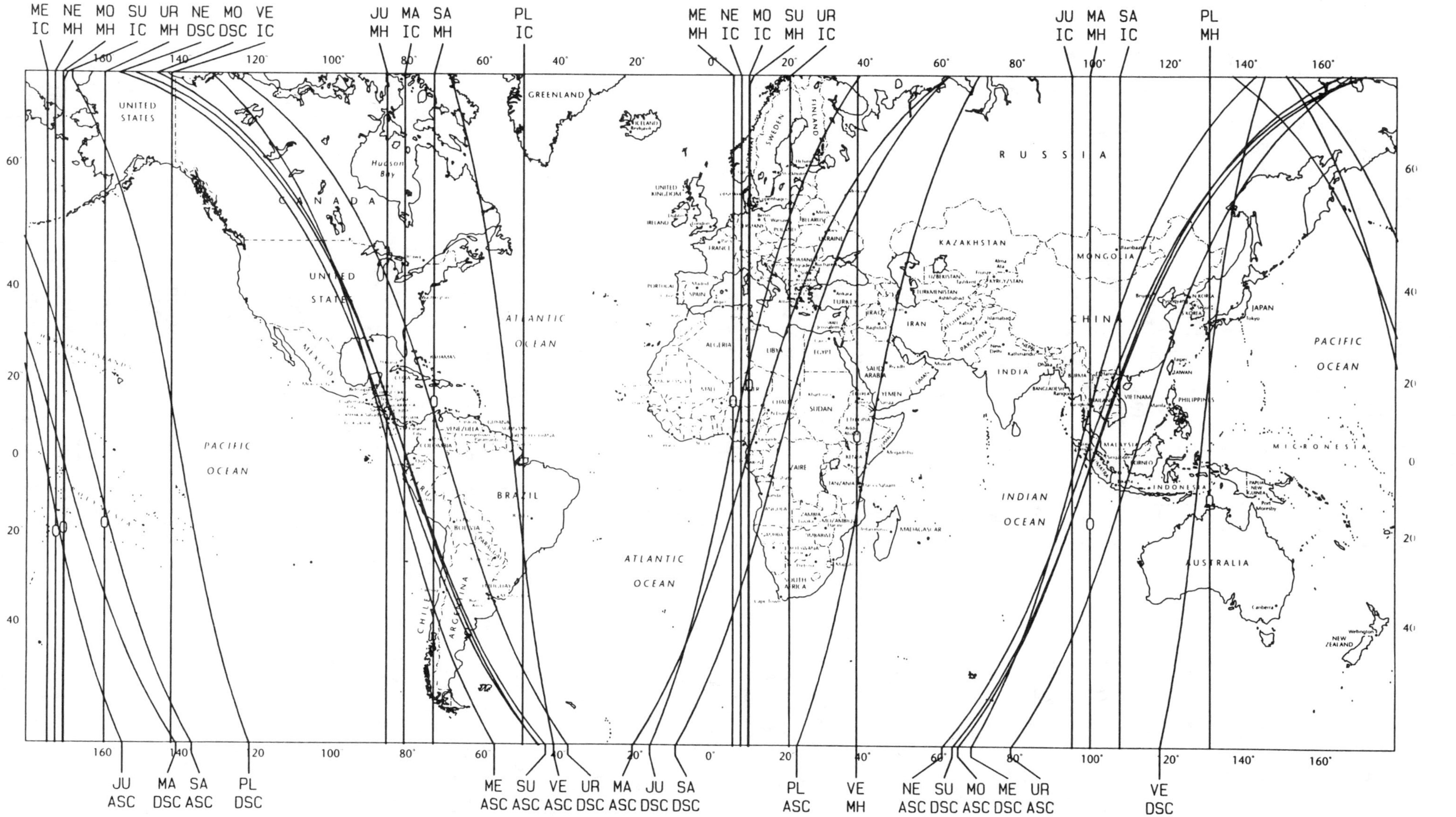

Lunar Eclipse

July 28, 1999 11:25 AM GMT

04/16/96

Lunar Eclipse
July 28, 1999 11:25 AM GMT

ASTRO*CARTO*GRAPHY LATITUDE CROSSINGS

NORTH LATITUDES

69N44 VE/MH * JU/DSC
69N23 VE/IC * MO/DSC
69N06 ME/DSC * MA/DSC
68N59 UR/ASC * NE/ASC
68N38 VE/IC * SU/ASC
68N36 SU/DSC * UR/ASC
68N24 MO/ASC * SA/ASC
68N03 VE/DSC * MA/DSC
68N00 SA/ASC * NE/ASC
67N59 VE/IC * NE/DSC
67N55 MO/ASC * JU/ASC
67N51 SU/DSC * SA/ASC
67N38 SA/ASC * UR/ASC
67N29 JU/ASC * NE/ASC
67N17 SU/DSC * JU/ASC
66N52 JU/ASC * UR/ASC
66N11 MO/ASC * MA/DSC
65N54 MA/DSC * NE/ASC
65N39 SU/DSC * MA/DSC
65N01 PL/MH * ME/DSC
64N55 MA/DSC * UR/ASC
62N56 MO/ASC * VE/DSC
62N32 VE/DSC * NE/ASC
62N27 UR/MH * PL/DSC
61N33 SU/DSC * VE/DSC
61N13 Anchorage

59N20 Stockholm
58N59 VE/MH * MA/ASC
58N18 MO/ASC * NE/ASC
57N15 PL/MH * NE/ASC
57N10 PL/MH * MO/ASC
55N57 PL/MH * SU/DSC
55N45 Moscow
55N45 VE/DSC * UR/ASC
55N41 Copenhagen
54N18 VE/MH * SA/DSC
53N23 UR/IC * JU/DSC
52N30 Berlin
51N30 London

49N14 Vancouver
48N50 Paris
48N25 PL/MH * UR/ASC
47N36 Seattle
46N06 MO/DSC * ME/ASC
45N32 Montreal
44N59 Minneapolis
43N39 Toronto
43N20 MA/DSC * JU/ASC
42N22 Boston
42N20 Detroit
42N12 SU/ASC * ME/ASC
42N01 JU/MH * UR/DSC
41N54 Rome
41N53 Chicago
40N43 New York

39N44 Denver
38N53 Washington
38N37 St. Louis
37N47 San Francisco
37N30 ME/IC * SA/ASC
36N33 SA/IC * ME/DSC
36N10 Las Vegas
35N40 Tokyo
34N04 Los Angeles
33N50 UR/IC * MA/ASC
33N45 Atlanta
33N27 Phoenix
32N47 Dallas
32N29 MA/IC * UR/DSC
32N13 NE/MH * SA/ASC
30N03 Cairo

CENTRAL LATITUDES

29N58 New Orleans
29N46 Houston
28N39 Delhi
26N28 MO/MH * SA/ASC
25N52 SU/IC * SA/ASC
25N47 Miami
24N58 SA/IC * NE/ASC
24N31 SU/MH * JU/DSC
23N45 MO/IC * JU/DSC
23N26 ME/ASC * NE/DSC
22N34 Calcutta
22N16 PL/MH * VE/DSC
22N16 Hong Kong
21N19 Honolulu
21N11 SA/IC * MO/ASC

19N55 SA/IC * SU/DSC
19N26 Mexico City
15N12 NE/IC * JU/DSC
15N10 JU/MH * MO/DSC
15N06 JU/MH * SU/ASC
14N35 Manila
13N40 MO/DSC * SU/ASC
13N40 VE/IC * PL/DSC
13N21 MA/MH * ME/DSC
13N01 UR/IC * SA/DSC
12N20 ME/IC * MA/DSC
11N04 SA/MH * UR/DSC
10N30 Caracas

9N02 JU/MH * NE/DSC
6N27 Lagos
5N28 ME/MH * JU/DSC
5N25 NE/MH * MA/DSC
4N30 MA/MH * NE/ASC
4N11 JU/MH * ME/ASC
2N03 SU/MH * MA/ASC
1N43 MA/IC * SU/ASC
1N23 MO/IC * MA/ASC
1N13 MA/IC * MO/DSC

1S13 MA/MH * MO/ASC
1S23 MO/MH * MA/DSC
1S43 MA/MH * SU/DSC
2S03 SU/IC * MA/DSC
4S11 JU/IC * ME/DSC
4S30 MA/IC * NE/DSC
5S25 NE/IC * MA/ASC
5S28 ME/IC * JU/ASC
9S02 JU/IC * NE/ASC

11S04 SA/IC * UR/ASC
12S20 ME/MH * MA/ASC
13S01 UR/MH * SA/ASC
13S21 MA/IC * ME/ASC
13S40 VE/MH * PL/ASC
13S40 MO/ASC * SU/DSC
15S06 JU/IC * SU/DSC
15S10 JU/IC * MO/ASC
15S12 NE/MH * JU/ASC
19S55 SA/MH * SU/DSC

21S11 SA/MH * MO/DSC
22S16 PL/IC * VE/ASC
22S54 Rio De Janeiro
23S26 ME/DSC * NE/ASC
23S45 MO/MH * JU/ASC
24S31 SU/IC * JU/ASC
24S58 SA/MH * NE/DSC
25S52 SU/MH * SA/DSC
26S12 Johannesburg
26S28 MO/IC * SA/DSC

32S13 NE/IC * SA/DSC

SOUTH LATITUDES

32S29 MA/MH * UR/ASC
33S50 UR/MH * MA/DSC
33S55 Sydney
34S36 Buenos Aires
36S33 SA/MH * ME/ASC
36S52 Auckland
37S30 ME/MH * SA/DSC
37S50 Melbourne

42S01 JU/IC * UR/ASC
42S12 SU/DSC * ME/DSC
43S20 MA/ASC * JU/DSC
46S06 MO/ASC * ME/DSC
48S25 PL/IC * UR/DSC

53S23 UR/MH * JU/ASC
54S18 VE/IC * SA/ASC
55S45 VE/ASC * UR/DSC
55S57 PL/IC * SU/ASC
57S10 PL/IC * MO/DSC
57S15 PL/IC * NE/DSC
58S18 MO/DSC * NE/DSC
58S59 VE/IC * MA/DSC

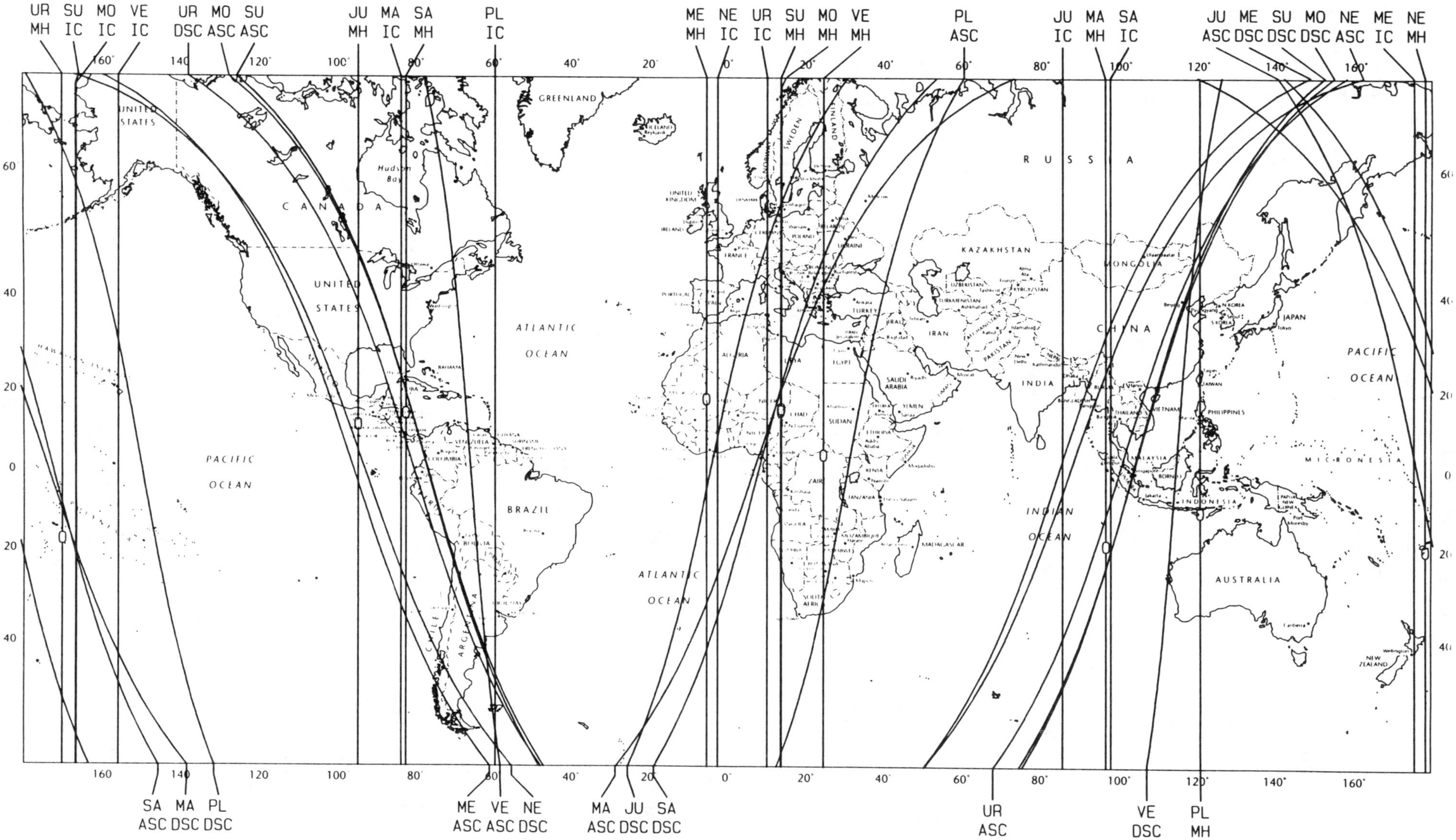

Solar Eclipse

August 11, 1999 11:09 AM GMT

04/16/96

04/16/96

Solar Eclipse
August 11, 1999 11:09 AM GMT

ASTRO*CARTO*GRAPHY LATITUDE CROSSINGS

NORTH LATITUDES

69N58	SU/IC * NE/DSC
69N57	MO/IC * NE/DSC
69N55	PL/MH * MA/DSC
69N18	VE/IC * ME/ASC
69N15	ME/DSC * SA/ASC
69N04	VE/DSC * MA/DSC
68N59	UR/ASC * NE/ASC
68N54	ME/DSC * JU/ASC
68N32	VE/IC * NE/DSC
68N18	SU/DSC * SA/ASC
68N11	MO/DSC * NE/ASC
68N01	SA/ASC * NE/ASC
67N56	MO/DSC * SA/ASC
67N40	SA/ASC * UR/ASC
67N36	SU/DSC * JU/ASC
67N30	JU/ASC * NE/ASC
67N19	SU/DSC * NE/ASC
67N10	MO/DSC * JU/ASC
66N56	ME/DSC * MA/DSC
66N54	JU/ASC * UR/ASC
66N35	VE/MH * JU/DSC
65N47	MA/DSC * NE/ASC
65N38	MA/ASC * PL/ASC
65N18	SU/DSC * MA/DSC
64N58	MO/DSC * MA/DSC
64N54	MA/DSC * UR/ASC
63N35	MO/DSC * UR/ASC
62N44	UR/MH * PL/DSC
61N55	ME/DSC * VE/DSC
61N13	Anchorage
60N50	PL/MH * ME/DSC
60N24	ME/ASC * NE/DSC
59N20	Stockholm
58N24	SU/IC * PL/DSC
58N11	MO/IC * PL/DSC
57N47	VE/DSC * NE/ASC
57N37	SU/DSC * UR/ASC
57N19	PL/MH * NE/ASC
56N03	MO/MH * JU/DSC
55N50	SU/MH * JU/DSC
55N45	Moscow
55N41	Copenhagen
55N01	PL/MH * VE/DSC
54N44	MA/DSC * JU/ASC
52N30	Berlin
51N30	London
50N08	UR/IC * JU/DSC
49N14	Vancouver
48N50	Paris
48N43	PL/MH * UR/ASC
48N18	JU/MH * SU/ASC
47N40	VE/MH * SA/DSC
47N38	JU/MH * MO/ASC
47N36	Seattle
46N59	PL/MH * SU/DSC
46N26	VE/DSC * UR/ASC
45N50	PL/MH * MO/DSC
45N32	Montreal
44N59	Minneapolis
43N39	Toronto
43N19	SU/DSC * VE/DSC
42N29	VE/MH * MA/ASC
42N22	Boston
42N20	Detroit
41N54	Rome
41N53	Chicago
41N43	MO/DSC * VE/DSC
41N19	ME/IC * SA/ASC
40N43	New York
39N44	Denver
39N05	JU/MH * UR/DSC

CENTRAL LATITUDES

38N53	Washington
38N37	St. Louis
37N47	San Francisco
36N10	Las Vegas
35N40	Tokyo
34N53	SA/IC * ME/DSC
34N39	NE/MH * SA/ASC
34N04	Los Angeles
33N45	Atlanta
33N27	Phoenix
32N47	Dallas
32N21	MA/MH * ME/DSC
31N52	ME/IC * MA/DSC
30N47	VE/IC * PL/DSC
30N03	Cairo
29N58	New Orleans
29N46	Houston
28N39	Delhi
27N08	SA/IC * NE/ASC
25N47	Miami
25N24	NE/MH * MA/DSC
24N20	MA/MH * NE/ASC
23N41	MA/IC * SU/ASC
23N29	MA/IC * MO/ASC
22N34	Calcutta
22N16	Hong Kong
21N20	MO/MH * SA/DSC
21N19	Honolulu
20N49	SU/MH * SA/DSC
20N12	MO/MH * MA/ASC
19N50	SA/MH * SU/ASC
19N48	SU/MH * MA/ASC
19N45	SA/MH * MO/ASC
19N26	Mexico City
16N47	MO/ASC * SU/ASC
16N07	MA/ASC * SA/DSC
14N35	Manila
11N07	MA/IC * UR/DSC
10N30	Caracas
10N16	UR/IC * MA/ASC
8N58	NE/IC * JU/DSC
8N32	UR/IC * SA/DSC
7N14	SA/MH * UR/DSC
6N27	Lagos
5N22	JU/MH * NE/DSC
4N39	ME/IC * JU/ASC
2N58	JU/IC * ME/DSC
2S58	JU/MH * ME/ASC
4S39	ME/MH * JU/DSC
5S22	JU/IC * NE/ASC
7S14	SA/IC * UR/ASC
8S32	UR/MH * SA/ASC
8S58	NE/MH * JU/ASC
10S16	UR/MH * MA/DSC
11S07	MA/MH * UR/ASC
16S07	MA/DSC * SA/ASC
16S47	MO/DSC * SU/DSC
19S45	SA/IC * MO/DSC
19S48	SU/IC * MA/DSC
19S50	SA/IC * SU/DSC
20S12	MO/IC * MA/DSC
20S49	SU/IC * SA/ASC
21S20	MO/IC * SA/ASC
22S54	Rio De Janeiro
23S29	MA/MH * MO/DSC
23S41	MA/MH * SU/DSC
24S20	MA/IC * NE/DSC
25S24	NE/IC * MA/ASC
26S12	Johannesburg

SOUTH LATITUDES

27S08	SA/MH * NE/DSC
30S47	VE/MH * PL/ASC
31S52	ME/MH * MA/ASC
32S21	MA/IC * ME/ASC
33S55	Sydney
34S36	Buenos Aires
34S39	NE/IC * SA/DSC
34S53	SA/MH * ME/ASC
36S52	Auckland
37S50	Melbourne
39S05	JU/IC * UR/ASC
41S19	ME/MH * SA/DSC
41S43	MO/ASC * VE/ASC
42S29	VE/IC * MA/DSC
43S19	SU/ASC * VE/ASC
45S50	PL/IC * MO/ASC
46S26	VE/ASC * UR/DSC
46S59	PL/IC * SU/ASC
47S38	JU/IC * MO/DSC
47S40	VE/IC * SA/ASC
48S18	JU/IC * SU/DSC
48S43	PL/IC * UR/DSC
50S08	UR/MH * JU/ASC
54S44	MA/ASC * JU/DSC
55S01	PL/IC * VE/ASC
55S50	SU/IC * JU/ASC
56S03	MO/IC * JU/ASC
57S19	PL/IC * NE/DSC
57S37	SU/ASC * UR/DSC
57S47	VE/ASC * NE/DSC
58S11	MO/MH * PL/ASC
58S24	SU/MH * PL/ASC

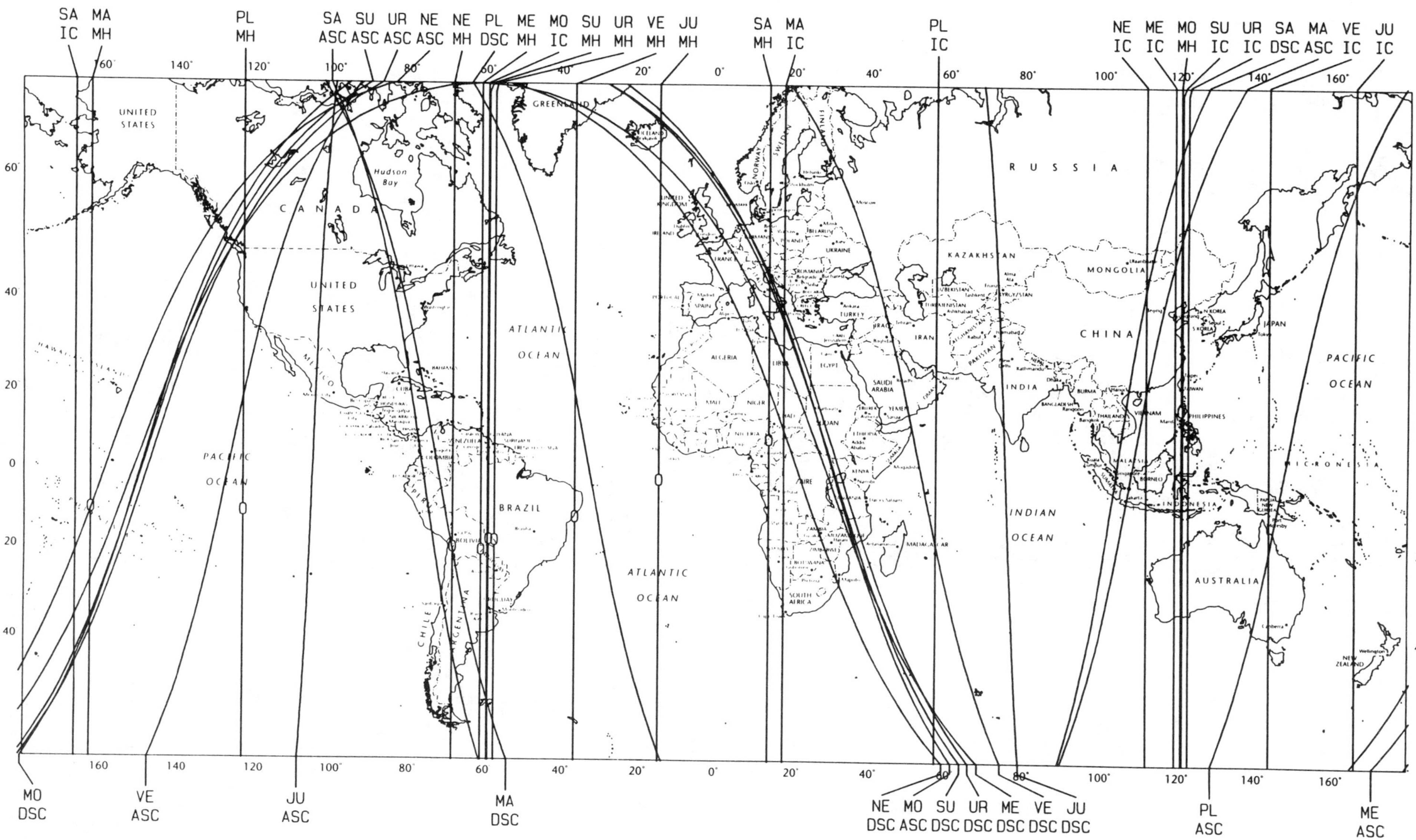

Lunar Eclipse

January 31, 1999 4:07 PM GMT

04/16/96

Lunar Eclipse
January 31, 1999 4:07 PM GMT

ASTRO*CARTO*GRAPHY LATITUDE CROSSINGS

NORTH LATITUDES

69N57 MO/IC * ME/DSC
69N56 SU/MH * ME/DSC
69N56 ME/ASC * PL/DSC
69N55 UR/MH * ME/DSC
69N48 NE/MH * ME/ASC
69N32 ME/DSC * NE/DSC
69N15 JU/ASC * SA/ASC
69N07 UR/ASC * NE/ASC
69N02 MO/DSC * SA/ASC
69N00 SU/ASC * VE/ASC
68N57 MO/DSC * JU/ASC
68N46 MO/DSC * MA/DSC
68N28 MA/DSC * JU/ASC
68N28 VE/ASC * NE/ASC
68N27 ME/MH * PL/DSC
68N10 VE/MH * ME/DSC
68N07 SU/ASC * SA/ASC
68N02 SA/ASC * NE/ASC
67N54 SU/ASC * MA/DSC
67N52 MA/DSC * NE/ASC
67N45 VE/ASC * UR/ASC
67N45 SU/ASC * NE/ASC
67N38 JU/ASC * NE/ASC
67N35 SU/ASC * JU/ASC
67N34 SA/ASC * UR/ASC
67N31 VE/MH * NE/DSC
67N31 JU/MH * MO/ASC
67N27 VE/ASC * SA/ASC
67N23 MA/DSC * UR/ASC
67N20 MO/IC * PL/DSC
67N09 VE/ASC * MA/DSC
67N07 SU/MH * PL/DSC
66N45 JU/ASC * UR/ASC
66N38 JU/MH * UR/DSC
66N24 JU/MH * SU/DSC
66N04 ME/ASC * SA/ASC
65N58 ME/ASC * MA/DSC
65N56 UR/MH * PL/DSC
65N19 VE/ASC * JU/ASC
65N10 UR/IC * SA/DSC
64N45 ME/ASC * JU/ASC
64N26 ME/ASC * VE/ASC
64N13 MO/DSC * NE/ASC
63N44 MA/DSC * SA/ASC
63N18 SU/IC * SA/DSC
62N55 MO/MH * SA/DSC
62N11 JU/MH * ME/DSC
61N13 Anchorage
60N29 ME/IC * SA/DSC

59N20 Stockholm
59N19 JU/MH * NE/DSC
58N55 PL/MH * NE/ASC
58N45 JU/IC * PL/ASC
56N37 PL/MH * MO/DSC
55N45 Moscow
55N41 Copenhagen
54N46 PL/MH * SU/ASC
54N38 UR/IC * MA/ASC
52N51 PL/MH * UR/ASC
52N41 MO/ASC * UR/DSC
52N30 Berlin
52N24 PL/MH * ME/ASC
51N39 SU/IC * MA/ASC
51N30 London
51N03 MO/MH * MA/ASC

49N41 ME/ASC * UR/ASC
49N14 Vancouver
48N50 Paris
47N36 Seattle
47N09 ME/IC * MA/ASC
45N32 Montreal
45N16 SA/MH * UR/DSC

CENTRAL LATITUDES

44N59 Minneapolis
44N30 SA/MH * MO/ASC
43N39 Toronto
43N17 SA/MH * SU/DSC
42N22 Boston
42N20 NE/IC * SA/DSC
42N20 Detroit
41N54 Rome
41N53 Chicago
40N43 New York

39N44 Denver
38N53 Washington
38N42 MA/IC * UR/DSC
38N37 St. Louis
37N47 San Francisco
37N06 MA/IC * MO/ASC
36N10 Las Vegas
36N05 MA/IC * SU/DSC
35N40 Tokyo
35N39 SA/MH * ME/DSC
34N04 Los Angeles
33N45 Atlanta
33N27 Phoenix
32N48 SU/ASC * ME/ASC
32N47 Dallas
30N03 Cairo

29N58 New Orleans
29N46 Houston
28N39 Delhi
28N04 MA/IC * ME/DSC
25N47 Miami
22N34 Calcutta
22N32 MO/DSC * ME/ASC
22N16 Hong Kong
21N19 Honolulu
21N00 SA/MH * NE/DSC
20N11 NE/IC * MA/ASC

19N58 VE/MH * PL/DSC
19N26 Mexico City
18N22 PL/MH * VE/ASC
15N38 MO/ASC * SU/DSC
14N35 Manila
10N41 MA/IC * NE/DSC
10N30 Caracas

6N27 Lagos

10S41 MA/MH * NE/ASC
15S38 MO/DSC * SU/ASC
18S22 PL/IC * VE/DSC
19S58 VE/IC * PL/ASC

20S11 NE/MH * MA/DSC
21S00 SA/IC * NE/ASC
22S32 MO/ASC * ME/DSC
22S54 Rio De Janeiro
26S12 Johannesburg
28S04 MA/MH * ME/ASC

32S48 SU/DSC * ME/DSC
33S55 Sydney
34S36 Buenos Aires
35S39 SA/IC * ME/ASC
36S05 MA/MH * SU/ASC
36S52 Auckland
37S06 MA/MH * MO/DSC
37S50 Melbourne
38S42 MA/MH * UR/ASC

42S20 NE/MH * SA/ASC
43S17 SA/IC * SU/ASC
44S30 SA/IC * MO/DSC

SOUTH LATITUDES

45S16 SA/IC * UR/ASC
47S09 ME/MH * MA/DSC
49S41 ME/DSC * UR/DSC

51S03 MO/IC * MA/DSC
51S39 SU/MH * MA/DSC
52S24 PL/IC * ME/DSC
52S41 MO/DSC * UR/ASC
52S51 PL/IC * UR/DSC
54S38 UR/MH * MA/DSC
54S46 PL/IC * SU/DSC
56S37 PL/IC * MO/ASC
58S45 JU/MH * PL/DSC
58S55 PL/IC * NE/DSC
59S19 JU/IC * NE/ASC

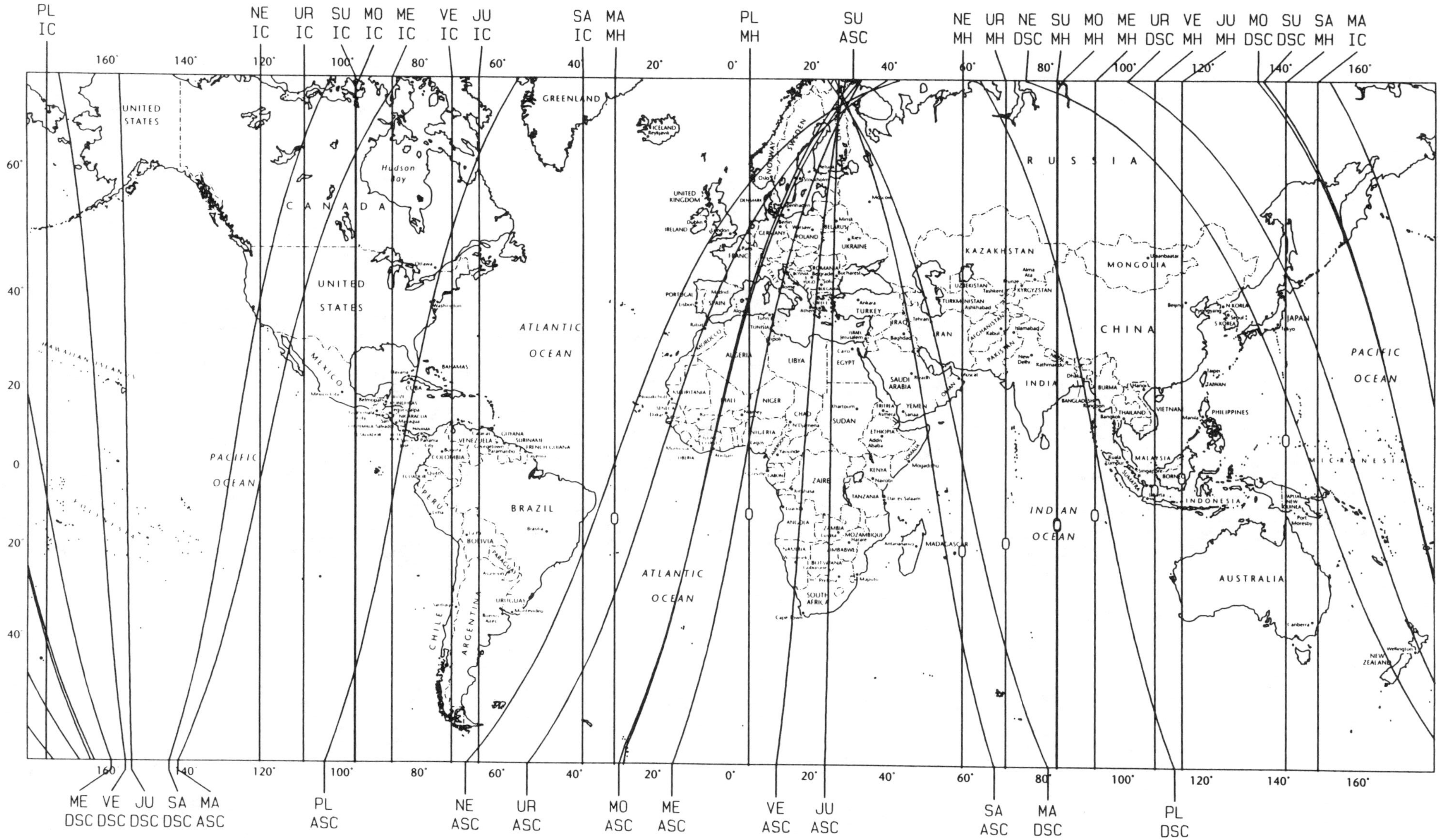

Solar Eclipse

February 16, 1999 6:39 AM GMT

05/21/97

05/21/97

Solar Eclipse
February 16, 1999 6:39 AM GMT

ASTRO*CARTO*GRAPHY LATITUDE CROSSINGS

NORTH LATITUDES

69N24	JU/ASC * SA/ASC
69N06	UR/ASC * NE/ASC
69N02	SU/MH * NE/DSC
69N00	MO/MH * NE/DSC
68N53	ME/IC * MA/ASC
68N51	SU/ASC * VE/ASC
68N45	SU/ASC * SA/ASC
68N39	MA/DSC * JU/ASC
68N39	VE/ASC * SA/ASC
68N27	SU/ASC * MA/DSC
68N24	VE/MH * UR/DSC
68N16	MO/ASC * SA/ASC
68N14	SU/ASC * JU/ASC
68N13	ME/ASC * NE/ASC
68N12	VE/ASC * MA/DSC
68N03	MO/ASC * MA/DSC
68N01	SA/ASC * NE/ASC
67N55	MA/DSC * NE/ASC
67N49	MO/ASC * VE/ASC
67N45	VE/ASC * NE/ASC
67N44	ME/ASC * SA/ASC
67N44	MO/ASC * NE/ASC
67N36	JU/ASC * NE/ASC
67N34	SA/ASC * UR/ASC
67N33	ME/ASC * MA/DSC
67N28	MA/DSC * UR/ASC
67N24	MO/ASC * JU/ASC
67N20	ME/ASC * UR/ASC
67N11	ME/MH * NE/DSC
67N10	SA/MH * SU/DSC
67N09	SU/ASC * NE/ASC
66N52	VE/ASC * UR/ASC
66N44	JU/ASC * UR/ASC
66N39	MA/DSC * SA/ASC
66N31	SA/MH * MO/DSC
66N13	JU/MH * UR/DSC
65N52	ME/ASC * VE/ASC
65N50	ME/ASC * JU/ASC
65N44	VE/ASC * JU/ASC
65N40	MO/ASC * UR/ASC
65N30	UR/MH * PL/DSC
64N16	SU/ASC * UR/ASC
63N52	UR/IC * SA/DSC
62N20	MO/IC * MA/ASC
62N12	SU/IC * MA/ASC
62N11	JU/IC * PL/ASC
61N54	VE/MH * NE/DSC
61N13	Anchorage
61N06	MA/IC * SU/DSC
60N23	MA/IC * MO/DSC
59N20	Stockholm
58N55	PL/MH * NE/ASC
57N57	JU/MH * NE/DSC
55N45	Moscow
55N41	Copenhagen
52N39	PL/MH * UR/ASC
52N30	Berlin
51N53	VE/IC * PL/ASC
51N30	London
49N14	Vancouver
48N50	Paris
47N36	Seattle
45N48	SU/MH * PL/DSC
45N34	SA/MH * UR/DSC
45N32	Montreal
45N24	MO/MH * PL/DSC
44N59	Minneapolis
43N39	Toronto
42N22	Boston
42N20	Detroit
41N54	Rome
41N53	Chicago

CENTRAL LATITUDES

40N56	PL/MH * SU/ASC
40N48	UR/IC * MA/ASC
40N43	New York
39N44	Denver
39N32	PL/MH * MO/ASC
39N16	NE/IC * SA/DSC
38N53	Washington
38N37	St. Louis
37N47	San Francisco
36N10	Las Vegas
35N40	Tokyo
34N04	Los Angeles
33N45	Atlanta
33N27	Phoenix
32N47	Dallas
30N16	MA/IC * UR/DSC
30N03	Cairo
29N58	New Orleans
29N46	Houston
28N39	Delhi
25N47	Miami
22N34	Calcutta
22N16	Hong Kong
21N19	Honolulu
20N05	SA/MH * NE/DSC
19N26	Mexico City
18N30	MO/DSC * SU/DSC
14N35	Manila
10N30	Caracas
8N13	PL/MH * ME/ASC
8N00	ME/MH * PL/DSC
6N27	Lagos
3N28	NE/MH * MA/DSC
2N06	MA/MH * NE/ASC
2S06	MA/IC * NE/DSC
3S28	NE/IC * MA/ASC
8S00	ME/IC * PL/ASC
8S13	PL/IC * ME/DSC
18S30	MO/ASC * SU/ASC
20S05	SA/IC * NE/ASC
22S54	Rio De Janeiro
26S12	Johannesburg
30S16	MA/MH * UR/ASC
33S55	Sydney
34S36	Buenos Aires
36S52	Auckland
37S50	Melbourne
39S16	NE/MH * SA/ASC
39S32	PL/IC * MO/DSC
40S48	UR/MH * MA/DSC
40S56	PL/IC * SU/DSC
45S24	MO/IC * PL/ASC
45S34	SA/IC * UR/ASC
45S48	SU/IC * PL/ASC
51S53	VE/MH * PL/DSC
52S39	PL/IC * UR/DSC
57S57	JU/IC * NE/ASC
58S55	PL/IC * NE/DSC

SOUTH LATITUDES

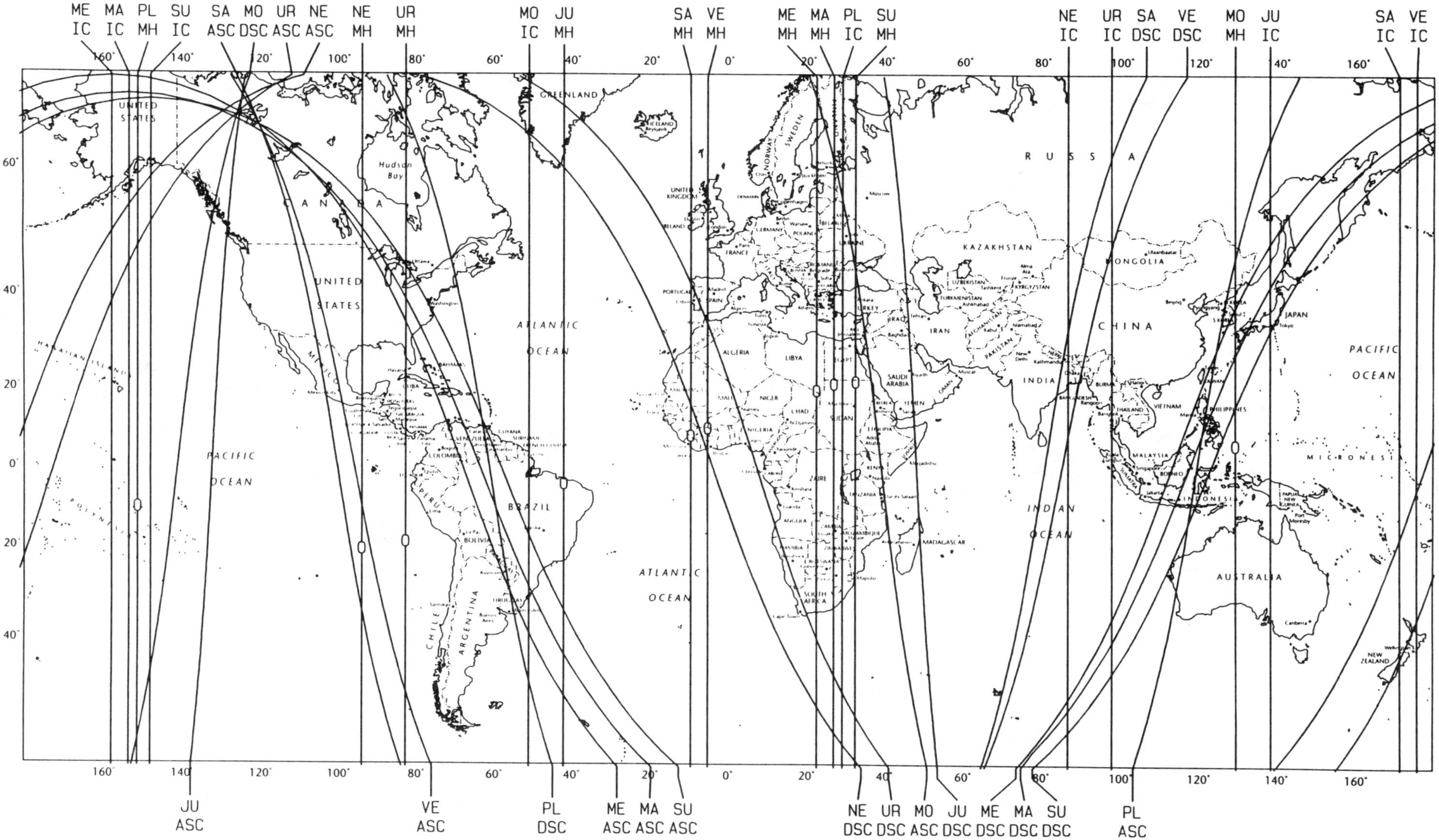

04/17/96

Mars Opp Pluto

June 2, 1998 9: 45 AM GMT

04/17/96

Mars Opp Pluto
June 2, 1998 9:45 AM GMT

ASTRO*CARTO*GRAPHY LATITUDE CROSSINGS

NORTH LATITUDES

69N51 MA/IC * ME/ASC
69N47 PL/MH * ME/ASC
69N38 SU/IC * ME/ASC
69N32 MO/IC * UR/DSC
69N17 UR/ASC * NE/ASC
68N50 JU/ASC * SA/ASC
68N38 MO/DSC * SA/ASC
68N20 PL/MH * MA/ASC
68N17 ME/IC * MA/DSC
68N16 VE/ASC * JU/ASC
68N15 SU/IC * MA/ASC
68N12 MO/DSC * VE/ASC
68N03 ME/MH * MO/ASC
68N01 SA/ASC * NE/ASC
67N53 VE/ASC * NE/ASC
67N53 MO/DSC * JU/ASC
67N47 PL/MH * SU/DSC
67N44 MA/IC * SU/DSC
67N43 JU/ASC * NE/ASC
67N42 MO/DSC * NE/ASC
67N33 VE/IC * ME/DSC
67N31 ME/IC * SU/DSC
67N29 SA/ASC * UR/ASC
67N20 VE/ASC * UR/ASC
67N19 SU/ASC * MA/ASC
67N19 JU/MH * UR/DSC
67N04 ME/ASC * NE/ASC
66N42 SU/ASC * NE/ASC
66N42 JU/ASC * UR/ASC
66N39 MO/DSC * ME/ASC
66N38 SA/IC * ME/DSC
66N36 ME/ASC * UR/ASC
66N34 MA/ASC * NE/ASC
66N32 ME/ASC * JU/ASC
66N28 MO/DSC * UR/ASC
66N19 SU/ASC * UR/ASC
66N17 UR/MH * PL/DSC
66N16 MO/DSC * SU/ASC
66N10 SU/ASC * JU/ASC
66N05 MA/ASC * UR/ASC
65N57 MO/DSC * MA/ASC
65N48 MA/ASC * JU/ASC
65N23 SU/ASC * ME/ASC
65N19 SU/ASC * VE/ASC
65N17 SU/ASC * SA/ASC
65N16 ME/ASC * VE/ASC
65N13 ME/ASC * SA/ASC
64N52 VE/IC * MA/DSC
64N49 VE/ASC * SA/ASC
64N25 MA/ASC * SA/ASC
64N25 MO/IC * NE/DSC
64N23 VE/ASC * MA/ASC
63N41 SA/IC * MA/DSC
63N32 JU/IC * PL/ASC
62N44 VE/IC * SU/DSC
61N41 UR/IC * SA/DSC
61N23 MA/MH * MO/ASC
61N13 SA/IC * SU/DSC
61N13 Anchorage
60N51 ME/ASC * MA/ASC
60N28 JU/MH * NE/DSC

59N20 Stockholm
59N07 SU/IC * NE/ASC
56N57 PL/MH * NE/ASC
56N28 PL/IC * MO/ASC
55N45 Moscow
55N41 Copenhagen
55N18 MA/IC * NE/ASC
55N16 NE/MH * SU/ASC
52N52 SU/IC * UR/ASC
52N30 Berlin
51N59 NE/MH * MA/ASC
51N30 London

CENTRAL LATITUDES

51N22 ME/IC * NE/ASC
50N16 NE/MH * ME/ASC

49N41 JU/IC * ME/DSC
49N14 Vancouver
49N06 PL/MH * UR/ASC
49N04 UR/IC * VE/DSC
48N50 Paris
47N36 Seattle
47N10 MO/MH * PL/ASC
46N10 MA/IC * UR/ASC
45N52 UR/MH * SU/ASC
45N32 Montreal
45N23 SU/MH * MO/ASC
44N59 Minneapolis
43N39 Toronto
42N58 SA/MH * UR/DSC
42N22 Boston
42N20 Detroit
42N12 JU/IC * MA/DSC
41N54 Rome
41N53 Chicago
40N43 New York

39N51 UR/MH * MA/ASC
39N44 Denver
39N00 ME/IC * UR/ASC
38N53 Washington
38N37 St. Louis
38N03 MO/MH * ME/DSC
37N47 San Francisco
36N10 Las Vegas
35N40 Tokyo
35N09 UR/MH * ME/ASC
34N58 VE/MH * UR/DSC
34N04 Los Angeles
33N45 Atlanta
33N27 Phoenix
33N26 NE/IC * SA/DSC
33N25 JU/IC * SU/DSC
32N47 Dallas
30N03 Cairo

29N58 New Orleans
29N46 Houston
29N11 ME/DSC * PL/ASC
28N39 Delhi
28N02 MO/MH * MA/DSC
25N47 Miami
22N34 Calcutta
22N16 Hong Kong
21N19 Honolulu

19N26 Mexico City
16N41 SA/MH * NE/DSC
16N00 MO/MH * SU/DSC
14N35 Manila
13N09 SU/ASC * PL/DSC
10N30 Caracas

9N11 MA/DSC * PL/ASC
8N36 NE/IC * VE/DSC
6N27 Lagos
4N44 VE/MH * NE/DSC

4S44 VE/IC * NE/ASC
8S36 NE/MH * VE/ASC
9S11 MA/ASC * PL/DSC

13S09 SU/DSC * PL/ASC
16S00 MO/IC * SU/ASC
16S41 SA/IC * NE/ASC

22S54 Rio De Janeiro
26S12 Johannesburg

SOUTH LATITUDES

28S02 MO/IC * MA/ASC
29S11 ME/ASC * PL/DSC

33S25 JU/MH * SU/ASC
33S26 NE/MH * SA/ASC
33S55 Sydney
34S36 Buenos Aires
34S58 VE/IC * UR/ASC
35S09 UR/IC * ME/DSC
36S52 Auckland
37S50 Melbourne
38S03 MO/IC * ME/ASC
39S00 ME/MH * UR/DSC
39S51 UR/IC * MA/DSC

42S12 JU/MH * MA/ASC
42S58 SA/IC * UR/ASC
45S23 SU/IC * MO/DSC
45S52 UR/IC * SU/DSC
46S10 MA/MH * UR/DSC
47S10 MO/IC * PL/DSC
49S04 UR/MH * VE/ASC
49S06 PL/IC * UR/DSC
49S41 JU/MH * ME/ASC

50S16 NE/IC * ME/DSC
51S22 ME/MH * NE/DSC
51S59 NE/IC * MA/DSC
52S52 SU/MH * UR/DSC
55S16 NE/IC * SU/DSC
55S18 MA/MH * NE/DSC
56S28 PL/MH * MO/DSC
56S57 PL/IC * NE/DSC
59S07 SU/MH * NE/DSC

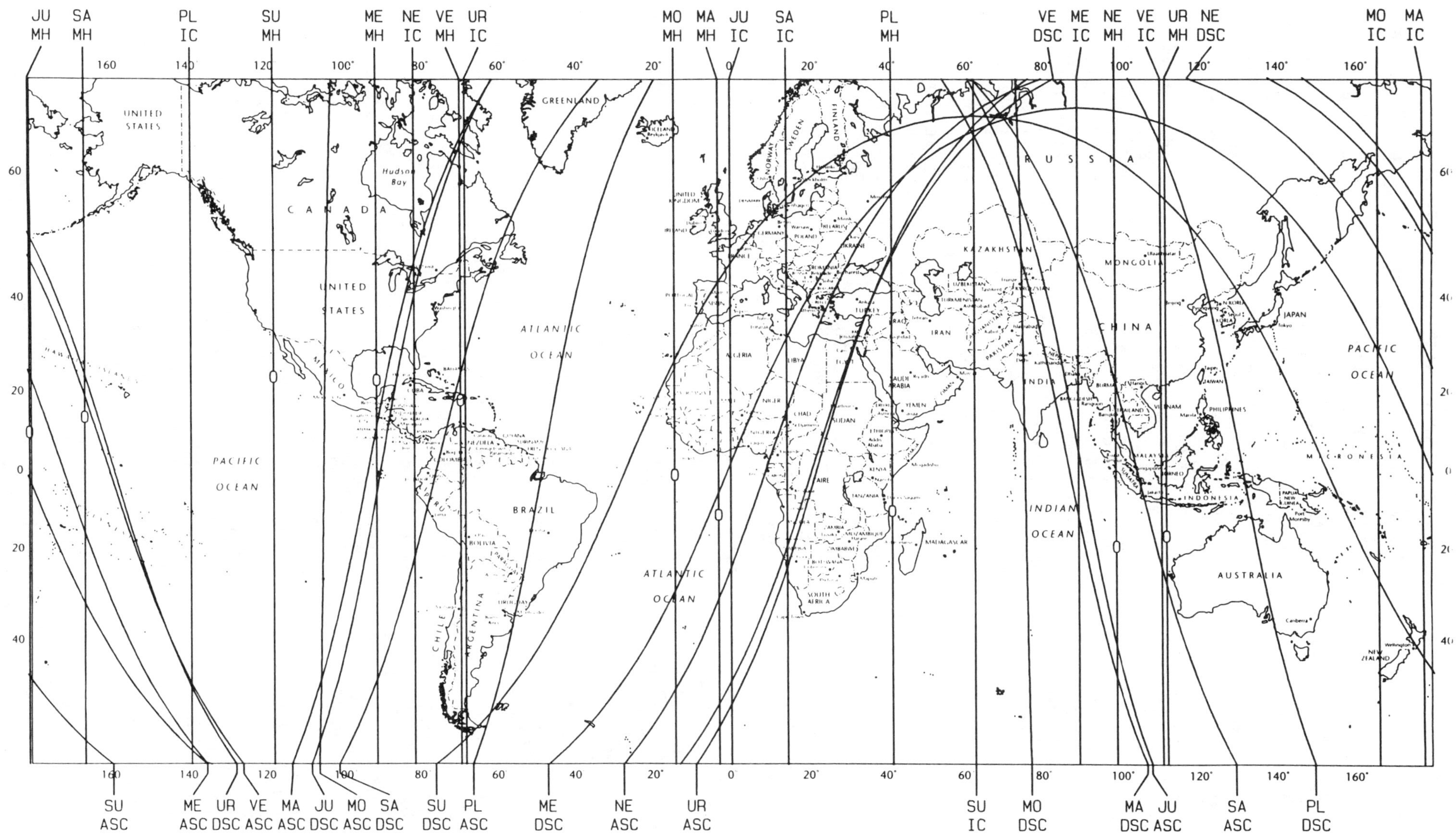

Summer Solstice

June 21, 1999 7:49 PM GMT

04/16/96

04/16/96

Summer Solstice
June 21, 1999 7:49 PM GMT

ASTRO*CARTO*GRAPHY LATITUDE CROSSINGS

NORTH LATITUDES

69N49 SU/IC * SA/ASC
69N16 MA/ASC * JU/DSC
69N15 MO/DSC * UR/ASC
69N09 MO/DSC * NE/ASC
69N02 SU/IC * JU/ASC
69N00 UR/ASC * NE/ASC
68N01 MO/DSC * VE/DSC
67N59 SA/ASC * NE/ASC
67N36 SA/ASC * UR/ASC
67N28 JU/ASC * NE/ASC
67N03 NE/MH * ME/ASC
66N54 SU/IC * NE/ASC
66N49 VE/DSC * SA/ASC
66N48 JU/ASC * UR/ASC
66N36 MO/DSC * ME/DSC
66N33 SU/DSC * NE/ASC
66N33 SU/ASC * UR/ASC
66N33 SU/DSC * MA/DSC
66N32 SU/ASC * JU/ASC
66N27 SU/ASC * VE/DSC
66N22 SU/ASC * SA/ASC
66N20 SU/ASC * ME/DSC
66N19 MA/DSC * NE/ASC
66N15 ME/DSC * SA/ASC
66N14 ME/ASC * PL/DSC
66N10 MO/DSC * SU/ASC
65N55 VE/DSC * JU/ASC
65N50 VE/IC * ME/ASC
65N45 ME/DSC * JU/ASC
65N39 UR/MH * ME/ASC
65N35 ME/DSC * VE/DSC
65N33 SU/IC * UR/ASC
65N18 UR/IC * MA/ASC
65N14 MA/DSC * UR/ASC
65N03 PL/MH * SU/DSC
65N01 SU/IC * ME/DSC
64N56 ME/DSC * MA/DSC
64N40 UR/IC * JU/DSC
64N36 SU/IC * MA/DSC
64N36 ME/DSC * UR/ASC
64N23 VE/DSC * MA/DSC
64N17 VE/MH * MA/ASC
64N17 SU/IC * VE/DSC
64N11 MO/DSC * SA/ASC
64N11 ME/IC * SU/ASC
63N29 VE/IC * PL/DSC
63N27 VE/MH * JU/DSC
63N16 MO/IC * UR/DSC
62N09 UR/MH * PL/DSC
61N47 NE/MH * SU/ASC
61N13 Anchorage
60N38 MO/IC * VE/ASC
60N03 ME/DSC * NE/ASC

59N20 Stockholm
58N12 PL/MH * ME/DSC
57N25 PL/MH * NE/ASC
57N07 VE/IC * SU/ASC
56N51 SA/IC * SU/DSC
56N49 MO/DSC * JU/ASC
56N30 UR/MH * SU/ASC
55N45 Moscow
55N41 Copenhagen
54N48 MA/IC * UR/DSC
52N46 SU/ASC * PL/DSC
52N30 Berlin
51N30 London
51N18 MA/IC * VE/ASC
51N16 JU/MH * UR/DSC

49N14 Vancouver
48N51 MO/IC * NE/DSC
48N50 Paris
48N26 PL/MH * UR/ASC

CENTRAL LATITUDES

47N44 PL/MH * VE/DSC
47N36 Seattle
47N27 JU/MH * VE/ASC
47N08 MO/DSC * MA/DSC
47N02 ME/IC * SA/ASC
46N54 NE/IC * MA/ASC
46N37 JU/IC * SU/DSC
45N32 Montreal
44N59 Minneapolis
43N39 Toronto
43N19 MA/MH * SU/DSC
42N22 Boston
42N20 Detroit
41N54 Rome
41N53 Chicago
41N01 NE/IC * JU/DSC
40N43 New York

39N44 Denver
38N53 Washington
38N46 VE/DSC * UR/ASC
38N37 St. Louis
37N47 San Francisco
36N10 Las Vegas
35N40 Tokyo
34N04 Los Angeles
33N45 Atlanta
33N27 Phoenix
32N47 Dallas
32N06 SA/IC * ME/DSC
31N02 MA/IC * NE/DSC
30N03 Cairo

29N58 New Orleans
29N46 Houston
29N15 MO/IC * ME/ASC
28N39 Delhi
28N13 UR/IC * SA/DSC
27N12 MO/MH * SU/DSC
25N47 Miami
24N18 VE/MH * SA/DSC
23N56 JU/MH * NE/DSC
23N44 SA/MH * UR/DSC
22N34 Calcutta
22N16 Hong Kong
21N23 NE/MH * SA/ASC
21N19 Honolulu

19N26 Mexico City
18N49 SA/MH * VE/ASC
15N29 SA/IC * NE/ASC
14N35 Manila
11N30 ME/MH * MA/ASC
10N30 Caracas

6N27 Lagos
5N24 MA/IC * ME/ASC
5N10 ME/IC * JU/ASC
2N11 JU/IC * ME/DSC

2S11 JU/MH * ME/ASC
5S10 ME/MH * JU/DSC
5S24 MA/MH * ME/DSC

11S30 ME/IC * MA/DSC
15S29 SA/MH * NE/DSC
18S49 SA/IC * VE/DSC

21S23 NE/IC * SA/DSC
22S54 Rio De Janeiro
23S44 SA/IC * UR/ASC
23S56 JU/IC * NE/ASC
24S18 VE/IC * SA/ASC
26S12 Johannesburg
27S12 MO/IC * SU/ASC

SOUTH LATITUDES

28S13 UR/MH * SA/ASC
29S15 MO/MH * ME/DSC

31S02 MA/MH * NE/ASC
32S06 SA/MH * ME/ASC
33S55 Sydney
34S36 Buenos Aires
36S52 Auckland
37S50 Melbourne
38S46 VE/ASC * UR/DSC

41S01 NE/MH * JU/ASC
43S19 MA/IC * SU/ASC
46S37 JU/MH * SU/ASC
46S54 NE/MH * MA/DSC
47S02 ME/MH * SA/DSC
47S08 MO/ASC * MA/ASC
47S27 JU/IC * VE/DSC
47S44 PL/IC * VE/ASC
48S26 PL/IC * UR/DSC
48S51 MO/MH * NE/ASC

51S16 JU/IC * UR/ASC
51S18 MA/MH * VE/DSC
52S46 SU/DSC * PL/ASC
54S48 MA/MH * UR/ASC
56S30 UR/IC * SU/DSC
56S49 MO/ASC * JU/DSC
56S51 SA/MH * SU/ASC
57S07 VE/MH * SU/DSC
57S25 PL/IC * NE/DSC
58S12 PL/IC * ME/ASC

Chapter 10/Astro*Carto*Graphy and Rectification

An Astro*Carto*Graphy map can be drawn only when an accurate birthtime is known. If the time of birth is correct within fifteen minutes, an Astro*Carto*Graphy map will still provide reasonably accurate information.

The Astro*Carto*Graphy lines move one degree every four minutes. If the time of birth is known within one hour, the lines move as follows:

4 minutes	1 degree	8 minutes	2 degrees
12 minutes	3 degrees	16 minutes	4 degrees
20 minutes	5 degrees	24 minutes	6 degrees
28 minutes	7 degrees	32 minutes	8 degrees
36 minutes	9 degrees	40 minutes	10 degrees
44 minutes	11 degrees	48 minutes	12 degrees
52 minutes	13 degrees	56 minutes	14 degrees
60 minutes	15 degrees		

This list shows that an Astro*Carto*Graphy map will not provide accurate information when the time of birth is more than thirty minutes in error.

There are many people who do not know their time of birth, so they are unable to benefit from Astro*Carto*Graphy. In Great Britain and other countries no time of birth is recorded on the birth certificate. The usual source is the mother or other relatives, but sometimes the mother can only remember that the birth was early morning or within a certain time range. Based on this information an astrologer can attempt to rectify the time of birth, a procedure done by examining important events that have taken place and finding the relevant astrological significators.

The Difficulty of Rectification

Rectification is difficult, especially when no time frame is known, and astrologers use various techniques to find one that gives accurate results. Only the **correct** time of birth will provide accurate information if the natal chart is rectified. If you ask five astrologers to rectify a natal chart, you may receive five different times of birth.

Astrology books say it is possible to tell a person's Ascendant from his or her appearance. This sounds easy, but in reality this approach is unsatisfactory. You can never tell if it is the Sun, Ascendant or Moon sign reflected in a person's appearance, and it is also difficult to look at a person and work out which sign is being expressed.

Considerations magazine reviewed the different techniques used to rectify natal charts. The magazine regularly features a rectification contest, giving astrologers the opportunity to work out the correct time of birth. It is research like this that will show which rectification techniques give the best results.

The object of this chapter is to examine the possibility of using Astro*Carto*Graphy to rectify a birth chart. Obviously, if a person has never travelled much in his or her life, there is no point using Astro*Carto*Graphy to rectify the time of birth.

If a person has experienced significant events connected to locations, it may be possible to use Astro*Carto*Graphy. When the exact birthtime is unknown the following information is needed to rectify the birth chart with Astro*Carto*Graphy.

a. Type of event—Based on the kind of event that occurred, decide which planetary influences would be the most appropriate.

b. Location of event—Search for times when the selected planetary influences were at the location.

c. Date of event—Having found a possible birthtime, examine the temporary influences to make sure the event is accurately described.

When more than one event is examined it is apparent when the rectified birthtime works for other locations. The more locations considered, the better to rectify the birthtime. If the approximate birthtime is known, it should be easier to use Astro*Carto*Graphy to rectify the natal chart, provided the person has experienced events at different locations.

First, draw an Astro*Carto*Graphy map for the approximate time given; then examine the planetary influences at the relevant locations to find out if the information is correct. It's possible that the rectified planetary influences are a few degrees away from the location. If a planetary line is eight degrees away from a location this can indicate that the birthtime should be rectified.

The planetary lines move one degree every four minutes. This means that moving a planetary line eight degrees will change the time of birth by thirty-two minutes. The thirty-two minutes needs to be added or subtracted from the approximate time that was given, and a new Astro*Carto*Graphy map drawn for the rectified time.

Is the difference in time added or subtracted?

Planetary line moves **westward** = plus

Approximate time	10:00 a.m.
lines are moved 8 degrees **westward**	+ 0: 32
Rectified time	10:32 a.m.

Planetary line moves **eastward** = minus

Approximate time	10:00 a.m.
lines are moved 8 degrees **eastward**	- 0:32
Rectified time	9:28 a.m.

The Midday Technique

When no time of birth is known a different approach is needed to try to rectify the natal chart with Astro*Carto*Graphy.

The natal chart is calculated for 12:00 noon, which is halfway through the day. Instead of drawing the usual Astro*Carto*Graphy map, the planetary lines are drawn on a transparent overlay. This overlay can then be placed on top of a world map to show the planetary influences for 12:00 noon. Obviously, the measurements for the world map and the Astro*Carto*Graphy overlay must be exactly the same.

With the Astro*Carto*Graphy midday overlay on top of the world map it is possible to try to rectify the natal chart. The overlay is moved slowly westward or eastward across the world map and, as this is done, the relevant locations are examined for the changing planetary influences. This is continued until an appropriate planetary influence appears at the location.

Once this is done other locations are examined while the overlay is in this position. If the planetary influences at other locations are not appropriate, then the rectification is incorrect. Only when the planetary influences at **all** the relevant locations agree with the known events and experiences is the rectification reliable.

According to Jim Lewis the rectification will suddenly fall into place when using this technique. It's a question of moving the overlay westward and eastward until you find the right combination.

Relocation Charts Technique

Without access to the marketed Astro*Carto*Graphy product it is impossible to use the midday technique to rectify the time of birth. There is another way.

Make a list of the locations where events or experiences have taken place. Then consider which planetary influences best describe what occurred. Then calculate relocation charts to put these planetary influences on the angles. This will give four different times for the natal chart as it is not yet known which angle is involved. These four times are then used for other locations to find out which one gives the best results.

Once the most appropriate birthtime is found, an Astro*Carto*Graphy map is drawn to make sure that the correct information is obtained. Finally, the rectified time is used to work out the temporary influences to check that there are no contradictions.

The Chartwheels computer program is a tremendous help when doing rectification. It is easy to rectify the times when any planet is moved to an angle, thus saving many calculations.

Example 1—Martin Luther King, Jr.

The time of birth for Martin Luther King, Jr. is quoted as "close to Noon" (statement from mother). The approximate time of birth makes the rectification easier.

Jim Lewis calculated an Astro*Carto*Graphy map for the given time. When the planetary lines were drawn on a map of the world he examined locations significant for Martin Luther King, Jr. After examining this Astro*Carto*Graphy map, it wasn't long before Jim Lewis realized he could rectify the time of birth.

When the planetary lines were moved approximately ten degrees **eastward** they accurately described what occurred at certain locations. As a result of this adjustment Jim Lewis arrived at a rectified time of 11:21 a.m. The new Astro*Carto*Graphy map shows these planetary influences:

PL IC at Memphis

This was where he was assassinated in 1968. (John F. Kennedy had PL MH in Dallas where he met the same fate).

At the given time of "close to noon," PL IC is the only planetary line that explains why Martin Luther King, Jr. was assassinated in Memphis.

JU ASC at Atlanta

At Atlanta, Georgia, he became a Baptist minister and soon was leader of the civil rights movement.

SU MH at Washington

In 1963, he made his famous "I have a dream" speech in Washington. More than 200,000 people took part in this peaceful civil rights demonstration.

A few years later Congress decided to make Martin Luther King's birthday a national holiday in America.

JU MH at Stockholm

At Stockholm he was awarded the Nobel Peace prize in recognition of his leadership in the peaceful fight for civil rights.

Example 2—Ronald Reagan

Ronald Reagan's birthtime is unknown, so for many years astrologers have attempted to rectify his natal chart. There are more than thirty rectified times of birth for him! This shows the serious problem faced by astrologers in trying to find the correct birthtime when it is unknown.

A few years ago Jim Lewis sent me a copy of the Astro*Carto*Graphy map that he had rectified for Ronald Reagan. He had tried maps for every possible time, but the one for 8:25 a.m. CST was startlingly accurate.

The SU ASC line is nearest Los Angeles, which indicates acting. He was the president of the Screen Actors Guild there. MA MH is at Washington where he survived a gunshot wound in an assassination attempt. At Washington he has also had two surgical operations.

During the time that Ronald Reagan was president he supported the Nicaraguan Contra Rebels who were trying to overthrow the Marxist government. This is shown by PL IC at Nicaragua. At Iran Ronald Reagan has JU IC, MO MH and SU DSC which suggests opportunity. Reagan's election was helped by the hostage crisis in Iran that Carter couldn't resolve.

In 1986 the "Irangate" scandal surfaced, linking Iran and Nicaragua. It started when an American hostage was released after eighteen months in captivity, and a magazine published the information that the United States

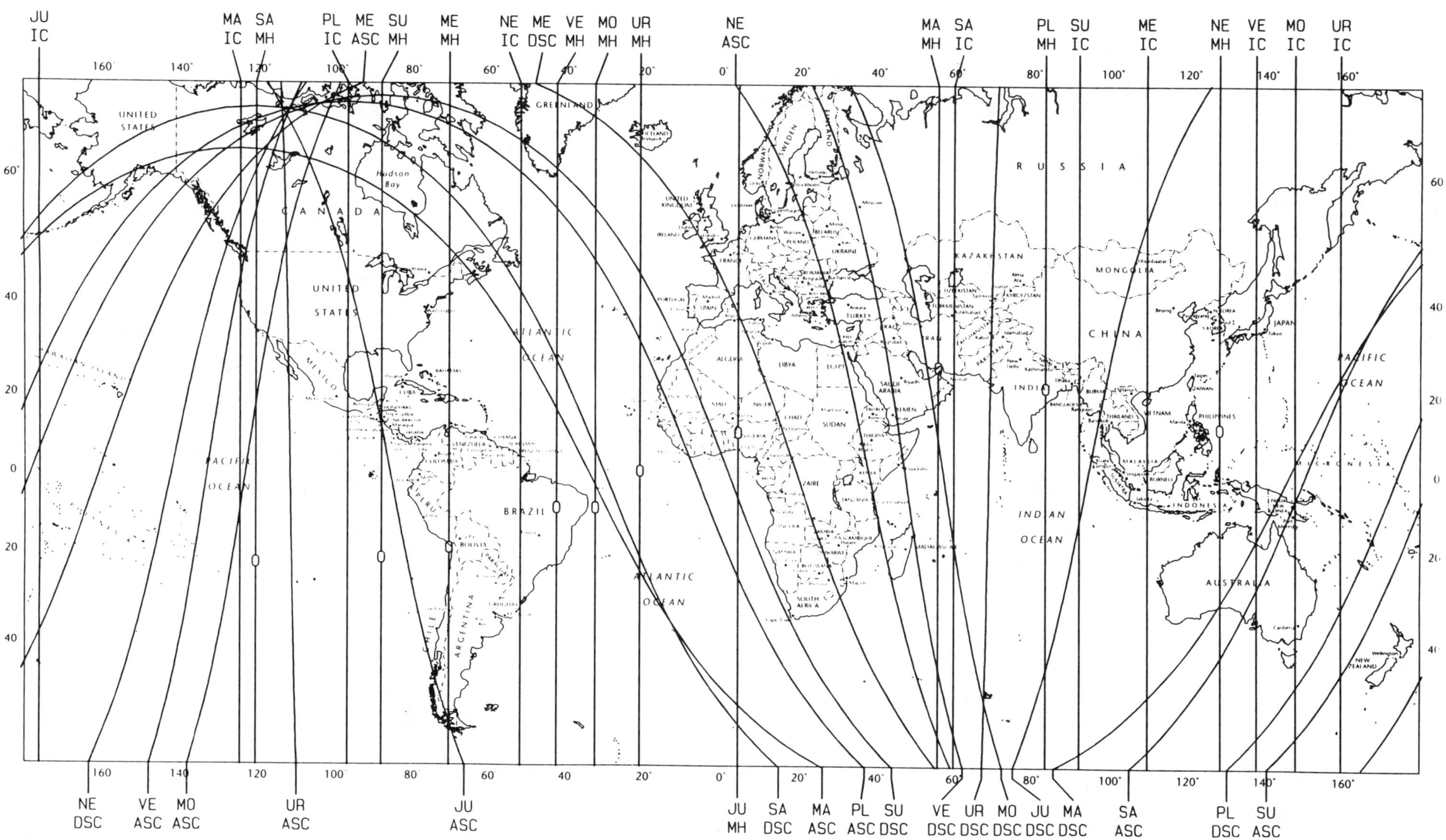

Martin Luther King Jr

January 15, 1929 12:00 PM CST

04/16/96

04/16/96

Martin Luther King Jr
January 15, 1929 11:21 AM CST

ASTRO*CARTO*GRAPHY LATITUDE CROSSINGS

NORTH LATITUDES

69N32	JU/MH * NE/ASC
69N26	PL/IC * ME/ASC
69N08	VE/MH * ME/DSC
68N50	MO/ASC * ME/ASC
68N40	PL/IC * SU/ASC
68N25	MO/ASC * SU/ASC
68N14	MO/ASC * PL/DSC
68N12	SU/ASC * ME/ASC
68N09	ME/ASC * PL/DSC
68N07	SU/ASC * PL/DSC
68N07	VE/ASC * NE/DSC
68N04	SU/MH * PL/ASC
67N57	ME/MH * SU/DSC
67N45	MA/IC * SA/ASC
67N44	JU/ASC * UR/ASC
67N40	SA/DSC * UR/ASC
67N39	JU/ASC * SA/DSC
67N36	VE/ASC * PL/DSC
67N35	SA/DSC * NE/DSC
67N35	SA/DSC * PL/DSC
67N34	NE/DSC * PL/DSC
67N34	VE/ASC * SA/DSC
67N27	SU/ASC * SA/DSC
67N27	JU/ASC * PL/DSC
67N24	UR/ASC * PL/DSC
67N16	ME/ASC * SA/DSC
67N06	JU/ASC * NE/DSC
67N05	SU/ASC * VE/ASC
67N01	SU/ASC * JU/ASC
67N00	MO/MH * ME/DSC
66N58	SU/ASC * NE/DSC
66N53	VE/ASC * JU/ASC
66N52	SU/ASC * UR/ASC
66N48	MO/ASC * SA/DSC
66N35	UR/ASC * NE/DSC
66N31	SA/MH * PL/DSC
66N05	ME/MH * PL/ASC
66N03	ME/ASC * JU/ASC
66N00	PL/IC * SA/DSC
65N50	VE/ASC * UR/ASC
65N48	MA/IC * PL/DSC
65N29	SA/MH * SU/ASC
65N21	ME/ASC * UR/ASC
65N06	ME/ASC * VE/ASC
64N27	NE/IC * SU/DSC
64N27	MA/IC * SU/ASC
64N17	SU/MH * SA/DSC
64N00	ME/ASC * NE/DSC
63N19	SU/ASC * MA/DSC
63N19	SA/MH * MA/ASC
63N15	ME/ASC * MA/ASC
63N13	MA/ASC * NE/DSC
63N06	VE/ASC * MA/ASC
62N56	MA/ASC * UR/ASC
62N55	MA/DSC * PL/DSC
62N39	UR/MH * ME/DSC
62N31	SA/MH * ME/ASC
62N18	MA/ASC * JU/ASC
62N13	MO/ASC * MA/ASC
61N52	MO/ASC * JU/ASC
61N13	Anchorage
61N13	VE/MH * SU/DSC
60N51	SA/MH * NE/DSC
60N45	NE/IC * PL/ASC
60N30	PL/IC * MA/ASC
60N23	MA/IC * ME/ASC
59N20	Stockholm
58N10	SU/MH * MA/ASC
57N48	ME/MH * SA/DSC
56N05	MO/MH * SU/DSC
55N57	VE/MH * PL/ASC
55N50	MA/IC * NE/DSC
55N45	Moscow

CENTRAL LATITUDES

55N41	Copenhagen
55N34	SA/MH * VE/ASC
54N12	JU/IC * SA/ASC
53N52	ME/IC * JU/DSC
52N30	Berlin
51N30	London
51N08	JU/IC * MA/DSC
50N51	MO/ASC * UR/ASC
49N46	ME/MH * MA/ASC
49N14	Vancouver
48N50	Paris
48N15	MO/MH * PL/ASC
47N36	Seattle
45N32	Montreal
45N30	UR/MH * SU/DSC
44N59	Minneapolis
44N54	MA/IC * VE/ASC
44N49	PL/IC * JU/ASC
43N39	Toronto
42N22	Boston
42N20	Detroit
42N19	NE/IC * SA/DSC
41N54	Rome
41N53	Chicago
40N43	New York
39N44	Denver
38N53	Washington
38N37	St. Louis
37N48	JU/MH * ME/DSC
37N47	San Francisco
36N10	Las Vegas
35N40	Tokyo
35N29	MA/DSC * SA/ASC
34N04	Los Angeles
33N45	Atlanta
33N27	Phoenix
32N47	Dallas
32N18	UR/MH * PL/ASC
31N21	NE/IC * MA/ASC
30N03	Cairo
29N58	New Orleans
29N46	Houston
28N39	Delhi
28N07	VE/MH * SA/DSC
25N47	Miami
25N28	JU/IC * PL/DSC
24N06	UR/IC * MA/DSC
22N34	Calcutta
22N16	Hong Kong
21N19	Honolulu
20N49	UR/IC * SA/ASC
19N26	Mexico City
18N53	SA/MH * MO/ASC
16N32	VE/MH * MA/ASC
14N35	Manila
12N15	SU/MH * JU/ASC
10N30	Caracas
7N16	MO/MH * SA/DSC
6N52	MA/MH * MO/DSC
6N27	Lagos
6N07	JU/IC * SU/ASC
2N06	MO/IC * MA/DSC
2S06	MO/MH * MA/ASC
6S07	JU/MH * SU/DSC
6S52	MA/IC * MO/ASC
7S16	MO/IC * SA/ASC
12S15	SU/IC * JU/DSC
16S32	VE/IC * MA/DSC

SOUTH LATITUDES

18S53	SA/IC * MO/DSC
20S49	UR/MH * SA/DSC
22S54	Rio De Janeiro
24S06	UR/MH * MA/ASC
25S28	JU/MH * PL/ASC
26S12	Johannesburg
28S07	VE/IC * SA/ASC
31S21	NE/MH * MA/DSC
32S18	UR/IC * PL/DSC
33S55	Sydney
34S36	Buenos Aires
35S29	MA/ASC * SA/DSC
36S52	Auckland
37S48	JU/IC * ME/ASC
37S50	Melbourne
42S19	NE/MH * SA/ASC
44S49	PL/MH * JU/DSC
44S54	MA/MH * VE/DSC
45S30	UR/IC * SU/ASC
48S15	MO/IC * PL/DSC
49S46	ME/IC * MA/DSC
50S51	MO/DSC * UR/DSC
51S08	JU/MH * MA/ASC
53S52	ME/MH * JU/ASC
54S12	JU/MH * SA/DSC
55S34	SA/IC * VE/DSC
55S50	MA/MH * NE/ASC
55S57	VE/IC * PL/DSC
56S05	MO/IC * SU/ASC
57S48	ME/IC * SA/ASC
58S10	SU/IC * MA/DSC

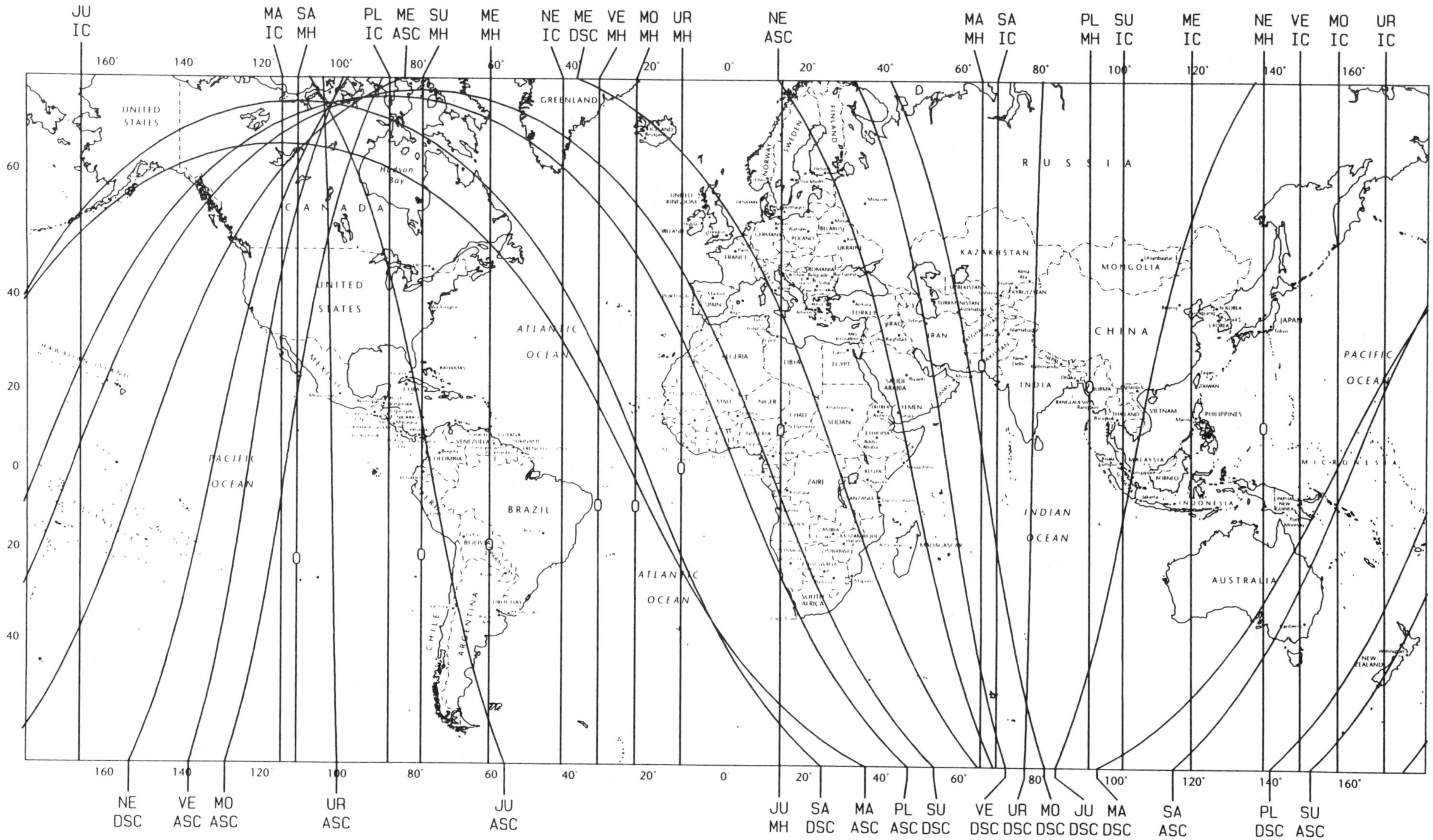

04/16/96

Martin Luther King Jr

January 15, 1929 11:21 AM CST

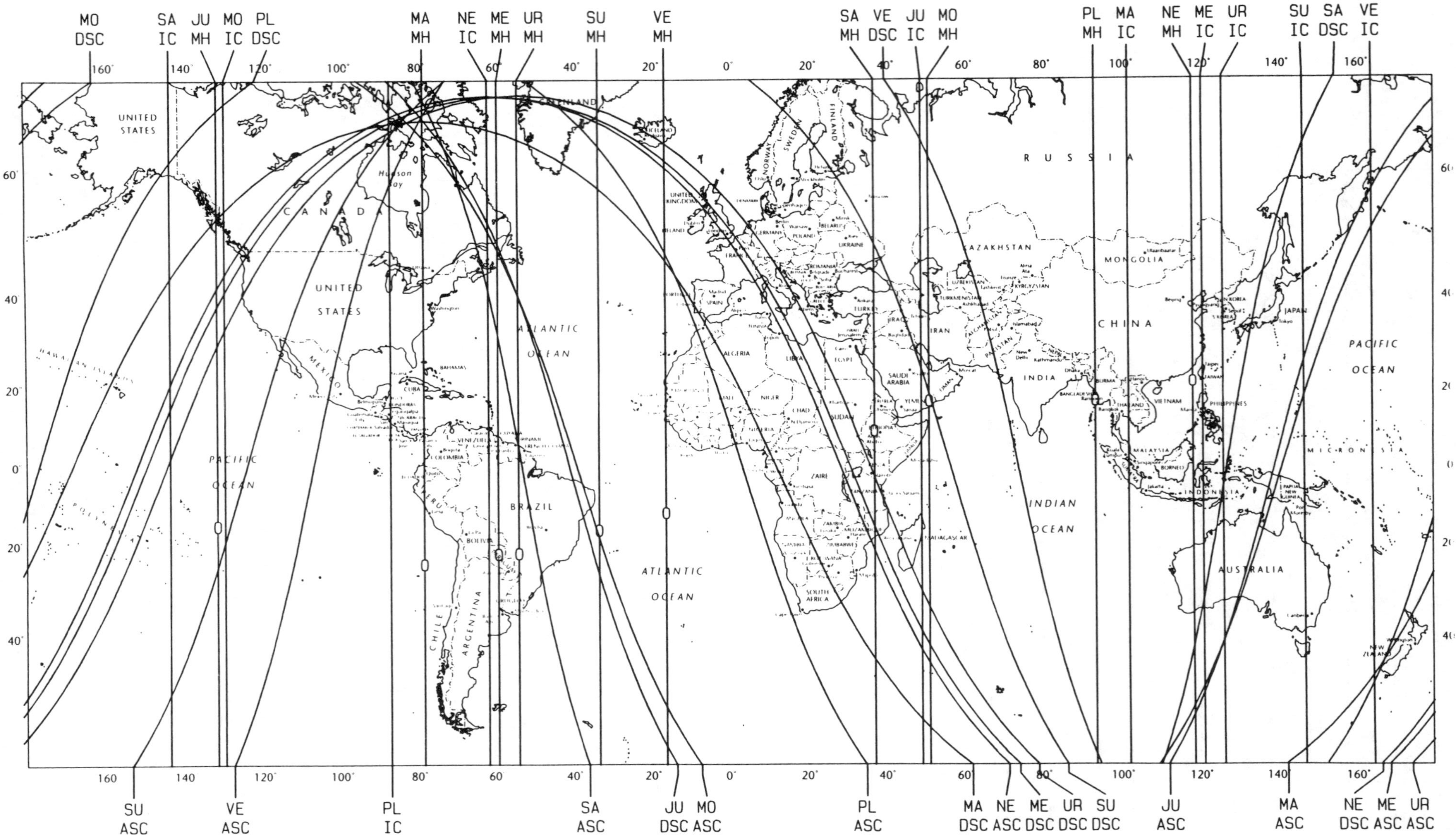

04/16/96

Ronald Reagan

February 6, 1911 8:25 AM CST

04/16/96

Ronald Reagan
February 6, 1911 8:25 AM CST

ASTRO*CARTO*GRAPHY LATITUDE CROSSINGS

NORTH LATITUDES

69N57	SU/ASC * VE/ASC
69N45	PL/IC * JU/DSC
68N35	ME/MH * NE/ASC
68N35	UR/ASC * NE/ASC
68N34	ME/MH * UR/ASC
68N29	NE/IC * UR/ASC
68N28	NE/IC * ME/ASC
68N28	UR/DSC * PL/ASC
68N28	ME/ASC * UR/ASC
68N25	UR/MH * NE/ASC
68N25	UR/MH * ME/DSC
68N23	ME/DSC * NE/ASC
68N19	MA/MH * SA/ASC
68N16	SU/ASC * SA/ASC
68N15	MA/MH * SU/ASC
67N57	VE/ASC * NE/DSC
67N55	MO/IC * PL/DSC
67N50	SA/ASC * NE/DSC
67N47	ME/DSC * PL/ASC
67N46	MA/MH * NE/DSC
67N45	PL/IC * MO/ASC
67N33	NE/ASC * PL/ASC
67N33	SU/ASC * NE/DSC
67N32	ME/ASC * VE/ASC
67N30	ME/ASC * SA/ASC
67N30	JU/DSC * NE/DSC
67N27	VE/ASC * SA/ASC
67N26	SU/ASC * JU/DSC
67N23	MA/MH * ME/ASC
67N21	SU/MH * UR/DSC
67N16	JU/MH * PL/DSC
67N10	ME/ASC * JU/DSC
67N04	SA/ASC * UR/ASC
66N58	MO/ASC * NE/DSC
66N52	SU/ASC * ME/ASC
66N51	VE/ASC * UR/ASC
66N49	MA/MH * UR/ASC
66N42	JU/DSC * UR/ASC
66N42	MA/MH * VE/ASC
66N37	MO/ASC * ME/ASC
66N36	PL/IC * NE/DSC
66N31	VE/ASC * JU/DSC
66N26	MO/ASC * SU/ASC
66N23	MA/MH * JU/DSC
66N23	SU/MH * ME/DSC
66N11	MA/DSC * JU/DSC
66N11	VE/ASC * MA/ASC
66N09	MA/DSC * SA/ASC
66N08	MA/ASC * UR/ASC
66N08	MO/ASC * MA/ASC
66N06	MO/ASC * UR/ASC
66N05	SU/ASC * MA/ASC
66N03	PL/IC * ME/ASC
66N02	SU/MH * NE/ASC
65N56	PL/IC * MA/ASC
65N54	ME/ASC * MA/ASC
65N31	MA/ASC * NE/DSC
65N29	SU/ASC * UR/ASC
65N17	MO/ASC * VE/ASC
65N17	NE/IC * MA/DSC
65N05	PL/IC * UR/ASC
64N58	ME/MH * MA/DSC
64N39	PL/IC * SU/ASC
64N23	MA/MH * MO/ASC
64N11	UR/MH * MA/DSC
64N01	JU/IC * VE/DSC
63N43	VE/MH * UR/DSC
63N24	SU/MH * PL/ASC
63N22	SU/IC * SA/DSC
63N17	JU/DSC * SA/ASC
62N14	MO/MH * VE/DSC
61N44	VE/MH * ME/DSC
61N32	SA/IC * PL/DSC
61N13	Anchorage

CENTRAL LATITUDES

60N51	VE/MH * NE/ASC
59N20	Stockholm
59N10	MO/ASC * SA/ASC
58N48	PL/IC * VE/ASC
58N10	SU/MH * MA/DSC
57N39	VE/IC * JU/ASC
55N45	Moscow
55N41	Copenhagen
55N01	MO/IC * MA/ASC
53N51	JU/MH * MA/ASC
53N13	NE/IC * JU/DSC
52N30	Berlin
51N55	VE/IC * MO/DSC
51N50	NE/IC * MO/ASC
51N30	London
49N48	SA/MH * SU/DSC
49N48	ME/MH * JU/DSC
49N14	Vancouver
48N50	Paris
48N44	ME/MH * MO/ASC
48N13	VE/MH * PL/ASC
47N36	Seattle
46N55	VE/MH * MA/DSC
45N32	Montreal
44N59	Minneapolis
44N29	MO/IC * NE/DSC
44N28	MA/DSC * PL/ASC
43N47	SA/IC * MA/ASC
43N39	Toronto
42N22	Boston
42N20	Detroit
42N02	JU/MH * NE/DSC
41N54	Rome
41N53	Chicago
41N29	NE/IC * SA/ASC
41N11	MO/IC * ME/ASC
40N43	New York
40N41	MO/ASC * JU/DSC
40N30	UR/MH * MO/ASC
40N29	UR/MH * JU/DSC
39N44	Denver
38N53	Washington
38N37	St. Louis
38N26	JU/MH * ME/ASC
37N47	San Francisco
36N10	Las Vegas
35N40	Tokyo
34N04	Los Angeles
33N45	Atlanta
33N29	MO/IC * UR/ASC
33N27	Phoenix
32N47	Dallas
32N12	ME/MH * SA/ASC
30N03	Cairo
29N59	JU/MH * UR/ASC
29N58	New Orleans
29N46	Houston
28N39	Delhi
26N14	SU/IC * JU/ASC
25N47	Miami
24N39	JU/IC * SU/DSC
22N34	Calcutta
22N16	Hong Kong
21N19	Honolulu
20N48	SA/IC * NE/DSC
19N26	Mexico City
18N47	MO/MH * SU/DSC
17N41	SU/IC * MO/DSC
15N03	SA/IC * ME/ASC
14N35	Manila

SOUTH LATITUDES

10N30	Caracas
6N27	Lagos
5N19	UR/MH * SA/ASC
2N18	SA/IC * UR/ASC
2S18	SA/MH * UR/DSC
5S19	UR/IC * SA/DSC
15S03	SA/MH * ME/DSC
17S41	SU/MH * MO/ASC
18S47	MO/IC * SU/ASC
20S48	SA/MH * NE/ASC
22S54	Rio De Janeiro
24S39	JU/MH * SU/ASC
26S12	Johannesburg
26S14	SU/MH * JU/DSC
29S59	JU/IC * UR/DSC
32S12	ME/IC * SA/DSC
33S29	MO/MH * UR/DSC
33S55	Sydney
34S36	Buenos Aires
36S52	Auckland
37S50	Melbourne
38S26	JU/IC * ME/DSC
40S29	UR/IC * JU/ASC
40S30	UR/IC * MO/DSC
40S41	MO/DSC * JU/ASC
41S11	MO/MH * ME/DSC
41S29	NE/MH * SA/DSC
42S02	JU/IC * NE/ASC
43S47	SA/MH * MA/DSC
44S28	MA/ASC * PL/DSC
44S29	MO/MH * NE/ASC
46S55	VE/IC * MA/ASC
48S13	VE/IC * PL/DSC
48S44	ME/IC * MO/DSC
49S48	ME/IC * JU/ASC
49S48	SA/IC * SU/ASC
51S50	NE/MH * MO/DSC
51S55	VE/MH * MO/ASC
53S13	NE/MH * JU/ASC
53S51	JU/IC * MA/DSC
55S01	MO/MH * MA/DSC
57S39	VE/MH * JU/DSC
58S10	SU/IC * MA/ASC
58S48	PL/MH * VE/DSC
59S10	MO/DSC * SA/DSC

had arranged an arms-for-hostage deal with Iran. More than $30 million was made from the sale of arms to Iran which was then transferred to the Contra Rebels in Nicaragua.

At the time that "Irangate" was made public, Reagan had C*C*G PL IC in Iran. The A*C*G PL IC line is at Nicaragua and had moved all the way to Iran when the "Irangate" scandal erupted. So A*C*G PL IC (Nicaragua) and C*C*G PL IC (Iran) shows how both countries are connected.

Ronald Reagan has SA MH at Moscow, showing his hard-line approach when negotiating with Gorbachev. Other locations for Ronald Reagan are examined in *The Astro*Carto*Graphy Book of Maps*. Jim Lewis' rectification is impressive, especially the link between Nicaragua and Iran.

Example 3—Salman Rushdie

In 1990, when Jim Lewis lectured in London he explained about his rectification of Salman Rushdie's chart. Salman Rushdie was born June 19, 1947 in Bombay, birthtime unknown. His book, *The Satanic Verses*, created a great deal of unrest amongst the Moslem population.

On February 14, 1989, the Iranian leader Ayatollah Khomeini told Moslems that Salman Rushdie had been sentenced to death. Ever since that announcement Salman Rushdie has been in hiding.

Jim Lewis arrived at a rectified time of 11:57 p.m. GMT, June **18th**, 1947. An Astro*Carto*Graphy map for this time and date shows the following planetary influences:

SU IC at London

It was at London that Salmon Rushdie established his literary reputation. His sudden rise to fame as a result of a death threat is shown by the other planetary influences within orb of London. Salman Rushdie also has MO IC, UR IC and PL DSC.

PL IC at Tehran

This planetary line is appropriate for an assassination announcement from the Iranian leader.

UR ASC at Bombay

With UR ASC at his birthplace, he was seen as different from other Muslims.

JU ASC at Los Angeles

Jim Lewis was of the opinion that Los Angeles would be a good place for him to seek refuge.

On the day Khomeini sentenced Rushdie to death there were two significant C*C*G aspects. C*C*G PL DSC was at Tehran, thus activating the permanent influence from A*C*G PL IC. C*C*G MA ASC was near Bombay, which would have a worldwide effect. This shows that it could be a dangerous time for him.

Two years after the death sentence was issued, C*C*G MA was conjunct A*C*G UR. This was potentially a very dangerous period for Rushdie, especially when he made any public appearances.

Example 4—John Major

British astrologer Russell Grant learned that John Major was born between 2:00 a.m. and 4:30 a.m. BST on March 29, 1943 in South London. John Major became British prime minister on November 27, 1990. Apart from the historical cease fire announcement by the IRA on August 31, 1994, there is little to work with to attempt a rectification. But, because the birthtime is known to be within two and a half hours it is worthwhile to attempt rectification.

First, horoscopes for 2:00 a.m. and 4:30 a.m., set for London, were calculated. From these it was apparent that the MH and ASC had to be between these positions.

	2:00 a.m.	**4:30 a.m.**
MH	22 LIB 040	00 SAG 11
ASC	19 SAG 06	27 CAP 40

Looking at the transits when John Major became prime minister, UR was at 7 CAP 43 and PL at 18 SCO 22. This suggests that his ASC is approximately 7-8 degrees CAP, and his MH at 18-19 degrees SCO.

John Major must have significant planetary influences in Northern Ireland to explain his efforts to negotiate a peace agreement. It's possible that he has VE IC passing through Northern Ireland. Using the coordinates for Belfast, 3:25 a.m. BST puts VE exactly on the IC. This also puts MO ASC near Northern Ireland, so these planetary lines seem appropriate for trying to the end the conflict there.

Using the 3:25 a.m. time, the AstroMaps Hi-Res program provided fine tuning. Events examined were the progressions and transits for the time he became prime minister and when the IRA announced its cease fire.

On November 27, 1990, John Major had C*C*G VE DSC and C*C*G UR ASC at London. He had experienced a quick rise to prime minister, having only joined the cabinet in 1987. This is shown by C*C*G UR ASC at London. Peace and unity would prove to be important for John Major as indicated by C*C*G VE DSC exactly at London.

On August 31, 1994, the IRA agreed to a cease fire with the British government—a big breakthrough for John Major. On this day he had C*C*G JU MH passing through Britain.

I rectified his birthtime to 3:28 a.m. BST to bring C*C*G JU MH within one degree orb of Belfast. This shows the huge success he achieved in bringing peace to Northern Ireland. The 3:28 a.m. time kept C*C*G VE DSC and C*C*G UR ASC within one degree orb of London on November 27, 1990.

Using the rectified time of 3:28 a.m. BST, other locations on his A*C*G map were checked. JU DSC passing through Europe didn't seem appropriate as Britain has experienced many difficulties in the European Union. UR IC is near Berlin, which indicates that Britain objected to the leading role Germany has in the European Union. Britain felt threatened about losing its independence due to the strong push by the German government towards monetary union.

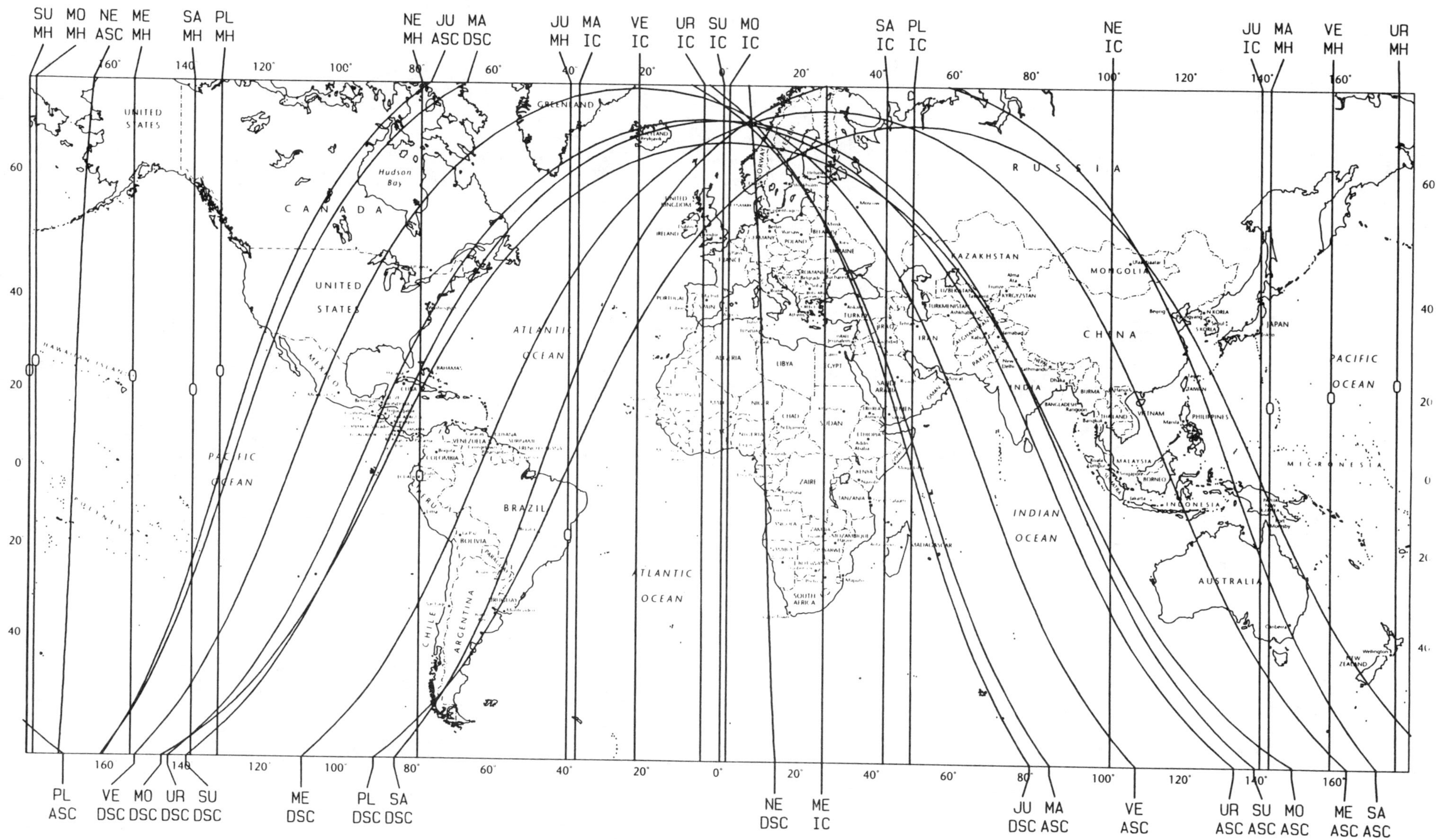

Salman Rushdie

June 18, 1947 11:57 PM GMT

04/16/96

04/16/96

Salman Rushdie
June 18, 1947 11:57 PM GMT

ASTRO*CARTO*GRAPHY LATITUDE CROSSINGS

NORTH LATITUDES

69N23 NE/MH * JU/ASC
69N20 UR/IC * MA/ASC
69N02 SU/IC * JU/DSC
68N56 MA/IC * VE/DSC
68N47 UR/IC * VE/ASC
68N42 JU/MH * VE/DSC
68N34 MO/IC * JU/DSC
68N12 VE/ASC * MA/ASC
68N10 SU/IC * VE/ASC
68N08 SU/IC * MA/ASC
67N56 MO/IC * VE/ASC
67N42 MO/IC * MA/ASC
67N16 VE/ASC * JU/DSC
67N10 NE/MH * MA/DSC
66N57 VE/ASC * NE/DSC
66N52 VE/ASC * SA/DSC
66N51 SA/IC * ME/ASC
66N46 SA/DSC * NE/DSC
66N42 JU/DSC * SA/DSC
66N37 ME/DSC * VE/ASC
66N37 JU/DSC * NE/DSC
66N36 SU/IC * UR/ASC
66N35 SU/DSC * UR/ASC
66N35 MO/IC * SU/ASC
66N32 MO/IC * UR/ASC
66N30 UR/IC * SU/DSC
66N30 SU/ASC * MA/ASC
66N28 SU/ASC * SA/DSC
66N28 ME/DSC * JU/DSC
66N26 ME/DSC * NE/DSC
66N25 SU/ASC * ME/DSC
66N25 SU/ASC * NE/DSC
66N24 SU/ASC * JU/DSC
66N23 MA/ASC * SA/DSC
66N19 SA/DSC * UR/ASC
66N17 MA/ASC * UR/ASC
66N16 ME/DSC * UR/ASC
66N16 ME/DSC * MA/ASC
66N15 SU/ASC * VE/ASC
66N11 ME/ASC * PL/DSC
66N11 UR/ASC * NE/DSC
66N05 JU/DSC * UR/ASC
66N03 ME/DSC * SA/DSC
66N02 SA/IC * PL/DSC
65N57 PL/IC * ME/ASC
65N47 MA/ASC * NE/DSC
65N44 VE/IC * UR/DSC
65N37 VE/ASC * UR/ASC
65N33 MO/IC * ME/DSC
65N16 SU/IC * ME/DSC
65N07 MO/IC * SA/DSC
64N57 VE/IC * SU/DSC
64N34 SU/IC * SA/DSC
64N25 ME/IC * PL/DSC
64N17 SU/ASC * PL/DSC
64N15 SU/IC * MO/DSC
64N14 MO/DSC * SA/DSC
64N10 UR/IC * ME/DSC
64N08 ME/IC * SU/ASC
64N08 MO/ASC * NE/DSC
64N06 UR/IC * MO/DSC
64N06 MO/DSC * ME/DSC
63N53 MO/ASC * MA/ASC
63N51 UR/ASC * PL/DSC
63N46 MO/ASC * JU/DSC
63N13 VE/ASC * PL/DSC
63N06 ME/IC * UR/ASC
63N05 MO/ASC * PL/DSC
62N55 MA/IC * UR/DSC
62N51 MO/ASC * VE/ASC
62N32 UR/IC * SA/DSC
62N16 JU/MH * UR/DSC
62N14 VE/IC * MO/DSC
62N10 JU/DSC * PL/DSC

CENTRAL LATITUDES

62N00 MA/ASC * PL/DSC
61N58 ME/IC * MO/ASC
61N20 MA/IC * SU/DSC
61N13 Anchorage
60N35 ME/IC * VE/ASC
60N32 JU/MH * SU/DSC

59N36 NE/DSC * PL/DSC
59N33 SA/IC * SU/ASC
59N20 Stockholm
58N06 MA/IC * MO/DSC
57N59 VE/IC * ME/DSC
57N37 MO/ASC * UR/ASC
57N26 SA/IC * MO/ASC
57N23 SA/IC * UR/ASC
57N12 JU/MH * MO/DSC
56N37 MO/IC * PL/DSC
56N35 MA/ASC * JU/DSC
56N17 PL/IC * SU/ASC
56N04 NE/MH * VE/DSC
55N57 NE/IC * SA/ASC
55N46 SU/IC * PL/DSC
55N45 Moscow
55N41 Copenhagen
54N29 NE/IC * PL/ASC
54N14 PL/IC * MO/ASC
53N43 ME/IC * MA/ASC
53N23 ME/IC * JU/DSC
53N19 PL/IC * UR/ASC
52N41 UR/IC * PL/DSC
52N30 Berlin
51N30 London
50N41 VE/IC * SA/DSC

49N44 SA/ASC * PL/ASC
49N14 Vancouver
49N02 SA/IC * VE/ASC
48N50 Paris
47N36 Seattle
46N27 MA/IC * ME/DSC
45N32 Montreal
44N59 Minneapolis
43N52 JU/MH * ME/DSC
43N39 Toronto
42N22 Boston
42N20 Detroit
41N54 Rome
41N53 Chicago
40N43 New York
40N16 PL/IC * VE/ASC

39N44 Denver
38N53 Washington
38N37 St. Louis
37N47 San Francisco
36N10 Las Vegas
35N40 Tokyo
35N29 VE/IC * PL/DSC
34N04 Los Angeles
33N45 Atlanta
33N27 Phoenix
33N17 NE/MH * UR/DSC
32N47 Dallas
32N46 NE/IC * ME/ASC
30N03 Cairo

29N58 New Orleans
29N46 Houston
28N39 Delhi
28N35 MO/ASC * SU/ASC
28N10 SA/IC * MA/ASC
25N55 MA/IC * SA/DSC
25N47 Miami
24N29 NE/MH * SU/DSC
23N47 SA/IC * JU/DSC

SOUTH LATITUDES

22N34 Calcutta
22N16 Hong Kong
21N19 Honolulu
20N18 JU/MH * SA/DSC

19N26 Mexico City
19N25 NE/MH * MO/DSC
14N35 Manila
10N30 Caracas

8N48 PL/IC * MA/ASC
6N27 Lagos
6N21 MA/IC * PL/DSC
1N34 PL/IC * JU/DSC
1N02 JU/MH * PL/DSC

1S02 JU/IC * PL/ASC
1S34 PL/MH * JU/ASC
6S21 MA/MH * PL/ASC
8S48 PL/MH * MA/DSC

19S25 NE/IC * MO/ASC

20S18 JU/IC * SA/ASC
22S54 Rio De Janeiro
23S47 SA/MH * JU/ASC
24S29 NE/IC * SU/ASC
25S55 MA/MH * SA/ASC
26S12 Johannesburg
28S10 SA/MH * MA/DSC
28S35 MO/DSC * SU/DSC

32S46 NE/MH * ME/DSC
33S17 NE/IC * UR/ASC
33S55 Sydney
34S36 Buenos Aires
35S29 VE/MH * PL/ASC
36S52 Auckland
37S50 Melbourne

40S16 PL/MH * VE/DSC
43S52 JU/IC * ME/ASC
46S27 MA/MH * ME/ASC
49S02 SA/MH * VE/DSC
49S44 SA/DSC * PL/DSC

50S41 VE/MH * SA/ASC
52S41 UR/MH * PL/ASC
53S19 PL/MH * UR/DSC
53S23 ME/MH * JU/ASC
53S43 ME/MH * MA/DSC
54S14 PL/MH * MO/DSC
54S29 NE/MH * PL/DSC
55S46 SU/MH * PL/ASC
55S57 NE/MH * SA/DSC
56S04 NE/IC * VE/ASC
56S17 PL/MH * SU/DSC
56S35 MA/DSC * JU/ASC
56S37 MO/MH * PL/ASC
57S12 JU/IC * MO/ASC
57S23 SA/MH * UR/DSC
57S26 SA/MH * MO/DSC
57S37 MO/DSC * UR/DSC
57S59 VE/MH * ME/ASC

Salman Rushdie
June 18, 1947 11:57 PM GMT

ASTRO*CARTO*GRAPHY LATITUDE CROSSINGS

NORTH LATITUDES	CENTRAL LATITUDES	SOUTH LATITUDES
		58S06 MA/MH * MO/ASC
		59S33 SA/MH * SU/DSC
		59S36 NE/ASC * PL/ASC

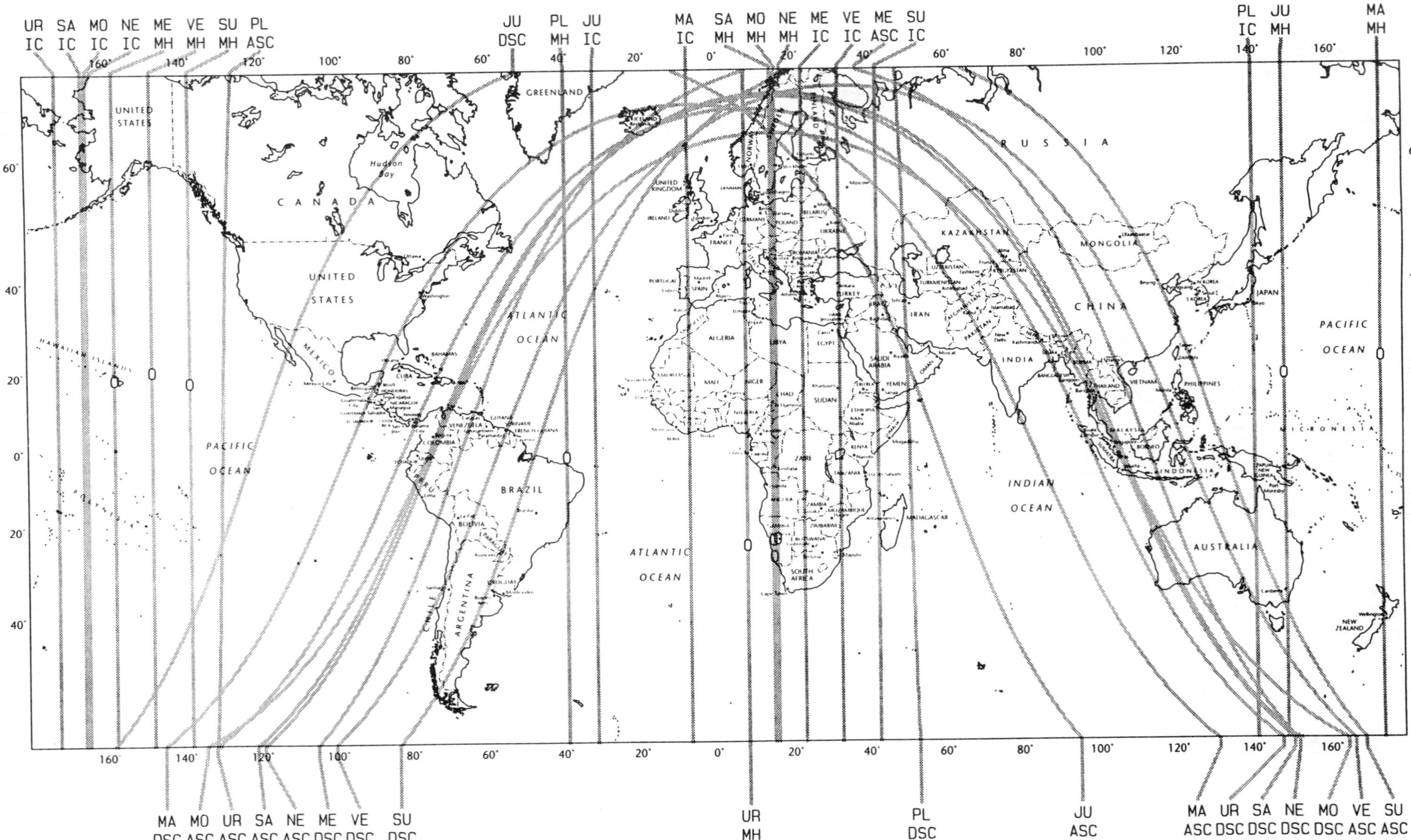

CYCLO*CARTO*GRAPHY

04/17/96

SU, MO, ME, VE AND MA ARE PROGRESSED
JU, SA, UR, NE AND PL ARE TRANSITING

Salman Rushdie

PROG	7	30	47	15:52 GMT
TRANSIT	2	14	89	0:00 GMT
NATAL	6	18	47	23:57 GMT

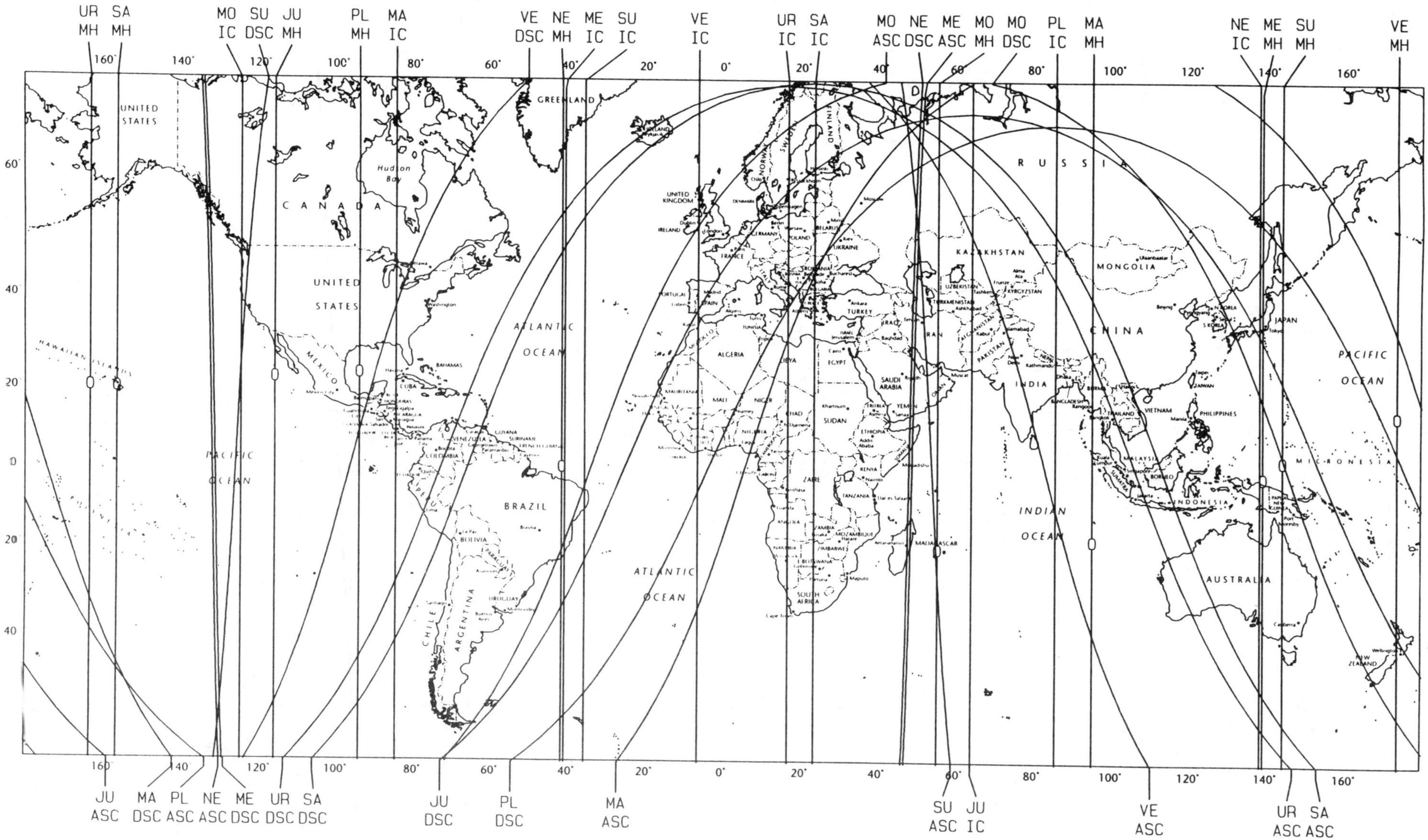

John Major

March 29, 1943 3:28 AM GDT

05/21/97

05/21/97

John Major
March 29, 1943 3:28 AM GDT

ASTRO*CARTO*GRAPHY LATITUDE CROSSINGS

NORTH LATITUDES

69N57	JU/IC * MA/ASC
69N44	UR/IC * SA/DSC
69N42	MO/ASC * VE/ASC
69N37	SA/DSC * UR/DSC
69N30	SA/IC * UR/ASC
69N26	MO/ASC * SA/ASC
69N03	VE/ASC * SA/ASC
68N56	MO/ASC * UR/ASC
68N17	SU/ASC * SA/ASC
68N00	VE/IC * UR/DSC
67N51	MO/MH * MA/ASC
67N42	PL/IC * MO/DSC
67N35	SA/ASC * NE/DSC
67N26	ME/ASC * SA/ASC
67N22	SA/IC * MO/ASC
67N08	MA/ASC * SA/ASC
67N07	VE/ASC * UR/ASC
67N02	VE/IC * SA/DSC
66N58	NE/IC * MA/DSC
66N51	MO/MH * JU/DSC
66N48	JU/DSC * SA/ASC
66N43	SU/ASC * UR/ASC
66N42	ME/MH * MA/DSC
66N38	MO/MH * SA/ASC
66N36	ME/ASC * MA/ASC
66N35	ME/ASC * JU/DSC
66N34	MA/ASC * JU/DSC
66N30	JU/DSC * NE/DSC
66N20	JU/DSC * UR/ASC
66N16	MA/ASC * NE/DSC
66N10	SU/ASC * JU/DSC
66N09	VE/ASC * JU/DSC
66N03	SU/ASC * VE/ASC
66N03	MA/ASC * UR/ASC
65N57	JU/ASC * PL/DSC
65N54	UR/ASC * NE/DSC
65N47	MO/DSC * PL/ASC
65N41	ME/ASC * UR/ASC
65N41	MA/MH * PL/ASC
65N35	UR/IC * MO/ASC
65N32	PL/IC * JU/ASC
65N27	VE/ASC * MA/ASC
65N15	MA/MH * MO/DSC
65N05	SU/ASC * MA/ASC
64N43	SU/MH * MA/DSC
64N32	MO/MH * UR/ASC
64N22	JU/IC * PL/DSC
64N20	VE/ASC * NE/DSC
64N19	SA/ASC * PL/DSC
64N13	JU/IC * SA/ASC
63N50	ME/ASC * VE/ASC
63N41	MA/MH * JU/ASC
63N25	UR/ASC * PL/DSC
62N40	MO/MH * PL/DSC
62N13	VE/ASC * PL/DSC
61N29	ME/ASC * PL/DSC
61N14	NE/DSC * PL/DSC
61N13	JU/IC * UR/ASC
61N13	Anchorage
60N52	SA/IC * JU/DSC
60N49	MO/MH * VE/ASC
60N47	SU/ASC * PL/DSC
59N20	Stockholm
59N03	MO/DSC * JU/ASC
58N35	SU/IC * UR/DSC
57N58	UR/IC * JU/DSC
55N45	Moscow
55N41	Copenhagen
55N21	SU/ASC * NE/DSC
55N13	ME/IC * UR/DSC
54N38	NE/MH * UR/DSC
54N12	SU/IC * SA/DSC
53N55	NE/IC * PL/ASC

CENTRAL LATITUDES

53N25	ME/MH * PL/ASC
52N56	VE/IC * MO/ASC
52N30	Berlin
51N41	PL/IC * SA/ASC
51N32	JU/IC * VE/ASC
51N30	London
51N12	SU/ASC * ME/ASC
49N47	SU/MH * PL/ASC
49N32	ME/IC * SA/DSC
49N14	Vancouver
48N50	Paris
48N42	NE/MH * SA/DSC
48N20	MA/ASC * PL/DSC
47N36	Seattle
46N09	SA/IC * PL/DSC
45N32	Montreal
44N59	Minneapolis
44N54	SA/IC * MA/ASC
43N43	PL/IC * UR/ASC
43N39	Toronto
42N22	Boston
42N20	Detroit
41N54	Rome
41N53	Chicago
40N43	New York
40N18	MA/MH * SA/ASC
39N44	Denver
38N53	Washington
38N37	St. Louis
38N31	UR/IC * PL/DSC
38N31	MA/IC * VE/DSC
37N54	VE/IC * JU/DSC
37N47	San Francisco
36N10	Las Vegas
35N40	Tokyo
34N04	Los Angeles
33N45	Atlanta
33N43	VE/MH * MA/DSC
33N27	Phoenix
33N06	NE/IC * JU/ASC
32N47	Dallas
32N02	UR/IC * MA/ASC
31N49	ME/MH * JU/ASC
30N03	Cairo
29N58	New Orleans
29N46	Houston
28N39	Delhi
27N45	MA/MH * UR/ASC
25N47	Miami
25N10	MO/IC * SU/DSC
22N34	Calcutta
22N32	SU/MH * JU/ASC
22N16	Hong Kong
21N19	Honolulu
19N46	NE/IC * MO/DSC
19N26	Mexico City
17N47	ME/MH * MO/DSC
14N35	Manila
10N30	Caracas
9N26	PL/MH * VE/DSC
6N27	Lagos
5N34	VE/MH * PL/ASC
3N58	SU/MH * MO/DSC
3S58	SU/IC * MO/ASC
5S34	VE/IC * PL/DSC
9S26	PL/IC * VE/ASC
17S47	ME/IC * MO/ASC
19S46	NE/MH * MO/ASC

SOUTH LATITUDES

22S32	SU/IC * JU/DSC
22S54	Rio De Janeiro
25S10	MO/MH * SU/ASC
26S12	Johannesburg
27S45	MA/IC * UR/DSC
31S49	ME/IC * JU/DSC
32S02	UR/MH * MA/DSC
33S06	NE/MH * JU/DSC
33S43	VE/IC * MA/ASC
33S55	Sydney
34S36	Buenos Aires
36S52	Auckland
37S50	Melbourne
37S54	VE/MH * JU/ASC
38S31	MA/MH * VE/ASC
38S31	UR/MH * PL/ASC
40S18	MA/IC * SA/DSC
43S43	PL/MH * UR/DSC
44S54	SA/MH * MA/DSC
46S09	SA/MH * PL/ASC
48S20	MA/DSC * PL/ASC
48S42	NE/IC * SA/ASC
49S32	ME/MH * SA/ASC
49S47	SU/IC * PL/DSC
51S12	SU/DSC * ME/DSC
51S32	JU/MH * VE/DSC
51S41	PL/MH * SA/DSC
52S56	VE/MH * MO/DSC
53S25	ME/IC * PL/DSC
53S55	NE/MH * PL/DSC
54S12	SU/MH * SA/ASC
54S38	NE/IC * UR/ASC
55S13	ME/MH * UR/ASC
55S21	SU/DSC * NE/ASC
57S58	UR/MH * JU/ASC
58S35	SU/MH * UR/ASC
59S03	MO/ASC * JU/DSC

When John Major became prime minister he had C*C*G PL MH in Bonn and C*C*G SA ASC near to Berlin. This shows that he experienced many difficulties as a result of Germany's dominant role in the European Union.

In 1991, he had C*C*G UR ASC and C*C*G NE ASC in Brussels—important influences for Britains role in Europe. At the time of the UR and NE conjunctions in 1993, he had C*C*G PL MH in Berlin and C*C*G UR ASC and C*C*G NE ASC passing through Germany. These temporary influences reiterate the difficulties that John Major experienced in what he saw as a German dominated European Union.

Although he has A*C*G JU DSC passing through Europe, the long-term C*C*G influences from PL MH, UR ASC and NE ASC tend to be more dominant.

Example 5—Mikhail Gorbachev

There are at least five times of birth for Mikhail Gorbachev and, to complicate matters, different locations have been quoted for his birthplace. Then there is a problem with the many time zones in the former Soviet Union, an area where there is some doubt about the accuracy of the information. Trying to rectify Gorbachev's natal chart under these circumstances is extremely difficult, so a different approach is needed. Because of these concerns, it is best not to consider place of birth, time zones and time changes, but to use GMT.

During the time that Gorbachev was the Soviet leader, significant changes took place in Europe. The opening of the Berlin Wall and the breakdown of communism in Eastern European countries brought an end to the Cold War.

First, different relocation charts calculated for Berlin were tried to determine the times when various planets were conjunct the angles. These times were noted and the AstroMaps Hi-Res program was used to set up Astro*Carto*Graphy maps. One time provided astonishingly accurate information at various locations. Cyclo*Carto*Graphy was then used to fine tune Gorbachev's birthtime.

My rectified time of birth for Gorbachev is 7:12 p.m. GMT, which results in an impressive Astro*Carto*Graphy map.

JU MH passes through Berlin and is near Prague. His greatest successes were the opening of the Berlin Wall and the reunification of Germany. In June 1989, Gorbachev visited West Germany where he was given an extraordinary welcome. During this visit there was "Gorbymania" everywhere he went. He announced that one day the Berlin Wall could disappear.

This JU MH line is near Malta where Gorbachev and Bush announced the start of a new era between the two superpowers. It was at Malta that the end of the Cold War was declared—another success for Gorbachev.

In 1990, Gorbachev was awarded the Nobel Peace Prize. The announcement was made in Oslo near Gorbachev's JU MH line. This is the same line that Martin Luther King, Jr. had when he won the Nobel Peace Prize.

JU MH and UR DSC have an influence in former Yugoslavia where Slovenia, Croatia and Bosnia-Herzegovina declared themselves independent countries. The declaration of independence met with Serbian disapproval, resulting in a civil war. This is shown by PL MH and SA IC near Belgrade.

PL MH, SA IC and UR DSC are powerful at the Baltics. Gorbachev tried to prevent the Baltic republics from becoming independent countries by sending in Soviet troops. Lithuania, where UR DSC is exact, was the first Baltic republic to declare independence.

MA MH is near Moscow where Gorbachev battled with great determination to put his plans into action. He met opposition from the old regime which was not used to a Soviet leader wanting to make sweeping changes. Soviet hardliners did not agree with Gorbachev's policies, so they decided to plan a coup to topple him.

Gorbachev has UR MH, VE DSC and MA ASC near Washington. America quickly realized that Gorbachev was different from other Soviet leaders. Gorbachev was seen as a leader determined to reform the Soviet Union, and one who was prepared to find peace with America after years of animosity.

At Beijing, China, where he made a historic visit in May 1989, Gorbachev has JU DSC. He was greeted by "Gorbymania" and inspired half a million people to occupy Tiananmen Square to call for democracy.

Looking at Cyclo*Carto*Graphy for these events, the temporary influences give supporting evidence for the rectified time of birth.

On the day that the Berlin Wall was opened Gorbachev had C*C*G JU MH, C*C*G SA IC and C*C*G NE IC at Berlin. This shows the dissolving (NE) of barriers (SA) in a spectacular (JU) way. These three C*C*G planets were significant on the Astro*Carto*Graphy map for the Berlin Wall.

On October 15, 1990, it was announced in Oslo that Gorbachev had won the Nobel Peace Prize. That day, Gorbachev had C*C*G VE DSC exact at Oslo.

When Gorbachev became the Soviet leader he had C*C*G PL DSC within two degrees orb of Moscow. This is a long-term temporary influence which shows he would try to transform the Soviet Union during the time he was leader.

The coup to remove Gorbachev as the Soviet leader took place when C*C*G SA IC was at Moscow. Although Gorbachev was quickly restored as Soviet president he had limited power and resigned four months later.

On December 3, 1989, Gorbachev and Bush met in Malta to discuss wide-ranging arms cuts. It was also on this day that the Cold War ended. Gorbachev had C*C*G NE IC exact at Malta during the meeting. This could indicate that Gorbachev had to make some sacrifices to sign the agreement.

Gorbachev's visit to Beijing in May 1989 was at a time when the Chinese people were demonstrating for greater democracy. After his visit his C*C*G UR DSC moved within orb of Beijing where the move for democracy gained greater momentum.

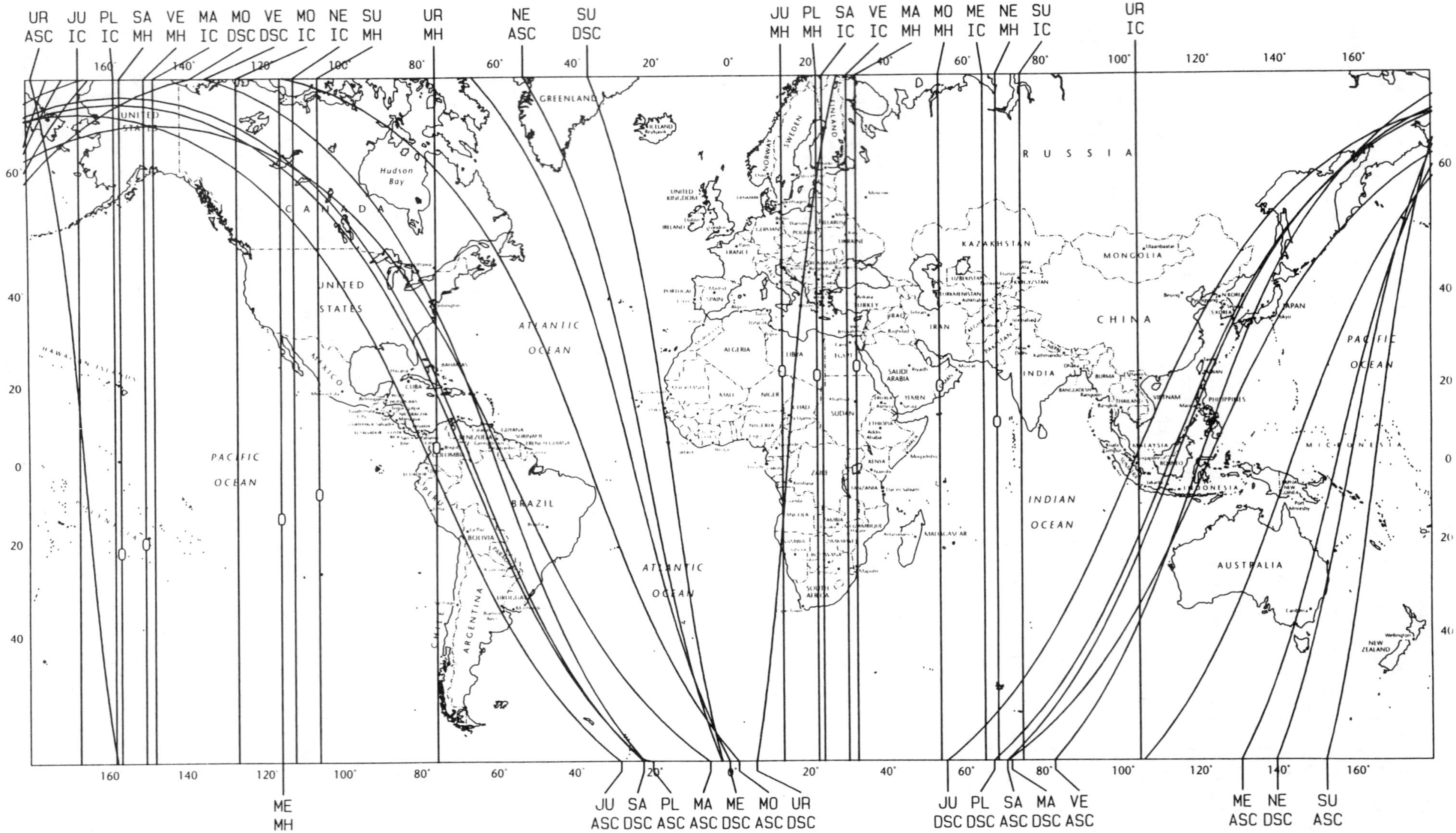

Mikhail Gorbachev

March 2, 1931 7:12 PM GMT

04/16/96

04/16/96

Mikhail Gorbachev
March 2, 1931 7:12 PM GMT

ASTRO*CARTO*GRAPHY LATITUDE CROSSINGS

NORTH LATITUDES

69N52 JU/IC * VE/ASC
69N47 NE/IC * MO/ASC
69N16 VE/ASC * NE/DSC
69N07 SU/MH * MO/ASC
69N05 MA/IC * MO/DSC
69N03 MO/IC * VE/DSC
68N57 JU/IC * ME/ASC
68N46 VE/MH * MO/DSC
68N38 SU/ASC * VE/ASC
68N24 VE/ASC * UR/ASC
68N22 PL/IC * SA/ASC
68N20 MO/DSC * SA/DSC
68N16 VE/MH * SA/DSC
68N09 MA/IC * SA/DSC
68N03 JU/IC * SA/ASC
67N52 ME/ASC * SA/ASC
67N41 SA/MH * PL/ASC
67N41 MO/DSC * PL/ASC
67N41 SA/MH * MO/DSC
67N40 SU/ASC * UR/ASC
67N31 VE/MH * PL/ASC
67N27 JU/IC * PL/DSC
67N23 MA/IC * PL/ASC
67N22 PL/IC * MO/DSC
67N16 SA/ASC * NE/DSC
67N11 ME/ASC * PL/DSC
67N03 SA/ASC * UR/ASC
66N54 SU/ASC * SA/ASC
66N52 ME/MH * VE/DSC
66N38 NE/DSC * PL/DSC
66N38 ME/ASC * JU/DSC
66N36 MO/DSC * JU/ASC
66N34 UR/ASC * PL/DSC
66N27 PL/IC * JU/ASC
66N25 JU/DSC * NE/DSC
66N24 JU/DSC * UR/ASC
66N21 SA/MH * JU/ASC
66N20 SU/ASC * PL/DSC
66N19 UR/ASC * NE/DSC
66N16 SU/ASC * JU/DSC
66N12 JU/DSC * PL/DSC
65N57 NE/IC * VE/DSC
65N47 VE/MH * JU/ASC
65N35 MA/DSC * JU/ASC
65N34 VE/MH * MA/DSC
65N32 JU/DSC * SA/ASC
65N30 MA/IC * JU/ASC
65N27 MO/IC * SA/DSC
65N20 SA/MH * MA/DSC
65N15 PL/IC * MA/DSC
64N57 JU/IC * MO/DSC
64N22 MO/IC * PL/ASC
64N21 VE/ASC * JU/DSC
64N20 JU/IC * MA/DSC
64N16 ME/ASC * UR/ASC
64N02 MO/IC * MA/ASC
63N53 MO/DSC * MA/DSC
63N48 SU/MH * VE/DSC
63N25 MA/ASC * PL/ASC
63N01 SA/IC * UR/DSC
62N57 MA/DSC * UR/ASC
62N17 ME/MH * SA/DSC
62N12 ME/ASC * MA/DSC
61N57 MO/DSC * UR/ASC
61N43 ME/MH * MA/ASC
61N15 MA/DSC * NE/DSC
61N13 SU/ASC * MA/DSC
61N13 Anchorage
60N59 NE/IC * SA/DSC
60N54 ME/MH * PL/ASC
60N52 SU/ASC * NE/DSC
60N45 NE/IC * MA/ASC
60N42 VE/ASC * SA/ASC
60N33 UR/MH * MO/ASC

CENTRAL LATITUDES

60N33 MO/IC * JU/ASC
60N07 MA/ASC * SA/DSC

59N46 VE/ASC * PL/DSC
59N29 NE/IC * PL/ASC
59N20 Stockholm
58N49 PL/MH * UR/DSC
58N31 SU/MH * MA/ASC
57N56 SU/MH * SA/DSC
56N15 SA/ASC * PL/DSC
56N15 MO/DSC * ME/ASC
56N11 SU/MH * PL/ASC
56N07 MO/DSC * SU/ASC
55N56 SU/ASC * ME/ASC
55N45 Moscow
55N41 Copenhagen
55N17 ME/MH * JU/ASC
54N25 MO/DSC * NE/DSC
53N08 NE/IC * JU/ASC
52N30 Berlin
51N30 London

49N14 Vancouver
48N50 Paris
48N15 ME/ASC * NE/DSC
48N08 SU/MH * JU/ASC
47N36 Seattle
45N32 Montreal
44N59 Minneapolis
43N39 Toronto
42N22 Boston
42N20 Detroit
41N54 Rome
41N53 Chicago
40N43 New York

39N44 Denver
38N53 Washington
38N37 St. Louis
37N47 San Francisco
37N25 UR/MH * VE/DSC
36N10 Las Vegas
35N40 Tokyo
34N08 UR/MH * MA/ASC
34N04 Los Angeles
33N45 Atlanta
33N27 Phoenix
32N47 Dallas
30N03 Cairo

29N58 New Orleans
29N46 Houston
28N39 Delhi
25N47 Miami
22N34 Calcutta
22N17 UR/MH * SA/DSC
22N16 Hong Kong
21N29 VE/DSC * MA/ASC
21N19 Honolulu

19N26 Mexico City
18N25 UR/MH * PL/ASC
14N35 Manila
14N09 JU/IC * UR/ASC
10N30 Caracas

6N27 Lagos
2N46 UR/IC * JU/DSC

2S46 UR/MH * JU/ASC

14S09 JU/MH * UR/DSC
18S25 UR/IC * PL/DSC

21S29 VE/ASC * MA/DSC

SOUTH LATITUDES

22S17 UR/IC * SA/ASC
22S54 Rio De Janeiro
26S12 Johannesburg

33S55 Sydney
34S08 UR/IC * MA/DSC
34S36 Buenos Aires
36S52 Auckland
37S25 UR/IC * VE/ASC
37S50 Melbourne

48S08 SU/IC * JU/DSC
48S15 ME/DSC * NE/ASC

53S08 NE/MH * JU/DSC
54S25 MO/ASC * NE/ASC
55S17 ME/IC * JU/DSC
55S56 SU/DSC * ME/DSC
56S07 MO/ASC * SU/DSC
56S11 SU/IC * PL/DSC
56S15 MO/ASC * ME/DSC
56S15 SA/DSC * PL/ASC
57S56 SU/IC * SA/ASC
58S31 SU/IC * MA/DSC
58S49 PL/IC * UR/ASC
59S29 NE/MH * PL/DSC
59S46 VE/DSC * PL/ASC

Example 6—United States Declaration of Independence

The time for this chart continually creates controversy among astrologers, the favorites being 4:47 p.m. LMT and 2:13 a.m. LMT. Jim Lewis used a different chart for the United States—11:00 a.m. LMT, July 6, 1775, when war was declared. This chart is featured in the *Astro*Carto*Graphy Book of Maps* showing U.S. foreign relations, especially those connected to military activities.

In *The Book of World Horoscopes* Nicholas Campion wrote many pages about the Declaration of Independence chart. After examining all the documentary evidence he suggested that 5:10 p.m. was the best of all the possibilities for the national horoscope of the United States.

With all this controversy, it is a good opportunity to use Astro*Carto*Graphy to try to rectify the Declaration of Independence chart. After calculating numerous relocation charts for various locations I arrived at a time that gives accurate information: 4:05 p.m. LMT, July 4, 1776. The Astro*Carto*Graphy map shows these planetary influences:

PL IC passes through California where many earthquakes have caused mass destruction. The area is prone to earthquakes as it lies on the San Andreas Fault.

In South America NE MH and MA DSC are near Colombia. In recent years the American government has been fighting a drugs (NE) war (MA) in this part of the world.

In the Middle East PL MH and ME IC pass through Iran. The combination of these planetary lines shows the terrorism and dictatorial regime that America has had to deal with in Iran.

SU IC and SA DSC at Iraq is where the U.S. shows its authority and control. It also indicates that the U.S. would not have easy relations with Iraq.

VE IC and JU IC are near Israel. These planetary influences are appropriate for the active role America has in finding peace in the Middle East.

SA IC, ME ASC and PL DSC are within orb of Hiroshima. A few days before an atomic bomb was dropped on Hiroshima the Japanese were told to surrender or face utter destruction. The same PL DSC line passes exactly through Pyongyang, the capital of North Korea. There was recently a crisis between the U.S. and North Korea about the diversion of plutonium into making bombs.

MA ASC and NE IC are near Viet Nam, site of an embarrassing war for the U.S. More than 45,000 Americans were killed in a war that cost $100 billion.

SA DSC, JU IC and VE IC are at Moscow. For many years SA DSC was the most dominant influence at Moscow, shown by America's fear of the Soviet Union. After the fall of Communism in Eastern Europe foreign relations between America and the Soviet Union improved. This was helped in a big way by the economic aid that America provides.

On the day of the first atomic bomb test in Mew Mexico C*C*G PL MH was **exact** at Alamogordo.

On January 17, 1991 the Gulf War, a big military success for the U.S., started. At the time Desert Storm was put into action C*C*G MA DSC was **exact** at Baghdad. Also, C*C*G NE MH was near Kuwait; hence a war (MA) with Iraq connected with oil (NE).

These temporary influences give convincing evidence for the rectified time of 4:05 p.m. LMT, but more research needs to be done to confirm it.

Looking at the six examples presented in this chapter, rectification with Astro*Carto*Graphy looks easy, but it is not as easy as it looks. It involves a great deal of testing to find a birthtime that fits all the facts. It's like doing detective work using clues and information to try to solve a case.

Once a suitable time of birth is found it is essential that Cyclo*Carto*Graphy is used to examine important events. This shows where small adjustments need to be made.

Usually, a rectification with Astro*Carto*Graphy gives the approximate time of birth, and Cyclo*Carto*Graphy is used for fine tuning. If Cyclo*Carto*Graphy doesn't give additional information for any of the events it is possible the angles are wrong on the Astro*Carto*Graphy map.

If you feel sure there is a VE line at a location, find the times when VE is on each angle. Draw Astro*Carto*Graphy maps for these times, and then check the planetary influences at other locations.

When it is impossible to find a suitable time of birth with Astro*Carto*Graphy, it's possible Cyclo*Carto*Graphy is strongly involved.

There is no sure way to rectify a natal chart with Astro*Carto*Graphy, but the technique, once mastered, does provide an interesting way of using events and experiences to rectify a birthtime.

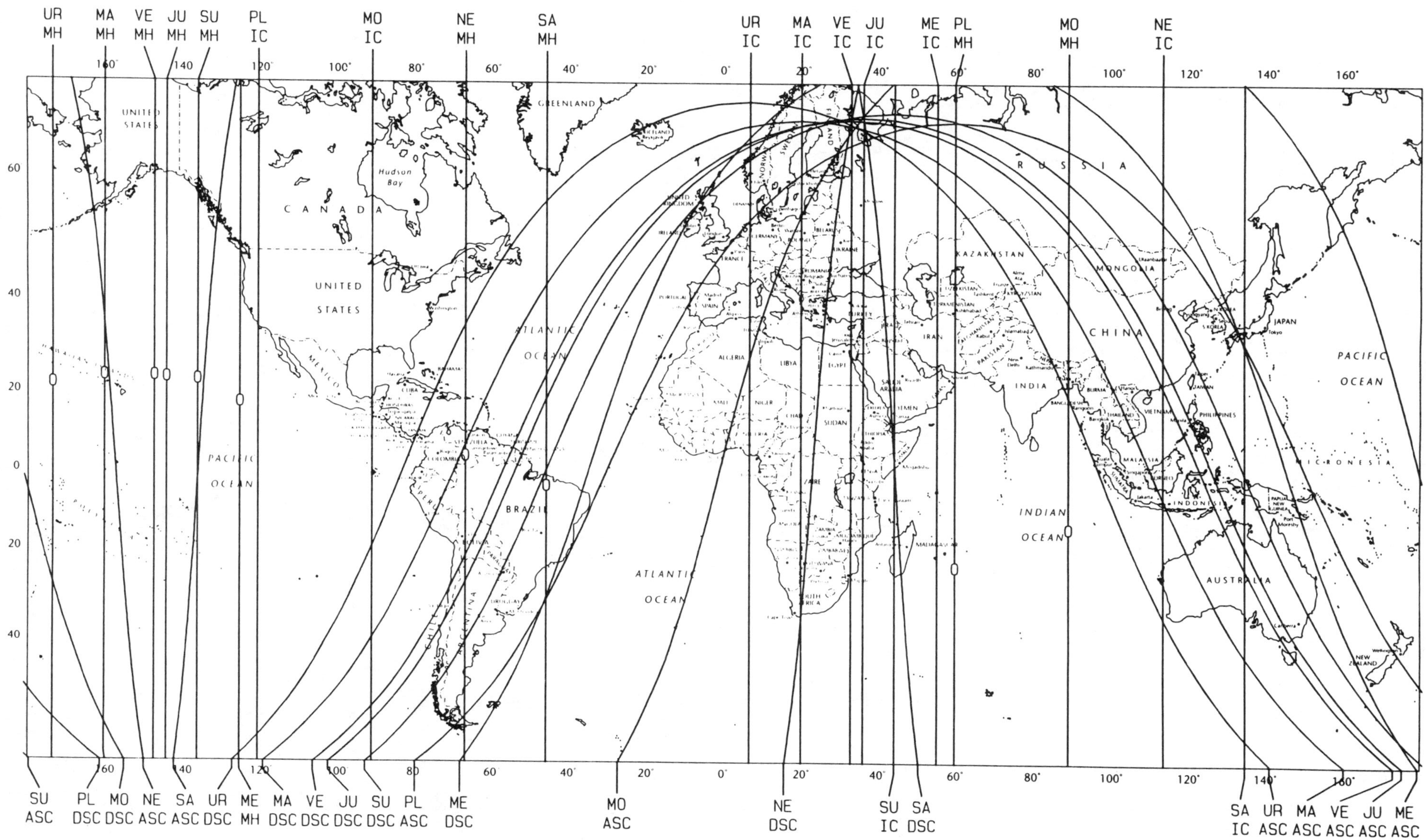

USA Dec. of Indep

July 4, 1776 4:05 PM LMT

04/16/96

04/16/96

USA Dec. of Indep
July 4, 1776 4:05 PM LMT

ASTRO*CARTO*GRAPHY LATITUDE CROSSINGS

NORTH LATITUDES

69N51 SA/DSC * NE/DSC
69N11 MO/MH * ME/ASC
68N59 MA/IC * ME/DSC
67N59 ME/DSC * UR/ASC
67N43 MA/IC * UR/ASC
67N00 MO/ASC * SU/DSC
67N00 JU/IC * SU/DSC
66N59 SU/DSC * SA/DSC
66N58 JU/IC * MO/ASC
66N54 MO/ASC * SA/DSC
66N51 VE/IC * NE/DSC
66N49 ME/IC * SU/ASC
66N48 VE/IC * SU/DSC
66N48 SU/DSC * NE/DSC
66N44 JU/DSC * SA/DSC
66N44 MO/ASC * JU/DSC
66N42 VE/IC * JU/DSC
66N42 JU/DSC * NE/DSC
66N39 SU/DSC * JU/DSC
66N36 JU/DSC * UR/ASC
66N34 JU/IC * SA/DSC
66N34 SU/DSC * UR/ASC
66N32 SU/IC * JU/ASC
66N28 VE/DSC * NE/DSC
66N27 MO/ASC * VE/ASC
66N27 VE/DSC * UR/ASC
66N26 JU/IC * VE/ASC
66N26 VE/ASC * SA/DSC
66N25 PL/MH * SU/ASC
66N24 SU/DSC * VE/DSC
66N21 MA/ASC * JU/DSC
66N19 VE/DSC * MA/ASC
66N17 SU/ASC * PL/DSC
66N17 SU/DSC * MA/ASC
66N14 ME/IC * PL/ASC
66N08 ME/DSC * MA/DSC
66N07 UR/ASC * NE/DSC
66N04 SU/IC * VE/ASC
66N03 VE/IC * UR/ASC
66N02 JU/ASC * PL/ASC
66N02 VE/DSC * JU/DSC
65N56 MA/IC * VE/DSC
65N56 MA/ASC * NE/DSC
65N55 MA/IC * JU/DSC
65N54 VE/IC * MA/ASC
65N53 MO/ASC * UR/ASC
65N51 UR/IC * MA/DSC
65N50 MO/ASC * MA/ASC
65N48 VE/ASC * PL/ASC
65N46 MA/ASC * UR/ASC
65N36 JU/IC * MA/ASC
65N34 MA/ASC * SA/DSC
65N33 ME/IC * JU/ASC
65N29 SU/IC * PL/ASC
65N27 JU/IC * UR/ASC
65N21 VE/IC * MO/ASC
65N21 SA/DSC * UR/ASC
65N18 MA/IC * SU/DSC
65N04 MA/ASC * PL/ASC
64N52 PL/MH * JU/ASC
64N51 MO/ASC * NE/DSC
64N50 ME/IC * VE/ASC
64N50 UR/ASC * PL/ASC
64N42 UR/IC * ME/DSC
64N31 SA/DSC * PL/ASC
64N30 SU/IC * MA/ASC
64N24 JU/IC * PL/ASC
64N08 UR/IC * VE/DSC
64N00 PL/MH * VE/ASC
63N49 VE/IC * PL/ASC
63N46 UR/IC * JU/DSC
63N41 ME/DSC * VE/DSC
63N32 NE/DSC * PL/ASC
63N24 SU/IC * UR/ASC

CENTRAL LATITUDES

63N17 MO/MH * PL/DSC
62N59 MO/ASC * PL/ASC
62N49 ME/DSC * JU/DSC
62N10 UR/IC * SU/DSC
62N00 ME/IC * MA/ASC
61N13 Anchorage
60N31 PL/MH * MA/ASC
60N21 MA/IC * PL/ASC

59N20 Stockholm
59N17 MO/MH * SU/ASC
59N10 NE/IC * ME/ASC
59N07 ME/IC * UR/ASC
56N42 SA/MH * UR/DSC
56N33 PL/MH * UR/ASC
56N28 SU/DSC * ME/DSC
55N45 Moscow
55N41 Copenhagen
54N39 MA/IC * MO/ASC
54N30 MO/MH * JU/ASC
53N58 UR/IC * PL/ASC
53N36 NE/IC * PL/DSC
52N30 Berlin
52N03 MO/MH * VE/ASC
51N30 London

49N14 Vancouver
48N50 Paris
47N36 Seattle
45N32 Montreal
44N59 Minneapolis
43N39 Toronto
43N04 SA/MH * MA/DSC
42N22 Boston
42N20 Detroit
41N54 Rome
41N53 Chicago
40N43 New York
40N19 NE/IC * SU/ASC

39N44 Denver
39N37 MO/MH * MA/ASC
38N53 Washington
38N37 St. Louis
37N47 San Francisco
36N10 Las Vegas
35N40 Tokyo
35N33 MA/MH * NE/ASC
35N10 NE/MH * UR/DSC
34N04 Los Angeles
33N45 Atlanta
33N27 Phoenix
32N47 Dallas
32N34 ME/ASC * PL/DSC
31N18 SA/IC * PL/DSC
30N45 SA/IC * ME/ASC
30N03 Cairo

29N58 New Orleans
29N46 Houston
28N39 Delhi
28N01 UR/IC * MO/ASC
27N29 NE/IC * JU/ASC
25N47 Miami
24N11 SA/MH * VE/DSC
22N34 Calcutta
22N16 Hong Kong
21N20 NE/IC * VE/ASC
21N19 Honolulu

19N26 Mexico City
18N52 MO/MH * UR/ASC
18N16 SA/MH * JU/DSC
14N35 Manila
10N30 Caracas

SOUTH LATITUDES

6N48 NE/MH * MA/DSC
6N27 Lagos
2N50 SU/IC * SA/DSC
0N25 SA/MH * SU/DSC

0S25 SA/IC * SU/ASC
2S50 SU/MH * SA/ASC
6S48 NE/IC * MA/ASC

18S16 SA/IC * JU/ASC
18S52 MO/IC * UR/DSC

21S20 NE/MH * VE/DSC
22S54 Rio De Janeiro
24S11 SA/IC * VE/ASC
26S12 Johannesburg
27S29 NE/MH * JU/DSC
28S01 UR/MH * MO/DSC

30S45 SA/MH * ME/DSC
31S18 SA/MH * PL/ASC
32S34 ME/DSC * PL/ASC
33S55 Sydney
34S36 Buenos Aires
35S10 NE/IC * UR/ASC
35S33 MA/IC * NE/DSC
36S52 Auckland
37S50 Melbourne
39S37 MO/IC * MA/DSC

40S19 NE/MH * SU/DSC
43S04 SA/IC * MA/ASC

52S03 MO/IC * VE/DSC
53S36 NE/MH * PL/ASC
53S58 UR/MH * PL/DSC
54S30 MO/IC * JU/DSC
54S39 MA/MH * MO/DSC
56S28 SU/ASC * ME/ASC
56S33 PL/IC * UR/DSC
56S42 SA/IC * UR/ASC
59S07 ME/MH * UR/DSC
59S10 NE/MH * ME/DSC
59S17 MO/IC * SU/DSC

Chapter 11/Personal Experiences

Whenever I read about a new astrological technique I try it on my own chart to see if it gives accurate information. My first reaction to Astro*Carto*Graphy was one of utter amazement as it accurately described what I had experienced at locations.

When I lecture on Astro*Carto*Graphy I explain about my experiences as they are impressive. Afterwards, astrologers tell me that those experiences were the most interesting part of the lecture. Because of this enthusiastic feedback, I thought it would be a good idea to include them in this book.

How I Started Studying Astrology

In 1982, my mother bought an astrology book which explained how to calculate and interpret a horoscope, but she was unable to understand the mathematics involved. As mathematics was my best subject at school I decided to read the book to see if I could understand it. It took me hours to calculate my own horoscope.

Then I interpreted my horoscope and was impressed with the results. I did the same for my mother and sister and they confirmed that the interpretation was accurate.

Looking at my own chart I saw that I would start a new job in January or February 1983. At the time I was unemployed, so I was curious to know if a job would materialize then. Much to my surprise I started a new job on February 4, 1983. It was then that I decided I wanted to learn more about astrology.

One month later I joined the Leicester Astrological Society. At the first few meetings I didn't understand much; it was like listening to a foreign language! A member of the society recommended that I study with the Faculty of Astrological Studies, so in April 1983 I enrolled in its correspondence course. The good thing about the Faculty of Astrological Studies is that it teaches many different ways astrology can be used.

How I Became Interested in Astro*Carto*Graphy

In 1982, I was in Phoenix, Arizona for two weeks holiday. During that time I felt like a completely different person, which I couldn't understand. After returning from Phoenix I felt depressed, and kept thinking about the fantastic time I had there and how I felt at home.

At the time I was unemployed, and decided that if I didn't get a job at the beginning of 1983, I would emigrate to Phoenix. Even when I started the new job I had plans to eventually relocate to Phoenix.

In September 1983, I purchased my own Astro*Carto*Graphy map after seeing an advertisement. I was shocked when it showed that Phoenix was one of the best places in the world for me! Two years later I purchased my own Cyclo*Carto*Graphy overlay, but since it was for 1986, it wasn't until later that it proved accurate.

In March 1986, I attended a weekend seminar on Astro*Carto*Graphy held by Jim Lewis. What I learned that weekend changed my life. After the seminar I took Jim Lewis' exam on Astro*Carto*Graphy, and a few weeks later I was the first astrologer in Britain to become a Certified Astro*Carto*Graphy Interpreter.

In 1987, Jim Lewis asked me to be his United Kingdom representative, and I was happy to accept, especially as I was using Astro*Carto*Graphy beneficially in my own life. Soon after, I started giving lectures on the subject, and was travelling all over Britain, visiting more than thirty astrology groups.

As a representative for Astro*Carto*Graphy I continually receive feedback from satisfied customers who write about their own experiences at locations. I have also done numerous Astro*Carto*Graphy interpretations for clients who required specialized advice about relocating. It was this kind of involvement with Astro*Carto*Graphy that helped me develop my interest.

Example 1—Holiday in Phoenix

During the two weeks I was in Phoenix I felt incredibly happy with life. I found it so easy to get along with people in Phoenix, something I had never experienced before. After returning to England I kept thinking about my holiday in Phoenix. I was convinced I would be successful and happy in life if I lived there, and had already arranged accommodations through contacts I made while on holiday.

I contacted the American Embassy in London for the necessary application forms and information. But, because my financial situation was in a bad state, I decided to wait a while before emigrating to America. Although I started a new job in February 1983, I assumed it would only be for a short while until I had saved enough money.

This was taking place at a time when I had no knowledge of Astro*Carto*Graphy. All I knew was that Phoenix was a place where I would be happy. A few months later I obtained my own Astro*Carto*Graphy map, and was amazed at what I discovered.

I have JU MH passing **exactly** through Phoenix! In the interpretation booklet provided with the map it said that a location on a JU MH is one of the best places to be. I also have UR MH near Phoenix where I would be able to express individuality in my career and social standing. The

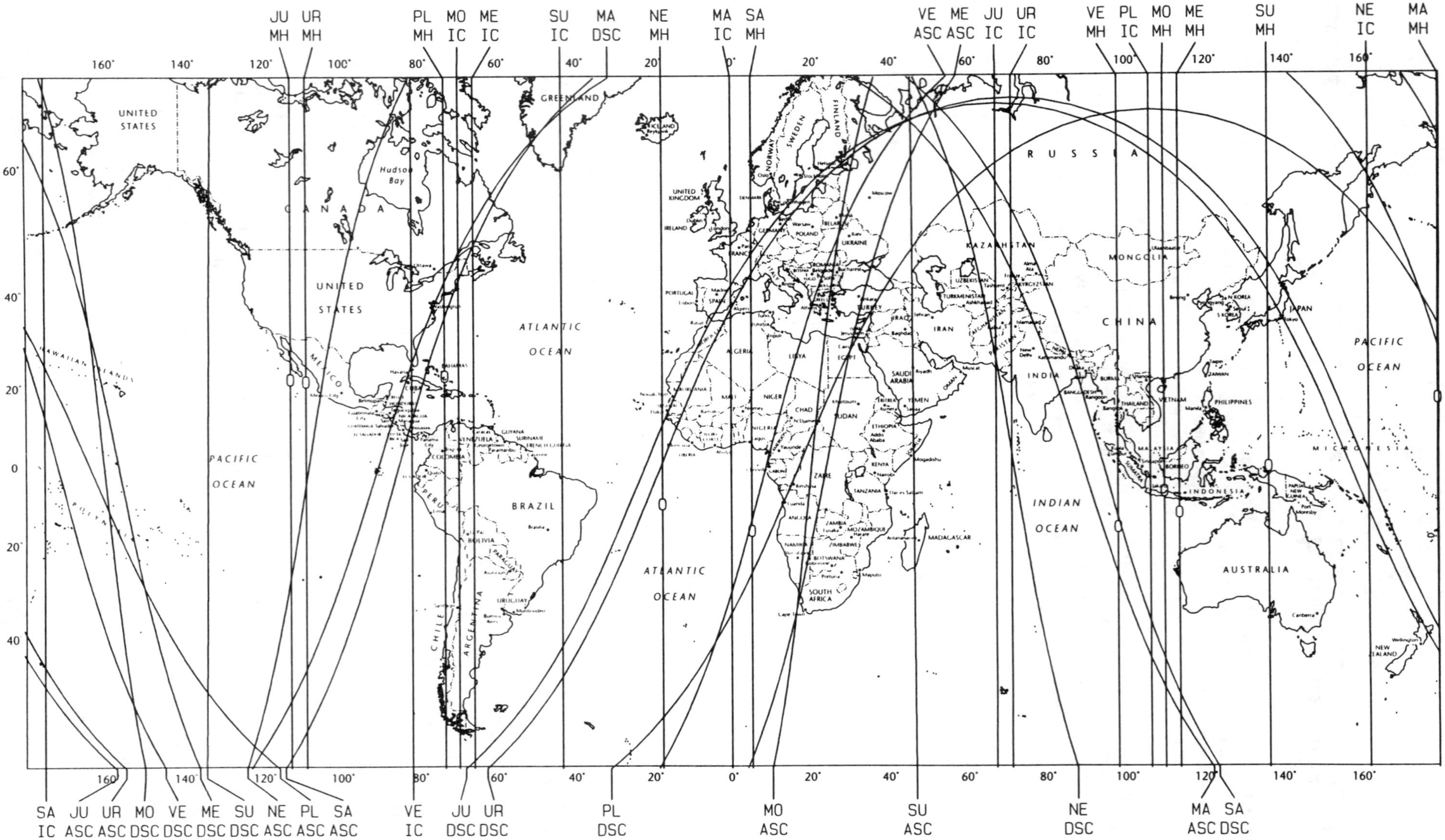

David Meadows

March 22, 1955 3:00 AM GMT

05/21/97

05/21/97

David Meadows
March 22, 1955 3:00 AM GMT

ASTRO*CARTO*GRAPHY LATITUDE CROSSINGS

NORTH LATITUDES

69N45 SU/ASC * SA/DSC
69N14 SA/DSC * NE/DSC
68N56 MO/ASC * MA/ASC
68N52 SA/IC * MO/DSC
68N22 VE/ASC * SA/DSC
68N03 JU/IC * UR/DSC
67N58 VE/ASC * NE/DSC
67N30 UR/IC * JU/ASC
67N26 ME/ASC * SA/DSC
67N12 SU/IC * SA/ASC
67N06 SA/DSC * UR/DSC
66N58 JU/DSC * SA/DSC
66N54 ME/ASC * UR/DSC
66N51 SU/IC * MA/DSC
66N48 MO/MH * PL/ASC
66N46 ME/ASC * JU/DSC
66N43 UR/DSC * NE/DSC
66N40 ME/MH * PL/ASC
66N39 JU/DSC * NE/DSC
66N37 VE/MH * PL/DSC
66N31 SU/ASC * VE/ASC
66N23 UR/ASC * PL/DSC
66N20 JU/DSC * UR/DSC
66N17 ME/ASC * NE/DSC
66N05 VE/ASC * MA/ASC
66N02 SU/ASC * JU/DSC
65N58 SU/ASC * UR/DSC
65N57 MA/ASC * JU/DSC
65N55 JU/ASC * PL/DSC
65N54 MA/ASC * UR/DSC
65N49 VE/ASC * JU/DSC
65N44 SU/ASC * MA/ASC
65N43 VE/MH * UR/ASC
65N39 VE/ASC * UR/DSC
65N03 MA/DSC * SA/ASC
64N31 ME/ASC * MA/ASC
64N17 VE/MH * JU/ASC
64N08 MA/MH * ME/DSC
63N47 PL/IC * UR/ASC
63N46 SU/MH * PL/ASC
63N03 MO/ASC * JU/DSC
62N43 MO/MH * UR/ASC
62N21 UR/IC * PL/DSC
62N18 MO/ASC * UR/DSC
61N56 PL/IC * JU/ASC
61N53 SU/ASC * ME/ASC
61N23 ME/MH * UR/ASC
61N13 Anchorage
61N06 JU/IC * PL/DSC
60N51 MA/ASC * NE/DSC
60N38 MO/MH * JU/ASC
60N23 SA/DSC * PL/DSC

59N20 Stockholm
59N12 NE/IC * VE/DSC
59N01 ME/MH * JU/ASC
58N47 JU/IC * SA/DSC
58N36 SA/IC * ME/DSC
57N57 MA/ASC * PL/DSC
57N03 NE/DSC * PL/DSC
55N45 Moscow
55N41 Copenhagen
54N57 UR/IC * SA/DSC
53N02 NE/IC * PL/ASC
52N53 ME/IC * MA/DSC
52N30 Berlin
51N30 London
51N04 MO/ASC * VE/ASC
50N26 JU/IC * MA/ASC

49N33 ME/IC * SA/ASC
49N14 Vancouver
49N00 SU/ASC * PL/DSC
48N50 Paris

CENTRAL LATITUDES

48N24 MO/IC * MA/DSC
48N06 SA/MH * JU/DSC
47N36 Seattle
46N22 SU/MH * UR/ASC
45N32 Montreal
45N03 UR/IC * MA/ASC
44N59 Minneapolis
44N42 SA/MH * UR/DSC
43N39 Toronto
43N27 MO/IC * SA/ASC
43N13 PL/MH * MA/DSC
42N25 MA/IC * JU/DSC
42N22 Boston
42N20 Detroit
41N54 Rome
41N53 Chicago
40N54 SU/MH * JU/ASC
40N43 New York
40N16 VE/DSC * PL/ASC

39N44 Denver
38N53 Washington
38N37 St. Louis
37N51 MA/IC * UR/DSC
37N47 San Francisco
36N15 PL/MH * SA/ASC
36N10 Las Vegas
35N40 Tokyo
34N26 MA/MH * PL/ASC
34N04 Los Angeles
33N53 MO/DSC * ME/DSC
33N45 Atlanta
33N27 Phoenix
32N47 Dallas
30N03 Cairo

29N58 New Orleans
29N54 MA/MH * VE/DSC
29N46 Houston
28N49 ME/ASC * PL/DSC
28N39 Delhi
26N58 JU/IC * NE/DSC
25N58 VE/IC * MA/DSC
25N47 Miami
25N34 SA/IC * PL/ASC
22N34 Calcutta
22N16 Hong Kong
21N19 Honolulu

19N26 Mexico City
14N35 Manila
13N01 SA/IC * VE/DSC
12N21 VE/IC * SA/ASC
10N59 NE/MH * JU/DSC
10N40 MO/ASC * PL/DSC
10N30 Caracas

6N27 Lagos
4N00 UR/IC * NE/DSC
1N34 NE/MH * UR/DSC

1S34 NE/IC * UR/ASC
4S00 UR/MH * NE/ASC

10S40 MO/DSC * PL/ASC
10S59 NE/IC * JU/ASC
12S21 VE/MH * SA/DSC
13S01 SA/MH * VE/ASC

22S54 Rio De Janeiro
25S34 SA/MH * PL/DSC
25S58 VE/MH * MA/ASC
26S12 Johannesburg
26S58 JU/MH * NE/ASC
28S49 ME/DSC * PL/ASC

SOUTH LATITUDES

29S54 MA/IC * VE/ASC

33S53 MO/ASC * ME/ASC
33S55 Sydney
34S26 MA/IC * PL/DSC
34S36 Buenos Aires
36S15 PL/IC * SA/DSC
36S52 Auckland
37S50 Melbourne
37S51 MA/MH * UR/ASC

40S16 VE/ASC * PL/DSC
40S54 SU/IC * JU/DSC
42S25 MA/MH * JU/ASC
43S13 PL/IC * MA/ASC
43S27 MO/MH * SA/DSC
44S42 SA/IC * UR/ASC
45S03 UR/MH * MA/DSC
46S22 SU/IC * UR/DSC
48S06 SA/IC * JU/ASC
48S24 MO/MH * MA/ASC
49S00 SU/DSC * PL/ASC
49S33 ME/MH * SA/DSC

50S26 JU/MH * MA/DSC
51S04 MO/DSC * VE/DSC
52S53 ME/MH * MA/ASC
53S02 NE/MH * PL/DSC
54S57 UR/MH * SA/ASC
57S03 NE/ASC * PL/ASC
57S57 MA/DSC * PL/ASC
58S36 SA/MH * ME/ASC
58S47 JU/MH * SA/ASC
59S01 ME/IC * JU/DSC
59S12 NE/MH * VE/ASC

combination of JU and UR showed that Phoenix would be an exciting place for me and one where I experience extraordinary good luck. After learning this, I made up my mind that I would definitely move to Phoenix. But I ended up moving to Munich!

Why did I not move to Phoenix after obtaining my Astro*Carto*Graphy map? Two years later an event took place that was to completely change my life.

While I was in Phoenix I had a terrifying experience from which it took me a couple of weeks to recover. I was riding a horse in the desert when it suddenly bolted. Being a beginner there was nothing I could do to bring the horse under control. I bounced up and down on the saddle and had difficulty holding on. Eventually, I was thrown off the horse, but as this happened my foot got stuck in the stirrup. I was dragged along the ground for a quite a while before I could free my foot. I couldn't walk properly afterwards from all the pain, so I returned home, still recovering from the experience. At Phoenix I have the following latitude crossings:

Crossings:	MO DSC/ME DSC
	MA MH/PL ASC
	PL MH/SA ASC

The MA/PL crossing seems appropriate for this accident and shows that I needed to be careful how I used my excess energy.

The temporary influences I experienced at the time I wanted to move to Phoenix show that I did not emigrate.

A*C*G	**Age 27**	**C*C*G**
SU 00 ARI 40		MO 02 PIS 11
PL 00 VIR 51		MA 02 GEM 54
MO 04 PIS 23		SA 15 LIB 16
ME 08 PIS 10		ME 20 ARI 59
MA 13 TAU 49		VE 23 PIS 39
SA 19 SCO 02		NE 25 SAG 20
JU 21 CAN 37		SU 25 ARI 34
VE 22 AQU 33		PL 28 LIB 45
UR 25 CAN 36		JU 28 LIB 58
NE 26 LIB 14		UR 29 SCO 35

1. C*C*G ME EQU A*C*G JU
2. C*C*G SU EQU A*C*G UR (orb 0 02)
3. C*C*G SU = A*C*G NE (app)

A*C*G	**(24 54)**	**Displaced C*C*G**
SU 00 ARI 40		NE 00 SAG 26
PL 00 VIR 51		PL 03 LIB 51
MO 04 PIS 23		JU 04 LIB 04
ME 08 PIS 10		UR 04 SCO 41
MA 13 TAU 49		MO 07 AQU 17
SA 19 SCO 02		MA 08 TAU 00
JU 21 CAN 37		SA 20 VIR 22
VE 22 AQU 33		ME 26 PIS 05
UR 25 CAN 36		VE 28 AQU 45
NE 26 LIB 14		

A*C*G	**(24 54)**	**Solar arc A*C*G**
SU 00 ARI 40		ME 03 ARI 04
PL 00 VIR 51		MA 08 GEM 43
MO 04 PIS 23		SA 13 SAG 56
ME 08 PIS 10		JU 16 LEO 31
MA 13 TAU 49		VE 17 PIS 27
SA 19 SCO 02		UR 20 LEO 30
JU 21 CAN 37		NE 21 SCO 08
VE 22 AQU 33		PL 25 VIR 45
UR 25 CAN 36		MO 29 PIS 17
NE 26 LIB 14		

4. solar arc MA EQU A*C*G EQU A*C*G ME

These are calculated for June 17, 1982, the day I had the accident in Phoenix.

The C*C*G SU EQU A*C*G UR aspect was exact, which confirms that I experienced new things in life. This aspect also has a connection with accidents (Ebertin).

C*C*G SU = A*C*G NE is a strong indicator that I did not know which direction my life was going. It was a most confusing time in my life; I felt indecisive about my future. As it was an applying aspect it became more powerful after I returned from Phoenix.

C*C*G ME EQU A*C*G JU indicated that I would be overly optimistic about my plans to move to Phoenix.

My Astro*Carto*Graphy solar return map for the year I visited Phoenix is amazing as it reiterates the information obtained from Cyclo*Carto*Graphy.

I have SA MH passing **exactly** through Leicester, which shows the problems I experienced careerwise. Also within orb of Leicester were MA MH and NE ASC, confirming that my plans to emigrate would not materialize.

Close to Phoenix are NE IC and SU DSC, which reiterate the influence from C*C*G SU = A*C*G NE. The confusion and uncertainty at this time was clearly connected to Phoenix.

There is no doubt that Phoenix is a beneficial location for me as shown by my Astro*Carto*Graphy map. But when an important move is planned, the temporary influences need to be considered to find out if the timing is good.

Example 2—Visit to Dresden

In 1970, I joined a penfriend club and started writing to three people in Germany. A few years later I obtained more penfriends, bringing the total to about twenty. I found it interesting to write to people from different parts of the world, but after a while some penfriends decided to stop writing, probably due to lack of interest.

Only one of the penfriends I started writing to in 1970 remained. She lived in Dresden, East Germany, and wrote good English. In 1985, she invited me to visit; because she lived in communist East Germany she was not allowed to visit England. I thought it would be unusual to visit a communist country, so I accepted the invitation and started arrangements to obtain a visa.

At the time I had no idea that visiting Dresden would change my life. For me it was an opportunity to find out how people really lived in an East European country. While I was in East Germany my penfriend and I got along well with each other. Although communication was sometimes difficult between us, it was not a problem. It was quite unusual communicating with an English-German dictionary! I felt happy and relaxed with my penfriend in Dresden. It was only when I returned to England that I realized how much I missed her.

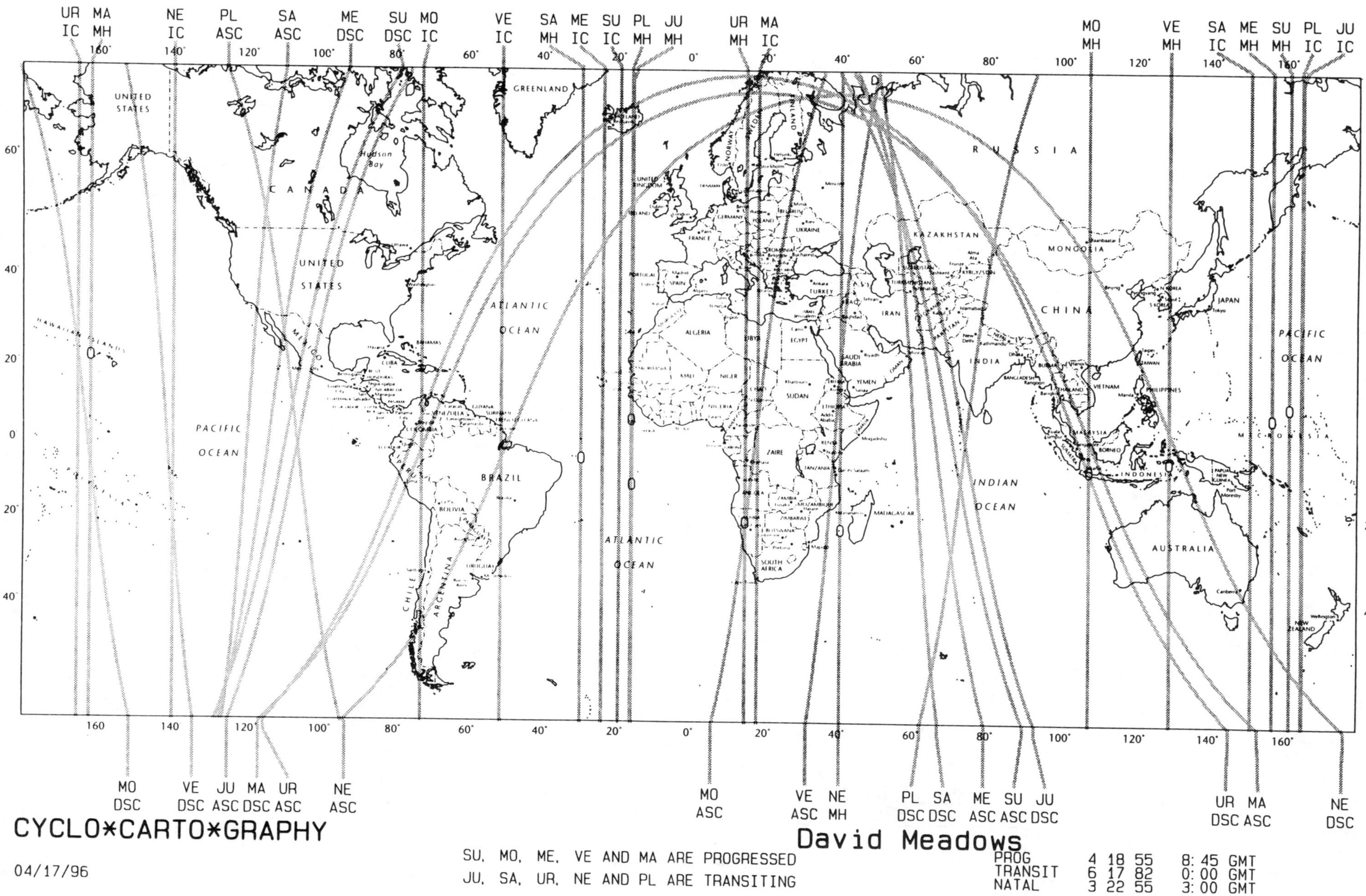

CYCLO*CARTO*GRAPHY

04/17/96

SU, MO, ME, VE AND MA ARE PROGRESSED
JU, SA, UR, NE AND PL ARE TRANSITING

David Meadows

PROG	4 18 55	8: 45 GMT
TRANSIT	6 17 82	0: 00 GMT
NATAL	3 22 55	3: 00 GMT

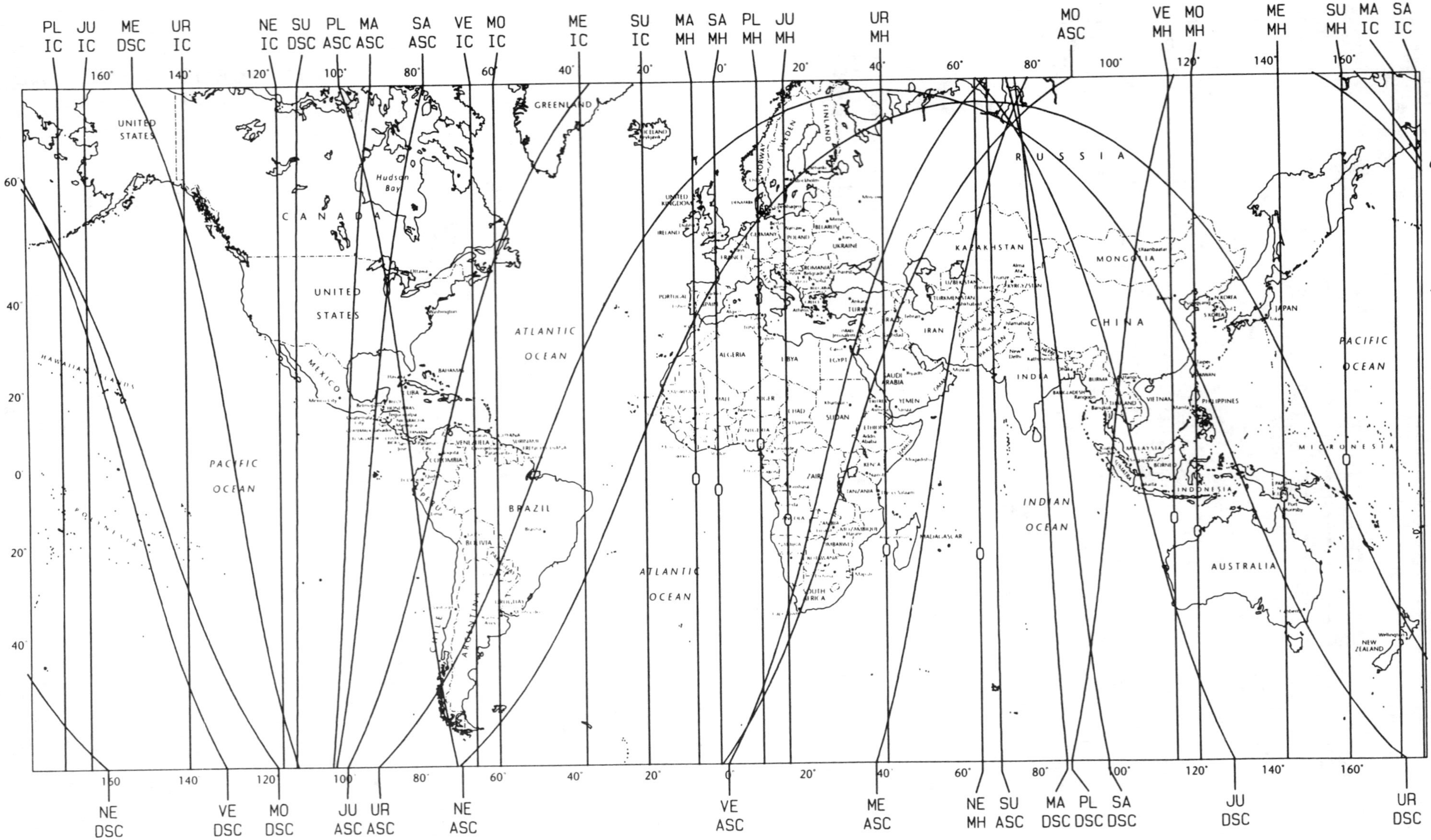

David Meadows SR

March 22, 1982 1:27 AM GMT

04/16/96

04/16/96

David Meadows SR
March 22, 1982 1:27 AM GMT

ASTRO*CARTO*GRAPHY LATITUDE CROSSINGS

NORTH LATITUDES

69N34	VE/ASC * JU/DSC
69N02	NE/MH * VE/ASC
68N56	ME/IC * JU/ASC
68N51	SU/ASC * JU/DSC
68N24	VE/MH * PL/DSC
68N18	SU/MH * MO/DSC
67N57	ME/ASC * MA/DSC
67N51	VE/ASC * NE/ASC
67N51	SU/ASC * NE/DSC
67N50	UR/DSC * NE/ASC
67N48	JU/DSC * NE/DSC
67N44	VE/ASC * UR/DSC
67N35	SA/DSC * NE/DSC
67N33	ME/ASC * NE/DSC
67N30	MA/DSC * NE/DSC
67N18	NE/MH * UR/DSC
67N16	MO/ASC * NE/DSC
67N11	ME/ASC * SA/DSC
67N07	JU/MH * UR/ASC
66N45	SU/ASC * UR/DSC
66N19	MO/ASC * MA/DSC
66N05	MO/ASC * SA/DSC
65N59	UR/MH * NE/ASC
65N57	ME/ASC * JU/DSC
65N47	PL/MH * UR/ASC
65N38	ME/ASC * UR/DSC
65N35	MA/IC * VE/DSC
65N34	MO/ASC * JU/DSC
65N28	MO/ASC * UR/DSC
65N17	JU/DSC * UR/DSC
64N52	SA/DSC * UR/DSC
64N51	MO/ASC * ME/ASC
64N45	MA/DSC * UR/DSC
63N52	JU/DSC * SA/DSC
63N47	MA/DSC * JU/DSC
63N30	MA/DSC * SA/DSC
63N29	MA/IC * MO/DSC
62N47	MO/ASC * SU/ASC
62N30	SA/MH * UR/ASC
61N52	MA/ASC * PL/ASC
61N35	SA/IC * VE/DSC
61N13	Anchorage
60N51	NE/MH * MO/ASC
60N09	SA/IC * MO/DSC
60N08	NE/DSC * PL/DSC
59N59	MA/MH * UR/ASC
59N20	Stockholm
59N17	SU/ASC * ME/ASC
58N00	JU/MH * NE/ASC
57N40	VE/MH * NE/DSC
55N45	Moscow
55N41	Copenhagen
55N02	MO/DSC * VE/DSC
54N06	UR/IC * ME/DSC
54N01	MO/MH * NE/DSC
53N53	PL/MH * NE/ASC
52N31	NE/MH * ME/ASC
52N30	Berlin
51N30	London
51N14	SU/IC * UR/ASC
50N18	PL/IC * MO/DSC
49N14	Vancouver
48N50	Paris
48N33	PL/IC * VE/DSC
48N23	MO/IC * JU/ASC
47N53	UR/DSC * PL/DSC
47N53	UR/MH * VE/ASC
47N36	Seattle
45N32	Montreal
44N59	Minneapolis
43N50	SA/MH * NE/ASC
43N39	Toronto

CENTRAL LATITUDES

42N51	SA/ASC * PL/ASC
42N22	Boston
42N20	Detroit
41N54	Rome
41N53	Chicago
40N43	New York
40N19	JU/IC * MO/DSC
39N44	Denver
38N53	Washington
38N37	St. Louis
37N47	San Francisco
36N10	Las Vegas
36N01	MA/MH * NE/ASC
35N40	Tokyo
35N30	VE/MH * UR/DSC
34N54	VE/IC * JU/ASC
34N04	Los Angeles
33N45	Atlanta
33N42	JU/IC * VE/DSC
33N27	Phoenix
32N47	Dallas
30N03	Cairo
29N58	New Orleans
29N46	Houston
29N26	ME/IC * UR/ASC
28N47	UR/MH * MO/ASC
28N39	Delhi
26N34	ME/MH * NE/DSC
25N47	Miami
24N13	MO/MH * UR/DSC
22N34	Calcutta
22N16	Hong Kong
21N19	Honolulu
19N26	Mexico City
18N45	JU/DSC * PL/DSC
14N35	Manila
10N30	Caracas
10N15	SU/IC * NE/ASC
6N27	Lagos
10S15	SU/MH * NE/DSC
18S45	JU/ASC * PL/ASC
22S54	Rio De Janeiro
24S13	MO/IC * UR/ASC
26S12	Johannesburg
26S34	ME/IC * NE/ASC
28S47	UR/IC * MO/DSC
29S26	ME/MH * UR/DSC
33S42	JU/MH * VE/ASC
33S55	Sydney
34S36	Buenos Aires
34S54	VE/MH * JU/DSC
35S30	VE/IC * UR/ASC
36S01	MA/IC * NE/DSC
36S52	Auckland
37S50	Melbourne
40S19	JU/MH * MO/ASC
42S51	SA/DSC * PL/DSC
43S50	SA/IC * NE/DSC
47S53	UR/IC * VE/DSC
47S53	UR/ASC * PL/ASC
48S23	MO/MH * JU/DSC
48S33	PL/MH * VE/ASC
50S18	PL/MH * MO/ASC
51S14	SU/MH * UR/DSC
52S31	NE/IC * ME/DSC
53S53	PL/IC * NE/DSC

SOUTH LATITUDES

54S01	MO/IC * NE/ASC
54S06	UR/MH * ME/ASC
55S02	MO/ASC * VE/ASC
57S40	VE/IC * NE/ASC
58S00	JU/IC * NE/DSC
59S17	SU/DSC * ME/DSC
59S59	MA/IC * UR/DSC

When I examined my Astro*Carto*Graphy map I was amazed at the planetary influences at Dresden—JU DSC and UR DSC within orb. These planetary influences indicate an unusual and beneficial relationship at Dresden.

Dresden is located at 51N03 latitude. At 51N04 I have a MO ASC/VE ASC latitude crossing which has a powerful effect at Dresden. In the Astro*Carto*Graphy interpretation booklet I read that the MO/VE crossing is where a man can find a glamorous wife. You can imagine what I thought when I read this!

When I examined the synastry between our natal charts I was also astonished. There were many harmonious aspects between our charts which showed that we get along well with each other. Astrology was showing that I had found my future marriage partner in Dresden, but I was still uncertain whether it would be possible for us to get married as she lived behind the Iron Curtain. Also, I didn't know if my penfriend felt the same way about me. The temporary influences in effect when I visited Dresden explain what I experienced.

A*C*G	**Age 30**	**C*C*G**
SU 00 ARI 40		NE 00 CAP 57
PL 00 VIR 51		MA 05 GEM 13
MO 04 PIS 23		PL 06 SCO 31
ME 08 PIS 10		JU 10 AQU 00
MA 13 TAU 49		UR 13 SAG 00
SA 19 SCO 02		MO 14 ARI 43
JU 21 CAN 37		SA 22 SCO 09
VE 22 AQU 33		ME 27 ARI 11
UR 25 CAN 36		VE 27 PIS 17
NE 26 LIB 14		SU 28 ARI 37

1. C*C*G NE SQU A*C*G SU
2. C*C*G MA EQU A*C*G MO
3. C*C*G SA EQU A*C*G VE
4. C*C*G ME = A*C*G NE
5. Prog. birthplace = A*C*G VE ASC (exact)

A*C*G	**(27 57)**	**Displaced C*C*G**
SU 00 ARI 40		NE 03 SAG 00
PL 00 VIR 51		MA 07 TAU 16
MO 04 PIS 23		PL 08 LIB 34
ME 08 PIS 10		JU 12 PIS 03
MA 13 TAU 49		UR 15 SCO 03
SA 19 SCO 02		MO 16 PIS 46
JU 21 CAN 37		SA 24 LIB 12
VE 22 AQU 33		ME 29 PIS 14
UR 25 CAN 36		VE 29 AQU 20
NE 26 LIB 14		

A*C*G	**(27 57)**	**Solar arc A*C*G**
SU 00 ARI 40		MO 02 ARI 20
PL 00 VIR 51		ME 06 ARI 07
MO 04 PIS 23		MA 11 GEM 46
ME 08 PIS 10		SA 16 SAG 59
MA 13 TAU 49		JU 19 SCO 34
SA 19 SCO 02		VE 20 PIS 30
JU 21 CAN 37		UR 23 LEO 33
VE 22 AQU 33		NE 24 SCO 11
UR 25 CAN 36		PL 28 VIR 48
NE 26 LIB 14		

5. solar arc JU = A*C*G SA
6. solar arc UR EQU A*C*G VE

Prog. birthplace was exactly on my A*C*G VE ASC which shows the start of a relationship. Although we had been writing to each other for many years it was the first time we had met.

A new relationship is reiterated by solar arc UR EQU A*C*G VE. But C*C*G SA EQU A*C*G VE shows that the relationship would not be an easy one. I had to overcome many obstacles and difficulties for the relationship to develop.

C*C*G NE EQU A*C*G SU and A*C*G ME = A*C*G NE explain why I was reluctant to make any kind of commitment in the relationship. This was not easy as C*C*G MA EQU MO shows impatience and wanting to make quick decisions.

My solar return map for the year I visited Dresden is relevant. JU IC and VE DSC were near Dresden, which indicates a successful partnership. PL ASC was also within orb of Dresden, showing that the partnership would change my life.

It took me a year and a half to make the decision that I wanted to marry my penfriend in Dresden. It was my knowledge of Astro*Carto*Graphy that helped me make this decision.

Example 3—Disastrous Holiday in Poland

After visiting my penfriend in East Germany I had the opportunity to accept another invitation, this time to visit a penfriend in Poland. As I had two penfriends in Poland I decided to visit both of them.

In 1985, I ordered my first Cyclo*Carto*Graphy overlay calculated for January 1, 1986. Although I have no Astro*Carto*Graphy planetary lines within orb of Poland the Cyclo*Carto*Graphy overlay showed significant information. I had C*C*G MA IC and C*C*G SA MH exact at Warsaw. After the Chernobyl disaster the Foreign Office in London advised people not to visit Poland, and it was only six weeks before my planned visit that the Foreign Office announced it was now safe to travel there. I thought these C*C*G planetary influences showed the problems in arranging a visit to Poland.

As my visit was planned for Poznan and Zielona Gora I assumed that C*C*G MA IC and C*C*G SA MH were not that important. My flight was to Warsaw, so maybe it would be delayed. I arrived in Warsaw exactly as scheduled; the penfriend from Poznan was waiting for me. She thought it would be a good idea to see something of Warsaw before travelling to Poznan, and suggested that we stayed overnight there.

The next day we went sightseeing in Warsaw, seeing a statue of Chopin at Lazienki Park and an interesting palace. When we returned to her car we saw that it had been broken into. My luggage had been stolen, and all I had left was the clothing I was wearing. Fortunately, I had my passport, travel documents and wallet in my pockets; otherwise I would have been in a very serious situation.

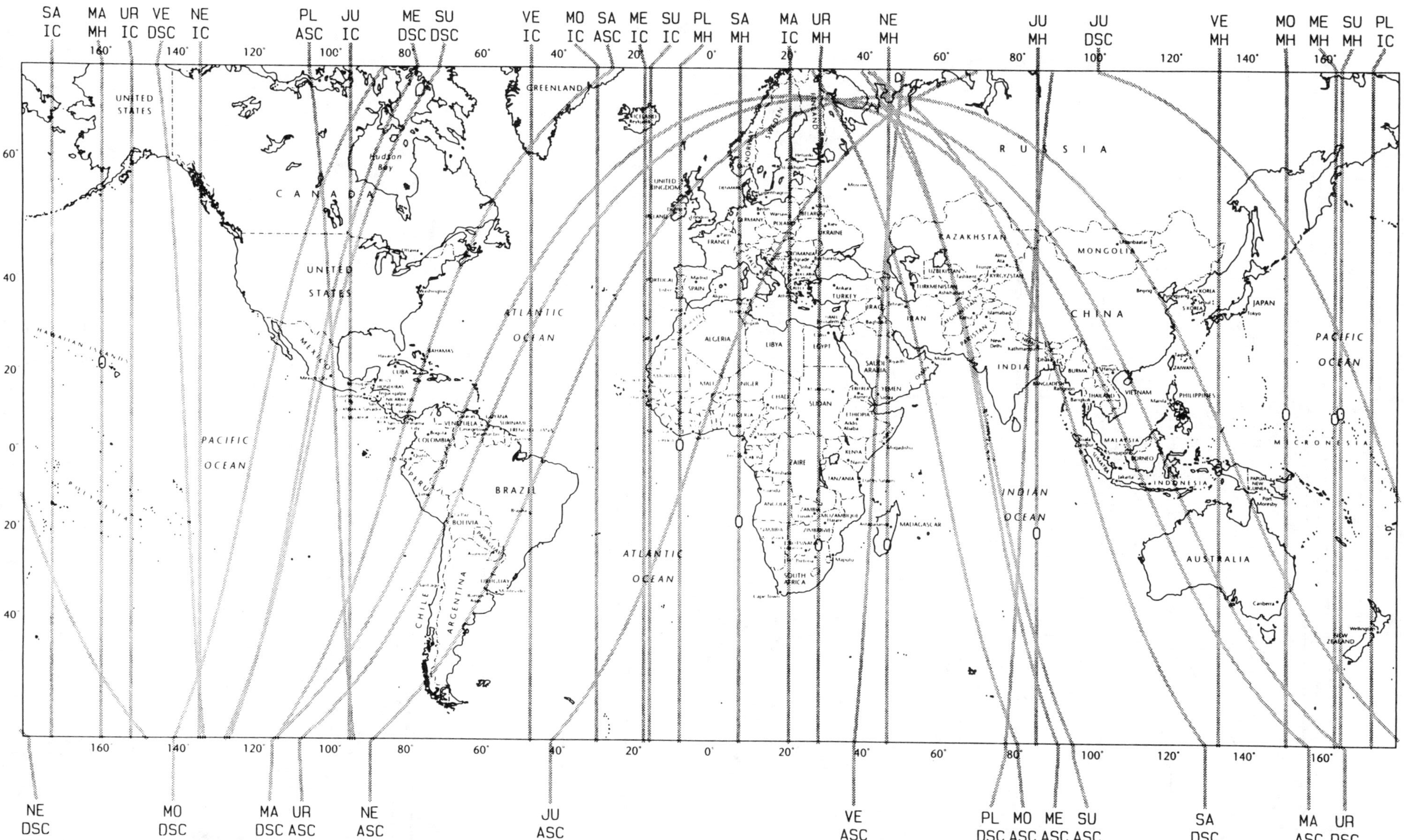

CYCLO*CARTO*GRAPHY

04/17/96

SU, MO, ME, VE AND MA ARE PROGRESSED
JU, SA, UR, NE AND PL ARE TRANSITING

David Meadows

PROG	4 21 55	15:05 GMT
TRANSIT	9 21 85	0:00 GMT
NATAL	3 22 55	3:00 GMT

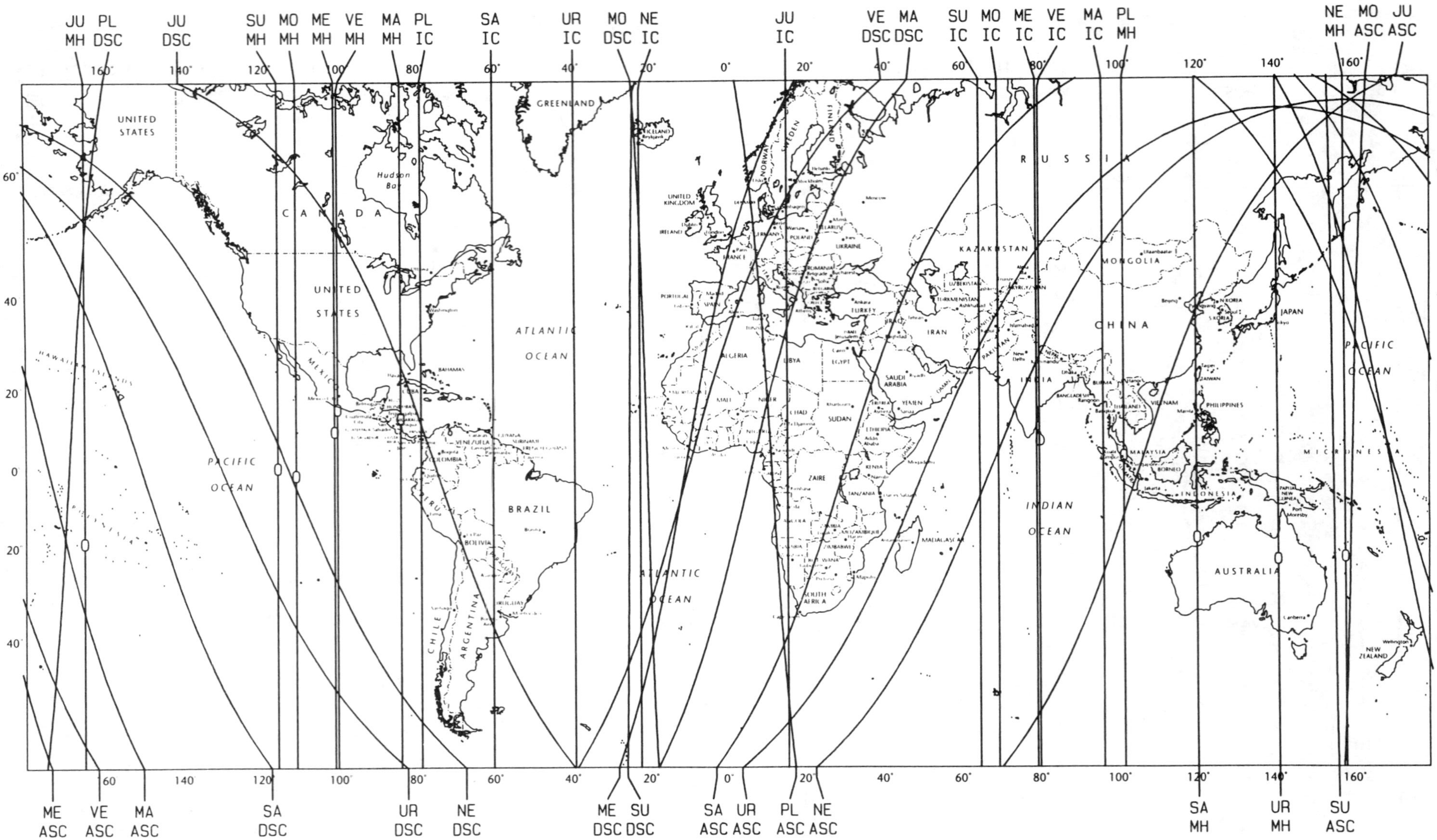

David Meadows SR

March 21, 1985 7:46 PM GMT

04/16/96

ASTRO*CARTO*GRAPHY LATITUDE CROSSINGS

NORTH LATITUDES

69N18	SU/ASC * SA/DSC
68N34	MO/ASC * JU/ASC
68N30	JU/IC * ME/DSC
68N20	NE/MH * SA/DSC
67N59	MO/DSC * SU/DSC
67N45	JU/ASC * SA/DSC
67N44	JU/ASC * NE/DSC
67N44	SA/DSC * NE/DSC
67N42	SU/ASC * NE/ASC
67N41	VE/IC * SA/ASC
67N39	MO/ASC * NE/DSC
67N37	MA/ASC * NE/ASC
67N27	ME/IC * SA/ASC
67N14	ME/ASC * NE/ASC
67N12	NE/MH * JU/ASC
67N04	UR/DSC * NE/ASC
67N01	ME/ASC * UR/DSC
66N50	UR/MH * NE/ASC
66N46	MO/ASC * SA/DSC
66N43	VE/ASC * UR/ASC
66N35	MA/ASC * UR/DSC
66N33	SU/ASC * UR/DSC
66N26	JU/ASC * UR/DSC
66N17	SU/ASC * MA/ASC
66N10	NE/MH * UR/DSC
66N08	MA/ASC * JU/ASC
66N03	SU/ASC * JU/ASC
65N41	SA/MH * UR/ASC
65N37	VE/ASC * NE/ASC
65N30	MO/ASC * UR/DSC
64N20	SA/DSC * UR/DSC
64N02	ME/ASC * JU/ASC
63N47	NE/MH * MA/ASC
63N40	MO/IC * SA/ASC
62N59	SU/MH * JU/DSC
62N41	SA/MH * NE/ASC
62N36	JU/MH * NE/DSC
61N52	NE/DSC * PL/DSC
61N23	PL/MH * UR/ASC
61N13	Anchorage
61N08	SU/IC * SA/ASC
60N31	MO/MH * JU/DSC
60N26	UR/MH * VE/ASC
60N15	VE/ASC * JU/ASC
60N07	MO/ASC * MA/ASC
60N06	UR/MH * JU/ASC
59N21	MA/IC * UR/ASC
59N20	Stockholm
58N52	ME/DSC * PL/ASC
57N03	JU/IC * VE/DSC
56N25	SU/ASC * ME/ASC
55N45	Moscow
55N41	Copenhagen
53N47	JU/MH * UR/DSC
53N42	PL/MH * NE/ASC
53N27	UR/DSC * PL/DSC
53N01	ME/MH * JU/DSC
52N33	NE/IC * MO/DSC
52N30	Berlin
52N15	VE/MH * JU/DSC
51N30	London
50N13	JU/MH * PL/DSC
49N29	MA/IC * NE/ASC
49N22	NE/MH * ME/ASC
49N14	Vancouver
49N13	VE/IC * UR/ASC
48N50	Paris
48N32	ME/IC * UR/ASC
47N42	VE/DSC * PL/ASC
47N36	Seattle
45N32	Montreal
44N59	Minneapolis

CENTRAL LATITUDES

44N12	SU/ASC * VE/ASC
43N39	Toronto
42N22	Boston
42N20	Detroit
41N54	Rome
41N53	Chicago
40N43	New York
40N40	SA/DSC * PL/DSC
40N36	MO/ASC * ME/ASC
39N44	Denver
38N53	Washington
38N37	JU/MH * SA/DSC
38N37	St. Louis
37N47	San Francisco
37N46	NE/MH * VE/ASC
37N19	MO/IC * UR/ASC
37N08	SA/MH * JU/ASC
36N10	Las Vegas
35N40	Tokyo
34N11	JU/IC * MA/DSC
34N04	Los Angeles
33N45	Atlanta
33N27	Phoenix
32N47	Dallas
31N11	MO/ASC * VE/ASC
30N03	Cairo
29N58	SU/IC * UR/ASC
29N58	New Orleans
29N46	Houston
28N39	Delhi
27N45	VE/IC * NE/ASC
26N18	ME/IC * NE/ASC
25N47	Miami
25N45	MA/MH * JU/DSC
22N34	Calcutta
22N16	Hong Kong
21N19	Honolulu
19N26	Mexico City
17N18	MA/DSC * PL/ASC
14N35	Manila
11N47	PL/IC * JU/DSC
10N30	Caracas
7N18	ME/ASC * VE/ASC
6N49	SU/MH * NE/DSC
6N27	Lagos
4N35	MO/IC * NE/ASC
4S35	MO/MH * NE/DSC
6S49	SU/IC * NE/ASC
7S18	ME/DSC * VE/DSC
11S47	PL/MH * JU/ASC
17S18	MA/ASC * PL/DSC
22S54	Rio De Janeiro
25S45	MA/IC * JU/ASC
26S12	Johannesburg
26S18	ME/MH * NE/DSC
27S45	VE/MH * NE/DSC
29S58	SU/MH * UR/DSC
31S11	MO/DSC * VE/DSC
33S55	Sydney
34S11	JU/MH * MA/ASC
34S36	Buenos Aires
36S52	Auckland
37S08	SA/IC * JU/DSC
37S19	MO/MH * UR/DSC
37S46	NE/IC * VE/DSC
37S50	Melbourne
38S37	JU/IC * SA/ASC

SOUTH LATITUDES

40S36	MO/DSC * ME/DSC
40S40	SA/ASC * PL/ASC
44S12	SU/DSC * VE/DSC
47S42	VE/ASC * PL/DSC
48S32	ME/MH * UR/DSC
49S13	VE/MH * UR/DSC
49S22	NE/IC * ME/DSC
49S29	MA/MH * NE/DSC
50S13	JU/IC * PL/ASC
52S15	VE/IC * JU/ASC
52S33	NE/MH * MO/ASC
53S01	ME/IC * JU/ASC
53S27	UR/ASC * PL/ASC
53S42	PL/IC * NE/DSC
53S47	JU/IC * UR/ASC
56S25	SU/DSC * ME/DSC
57S03	JU/MH * VE/ASC
58S52	ME/ASC * PL/DSC
59S21	MA/MH * UR/DSC

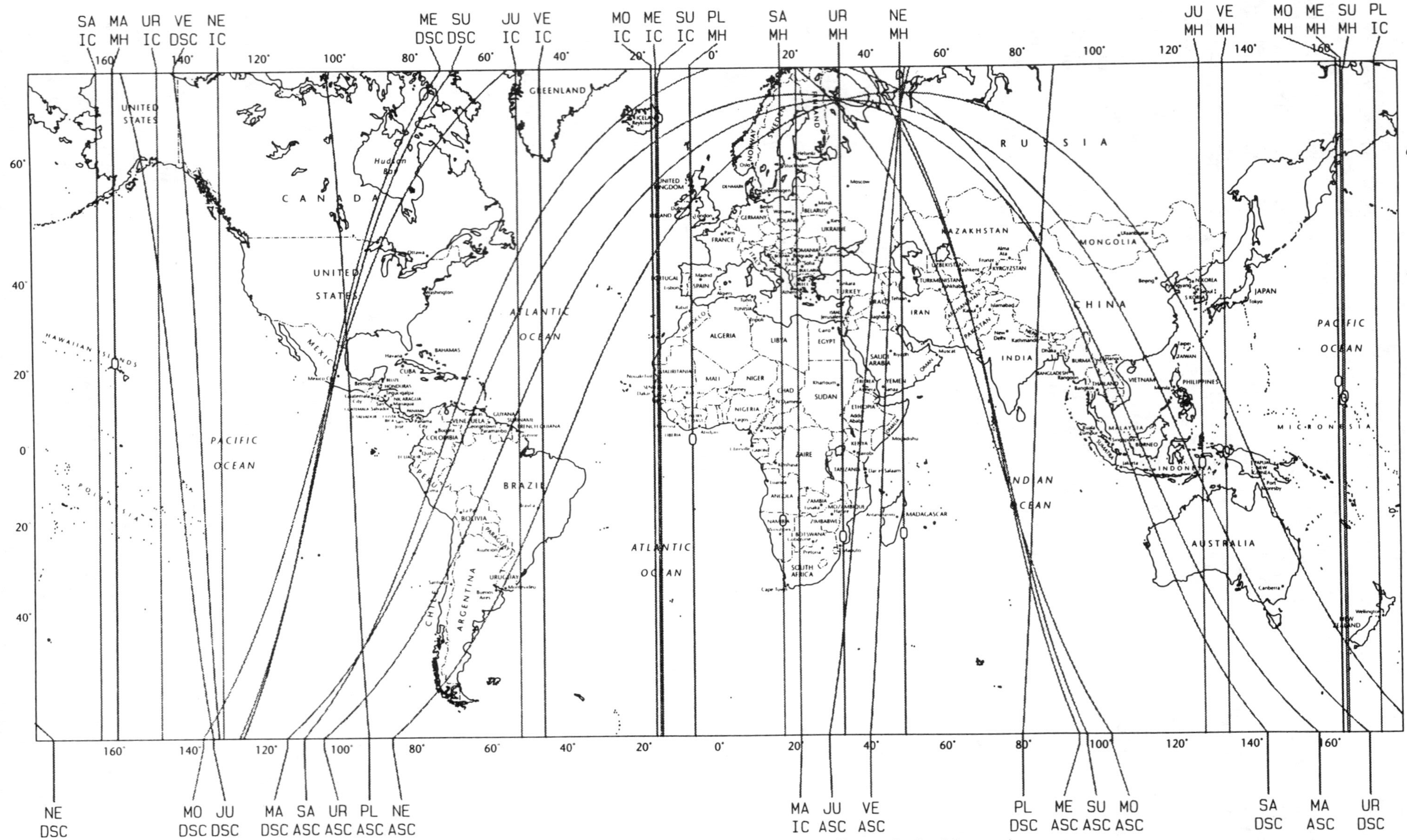

CYCLO*CARTO*GRAPHY

05/21/97

SU, MO, ME, VE AND MA ARE PROGRESSED
JU, SA, UR, NE AND PL ARE TRANSITING

David Meadows

PROG	4 22 55	12:54 GMT
TRANSIT	8 19 86	0:00 GMT
NATAL	3 22 55	3:00 GMT

We went to a police station to report the theft, but nobody could speak English. My penfriend couldn't speak much English either.

The British Embassy in Warsaw said I should contact my travel insurance company and ask that they send me some money so that I could buy some new clothes. But, they said, I should insist they send me either pounds or American dollars.

The Polish government would not allow my travel insurance company to send me hard currency. I could only receive Polish money, which I didn't want as it wasn't worth much. How I managed to survive two weeks in Poland under these conditions I do not know. It was the worst holiday I had ever experienced and I was glad to return to England.

I thought things would improve when I went to Zielona Gora to visit my other penfriend, but he lived on a farm with his parents where living conditions were like going back fifty years in time. At least this penfriend spoke very good English. We visited a town called Wroclaw and stayed in a youth hostel. While there he introduced me to one of his acquaintances.

This was an unbelievable experience for me. When I met her it was similar to the experience I had in Dresden! My penfriend had to translate for us as she couldn't speak English. Although we were immediately attracted to each other, nothing developed between us. This was probably due to our inability to communicate.

I was convinced there was some kind of connection between Dresden and Wroclaw, and when I looked in my atlas I saw that Wroclaw was at 51N06, exactly the same latitude as Dresden. This meant that my MO ASC/VE ASC latitude crossing at 51N04 also manifested at Wroclaw. I then thought that if I travelled along this latitude I would meet many different women!

On the day my luggage was stolen in Warsaw I had some significant temporary influences in effect:

A*C*G	**Age 31**	**C*C*G**
SU 00 ARI 40		SA 01 SAG 30
PL 00 VIR 51		NE 03 CAP 30
MO 04 PIS 23		MA 05 GEM 22
ME 08 PIS 10		PL 08 SCO 00
MA 13 TAU 49		UR 17 SAG 22
SA 19 SCO 02		JU 22 PIS 03
JU 21 CAN 37		MO 27 ARI 49
VE 22 AQU 33		VE 28 PIS 17
UR 25 CAN 36		ME 28 ARI 58
NE 26 LIB 14		SU 29 ARI 28

1. C*C*G SA EQU A*C*G PL

2. Prog. Warsaw = A*C*G PL DSC

3. C*C*G MA EQU A*C*G MO

3. C*C*G MA IC = Warsaw

4. C*C*G SA MH = Poznan

5. Prog. Warsaw = C*C*G NE MH

A*C*G	**(28 48)**	**Displaced C*C*G**
SU 00 ARI 40		ME 00 ARI 10
PL 00 VIR 51		SA 02 SCO 42
MO 04 PIS 23		NE 04 SAG 42
ME 08 PIS 10		MA 06 TAU 34
MA 13 TAU 49		PL 09 LIB 12
SA 19 SCO 02		UR 18 SCO 34
JU 21 CAN 37		JU 23 AQU 15
VE 22 AQU 33		MO 29 PIS 01
UR 25 CAN 36		VE 29 AQU 29
NE 26 LIB 14		

6. Disp. C*C*G ME = A*C*G SU

7. Disp. C*C*G UR = A*C*G SA

8. Disp. C*C*G JU = A*C*G VE

A*C*G	**(28 48)**	**Solar arc A*C*G**
SU 00 ARI 40		MO 03 ARI 11
PL 00 VIR 51		ME 06 ARI 58
MO 04 PIS 23		MA 12 GEM 37
ME 08 PIS 10		SA 17 SAG 50
MA 13 TAU 49		JU 20 LEO 25
SA 19 SCO 02		VE 21 PIS 21
JU 21 CAN 37		UR 24 LEO 24
VE 22 AQU 33		NE 25 SCO 02
UR 25 CAN 36		PL 29 VIR 39
NE 26 LIB 14		

Although C*C*G SA MH had moved away from Warsaw, it was now exact at Poznan! My penfriend lived in Poznan and my luggage was stolen from her car.

C*C*G MA IC was at Warsaw when I arrived and soon experienced a challenging situation. I needed all my will power and determination to get through the struggle for survival.

To make matters more intense, prog. Warsaw was on my A*C*G PL DSC and C*C*G NE MH. These combined planetary influences show the hardship and trauma I experienced as a result of crime.

C*C*G SA EQU A*C*G PL reiterated the temporary influences that were dominant in Poland. Disp. C*C*G UR = A*C*G SA also indicated that things were not straightforward in my life.

Disp. C*C*G JU = A*C*G VE was appropriate for my visit to Wroclaw where my MO ASC/VE ASC latitude crossing was triggered. With all the other temporary influences in Poland it is easy to see why nothing developed from this relationship.

On my solar return map PL MH and NE ASC were within orb of Poland. These planetary lines had more influence in Poznan and Zielona Gora where I stayed.

Solar return PL MH and NE ASC reiterate prog. Warsaw = A*C*G PL DSC and C*C*G NE MH. As a result, these two planetary influences had a powerful effect when I was in Warsaw.

Example 4—Visit to Hungary

Now that I was sure about marrying my penfriend in Dresden, I suggested we go to Hungary for a holiday. She was limited in where she could travel, but two weeks in Hungary was not a problem. We spent one week at Lake Balaton and the other in Budapest.

Near the end of the holiday we became engaged, not knowing when we would be allowed to marry. We needed

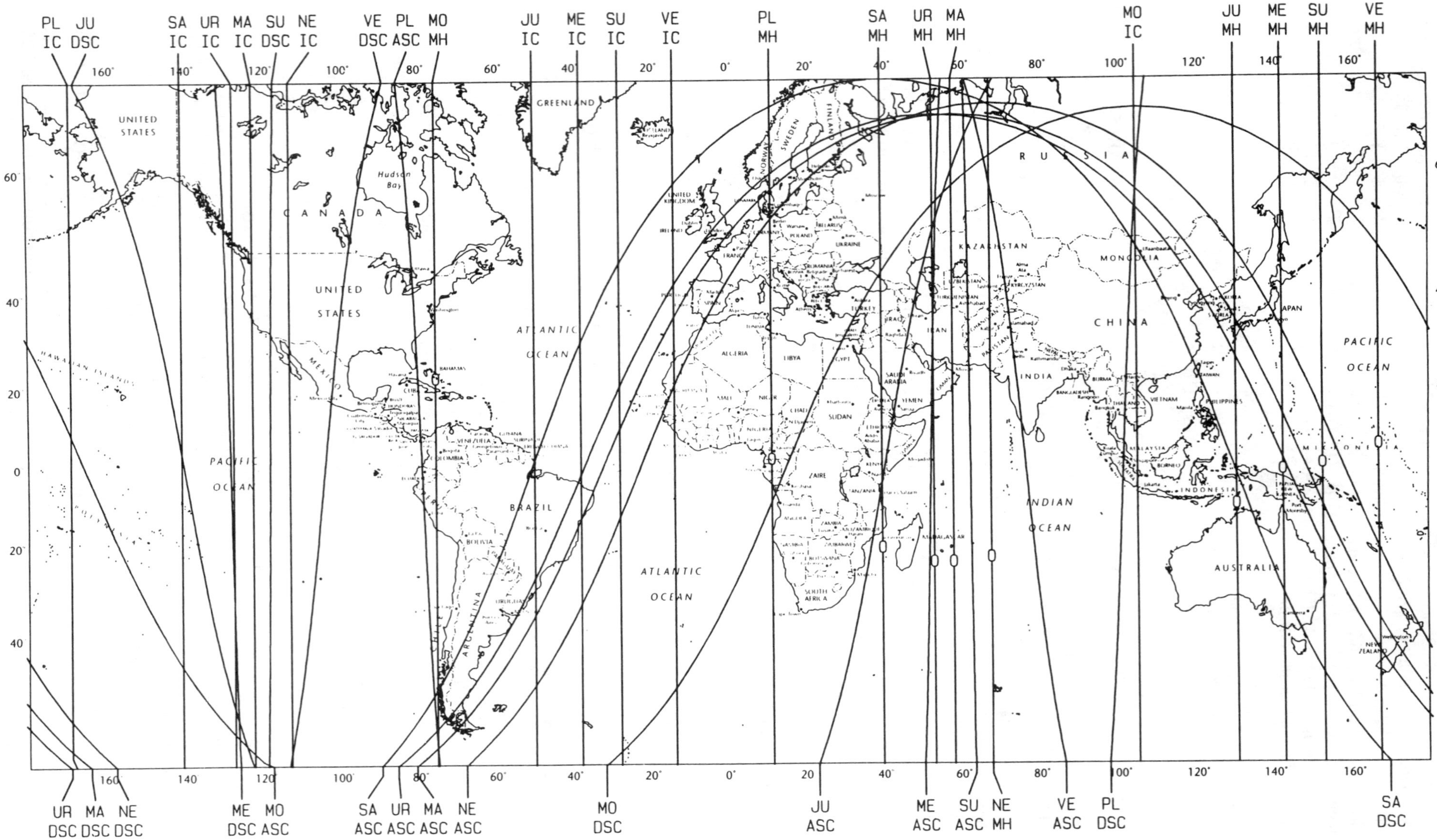

David Meadows SR

March 22, 1986 1:56 AM GMT

04/16/96

David Meadows SR
March 22, 1986 1:56 AM GMT

ASTRO*CARTO*GRAPHY LATITUDE CROSSINGS

NORTH LATITUDES

69N31	UR/MH * SA/DSC
69N21	NE/MH * JU/ASC
69N20	ME/ASC * SA/DSC
69N04	MA/MH * SA/DSC
68N40	SU/ASC * SA/DSC
68N29	VE/ASC * SA/DSC
68N00	JU/ASC * SA/DSC
67N46	SA/DSC * NE/ASC
67N45	JU/ASC * NE/ASC
67N42	VE/ASC * NE/ASC
67N40	NE/MH * SA/DSC
67N39	SU/ASC * NE/ASC
67N35	PL/MH * SA/ASC
67N29	MA/MH * NE/ASC
67N17	ME/ASC * NE/ASC
67N13	MO/ASC * PL/DSC
67N06	UR/MH * NE/ASC
66N41	ME/ASC * UR/DSC
66N41	MA/ASC * UR/DSC
66N40	ME/ASC * MA/ASC
66N39	SU/ASC * MA/DSC
66N39	UR/ASC * NE/ASC
66N38	MA/MH * UR/DSC
66N37	UR/MH * MA/ASC
66N36	VE/ASC * JU/ASC
66N35	MA/DSC * JU/ASC
66N35	VE/ASC * MA/DSC
66N29	SU/ASC * UR/DSC
66N24	NE/MH * MA/DSC
66N21	MO/DSC * NE/DSC
66N21	JU/ASC * UR/DSC
66N19	VE/ASC * UR/DSC
66N11	MA/ASC * NE/ASC
66N08	SA/MH * UR/ASC
66N01	NE/MH * UR/DSC
65N36	SA/MH * MA/ASC
65N32	MA/DSC * SA/DSC
65N11	SA/MH * NE/ASC
65N06	MO/DSC * MA/DSC
65N06	JU/MH * MO/ASC
64N55	MO/DSC * SA/DSC
64N41	MO/DSC * UR/DSC
64N26	SU/ASC * JU/ASC
63N43	SA/DSC * UR/DSC
62N44	MO/IC * NE/DSC
62N30	NE/DSC * PL/DSC
62N12	ME/MH * MO/ASC
62N08	NE/MH * MO/DSC
61N44	MO/DSC * VE/ASC
61N13	Anchorage
60N19	MA/MH * JU/ASC
60N04	PL/MH * UR/ASC
59N59	MO/DSC * SU/ASC
59N20	Stockholm
58N46	VE/IC * SA/ASC
58N37	NE/MH * VE/ASC
58N21	MA/MH * MO/DSC
58N13	SU/MH * MO/ASC
57N57	PL/MH * MA/ASC
57N44	MO/IC * PL/DSC
57N41	MA/DSC * PL/DSC
57N41	MO/IC * MA/DSC
57N21	MO/DSC * JU/ASC
56N18	MO/DSC * ME/ASC
55N45	Moscow
55N43	UR/MH * MO/DSC
55N41	Copenhagen
55N13	UR/DSC * PL/DSC
55N02	MO/IC * UR/DSC
53N44	PL/MH * NE/ASC
53N37	ME/ASC * JU/ASC
52N30	Berlin
51N52	UR/MH * JU/ASC

CENTRAL LATITUDES

51N30	London
49N55	SA/DSC * PL/DSC
49N14	Vancouver
48N58	MO/IC * SA/DSC
48N50	Paris
48N41	VE/MH * MO/ASC
48N11	JU/MH * NE/DSC
47N36	Seattle
46N44	SU/IC * SA/ASC
45N32	Montreal
44N59	Minneapolis
44N51	SA/MH * MO/DSC
43N39	Toronto
42N48	VE/IC * UR/ASC
42N22	Boston
42N20	Detroit
41N54	Rome
41N53	Chicago
40N43	New York
39N44	Denver
38N53	Washington
38N37	St. Louis
37N47	San Francisco
36N29	VE/IC * MA/ASC
36N10	Las Vegas
35N40	Tokyo
35N03	JU/MH * MA/DSC
34N04	Los Angeles
33N45	Atlanta
33N27	Phoenix
32N47	Dallas
32N46	ME/MH * NE/DSC
32N15	UR/MH * ME/ASC
30N58	ME/IC * SA/ASC
30N03	Cairo
29N58	New Orleans
29N46	Houston
28N39	Delhi
26N59	JU/MH * UR/DSC
25N47	Miami
22N34	Calcutta
22N16	Hong Kong
21N19	Honolulu
20N53	VE/IC * NE/ASC
20N44	SU/IC * UR/ASC
19N26	Mexico City
14N35	Manila
12N52	ME/MH * MA/DSC
12N22	SU/MH * NE/DSC
10N30	Caracas
10N14	SU/IC * MA/ASC
8N34	PL/IC * MO/ASC
6N27	Lagos
3N49	SA/IC * JU/DSC
1N48	JU/IC * SA/ASC
1N42	ME/MH * UR/DSC
1S42	ME/IC * UR/ASC
1S48	JU/MH * SA/DSC
3S49	SA/MH * JU/ASC
8S34	PL/MH * MO/DSC
10S14	SU/MH * MA/DSC
12S22	SU/IC * NE/ASC
12S52	ME/IC * MA/ASC
20S44	SU/MH * UR/DSC
20S53	VE/MH * NE/DSC
22S54	Rio De Janeiro
26S12	Johannesburg

SOUTH LATITUDES

26S59	JU/IC * UR/ASC
30S58	ME/MH * SA/DSC
32S15	UR/IC * ME/DSC
32S46	ME/IC * NE/ASC
33S55	Sydney
34S36	Buenos Aires
35S03	JU/IC * MA/ASC
36S29	VE/MH * MA/DSC
36S52	Auckland
37S50	Melbourne
42S48	VE/MH * UR/DSC
44S51	SA/IC * MO/ASC
46S44	SU/MH * SA/DSC
48S11	JU/IC * NE/ASC
48S41	VE/IC * MO/DSC
48S58	MO/MH * SA/ASC
49S55	SA/ASC * PL/ASC
51S52	UR/IC * JU/DSC
53S37	ME/DSC * JU/DSC
53S44	PL/IC * NE/DSC
55S02	MO/MH * UR/ASC
55S13	UR/ASC * PL/ASC
55S43	UR/IC * MO/ASC
56S18	MO/ASC * ME/DSC
57S21	MO/ASC * JU/DSC
57S41	MO/MH * MA/ASC
57S41	MA/ASC * PL/ASC
57S44	MO/MH * PL/ASC
57S57	PL/IC * MA/DSC
58S13	SU/IC * MO/DSC
58S21	MA/IC * MO/ASC
58S37	NE/IC * VE/DSC
58S46	VE/MH * SA/DSC
59S59	MO/ASC * SU/DSC

to obtain permission from the East German authorities, which would not be easy.

On my Astro*Carto*Graphy map I have VE ASC near Budapest; it seemed a good place for the engagement.

A*C*G	Age 32	C*C*G
SU 00 ARI 40		SU 00 TAU 25
PL 00 VIR 51		ME 00 TAU 59
MO 04 PIS 23		NE 05 CAP 46
ME 08 PIS 10		MA 06 GEM 36
MA 13 TAU 49		PL 10 SCO 24
SA 19 SCO 02		MO 13 TAU 25
JU 21 CAN 37		SA 13 SAG 27
VE 22 AQU 33		UR 22 SAG 04
UR 25 CAN 36		JU 28 ARI 04
NE 26 LIB 14		VE 29 PIS 25

1. C*C*G MO = A*C*G MA
2. Prog. Budapest = C*C*G VE ASC

A*C*G	(29 45)	Displaced C*C*G
SU 00 ARI 40		ME 01 ARI 14
PL 00 VIR 51		NE 06 SAG 01
MO 04 PIS 23		MA 06 TAU 51
ME 08 PIS 10		PL 10 LIB 39
MA 13 TAU 49		MO 13 ARI 40
SA 19 SCO 02		SA 13 SCO 42
JU 21 CAN 37		UR 22 SCO 19
VE 22 AQU 33		JU 28 PIS 19
UR 25 CAN 36		VE 29 AQU 40
NE 26 LIB 14		

3. Disp. C*C*G ME = A*C*G SU
4. Disp. C*C*G SA = A*C*G MA
5. Disp. C*C*G UR = A*C*G VE

A*C*G	(29 45)	Solar arc A*C*G
SU 00 ARI 40		PL 00 LIB 36
PL 00 VIR 51		MO 04 ARI 08
MO 04 PIS 23		ME 07 ARI 55
ME 08 PIS 10		MA 13 GEM 34
MA 13 TAU 49		SA 18 SAG 47
SA 19 SCO 02		JU 21 LEO 22
JU 21 CAN 37		VE 22 PIS 28
VE 22 AQU 33		UR 25 LEO 21
UR 25 CAN 36		NE 25 SCO 59
NE 26 LIB 14		

6. solar arc PL = A*C*G SU

On the day I became engaged I had prog. Budapest = C*C*G VE ASC. So it was a good time for that to happen. C*C*G MO = A*C*G MA indicated that I would express my feelings more easily while I was in Hungary. Disp. C*C*G UR = A*C*G VE showed the change in our relationship. We were no longer penfriends after writing to each other for fifteen years.

Disp. C*C*G SA = A*C*G MA indicated I had many obstacles to overcome before we married. Solar arc PL = A*C*G SU suggested that my life would change as a result of becoming engaged.

The solar return map for the year that I became engaged is impressive as it explains what I experienced. Within orb of Budapest were MO DSC, VE MH, SA DSC and PL DSC. The VE MH and SA DSC were emphasized as they cross each other on the line of latitude for Budapest.

These planetary energies at Budapest are a clear indication of my engagement and not knowing when we could get married. We had a battle with the authorities (PL DSC) ahead of us, as well as many obstacles (SA DSC) to overcome before we could marry (MO DSC and VE MH). The emotional power struggle with the East German authorities is shown by MO DSC and PL DSC exact at Dresden.

Example 5—Wedding in Dresden

As expected, the East German authorities tried everything possible to stop us from marrying. The British Embassy in East Berlin was doing all it could to help me obtain permission, and the Foreign Office in London was also putting pressure on the East German authorities. My local MP also became involved in the power struggle.

In March 1988, I calculated my solar return Astro*Carto*Graphy map for my thirty-third birthday. From this solar return map I predicted we would marry before my thirty-fourth birthday. I had MO MH, VE MH and PL IC all near Dresden, and was convinced this indicated marriage. PL IC passed exactly through East Berlin where our application for marriage was being processed. I couldn't help but think some kind of miracle would happen in East Berlin. I told my fiancee this information and she thought it impossible to marry so quickly.

I visited Dresden at Christmas 1988 just after the East German government had turned down another application. With only three months to my thirty-fourth birthday, I didn't see how my prediction could come true. We decided to write a letter to the East German leader, Erich Honecker. I wanted to type the letter back in England so I could make a few copies of it.

When I was returning to England I was stopped at the East German border. At Checkpoint Charlie I had to go inside a small room and empty all my pockets. The border official saw the drafted letter to the East German leader and disappeared out of the room with it. After what seemed a long time he returned and said I could leave. There is no doubt he made a copy of the letter and passed it on.

Much to our amazement, on January 18, 1989 we were given permission to marry, but all my documents had to be re-translated as they were more than six months old. No date could be arranged for the wedding until all the translation work had been done. During this time I told my family it would be best to arrange visas to visit East Germany, but they couldn't understand this as no wedding date had been arranged. It usually took four to five weeks to obtain an East German visa and it was valid for six months. There was still the chance that I could marry before my thirty-fourth birthday.

On February 28, 1989, my fiancee was told she could go to the registry office to arrange a date for the wedding. She went there and was told only one date was available: March 18, 1989. There was a choice of 9:00 a.m. or 10:00

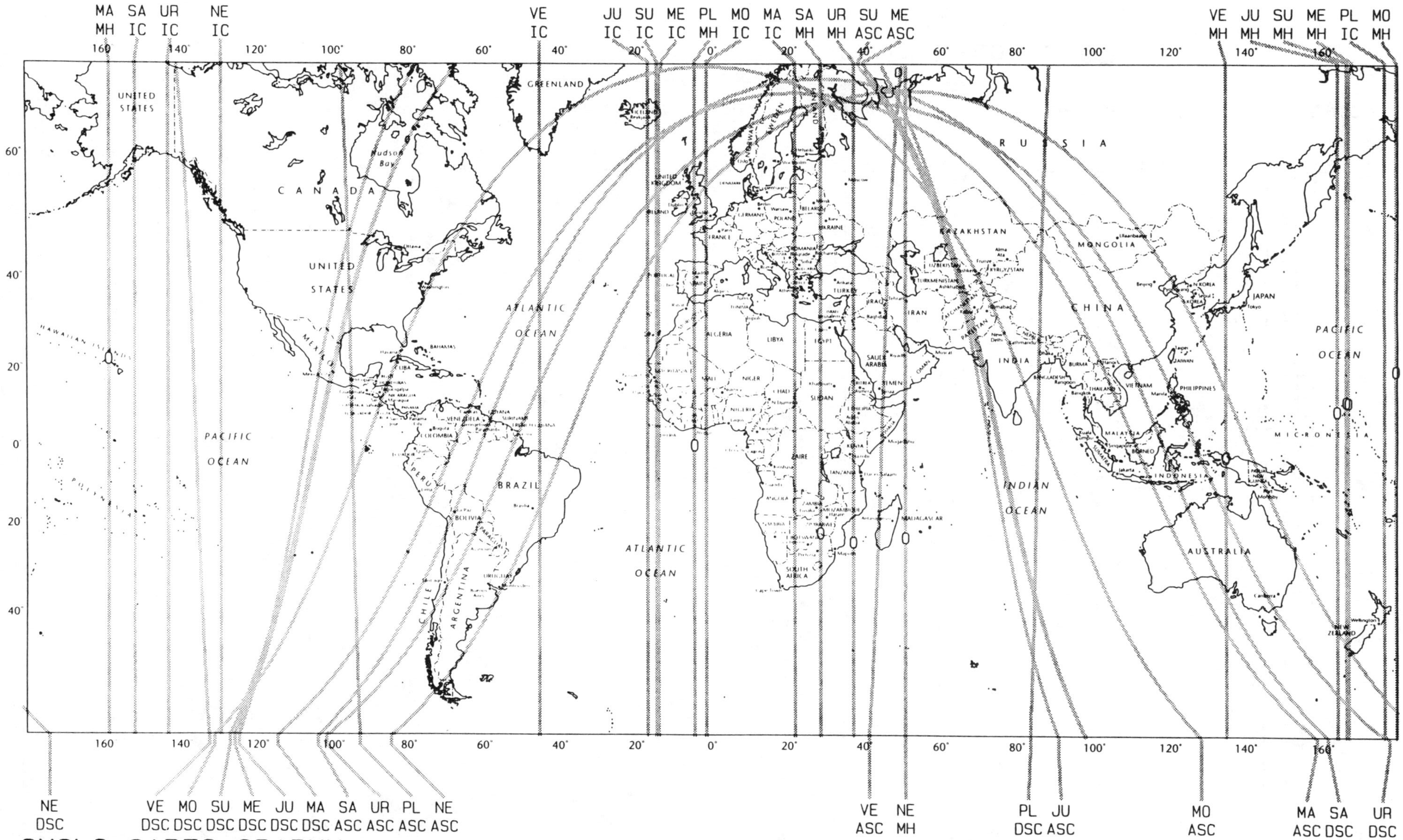

CYCLO*CARTO*GRAPHY

04/17/96

David Meadows

SU, MO, ME, VE AND MA ARE PROGRESSED
JU, SA, UR, NE AND PL ARE TRANSITING

PROG	4	23	55	13:28 GMT
TRANSIT	8	28	87	0:00 GMT
NATAL	3	22	55	3:00 GMT

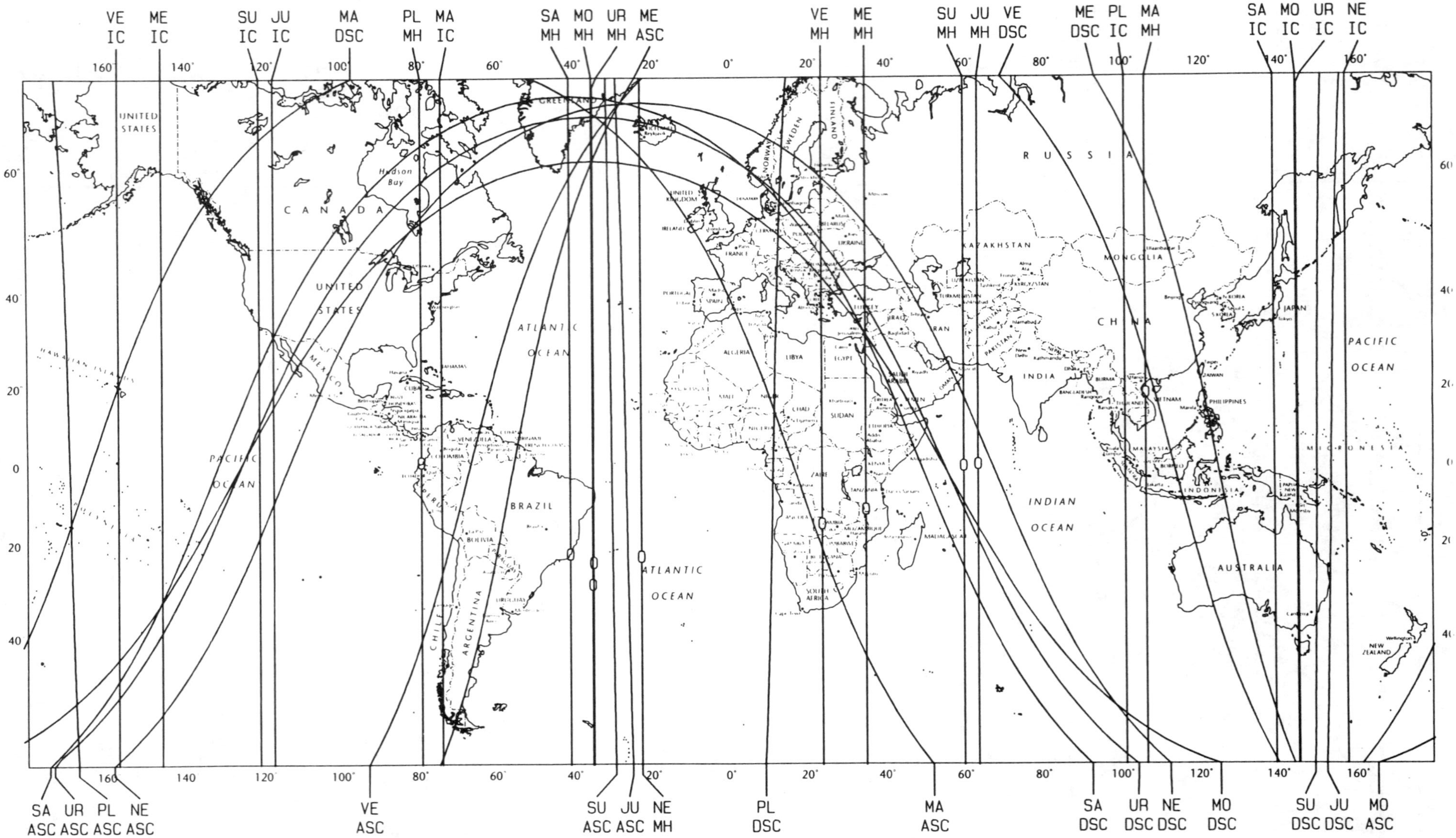

David Meadows SR

March 22, 1987 8:05 AM GMT

04/16/96

David Meadows SR
March 22, 1987 8:05 AM GMT

ASTRO*CARTO*GRAPHY LATITUDE CROSSINGS

NORTH LATITUDES

69N46 NE/MH * VE/ASC
68N25 MA/ASC * SA/ASC
68N19 MO/MH * SA/DSC
68N19 UR/MH * SA/DSC
68N17 SA/MH * MA/ASC
68N09 SU/ASC * SA/DSC
67N58 JU/ASC * SA/DSC
67N53 ME/ASC * SA/DSC
67N53 VE/ASC * SA/DSC
67N53 ME/ASC * VE/ASC
67N47 SA/DSC * NE/ASC
67N44 ME/ASC * NE/ASC
67N44 VE/ASC * NE/ASC
67N43 JU/ASC * NE/ASC
67N35 SU/ASC * NE/ASC
67N26 VE/ASC * JU/ASC
67N24 NE/MH * SA/DSC
67N22 UR/MH * NE/ASC
67N21 MO/MH * NE/ASC
67N14 MA/ASC * NE/ASC
67N12 ME/ASC * JU/ASC
66N50 MO/MH * MA/ASC
66N47 SA/MH * NE/ASC
66N46 UR/MH * MA/ASC
66N28 MO/MH * UR/ASC
66N28 MA/ASC * UR/DSC
66N25 SU/ASC * UR/DSC
66N25 VE/ASC * UR/DSC
66N23 ME/ASC * UR/DSC
66N21 SA/MH * UR/ASC
66N20 JU/ASC * UR/DSC
66N09 SU/ASC * VE/ASC
66N04 UR/ASC * NE/ASC
65N59 NE/MH * UR/DSC
65N49 VE/ASC * MA/ASC
65N37 SU/ASC * MA/ASC
65N29 ME/ASC * MA/ASC
65N24 JU/IC * MA/DSC
65N11 SU/ASC * ME/ASC
64N49 MA/IC * SA/ASC
64N47 PL/IC * ME/DSC
64N34 MA/ASC * JU/ASC
64N02 UR/MH * VE/ASC
64N00 SU/IC * MA/DSC
63N54 MO/MH * VE/ASC
63N21 PL/MH * SA/ASC
63N09 NE/DSC * PL/DSC
62N03 NE/MH * MA/ASC
61N34 UR/MH * ME/ASC
61N30 MO/DSC * ME/ASC
61N30 UR/MH * MO/DSC
61N28 MO/ASC * VE/ASC
61N27 MO/DSC * SU/ASC
61N23 SA/MH * MO/ASC
61N20 MO/DSC * JU/ASC
61N19 MO/MH * ME/ASC
61N13 Anchorage
60N56 NE/MH * MO/DSC
60N50 MA/IC * UR/ASC
60N40 MO/DSC * MA/ASC
60N28 SA/DSC * UR/DSC

59N27 SA/MH * VE/ASC
59N26 VE/MH * NE/DSC
59N20 Stockholm
59N03 MA/MH * ME/DSC
58N49 PL/MH * UR/ASC
57N08 MA/IC * NE/ASC
56N57 UR/DSC * PL/DSC
56N09 SA/DSC * PL/DSC
55N45 Moscow
55N41 Copenhagen
55N15 MA/IC * MO/ASC
53N49 PL/MH * NE/ASC

CENTRAL LATITUDES

53N04 PL/MH * MO/ASC
52N52 SA/MH * ME/ASC
52N50 ME/MH * NE/DSC
52N30 Berlin
51N44 MO/ASC * NE/ASC
51N30 London
51N03 MO/DSC * PL/DSC
50N12 VE/MH * UR/DSC

49N14 Vancouver
48N50 Paris
47N36 Seattle
47N29 VE/MH * SA/DSC
45N32 Montreal
44N59 Minneapolis
43N44 VE/MH * MO/DSC
43N39 Toronto
43N11 ME/IC * MA/DSC
42N22 Boston
42N20 Detroit
41N54 Rome
41N53 Chicago
40N43 New York
40N18 PL/IC * VE/DSC

39N44 Denver
38N53 Washington
38N38 ME/MH * UR/DSC
38N37 St. Louis
37N47 San Francisco
36N10 Las Vegas
35N40 Tokyo
34N04 Los Angeles
33N45 Atlanta
33N27 Phoenix
32N57 MO/DSC * SA/DSC
32N47 Dallas
32N25 ME/MH * MO/DSC
32N11 ME/MH * SA/DSC
32N00 JU/IC * SA/ASC
30N03 Cairo

29N58 New Orleans
29N46 Houston
28N39 Delhi
26N13 MA/MH * VE/DSC
25N47 Miami
25N02 SU/IC * SA/ASC
22N34 Calcutta
22N16 Hong Kong
21N19 Honolulu
20N30 VE/IC * MA/DSC

19N26 Mexico City
18N26 JU/IC * UR/ASC
17N41 SU/MH * NE/DSC
15N20 JU/IC * MO/ASC
14N35 Manila
14N25 MA/ASC * PL/DSC
10N38 SU/IC * UR/ASC
10N30 Caracas

9N16 JU/MH * NE/DSC
8N56 SU/IC * MO/ASC
6N27 Lagos
1N58 MO/ASC * UR/ASC

1S58 MO/DSC * UR/DSC
8S56 SU/MH * MO/DSC
9S16 JU/IC * NE/ASC

10S38 SU/MH * UR/DSC
14S25 MA/DSC * PL/ASC
15S20 JU/MH * MO/DSC
17S41 SU/IC * NE/ASC

SOUTH LATITUDES

18S26 JU/MH * UR/DSC

20S30 VE/MH * MA/ASC
22S54 Rio De Janeiro
25S02 SU/MH * SA/DSC
26S12 Johannesburg
26S13 MA/IC * VE/ASC

32S00 JU/MH * SA/DSC
32S11 ME/IC * SA/ASC
32S25 ME/IC * MO/ASC
32S57 MO/ASC * SA/ASC
33S55 Sydney
34S36 Buenos Aires
36S52 Auckland
37S50 Melbourne
38S38 ME/IC * UR/ASC

40S18 PL/MH * VE/ASC
43S11 ME/MH * MA/ASC
43S44 VE/IC * MO/ASC
47S29 VE/IC * SA/ASC

50S12 VE/IC * UR/ASC
51S03 MO/ASC * PL/ASC
51S44 MO/DSC * NE/DSC
52S50 ME/IC * NE/ASC
52S52 SA/IC * ME/DSC
53S04 PL/IC * MO/DSC
53S49 PL/IC * NE/DSC
55S15 MA/MH * MO/DSC
56S09 SA/ASC * PL/ASC
56S57 UR/ASC * PL/ASC
57S08 MA/MH * NE/DSC
58S49 PL/IC * UR/DSC
59S03 MA/IC * ME/ASC
59S26 VE/IC * NE/ASC
59S27 SA/IC * VE/DSC

a.m. As my family and I had applied earlier for our East German visas we had no problem marrying at short notice.

The marriage took place four days before my thirty-fourth birthday. Yet three months before then it was hard to believe my prediction would come true.

A*C*G	Age 33	C*C*G
SU 00 ARI 40		VE 01 ARI 09
PL 00 VIR 51		SU 01 TAU 53
MO 04 PIS 23		ME 04 TAU 06
ME 08 PIS 10		UR 05 CAP 36
MA 13 TAU 49		MA 07 GEM 43
SA 19 SCO 02		MO 08 GEM 20
JU 21 CAN 37		NE 13 CAP 10
VE 22 AQU 33		SA 13 CAP 57
UR 25 CAN 36		PL 17 SCO 09
NE 26 LIB 14		JU 29 TAU 06

1). C*C*G VE = A*C*G SU

2). C*C*G MA EQU A*C*G ME

3). C*C*G MO EQU A*C*G ME

4). C*C*G JU IC at Dresden (orb $1\frac{1}{2}°$)

A*C*G	(31 13)	Displaced C*C*G
SU 00 ARI 40		ME 02 ARI 53
PL 00 VIR 51		UR 04 SAG 23
MO 04 PIS 23		MA 06 TAU 30
ME 08 PIS 10		MO 07 TAU 07
MA 13 TAU 49		NE 11 SAG 57
SA 19 SCO 02		SA 12 SAG 44
JU 21 CAN 37		PL 15 LIB 56
VE 22 AQU 33		JU 27 ARI 53
UR 25 CAN 36		VE 29 AQU 56
NE 26 LIB 14		

5. Disp. C*C*G VE = A*C*G PL

A*C*G	(31 13)	Solar arc A*C*G
SU 00 ARI 40		PL 02 LIB 04
PL 00 VIR 51		MO 05 ARI 36
MO 04 PIS 23		ME 09 ARI 23
ME 08 PIS 10		MA 15 GEM 02
MA 13 TAU 49		SA 20 SAG 15
SA 19 SCO 02		JU 22 LEO 50
JU 21 CAN 37		VE 23 PIS 46
VE 22 AQU 33		UR 26 LEO 49
UR 25 CAN 36		NE 27 SCO 27
NE 26 LIB 14		

6. solar arc NE MH = Dresden

7. solar arc JU = A*C*G VE

Other temporary influences showed that I would marry at that time. C*C*G VE = A*C*G SU is a classic "marriage" aspect.

C*C*G JU IC was near to Dresden on the day I married. This shows the good luck I experienced to be able to marry on that day. It was reiterated by solar arc JU = A*C*G VE.

Solar arc NE MH = Dresden indicated that it was like a dream come true. After waiting so long to obtain permission, it was hard to believe it was real. The impossible marriage can also be seen from disp. C*C*G VE = A*C*G PL.

C*C*G MA and C*C*G MO EQU A*C*G ME is appropriate for my anger and determination to achieve what I wanted. The next move would have been to take my case to the United Nations.

Solar return JU MH at Leicester showed that it would be a successful year for me. This was also when I started to travel throughout Great Britain lecturing on Astro*Carto*Graphy.

Example 6—Martina Leaving Dresden

I had to return to England on my own. My wife had to apply for an exit visa to leave East Germany, not knowing how long it would take.

After my marriage prediction came true, my wife was convinced of the validity of Astro*Carto*Graphy. She asked me when would be a good time to apply for an exit visa to leave East Germany. I examined her Cyclo*Carto*Graphy planets and saw that JU MH was at Dresden the end of April 1989, and told her to tell the authorities that she wanted to leave East Germany then.

There were many formalities to complete, but on April 25, 1989, my wife was given an exit visa valid until the end of the month. The next day she visited the British Embassy in East Berlin to collect an entry visa.

A*C*G	Age 32	C*C*G
ME 02 LIB 43		UR 05 CAP 37
PL 04 VIR 41		MO 07 VIR 26
UR 07 LEO 13		JU 07 GEM 44
SU 08 VIR 31		SU 07 LIB 56
JU 12 VIR 45		NE 13 CAP 17
MO 19 GEM 19		SA 15 CAP 00
VE 22 CAN 11		PL 16 SCO 16
MA 24 PIS 26		MA 16 PIS 43
SA 25 SCO 04		VE 27 LEO 25
NE 26 LIB 57		ME 28 VIR 15

1. C*C*G JU MH = Dresden

2. Prog. Dresden = C*C*G UR IC

3. C*C*G JU EQU A*C*G SU

4. C*C*G MO = A*C*G SU (app)

A*C*G	(29 25)	Displaced C*C*G
ME 02 LIB 43		UR 06 SAG 12
PL 04 VIR 41		MO 08 CAN 01
UR 07 LEO 13		JU 08 TAU 19
SU 08 VIR 31		NE 13 SAG 52
JU 12 VIR 45		SA 15 SAG 35
MO 19 GEM 19		PL 16 LIB 51
VE 22 CAN 11		MA 17 AQU 18
MA 24 PIS 26		VE 28 CAN 00
SA 25 SCO 04		ME 28 LEO 50
NE 26 LIB 57		

A*C*G	(29 25)	Solar arc A*C*G
ME 02 LIB 43		ME 02 SCO 08
PL 04 VIR 41		PL 04 LIB 06
UR 07 LEO 13		UR 06 VIR 38
SU 08 VIR 31		JU 12 LIB 10
JU 12 VIR 45		MO 18 CAN 44
MO 19 GEM 19		VE 21 LEO 36
VE 22 CAN 11		MA 23 ARI 51
MA 24 PIS 26		SA 24 SAG 29

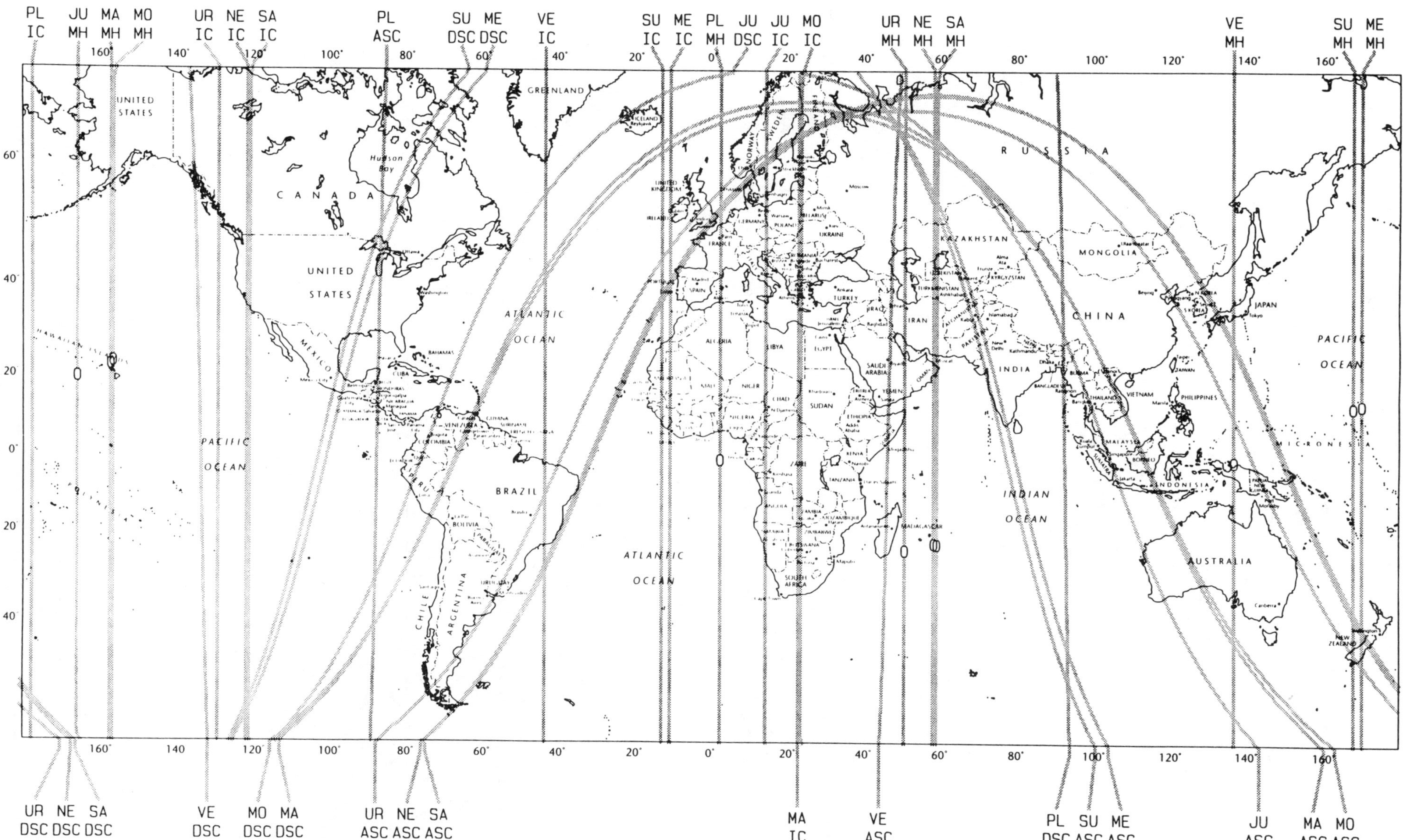

CYCLO*CARTO*GRAPHY

04/17/96

SU, MO, ME, VE AND MA ARE PROGRESSED
JU, SA, UR, NE AND PL ARE TRANSITING

David Meadows

PROG	4 25 55	2:48 GMT
TRANSIT	3 18 89	0:00 GMT
NATAL	3 22 55	3:00 GMT

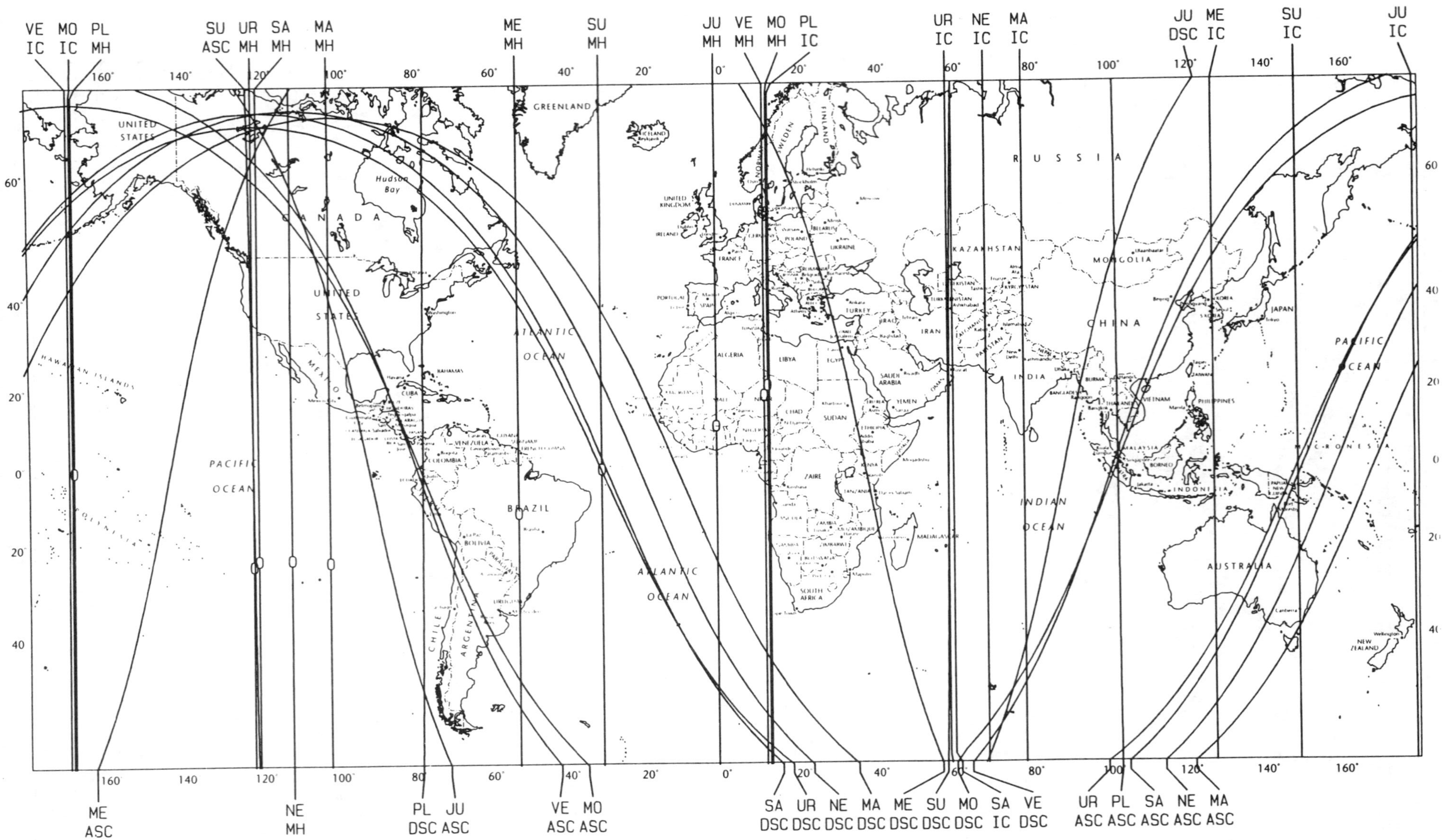

David Meadows SR

March 21, 1988 2: 11 PM GMT

04/16/96

David Meadows SR
March 21, 1988 2:11 PM GMT

ASTRO*CARTO*GRAPHY LATITUDE CROSSINGS

NORTH LATITUDES

69N45 NE/MH * ME/ASC
68N36 PL/MH * MO/ASC
68N36 VE/IC * MO/DSC
68N26 SU/ASC * JU/ASC
68N03 JU/IC * MO/DSC
67N53 ME/ASC * NE/ASC
67N45 UR/MH * JU/ASC
67N43 SA/DSC * NE/DSC
67N42 JU/ASC * SA/ASC
67N42 UR/MH * SA/ASC
67N41 SU/ASC * SA/ASC
67N41 SA/MH * NE/ASC
67N38 ME/ASC * SA/DSC
67N37 JU/ASC * NE/ASC
67N37 MA/MH * NE/DSC
67N37 UR/MH * NE/ASC
67N32 SU/ASC * NE/ASC
67N29 NE/MH * SA/DSC
67N10 MA/ASC * SA/DSC
67N09 MA/DSC * NE/DSC
67N07 VE/ASC * SA/ASC
66N57 NE/MH * MA/ASC
66N54 SA/MH * JU/ASC
66N39 MA/MH * SA/DSC
66N35 VE/ASC * NE/ASC
66N33 ME/ASC * MA/ASC
66N29 VE/MH * ME/DSC
66N22 SU/ASC * UR/ASC
66N22 SA/MH * UR/DSC
66N21 JU/ASC * UR/DSC
66N21 MA/ASC * UR/DSC
66N19 ME/ASC * UR/DSC
66N17 MA/ASC * JU/ASC
66N16 MO/ASC * SA/ASC
66N10 SA/MH * MA/ASC
66N07 VE/ASC * UR/ASC
66N04 NE/MH * UR/DSC
66N00 UR/MH * MA/ASC
65N52 SU/ASC * MA/ASC
65N45 MO/MH * ME/DSC
65N37 MO/ASC * NE/ASC
65N31 MO/ASC * UR/ASC
65N28 ME/ASC * JU/ASC
65N25 PL/IC * ME/DSC
65N14 MA/DSC * PL/DSC
65N13 UR/ASC * NE/ASC
65N06 VE/ASC * MA/ASC
65N05 MA/MH * UR/DSC
64N58 SU/IC * VE/DSC
64N07 MO/ASC * MA/ASC
63N49 NE/DSC * PL/DSC
63N41 SA/MH * ME/ASC
63N15 SU/ASC * VE/ASC
62N52 UR/MH * VE/ASC
62N42 ME/ASC * VE/ASC
62N21 UR/MH * ME/ASC
62N15 SA/MH * VE/ASC
61N43 SU/IC * MO/DSC
61N28 SU/ASC * ME/ASC
61N13 Anchorage
60N51 MO/ASC * ME/ASC
60N48 SA/DSC * PL/DSC
60N37 MO/ASC * SU/ASC
60N15 UR/MH * MO/ASC

59N37 SA/MH * MO/ASC
59N20 Stockholm
59N16 NE/MH * JU/ASC
58N42 PL/MH * SA/ASC
58N39 UR/DSC * PL/DSC
58N29 MO/IC * SA/ASC
58N01 VE/IC * SA/ASC
57N40 PL/MH * UR/ASC
57N29 ME/MH * MA/DSC

CENTRAL LATITUDES

57N28 MO/IC * UR/ASC
57N17 NE/MH * VE/ASC
57N00 VE/IC * UR/ASC
55N45 Moscow
55N41 Copenhagen
54N43 VE/ASC * JU/ASC
54N36 NE/MH * MO/ASC
53N57 PL/MH * NE/ASC
53N38 MO/IC * NE/ASC
52N57 VE/IC * NE/ASC
52N30 Berlin
52N01 ME/MH * NE/DSC
51N55 ME/IC * VE/DSC
51N30 London

49N22 JU/IC * SA/ASC
49N19 MO/ASC * JU/ASC
49N14 Vancouver
49N01 JU/DSC * PL/ASC
48N50 Paris
48N43 JU/IC * UR/ASC
48N12 MA/MH * VE/ASC
47N36 Seattle
47N23 ME/IC * MO/DSC
45N41 MA/MH * MO/ASC
45N32 Montreal
44N59 Minneapolis
44N14 PL/MH * MA/ASC
43N45 MO/IC * MA/ASC
43N39 Toronto
43N25 ME/MH * SA/DSC
42N40 VE/IC * MA/ASC
42N22 Boston
42N20 Detroit
41N54 Rome
41N53 Chicago
41N08 MA/MH * JU/ASC
40N43 New York
40N10 JU/IC * NE/ASC

39N57 ME/MH * UR/DSC
39N44 Denver
38N53 Washington
38N39 SU/MH * MA/DSC
38N38 SA/ASC * UR/ASC
38N37 St. Louis
37N47 San Francisco
36N10 Las Vegas
35N40 Tokyo
34N04 Los Angeles
33N45 Atlanta
33N27 Phoenix
32N47 Dallas
30N03 Cairo

29N58 New Orleans
29N46 Houston
28N39 Delhi
25N47 Miami
22N59 JU/IC * MA/ASC
22N45 SU/MH * NE/DSC
22N34 Calcutta
22N16 Hong Kong
21N19 Honolulu

19N26 Mexico City
19N12 MO/ASC * VE/ASC
14N35 Manila
10N30 Caracas

6N27 Lagos
4N20 UR/MH * SU/ASC
3N54 VE/DSC * PL/ASC
3N11 SU/MH * SA/DSC
1N05 MO/DSC * PL/ASC

SOUTH LATITUDES

0N05 SU/IC * UR/ASC

0S05 SU/MH * UR/DSC
1S05 MO/ASC * PL/DSC
3S11 SU/IC * SA/ASC
3S54 VE/ASC * PL/DSC
4S20 UR/IC * SU/DSC

19S12 MO/DSC * VE/DSC

22S45 SU/IC * NE/ASC
22S54 Rio De Janeiro
22S59 JU/MH * MA/DSC
26S12 Johannesburg

33S55 Sydney
34S36 Buenos Aires
36S52 Auckland
37S50 Melbourne
38S38 SA/DSC * UR/DSC
38S39 SU/IC * MA/ASC
39S57 ME/IC * UR/ASC

40S10 JU/MH * NE/DSC
41S08 MA/IC * JU/DSC
42S40 VE/MH * MA/DSC
43S25 ME/IC * SA/ASC
43S45 MO/MH * MA/DSC
44S14 PL/IC * MA/DSC
45S41 MA/IC * MO/DSC
47S23 ME/MH * MO/ASC
48S12 MA/IC * VE/DSC
48S43 JU/MH * UR/DSC
49S01 JU/ASC * PL/DSC
49S19 MO/DSC * JU/DSC
49S22 JU/MH * SA/DSC

51S55 ME/MH * VE/ASC
52S01 ME/IC * NE/ASC
52S57 VE/MH * NE/DSC
53S38 MO/MH * NE/DSC
53S57 PL/IC * NE/DSC
54S36 NE/IC * MO/DSC
54S43 VE/DSC * JU/DSC
57S00 VE/MH * UR/DSC
57S17 NE/IC * VE/DSC
57S28 MO/MH * UR/DSC
57S29 ME/IC * MA/ASC
57S40 PL/IC * UR/DSC
58S01 VE/MH * SA/DSC
58S29 MO/MH * SA/DSC
58S39 UR/ASC * PL/ASC
58S42 PL/IC * SA/DSC
59S16 NE/IC * JU/DSC
59S37 SA/IC * MO/DSC

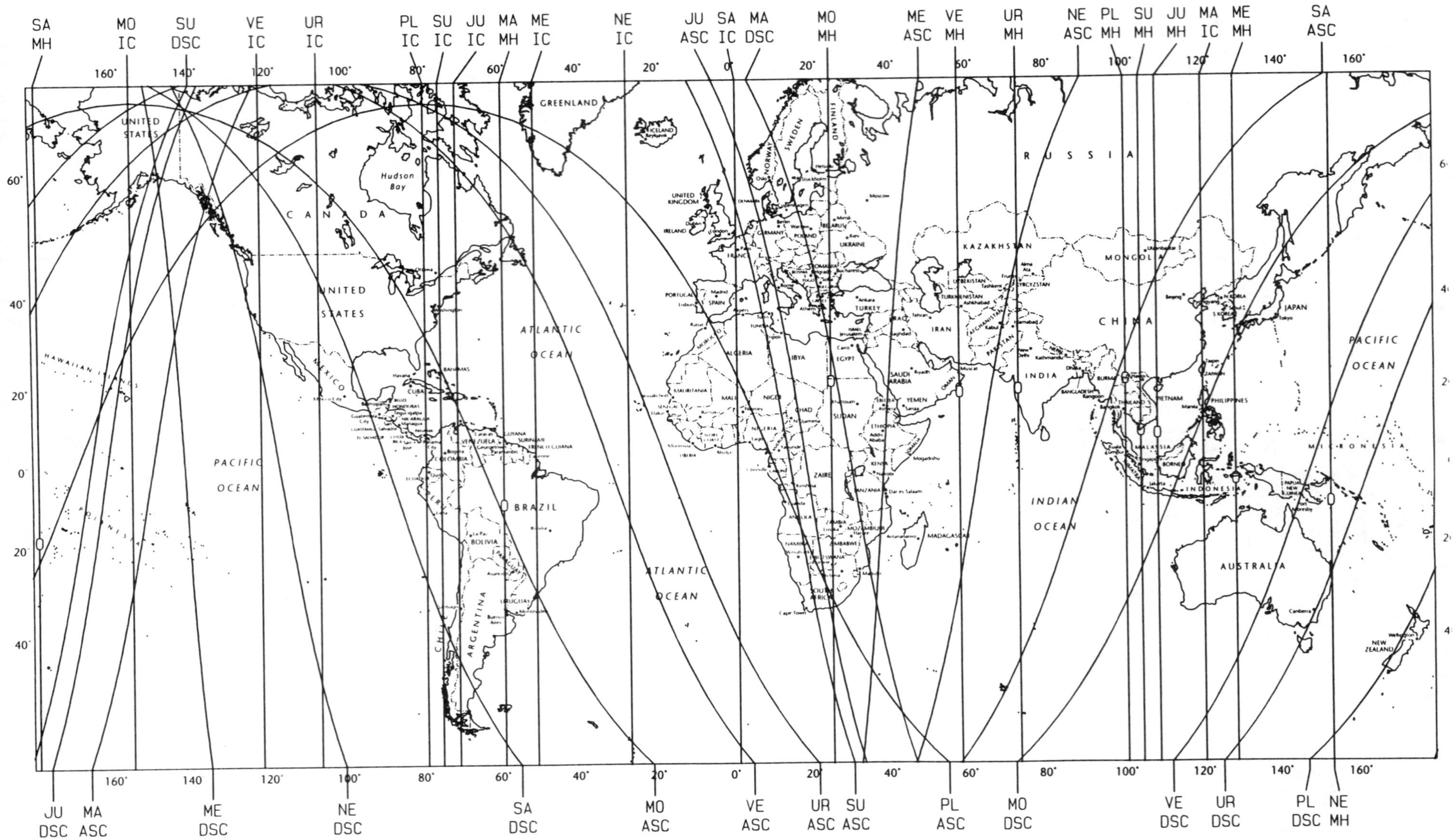

Martina Meadows

August 30, 1956 6:00 AM CET

05/21/97

05/21/97

Martina Meadows
August 30, 1956 6:00 AM CET

ASTRO*CARTO*GRAPHY LATITUDE CROSSINGS

NORTH LATITUDES

69N47	VE/DSC * NE/DSC
69N44	VE/IC * UR/DSC
69N32	MA/ASC * UR/DSC
69N22	VE/DSC * SA/DSC
68N42	ME/DSC * VE/DSC
68N32	SU/DSC * NE/DSC
68N22	MO/ASC * VE/DSC
68N19	MO/ASC * ME/DSC
68N19	SU/DSC * SA/DSC
68N08	UR/ASC * PL/DSC
68N06	PL/IC * UR/ASC
68N05	SU/IC * PL/ASC
68N05	SA/DSC * NE/DSC
67N56	JU/IC * PL/ASC
67N56	MO/ASC * SU/DSC
67N51	VE/ASC * PL/DSC
67N50	MO/IC * VE/DSC
67N49	JU/DSC * SA/DSC
67N47	MO/ASC * NE/DSC
67N44	MO/ASC * JU/DSC
67N41	MO/ASC * SA/DSC
67N33	JU/DSC * NE/DSC
67N28	MO/ASC * UR/DSC
67N20	SU/IC * UR/ASC
67N18	SA/DSC * UR/DSC
66N59	UR/DSC * NE/DSC
66N54	MA/MH * PL/ASC
66N34	JU/DSC * UR/DSC
66N30	SA/MH * MO/DSC
66N20	JU/IC * UR/ASC
66N06	MO/ASC * MA/ASC
65N40	UR/IC * PL/DSC
65N35	SU/DSC * UR/DSC
65N33	ME/IC * PL/ASC
65N06	PL/IC * VE/ASC
64N44	VE/IC * MO/ASC
64N23	ME/DSC * UR/DSC
64N12	MA/ASC * SA/DSC
64N08	SA/IC * JU/ASC
63N38	SU/IC * VE/ASC
63N15	MO/ASC * PL/DSC
62N31	ME/MH * SA/ASC
62N17	MA/MH * UR/ASC
62N06	SU/DSC * ME/DSC
61N59	MO/IC * UR/DSC
61N41	JU/IC * VE/ASC
61N26	VE/IC * PL/DSC
61N13	Anchorage
60N46	SA/DSC * PL/DSC
59N50	MA/ASC * NE/DSC
59N41	VE/IC * SA/DSC
59N25	UR/IC * MO/ASC
59N20	Stockholm
58N48	ME/DSC * JU/DSC
57N48	ME/IC * UR/ASC
57N47	SA/MH * VE/DSC
57N04	MA/IC * SA/ASC
57N01	NE/MH * MO/DSC
56N47	NE/DSC * PL/DSC
56N44	NE/IC * PL/ASC
55N45	Moscow
55N41	Copenhagen
55N41	SA/IC * SU/ASC
53N41	MA/MH * VE/ASC
52N59	MA/ASC * PL/DSC
52N30	Berlin
51N30	London
49N43	MO/IC * SU/DSC
49N14	Vancouver
48N50	Paris
47N36	Seattle
47N26	UR/MH * NE/ASC

CENTRAL LATITUDES

45N48	ME/DSC * PL/DSC
45N32	Montreal
44N59	Minneapolis
44N14	ME/IC * VE/ASC
44N00	UR/IC * SA/DSC
43N43	JU/MH * SA/ASC
43N39	Toronto
42N22	Boston
42N20	Detroit
41N54	Rome
41N53	Chicago
40N43	New York
40N42	SA/MH * UR/DSC
39N44	Denver
38N53	Washington
38N37	St. Louis
37N51	MO/IC * JU/DSC
37N47	San Francisco
36N13	SU/MH * SA/ASC
36N10	Las Vegas
35N40	Tokyo
34N04	Los Angeles
33N45	Atlanta
33N27	Phoenix
32N48	ME/DSC * MA/ASC
32N47	Dallas
32N33	PL/IC * MO/ASC
32N13	MO/IC * PL/DSC
30N19	ME/MH * MO/DSC
30N03	Cairo
29N58	New Orleans
29N46	Houston
28N52	MO/MH * MA/DSC
28N44	JU/DSC * PL/DSC
28N39	Delhi
27N42	PL/MH * SA/ASC
26N54	VE/IC * NE/DSC
26N33	NE/IC * UR/ASC
25N47	Miami
25N18	SU/IC * MO/ASC
22N34	Calcutta
22N33	SA/IC * PL/ASC
22N16	Hong Kong
21N19	Honolulu
19N26	Mexico City
16N06	JU/IC * MO/ASC
15N27	SU/DSC * PL/DSC
14N35	Manila
13N38	NE/MH * VE/DSC
12N41	MA/IC * MO/DSC
10N30	Caracas
6N27	Lagos
12S41	MA/MH * MO/ASC
13S38	NE/IC * VE/ASC
15S27	SU/ASC * PL/ASC
16S06	JU/MH * MO/DSC
22S33	SA/MH * PL/DSC
22S54	Rio De Janeiro
25S18	SU/MH * MO/DSC
26S12	Johannesburg
26S33	NE/MH * UR/DSC
26S54	VE/MH * NE/ASC
27S42	PL/IC * SA/DSC
28S44	JU/ASC * PL/ASC
28S52	MO/IC * MA/ASC
30S19	ME/IC * MO/ASC
32S13	MO/MH * PL/ASC
32S33	PL/MH * MO/DSC

SOUTH LATITUDES

32S48	ME/ASC * MA/DSC
33S55	Sydney
34S36	Buenos Aires
36S13	SU/IC * SA/DSC
36S52	Auckland
37S50	Melbourne
37S51	MO/MH * JU/ASC
40S42	SA/IC * UR/ASC
43S43	JU/IC * SA/DSC
44S00	UR/MH * SA/ASC
44S14	ME/MH * VE/DSC
45S48	ME/ASC * PL/ASC
47S26	UR/IC * NE/DSC
49S43	MO/MH * SU/ASC
52S59	MA/DSC * PL/ASC
53S41	MA/IC * VE/DSC
55S41	SA/MH * SU/DSC
56S44	NE/MH * PL/DSC
56S47	NE/ASC * PL/ASC
57S01	NE/IC * MO/ASC
57S04	MA/MH * SA/DSC
57S47	SA/IC * VE/ASC
57S48	ME/MH * UR/DSC
58S48	ME/ASC * JU/ASC
59S25	UR/MH * MO/DSC
59S41	VE/MH * SA/ASC
59S50	MA/DSC * NE/ASC

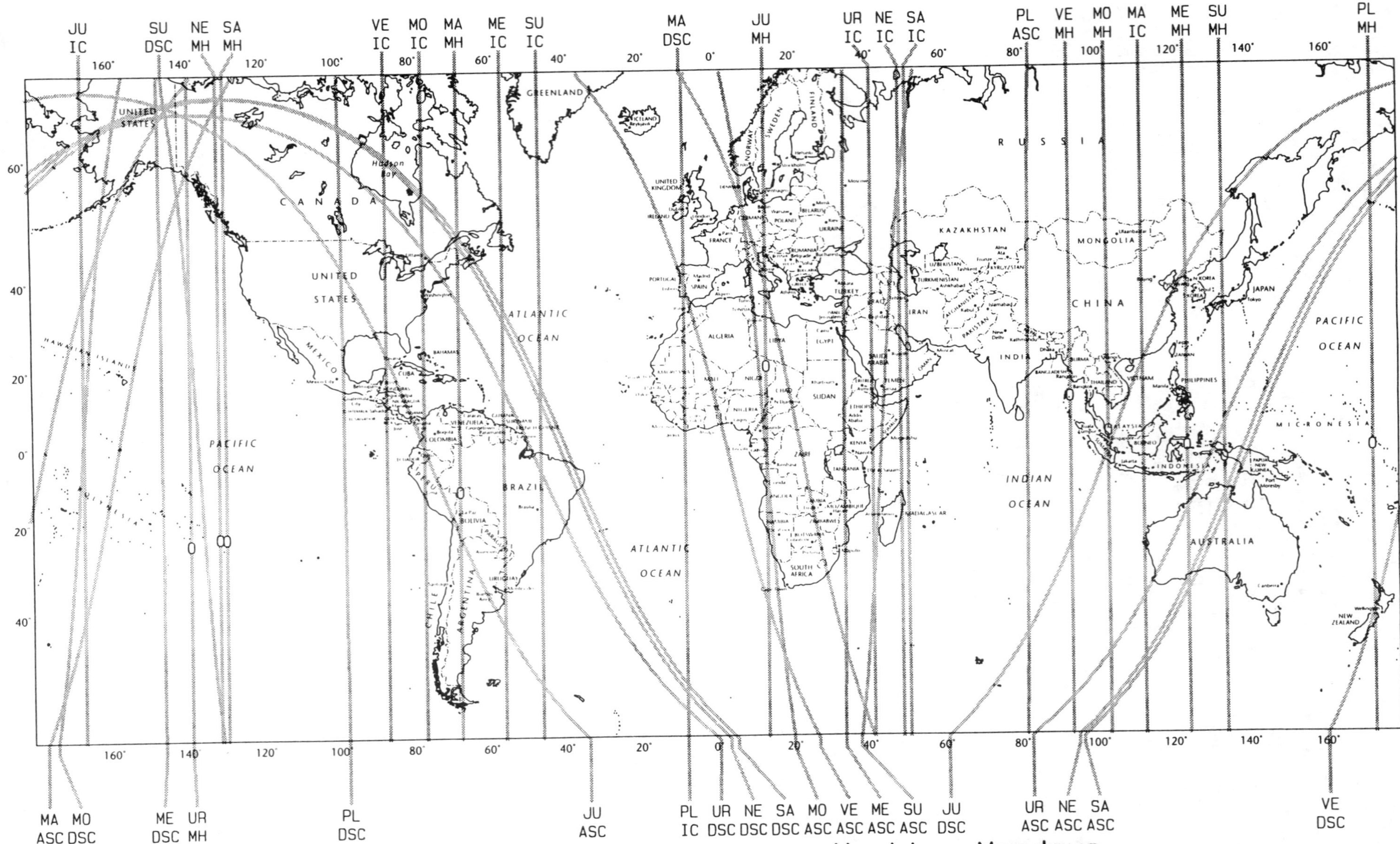

CYCLO*CARTO*GRAPHY

04/17/96

SU, MO, ME, VE AND MA ARE PROGRESSED
JU, SA, UR, NE AND PL ARE TRANSITING

Martina Meadows

PROG	10	1	56	20:59 GMT
TRANSIT	4	30	89	0:00 GMT
NATAL	8	30	56	5:00 GMT

SA 25 SCO 04	NE 26 SCO 22
NE 26 LIB 57	

5. solar arc SA EQU A*C*G MA

My wife left East Germany on April 29, 1989 when her C*C*G JU MH was exactly at Dresden. Another indication for breaking away from East Germany was prog. Dresden = C*C*G UR IC.

The new beginning in her life was shown by the applying C*C*G MO = A*C*G SU. C*C*G JU EQU A*C*G SU, indicating the optimism she had about starting a new life in the west.

The solar return map shows that PL DSC was within orb of both Dresden and Leicester. This is a clear indication that her life would be transformed by moving from Dresden to Leicester.

My wife almost didn't make it to England as she didn't have a transit visa to travel through Holland. Near the West German/Dutch border she nearly had to leave the train. The West German border official told her it would be best for her to leave the train before it reached Holland; then she could live in West Germany as a refugee.

The Dutch border official said a person married to a foreigner has dual nationality. When my wife told him I was waiting at Hook of Holland he decided to give her a transit visa free of charge.

The solar return PL ASC passed exactly through the West German/Dutch border where the journey nearly ended. The East German authorities were more concerned that my wife had an entry visa for Britain. They didn't mention anything about a transit visa for Holland. Maybe they assumed she was travelling to Britain by plane.

Example 7—Martina in Leicester

Before my wife left Dresden I told her she would probably find it difficult to settle in England. She has SA IC and SU ASC near Leicester on her Astro*Carto*Graphy map, so I knew she wouldn't find it easy living there. I was prepared to help her get through this difficult adjustment.

Martina said that anywhere had to be better than living in East Germany, and she would have more freedom living in England.

Solar arc SA EQU A*C*G MA also indicated that she would experience some difficulties living in England.

Martina had only been in England for a few days when she realized it would take a long time to settle in. Living in England was much different from what she had expected.

When she received the first letter from her parents she broke down in tears. She didn't know when she would see her family in Dresden again as she was not allowed to visit East Germany. When the tears started happening every day, she became a big worry for me.

On her solar return map Martina had PL ASC and MO DSC within orb of Leicester. This explains the traumatic time she went through emotionally.

Example 8—Martina Moves to Munich

In January 1990, Martina started a job with a German company in Leicester, which made her happy. This was a good opportunity for her to start to settle down in England. I was unemployed and found the job search difficult. We were struggling financially even though Martina had a job, and kept hoping our situation would improve as we were slowly getting into more debt. Martina still wasn't completely settled in England as she didn't feel secure in life. She also felt lonely as she found it difficult to get along with people.

On 11th April 11, 1991, something happened that Martina never recovered from. Just before she left work the boss told her he had some bad news. The company had to lay off eight employees and her name was on the list. The layoff was immediate, so she returned home without a job. Martina was shocked that she had lost her job so unexpectedly and within the space of five minutes. Something like this had never happened to her before. Because she had not worked in England for three years she was unable to claim unemployment benefit. This depressed her and she couldn't see how we could manage financially.

It was obvious that Martina would never be happy living in England. Moving away from the powerful influence of SA IC would be the best thing for her. With JU ASC passing through Germany I knew that Martina would be much more happier there. I realized that such a move would be a big change for me, but I couldn't see another choice.

We decided to move to Munich for two reasons. First, Astro*Carto*Graphy showed Munich to be a good place for both of us. Martina has JU ASC near Munich with SU ASC also within orb. At Munich I have UR DSC and JU DSC. Martina's sister lived in Munich, so she would be able to help us. We decided that Martina would move and stay with her sister. Once she had her own accommodations, I would join her.

The JU ASC line had an immediate effect when Martina arrived in Munich. Within a few days of arriving she had a new job, and shortly after she had a second job. The biggest problem was in finding accommodations in Munich, but Martina advertised in the local newspaper and easily found something. Even she couldn't believe she was so successful in a short space of time. She had become used to the SA IC influence in England and thought it would take a long time to rebuild her self-confidence.

A*C*G	**Age 34**	**C*C*G**
ME 02 LIB 43		MO 05 LIB 06
PL 04 VIR 41		SA 09 AQU 19
UR 07 LEO 13		SU 09 LIB 48
SU 08 VIR 31		JU 09 LEO 58
JU 12 VIR 45		UR 14 CAP 34
MO 19 GEM 19		MA 16 PIS 27
VE 22 CAN 11		NE 17 CAP 47
MA 24 PIS 26		PL 20 SCO 28
SA 25 SCO 04		ME 27 VIR 38
NE 26 LIB 57		VE 29 LEO 42

1. C*C*G MA DSC = Munich

A*C*G	**(31 17)**	**Displaced C*C*G**
ME 02 LIB 43		MO 03 VIR 49
PL 04 VIR 41		SA 08 CAP 02

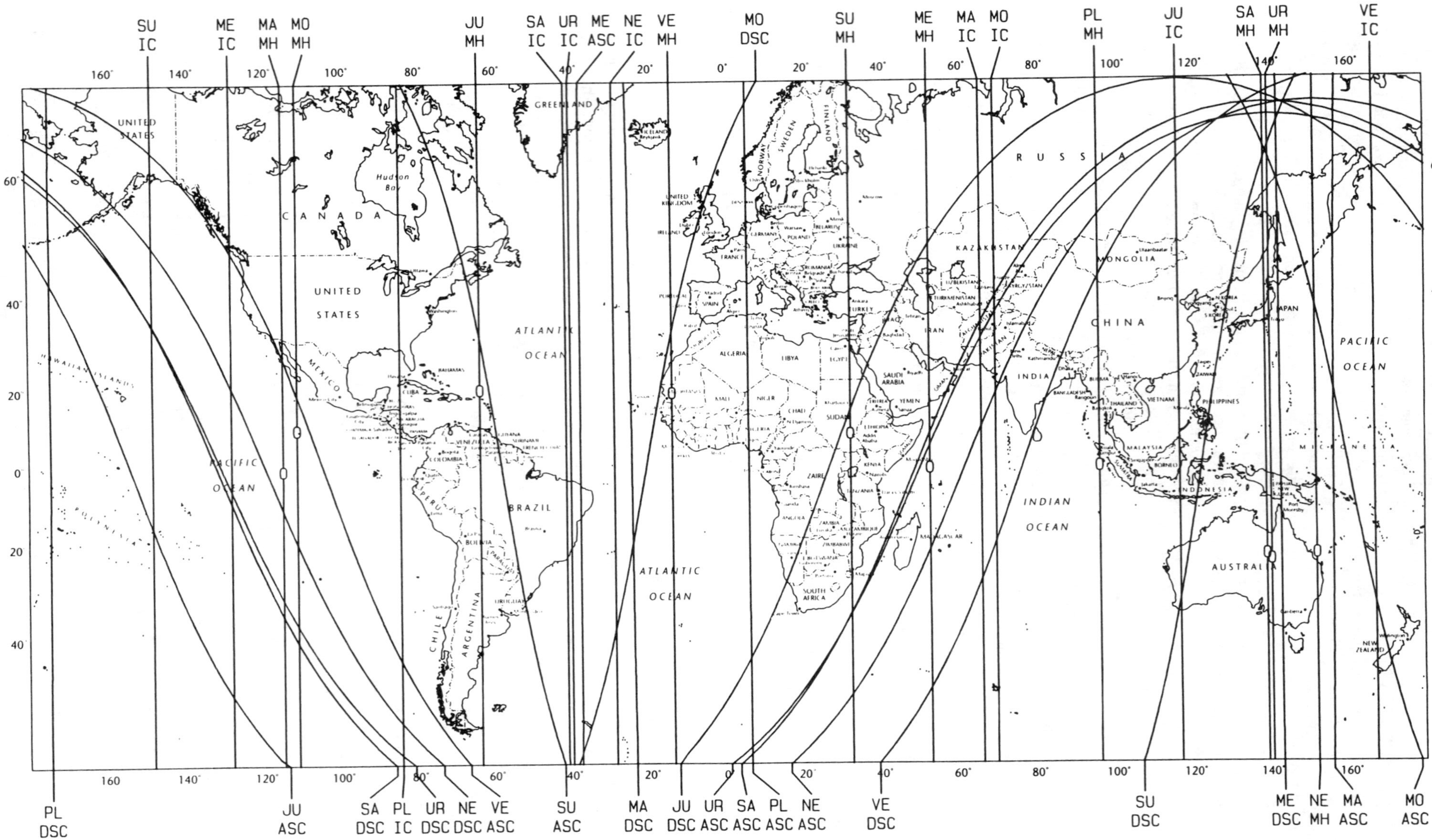

Martina Meadows SR

August 30, 1988 9:50 AM GMT

04/16/96

04/16/96

Martina Meadows SR
August 30, 1988 9:50 AM GMT

ASTRO*CARTO*GRAPHY LATITUDE CROSSINGS

NORTH LATITUDES

69N41 SU/DSC * VE/DSC
69N35 VE/ASC * PL/DSC
69N11 MO/ASC * JU/ASC
69N04 ME/DSC * VE/DSC
68N42 UR/MH * VE/DSC
68N33 PL/MH * JU/DSC
68N30 SA/MH * VE/DSC
68N24 VE/DSC * JU/ASC
68N20 SA/MH * JU/ASC
68N10 UR/MH * JU/ASC
68N02 MO/DSC * PL/ASC
67N47 ME/DSC * JU/ASC
67N39 MA/ASC * NE/DSC
67N35 UR/MH * SA/DSC
67N35 SU/DSC * NE/ASC
67N33 ME/DSC * SA/DSC
67N33 JU/ASC * NE/ASC
67N32 SA/DSC * NE/ASC
67N31 VE/DSC * SA/ASC
67N31 MO/ASC * SA/ASC
67N31 JU/ASC * SA/DSC
67N30 SU/DSC * SA/DSC
67N30 MO/ASC * VE/DSC
67N29 ME/DSC * NE/ASC
67N28 SU/DSC * JU/ASC
67N20 UR/MH * NE/ASC
67N15 SA/MH * NE/ASC
67N06 NE/MH * SA/DSC
67N03 VE/IC * NE/DSC
66N51 MO/ASC * NE/ASC
66N35 MA/ASC * SA/DSC
66N22 SA/MH * UR/ASC
66N21 ME/DSC * UR/DSC
66N21 SU/DSC * UR/DSC
66N18 MO/ASC * UR/ASC
65N59 JU/IC * SA/ASC
65N59 JU/ASC * UR/DSC
65N57 NE/MH * UR/DSC
65N55 VE/DSC * NE/ASC
65N53 NE/MH * JU/ASC
65N52 VE/DSC * UR/ASC
65N51 UR/ASC * NE/ASC
65N43 SU/DSC * ME/DSC
65N28 MA/ASC * UR/DSC
65N06 VE/IC * SA/DSC
64N31 JU/IC * UR/ASC
64N22 MA/ASC * JU/ASC
64N01 VE/IC * UR/DSC
63N31 NE/DSC * PL/DSC
63N30 JU/IC * NE/ASC
63N17 SU/IC * VE/ASC
63N03 UR/MH * SU/DSC
63N01 SA/MH * MO/ASC
62N20 MO/ASC * SU/DSC
61N41 UR/MH * MO/ASC
61N38 JU/IC * VE/DSC
61N36 MO/IC * JU/DSC
61N28 SA/MH * SU/DSC
61N13 Anchorage
60N38 PL/MH * SA/ASC
60N25 VE/IC * JU/ASC

59N48 MA/IC * JU/DSC
59N20 Stockholm
58N43 SA/DSC * PL/DSC
58N40 PL/MH * UR/ASC
58N32 MO/ASC * ME/DSC
57N46 UR/DSC * PL/DSC
55N45 Moscow
55N41 Copenhagen
54N12 PL/MH * NE/ASC
52N30 Berlin
51N30 London

CENTRAL LATITUDES

49N59 SU/IC * NE/DSC
49N56 ME/MH * JU/DSC
49N14 Vancouver
48N58 ME/IC * VE/ASC
48N50 Paris
47N36 Seattle
45N32 Montreal
44N59 Minneapolis
44N00 PL/MH * VE/DSC
43N40 JU/ASC * PL/DSC
43N39 Toronto
42N22 Boston
42N20 Detroit
41N54 Rome
41N53 Chicago
40N43 New York
40N27 MO/IC * SA/ASC

39N44 Denver
38N53 Washington
38N37 St. Louis
38N02 NE/MH * MO/ASC
37N55 SA/DSC * UR/DSC
37N47 San Francisco
37N12 MO/IC * UR/ASC
36N10 Las Vegas
35N40 Tokyo
35N14 MA/IC * SA/ASC
35N01 SU/IC * UR/DSC
34N49 SU/IC * SA/DSC
34N04 Los Angeles
33N45 Atlanta
33N30 VE/MH * MO/DSC
33N27 Phoenix
32N47 Dallas
32N02 JU/MH * SU/ASC
31N48 MA/IC * UR/ASC
30N03 Cairo

29N58 New Orleans
29N46 Houston
28N39 Delhi
27N25 MA/MH * VE/ASC
25N47 Miami
22N34 Calcutta
22N16 Hong Kong
21N19 Honolulu
20N37 ME/IC * NE/DSC

19N26 Mexico City
18N57 MO/ASC * MA/ASC
18N53 MO/IC * NE/ASC
18N42 MO/MH * VE/ASC
14N51 SU/MH * JU/DSC
14N35 Manila
10N36 MA/IC * NE/ASC
10N30 Caracas

8N50 ME/MH * SA/ASC
6N27 Lagos
5N40 ME/MH * UR/ASC

5S40 ME/IC * UR/DSC
8S50 ME/IC * SA/DSC

10S36 MA/MH * NE/DSC
14S51 SU/IC * JU/ASC
18S42 MO/IC * VE/DSC
18S53 MO/MH * NE/DSC
18S57 MO/DSC * MA/DSC

20S37 ME/MH * NE/ASC
22S54 Rio De Janeiro
26S12 Johannesburg
27S25 MA/IC * VE/DSC

SOUTH LATITUDES

31S48 MA/MH * UR/DSC
32S02 JU/IC * SU/DSC
33S30 VE/IC * MO/ASC
33S55 Sydney
34S36 Buenos Aires
34S49 SU/MH * SA/ASC
35S01 SU/MH * UR/ASC
35S14 MA/MH * SA/DSC
36S52 Auckland
37S12 MO/MH * UR/DSC
37S50 Melbourne
37S55 SA/ASC * UR/ASC
38S02 NE/IC * MO/DSC

40S27 MO/MH * SA/DSC
43S40 JU/DSC * PL/ASC
44S00 PL/IC * VE/ASC
48S58 ME/MH * VE/DSC
49S56 ME/IC * JU/ASC
49S59 SU/MH * NE/ASC

54S12 PL/IC * NE/DSC
57S46 UR/ASC * PL/ASC
58S32 MO/DSC * ME/ASC
58S40 PL/IC * UR/DSC
58S43 SA/ASC * PL/ASC
59S48 MA/MH * JU/ASC

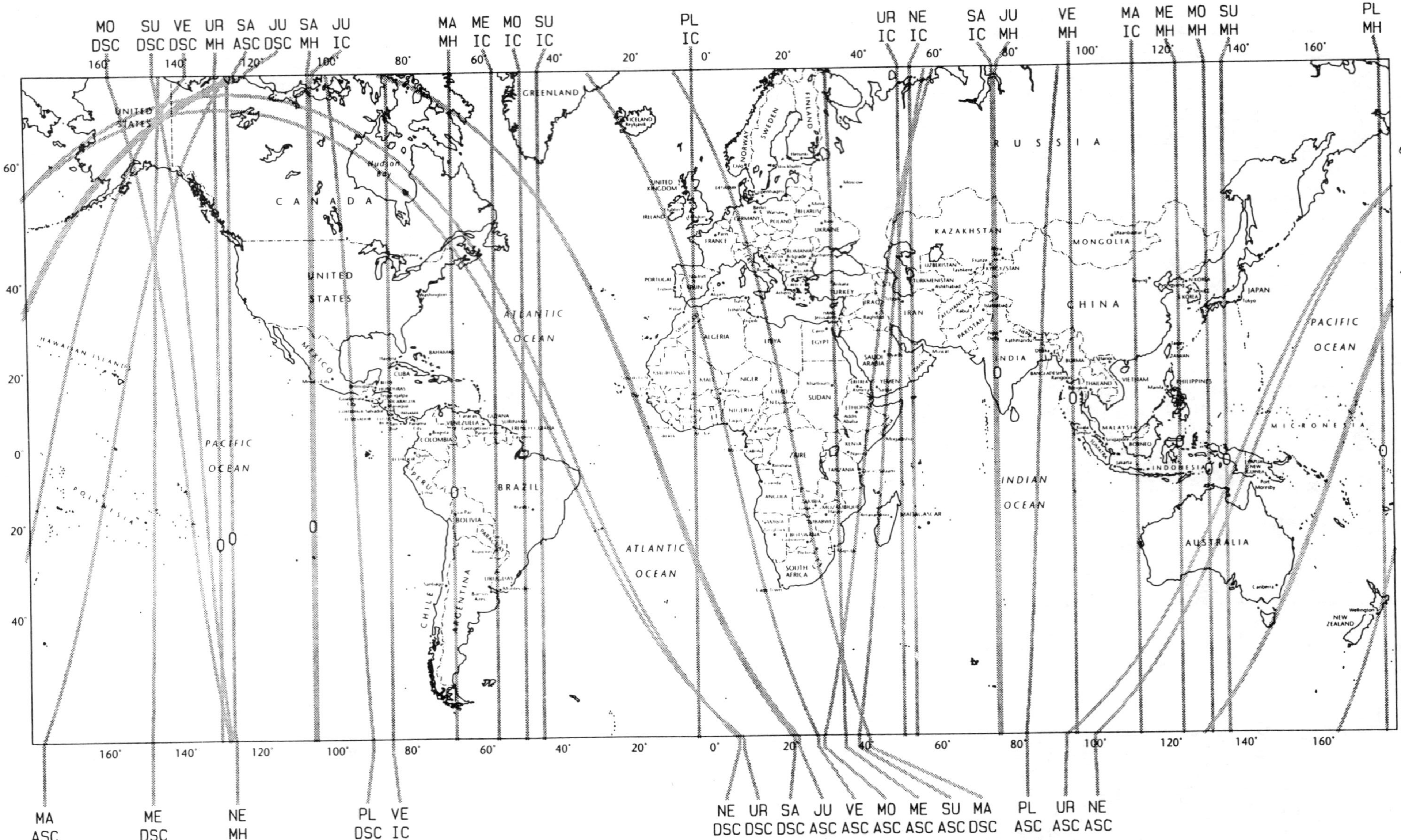

CYCLO*CARTO*GRAPHY

04/17/96

SU, MO, ME, VE AND MA ARE PROGRESSED
JU, SA, UR, NE AND PL ARE TRANSITING

Martina Meadows

PROG	10 3 56	22:20 GMT
TRANSIT	5 21 91	0:00 GMT
NATAL	8 30 56	5:00 GMT

UR 07 LEO 13
SU 08 VIR 31
JU 12 VIR 45
MO 19 GEM 19
VE 22 CAN 11
MA 24 PIS 26
SA 25 SCO 04
NE 26 LIB 57

JU 08 CAN 41
UR 13 SAG 17
MA 15 AQU 10
NE 16 SAG 30
PL 19 LIB 11
ME 26 LEO 21
VE 28 CAN 25

2. Disp. C*C*G MO = A*C*G PL

A*C*G	**(31 17)**	**Solar arc A*C*G**
ME 02 LIB 43		ME 04 SCO 00
PL 04 VIR 41		PL 05 LIB 58
UR 07 LEO 13		UR 08 VIR 30
SU 08 VIR 31		JU 14 VIR 02
JU 12 VIR 45		MO 20 CAN 36
MO 19 GEM 19		VE 23 LEO 28
VE 22 CAN 11		MA 25 ARI 43
MA 24 PIS 26		MA 26 SAG 21
SA 25 SCO 04		NE 28 SCO 14
NE 26 LIB 57		

3. solar arc UR = A*C*G SU

On the day Martina moved to Munich she had solar arc UR **exactly** on her A*C*G SU (one minute orb). This aspect shows that it was a good time for Martina to make changes in her life.

C*C*G MA DSC at Munich gave her the energy and determination to achieve what she aimed for. She pushed herself to the limit by taking on two jobs to save as much money as possible.

Disp. C*C*G MO = A*C*G PL also motivated her emotionally to attain the goal of finding her own accommodations. She knew that when she had achieved this I would move to Munich.

Solar return PL DSC and VE IC are within orb of Munich where finding a place to live transformed her life. Moving to Munich on her own, she missed me immensely.

Within orb of Leicester were solar return JU IC and MA ASC. These planetary influences would activate the A*C*G SA IC line that is always at Leicester, but Martina no longer had the willpower to overcome the struggle for survival in England.

This example shows how quickly life can change by moving to a new location. It is also important to consider the temporary influences at the time to make sure that the move will be successful.

Example 9—David Moves to Munich

After Martina had been in Munich just over five months she was able to move into her own accommodations. Half of this time was spent waiting for the rented accommodations to become available.

A*C*G	**Age 36**	**C*C*G**
SU 00 ARI 40		SA 03 AQU 08
PL 00 VIR 51		VE 04 ARI 03
MO 04 PIS 23		SU 04 TAU 21
ME 08 PIS 10		ME 09 TAU 27
MA 13 TAU 49		MA 09 GEM 36
SA 19 SCO 02		JU 11 VIR 22
JU 21 CAN 37		UR 11 CAP 33
VE 22 AQU 33		NE 15 CAP 29
UR 25 CAN 36		MO 19 CAN 26
NE 26 LIB 14		PL 21 SCO 14

A*C*G	**(33 41)**	**Displaced C*C*G**
SU 00 ARI 40		VE 00 PIS 22
PL 00 VIR 51		ME 05 ARI 46
MO 04 PIS 23		MA 05 TAU 55
ME 08 PIS 10		JU 07 LEO 41
MA 13 TAU 49		UR 07 SAG 52
SA 19 SCO 02		NE 11 SAG 48
JU 21 CAN 37		MO 15 GEM 45
VE 22 AQU 33		PL 17 LIB 33
UR 25 CAN 36		SA 29 SAG 27
NE 26 LIB 14		

1. Disp. C*C*G VE = A*C*G PL

A*C*G	**(33 41)**	**Solar arc A*C*G**
SU 00 ARI 40		PL 04 LIB 32
PL 00 VIR 51		MO 08 ARI 04
MO 04 PIS 23		ME 11 ARI 51
ME 08 PIS 10		MA 17 GEM 30
MA 13 TAU 49		SA 22 SAG 43
SA 19 SCO 02		JU 25 LEO 18
JU 21 CAN 37		VE 26 PIS 14
VE 22 AQU 33		UR 29 LEO 17
UR 25 CAN 36		NE 29 SCO 55
NE 26 LIB 14		

2. solar arc NE EQU A*C*G PL

I had two and a half months to prepare the move to Munich, during which time I was busy packing boxes. At the time I had C*C*G PL EQU A*C*G SA, which was the preparation for a move into the unknown.

On the day I moved to Munich I had disp. C*C*G VE = A*C*G PL, confirming that we could now make a new start in life together. After what we had been through our love for each other was stronger than ever.

Solar arc NE EQU A*C*G PL manifested as the uncertainty about how things would work out for me in Munich. As I couldn't speak German I didn't know what I would be able to do careerwise.

Solar return SA MH passes exactly through Leicester with JU IC and MO ASC not too far away. These planetary influences confirm how unsettled I felt in life, especially after Martina's job ended.

Solar return PL DSC was near Munich (orb about one degree) where I would make a new start in life. For the past few years I had noticed a solar return PL line passing through Germany, which must have been telling me something. Solar return MA ASC was also near Munich, which shows that I was determined to overcome any difficulties I experienced there. The combination of PL DSC and MA ASC indicated the challenge I faced to start a new life in Munich.

At Munich there is also an influence from MO ASC, so Martina and I have the same three planetary influences there, but shown in a different way.

Martina had C*C*G MA DSC at Munich and disp. C*C*G MO = A*C*G PL (anywhere aspect). I had PL

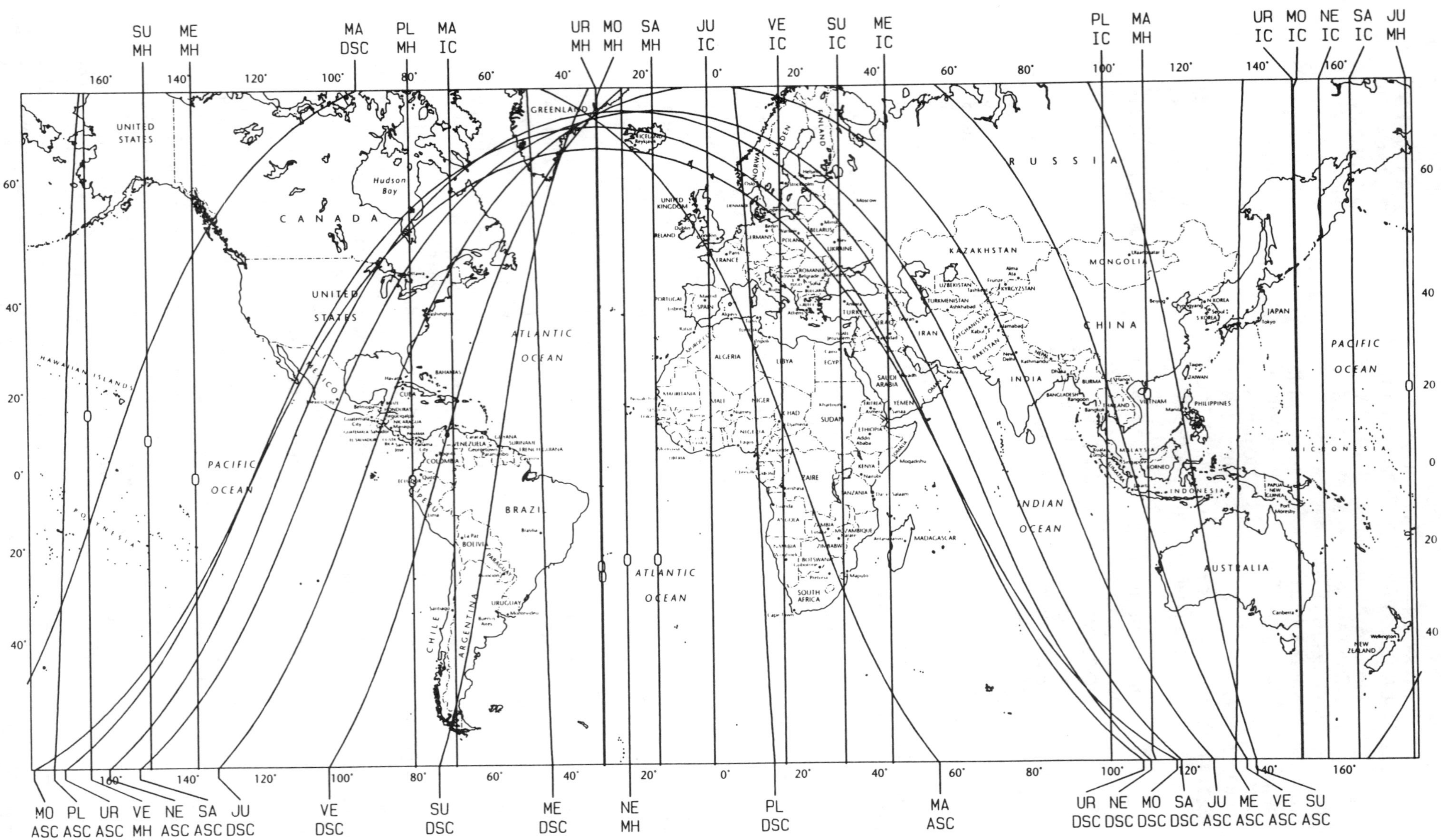

Martina Meadows SR

August 30, 1990 9:53 PM GMT

04/16/96

Martina Meadows SR
August 30, 1990 9:53 PM GMT

ASTRO*CARTO*GRAPHY LATITUDE CROSSINGS

NORTH LATITUDES

69N59 JU/ASC * PL/DSC
69N57 NE/MH * VE/DSC
69N32 SA/MH * JU/DSC
69N07 VE/IC * JU/ASC
68N41 NE/MH * JU/DSC
68N10 SU/DSC * MA/ASC
67N53 SA/ASC * NE/DSC
67N52 JU/DSC * NE/ASC
67N51 MO/MH * VE/DSC
67N49 MO/MH * NE/ASC
67N49 VE/DSC * NE/ASC
67N48 UR/MH * NE/ASC
67N46 UR/MH * VE/DSC
67N45 SA/MH * NE/DSC
67N45 NE/MH * SA/ASC
67N41 MA/ASC * NE/ASC
67N38 MO/MH * JU/DSC
67N35 UR/MH * JU/DSC
67N31 SU/DSC * NE/ASC
67N24 VE/DSC * JU/DSC
67N22 MA/ASC * JU/DSC
67N21 VE/DSC * MA/ASC
67N18 JU/IC * SA/DSC
67N18 MO/MH * SA/ASC
67N17 UR/MH * SA/ASC
67N13 MA/ASC * SA/ASC
67N11 PL/IC * SU/ASC
67N07 VE/DSC * SA/ASC
67N03 UR/MH * MA/ASC
66N59 MO/MH * MA/ASC
66N59 JU/DSC * SA/ASC
66N34 SU/DSC * SA/ASC
66N26 SU/IC * JU/ASC
66N23 JU/IC * NE/DSC
66N21 MO/MH * UR/DSC
66N21 SA/DSC * PL/DSC
66N20 MA/ASC * UR/DSC
66N19 VE/DSC * UR/ASC
66N14 JU/DSC * UR/ASC
66N13 SU/DSC * UR/ASC
66N13 NE/MH * UR/DSC
66N04 ME/DSC * NE/ASC
66N04 SA/ASC * UR/ASC
66N03 SU/DSC * JU/DSC
65N40 SA/MH * UR/DSC
65N22 ME/DSC * UR/ASC
64N51 NE/DSC * PL/DSC
64N50 NE/MH * MA/ASC
64N30 ME/DSC * SA/ASC
64N15 VE/IC * SA/DSC
63N59 UR/MH * MO/ASC
63N50 NE/MH * MO/DSC
63N43 MO/ASC * VE/DSC
63N41 MO/ASC * SU/DSC
63N41 MO/DSC * MA/ASC
63N28 JU/IC * UR/DSC
63N20 SU/DSC * VE/DSC
63N16 SA/MH * MO/DSC
63N08 UR/ASC * NE/ASC
63N04 MO/ASC * JU/DSC
62N55 MO/ASC * ME/DSC
62N34 ME/DSC * JU/DSC
62N28 ME/IC * JU/ASC
61N54 VE/IC * NE/DSC
61N13 Anchorage
61N12 SA/MH * MA/ASC
61N04 UR/DSC * PL/DSC
60N57 JU/IC * MO/DSC
60N50 MA/IC * UR/ASC
60N45 MO/ASC * SA/ASC
60N11 MA/IC * NE/ASC

59N20 Stockholm
58N40 SU/IC * SA/DSC

CENTRAL LATITUDES

58N18 MO/DSC * PL/DSC
58N02 MA/IC * MO/ASC
57N39 ME/DSC * VE/DSC
57N08 VE/IC * UR/DSC
56N18 PL/MH * UR/ASC
56N09 MA/IC * SA/ASC
55N45 Moscow
55N41 Copenhagen
54N21 VE/IC * MO/DSC
54N19 PL/MH * NE/ASC
54N02 SU/IC * NE/DSC
53N15 PL/MH * MO/ASC
52N30 Berlin
51N30 London
50N57 MA/MH * SU/ASC
50N42 ME/IC * SA/DSC
50N28 SU/DSC * ME/DSC

49N46 ME/MH * MA/DSC
49N33 JU/IC * MA/ASC
49N17 MO/ASC * NE/ASC
49N14 Vancouver
48N50 Paris
47N46 PL/MH * SA/ASC
47N36 Seattle
47N09 MA/IC * JU/DSC
46N17 SU/IC * UR/DSC
45N32 Montreal
44N59 Minneapolis
43N39 Toronto
43N23 SU/IC * MO/DSC
42N36 ME/IC * NE/DSC
42N22 Boston
42N20 Detroit
41N54 Rome
41N53 Chicago
40N43 New York

39N44 Denver
38N53 Washington
38N37 St. Louis
37N47 San Francisco
36N10 Las Vegas
35N40 Tokyo
34N04 Los Angeles
33N45 Atlanta
33N27 Phoenix
32N47 Dallas
31N39 MA/ASC * PL/DSC
30N52 ME/IC * UR/DSC
30N28 PL/MH * JU/DSC
30N03 Cairo

29N59 SU/MH * MA/DSC
29N58 New Orleans
29N46 Houston
28N39 Delhi
28N33 ME/IC * MO/DSC
25N47 Miami
22N34 Calcutta
22N16 Hong Kong
21N35 PL/IC * VE/ASC
21N19 Honolulu

19N26 Mexico City
15N22 MA/IC * VE/DSC
14N35 Manila
12N54 VE/IC * MA/ASC
10N30 Caracas

6N27 Lagos
4N26 MO/DSC * UR/DSC

4S26 MO/ASC * UR/ASC

SOUTH LATITUDES

12S54 VE/MH * MA/DSC
15S22 MA/MH * VE/ASC

21S35 PL/MH * VE/DSC
22S54 Rio De Janeiro
26S12 Johannesburg
28S33 ME/MH * MO/ASC
29S59 SU/IC * MA/ASC

30S28 PL/IC * JU/ASC
30S52 ME/MH * UR/ASC
31S39 MA/DSC * PL/ASC
33S55 Sydney
34S36 Buenos Aires
36S52 Auckland
37S50 Melbourne

42S36 ME/MH * NE/ASC
43S23 SU/MH * MO/ASC
46S17 SU/MH * UR/ASC
47S09 MA/MH * JU/ASC
47S46 PL/IC * SA/DSC
49S17 MO/DSC * NE/DSC
49S33 JU/MH * MA/DSC
49S46 ME/IC * MA/ASC

50S28 SU/ASC * ME/ASC
50S42 ME/MH * SA/ASC
50S57 MA/IC * SU/DSC
53S15 PL/IC * MO/DSC
54S02 SU/MH * NE/ASC
54S19 PL/IC * NE/DSC
54S21 VE/MH * MO/ASC
56S09 MA/MH * SA/DSC
56S18 PL/IC * UR/DSC
57S08 VE/MH * UR/ASC
57S39 ME/ASC * VE/ASC
58S02 MA/MH * MO/DSC
58S18 MO/ASC * PL/ASC
58S40 SU/MH * SA/ASC

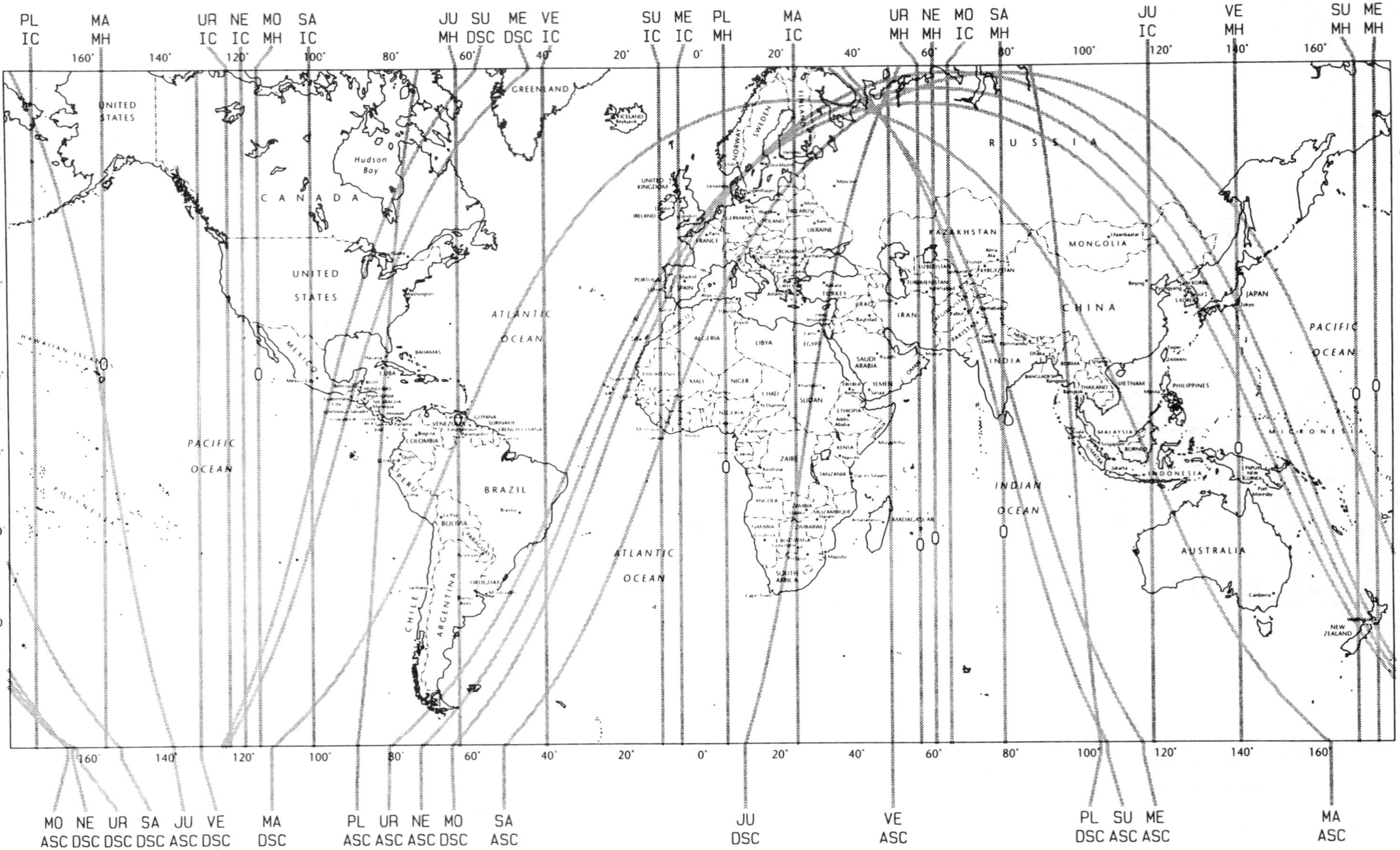

CYCLO*CARTO*GRAPHY

04/17/96

SU, MO, ME, VE AND MA ARE PROGRESSED
JU, SA, UR, NE AND PL ARE TRANSITING

David Meadows

PROG	4 27 55	17:45 GMT
TRANSIT	11 1 91	0:00 GMT
NATAL	3 22 55	3:00 GMT

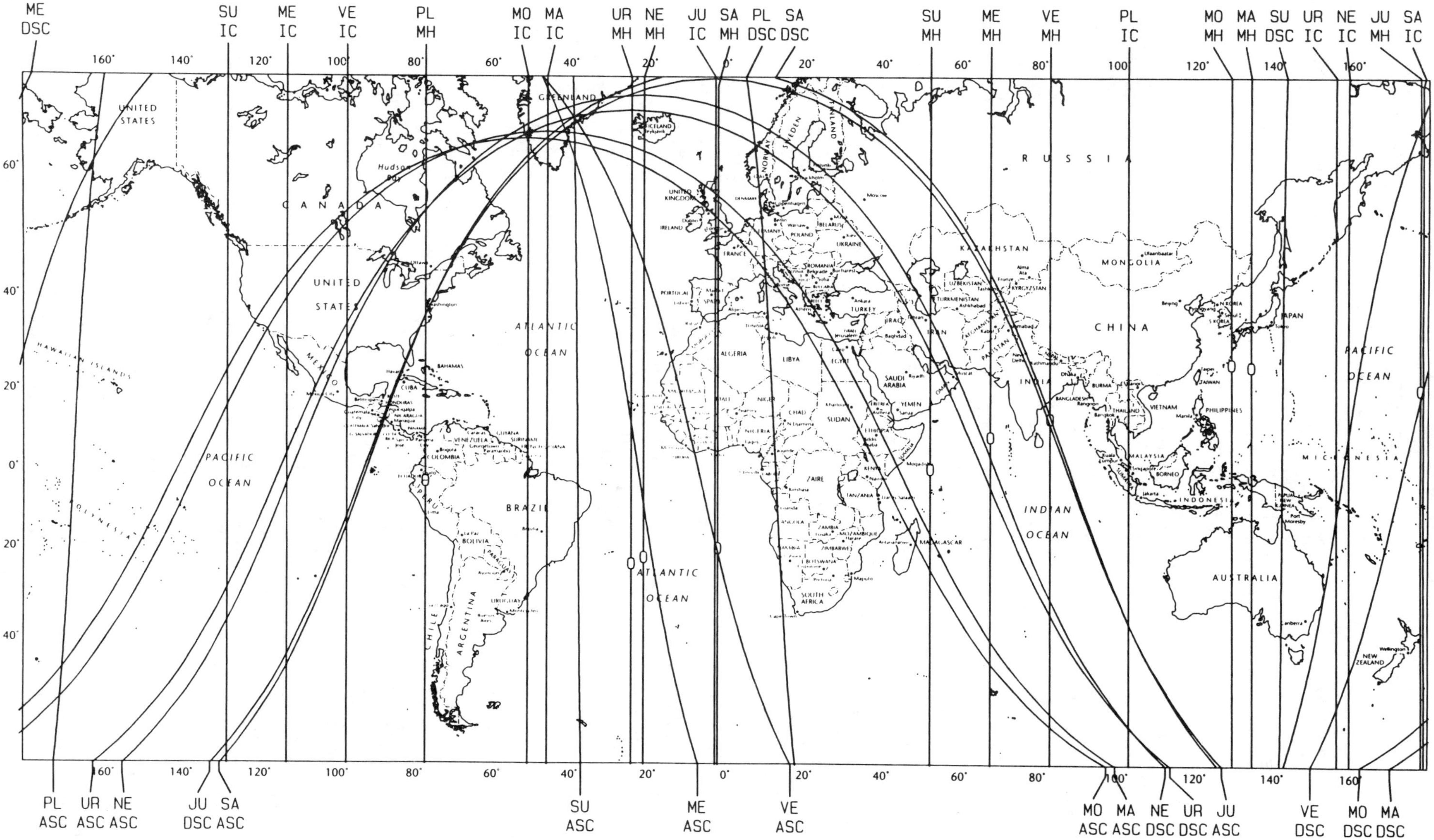

David Meadows SR

March 22, 1991 8:35 AM GMT

04/16/96

04/16/96

David Meadows SR
March 22, 1991 8:35 AM GMT

ASTRO*CARTO*GRAPHY LATITUDE CROSSINGS

NORTH LATITUDES

69N54 SA/MH * JU/ASC
69N43 JU/ASC * PL/DSC
69N35 NE/MH * SA/ASC
69N30 MA/IC * VE/ASC
69N11 UR/MH * SA/ASC
68N58 MA/IC * ME/ASC
68N55 NE/MH * JU/DSC
68N32 UR/MH * JU/DSC
68N21 UR/MH * NE/ASC
68N18 JU/DSC * NE/ASC
68N04 SA/ASC * NE/ASC
67N23 SU/ASC * NE/ASC
67N20 JU/IC * NE/DSC
67N15 SA/MH * NE/DSC
67N11 VE/ASC * NE/ASC
66N52 NE/MH * UR/DSC
66N50 ME/ASC * NE/ASC
66N30 JU/DSC * UR/ASC
66N27 MA/IC * NE/ASC
66N22 SA/IC * ME/DSC
66N16 SU/ASC * SA/ASC
66N14 VE/ASC * SA/ASC
66N12 SU/ASC * UR/ASC
66N12 VE/ASC * UR/ASC
66N11 SA/ASC * UR/ASC
66N11 SU/ASC * VE/ASC
65N53 NE/DSC * PL/DSC
65N51 ME/ASC * UR/ASC
65N47 VE/ASC * JU/DSC
65N42 JU/MH * ME/DSC
65N38 SU/ASC * JU/DSC
65N30 MO/IC * NE/ASC
65N25 ME/ASC * SA/ASC
65N23 MA/IC * UR/ASC
65N23 JU/IC * UR/DSC
65N17 SA/MH * UR/DSC
64N55 ME/ASC * JU/DSC
64N41 MA/ASC * SA/ASC
64N39 MA/ASC * JU/DSC
64N37 MA/DSC * UR/ASC
64N37 ME/ASC * MA/ASC
64N36 MO/IC * MA/DSC
64N32 MO/IC * UR/ASC
64N32 SU/ASC * MA/ASC
64N23 VE/ASC * MA/ASC
64N22 MA/DSC * NE/ASC
64N06 MA/IC * SA/ASC
63N57 MO/DSC * UR/ASC
63N55 MO/ASC * SA/ASC
63N54 MA/IC * MO/ASC
63N51 MO/ASC * JU/DSC
63N50 MO/DSC * NE/ASC
63N36 MO/ASC * ME/ASC
63N30 MO/ASC * SU/ASC
63N28 MA/IC * JU/DSC
63N26 UR/DSC * PL/DSC
63N04 UR/MH * MA/ASC
63N03 MO/ASC * VE/ASC
62N55 MO/DSC * MA/DSC
62N32 NE/MH * MA/ASC
62N00 MO/IC * SA/ASC
61N38 SU/ASC * ME/ASC
61N25 UR/MH * MO/ASC
61N24 PL/MH * MO/DSC
61N23 MO/IC * JU/DSC
61N14 VE/DSC * PL/ASC
61N13 Anchorage
60N59 PL/MH * MA/DSC
60N42 NE/MH * MO/ASC

59N20 Stockholm
58N54 SU/MH * SA/DSC
57N29 SU/MH * JU/ASC
57N10 JU/IC * MA/ASC

CENTRAL LATITUDES

56N53 SA/MH * MA/ASC
55N45 Moscow
55N41 Copenhagen
55N35 UR/ASC * NE/ASC
54N45 PL/MH * UR/ASC
54N41 VE/IC * MO/DSC
54N39 PL/MH * NE/ASC
53N57 JU/IC * MO/ASC
53N36 SA/MH * MO/ASC
52N48 VE/IC * MA/DSC
52N30 Berlin
51N41 UR/MH * VE/ASC
51N30 London
50N24 MA/ASC * PL/DSC

49N14 Vancouver
48N50 Paris
47N36 Seattle
46N05 NE/MH * VE/ASC
45N32 Montreal
45N03 MO/ASC * PL/DSC
44N59 Minneapolis
44N33 ME/MH * SA/DSC
44N07 ME/IC * MO/DSC
43N39 Toronto
42N33 ME/MH * JU/ASC
42N22 Boston
42N20 Detroit
41N54 Rome
41N53 Chicago
40N43 New York

39N50 ME/IC * MA/DSC
39N44 Denver
38N53 Washington
38N37 St. Louis
38N20 JU/DSC * SA/ASC
37N47 San Francisco
36N25 PL/MH * JU/DSC
36N19 PL/MH * SA/ASC
36N10 Las Vegas
35N59 SU/MH * NE/DSC
35N40 Tokyo
34N39 VE/IC * UR/ASC
34N04 Los Angeles
33N45 Atlanta
33N27 Phoenix
32N47 Dallas
31N23 VE/IC * NE/ASC
30N03 Cairo

29N58 New Orleans
29N46 Houston
28N51 SU/MH * UR/DSC
28N39 Delhi
25N47 Miami
24N28 SU/IC * MO/DSC
22N34 Calcutta
22N16 Hong Kong
21N19 Honolulu
20N26 SA/IC * VE/DSC

19N26 Mexico City
17N55 JU/MH * VE/DSC
16N03 SU/IC * MA/DSC
14N35 Manila
13N50 VE/MH * SA/DSC
11N54 UR/MH * ME/ASC
11N38 VE/MH * JU/ASC
10N30 Caracas

8N36 NE/IC * ME/DSC
6N27 Lagos
4N17 ME/IC * UR/ASC
3N19 ME/MH * NE/DSC

SOUTH LATITUDES

3S19 ME/IC * NE/ASC
4S17 ME/MH * UR/DSC
8S36 NE/MH * ME/ASC

11S38 VE/IC * JU/DSC
11S54 UR/IC * ME/DSC
13S50 VE/IC * SA/ASC
16S03 SU/MH * MA/ASC
17S55 JU/IC * VE/ASC

20S26 SA/MH * VE/ASC
22S54 Rio De Janeiro
24S28 SU/MH * MO/ASC
26S12 Johannesburg
28S51 SU/IC * UR/ASC

31S23 VE/MH * NE/DSC
33S55 Sydney
34S36 Buenos Aires
34S39 VE/MH * UR/DSC
35S59 SU/IC * NE/ASC
36S19 PL/IC * SA/DSC
36S25 PL/IC * JU/ASC
36S52 Auckland
37S50 Melbourne
38S20 JU/ASC * SA/DSC
39S50 ME/MH * MA/ASC

42S33 ME/IC * JU/DSC
44S07 ME/MH * MO/ASC
44S33 ME/IC * SA/ASC
45S03 MO/DSC * PL/ASC
46S05 NE/IC * VE/DSC

50S24 MA/DSC * PL/ASC
51S41 UR/IC * VE/DSC
52S48 VE/MH * MA/ASC
53S36 SA/IC * MO/DSC
53S57 JU/MH * MO/DSC
54S39 PL/IC * NE/DSC
54S41 VE/MH * MO/ASC
54S45 PL/IC * UR/DSC
55S35 UR/DSC * NE/DSC
56S53 SA/IC * MA/DSC
57S10 JU/MH * MA/DSC
57S29 SU/IC * JU/DSC
58S54 SU/IC * SA/ASC

DSC, MA ASC and MO ASC at Munich on my solar return map.

When I moved to Munich, C*C*G PL MH was just starting to pass through Germany. One year later C*C*G PL MH had reached Munich and had an effect for two years.

It was C*C*G PL MH at Munich that helped me start a new career, one which I had previously never considered. However, when C*C*G PL MH first arrived at Munich I went through a traumatic time in my life. I was working for an agency that had arranged my employment at a computer company for six months. If this company was impressed with me, a transfer could be arranged from the agency I worked for.

My job was accounts clerk, which involved a great amount of work. I was determined to make a good impression, so I decided to work long hours. After five months I had worked more than 150 hours overtime, and the manager wanted to employ me on a permanent basis after the six months contract expired. It was just a matter of sorting out the formalities between my agency and the computer company.

After I was offered a full-time position at the computer company I realized I had not been paid for all the overtime. I asked my agency to pay this overtime with the last salary as I would be leaving in two weeks time. It was written in my work contract that I would be paid for overtime, so I was entitled to it.

C*C*G PL MH had just reached Munich for the first time when I was offered a full-time position with this computer company. I assumed this would be the start of a new career in Munich. Something unexpected happened next: My agency refused to pay me for the overtime and said I would have to take time off. This wasn't possible as I would be starting a new job in a few days time.

Then quite unexpectedly I was told that the computer company no longer wanted me, so I would be staying with the agency. But the computer company didn't know anything about this, so they made a strong protest to my agency. I was convinced my agency had blocked the transfer to the computer company, and as a result I was determined to be paid for the overtime.

After the new job with the computer company fell through my agency had no work for me. I was told I would have to use my overtime as holiday until they found work for me. It was obvious that my agency had found a way to avoid paying me for overtime. I had it in writing that I would be paid, so I decided to put the matter in the hands of a solicitor.

The solicitor wrote a letter to my agency requesting that they pay me for the overtime. Shortly after my agency received this letter I was given notice that my contract would be terminated. As a result I took my agency to court for unfair dismissal.

On the day the court case took place, C*C*G PL MH was exact (orb three minutes) at Munich. I won the case and was awarded compensation plus the outstanding payment for overtime.

A month before I appeared in court, C*C*G PL MH moved back within orb of Munich. At this time I started an accountancy course which lasted six months. It was paid for by the employment office to help me obtain a new job.

C*C*G PL MH was at Munich during the time I took the course. Despite some language difficulties I completed it with extraordinary success, and after six months I was a qualified accountant, something that would have been impossible to achieve in England. The fact that I did so in a foreign language makes it all the more miraculous.

The subject that interested me the most in this course was tax. After I had qualified as an accountant I decided to get a job doing tax returns. Four months after the course finished I started a new job as a tax accountant.

In England I never imagined I would be specializing in taxes for a career. But C*C*G PL MH at Munich helped me start a new career that has transformed my life.

A career specializing in tax is appropriate as PL is the ruler of SCO—other people's money.

I'm now happy living in Munich and would never move back to England. When I left England it was a move into the unknown as I didn't know what to expect. It just goes to show what a difference a location can make.

Example 10—David/Martina (Relationship)

An Astro*Carto*Graphy map drawn for our relationship provides accurate information about what we have experienced together.

We first met in East Berlin, so this is the birth location of the relationship chart. We had been writing to each other for fifteen years before we first met, which is shown in our composite chart.

The composite chart shows the start of the relationship, which in our case was as penfriends. This is confirmed by the composite grand trine of ME, MA and JU.

Using the relationship chart the planetary influences at Berlin show how things developed between us after we first met:

Lines:	JU IC
	PL IC
	VE DSC
Crossings:	JU IC/VE DSC
	SA MH/VE ASC
	PL IC/VE DSC
	MO DSC/PL DSC
	VE MH/NE DSC
	SU DSC/ME DSC
	ME MH/JU DSC
	NE MH/ME ASC
	MO IC/PL ASC
	MA IC/PL ASC

JU IC and PL IC are near Berlin with VE DSC not too far away. These planetary energies manifested as a successful (JU IC) and life-changing (PL IC) relationship that resulted in marriage (VE DSC).

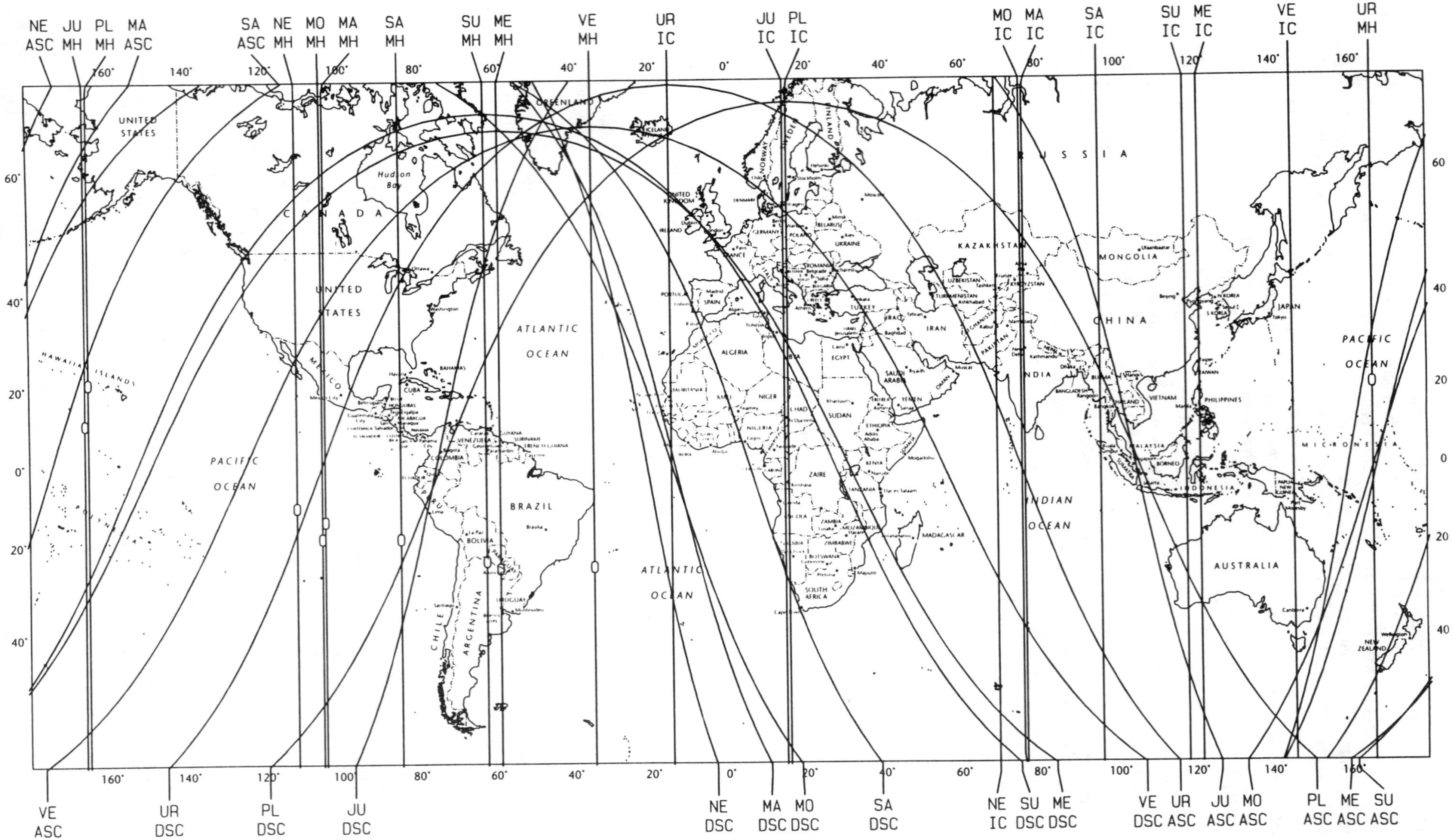

David/Martina

December 10, 1955 4:00 PM GMT

04/16/96

04/16/96

David/Martina
December 10, 1955 4:00 PM GMT

ASTRO*CARTO*GRAPHY LATITUDE CROSSINGS

NORTH LATITUDES

69N56	MA/DSC * SA/DSC
68N31	VE/MH * UR/DSC
68N13	SA/DSC * NE/DSC
68N02	JU/IC * PL/DSC
67N55	UR/ASC * PL/DSC
67N42	JU/DSC * SA/DSC
67N19	JU/DSC * NE/DSC
67N13	SA/DSC * UR/DSC
67N07	MO/DSC * SU/ASC
67N05	SU/MH * MO/DSC
67N04	ME/MH * SU/DSC
66N58	JU/IC * UR/ASC
66N53	UR/DSC * NE/DSC
66N46	PL/IC * UR/ASC
66N42	MO/IC * JU/ASC
66N30	JU/DSC * UR/DSC
66N29	MA/DSC * UR/DSC
66N28	MA/DSC * JU/DSC
66N22	SU/DSC * UR/DSC
66N20	SU/DSC * JU/DSC
66N17	SU/DSC * MA/DSC
66N11	MA/IC * JU/ASC
65N59	SU/DSC * NE/DSC
65N50	ME/MH * MO/DSC
65N41	VE/ASC * SA/DSC
65N37	SU/DSC * VE/ASC
65N31	SA/MH * SU/ASC
65N30	VE/ASC * NE/DSC
65N24	VE/ASC * MA/DSC
65N23	MO/DSC * ME/DSC
65N20	SU/MH * ME/ASC
65N13	ME/DSC * UR/DSC
65N06	VE/ASC * JU/DSC
65N04	ME/DSC * VE/ASC
65N02	ME/DSC * JU/DSC
64N56	UR/IC * PL/DSC
64N54	SU/DSC * SA/DSC
64N36	ME/DSC * MA/DSC
64N35	VE/DSC * PL/DSC
64N32	VE/MH * SU/DSC
64N28	VE/ASC * UR/DSC
64N25	ME/DSC * NE/DSC
64N23	UR/IC * VE/DSC
64N19	MO/DSC * VE/ASC
64N12	PL/MH * MA/ASC
64N12	MO/DSC * UR/DSC
64N03	VE/MH * SA/DSC
63N43	ME/MH * VE/ASC
63N32	JU/MH * MA/ASC
63N21	VE/MH * ME/DSC
63N05	SA/MH * ME/ASC
63N02	SU/MH * VE/ASC
62N54	ME/DSC * SA/DSC
62N47	ME/MH * UR/DSC
62N32	MO/DSC * JU/DSC
62N02	SU/DSC * PL/DSC
61N31	ME/DSC * PL/DSC
61N17	SU/MH * UR/DSC
61N14	MA/DSC * NE/DSC
61N13	Anchorage
60N41	SA/DSC * PL/DSC
60N39	MA/MH * SU/ASC
60N22	MO/MH * SU/ASC
59N20	Stockholm
58N00	UR/IC * SU/DSC
57N58	PL/MH * MO/ASC
57N43	VE/MH * PL/DSC
57N31	UR/IC * ME/DSC
57N28	NE/MH * SU/ASC
57N22	MA/DSC * PL/DSC
57N12	JU/MH * MO/ASC
57N07	MA/MH * ME/ASC
56N53	NE/DSC * PL/DSC

CENTRAL LATITUDES

56N45	MO/MH * ME/ASC
56N39	VE/MH * MA/DSC
55N53	NE/IC * PL/ASC
55N45	Moscow
55N41	Copenhagen
55N17	JU/IC * VE/DSC
55N10	SA/MH * VE/ASC
54N46	PL/IC * VE/DSC
54N16	MO/DSC * PL/DSC
54N12	VE/MH * NE/DSC
53N51	SU/DSC * ME/DSC
53N38	ME/MH * JU/DSC
53N16	NE/MH * ME/ASC
52N30	Berlin
51N30	London
51N11	MO/IC * PL/ASC
50N30	MA/IC * PL/ASC
49N14	Vancouver
48N50	Paris
48N00	UR/IC * SA/DSC
47N52	VE/MH * MO/DSC
47N36	Seattle
47N00	SU/MH * JU/DSC
45N32	Montreal
44N59	Minneapolis
43N39	Toronto
43N25	SA/MH * UR/DSC
42N22	Boston
42N20	Detroit
41N54	Rome
41N53	Chicago
40N43	New York
39N44	Denver
38N53	Washington
38N37	St. Louis
38N33	MA/MH * VE/ASC
38N02	MO/DSC * NE/DSC
37N47	San Francisco
37N33	MO/MH * VE/ASC
36N47	SA/IC * JU/ASC
36N10	Las Vegas
35N40	Tokyo
34N04	Los Angeles
33N45	Atlanta
33N41	ME/MH * PL/DSC
33N27	Phoenix
32N47	Dallas
31N50	JU/IC * ME/DSC
31N35	UR/MH * NE/ASC
30N23	PL/IC * ME/DSC
30N03	Cairo
29N58	New Orleans
29N46	Houston
28N44	PL/MH * SA/ASC
28N39	Delhi
28N02	JU/IC * SU/DSC
27N55	NE/MH * VE/ASC
27N21	SU/MH * PL/DSC
26N25	JU/MH * SA/ASC
26N18	PL/IC * SU/DSC
25N47	Miami
23N13	SA/IC * PL/ASC
22N34	Calcutta
22N16	Hong Kong
21N19	Honolulu
19N26	Mexico City
15N58	NE/IC * UR/ASC
14N35	Manila
10N30	Caracas
9N58	MO/ASC * MA/ASC

SOUTH LATITUDES

6N27	Lagos
4N54	UR/IC * MA/DSC
4N52	JU/ASC * PL/ASC
3N08	MA/MH * UR/DSC
1N13	UR/IC * MO/DSC
1N02	MO/MH * UR/DSC
1S02	MO/IC * UR/ASC
1S13	UR/MH * MO/ASC
3S08	MA/IC * UR/ASC
4S52	JU/DSC * PL/DSC
4S54	UR/MH * MA/ASC
9S58	MO/DSC * MA/DSC
15S58	NE/MH * UR/DSC
22S54	Rio De Janeiro
23S13	SA/MH * PL/DSC
26S12	Johannesburg
26S18	PL/MH * SU/ASC
26S25	JU/IC * SA/DSC
27S21	SU/IC * PL/ASC
27S55	NE/IC * VE/DSC
28S02	JU/MH * SU/ASC
28S44	PL/IC * SA/DSC
30S23	PL/MH * ME/ASC
31S35	UR/IC * NE/DSC
31S50	JU/MH * ME/ASC
33S41	ME/IC * PL/ASC
33S55	Sydney
34S36	Buenos Aires
36S47	SA/MH * JU/DSC
36S52	Auckland
37S33	MO/IC * VE/DSC
37S50	Melbourne
38S02	MO/ASC * NE/ASC
38S33	MA/IC * VE/DSC
43S25	SA/IC * UR/ASC
47S00	SU/IC * JU/ASC
47S52	VE/IC * MO/ASC
48S00	UR/MH * SA/ASC
50S30	MA/MH * PL/DSC
51S11	MO/MH * PL/DSC
53S16	NE/IC * ME/DSC
53S38	ME/IC * JU/ASC
53S51	SU/ASC * ME/ASC
54S12	VE/IC * NE/ASC
54S16	MO/ASC * PL/ASC
54S46	PL/MH * VE/ASC
55S10	SA/IC * VE/DSC
55S17	JU/MH * VE/ASC
55S53	NE/MH * PL/DSC
56S39	VE/IC * MA/ASC
56S45	MO/IC * ME/DSC
56S53	NE/ASC * PL/ASC
57S07	MA/IC * ME/DSC
57S12	JU/IC * MO/DSC
57S22	MA/ASC * PL/ASC
57S28	NE/IC * SU/DSC
57S31	UR/MH * ME/ASC
57S43	VE/IC * PL/ASC

David/Martina
December 10, 1955 4:00 PM GMT

ASTRO*CARTO*GRAPHY LATITUDE CROSSINGS

NORTH LATITUDES	CENTRAL LATITUDES	SOUTH LATITUDES
		57S58 PL/IC * MO/DSC
		58S00 UR/MH * SU/ASC

As we wanted to marry in a Communist country it could have taken years to sort out all the red tape. The fact that we married within one and a half years was really miraculous.

There are ten latitude crossings within orb of Berlin which will give additional information to the influence from JU IC, PL IC and VE DSC.

JU/VE and PL/VE reiterate the JU IC, PL IC and VE DSC planetary lines, so they will have a powerful effect.

SA/VE indicates some difficulties to overcome before marriage. It also shows that we experienced long periods of separation due to the political situation.

The two MO/PL crossings are appropriate for the powerful feelings we had for each other. This manifested when we were together or in our letters to each other.

MA/PL shows the strong determination we had to battle with the authorities to obtain permission to get married.

VE/NE manifested as the uncertainty we both experienced when I first visited Dresden. During the visit we felt strongly attracted to each other, but kept everything on a friendship basis. It was only after I returned to England that we realized it was more than friendship.

ME/NE indicated the initial doubts we had about what type of relationship was developing between us. We both were confused about why we got along so well during the visit.

SU/ME and ME/JU shows the good communication between us despite the language difficulties we experienced. But as our relationship developed we understood each other very well.

JU IC, PL IC and VE DSC also have an influence at Dresden, which is not far from Berlin. These planetary energies show that Dresden was a good location for us to get married.

A*C*G	**Age 33**	**C*C*G**
JU 03 VIR 46		UR 05 CAP 36
UR 04 LEO 16		NE 13 CAP 10
MO 04 SCO 39		ME 13 AQU 14
PL 04 VIR 43		SA 13 CAP 57
MA 05 SCO 26		PL 17 SCO 09
VE 14 CAP 23		MO 20 CAP 51
SU 16 SAG 46		SU 23 CAP 28
ME 20 SAG 19		MA 27 SCO 09
SA 24 SCO 46		VE 27 AQU 10
NE 28 LIB 11		JU 29 TAU 06

1. C*C*G SA = A*C*G VE
2. Prog. Berlin = A*C*G UR ASC
3. C*C*G UR DSC = Dresden (exact)

A*C*G	**(36 42)**	**Displaced C*C*G**
JU 03 VIR 46		NE 06 SAG 28
UR 04 LEO 16		ME 06 CAP 32
MO 04 SCO 39		SA 07 SAG 15
PL 04 VIR 43		PL 10 LIB 27
MA 05 SCO 26		MO 14 SAG 09
VE 14 CAP 23		MA 20 LIB 27
SU 16 SAG 46		VE 20 CAP 28
ME 20 SAG 19		JU 22 ARI 24
SA 24 SCO 46		UR 28 SCO 54
NE 28 LIB 11		

A*C*G	**(36 42)**	**Solar arc A*C*G**
JU 03 VIR 46		SA 01 CAP 28
UR 04 LEO 16		NE 04 SAG 53
MO 04 SCO 39		JU 10 LIB 28
PL 04 VIR 43		UR 10 VIR 58
MA 05 SCO 26		MO 11 SAG 21
VE 14 CAP 23		PL 11 LIB 25
SU 16 SAG 46		MA 12 SAG 08
ME 20 SAG 19		VE 21 AQU 05
SA 24 SCO 46		ME 27 CAP 01
NE 28 LIB 11		

4. solar arc NE EQU A*C*G PL

On the day we married, C*C*G UR DSC was exact at Dresden. This shows the sudden change of circumstances in our relationship and the quick arrangements we had to make for the wedding.

Prog. Berlin = A*C*G UR ASC manifested as the unexpected change of mind by the East German authorities in giving us permission to marry. This sudden breakthrough was a shock for us as three weeks earlier our marriage application had been turned down.

C*C*G SA = A*C*G VE was an excellent aspect for marriage. This shows a firm commitment for a long-lasting marriage where there is stability, loyalty and faithfulness.

Solar return PL ASC was near Berlin the year we married. Just over one month after this solar return took effect the East German authorities gave us permission to marry. PL ASC at Berlin indicated a miracle would take place there to enable us to marry before my birthday.

April 29, 1989	C*C*G UR DSC = Dresden (exact)
May 21, 1991	C*C*G VE MH = Munich
May 21, 1991	C*C*G PL DSC = Leicester (exact)
November 1, 1991	C*C*G PL DSC = Leicester
November 1, 1991	C*C*G VE MH = Munich (exact)

On the day that Martina left East Germany, C*C*G UR DSC was once again exact at Dresden. C*C*G UR DSC had returned to exactly the same place it occupied on the day we married. With C*C*G UR DSC again exact at Dresden it brought changes for us. This time Martina was allowed to travel to England so that we could start married life together.

When Martina moved to Munich C*C*G PL DSC was exact at Leicester. After experiencing many difficulties in life living in Leicester we decided to make a new beginning. C*C*G PL DSC at Leicester showed that it was a good time to make a new start in life.

C*C*G VE MH was within orb of Munich when she moved there. When Martina arrived in Munich her priority was to earn as much money as possible so that she could obtain her own accommodations.

When I moved to Germany a few months later C*C*G VE MH was exact at Munich. After I was a short time in Munich we had resolved the serious financial difficulties we were experiencing in England.

The relationship chart map shows that SU DSC and ME

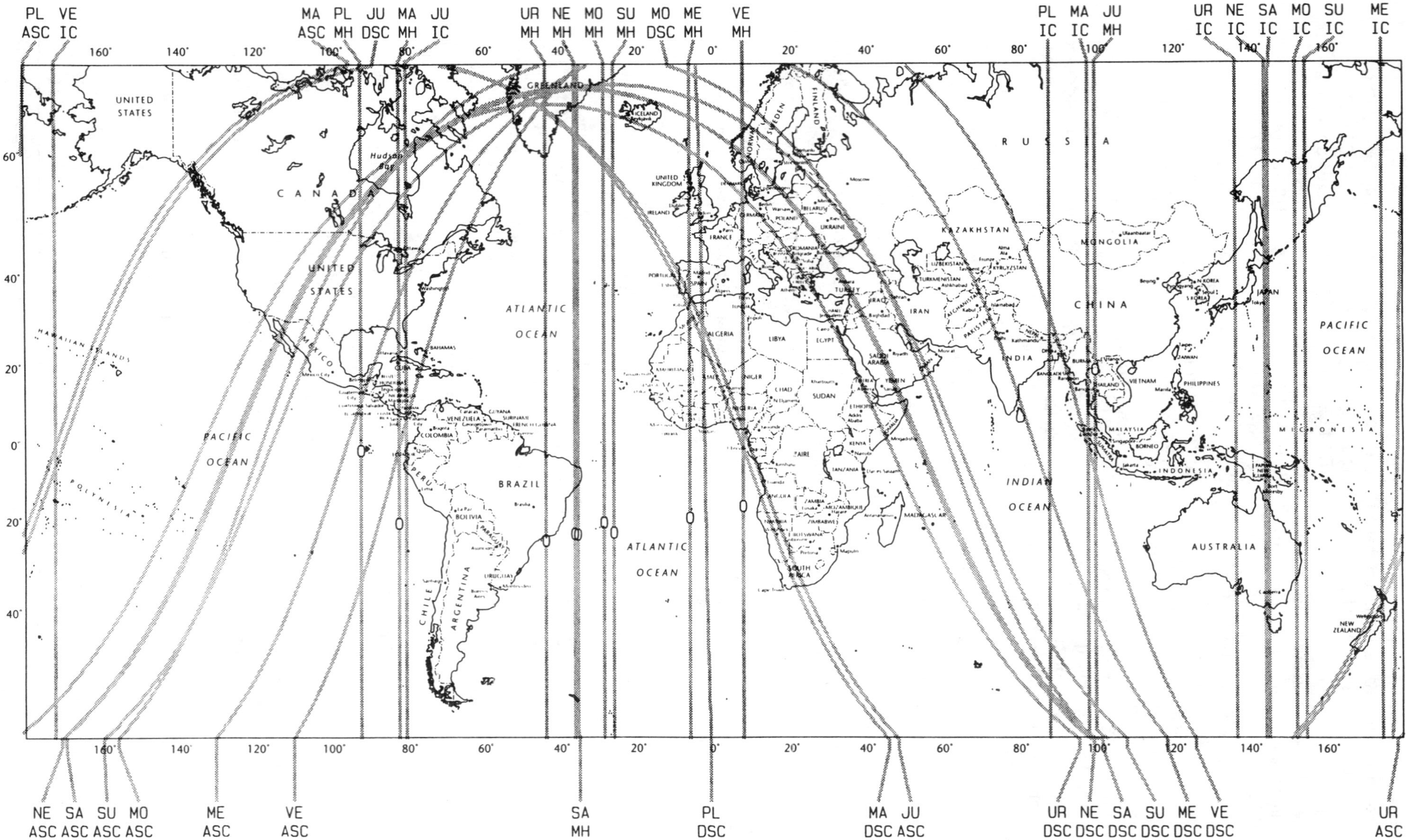

CYCLO*CARTO*GRAPHY

04/17/96

SU, MO, ME, VE AND MA ARE PROGRESSED
JU, SA, UR, NE AND PL ARE TRANSITING

David/Martina

PROG	1	12	56	22:31 GMT
TRANSIT	3	18	89	0:00 GMT
NATAL	12	10	55	16:00 GMT

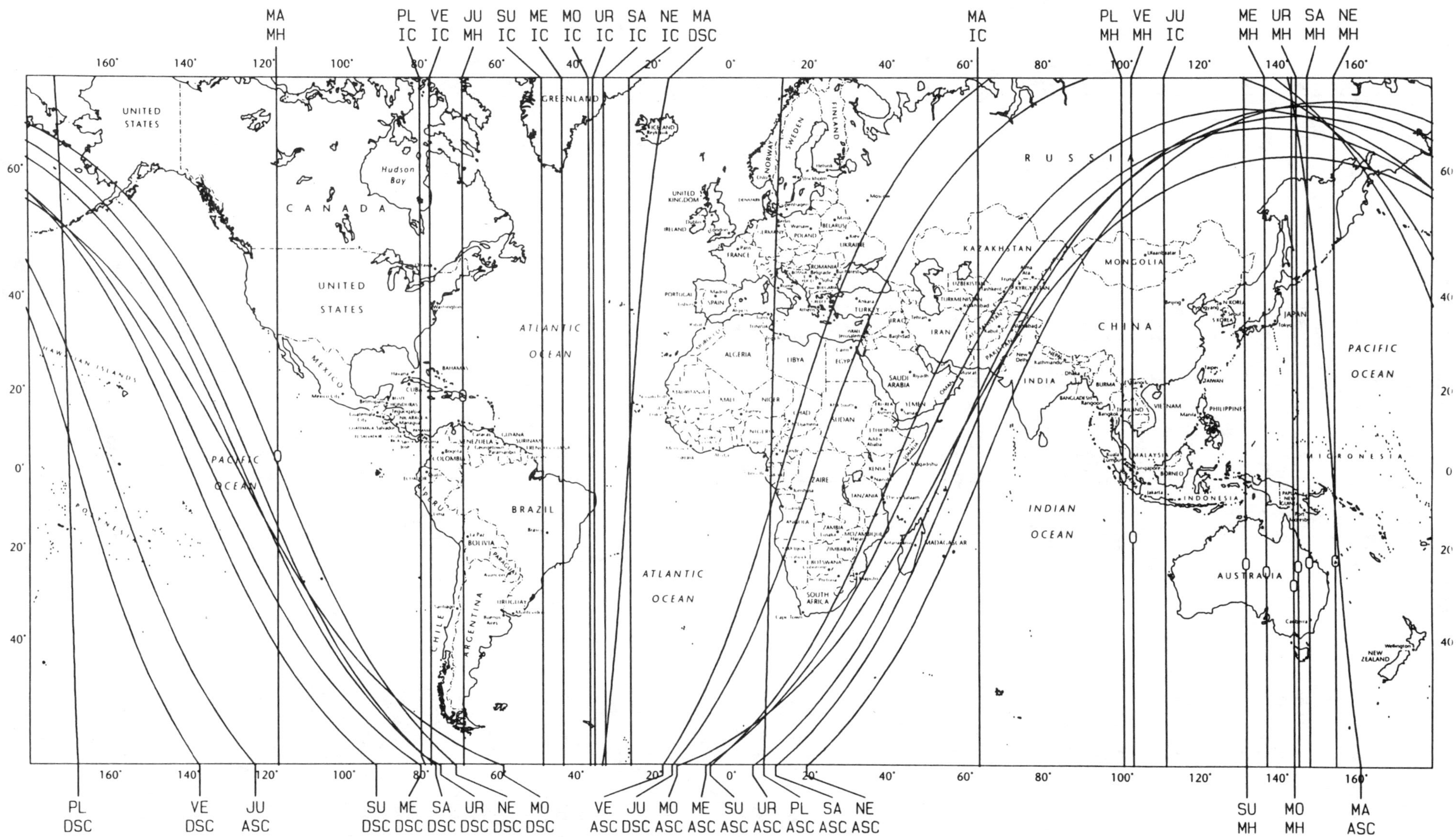

David/Martina SR

December 10, 1988 3:04 AM GMT

04/16/96

04/16/96

David/Martina SR
December 10, 1988 3:04 AM GMT

ASTRO*CARTO*GRAPHY LATITUDE CROSSINGS

NORTH LATITUDES

69N53	SU/MH * JU/ASC
69N13	MA/IC * VE/ASC
69N10	ME/MH * JU/ASC
68N52	VE/DSC * MA/ASC
68N47	MO/MH * VE/DSC
68N26	UR/MH * VE/DSC
68N17	MO/MH * MA/ASC
67N52	MO/MH * JU/ASC
67N48	MA/ASC * JU/ASC
67N39	SA/MH * NE/ASC
67N38	VE/DSC * NE/ASC
67N36	UR/MH * JU/ASC
67N31	JU/ASC * NE/ASC
67N31	SA/MH * VE/DSC
67N30	UR/MH * NE/ASC
67N28	MA/ASC * NE/ASC
67N26	MO/MH * NE/ASC
67N20	VE/DSC * SA/DSC
67N20	JU/ASC * SA/ASC
67N19	UR/MH * SA/ASC
67N18	MA/ASC * SA/ASC
67N17	MO/MH * SA/ASC
67N12	NE/MH * SA/DSC
67N10	SA/ASC * NE/ASC
66N59	ME/MH * SU/DSC
66N59	SU/DSC * SA/ASC
66N58	ME/MH * SA/ASC
66N57	SU/DSC * NE/ASC
66N54	SA/MH * JU/ASC
66N49	ME/MH * NE/ASC
66N37	MO/MH * SU/DSC
66N33	SU/DSC * MA/ASC
66N31	SU/MH * SA/ASC
66N31	UR/MH * SU/DSC
66N20	MA/ASC * UR/ASC
66N20	MO/MH * UR/ASC
66N20	SU/DSC * UR/DSC
66N19	SA/MH * UR/DSC
66N16	JU/ASC * UR/DSC
66N15	SA/MH * SU/DSC
66N13	VE/DSC * UR/DSC
66N08	ME/MH * UR/ASC
66N08	SU/MH * NE/ASC
66N03	NE/MH * UR/DSC
65N48	SU/DSC * JU/ASC
65N47	UR/MH * MA/ASC
65N46	SU/MH * UR/ASC
65N42	JU/IC * SU/ASC
65N40	SU/DSC * VE/DSC
65N23	NE/MH * SU/DSC
65N13	UR/ASC * NE/ASC
65N02	SU/MH * ME/ASC
65N00	VE/DSC * JU/ASC
64N58	MO/MH * ME/DSC
64N55	UR/MH * ME/DSC
64N54	ME/DSC * MA/ASC
64N51	NE/MH * JU/ASC
64N49	NE/MH * VE/DSC
64N44	SA/MH * ME/DSC
64N30	ME/ASC * NE/ASC
64N15	VE/MH * SU/ASC
64N05	NE/MH * ME/DSC
64N02	ME/ASC * UR/ASC
63N51	NE/DSC * PL/DSC
63N50	ME/DSC * VE/DSC
63N40	PL/MH * SU/ASC
63N40	ME/DSC * JU/ASC
63N13	ME/ASC * SA/ASC
63N02	MA/IC * JU/DSC
62N48	JU/IC * ME/ASC
62N34	JU/IC * SA/ASC
62N14	JU/IC * UR/ASC
61N53	UR/MH * MO/DSC
61N52	MO/DSC * MA/ASC

CENTRAL LATITUDES

61N50	SA/MH * MO/DSC
61N43	ME/MH * MO/ASC
61N29	NE/MH * MO/DSC
61N21	SU/MH * MO/ASC
61N13	SA/DSC * PL/DSC
61N13	Anchorage
60N49	VE/MH * ME/ASC
60N48	JU/IC * NE/ASC
60N44	MO/DSC * VE/DSC
60N18	MO/DSC * JU/ASC
60N03	PL/MH * ME/ASC
59N32	VE/MH * SA/ASC
59N29	VE/MH * UR/ASC
59N20	Stockholm
59N02	SA/ASC * UR/ASC
58N52	UR/DSC * PL/DSC
58N27	PL/MH * UR/ASC
58N23	PL/MH * SA/ASC
57N42	SU/DSC * ME/DSC
57N39	JU/IC * MO/ASC
56N49	VE/MH * NE/ASC
56N35	SA/MH * MA/ASC
55N45	Moscow
55N41	Copenhagen
55N17	PL/MH * NE/ASC
54N47	VE/MH * MO/ASC
54N26	MO/DSC * SU/DSC
53N42	PL/MH * MO/ASC
53N02	ME/DSC * PL/DSC
52N54	MO/DSC * PL/DSC
52N30	Berlin
52N23	MO/DSC * ME/DSC
52N01	SU/DSC * PL/DSC
51N30	London
50N12	MO/ASC * NE/ASC
49N14	Vancouver
48N50	Paris
47N36	Seattle
45N32	Montreal
44N59	Minneapolis
43N39	Toronto
42N22	Boston
42N20	Detroit
41N54	Rome
41N53	Chicago
41N30	MA/IC * SU/ASC
40N43	New York
39N44	Denver
38N53	Washington
38N37	St. Louis
37N47	San Francisco
36N10	Las Vegas
35N40	Tokyo
34N04	Los Angeles
33N45	Atlanta
33N27	Phoenix
32N47	Dallas
31N57	MA/IC * ME/ASC
30N12	JU/ASC * PL/DSC
30N03	Cairo
29N58	New Orleans
29N46	Houston
29N36	MO/ASC * SA/ASC
28N39	Delhi
25N47	Miami
22N34	Calcutta
22N16	Hong Kong
21N19	Honolulu
19N26	Mexico City
19N09	MA/IC * UR/ASC

SOUTH LATITUDES

17N50	MA/IC * MO/ASC
14N35	Manila
13N57	MA/IC * SA/ASC
11N42	MO/ASC * UR/ASC
10N30	Caracas
9N07	VE/DSC * PL/DSC
8N58	NE/IC * MA/DSC
6N27	Lagos
1N31	MA/MH * NE/DSC
1S31	MA/IC * NE/ASC
8S58	NE/MH * MA/ASC
9S07	VE/ASC * PL/ASC
11S42	MO/DSC * UR/DSC
13S57	MA/MH * SA/DSC
17S50	MA/MH * MO/DSC
19S09	MA/MH * UR/DSC
22S54	Rio De Janeiro
26S12	Johannesburg
29S36	MO/DSC * SA/DSC
30S12	JU/DSC * PL/ASC
31S57	MA/MH * ME/DSC
33S55	Sydney
34S36	Buenos Aires
36S52	Auckland
37S50	Melbourne
41S30	MA/MH * SU/DSC
50S12	MO/DSC * NE/DSC
52S01	SU/ASC * PL/ASC
52S23	MO/ASC * ME/ASC
52S54	MO/ASC * PL/ASC
53S02	ME/ASC * PL/ASC
53S42	PL/IC * MO/DSC
54S26	MO/ASC * SU/ASC
54S47	VE/IC * MO/DSC
55S17	PL/IC * NE/DSC
56S35	SA/IC * MA/DSC
56S49	VE/IC * NE/DSC
57S39	JU/MH * MO/DSC
57S42	SU/ASC * ME/ASC
58S23	PL/IC * SA/DSC
58S27	PL/IC * UR/DSC
58S52	UR/ASC * PL/ASC
59S02	SA/DSC * UR/DSC
59S29	VE/IC * UR/DSC
59S32	VE/IC * SA/DSC

DSC are near Leicester. These planetary influences do not show the many difficulties we experienced in life residing in Leicester.

The combination of SU DSC and ME DSC has more to do with the communication between us. But during the time that we lived in Leicester, C*C*G PL DSC had a powerful influence. Most of the time C*C*G PL DSC was at or near Leicester, which manifested as intense experiences for us.

The Astro*Carto*Graphy relationship chart map provides accurate information about what two people experience together, but it is also important to apply the techniques to find the temporary influences on an Astro*Carto*Graphy relationship chart map.

Example 11—David/Martina (First Meeting)

When I visited Dresden for the first time we had arranged to meet at Checkpoint Charlie in East Berlin. In case we missed each other at the border we had a second meeting point inside a well-known hotel.

I went through Checkpoint Charlie without any problem and arrived in East Berlin just before 2:30 p.m. We had arranged to meet between 2:00 p.m. and 3:00 p.m. at the border, so I expected Martina to be waiting for me.

Martina wasn't there when I arrived. I was totally confused when I looked at my map because I wasn't on the street I should have been.

At 3:00 p.m. there was still no sign of Martina, so I decided to make my way to the second meeting point. I was looking at my map when a couple asked me if I needed some help. I couldn't speak German, so I said the name of the hotel. They offered to drive me to this hotel and I gladly accepted.

I arrived at the hotel a few minutes later and asked if there were any message for me. The receptionist couldn't find anything and said I should wait inside the hotel. I waited near the entrance so that I would see Martina when she arrived. At 3:52 p.m. Martina arrived at the hotel and thought I was angry because I looked at my watch. I did so to make a note of the time when we first met.

It turned out that the East German government was reconstructing the border at Checkpoint Charlie, which meant I arrived at a different border crossing. Martina was waiting at the usual Checkpoint Charlie crossing on the East German side.

A chart calculated for our first meeting shows the confusion we experienced at Checkpoint Charlie. The exact MO CJN NE was SQU SU CJN ME, which explains why things did not go as planned.

At the time I arrived in East Berlin UR was CJN the ASC. I was shocked that Martina wasn't waiting for me as we had planned everything carefully.

The MO CJN NE SQU SU CJN ME also had a strong effect after we met. Communication between us was often confusing due to the different languages.

After I returned to England there was a great deal of uncertainty about how things would develop between us, so this square aspect was dominant before and after we first met.

An Astro*Carto*Graphy map of our first meeting chart has MO ASC, PL MH and NE ASC within orb of Berlin. The confusing circumstances under which we met are reiterated by MO ASC and NE ASC. During the week that I stayed in East Germany we were unsure about our feelings for each other, but PL MH showed that my visit to East Germany would end up changing our lives.

NE ASC is within orb of Leicester showing the uncertainty of whether Martina would be allowed to move to England. People I knew in England thought I was deceiving myself about wanting to marry in East Germany.

A*C*G	**Age 2**	**C*C*G**
MO 00 CAP 02		SU 01 LIB 47
NE 00 CAP 57		ME 03 LIB 59
VE 01 VIR 47		UR 05 CAP 36
PL 06 SCO 31		VE 05 VIR 53
MA 09 VIR 28		MA 11 VIR 32
JU 10 AQU 00		NE 13 CAP 10
UR 13 SAG 00		SA13 CAP 57
SA 22 SCO 09		PL 17 SCO 09
ME 28 VIR 18		MO 20 AQU 38
SU 28 VIR 39		JU 29 TAU 06

1. C*C*G SU EQU A*C*G NE (sep)
2. C*C*G NE ASC = Dresden
3. C*C*G UR ASC = Berlin

A*C*G	**(3 08)**	**Displaced C*C*G**
MO 00 CAP 02		ME 00 LIB 51
NE 00 CAP 57		UR 02 CAP 28
VE 01 VIR 47		VE 02 VIR 45
PL 06 SCO 31		MA 08 VIR 24
MA 09 VIR 28		NE 10 CAP 02
JU 10 AQU 00		SA 10 CAP 49
UR 13 SAG 00		PL 14 SCO 01
SA 22 SCO 09		MO 17 AQU 30
ME 28 VIR 18		JU 25 TAU 58
SU 28 VIR 39		

4. Disp. C*C*G VE = A*C*G VE
5. Disp. C*C*G MA = A*C*G MA

A*C*G	**(3 08)**	**Solar arc A*C*G**
MO 00 CAP 02		ME 01 LIB 26
NE 00 CAP 57		MO 03 CAP 10
VE 01 VIR 47		NE 04 CAP 05
PL 06 SCO 31		VE 04 VIR 55
MA 09 VIR 28		PL 09 SCO 39
JU 10 AQU 00		MA 12 VIR 36
UR 13 SAG 00		JU 13 AQU 08
SA 22 SCO 09		UR 16 SAG 08
ME 28 VIR 18		SA 25 SCO 17
SU 28 VIR 39		

6. solar arc ME EQU A*C*G NE
7. solar arc PL EQU A*C*G JU
8. solar arc MA EQU A*C*G UR

01.11.91 - C*C*G UR ASC = Munich

On the day we married, C*C*G NE ASC was just over one degree from Dresden, so this influence still had some

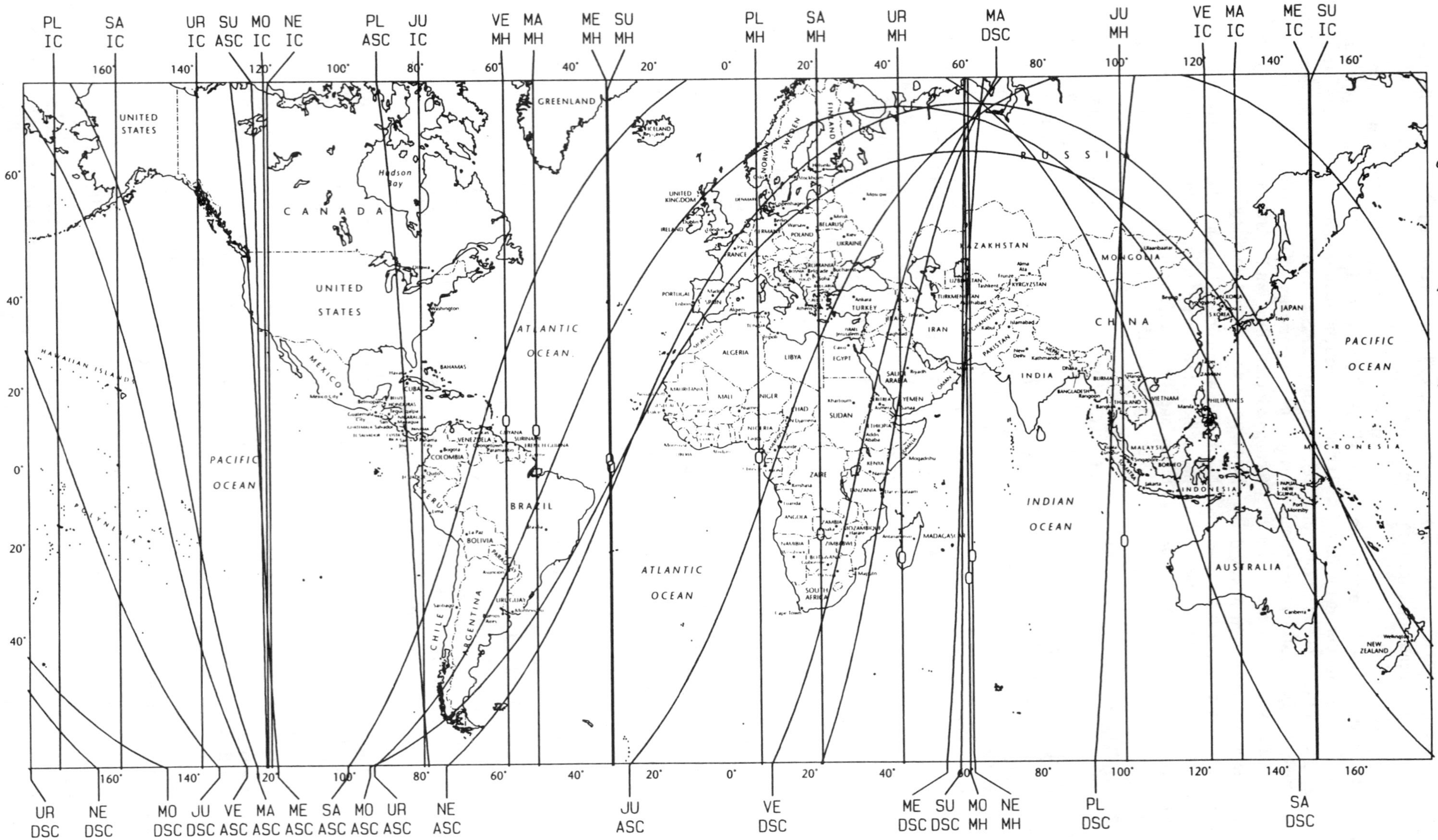

David/Martina First Meeting

September 21, 1985 1:52 PM GMT

04/16/96

04/16/96

David/Martina First Meeting
September 21, 1985 1:52 PM GMT

ASTRO*CARTO*GRAPHY LATITUDE CROSSINGS

NORTH LATITUDES

69N21 VE/IC * JU/DSC
68N56 MO/MH * SA/DSC
68N54 SU/DSC * SA/DSC
68N41 NE/MH * SA/DSC
68N08 VE/DSC * MA/DSC
68N06 MA/IC * JU/DSC
67N43 ME/DSC * SA/DSC
67N40 SU/DSC * NE/ASC
67N40 MO/MH * NE/ASC
67N39 VE/DSC * SA/DSC
67N38 ME/DSC * NE/DSC
67N38 VE/DSC * NE/DSC
67N38 SA/DSC * NE/DSC
67N38 MA/DSC * NE/DSC
67N36 MA/DSC * SA/DSC
67N32 JU/ASC * NE/DSC
67N30 ME/DSC * VE/DSC
67N17 UR/DSC * NE/ASC
67N11 ME/DSC * MA/DSC
67N07 JU/ASC * SA/DSC
67N01 MO/MH * SU/DSC
66N39 UR/MH * NE/ASC
66N35 ME/DSC * JU/ASC
66N31 SU/DSC * UR/DSC
66N31 MO/MH * UR/DSC
66N24 NE/MH * UR/DSC
66N21 MA/DSC * JU/ASC
66N16 VE/DSC * UR/DSC
66N12 JU/ASC * UR/DSC
66N10 MA/DSC * UR/DSC
66N02 ME/DSC * UR/DSC
66N02 SA/MH * UR/ASC
65N55 VE/DSC * JU/ASC
65N44 NE/MH * JU/ASC
65N31 NE/MH * VE/DSC
65N27 MO/MH * JU/ASC
65N25 SU/DSC * JU/ASC
64N53 MO/MH * VE/DSC
64N47 SU/DSC * VE/DSC
64N26 SA/DSC * UR/DSC
64N26 NE/MH * MA/DSC
63N29 MO/MH * MA/DSC
63N15 SU/DSC * MA/DSC
62N55 SU/MH * SA/ASC
62N43 ME/MH * SA/ASC
62N40 PL/MH * UR/ASC
62N34 MO/ASC * SU/DSC
62N34 MO/ASC * MA/DSC
62N34 NE/MH * MO/DSC
62N33 MO/DSC * ME/DSC
62N32 MO/ASC * VE/DSC
62N31 PL/IC * VE/ASC
62N20 MO/ASC * JU/ASC
62N15 ME/IC * JU/DSC
62N13 SA/MH * NE/ASC
62N08 JU/MH * NE/DSC
62N05 SU/IC * JU/DSC
61N38 NE/DSC * PL/DSC
61N31 UR/MH * MO/ASC
61N20 MO/DSC * SA/DSC
61N13 Anchorage

59N34 MO/DSC * UR/DSC
59N29 SA/IC * MA/ASC
59N20 Stockholm
57N16 UR/MH * JU/ASC
56N41 SA/MH * MO/ASC
55N54 JU/MH * MO/DSC
55N45 Moscow
55N44 MO/DSC * PL/DSC
55N41 Copenhagen
54N47 PL/MH * NE/ASC
52N40 JU/MH * UR/DSC
52N39 UR/DSC * PL/DSC

CENTRAL LATITUDES

52N37 JU/MH * PL/DSC
52N30 Berlin
51N30 London

49N53 VE/IC * NE/DSC
49N14 Vancouver
48N53 PL/MH * MO/ASC
48N50 Paris
47N36 Seattle
47N28 NE/MH * ME/DSC
45N32 Montreal
44N59 Minneapolis
44N22 MA/MH * SA/ASC
43N39 Toronto
42N23 VE/IC * MO/DSC
42N22 Boston
42N20 Detroit
41N54 Rome
41N53 Chicago
41N44 MA/IC * NE/DSC
41N15 UR/MH * VE/DSC
40N43 New York

39N44 Denver
38N53 Washington
38N37 St. Louis
37N47 San Francisco
37N41 SA/DSC * PL/DSC
36N58 SA/IC * VE/ASC
36N10 Las Vegas
35N40 Tokyo
35N35 MO/MH * ME/DSC
34N42 JU/MH * SA/DSC
34N07 MA/IC * MO/DSC
34N04 Los Angeles
33N45 Atlanta
33N27 Phoenix
32N59 SU/MH * UR/ASC
32N47 Dallas
32N24 ME/MH * UR/ASC
31N02 SA/MH * JU/ASC
30N03 Cairo

29N58 New Orleans
29N46 Houston
28N48 VE/MH * SA/ASC
28N39 Delhi
25N47 Miami
25N06 VE/IC * UR/DSC
22N34 Calcutta
22N16 Hong Kong
21N19 Honolulu

19N26 Mexico City
19N26 UR/MH * MA/DSC
14N35 Manila
10N53 SU/DSC * ME/DSC
10N30 Caracas

9N50 PL/IC * JU/DSC
8N28 MA/IC * UR/DSC
8N21 MO/ASC * NE/ASC
6N27 Lagos
6N26 ME/IC * NE/DSC
5N35 SU/IC * NE/DSC
3N21 ME/IC * MO/DSC
2N40 SU/IC * MO/DSC

2S40 SU/MH * MO/ASC
3S21 ME/MH * MO/ASC
5S35 SU/MH * NE/ASC
6S26 ME/MH * NE/ASC
8S21 MO/DSC * NE/DSC
8S28 MA/MH * UR/ASC
9S50 PL/MH * JU/ASC

SOUTH LATITUDES

10S53 SU/ASC * ME/ASC
19S26 UR/IC * MA/ASC

22S54 Rio De Janeiro
25S06 VE/MH * UR/ASC
26S12 Johannesburg
28S48 VE/IC * SA/DSC

31S02 SA/IC * JU/DSC
32S24 ME/IC * UR/DSC
32S59 SU/IC * UR/DSC
33S55 Sydney
34S07 MA/MH * MO/ASC
34S36 Buenos Aires
34S42 JU/IC * SA/ASC
35S35 MO/IC * ME/ASC
36S52 Auckland
36S58 SA/MH * VE/DSC
37S41 SA/ASC * PL/ASC
37S50 Melbourne

41S15 UR/IC * VE/ASC
41S44 MA/MH * NE/ASC
42S23 VE/MH * MO/ASC
44S22 MA/IC * SA/DSC
47S28 NE/IC * ME/ASC
48S53 PL/IC * MO/DSC
49S53 VE/MH * NE/ASC

52S37 JU/IC * PL/ASC
52S39 UR/ASC * PL/ASC
52S40 JU/IC * UR/ASC
54S47 PL/IC * NE/DSC
55S44 MO/ASC * PL/ASC
55S54 JU/IC * MO/ASC
56S41 SA/IC * MO/DSC
57S16 UR/IC * JU/DSC
59S29 SA/MH * MA/DSC
59S34 MO/ASC * UR/ASC

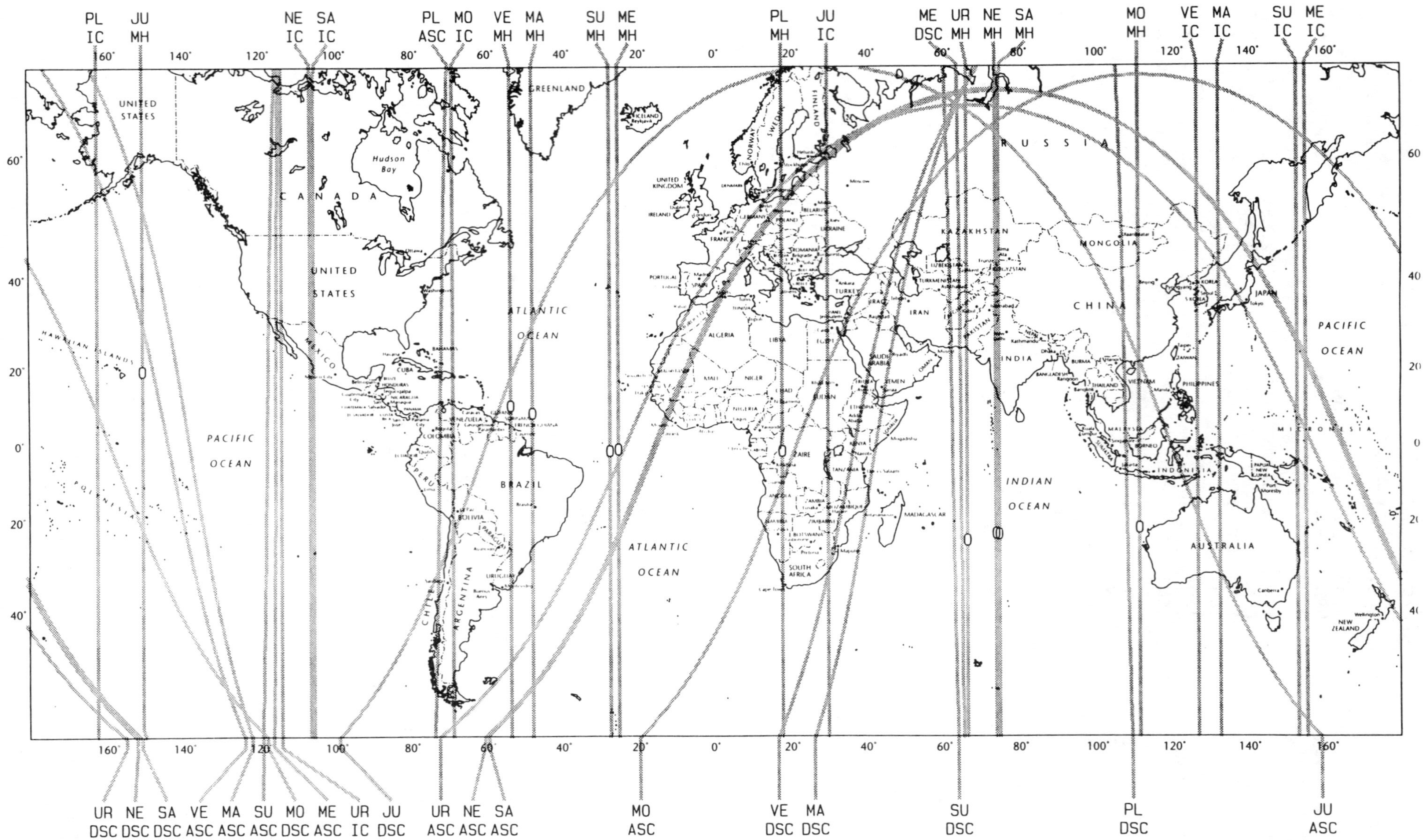

CYCLO*CARTO*GRAPHY

04/17/96

SU, MO, ME, VE AND MA ARE PROGRESSED
JU, SA, UR, NE AND PL ARE TRANSITING

David/Martina First Meeting

PROG	9 25 85	1:35 GMT
TRANSIT	3 18 89	0:00 GMT
NATAL	9 21 85	13:52 GMT

effect after the wedding. This was the uncertainty in not knowing when Martina would be allowed to leave Dresden.

C*C*G SU EQU A*C*G NE was in effect when we became engaged, indicating the uncertain time we had ahead of us. When Martina left Dresden this aspect had nearly moved out of orb.

Solar arc ME EQU A*C*G NE reiterates the doubts we experienced about our plans to marry. We had followed the correct procedure in our application and couldn't understand why the authorities refused to give us the permission.

C*C*G UR ASC at Berlin manifested as the sudden and unexpected decision that we could marry. Our application had been turned down less than a month before the authorities changed their minds. This was reiterated by solar arc PL EQU A*C*G JU.

Solar arc MA EQU A*C*G UR was appropriate for making the arrangements for the wedding at short notice. It also showed the move towards freedom and independence.

The solar return map for the year of our wedding has VE MH close to Dresden with PL ASC also within orb. This was a good indicator of marriage against all the odds.

MO IC and ME ASC are within orb of Leicester and need to be considered with A*C*G NE ASC. This indicated unclear thoughts and feelings about living in Leicester.

This is the only Astro*Carto*Graphy map I've done for the exact time when two people first met each other. Unless a note is made of the exact time for the first meeting nobody will be able to remember the details. Although our first meeting Astro*Carto*Graphy map provides accurate information about what we experienced, it should not be used on its own.

Had I only examined the first meeting Astro*Carto*Graphy map I would have interpreted the powerful NE influence as disappointment. The first meeting solar return map showed the possibility of marriage, but the other NE temporary influences were more powerful. When this first meeting Astro*Carto*Graphy map is used with the other synastry techniques it is possible to interpret it more effectively.

Example 12—David/Martina (Wedding)

For an important event like marriage I would have preferred to elect the date and time. But we had no choice and were only offered one date.

It was a miracle that we could marry, so we accepted the date offered. Had we refused this date, it would have meant reapplying.

The most powerful aspect in the marriage chart is MO SQU PL, which has a powerful influence in England. MO IC and PL DSC are near Leicester, which showed the emotionally traumatic life we experienced in England.

UR DSC is within orb of Dresden, indicating the change of circumstances in our relationship. It also explained that we had more freedom and independence after marrying.

A*C*G	**Age 2**	**C*C*G (Martina)**
MA 02 GEM 02		MA 03 GEM 27
UR 05 CAP 36		SA 09 AQU 19
NE 13 CAP 10		JU 09 LEO 58
SA 13 CAP 57		MO 11 VIR 07
ME 14 PIS 39		UR 14 CAP34
MO 16 LEO 45		NE 17 CAP 47
PL 17 SCO 09		ME 18 PIS 08
VE 24 PIS 25		PL 20 SCO 28
SU 27 PIS 56		VE 26 PIS 54
JU 29 TAU 06		SU 29 PIS 55

1. C*C*G UR EQU A*C*G SA

A*C*G	**Age 2**	**C*C*G (David)**
MA 02 GEM 02		SU 00 ARI 20
UR 05 CAP 36		SA 03 AQU 08
NE 13 CAP 10		MA 03 GEM 45
SA 13 CAP 57		JU 11 VIR 22
ME 14 PIS 39		UR 11 CAP 33
MO 16 LEO 45		NE 15 CAP 29
PL 17 SCO 09		MO 15 VIR 53
VE 24 PIS 25		ME 18 PIS 51
SU 27 PIS 56		PL 21 SCO 14
JU 29 TAU 06		VE 27 PIS 25

1. C*C*G VE = A*C*G SU

When Martina moved to Munich there was C*C*G UR EQU A*C*G SA, which indicated difficult changes were taking place within our marriage. When I joined Martina in Munich there was C*C*G VE = A*C*G SU.

A solar return map for our second wedding anniversary shows which planetary influences the wedding chart had when we moved to Munich.

PL ASC at Munich reiterates the information obtained from our individual solar return maps. We both had a solar return PL DSC passing through Germany near Munich when the move took place.

MO DSC and VE DSC within orb of Leicester showed that there was a strong feeling of love in the marriage. It was difficult to be apart when Martina moved to Munich on her own, but we had experienced much worse waiting to marry.

The marriage chart Astro*Carto*Graphy map does show some interesting and relevant information, but this map should not be used on its own to discover what a couple will experience together.

These examples show how I've changed my life with Astro*Carto*Graphy. Not all the events were happy ones, but at least Astro*Carto*Graphy explained why this was so.

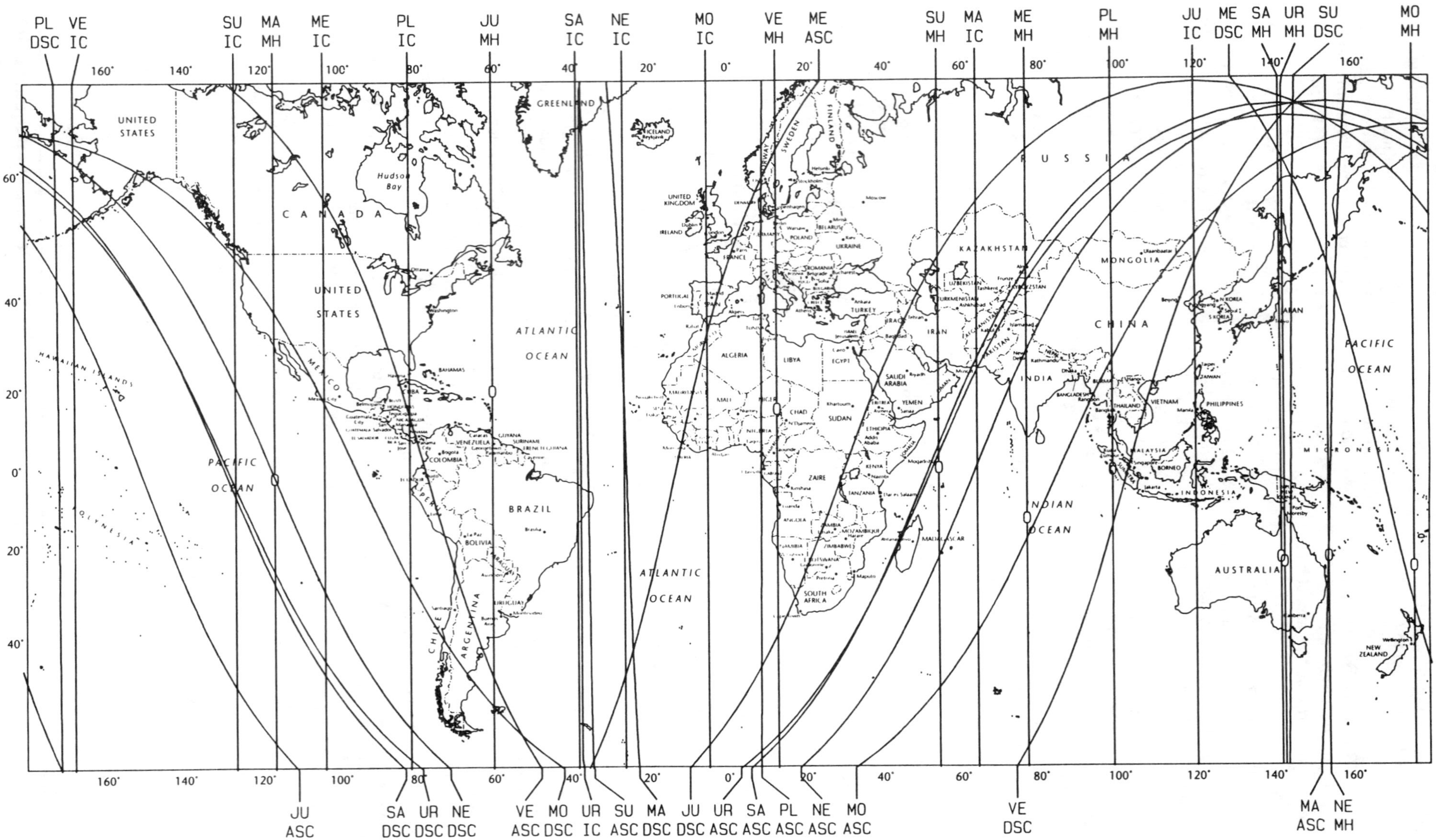

David/Martina 1st Meeting SR

September 21, 1988 8:14 AM GMT

04/16/96

ASTRO*CARTO*GRAPHY LATITUDE CROSSINGS

NORTH LATITUDES

69N29	SU/IC * VE/ASC
69N19	ME/DSC * JU/ASC
68N20	PL/MH * JU/DSC
68N16	SA/MH * JU/ASC
68N09	UR/MH * JU/ASC
67N39	SU/DSC * JU/ASC
67N39	MA/ASC * NE/DSC
67N37	VE/DSC * JU/ASC
67N33	JU/ASC * NE/ASC
67N32	SU/DSC * VE/DSC
67N31	SU/DSC * NE/ASC
67N31	UR/MH * SA/DSC
67N31	VE/DSC * NE/ASC
67N29	SA/DSC * NE/ASC
67N29	VE/DSC * SA/DSC
67N28	SU/DSC * SA/DSC
67N26	JU/ASC * SA/DSC
67N23	ME/DSC * SA/ASC
67N20	UR/MH * NE/ASC
67N16	SA/MH * NE/ASC
67N04	NE/MH * SA/DSC
66N44	ME/DSC * NE/ASC
66N40	MA/ASC * SA/DSC
66N22	VE/DSC * UR/DSC
66N22	SA/MH * UR/ASC
66N21	SU/DSC * UR/DSC
66N16	ME/DSC * UR/ASC
66N13	UR/MH * VE/DSC
66N13	MO/MH * NE/DSC
66N01	NE/MH * JU/ASC
66N01	JU/IC * SA/ASC
65N58	NE/MH * UR/DSC
65N57	JU/ASC * UR/DSC
65N50	SA/MH * VE/DSC
65N50	MA/MH * VE/ASC
65N50	UR/ASC * NE/ASC
65N36	MA/ASC * UR/DSC
65N16	MO/DSC * NE/DSC
65N12	MO/ASC * SA/DSC
64N55	MA/ASC * JU/ASC
64N54	MO/ASC * UR/DSC
64N48	MO/DSC * PL/DSC
64N40	JU/IC * UR/ASC
64N36	ME/DSC * VE/DSC
64N29	MO/ASC * JU/ASC
64N19	MO/ASC * MA/ASC
64N18	VE/IC * MO/DSC
63N48	JU/IC * NE/ASC
63N45	NE/MH * MO/ASC
63N35	MO/MH * SA/DSC
63N30	NE/DSC * PL/DSC
63N17	VE/MH * ME/ASC
63N05	ME/MH * JU/DSC
62N36	SA/MH * ME/DSC
62N32	MO/MH * UR/DSC
62N08	MO/ASC * SU/DSC
62N06	VE/IC * NE/DSC
61N50	UR/MH * ME/DSC
61N30	MO/ASC * ME/DSC
61N25	UR/MH * MO/ASC
61N13	Anchorage
61N11	SA/MH * MO/ASC
60N35	PL/MH * SA/ASC
59N53	ME/ASC * PL/ASC
59N20	Stockholm
59N03	SU/DSC * ME/DSC
58N51	SA/DSC * PL/DSC
58N50	PL/MH * UR/ASC
57N50	UR/DSC * PL/DSC
57N24	ME/IC * VE/ASC
57N01	MO/MH * JU/ASC
56N35	MA/IC * JU/DSC
56N31	VE/IC * SA/DSC

CENTRAL LATITUDES

55N45	Moscow
55N41	Copenhagen
55N32	VE/IC * UR/DSC
55N28	MO/ASC * VE/DSC
54N35	PL/MH * NE/ASC
52N30	Berlin
51N30	London
51N14	JU/IC * MO/ASC
49N14	Vancouver
48N50	Paris
48N15	SU/IC * MO/DSC
47N58	SU/MH * JU/DSC
47N36	Seattle
46N21	JU/IC * VE/DSC
46N01	NE/MH * ME/DSC
45N52	ME/MH * SA/ASC
45N32	Montreal
44N59	Minneapolis
44N37	JU/ASC * PL/DSC
43N39	Toronto
43N20	ME/MH * UR/ASC
42N22	Boston
42N20	Detroit
41N54	Rome
41N53	Chicago
41N24	ME/DSC * MA/ASC
40N43	New York
39N44	Denver
38N53	Washington
38N37	St. Louis
38N28	VE/IC * JU/ASC
38N02	MA/MH * MO/DSC
37N47	San Francisco
36N10	Las Vegas
35N40	Tokyo
34N04	Los Angeles
33N45	Atlanta
33N27	Phoenix
33N16	MO/IC * ME/ASC
32N47	Dallas
31N45	SA/DSC * UR/DSC
30N03	Cairo
29N58	New Orleans
29N46	Houston
29N12	ME/MH * NE/ASC
28N39	Delhi
27N43	MA/IC * SA/ASC
27N12	PL/MH * MO/ASC
25N47	Miami
24N51	MA/IC * UR/ASC
22N34	Calcutta
22N16	Hong Kong
21N37	SU/IC * NE/DSC
21N19	Honolulu
19N26	Mexico City
18N40	NE/IC * MA/DSC
17N33	ME/IC * MO/DSC
14N35	Manila
14N12	PL/IC * VE/ASC
10N30	Caracas
6N27	Lagos
6N24	SU/MH * SA/ASC
4N05	SU/MH * UR/ASC
1N21	MA/IC * NE/ASC
1S21	MA/MH * NE/DSC
4S05	SU/IC * UR/DSC
6S24	SU/IC * SA/DSC
14S12	PL/MH * VE/DSC

SOUTH LATITUDES

17S33	ME/MH * MO/ASC
18S40	NE/MH * MA/ASC
21S37	SU/MH * NE/ASC
22S54	Rio De Janeiro
24S51	MA/MH * UR/DSC
26S12	Johannesburg
27S12	PL/IC * MO/DSC
27S43	MA/MH * SA/DSC
29S12	ME/IC * NE/DSC
31S45	SA/ASC * UR/ASC
33S16	MO/MH * ME/DSC
33S55	Sydney
34S36	Buenos Aires
36S52	Auckland
37S50	Melbourne
38S02	MA/IC * MO/ASC
38S28	VE/MH * JU/DSC
41S24	ME/ASC * MA/DSC
43S20	ME/IC * UR/DSC
44S37	JU/DSC * PL/ASC
45S52	ME/IC * SA/DSC
46S01	NE/IC * ME/ASC
46S21	JU/MH * VE/ASC
47S58	SU/IC * JU/ASC
48S15	SU/MH * MO/ASC
51S14	JU/MH * MO/DSC
54S35	PL/IC * NE/DSC
55S28	MO/DSC * VE/ASC
55S32	VE/MH * UR/ASC
56S31	VE/MH * SA/ASC
56S35	MA/MH * JU/ASC
57S01	MO/IC * JU/DSC
57S24	ME/MH * VE/DSC
57S50	UR/ASC * PL/ASC
58S50	PL/IC * UR/DSC
58S51	SA/ASC * PL/ASC
59S03	SU/ASC * ME/ASC
59S53	ME/DSC * PL/DSC

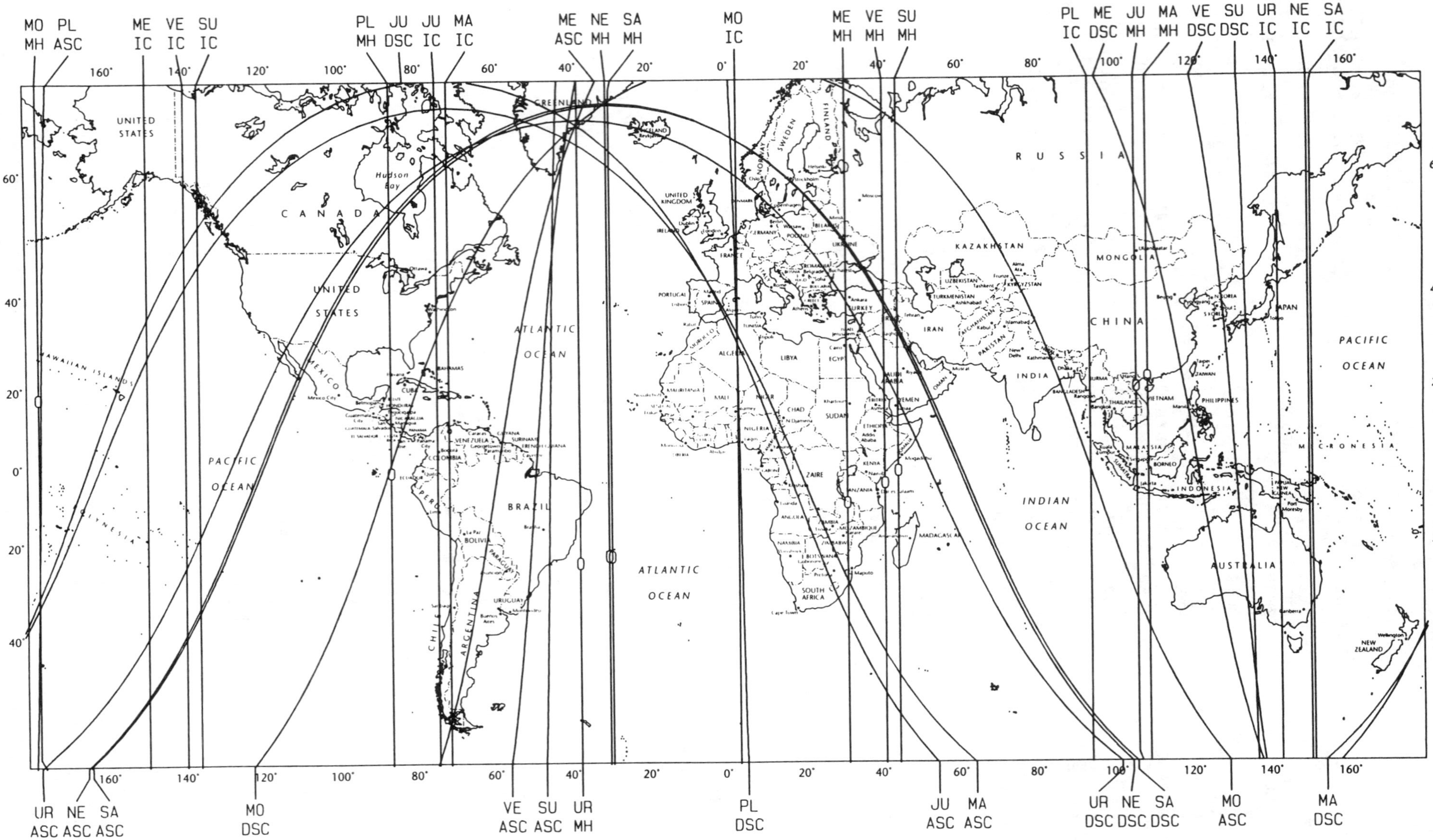

David/Martina Wedding

March 18, 1989 9:08 AM GMT

04/16/96

David/Martina Wedding
March 18, 1989 9:08 AM GMT

ASTRO*CARTO*GRAPHY LATITUDE CROSSINGS

NORTH LATITUDES

69N48 PL/MH * JU/DSC
69N13 ME/MH * MO/ASC
68N15 SA/MH * MO/DSC
68N04 NE/MH * MO/DSC
68N02 MO/DSC * NE/ASC
68N01 SA/MH * NE/DSC
67N56 ME/ASC * NE/ASC
67N51 NE/MH * SA/ASC
67N51 UR/MH * NE/ASC
67N51 MO/DSC * SA/ASC
67N47 VE/ASC * NE/ASC
67N43 ME/ASC * SA/ASC
67N40 JU/IC * MA/DSC
67N39 UR/MH * SA/ASC
67N34 VE/ASC * SA/ASC
67N30 SU/ASC * NE/ASC
67N25 JU/ASC * NE/ASC
67N15 SU/ASC * SA/ASC
67N15 JU/ASC * SA/ASC
67N13 SU/ASC * JU/ASC
67N03 VE/MH * MO/ASC
67N00 PL/MH * MA/DSC
66N30 MA/ASC * NE/ASC
66N25 ME/ASC * UR/DSC
66N25 MO/DSC * UR/DSC
66N24 VE/ASC * UR/ASC
66N24 JU/ASC * UR/ASC
66N24 VE/ASC * JU/ASC
66N22 MA/ASC * SA/ASC
66N19 SU/ASC * UR/ASC
66N14 NE/MH * UR/DSC
66N11 SA/MH * UR/DSC
66N08 UR/MH * ME/ASC
66N02 MA/ASC * UR/ASC
65N59 UR/MH * MO/DSC
65N59 SU/MH * MO/ASC
65N56 ME/ASC * JU/ASC
65N55 MO/DSC * JU/ASC
65N55 MO/DSC * ME/ASC
65N52 UR/MH * JU/ASC
65N07 SU/ASC * MA/ASC
65N07 MO/DSC * VE/ASC
64N52 SA/ASC * UR/ASC
64N41 MO/DSC * MA/ASC
64N36 SA/DSC * NE/DSC
64N29 SA/DSC * PL/DSC
64N29 VE/ASC * MA/ASC
64N29 NE/DSC * PL/DSC
64N15 ME/ASC * MA/ASC
64N10 MO/IC * SA/DSC
64N09 MO/IC * NE/DSC
63N54 UR/ASC * NE/ASC
63N49 MO/DSC * SU/ASC
63N47 UR/MH * MA/ASC
63N23 NE/MH * JU/ASC
63N04 SA/MH * JU/ASC
62N21 MA/IC * UR/ASC
61N58 ME/ASC * VE/ASC
61N49 MA/IC * NE/ASC
61N30 JU/IC * UR/ASC
61N26 NE/MH * MA/ASC
61N20 MA/IC * SA/ASC
61N13 Anchorage
61N08 SA/MH * MA/ASC
60N41 JU/IC * NE/ASC
60N16 UR/DSC * PL/DSC
60N09 JU/IC * SA/ASC

59N53 MO/IC * UR/DSC
59N25 JU/MH * ME/DSC
59N20 Stockholm
58N06 SU/ASC * ME/ASC
56N40 PL/MH * UR/ASC
55N45 Moscow

CENTRAL LATITUDES

55N41 Copenhagen
54N11 PL/MH * NE/ASC
54N03 MA/MH * ME/DSC
53N24 PL/MH * SA/ASC
53N15 SU/IC * JU/DSC
52N30 Berlin
51N30 London
50N16 ME/MH * SA/DSC

49N53 VE/IC * JU/DSC
49N48 ME/MH * NE/DSC
49N14 Vancouver
48N50 Paris
48N40 SU/ASC * VE/ASC
47N36 Seattle
46N48 SU/IC * MA/DSC
45N32 Montreal
44N59 Minneapolis
43N39 Toronto
43N31 MA/ASC * JU/ASC
42N52 VE/IC * MA/DSC
42N22 Boston
42N20 Detroit
41N54 Rome
41N53 Chicago
40N43 New York

39N44 Denver
39N25 VE/MH * SA/DSC
39N20 ME/MH * UR/DSC
38N53 Washington
38N37 St. Louis
38N33 VE/MH * NE/DSC
38N09 MA/IC * MO/DSC
37N47 San Francisco
36N38 ME/IC * JU/DSC
36N10 Las Vegas
35N40 Tokyo
34N08 SU/MH * SA/DSC
34N04 Los Angeles
33N45 Atlanta
33N27 Phoenix
33N04 SU/MH * NE/DSC
32N54 MA/ASC * PL/DSC
32N47 Dallas
32N42 MO/IC * MA/ASC
32N31 JU/IC * MO/DSC
30N51 JU/ASC * PL/DSC
30N42 MO/IC * JU/ASC
30N03 Cairo

29N58 New Orleans
29N46 Houston
28N39 Delhi
28N01 ME/IC * MA/DSC
26N51 MO/IC * PL/DSC
25N47 Miami
23N57 VE/MH * UR/DSC
22N34 Calcutta
22N16 Hong Kong
21N19 Honolulu

19N26 Mexico City
16N58 SU/MH * UR/DSC
14N35 Manila
10N30 Caracas

6N27 Lagos
1N14 PL/MH * MO/DSC

1S14 PL/IC * MO/ASC

16S58 SU/IC * UR/ASC

22S54 Rio De Janeiro

SOUTH LATITUDES

23S57 VE/IC * UR/ASC
26S12 Johannesburg
26S51 MO/MH * PL/ASC
28S01 ME/MH * MA/ASC

30S42 MO/MH * JU/DSC
30S51 JU/DSC * PL/ASC
32S31 JU/MH * MO/ASC
32S42 MO/MH * MA/DSC
32S54 MA/DSC * PL/ASC
33S04 SU/IC * NE/ASC
33S55 Sydney
34S08 SU/IC * SA/ASC
34S36 Buenos Aires
36S38 ME/MH * JU/ASC
36S52 Auckland
37S50 Melbourne
38S09 MA/MH * MO/ASC
38S33 VE/IC * NE/ASC
39S20 ME/IC * UR/ASC
39S25 VE/IC * SA/ASC

42S52 VE/MH * MA/ASC
43S31 MA/DSC * JU/DSC
46S48 SU/MH * MA/ASC
48S40 SU/DSC * VE/DSC
49S48 ME/IC * NE/ASC
49S53 VE/MH * JU/ASC

50S16 ME/IC * SA/ASC
53S15 SU/MH * JU/ASC
53S24 PL/IC * SA/DSC
54S03 MA/IC * ME/ASC
54S11 PL/IC * NE/DSC
56S40 PL/IC * UR/DSC
58S06 SU/DSC * ME/DSC
59S25 JU/IC * ME/ASC
59S53 MO/MH * UR/ASC

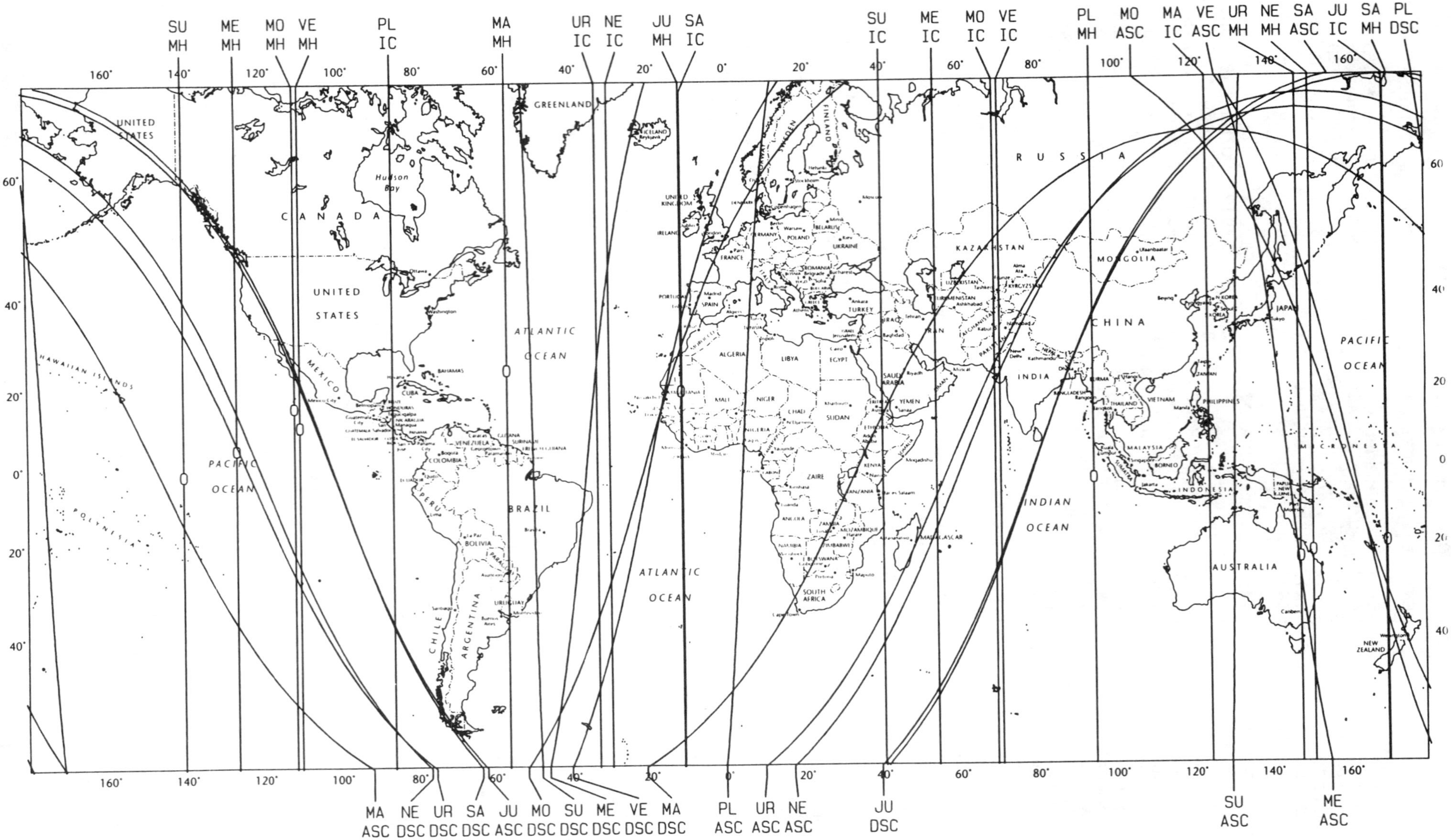

David/Martina Wedding SR

March 18, 1991 9: 24 PM GMT

04/16/96

04/16/96

David/Martina Wedding SR
March 18, 1991 9:24 PM GMT

ASTRO*CARTO*GRAPHY LATITUDE CROSSINGS

NORTH LATITUDES

69N55 SA/MH * JU/ASC
69N45 JU/ASC * PL/DSC
69N32 NE/MH * SA/ASC
69N09 UR/MH * SA/ASC
68N55 NE/MH * JU/DSC
68N37 ME/ASC * VE/ASC
68N36 VE/DSC * PL/ASC
68N32 UR/MH * JU/DSC
68N20 UR/MH * NE/ASC
68N18 JU/DSC * NE/ASC
68N04 SA/ASC * NE/ASC
67N24 SU/ASC * NE/ASC
67N18 JU/IC * NE/DSC
67N17 SA/MH * NE/DSC
67N15 VE/ASC * NE/ASC
67N06 ME/ASC * NE/ASC
66N52 NE/MH * UR/DSC
66N31 SU/ASC * VE/ASC
66N30 JU/DSC * UR/ASC
66N20 VE/ASC * SA/ASC
66N15 SU/ASC * SA/ASC
66N15 VE/ASC * UR/ASC
66N12 SU/ASC * UR/ASC
66N11 SA/ASC * UR/ASC
66N07 MA/IC * NE/ASC
66N03 ME/ASC * UR/ASC
65N52 NE/DSC * PL/DSC
65N51 VE/ASC * JU/DSC
65N51 ME/ASC * SA/ASC
65N35 SU/ASC * JU/DSC
65N21 JU/IC * UR/DSC
65N18 SA/MH * UR/DSC
65N17 ME/ASC * JU/DSC
65N14 MO/ASC * NE/ASC
65N06 MA/IC * UR/ASC
64N45 MA/DSC * UR/ASC
64N43 MO/ASC * MA/DSC
64N43 MA/ASC * SA/ASC
64N39 MA/ASC * JU/DSC
64N35 MA/DSC * NE/ASC
64N34 MO/ASC * UR/ASC
64N33 ME/ASC * MA/ASC
64N31 SU/ASC * MA/ASC
64N18 VE/ASC * MA/ASC
63N44 SU/ASC * ME/ASC
63N24 UR/DSC * PL/DSC
63N23 MA/IC * SA/ASC
63N13 MO/ASC * SA/ASC
63N00 MA/IC * MO/ASC
62N50 UR/MH * MA/ASC
62N49 MO/ASC * JU/DSC
62N42 MA/IC * JU/DSC
62N14 NE/MH * MA/ASC
61N33 PL/MH * MA/DSC
61N13 Anchorage
60N27 SU/MH * SA/DSC

59N50 MO/DSC * PL/ASC
59N29 SU/MH * JU/ASC
59N20 Stockholm
57N17 MO/ASC * SU/ASC
56N21 JU/IC * MA/ASC
56N14 SA/MH * MA/ASC
55N46 UR/ASC * NE/ASC
55N45 Moscow
55N41 Copenhagen
54N49 PL/MH * UR/ASC
54N43 PL/MH * NE/ASC
52N31 MO/ASC * ME/ASC
52N30 Berlin
51N36 VE/IC * MA/DSC
51N30 London
50N39 MO/IC * MA/DSC
50N00 ME/MH * SA/DSC

CENTRAL LATITUDES

49N14 Vancouver
48N57 MA/ASC * PL/DSC
48N50 Paris
48N47 ME/MH * JU/ASC
48N28 UR/MH * VE/ASC
47N36 Seattle
46N41 NE/IC * ME/DSC
45N32 Montreal
44N59 Minneapolis
43N39 Toronto
42N22 Boston
42N20 Detroit
41N54 Rome
41N53 Chicago
40N46 NE/MH * VE/ASC
40N43 New York
40N33 SU/MH * NE/DSC

39N44 Denver
38N53 Washington
38N37 St. Louis
37N47 San Francisco
36N50 PL/MH * SA/ASC
36N27 SA/IC * VE/DSC
36N19 PL/MH * JU/DSC
36N16 ME/IC * MA/DSC
36N10 Las Vegas
35N41 JU/MH * VE/DSC
35N40 Tokyo
35N03 UR/MH * MO/ASC
34N04 Los Angeles
33N53 SU/MH * UR/DSC
33N45 Atlanta
33N27 Phoenix
32N47 Dallas
30N45 SA/IC * MO/DSC
30N09 JU/MH * MO/DSC
30N03 Cairo

29N58 New Orleans
29N46 Houston
28N39 Delhi
28N17 VE/IC * UR/ASC
27N18 UR/IC * ME/DSC
27N16 NE/MH * MO/ASC
26N16 MO/MH * SA/DSC
25N47 Miami
25N44 MO/IC * UR/ASC
24N58 MO/MH * JU/ASC
23N50 VE/IC * NE/ASC
23N01 VE/MH * SA/DSC
22N34 Calcutta
22N16 Hong Kong
21N47 VE/MH * JU/ASC
21N19 Honolulu
20N51 MO/IC * NE/ASC

19N26 Mexico City
18N56 JU/DSC * SA/ASC
15N15 MO/DSC * VE/DSC
15N11 ME/MH * NE/DSC
14N35 Manila
13N40 SU/IC * MA/DSC
10N30 Caracas

7N00 ME/MH * UR/DSC
6N27 Lagos

7S00 ME/IC * UR/ASC

13S40 SU/MH * MA/ASC
15S11 ME/IC * NE/ASC
15S15 MO/ASC * VE/ASC
18S56 JU/ASC * SA/DSC

SOUTH LATITUDES

20S51 MO/MH * NE/DSC
21S47 VE/IC * JU/DSC
22S54 Rio De Janeiro
23S01 VE/IC * SA/ASC
23S50 VE/MH * NE/DSC
24S58 MO/IC * JU/DSC
25S44 MO/MH * UR/DSC
26S12 Johannesburg
26S16 MO/IC * SA/ASC
27S16 NE/IC * MO/DSC
27S18 UR/MH * ME/ASC
28S17 VE/MH * UR/DSC

30S09 JU/IC * MO/ASC
30S45 SA/MH * MO/ASC
33S53 SU/IC * UR/ASC
33S55 Sydney
34S36 Buenos Aires
35S03 UR/IC * MO/DSC
35S41 JU/IC * VE/ASC
36S16 ME/MH * MA/ASC
36S19 PL/IC * JU/ASC
36S27 SA/MH * VE/ASC
36S50 PL/IC * SA/DSC
36S52 Auckland
37S50 Melbourne

40S33 SU/IC * NE/ASC
40S46 NE/IC * VE/DSC
46S41 NE/MH * ME/ASC
48S28 UR/IC * VE/DSC
48S47 ME/IC * JU/DSC
48S57 MA/DSC * PL/ASC
50S00 ME/IC * SA/ASC

50S39 MO/MH * MA/ASC
51S36 VE/MH * MA/ASC
52S31 MO/DSC * ME/DSC
54S43 PL/IC * NE/DSC
54S49 PL/IC * UR/DSC
55S46 UR/DSC * NE/DSC
56S14 SA/IC * MA/DSC
56S21 JU/MH * MA/DSC
57S17 MO/DSC * SU/DSC
59S29 SU/IC * JU/DSC
59S50 MO/ASC * PL/DSC

Chapter 12/Astro*Carto*Graphy for Clients

When an official Astro*Carto*Graphy map is purchased it is supplied with an interpretation booklet written by Jim Lewis. For most people this booklet is sufficient to find out which planetary influences are at certain locations.

The advantage of Astro*Carto*Graphy is that it can be used by people who have no knowledge of astrology. The Astro*Carto*Graphy interpretation booklet gives meanings for all forty planetary lines and the forty-five different combinations of latitude crossings.

Once the basics of Astro*Carto*Graphy are understood the next step is to purchase a transparent Cyclo*Carto*Graphy overlay which is placed on top of the Astro*Carto*Graphy map. It shows where the progressed and transiting planets are on the Astro*Carto*Graphy map, and an interpretation booklet written by Jim Lewis is provided. This booklet explains how to apply the various Cyclo*Carto*Graphy techniques and obtain additional information about interpreting the Astro*Carto*Graphy map.

The Astro*Carto*Graphy and Cyclo*Carto*Graphy interpretation booklets are only available for customers who purchase the marketed maps and overlays. Names and addresses of companies that are authorized distributors of Astro*Carto*Graphy products and services are included in Appendix 1.

With some astrological computer programs now including Astro*Carto*Graphy it is possible for many astrologers to use this technique. As a result most astrologers see no reason why they should purchase an official Astro*Carto*Graphy map.

Many astrologers do not know how to interpret the maps, so I wasn't surprised to be asked to recommend a good book on the subject.

Other astrologers who can draw Astro*Carto*Graphy maps on a computer have some knowledge about how to interpret the planetary lines. Usually, they purchased their own Astro*Carto*Graphy maps and found the interpretation booklet provided good information about the technique.

When I give a lecture on Astro*Carto*Graphy I start by asking the audience two questions. The first is how many astrologers in the audience have knowledge of Astro*Carto*Graphy. Usually, about a quarter put up their hands. Then I ask how many have knowledge of Cyclo*Carto*Graphy, which always takes most of them by surprise. Sometimes no astrologers in the audience have ever heard about Cyclo*Carto*Graphy. Other times no more than two know what Cyclo*Carto*Graphy is, and that's usually because they have purchased their own overlay.

Quite a few astrologers use Astro*Carto*Graphy for clients who ask for advice about moving to a different location. As Astro*Carto*Graphy is available in some software it is easy to use these programs to draw maps, but when they advise clients by examining only the Astro*Carto*Graphy map they are making a serious mistake. It is essential that the temporary influences on the Astro*Carto*Graphy map be taken into consideration as they provide important additional information. But Cyclo*Carto*Graphy is not included in the computer programs that can draw Astro*Carto*Graphy maps.

In Chapter 5 I explained how the various Cyclo*Carto*Graphy techniques can be used without having to purchase an official overlay. Other techniques to find the temporary influences on an Astro*Carto*Graphy map were explained in Chapters 6 and 7. It is recommended that you have a thorough understanding of these chapters before using Astro*Carto*Graphy for clients.

Most clients who request an Astro*Carto*Graphy interpretation are unhappy in life at their present location and would like to move to a better place. But there are other ways in which Astro*Carto*Graphy can be used for clients.

I've received many requests for Astro*Carto*Graphy interpretations from businessmen who wanted to expand into the overseas market. They want to know which locations are most beneficial for them for expansion and when they should do so. The information that a businessman receives from an Astro*Carto*Graphy interpretation will tell him at which locations he'll have the most success. Where else could he obtain such detailed information about expanding his business abroad?

Most important is that it is not necessary to visit a location for the planetary influences there to have an effect. This is known as remote activation and can manifest in many different ways.

There are many possibilities for investing money, and by using Astro*Carto*Graphy it is possible to discover which countries and locations would be most beneficial. Having found a good place to invest, there is no need to travel there. Doing business at the location from far away is enough for the planetary lines there to have an influence. Even if the location is never visited during the time that money is invested there, the planetary energies will still be working.

Ten years ago I used my Astro*Carto*Graphy map to find a good place to invest money on a long-term basis. The value of my investment has increased **fivefold**, which works out at fifty percent profit per year. Even I couldn't imagine the huge success I would achieve with this investment.

The effects of planetary lines can also come to you via other people. Even if you have never visited the location where a person comes from, you will experience the planetary influences at that location. It's as if the person brings the planetary influences at a location to you. When you are with this person the planetary lines at the location he or she comes from will manifest in your life.

In Chapter 4 I gave an example of remote activation when I examined Matthew's Astro*Carto*Graphy map. He has never visited Bangkok in his life but still experienced the effects of VE ASC and MO IC at that location. He met a woman from Bangkok and they were attracted to each other. They met in England, so Matthew experienced the planetary influences at Bangkok via this woman.

The **birth** location of the other person is usually the one that is activated when you are together. The planetary lines at this person's place of birth will explain what you will experience in life.

It is also possible for other people to use your planetary lines at a location. If you live under a MA DSC line you could attract people who are aggressive, violent and bossy.

No matter which country you reside in you will have the opportunity to purchase many different products. Usually, you know only which country a product was made in, but most times this will be enough to use Astro*Carto*Graphy. Many countries are small in size, so the planetary lines will mostly have an effect at every location in a particular country.

If a product is made in America, Canada, Australia, China, etc., then there will be many planetary lines passing through the country. Unless you can find out the exact location, it is not possible to use Astro*Carto*Graphy.

Buying a car is something that most people take for granted. They do so more for how a car looks and performs than where it was produced. There are many reasons why a person decides to buy a certain type of car, however, I've yet to meet someone who has done so by using Astro*Carto*Graphy.

You probably haven't thought about the large range of foreign cars available and what effects they will have in your life. I strongly recommend that you examine your own Astro*Carto*Graphy map before deciding which car to buy.

In my life I have owned six different types of cars. The first two were British, which had more than their fair share of problems. This is confirmed by the MA IC and SA MH lines I have within orb of Britain. The best cars I've had for reliability and overall satisfaction were produced in Japan and Spain. In Japan I have SU MH, JU ASC and UR ASC, whereas JU DSC is powerful in Spain.

Other products that can be purchased by using Astro*Carto*Graphy are computers, stereo equipment, fax machines, clothes, food, etc. You will not be surprised that I bought a Japanese fax machine.

The Astro*Carto*Graphy planetary lines can also have an effect in your job. If you work for a foreign company, your Astro*Carto*Graphy map will have an effect by remote activation. You can find out interesting information when you examine the location where the company's head office is located.

Perhaps your boss is a foreigner, so again your Astro*Carto*Graphy map can be used to obtain further information. If there are any type of foreign connections in your job I recommend that you examine your own individual Astro*Carto*Graphy map.

One of the most unusual ways in which an Astro*Carto*Graphy map can be used is to obtain a talisman. There is a big business selling lucky charms, but how many people consider where these things come from?

There are many different reasons why people purchase lucky charms. Some would like to improve their financial situation, whereas others are looking for better luck in partnerships.

When an Astro*Carto*Graphy map is used to obtain a talisman you'll be amazed with the results. If you are interested in finding a partner, a talisman from a location on one of your VE lines will have a powerful influence.

In Chapter 11 I explained that I have a JU MH line exactly in Phoenix. When I was in Phoenix I purchased a western style tie which I thought nothing about at the time. Years later I noticed that whenever I wore this tie I always had good luck in my life! I then realized that using Astro*Carto*Graphy to obtain a talisman offered enormous potential for people to improve their lives.

Many of my clients have reported that past lives are shown on an Astro*Carto*Graphy map. I have no personal experience of Astro*Carto*Graphy and reincarnation, but it is certainly an interesting area.

Interpretation of dreams could also be helped by examining an Astro*Carto*Graphy map. If you dream about a certain place, the planetary influences there could show which potentials you should use in your life.

Maybe you've always wanted to visit a place but have never done so. People have significant planetary lines at their "dream holiday" locations.

Astro*Carto*Graphy can also be used to plan where to go for a holiday. Maybe you wonder why some holidays were terrific and others were terrible. Some people go to the same holiday destination year after year because they enjoy it there.

How to Approach Interpretations for Clients

After many years of doing Astro*Carto*Graphy interpretations for clients I have learned the most effective way to approach this type of astrology.

Clients usually request an Astro*Carto*Graphy interpretation because they are unhappy in life at their present location. These types of interpretations usually involve either relationships and/or career. Occasionally there are exceptions; for example, businessmen wishing to expand overseas or considering a job at a particular location. Surprisingly, I do not receive requests for interpretations from people who are interested in investments. I thought many people would ask for the best place to invest money or play a lottery.

There are basically four stages involved when doing an Astro*Carto*Graphy interpretation for a client:

1. Obtain background information from the client
2. Examine the Astro*Carto*Graphy map
3. Examine the Astro*Carto*Graphy temporary influences
4. Tell client outcome of the analysis

Obtain Background Information from Client

This may seem the obvious thing to do before beginning an interpretation, but if you are not careful you will miss the required vital information. I learned this fact the hard way when I gave irrelevant information to the client.

A client wanted to know where his best places in the world were. Without obtaining more information I examined his Astro*Carto*Graphy map for career and relationships, but he wanted to know where he could expand his business. As this client was happily married, the information about beneficial locations for love relationships was totally irrelevant.

Thus, when clients ask for their best place in the world, I ask: "The best place for what?" It is essential to find out why a person is looking for a good location so that you are in a better position to help. I also find it best to ask clients if they have any particular countries in mind. The answer you receive to this question will tell you if the client is prepared to travel a long distance to relocate.

There is no point examining the planetary lines in countries like Australia and New Zealand if the client wishes to remain in Europe. Other countries may be inappropriate due to the language or for other reasons. I usually ask clients if they can speak any foreign languages or if communication would be a problem in another country. As I speak to clients I make notes about what is important for the interpretation.

Even when I receive a requests for an interpretation by mail I telephone the clients to obtain as much background information as possible. Usually letters from clients requesting an interpretation do not give you all the information you need. By speaking to clients you obtain a clear picture of their circumstances and what they wish to achieve from the Astro*Carto*Graphy interpretation.

Examine the Astro*Carto*Graphy Map

Having obtained detailed information from the client it is time to examine the Astro*Carto*Graphy map. As you do so, put the client's natal chart next to it.

Also required in looking for locations on an Astro*Carto*Graphy map is a good atlas of the world. No expense should be spared in purchasing a top quality world atlas with an index of latitudes and longitudes.

I always start by examining the planetary influences at the location where the client lives. If the client is unhappy there, the Astro*Carto*Graphy map will confirm this. If the place of residence location shows that the client should be happy and successful, proceed with caution. In these situations I examine the natal chart to find out if the beneficial planetary influences at a location are afflicted. Then I examine the temporary influences affecting the place of residence location. If these do not explain why the client is unhappy in life at that location, the birth time is usually inaccurate.

I did an interpretation for a client who was unhappy and depressed at the location she lived. She had difficulty with relationships and wanted to know where she could find a partner. When I examined her Astro*Carto*Graphy map she had JU DSC exact where she was living. The influence from JU DSC did not explain why she unhappy where she resided. In her natal chart JU was not afflicted in any way. Also, she had no negative Astro*Carto*Graphy temporary influences at the place of residence location. It was then that I started to question her time of birth.

When I spoke to this client on the telephone she told me that she had lived at the location for more than ten years, and had always been unhappy there, so JU DSC was totally inappropriate. I then asked her if she was sure of her birth time. She replied that her birth time had been rectified with a pendulum by two Russian healers. She didn't know her time of birth, but assumed that it had been correctly rectified.

It is essential that you examine the client's place of residence location before you start the interpretation. It is also a good idea to examine other locations the client has lived to make sure the planetary influences provide the relevant information.

Having confirmed that the Astro*Carto*Graphy map is accurate, you can start the interpretation. Always keep in mind what the client wishes to achieve in life by relocating when you study the Astro*Carto*Graphy map.

How do you find the beneficial locations? The planetary lines on an Astro*Carto*Graphy map can be divided into two groups. Those that are beneficial and the others that should be avoided.

Beneficial lines:	JU MH, JU IC, JU ASC, JU DSC
	SU MH, SU IC, SU ASC, SU DSC
	VE MH, VE IC, VE ASC, VE DSC
	ME MH, ME IC, ME ASC, ME DSC
Lines to avoid:	SA MH, SA IC, SA ASC, SA DSC
	PL MH, PL IC, PL ASC, PL DSC
	NE MH, NE IC, NE ASC, NE DSC
	UR MH, UR IC, UR ASC, UR DSC
	MA MH, MA IC, MA ASC, MA DSC
	MO MH, MO IC, MO ASC, MO DSC

These are general guidelines as much depends on how the planets are aspected in the natal chart. That is why you should also study the natal chart to find out which lines to look for on the Astro*Carto*Graphy map.

If the client has a close T-square combination involving VE, SA, NE and MO, then VE lines should not be considered on the Astro*Carto*Graphy map. Any location on a VE line would activate this T-square, so this would not be a good place for partnerships. Instead, it would be better to examine locations on the JU DSC, SU DSC and ME DSC lines.

It could well be that the client has JU in close opposition aspect to SA in the natal chart. In this case the JU lines on the Astro*Carto*Graphy map would be next to SA, thus reducing the usual good luck from JU. If the declination of JU and SA are in narrow orb, then the ASC and DSC lines of JU and SA will be close together.

You will also find that some of the beneficial lines can be eliminated as they only pass through the ocean. Before you decide to forget a line that passes through the ocean do check if there are any small islands influenced, something it is easy to overlook.

Sometimes there will be two or more planetary lines having an influence at a location. Again, these can be divided into two groups, i.e. beneficial and those that should be avoided.

Beneficial planetary combinations:	SU/JU
	JU/MO
	JU/UR
	JU/PL
	SU/VE
	SU/JU
	ME/JU
	SU/ME
	SU/JU/UR
	SU/JU/PL
	SU/JU/UR/PL
	MO/JU/UR/PL
Planetary combinations to avoid:	SU/MA
	SU/SA
	SU/NE
	MA/PL
	MA/NE
	SU/MA/NE
	MA/NE/PL
	MA/UR
	SA/UR
	MA/SA/UR
	SU/SA/UR
	SU/SA/NE
	SU/SA/PL
	SA/UR/PL

These lists will give you an idea of what to look for and what you should avoid when examining locations on an Astro*Carto*Graphy map. Again, much depends on how the planets are aspected in the natal chart.

When you have examined the Astro*Carto*Graphy map in such a manner you will be able to see which planetary lines show the best possibilities. This is when you need to use a world atlas and the latitude crossings.

You may have read "west of line best" in the Astro*Carto*Graphy interpretation booklet. This suggests that there is a difference when a location is west or east of the planetary line. Jim Lewis told me that a location west of a planetary line has to do with internal realities and the planet is angular. When the location is east of a planetary line it involves external realities and the planet is cadent. See Diagram 19.

I have done my own research about locations west or east of a planetary line and found that there was hardly any difference. Sometimes one side of the line is better than the other to avoid the influence of other undesirable planetary lines.

Diagram 20 shows two locations having an influence from JU MH. Location A is west of the JU MH line, whereas location B is east. Location A is also influenced by NE IC which is not so good. Location B is near to VE MH.

Location B is much better as it is influenced by JU MH and VE MH. According to Jim Lewis' interpretation booklet it would be better when location B is west of JU MH and east of VE MH. In this particular case location B is the wrong side of JU MH and VE MH.

I have read Jim Lewis' interpretation booklet and made a list of whether west or east of line is best (see Diagram 21). Sometimes Jim Lewis writes that "east of line may be better," which is only a suggestion. Where Jim Lewis is not certain about which side of a line is best I have made an asterisk. For four of the planetary lines there is nothing written whether west or east of line is best.

By all means use this list for which side of a planetary line works best. But be prepared to use the other side of a line when it avoids the influence of other undesirable planetary lines.

If you are looking for locations concerning career, examine the MH lines first. Some of the other angular lines can also be considered for career, and sometimes I have found that there isn't much difference between the four lines for a planet. For example, Venus is connected to partnerships and relationships, so all the VE lines will have an effect in this area of life. DSC lines usually concern relationships, so these lines can also be considered.

After examining the beneficial planetary lines on the Astro*Carto*Graphy map you will have found particular cities or countries that look excellent for the client. Up to this stage you might not have not needed your atlas to find suitable locations, and you have also probably found numerous locations along or near the beneficial planetary lines.

The final stage of examining the Astro*Carto*Graphy map is the fine-tuning process to select those locations that are most beneficially influenced. This is achieved by using the latitude crossings on the Astro*Carto*Graphy map.

I find it best to first examine the list of latitude crossings and try to find where the beneficial crossings are located. Usually, about two or three latitude crossings will stand out as being positive influences.

Having located the latitudes where beneficial crossings occur I examine the Astro*Carto*Graphy map to find out if they activate any locations previously selected.

Then some adjustments may be needed in selecting

Diagram 19—Location West or East of Planetary Line

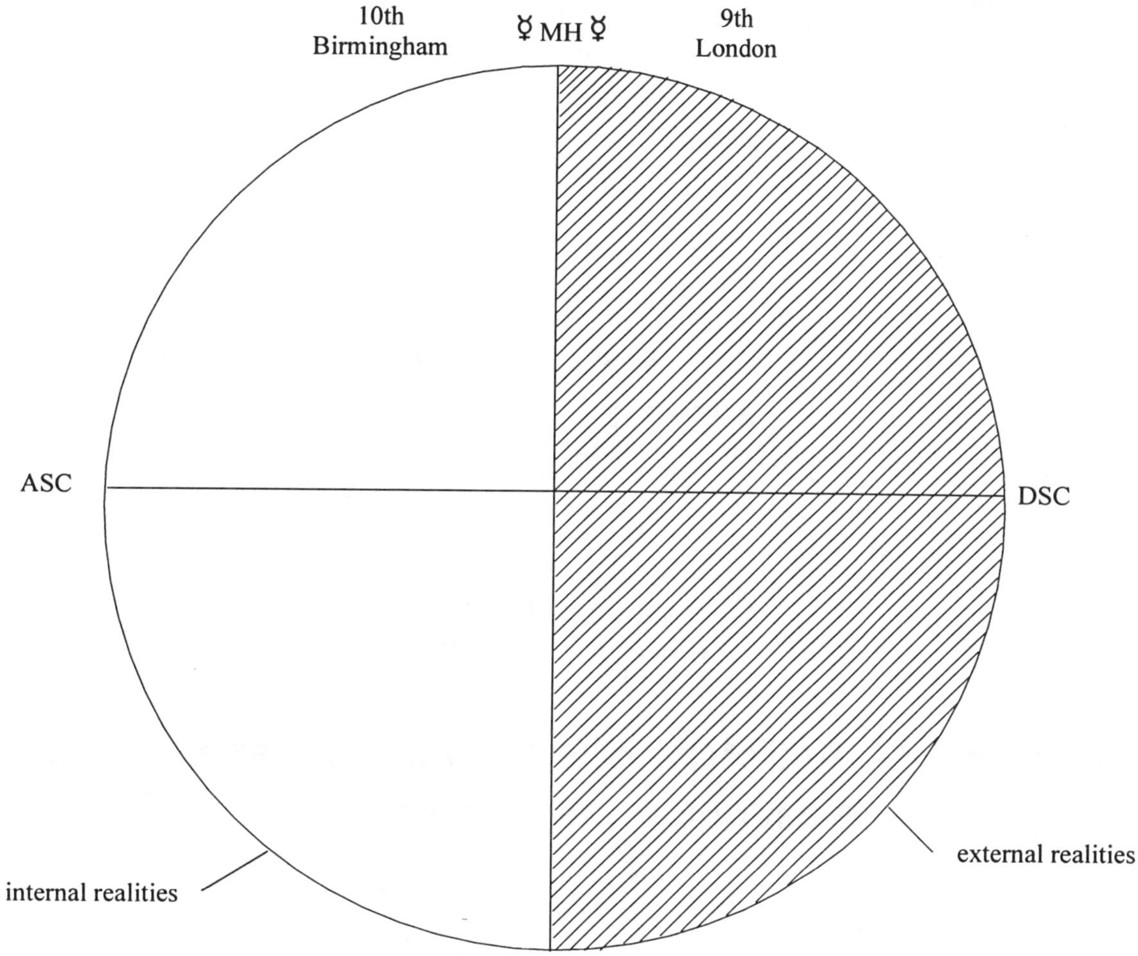

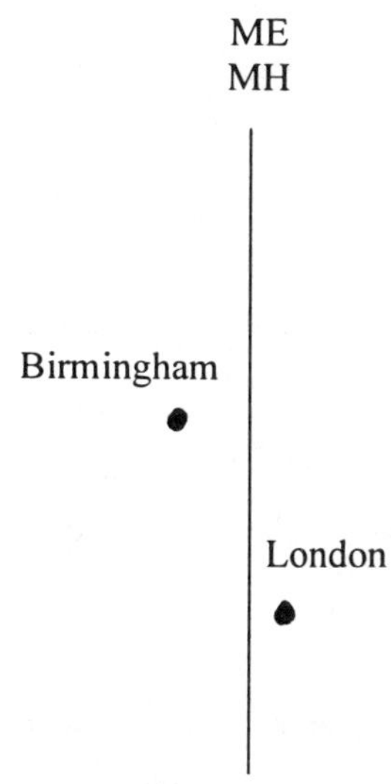

a) Location west of line—planet is angular
b) Location east of line—planet is cadent

Diagram 20—West or East of a Planetary Line

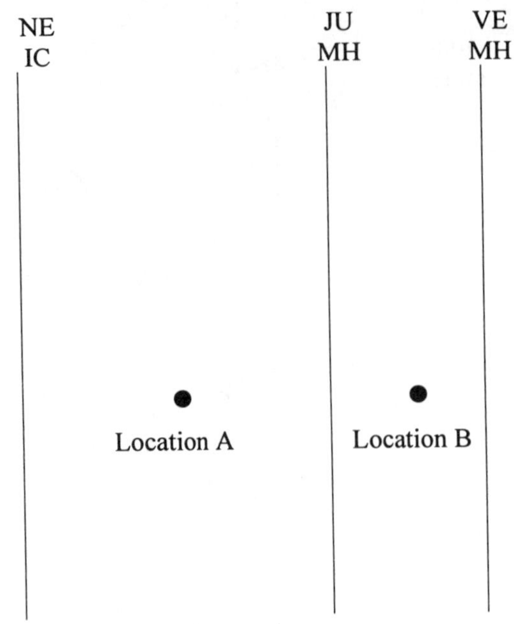

Location A = JU MH, NE IC
Location B = JU MH, VE MH

Diagram 21—Astro*Carto*Graphy Intepretation Booklet West or East of Planetary Line

	MH	ASC	IC	DSC
SU	West*	West	West	West
MO	East*	East*	1	West*
ME	West	West	East	East
VE	East	West	West	West
MA	West	West	East	2
JU	West	East	West	–
SA	–	East*	–	West
UR	West	East	East	West
NE	West	East*	East	East
PL	West	West	–	East*

(interior relaities) (exterior realities)

* = may be best (not certain)
1 = "things out of your control keep happening, especially west of this line"
2 = east of line problems are more self-originated than west
– = nothing written which side is best

suitable locations for the client. This is where an atlas is essential in fine-tuning locations along or near the beneficial planetary lines. The idea of this fine-tuning is to find a location that has favorable planetary lines and latitude crossings.

Sometimes it is not possible to find latitude crossings with positive planetary influences. In this case you need to examine the latitude crossings over a broad area and try to find where the minimum negative influences take place.

In countries like Sweden, Norway, Finland, etc. there are usually numerous latitude crossings which will tend to cancel out each other. When you find a location that has a long list of latitude crossings, ignore them.

The planetary lines at a location are more powerful than the latitude crossings, although when a latitude crossing is **exact** at a location it will have a more powerful influence.

Examine the Astro*Carto*Graphy Temporary Influences

Having found the most favorable locations for the client you now need to examine the temporary influences on the Astro*Carto*Graphy map. This is achieved by using Cyclo*Carto*Graphy, an Astro*Carto*Graphy solar return map and Astro*Carto*Graphy solar arc progressions. These techniques are explained in detail in Chapters 5, 6 and 7.

I find it best to write out a list of the temporary influences in two columns. One for the "anywhere aspects" and the other for "location aspects." Once you have used all the techniques to find the temporary influences you need to synthesize the information.

Usually, certain themes will stand out as you synthesize all the information for the temporary influences. Most important are the repetition of themes as they emphasize the trends that are likely to manifest during the time period under consideration.

The anywhere aspects will show the general influences that the person is likely to experience. Sometimes, from these aspects alone you can see that the client should wait a year or two before moving.

Suppose that a client has a T-square involving MO, MA, UR and NE with very close orbs. If C*C*G NE activated this T-square it would not be a good time for the client to relocate. The person's expectations would be too high and end in disappointment. It would be best for the client to wait until C*C*G NE has moved away from the T-square planets.

The Astro*Carto*Graphy map was used to select beneficial locations for the client. The temporary influences at these locations will show if the time is right to move there. If C*C*G MA, SA or NE is at one of the locations, then it is not a good time for the client to relocate, and it is better to wait until these temporary influences have moved away.

C*C*G PL at a location is an appropriate temporary influence when the person wants to make a new start in life. A move to a new country is usually a major change in life, so moving to a location where there is a C*C*G PL line can be a very positive influence.

C*C*G PL can be a difficult temporary influence when it is at the location you've lived for many years. When no major changes are made in life the person usually experiences the negative side of C*C*G PL.

C*C*G UR is also an appropriate temporary influence for making changes in life. By moving to a location where C*C*G UR is prominent it gives the person the chance to make the necessary changes.

As C*C*G JU moves very quickly across the Astro*Carto*Graphy map it usually doesn't stay long at a location. The exception is when JU is moving retrograde and when it is about to change direction. Then C*C*G JU will remain at a location for approximately two months.

When looking for a good time to move to a new location, the C*C*G JU can be used for fine-tuning. The time period that C*C*G JU is at the new location is what I usually look for. But it is important to take other factors into consideration when selecting the most appropriate time period for a person to move.

Cyclo*Carto*Graphy is the most important technique when examining the temporary influences. The Astro*Carto*Graphy solar return map and Astro*Carto*Graphy solar arc progressions give additional information to the Cyclo*Carto*Graphy temporary influences. Sometimes these auxiliary techniques provide repetition of themes to the Cyclo*Carto*Graphy temporary influences.

Most important is to examine the temporary influences at the place of residence location. These show if the person is prepared to relocate and what he or she will experience by remaining at the present location.

Temporary influences to the birth location are also significant. Even if the person has not lived at the birth location for many years the temporary influences will still have an effect. They will manifest as "anywhere aspects".

Tell Client Outcome of the Analysis

There are three different ways to tell the client what you have found in the Astro*Carto*Graphy analysis.

a. Consultation

b. Telephone

c. Written Report

The decision is made according to the client's circumstances, something known before you start the analysis. I ask the client for his or her preference.

Consultation

When I was living in Britain I had very few Astro*Carto*Graphy personal consultations for clients as most requested a written report. The British clients weren't prepared to travel a few hours to visit me, but since living in Munich and learning German, most of my interpretations are personal consultations.

With a personal consultation the initial approach is by telephone or letter. Once you have obtained the background information from the client, allow yourself plenty of time for the analysis. Do not make an appointment with the client at this stage.

After you have completed the analysis, contact the client to make an appointment for the consultation. Do not give any information about the analysis or you will end up doing a telephone interpretation.

When the client visits you for the consultation, don't rush straight into the interpretation. The client is obviously nervous about what you have found out, so give that person time to relax. Usually, ten or fifteen minutes general conservation with tea or coffee is sufficient to put the client at ease.

If you have a computer program with Astro*Carto*Graphy, have the client's map on the computer screen when you start the consultation. By doing so you can quickly examine the temporary influences for various dates if the client requests additional information.

Most important during the consultation is to be a good listener. Let the client speak and simply listen to what is said. When asked which locations are best for the client, present the information.

When explaining about temporary influences the client will ask if certain dates or months are good times. This is when you may need to use your computer to find out the answers. Clients are always amazed when you use an Astro*Carto*Graphy computer program to answer their questions.

After the client has left, make notes about what was discussed during the consultation. You should always make a photocopy of the client's Astro*Carto*Graphy map for future reference purposes. Most clients will contact you later to report back about their experiences. When they do, you can quickly refer to their Astro*Carto*Graphy map and your notes.

There are times when a client wants to make an appointment for a consultation at short notice. Provided I'm not completely booked I will try to arrange an appointment within the next few days. Usually this means I do not have enough time to do preparation work for the consultation, and under these circumstances I wait until the client arrives and see how things develop during the consultation. These types of consultations often turn out to be the most interesting as you do not know what kind of advice the client is seeking.

Telephone

Clients prefer consultations by telephone rather than those recorded on cassette. A telephone consultation is advantageous for the client as questions can be asked.

After obtaining the background information from the client, ask them to send their original Astro*Carto*Graphy map. Most clients who request an Astro*Carto*Graphy interpretation have their own map and need specialized advice. If the client does not possess an individual Astro*Carto*Graphy map, recommend that it be purchased.

At this stage I always ask for advance payment for the interpretation and, if necessary, for the Astro*Carto*Graphy map. There is nothing worse than having completed the interpretation, only to be faced with problems in getting the client to pay the invoice. Unfortunately, there are many people who do not pay invoices or are slow to send money. I tell clients I will start the interpretation when payment has been received.

When the analysis work is completed, make a photocopy of the client's Astro*Carto*Graphy map. Then return the original map to the client with the request that he or she contact you for an appointment.

The reason for this is that the client will be able to refer to the Astro*Carto*Graphy map during the telephone consultation. Whenever possible, make the appointment as soon as possible while the analysis is still fresh in your mind.

The telephone consultation is similar to the personal consultation except for the in-person contact. Again, it is best to let the client speak to you and listen to what is said. On average a telephone consultation lasts approximately one hour. During that time you have the client's Astro*Carto*Graphy map on the computer screen in case the client asks additional questions.

Written Report

Many of the Astro*Carto*Graphy interpretations I do are written reports. They take the most time to complete. As with telephone consultations I ask the client to send me the original Astro*Carto*Graphy map with advance payment. Again, I recommend that clients purchase their own individual Astro*Carto*Graphy map if they don't have one.

Once the analysis has been completed, write out the information obtained. It is not necessary to write a report of thirty or more pages. What the client needs can be said in 1,000 to 1,500 words, including the planetary influences and temporary influences at various locations. The client wants to know **where** and **when**, which you can explain with a four to five page report. Before sending the report to the client, make a copy for yourself, along with the client's Astro*Carto*Graphy map; then they will be available when the client calls again.

Two examples will illustrate how to find favorable locations on an Astro*Carto*Graphy map. Then the temporary influences will be explored.

Example 1—Client A

Nigel's Astro*Carto*Graphy map was featured in Chapter 2. It emphasizes some important points in interpretation. This is a fictitious map, so imagine that Nigel asked you to do an interpretation. He is currently living in Plymouth in southwestern England, and is unhappy. He wants to move so he can be more successful in his career. Presently he is a surveyor for an estate agency, but prefers to be a real estate salesman. Due to the recession he hasn't had the confidence to branch out into selling real estate.

Nigel would like to make a new start in life and move away from Britain. It is most important that he move to an English speaking country as he has difficulties learning foreign languages. Because it is May 1993, the

Cyclo*Carto*Graphy planets can be calculated for June 1, 1993

Examine Nigel's Astro*Carto*Graphy map to try to find favorable locations for him. Once you have done this, examine the temporary influences to find a good time for Nigel to move. It doesn't matter if you make mistakes at this stage. More important is that you learn from any mistakes and improve your ability to complete an Astro*Carto*Graphy interpretation. The more interpretations you do, the better you will be.

Now, let's compare notes to see how you've done with the interpretation. Did you also examine the temporary influences for Nigel to find a good time for him to move?

Many astrologers will have quickly examined locations on or near Nigel's JU MH line to find suitable places. This seems to be the natural place to start as JU MH passes through the United States and Canada.

First, examine Nigel's natal chart to find out how the planets aspect each other. When you do this you will see that Nigel has an afflicted JU which needs to be taken into consideration when examining his Astro*Carto*Graphy map.

Nigel has a T-square combination involving JU, SU, MA and SA, so the usual good luck from JU will be somewhat subdued. This means that if Nigel moves to a location on or near one of his JU lines it will have a reduced effect. It could take Nigel much longer than usual to begin to experience the effect of a JU line.

In Nigel's case, look for locations where JU is not the only planetary influence. In the United States he has JU MH with ME ASC which would help in finding a suitable location.

At Plymouth where Nigel presently lives he has MO ASC and NE DSC powerfully placed. In fact, Plymouth is virtually on his MO ASC line. MO lines are usually too emotional for most people, especially for men, so I tend to avoid these lines on their own.

Even though VE MH and JU DSC are within orb of Plymouth, MO ASC and NE DSC are the strongest planetary influences. This combination isn't easy for relating with other people, so it is clear why Nigel is unhappy. The VE MH/JU DSC latitude crossing reiterates these planetary lines that are within orb of Plymouth.

As SU is part of the T-square combination with MA, SA and JU, do not examine the SU lines on Nigel's Astro*Carto*Graphy map. The SU is near SA, another reason not to consider SU lines.

This leaves ME, VE and JU, especially when JU is in combination with ME, VE, UR and PL. Next I'll examine each of these lines, with emphasis on locations in English speaking countries.

ME IC passes through New Zealand, but is not suitable as NE ASC is nearby. Looking along the ME IC line, it passes through some South Pacific islands. New Caledonia looks most promising, especially Noumea as it falls on the JU IC/VE DSC latitude crossing. But New Caledonia is owned by France, which makes it inappropriate for relocation. It would, however, be a good place for a holiday.

ME MH has an influence in Ireland, but this can't be considered because NE DSC is also powerful.

ME DSC can be ignored as it passes through Russia. Other locations along this line are inappropriate or influenced by undesirable planetary lines.

ME ASC looks like the most promising line, especially the eastern United States and Canada where JU MH also has an influence. This is an area to keep in mind when the latitude crossings are examined.

VE IC passes through the Pacific Ocean, so this line can be ignored.

VE DSC can also be eliminated as it is near the SU DSC and SA ASC lines.

VE MH has an influence in southeastern England where the JU DSC line has a powerful influence. But MO ASC and NE DSC would also remain within orb of southeastern England, so he would not be able to move away from these influences. However, southeastern England would certainly be much better than Plymouth.

VE ASC passes through Costa Rica and Cuba, but both are inappropriate due to the Spanish language. Looking at this line in finer detail, there is Jamaica, Cayman Islands and the Bahamas which offer good possibilities. In these areas JU MH also has a strong influence, so the latitude crossings will help in finding out which places are best.

This is all that can be found on the Astro*Carto*Graphy map. But a JU/PL combination offers a very good possibility as they are powerful in Alaska, i.e. PL MH and JU ASC. Probably not many astrologers noticed that Alaska is beneficial for Nigel.

Looking at the latitude crossings I noticed that JU DSC/PL DSC is **exact** at Anchorage. This place is also exactly on Nigel's JU ASC line and strongly influenced by PL MH. Although there are many latitude crossings at Anchorage the exact JU DSC/PL DSC is significant. This double JU/PL influence would make Anchorage a lucky place for Nigel. The powerful JU/PL planetary influence tends to overrule the effects of the T-square combination involving JU.

Initially, I thought Toronto would be suitable for Nigel, but this location is exactly on the MA ASC/SA ASC latitude crossing. This is not a positive latitude crossing as it would emphasize the JU, SA, MA and SU T-square combination.

The list of latitude crossings shows that places about thirty-eight or thirty-nine degrees would be the best. The JU IC/PL ASC crossing at 39N33 is where I will try to find a location. With an atlas I chose Fairmont, West Virginia (39N29 80W10). It is two degrees away from JU MH and also influenced by ME ASC.

The Cayman Islands are on the NE MH/ME ASC latitude crossing, which I would prefer to avoid. At Kingston, Jamaica the NE MH/ME ASC latitude crossing has a wider orb, but still has some influence.

In the Bahamas, west of JU MH would be best at about twenty-three to twenty-four degrees north latitude. At

Andros Town (Andros Island) there is the MO MH/VE DSC latitude crossing (24N21). This place is within one degree orb of the JU MH line and also influenced by VE ASC.

To sum up the analysis of Nigel's Astro*Carto*Graphy map, his best places are Anchorage, Andros Town and Fairmont.

Having found three appropriate locations for Nigel it is essential to examine the temporary influences to find out a good time to move. This involves a great deal of work, but is needed to obtain a detailed picture of the temporary influences.

A*C*G	**Age 47**	**C*C*G**
MO 02 ARI 57		SA 02 PIS 48
NE 08 LIB 30		JU 04 LIB 56
UR 12 GEM 48		SU 08 PIS 24
VE 14 CAP 17		VE 14 PIS 29
PL 15 LEO 31		MA 15 CAN 50
SU 20 CAP 10		ME 20 PIS 06
SA 23 CAN 24		NE 22 CAP 20
JU 24 LIB 20		UR 23 CAP 30
MA 27 CAN 51		PL 24 SCO 56
ME 29 SAG 51		MO 25 SAG 34

1. C*C*G NE = A*C*G SA
2. C*C*G UR = A*C*G SA
3. C*C*G MA DSC = Kingston (Jamaica)
4. Prog. Plymouth = C*C*G MO DSC

A*C*G	**(48 14)**	**Displaced C*C*G**
MO 02 ARI 57		ME 01 AQU 52
NE 08 LIB 30		NE 04 SAG 06
UR 12 GEM 48		UR 05 SAG 16
VE 14 CAP 17		PL 06 LIB 42
PL 15 LEO 31		MO 07 SCO 20
SU 20 CAP 10		SA 14 CAP 34
SA 23 CAN 24		JU 16 LEO 42
JU 24 LIB 20		VE 26 CAP 15
MA 27 CAN 51		MA 27 TAU 36
ME 29 SAG 51		

5. Disp. C*C*G SA = A*C*G VE

A*C*G	**(48 14)**	**Solar arc A*C*G**
MO 02 ARI 57		UR 01 LEO 02
NE 08 LIB 30		VE 02 PIS 31
UR 12 GEM 48		PL 03 LIB 45
VE 14 CAP 17		SA 11 VIR 38
PL 15 LEO 31		JU 12 SAG 34
SU 20 CAP 10		MA 16 VIR 05
SA 23 CAN 24		ME 18 AQU 05
JU 24 LIB 20		MO 21 TAU 11
MA 27 CAN 51		NE 26 SCO 44
ME 29 SAG 51		

6. Solar arc PL = A*C*G MO
7. Solar arc JU = A*C*G UR
8. Solar arc VE ASC = birthplace
9. Solar arc SU ASC = Plymouth
10. Solar arc MO DSC = Anchorage

The anywhere aspects and the location aspects are:

Anywhere aspects: C*C*G UR = A*C*G SA
C*C*G NE = A*C*G SA
solar arc PL = A*C*G MO
solar arc JU = A*C*G JU
solar arc VE ASC = birthplace

Location aspects: C*C*G MA DSC = Kingston
Prog. Plymouth = C*C*G MO DSC
solar arc SU ASC = Plymouth
solar arc MO DSC = Anchorage

The Astro*Carto*Graphy solar return map for Nigel's forty-seventh birthday has the following information:

Anywhere aspects: T-Square combination
SU, ME, UR, NE, MA, JU
Birthplace =
SU MH, UR MH, NE MH, MA IC

Location aspects: Fairmont = SU ASC, UR ASC, NE ASC, MA DSC, ME ASC
Anchorage = PL ASC
Plymouth = SU MH, UR MH, NE MH, MA IC
Jamaica + Andros Town = MO DSC

The anywhere aspects show a strong emphasis on the natal T-square involving SU, MA, JU and SA. C*C*G UR and C*C*G NE both activate this T-square, so this would not be a good time to move. It would be best for Nigel to wait until C*C*G NE has moved away from the T-square before relocating.

This will be difficult for him because it will take three to four years before C*C*G NE has moved away from his T-square. But moving when C*C*G NE is activating the T-square will not work out as Nigel expects.

The solar return chart shows a repetition of themes as to what was obtained from the Cyclo*Carto*Graphy anywhere aspects. The solar return T-square also activates the natal T-square, thus emphasizing the importance of this combination of planets. The solar return has SU MH, UR MH, NE MH, MA IC at Plymouth and his birthplace, which shows that they are important temporary influences.

In this case there is really no need to examine the location aspects temporary influences as a move cannot be recommended. All I can suggest is that he consider southeastern England as a possibility where he has JU DSC and VE MH. But he should also remember that the C*C*G UR and C*C*G NE will have an effect in southeastern England.

What Nigel could do is visit Anchorage, Andros Town and Fairmont for a holiday to find out how things are for him in these places. Then, a few years later he could decide which place is best. I would also suggest that he obtain a talisman from Anchorage where he has JU ASC and PL MH. The combined energies of JU and PL usually bring extraordinary good luck, so a talisman would be a good way to pick up these energies.

Example 2—Client B

This Astro*Carto*Graphy map was also included in Chapter 2. Although it is fictitious, it is a good example as the temporary influences are interesting. This time it is possible to find a good time when the client can move to a new location.

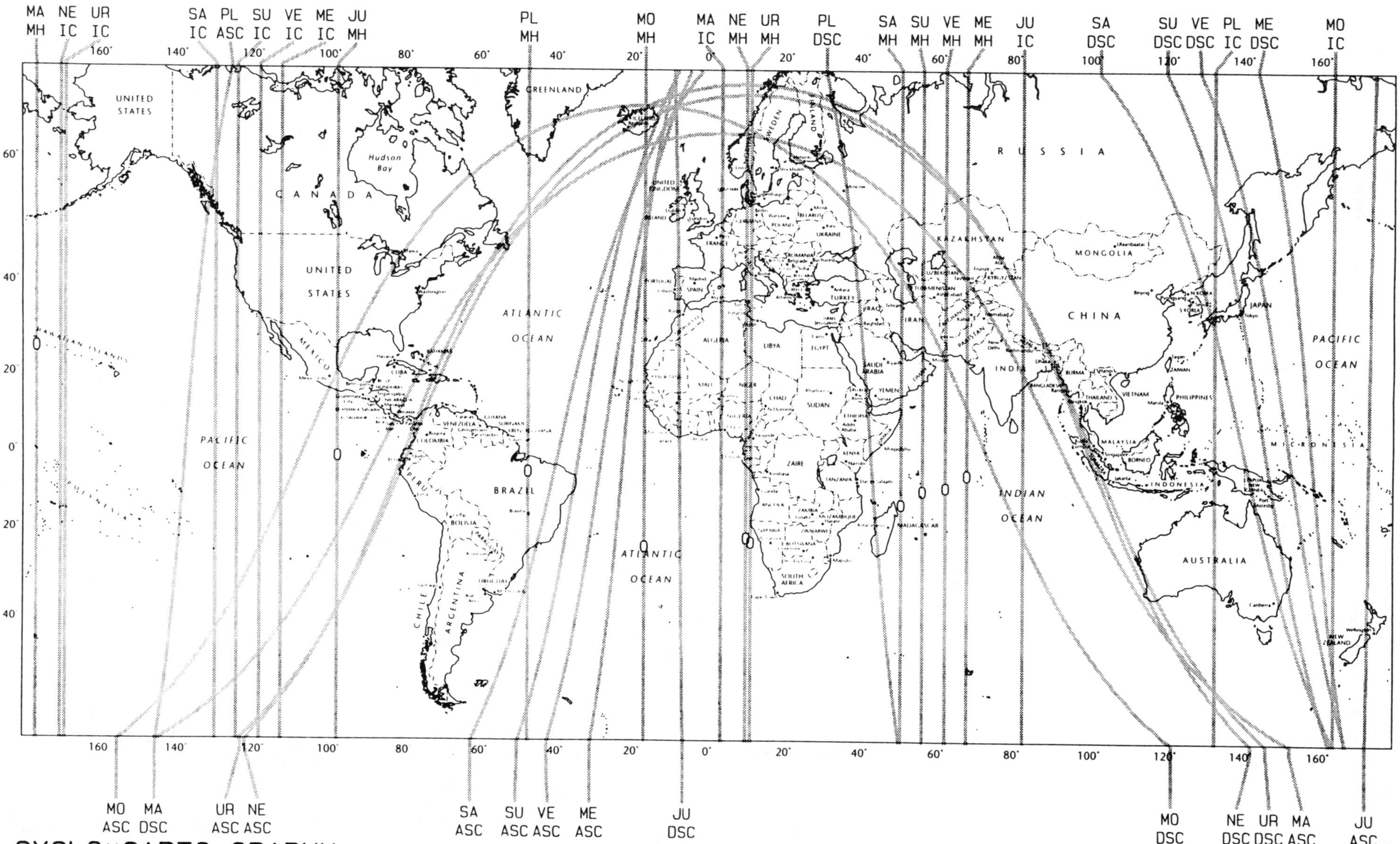

CYCLO*CARTO*GRAPHY

04/17/96

Nigel

SU, MO, ME, VE AND MA ARE PROGRESSED
JU, SA, UR, NE AND PL ARE TRANSITING

PROG	2	25	46	20:59 GMT
TRANSIT	6	1	93	0:00 GMT
NATAL	1	9	46	11:32 GMT

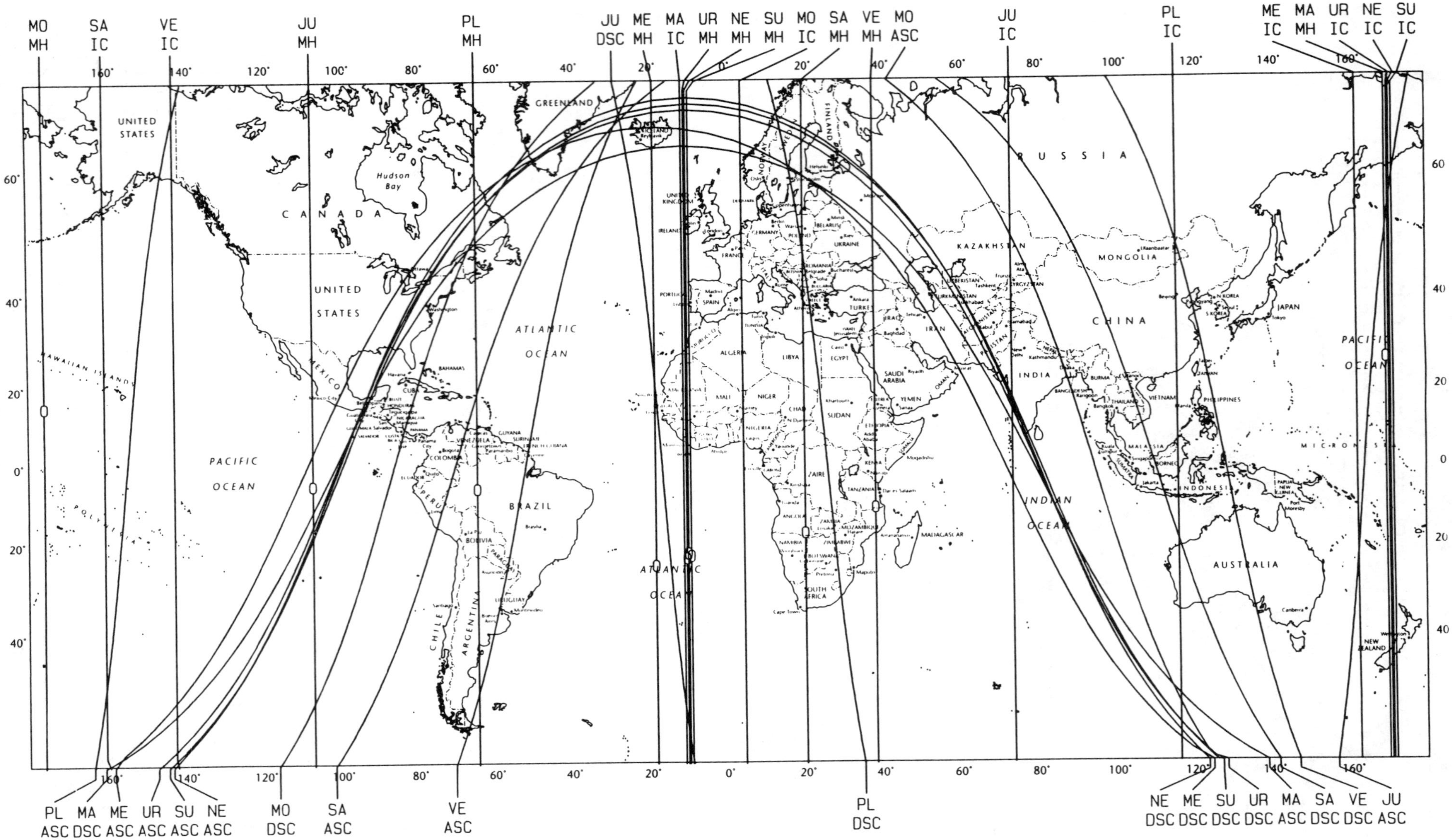

Nigel SR

January 9, 1993 12: 42 PM GMT

04/16/96

04/16/96

Nigel SR
January 9, 1993 12:42 PM GMT

ASTRO*CARTO*GRAPHY LATITUDE CROSSINGS

NORTH LATITUDES

69N10	ME/MH * SA/ASC
68N32	UR/MH * NE/ASC
68N32	SU/MH * NE/DSC
68N31	MA/IC * NE/ASC
68N19	ME/MH * NE/ASC
68N05	SA/ASC * NE/ASC
67N57	NE/MH * SU/ASC
67N57	UR/MH * SU/ASC
67N56	MA/IC * SU/ASC
67N54	MO/IC * NE/DSC
67N46	VE/ASC * NE/ASC
67N41	ME/MH * SU/ASC
67N37	JU/DSC * NE/ASC
67N23	NE/MH * UR/DSC
67N23	MA/IC * UR/ASC
67N23	SU/MH * UR/DSC
67N22	MO/IC * SU/DSC
67N11	SU/ASC * SA/ASC
67N11	ME/MH * UR/ASC
66N59	VE/ASC * JU/DSC
66N58	SU/ASC * JU/DSC
66N58	SU/ASC * VE/ASC
66N56	VE/IC * PL/ASC
66N45	NE/DSC * PL/DSC
66N42	MO/IC * UR/DSC
66N33	SA/ASC * UR/ASC
66N32	JU/DSC * UR/ASC
66N31	JU/DSC * SA/ASC
66N26	VE/ASC * UR/ASC
66N12	SU/DSC * PL/DSC
65N53	VE/ASC * SA/ASC
65N35	SA/MH * NE/DSC
65N31	MA/IC * ME/DSC
65N30	ME/ASC * JU/DSC
65N29	UR/MH * ME/DSC
65N28	NE/MH * ME/DSC
65N25	SU/MH * ME/DSC
65N21	UR/DSC * PL/DSC
65N20	ME/ASC * VE/ASC
65N17	ME/ASC * SA/ASC
65N06	SA/MH * SU/DSC
64N15	SA/MH * UR/DSC
63N55	MO/IC * ME/DSC
63N37	UR/MH * MA/ASC
63N37	NE/MH * MA/ASC
63N37	SU/MH * MA/ASC
63N26	ME/MH * MA/DSC
63N22	JU/IC * SA/DSC
63N20	ME/ASC * UR/ASC
62N55	MA/DSC * JU/DSC
62N47	MO/IC * MA/ASC
62N33	SU/ASC * ME/ASC
62N01	VE/ASC * MA/DSC
61N33	MO/DSC * NE/ASC
61N13	Anchorage
61N13	ME/DSC * PL/DSC
61N03	MA/DSC * SA/ASC
60N47	MO/DSC * ME/ASC
60N46	MA/ASC * PL/DSC
60N05	SA/MH * ME/DSC
59N57	SA/MH * MA/ASC
59N51	ME/ASC * NE/ASC
59N32	VE/MH * NE/DSC
59N31	ME/DSC * MA/ASC
59N26	SU/IC * JU/ASC
59N20	Stockholm
59N09	VE/MH * SU/DSC
58N56	MO/DSC * SU/ASC
58N31	MO/DSC * UR/ASC
57N53	VE/MH * UR/DSC
57N04	PL/MH * ME/ASC
56N45	NE/IC * JU/ASC
56N05	PL/MH * NE/ASC

CENTRAL LATITUDES

55N45	Moscow
55N41	Copenhagen
55N11	UR/IC * JU/ASC
54N49	PL/MH * SU/ASC
54N48	PL/MH * UR/ASC
54N24	SU/ASC * UR/ASC
52N51	VE/MH * MA/ASC
52N30	Berlin
51N53	MA/MH * JU/ASC
51N43	PL/MH * MO/DSC
51N30	London
51N05	JU/IC * MO/ASC
50N46	VE/MH * ME/DSC
50N27	PL/MH * MA/DSC
49N39	PL/IC * VE/DSC
49N14	Vancouver
49N05	MO/DSC * MA/DSC
48N50	Paris
48N39	SA/MH * PL/DSC
47N36	Seattle
45N32	Montreal
44N59	Minneapolis
44N58	SU/DSC * NE/DSC
43N39	Toronto
42N22	Boston
42N20	Detroit
41N54	Rome
41N53	Chicago
40N43	New York
39N44	Denver
38N53	Washington
38N37	St. Louis
37N47	San Francisco
36N10	Las Vegas
35N40	Tokyo
34N04	Los Angeles
33N45	Atlanta
33N27	Phoenix
32N47	Dallas
30N03	Cairo
29N58	New Orleans
29N46	Houston
28N39	Delhi
25N47	Miami
22N34	Calcutta
22N16	Hong Kong
22N10	ME/MH * JU/DSC
21N19	Honolulu
19N26	Mexico City
19N09	SU/ASC * MA/DSC
18N58	PL/MH * SA/ASC
17N33	JU/IC * SU/DSC
16N21	JU/IC * NE/DSC
15N56	UR/ASC * NE/ASC
14N38	JU/IC * UR/DSC
14N35	Manila
11N00	JU/IC * MA/ASC
10N30	MA/DSC * NE/ASC
10N30	Caracas
8N52	MA/DSC * UR/ASC
6N27	Lagos
3N54	JU/MH * ME/ASC
3S54	JU/IC * ME/DSC
8S52	MA/ASC * UR/DSC
10S30	MA/ASC * NE/DSC
11S00	JU/MH * MA/DSC
14S38	JU/MH * UR/ASC
15S56	UR/DSC * NE/DSC

SOUTH LATITUDES

16S21	JU/MH * NE/ASC
17S33	JU/MH * SU/ASC
18S58	PL/IC * SA/DSC
19S09	SU/DSC * MA/ASC
22S10	ME/IC * JU/ASC
22S54	Rio De Janeiro
26S12	Johannesburg
33S55	Sydney
34S36	Buenos Aires
36S52	Auckland
37S50	Melbourne
44S58	SU/ASC * NE/ASC
48S39	SA/IC * PL/ASC
49S05	MO/ASC * MA/ASC
49S39	PL/MH * VE/ASC
50S27	PL/IC * MA/ASC
50S46	VE/IC * ME/ASC
51S05	JU/MH * MO/DSC
51S43	PL/IC * MO/ASC
51S53	MA/IC * JU/DSC
52S51	VE/IC * MA/DSC
54S24	SU/DSC * UR/DSC
54S48	PL/IC * UR/DSC
54S49	PL/IC * SU/DSC
55S11	UR/MH * JU/DSC
56S05	PL/IC * NE/DSC
56S45	NE/MH * JU/DSC
57S04	PL/IC * ME/DSC
57S53	VE/IC * UR/ASC
58S31	MO/ASC * UR/DSC
58S56	MO/ASC * SU/DSC
59S09	VE/IC * SU/ASC
59S26	SU/MH * JU/DSC
59S31	ME/ASC * MA/DSC
59S32	VE/IC * NE/ASC
59S51	ME/DSC * NE/DSC
59S57	SA/IC * MA/DSC

Wendy gives you this background information for the Astro*Carto*Graphy interpretation: born in London, but moved to the Costa del Sol, Spain at an early age when her parents decided to leave Britain; went to school in Spain; and speaks Spanish fluently. Although she enjoys the climate in southern Spain she has experienced many difficulties in the area of partnerships. Relationships with men never seem to last long and she wonders if she will ever find the right partner.

At the beginning of 1983 Wendy was depressed after a partnership broke up just before Christmas. The man had made a big impact in her life and she thought he would be her future marriage partner. Wendy is nearly thirty years old and worries that time is running out to find the right partner. She is prepared to leave Spain if it means she will have more success in partnerships. You are asked to do an interpretation for her in July 1983.

Wendy's natal chart in right ascension has a T-square combination. This involves VE, UR, NE and the Moon's Node, which is a clear indication why partnerships are not easy for her. Relationships will tend to be short as she expects too much from the other person too soon and ends up disappointed.

VE and UR are conjunct so they will be next to each other on the Astro*Carto*Graphy map. As Wendy is looking for stability in relationships, locations where VE is dominant are not appropriate because UR is always involved.

The natal chart shows that Wendy has MA OPP JU, so these two planets will be next to each other on the Astro*Carto*Graphy map. This means that JU lines are not be suitable for Wendy because the opposition aspect is activated. Locations with MA and JU together in opposition would tend to make Wendy even more impatient in life. If MA and JU were conjunct they would probably manifest more positively for Wendy.

This leaves the SU and ME lines as being the most appropriate planetary lines for Wendy on her Astro*Carto*Graphy map. The DSC lines are the most important to consider, followed by the other angular lines.

At London Wendy has VE IC within eight degrees orb of influence. Normally, UR IC would be considered out of orb from London as it is more than eight degrees away, but with VE and UR being in a close conjunction (one degree orb), both can be considered as being within orb of London.

Mijas, Spain is powerfully influenced by VE IC, UR IC and SA DSC. With SA on the DSC angle it would tend to bring difficulties in the area of partnerships. VE IC and UR IC show that relationships don't seem to last all that long as problems soon turn up. This is reiterated by the VE ASC/UR ASC and VE IC/SA DSC crossings at Mijas. The MA MH/MO DSC crossing shows that Wendy is usually impatient when she is with other people. The Astro*Carto*Graphy map confirms why Wendy has had a difficult time in finding the right partner.

For Wendy, there are eight planetary lines to examine as only the SU and ME will be considered.

ME DSC passes through the United States, but it can't be considered as it is too near MA ASC, SA MH and NE MH.

ME ASC can be ignored as it passes through Russia. Other locations on or near this line are inappropriate.

ME IC is mostly in the Atlantic Ocean, so this line doesn't look promising. The Canary Islands are a possibility, but can't be considered due to the powerful influence from NE DSC. Iceland is also inappropriate due to the influence from SA DSC.

ME MH passes through the Pacific Ocean, but is too near the NE ASC line to be examined in more detail.

SU DSC provides much better chances of finding a suitable location for partnerships. This line is in the southwestern part of the United States, and southern California looks to be the best area. More important is that a location not be within orb of the MO MH line. In the natal chart the MO links up with the VE, UR, NE and Nodes T-square. Therefore, it is best to avoid the MO lines on the Astro*Carto*Graphy map.

SU IC has a strong influence in Brazil, especially at Rio de Janeiro, so this could be an excellent location for Wendy.

SU MH is in Australia, but Adelaide can't be considered due to the influence from PL ASC. Northern Australia would be better as it is out of orb of PL ASC. There is no well-known location in Australia on or near the SU MH line.

Another possibility for the SU MH line is Japan, where it is powerfully influenced. Perhaps Osaka would be a beneficial location for Wendy.

SU ASC passes through East Africa and then parts of Europe. West of the SU ASC line would be better in Europe so that there is no influence from MA DSC. Looking at locations on and west of the SU ASC line, Vienna, Austria could be a good possibility.

For Wendy, her SU lines provide the best possibilities in finding beneficial locations. As she is specifically interested in partnerships the SU DSC would appear to be the most appropriate line to examine. But at this stage I would not rule out the other SU lines as other factors need to be taken into consideration.

The list of latitude crossings for Wendy is now used to select locations along her SU lines. Locations at fifty-three degrees north latitude look the best as there are two JU/PL crossings. Otherwise, there doesn't appear to be a latitude that stands out as being beneficial.

Looking first at locations on or near her SU DSC line, San Diego and Las Vegas look good. Las Vegas is on the same latitude as Mijas, so Wendy would need to be aware of the tendency to rush into relationships. San Diego is influenced by VE MH/MO ASC, MA DSC/JU ASC, SA MH/ME DSC, MA MH/MO DSC and NE MH/JU DSC. These crossings look better for long-term relationships as the planetary influences are more balanced. Of the two locations I would choose San Diego as being better for partnerships.

At Rio de Janeiro, the latitude crossings are SA IC/UR ASC and MO IC/JU ASC. As both latitude crossings have more than a two degree orb, they will not have a powerful influence at Rio de Janeiro.

There are six latitude crossings at Osaka, Japan using a three degree orb. These are VE MH/MO ASC, MA DSC/JU ASC, SA MH/ME DSC, MA MH/MO DSC, NE MH/JU DSC and VE IC/SA DSC. This is virtually a combined influence of the crossings at San Diego and Las Vegas.

At Vienna Wendy has SA IC/PL ASC, SA MH/JU DSC, MO MH/SU DSC and MO ASC/NE ASC within two degrees orb. The latitude crossings can have an effect up to three degrees orb. (Sometimes I reduce this to two degrees to find out which are the strongest.) If you use a ruler for the latitude crossings you will find that the orb will be up to three degrees.

No matter where you look on Wendy's SU lines you will find some challenging latitude crossings. But don't forget that the planetary lines at a location are the most powerful influences. The latitude crossings will give additional information to how a planetary line at a location will be influenced.

The Cyclo*Carto*Graphy planets have been calculated for August 1, 1983.

A*C*G	**Age 29**	**C*C*G**
SA 01 SCO 50		PL00 SCO 58
JU 04 CAN 17		UR 03 SAG 14
MA 07 CAP 06		MO 07 SCO 28
ME 14 CAN 50		ME 11 CAN 47
MO 16 LIB 15		SU 18 CAN 55
SU 18 GEM 45		NE 26 SAG 35
NE 22 LIB 20		SA 27 LIB 20
UR 23 CAN 05		MA 27 SAG 48
VE 24 CAN 17		JU 29 SCO 06
PL 28 LEO 41		VE 29 LEO 13

1. C*C*G PL = A*C*G SA
2. C*C*G JU EQU A*C*G PL
3. C*C*G VE = A*C*G PL
4. Prog. birthplace = A*C*G JU ASC
5. Prog. Osaka = A*C*G ME MH
6. C*C*G JU DSC = birthplace

A*C*G	**(30 10)**	**Displaced C*C*G**
SA 01 SCO 50		PL00 LIB 48
JU 04 CAN 17		UR 03 SCO 04
MA 07 CAP 06		MO 07 LIB 18
ME 14 CAN 50		ME 11 GEM 37
MO 16 LIB 15		NE 26 SCO 25
SU 18 GEM 45		SA 27 VIR 10
NE 22 LIB 20		MA 27 SCO 38
UR 23 CAN 05		JU 28 LIB 56
VE 24 CAN 17		VE 29 CAN 03
PL 28 LEO 41		

A*C*G	**(30 10)**	**Solar arc A*C*G**
SA 01 SCO 50		SA 02 SAG 00
JU 04 CAN 17		JU 04 LEO 27
MA 07 CAP 06		MA 07 AQU 16
ME 14 CAN 50		ME 15 LEO 00
MO 16 LIB 15		MO 16 SCO 25
SU 18 GEM 45		NE 22 SCO 30
NE 22 LIB 20		UR 23 LEO 15
UR 23 CAN 05		VE 24 LEO 27
VE 24 CAN 17		PL 28 VIR 51
PL 28 LEO 41		

7. Solar arc MO DSC = birthplace (app)
8. Solar arc PL DSC = Rio de Janeiro

Having examined the temporary influences on Wendy's Astro*Carto*Graphy map it is best to make a list for the anywhere aspects and location aspects.

Anywhere aspects: C*C*G PL = A*C*G SA
C*C*G JU EQU A*C*G PL
C*C*G VE = A*C*G PL
Prog. birthplace = A*C*G JU ASC
C*C*G JU DSC = birthplace
Solar arc MO DSC = birthplace

Location aspects: Prog. Osaka = A*C*G ME MH
Solar arc PL DSC = Rio de Janeiro

The Astro*Carto*Graphy map for Wendy's twenty-ninth birthday has the following information:

Anywhere aspects: solar return JU = n. DSC
solar return UR = n. DSC
solar return IC = n. MA
solar return ME = n. ASC
solar return MH = n. ME
solar return MH = n. JU
solar return - SU CJN MO OPP NE

Location aspects: Mijas = NE IC
Vienna = PL ASC
San Diego = NE DSC, MO ASC
Las Vegas = NE DSC, MO ASC
Rio de Janeiro = ME MH, JU IC, VE ASC
Osaka = ME IC, VE DSC, JU MH

What these lists don't show is that Wendy is experiencing her Saturn return. The beginning of 1983 was especially difficult for her as she had the following temporary influences:

C*C*G SA = A*C*G SA
C*C*G MO = A*C*G SA

This is the same as saying pr. MO CJN n. SA and tr. SA CJN n. SA, but the C*C*G planets are calculated in right ascension. The Saturn return influence will last until about end of October 1983. Wendy should not move to a new location while the influences from her Saturn return are still effective.

PL is significant in the anywhere aspects temporary influences. First, C*C*G PL = A*C*G SA puts more emphasis on the Saturn return. What is interesting is that C*C*G PL = A*C*G SA only occurs in right ascension. If the usual natal chart and transits had been used for August 1, 1983, the orb between natal SA and transit PL would be more than six degrees. This shows the significance of using a right ascension horoscope with Astro*Carto*Graphy.

C*C*G VE = A*C*G PL is a separating aspect, but will still have an influence for about six months. During this time it is possible that Wendy could start a new relationship.

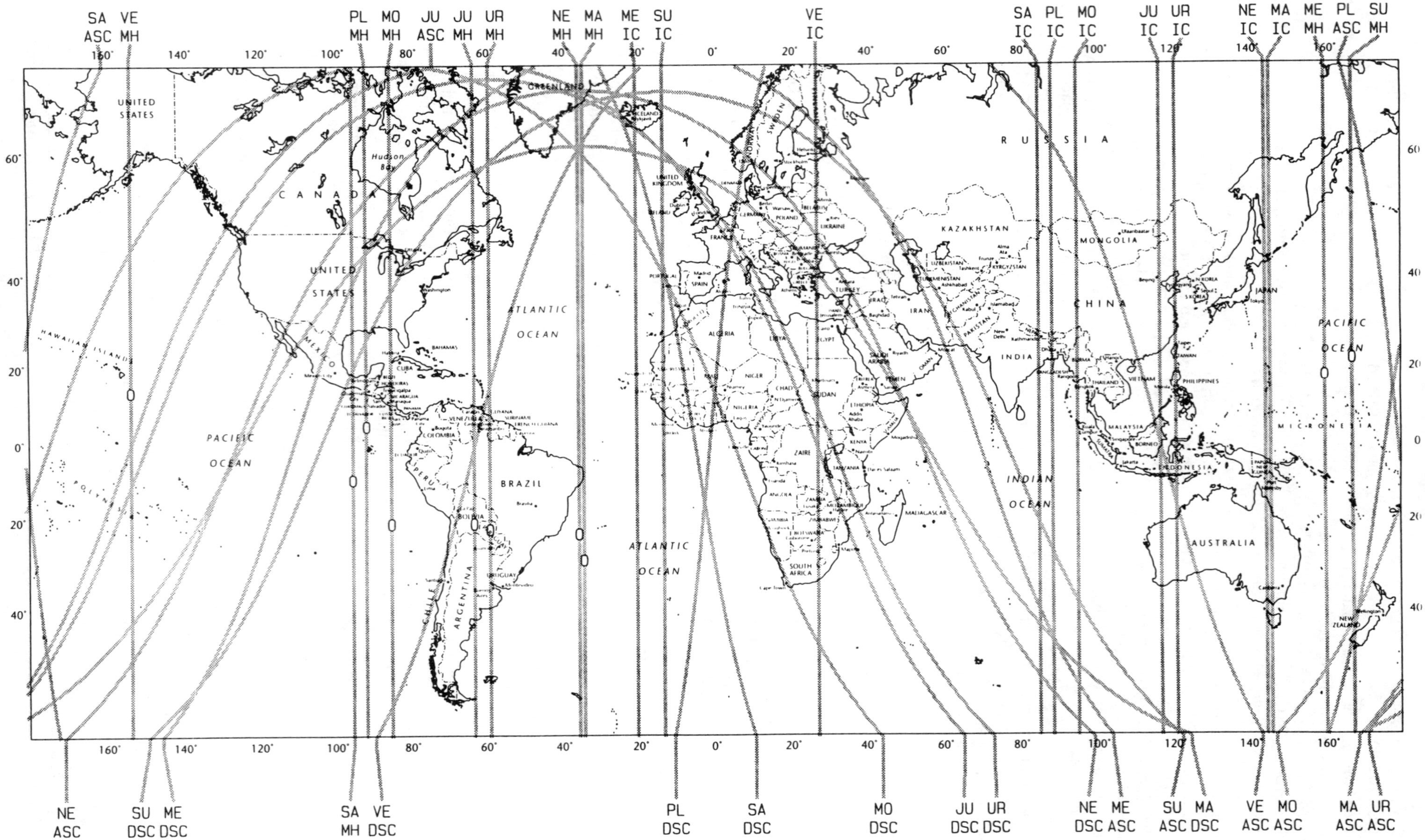

CYCLO*CARTO*GRAPHY

04/17/96

Wendy

SU, MO, ME, VE AND MA ARE PROGRESSED
JU, SA, UR, NE AND PL ARE TRANSITING

PROG	7 10 54	6: 11 GMT
TRANSIT	8 1 83	0: 00 GMT
NATAL	6 11 54	2: 50 GMT

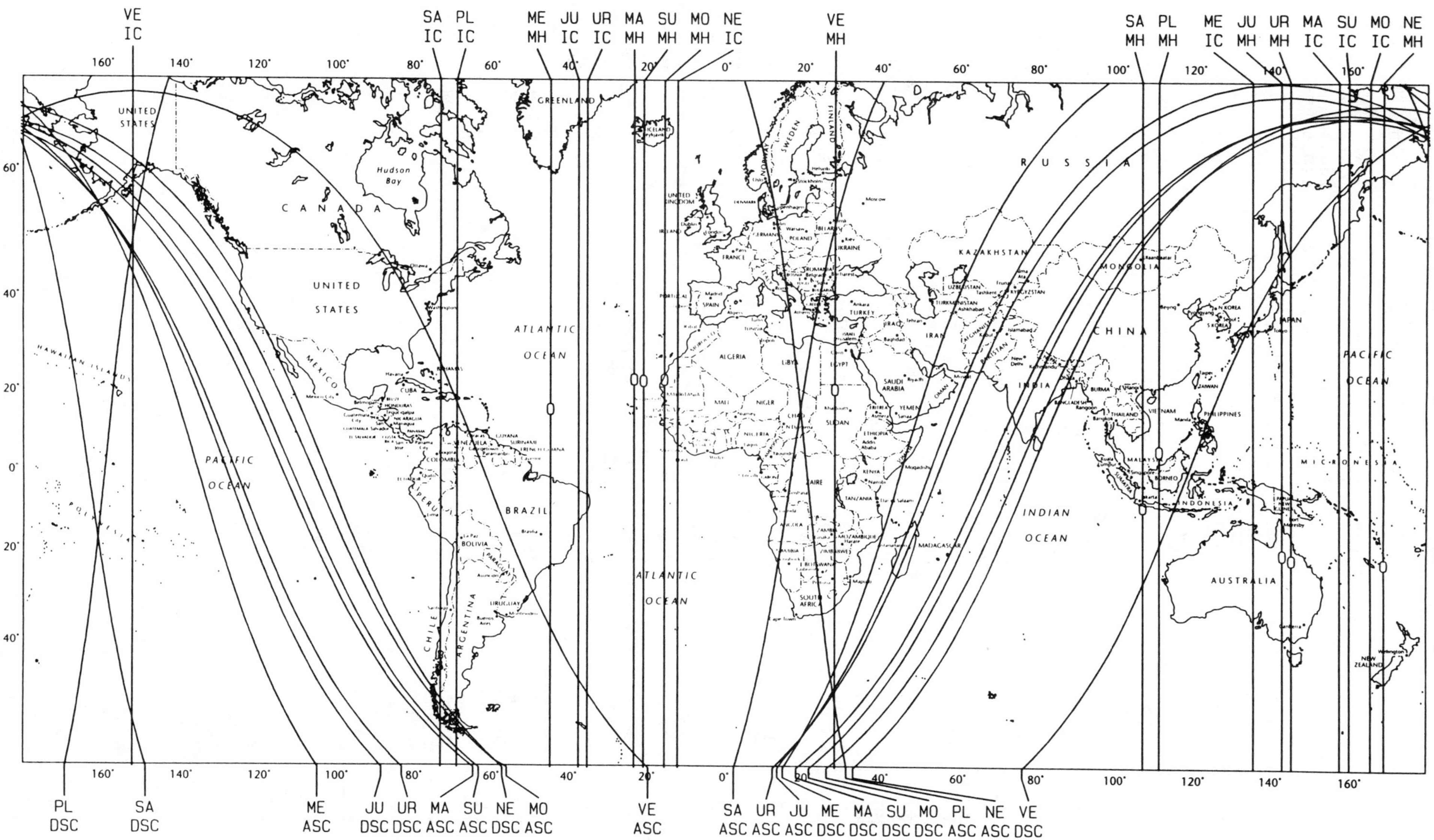

Wendy SR

June 11, 1983 1:21 PM GMT

04/16/96

04/16/96

Wendy SR
June 11, 1983 1:21 PM GMT

ASTRO*CARTO*GRAPHY LATITUDE CROSSINGS

NORTH LATITUDES

69N51 UR/MH * JU/DSC
69N43 ME/IC * JU/ASC
69N29 ME/ASC * SA/DSC
69N16 MA/IC * JU/DSC
69N03 SU/IC * JU/DSC
68N40 JU/MH * UR/ASC
68N40 VE/ASC * PL/DSC
68N25 ME/IC * UR/ASC
68N24 MO/IC * JU/DSC
68N15 MA/IC * UR/DSC
68N03 SU/IC * UR/DSC
67N54 NE/MH * JU/DSC
67N50 JU/DSC * NE/DSC
67N48 MO/IC * NE/ASC
67N45 UR/DSC * NE/ASC
67N38 SA/DSC * NE/DSC
67N36 SU/IC * NE/ASC
67N28 MO/IC * UR/DSC
67N28 MA/IC * NE/ASC
67N19 ME/ASC * NE/DSC
67N01 VE/DSC * NE/DSC
67N00 NE/MH * UR/DSC
66N55 MA/IC * SU/DSC
66N50 MO/IC * SU/ASC
66N47 PL/MH * JU/ASC
66N45 ME/ASC * VE/DSC
66N41 NE/MH * SU/ASC
66N37 NE/MH * MO/ASC
66N35 SU/IC * MA/ASC
66N35 MO/DSC * MA/ASC
66N34 SU/IC * MO/DSC
66N34 MO/ASC * SU/ASC
66N32 MO/ASC * UR/DSC
66N29 SU/ASC * UR/DSC
66N29 SU/DSC * NE/ASC
66N28 MA/IC * MO/DSC
66N25 MO/IC * MA/ASC
66N15 UR/MH * SU/DSC
66N14 MO/ASC * JU/DSC
66N12 NE/MH * MA/ASC
66N10 MA/DSC * NE/ASC
66N08 UR/MH * MA/DSC
66N06 UR/MH * NE/ASC
66N06 MO/ASC * SA/DSC
66N01 JU/MH * SU/DSC
66N01 MO/ASC * VE/DSC
65N56 JU/MH * MA/DSC
65N52 VE/DSC * SA/DSC
65N49 VE/DSC * JU/DSC
65N49 SU/ASC * VE/DSC
65N49 SU/ASC * JU/DSC
65N46 SA/MH * JU/ASC
65N46 SU/ASC * SA/DSC
65N43 JU/MH * NE/ASC
65N43 JU/DSC * SA/DSC
65N27 VE/DSC * UR/DSC
65N24 VE/DSC * MA/ASC
65N21 UR/MH * MO/DSC
65N19 SU/DSC * MA/DSC
65N17 MA/ASC * UR/DSC
65N05 MO/ASC * ME/ASC
65N02 ME/IC * MA/DSC
65N01 JU/MH * MO/DSC
65N00 MA/ASC * SA/DSC
65N00 ME/IC * SU/DSC
64N55 PL/MH * UR/ASC
64N45 SA/DSC * UR/DSC
64N17 MA/ASC * JU/DSC
64N09 ME/IC * NE/ASC
63N58 SU/ASC * ME/ASC
63N45 SA/MH * UR/ASC
63N38 ME/IC * MO/DSC
63N06 NE/MH * VE/DSC
62N15 ME/ASC * MA/ASC

CENTRAL LATITUDES

61N53 VE/IC * NE/DSC
61N51 MO/IC * VE/DSC
61N13 Anchorage
60N38 NE/DSC * PL/DSC

59N28 MO/DSC * NE/ASC
59N26 SU/IC * VE/DSC
59N20 Stockholm
59N09 VE/IC * MO/ASC
58N14 PL/MH * MA/DSC
58N10 MA/IC * VE/DSC
58N08 MO/ASC * PL/DSC
57N30 PL/MH * SU/DSC
56N58 VE/IC * SU/ASC
56N09 SA/MH * MA/DSC
56N06 SU/ASC * PL/DSC
55N57 ME/ASC * JU/DSC
55N45 Moscow
55N43 ME/ASC * UR/DSC
55N41 Copenhagen
55N13 VE/IC * MA/ASC
55N13 SA/MH * SU/DSC
55N02 JU/DSC * UR/DSC
54N34 MA/ASC * PL/DSC
54N06 PL/MH * MO/DSC
53N23 PL/MH * NE/ASC
53N13 VE/MH * SA/ASC
52N30 Berlin
51N30 London
51N11 SA/MH * MO/DSC
50N39 VE/IC * PL/DSC
50N02 SA/MH * NE/ASC

49N14 Vancouver
49N08 UR/DSC * PL/DSC
49N02 UR/MH * VE/DSC
48N50 Paris
48N46 VE/IC * UR/DSC
48N40 JU/DSC * PL/DSC
48N08 VE/IC * JU/DSC
47N36 Seattle
46N37 JU/MH * VE/DSC
46N25 ME/ASC * PL/DSC
45N32 Montreal
44N59 Minneapolis
44N49 VE/IC * ME/ASC
43N39 Toronto
42N22 Boston
42N20 Detroit
41N54 Rome
41N53 Chicago
40N43 New York

39N44 Denver
38N53 Washington
38N37 St. Louis
37N47 San Francisco
36N50 ME/IC * VE/DSC
36N10 Las Vegas
35N40 Tokyo
34N04 Los Angeles
33N45 Atlanta
33N27 Phoenix
32N47 Dallas
30N03 Cairo

29N58 New Orleans
29N46 Houston
28N39 Delhi
26N50 SA/IC * VE/ASC
25N47 Miami
22N34 Calcutta
22N16 Hong Kong
21N19 Honolulu

SOUTH LATITUDES

19N26 Mexico City
17N48 PL/IC * VE/ASC
16N17 SA/ASC * PL/ASC
14N35 Manila
10N30 Caracas

6N27 Lagos

16S17 SA/DSC * PL/DSC
17S48 PL/MH * VE/DSC

22S54 Rio De Janeiro
26S12 Johannesburg
26S50 SA/MH * VE/DSC

33S55 Sydney
34S36 Buenos Aires
36S50 ME/MH * VE/ASC
36S52 Auckland
37S50 Melbourne

44S49 VE/MH * ME/DSC
46S25 ME/DSC * PL/ASC
46S37 JU/IC * VE/ASC
48S08 VE/MH * JU/ASC
48S40 JU/ASC * PL/ASC
48S46 VE/MH * UR/ASC
49S02 UR/IC * VE/ASC
49S08 UR/ASC * PL/ASC

50S02 SA/IC * NE/DSC
50S39 VE/MH * PL/ASC
51S11 SA/IC * MO/ASC
53S13 VE/IC * SA/DSC
53S23 PL/IC * NE/DSC
54S06 PL/IC * MO/ASC
54S34 MA/DSC * PL/ASC
55S02 JU/ASC * UR/ASC
55S13 SA/IC * SU/ASC
55S13 VE/MH * MA/DSC
55S43 ME/DSC * UR/ASC
55S57 ME/DSC * JU/ASC
56S06 SU/DSC * PL/ASC
56S09 SA/IC * MA/ASC
56S58 VE/MH * SU/DSC
57S30 PL/IC * SU/ASC
58S08 MO/DSC * PL/ASC
58S10 MA/MH * VE/ASC
58S14 PL/IC * MA/ASC
59S09 VE/MH * MO/DSC
59S26 SU/MH * VE/ASC
59S28 MO/ASC * NE/DSC

C*C*G JU EQU A*C*G PL will be effective for a few weeks, so not too much importance can be put on this temporary influence. But during this time it could bring out the more positive side of the other PL temporary influences.

The remaining anywhere aspects place a strong emphasis on the place of birth as it is activated three different times. These are prog. birthplace = A*C*G JU ASC, C*C*G JU DSC = birthplace and solar arc MO DSC = birthplace. This suggests that Wendy will experience a more positive time in relationships, which will have an effect anywhere in the world. This will probably be more noticeable when the effects of the Saturn return have moved away.

Solar arc A*C*G PL DSC = Rio de Janeiro is an excellent temporary influence to have for starting a new partnership. Normally, A*C*G SU IC at Rio de Janeiro would have more influence in the home and family. But solar arc PL DSC at Rio de Janeiro will put more emphasis on relationships if Wendy moves there. Therefore, Rio de Janeiro can be considered an excellent place to meet a new partner.

Prog. Osaka = A*C*G ME MH indicates a good time to start a new career at this location, especially in the field of communications. This is reiterated by the natal SU in GEM. Although Osaka is more suited for career and position in society, she would meet interesting people there. This could also lead to the start of a new relationship, especially when considering the temporary influences at the birthplace.

I have left the solar return Astro*Carto*Graphy map until the end to show how I use it for an interpretation. Conjunction aspects between the natal chart angles and solar return planets are very important. Conjunction aspects between solar return chart angles for the place of birth and the natal planets should also be considered.

I find it best to draw a solar return chart with the natal planets on the outside. You can also draw a natal chart with the solar return planets on the outside. Then it will be easier to find the conjunction aspects between the angles and planets, using a five degree orb.

Looking at the solar return and natal charts, the aspects between angles and planets are:

Solar return JU = natal DSC

Solar return UR = natal DSC

Solar return ME = natal ASC

Solar return MH = natal ME

Solar return MH = natal JU

Solar return IC = natal MA

These aspects show that partnerships will be more beneficial for Wendy this year (solar return JU + UR = natal DSC). Career should also be much better for Wendy as solar return MH is at the midpoint of natal ME and JU. Perhaps there will be some arguments and disagreements within the family if she decides to leave Spain (solar return IC = natal MA).

A new beginning in life is shown by SU CJN MO in the solar return, but with both planets in OPP to NE Wendy shouldn't expect her life to automatically be more successful after moving to a new location. She needs to allow time for the new planetary influences at another location to have an effect in her life.

These solar return potential influences will have a better chance to manifest when Wendy is at a location where they are emphasized. Next, I examine locations on the Astro*Carto*Graphy solar return map.

At Mijas Wendy has NE IC just within orb, so it would only have a mild influence. It could manifest as her uncertainty about moving.

PL ASC at Vienna could indicate the opportunity to start a new life there. Wendy has no other temporary influences at Vienna, but the anywhere aspects shouldn't be forgotten.

NE DSC and MO ASC both have an influence at San Diego and Las Vegas, which are not good for a long-term relationship. She would be unable to see the reality of a new partnership until much later, when she is likely to end up disappointed.

Rio de Janeiro is influenced by ME MH, JU IC and VE ASC, which are beneficial planetary influences. With solar arc A*C*G PL DSC also at Rio de Janeiro this would be an excellent place to move to.

ME IC, VE DSC and JU MH are at Osaka, Japan, which are also beneficial planetary influences. VE DSC shows the potential of starting a new relationship, possibly marriage, with also success in her career.

During the first week in December 1983 tr. JU conjuncts her natal SU. At that time C*C*G JU will be at the locations where A*C*G SU is. As the effects of the Saturn return move away from Mijas in October 1983, a move at the beginning of December 1983 would be good.

After taking all the temporary influences into consideration, Rio de Janeiro and Osaka look to be the best places for Wendy. The best time is clearly the beginning of December 1983.

This example shows the importance of using all the techniques to obtain a complete picture of which planetary influences have an effect. Only then can you discover which location are the most beneficial.

Appendix

How to Calculate Lines for Astro*Carto*Graphy

Calculating the planetary lines for an Astro*Carto*Graphy map is a complicated and time-consuming process. It would take many hours to calculate all the lines and then draw them on a world map.

The MH and IC lines for the planets are not too difficult to calculate and draw on a map of the world. It is the ASC and DSC lines that require a great many calculations.

As the ASC and DSC lines curve on a world map their positions need to be calculated exactly. This can only be done when they are calculated for small increments of geographical latitude.

If increments of five degrees are used from 60S00 to 70N00, then up to **fifty-four** calculations are required for a planet's ASC and DSC lines. Ten planets that means 540 different calculations!

Once these calculations are completed the positions are plotted on a world map. Then all the points for a planet's ASC and DSC are joined to form a curving line.

This is easier to understand when you examine the following lists calculated for Uri Geller. From these lists it is possible to draw an Astro*Carto*Graphy map for Uri Geller. As an example I will use Uri Geller's chart and calculate the position of the Venus lines at 40N00 geographical latitude.

It is easier when the RAMC for **Greenwich** is used in calculating the positions of the planetary lines. Greenwich is at zero degrees geographical latitude, so the RAMC for Tel Aviv needs to be converted. RAMC at Tel Aviv = 122 45 (right ascension of the Midheaven).

Greenwich RAMC is calculated as follows:

RAMC at Tel Aviv	122 45
Tel Aviv (34E46)	<u>-34 46</u>*
RAMC at Greenwich	87 59

*east longitude (minus) west longitude (plus)

VE MH

RAMC at Greenwich	87 59
Right Ascension Venus	<u>-228 07</u>
(- = east) (+ = west)	-140 08
VE MH is at longitude	140E08

VE IC

VE MH	-140 08
(always + 180 00)	<u>+ 180 00</u>
(- = east) (+ = west)	+ 39 52
VE IC is at longitude	39W52

Ascensional Difference for Venus at 40N00 Latitude

VE (declination = 14S34)

Asc. Diff.	= arcsin (tan decl. x tan latitude)
	= arcsin (tan -14 34 x tan 40)
	= -12.59440088
	= -12°36"

VE ASC	Calculate Oblique Ascension for Venus
Oblique Ascension	= Right Ascension - Ascensional Difference
	= (228 07) - (-12 36)
	= 240 43
RAMC VE ASC	= Oblique Ascension - 90
	= 240 43 - 90
	= 150 43

Longitude VE ASC	
RAMC at Greenwich	= 87 59
RAMC VE ASC	= -150 43
Difference	= -62 44
VE ASC at 40N00	**= longitude 62E44**

VE DSC	Calculate Oblique Descension for Venus
Oblique Descension	= Right Ascension + Ascensional Difference
	= (228 07) + (-12 36)
	= 215 31
RAMC VE DSC	= Oblique Ascension + 90
	= 215 31 + 90
	= 305 31

Longitude VE DSC	
RAMC at Greenwich	= 87 59
RAMC VE DSC	= -305 31
Difference	= -217 32*
Subtract 360 00 for west	= - 360 00
Difference	= 142 28
VE DSC at 40N00	**= longitude 142W28**

* As this is over 180 degrees an adjustment needs to be made to obtain the longitude in west. The is achieved by subtracting 360.

Now you will understand why it is best that an Astro*Carto*Graphy map is drawn by a computer. Even with a scientific calculator it would take a long time to plot all the lines on a world map.

In Appendix 3 are details about three astrological computer programs I recommend for drawing Astro*Carto*Graphy maps. If you don't have a computer, then it is possible to order your own individual Astro*Carto*Graphy map. Astro*Carto*Graphy maps can be ordered from:

Astro Numeric Service
P.O. Box 336
Ashland OR 97520 USA
Phone: 541-488-1347
Fax: 541-488-2058
E-mail: info@astronumerics.com

Equinox
The Astrology Shop
78 Neal Street
Covent Garden
London WC2H 9PA England
Phone: (0171) 497-1001
Fax: (0171) 497-0344
E-mail: equinox@equinox.uk.com

How to Calculate Parans

In Chapter 2, I explained how a paran printout is used to find out which planets are exactly conjunct the angles in a natal chart. A paran printout can be calculated with the NOVA astrological computer program. If you do not have this computer program it is possible to calculate a paran printout with a scientific calculator.

The formula for calculating a paran printout is in *Essays in Astrology*, but is not easy to understand so I will try to explain the calculations. Uri Geller's natal chart will be used to explain how a paran printout is calculated. Once the formula is understood it is easy to do the calculations.

A scientific calculator is required to do the calculations for ASC and DSC. All that is needed is the planet's right ascension and declination.

MH = Right Ascension of planet
IC = Right Ascension of planet (*) + or - 180 degrees
*If Right Ascension < 180, then + 180
*If Right Ascension > 180 then - 180
ASC = Oblique Ascension - 90
DSC = Oblique Descension + 90

Using Uri Geller's Sun will be used to explain how to calculate the RAMC for the MH, IC, ASC and DSC.

MH

Right ascension of Sun = **267 17**

As no further calculations are required this is written in the paran printout grid.

IC

Right Ascension of Sun = 267 17

As Right Ascension > 180, then subtract 180

267 17 - 180 00 = **87 17**

Now you need to calculate the **Ascensional Difference** for the Sun. This is the only time that a scientific calculator is really needed for a planet.

Ascensional Difference = arcsin (tan decl. x tan lat)

For Uri Geller the calculation for his Sun is as follows:

Asc. Diff.	= arcsin (tan decl. x tan latitude)
	= arcsin (tan -23 25 x tan 32 04)
	= -15.74295666
ASC	Calculate Oblique Ascension
Oblique Ascension	= Right Ascension - Ascensional Difference
	= (267 17) - (-15.742295666) *
	= 283 01
ASC	= Oblique Ascension - 90
	= 283 01 - 90
	= 193 01
DSC	Calculate Oblique Descension
Oblique Descension Difference	= Right Ascension + Ascensional
	= (267 17) + (-15.74295666) *
	= 251 32
DSC	= Oblique Descension + 90
	= 251 32 + 90
	= 341 32

*-15.74295666 must be converted to degrees and minutes during the calculation.

From these calculations the R.A.M.C.'s for Sun are:

MH 267 17
IC 87 17

ASC 193 01
DSC 342 32
These are the same positions calculated by the NOVA program for Uri Geller. The calculation process is repeated for the other planets to complete the paran printout.

Although the calculations look complicated on paper they are easy to do with a scientific calculator.

Computer Software for Astro*Carto*Graphy

The computer program I have used for many years for Astro*Carto*Graphy is Astro*Maps Hi-Res, available from Matrix. In fact, this is the only computer program that has Jim Lewis' permission to use the Astro*Carto*Graphy trademark.

It is also possible to examine the Cyclo*Carto*Graphy planetary lines at locations with this program. This can only be achieved by first calculating the progressions, and afterwards doing the same for transits. It is not possible to combine the progressions and transits on one map.

For more information about the "Astro*Maps Hi-Res" program contact Matrix software at one of the following addresses:

Matrix Software
407 N. State Street
Big Rapids MI 49307 U.S.A.
Phone: 616-796-2483
Fax: 616-796-3060

Martin Davis
Library Wing
Abbey St. Bathans House
Abbey St. Bathans
Nr. Duns
Berwickshire TD11 3TX
United Kingdom
Phone: (01361) 840 340
Fax: (01361) 840 284
E-mail: martin@astral.demon.co.uk

For years I didn't consider other computer software featuring Astro*Carto*Graphy because only Matrix has permission to include it in their Astro*Maps Hi-Res program. There is no escaping the fact that Astro*Carto*Graphy is starting to be featured on more astrological computer software. Even when the Astro*Carto*Graphy trademark is not used the technique can be included under a different name.

There have been rapid developments with computer astrological software that features Astro*Carto*Graphy. Matrix's Win*Star program for Windows features Astro*Carto*Graphy. In my opinion, the DOS program Astro*Maps Hi-Res is much better for Astro*Carto*Graphy. Although there are many different country and continent maps available in both Win*Star and Astro*Maps Hi-Res, it is only possible to examine the Canary Islands on the world map. This makes things difficult when you want to examine the various techniques to find the temporary influences.

SolarMaps is an impressive program for Astro*Carto*Graphy that allows you to draw a closeup map of any part of the world for the planetary lines. Its just a matter of making a frame of the desired area and a closeup map is automatically drawn.

SolarMaps also has an option to include Chiron, the Moons Nodes and the asteroids on an Astro*Carto*Graphy map which is excellent for research purposes. Unlike the Matrix programs the ASC and DSC lines are easy to recognize on an Astro*Carto*Graphy map produced by SolarMaps.

Although SolarMaps includes an option for Cyclo*Carto*Graphy it is not the combination of progressions and transits that are drawn on an overlay. It is the same as the Matrix programs, i.e. either progressions or transits, but not both together.

For further information about the SolarMaps program contact one of the following suppliers at the following addresses:

Astrolabe
Box 1750
Brewster MA 02631 U.S.A.
Phone: 508-896-5081
Fax: 508-896-5189
E-mail: astrolabe@alabe.com

Roy Gillett Associates
32 Glynswood
Camberley
Surrey GU15 1HU, United Kingdom
Phone and fax: (01276) 683898
E-mail: roy.gillett@dial.pipex.com.uk.com

Esoteric Technologies Pty Ltd
P.O. Box 578
Magill
SA 5072
Australia
Phone and fax: (08) 8331 3057
E-mail: solarfire@dove.mtx.net.au

Uri Geller	Natal Chart	Astro-Mapping
Dec. 20,1946	Tel Aviv	2:0:0 AM EET
32N04'00"	8:10:59 ST	34E46'00"
Tropical		Geocentric

Sun	is on the upper Meridian at	179 E 18	is on the Nadir at	23 N 25
	is on the Zenith at	23 S 25	never sets at	66 S 35
	is on the lower Meridian at	0 W 42	never rises at	66 N 35

At	60 S 00	Su rises at long.	40 E 41	and sets at long.	42 W 05
At	55 S 00	Su rises at long.	51 E 05	and sets at long.	52 W 29
At	50 S 00	Su rises at long.	58 E 13	and sets at long.	59 W 37
At	45 S 00	Su rises at long.	63 E 38	and sets at long.	65 W 02
At	40 S 00	Su rises at long.	67 E 59	and sets at long.	69 W 23
At	35 S 00	Su rises at long.	71 E 38	and sets at long.	73 W 03
At	30 S 00	Su rises at long.	74 E 49	and sets at long.	76 W 13
At	25 S 00	Su rises at long.	77 E 39	and sets at long.	79 W 03
At	20 S 00	Su rises at long.	80 E 13	and sets at long.	81 W 38
At	15 S 00	Su rises at long.	82 E 38	and sets at long.	84 W 02
At	10 S 00	Su rises at long.	84 E 55	and sets at long.	86 W 19
At	5 S 00	Su rises at long.	87 E 07	and sets at long.	88 W 32
At	0 N 00	Su rises at long.	89 E 18	and sets at long.	90 W 42
At	5 N 00	Su rises at long.	91 E 28	and sets at long.	92 W 53
At	10 N 00	Su rises at long.	93 E 41	and sets at long.	95 W 05
At	15 N 00	Su rises at long.	95 E 58	and sets at long.	97 W 22
At	20 N 00	Su rises at long.	98 E 22	and sets at long.	99 W 47
At	25 N 00	Su rises at long.	100 E 57	and sets at long.	102 W 21
At	30 N 00	Su rises at long.	103 E 47	and sets at long.	105 W 11
At	35 N 00	Su rises at long.	106 E 57	and sets at long.	108 W 22
At	40 N 00	Su rises at long.	110 E 37	and sets at long.	112 W 01
At	45 N 00	Su rises at long.	114 E 58	and sets at long.	116 W 22
At	50 N 00	Su rises at long.	120 E 23	and sets at long.	121 W 47
At	55 N 00	Su rises at long.	127 E 31	and sets at long.	128 W 55
At	60 N 00	Su rises at long.	137 E 55	and sets at long.	139 W 19
At	65 N 00	Su rises at long.	157 E 35	and sets at long.	158 W 59

Moon	is on the upper Meridian at	139 E 21	is on the Nadir at	15 N 34
	is on the Zenith at	15 S 34	never sets at	74 S 26
	is on the lower Meridian at	40 W 39	never rises at	74 N 26

At	60 S 00	Mo rises at long.	20 E 29	and sets at long.	101 W 47
At	55 S 00	Mo rises at long.	25 E 53	and sets at long.	107 W 12
At	50 S 00	Mo rises at long.	29 E 57	and sets at long.	111 W 15
At	45 S 00	Mo rises at long.	33 E 10	and sets at long.	114 W 28
At	40 S 00	Mo rises at long.	35 E 49	and sets at long.	117 W 08
At	35 S 00	Mo rises at long.	38 E 06	and sets at long.	119 W 24
At	30 S 00	Mo rises at long.	40 E 05	and sets at long.	121 W 23
At	25 S 00	Mo rises at long.	41 E 53	and sets at long.	123 W 11
At	20 S 00	Mo rises at long.	43 E 32	and sets at long.	124 W 50
At	15 S 00	Mo rises at long.	45 E 04	and sets at long.	126 W 22

At	10 S 00	Mo rises at long.	46 E 32	and sets at long.	127 W 50
At	5 S 00	Mo rises at long.	47 E 57	and sets at long.	129 W 15
At	0 N 00	Mo rises at long.	49 E 21	and sets at long.	130 W 39
At	5 N 00	Mo rises at long.	50 E 45	and sets at long.	132 W 03
At	10 N 00	Mo rises at long.	52 E 10	and sets at long.	133 W 28
At	15 N 00	Mo rises at long.	53 E 38	and sets at long.	134 W 56
At	20 N 00	Mo rises at long.	55 E 10	and sets at long.	136 W 28
At	25 N 00	Mo rises at long	56 E 49	and sets at long.	138 W 07
At	30 N 00	Mo rises at long.	58 E 37	and sets at long.	139 W 55
At	35 N 00	Mo rises at long.	60 E 36	and sets at long.	141 W 54
At	40 N 00	Mo rises at long.	62 E 52	and sets at long.	144 W 11
At	45 N 00	Mo rises at long.	65 E 32	and sets at long.	146 W 50
At	50 N 00	Mo rises at long.	68 E 45	and sets at long.	150 W 03
At	55 N 00	Mo rises at long.	72 E 48	and sets at long.	154 W 07
At	60 N 00	Mo rises at long.	78 E 13	and sets at long.	159 W 31
At	65 N 00	Mo rises at long.	86 E 03	and sets at long.	167 W 22
At	70 N 00	Mo rises at long.	99 E 20	and sets at long.	179 E 22

Mercury	is on the upper Meridian at	159 E 45	is on the Nadir at	20 N 48
	is on the Zenith at	20 S 48	never sets at	69 S 12
	is on the lower Meridian at	20 W 15	never rises at	69 N 12

At	60 S 00	Me rises at long.	28 E 36	and sets at long.	69 W 05
At	55 S 00	Me rises at long.	36 E 53	and sets at long.	77 W 23
At	50 S 00	Me rises at long.	42 E 50	and sets at long.	83 W 19
At	45 S 00	Me rises at long.	47 E 25	and sets at long.	87 W 55
At	40 S 00	Me rises at long.	51 E 10	and sets at long.	91 W 39
At	35 S 00	Me rises at long.	54 E 19	and sets at long.	94 W 49
At	30 S 00	Me rises at long.	57 E 05	and sets at long.	97 W 35
At	25 S 00	Me rises at long.	59 E 33	and sets at long.	100 W 03
At	20 S 00	Me rises at long.	61 E 48	and sets at long.	102 W 18
At	15 S 00	Me rises at long.	63 E 54	and sets at long.	104 W 24
At	10 S 00	Me rises at long.	65 E 55	and sets at long.	106 W 24
At	5 S 00	Me rises at long.	67 E 51	and sets at long.	108 W 21
At	0 N 00	Me rises at long.	69 E 45	and sets at long.	110 W 15
At	5 N 00	Me rises at long.	71 E 39	and sets at long.	112 W 09
At	10 N 00	Me rises at long.	73 E 36	and sets at long.	114 W 05
At	15 N 00	Me rises at long.	75 E 36	and sets at long.	116 W 06
At	20 N 00	Me rises at long.	77 E 42	and sets at long.	118 W 12
At	25 N 00	Me rises at long.	79 E 57	and sets at long.	120 W 27
At	30 N 00	Me rises at long.	82 E 25	and sets at long.	122 W 55
At	35 N 00	Me rises at long.	85 E 11	and sets at long.	125 W 41
At	40 N 00	Me rises at long.	88 E 21	and sets at long.	128 W 50
At	45 N 00	Me rises at long.	92 E 05	and sets at long.	132 W 35
At	50 N 00	Me rises at long.	96 E 41	and sets at long.	137 W 10
At	55 N 00	Me rises at long.	102 E 37	and sets at long.	143 W 07
At	60 N 00	Me rises at long.	110 E 55	and sets at long.	151 W 24
At	65 N 00	Me rises at long.	124 E 20	and sets at long.	164 W 49

Venus	is on the upper Meridian at	140 E 08	is on the Nadir at	14 N 34
	is on the Zenith at	14 S 34	nevr sets at	75 S 26
	is on the lower Meridian at	39 W 52	never rises at	75 N 26

At	60 S 00	Ve rises at long.	23 E 22	and sets at long.	103 W 06
At	55 S 00	Ve rises at long.	28 E 20	and sets at long.	108 W 04
At	50 S 00	Ve rises at long.	32 E 05	and sets at long.	111 W 49
At	45 S 00	Ve rises at long.	35 E 04	and sets at long.	114 W 48
At	40 S 00	Ve rises at long.	37 E 32	and sets at long.	117 W 16
At	35 S 00	Ve rises at long.	39 E 39	and sets at long.	119 W 23
At	30 S 00	Ve rises at long.	41 E 30	and sets at long.	121 W 14
At	25 S 00	Ve rises at long.	43 E 10	and sets at long.	122 W 54
At	20 S 00	Ve rises at long.	44 E 42	and sets at long.	124 W 26
At	15 S 00	Ve rises at long.	46 E 08	and sets at long.	125 W 52
At	10 S 00	Ve rises at long.	47 E 30	and sets at long.	127 W 15
At	5 S 00	Ve rises at long.	48 E 50	and sets at long.	128 W 34
At	0 N 00	Ve rises at long.	50 E 08	and sets at long.	129 W 52
At	5 N 00	Ve rises at long.	51 E 26	and sets at long.	131 W 10
At	10 N 00	Ve rises at long.	52 E 45	and sets at long.	132 W 30
At	15 N 00	Ve rises at long.	54 E 08	and sets at long.	133 W 52
At	20 N 00	Ve rises at long.	55 E 34	and sets at long.	135 W 18
At	25 N 00	Ve rises at long.	57 E 06	and sets at long.	136 W 50
At	30 N 00	Ve rises at long.	58 E 46	and sets at long.	138 W 30
At	35 N 00	Ve rises at long.	60 E 37	and sets at long.	140 W 21
At	40 N 00	Ve rises at long.	62 E 44	and sets at long.	142 W 28
At	45 N 00	Ve rises at long.	65 E 12	and sets at long.	144 W 56
At	50 N 00	Ve rises at long.	68 E 11	and sets at long.	147 W 55
At	55 N 00	Ve rises at long.	71 E 56	and sets at long.	151 W 40
At	60 N 00	Ve rises at long.	76 E 54	and sets at long.	156 W 38
At	65 N 00	Ve rises at long.	84 E 01	and sets at long.	163 W 45
At	70 N 00	Ve rises at long.	95 E 43	and sets at long.	175 W 27

Mars	is on the upper Meridian at	175 W 51	is on the Nadir at	24 N 13
	is on the Zenith at	24 S 13	never sets at	65 S 47
	is on the lower Meridian at	4 E 09	never rises at	65 N 47

At	60 S 00	Ma rises at long.	42 E 59	and sets at long.	34 W 41
At	55 S 00	Ma rises at long.	54 E 11	and sets at long.	45 W 53
At	50 S 00	Ma rises at long.	61 E 44	and sets at long.	53 W 26
At	45 S 00	Ma rises at long.	67 E 25	and sets at long.	59 W 07
At	40 S 00	Ma rises at long.	71 E 59	and sets at long.	63 W 41
At	35 S 00	Ma rises at long.	75 E 48	and sets at long.	67 W 30
At	30 S 00	Ma rises at long.	79 E 06	and sets at long.	70 W 48
At	25 S 00	Ma rises at long.	82 E 03	and sets at long.	73 W 45
At	20 S 00	Ma rises at long.	84 E 44	and sets at long.	76 W 26
At	15 S 00	Ma rises at long.	87 E 14	and sets at long.	78 W 56
At	10 S 00	Ma rises at long.	89 E 36	and sets at long.	81 W 18
At	5 S 00	Ma rises at long.	91 E 54	and sets at long.	83 W 36
At	0 N 00	Ma rises at long.	94 E 09	and sets at long.	85 W 51

At	5 N 00	Ma rises at long.	96 E 24	and sets at long.	88 W 06
At	10 N 00	Ma rises at long.	98 E 42	and sets at long.	90 W 24
At	15 N 00	Ma rises at long.	101 E 04	and sets at long.	92 W 46
At	20 N 00	Ma rises at long.	103 E 34	and sets at long.	95 W 16
At	25 N 00	Ma rises at long.	106 E 15	and sets at long.	97 W 57
At	30 N 00	Ma rises at long.	109 E 12	and sets at long.	100 W 54
At	35 N 00	Ma rises at long.	112 E 30	and sets at long.	104 W 12
At	40 N 00	Ma rises at long.	116 E 19	and sets at long.	108 W 01
At	45 N 00	Ma rises at long.	120 E 53	and sets at long.	112 W 35
At	50 N 00	Ma rises at long.	126 E 34	and sets at long.	118 W 16
At	55 N 00	Ma rises at long.	134 E 07	and sets at long.	125 W 49
At	60 N 00	Ma rises at long.	145 E 19	and sets at long.	137 W 01
At	65 N 00	Ma rises at long.	168 E 50	and sets at long.	160 W 32
Jupiter	is on the upper Meridian at		138 E 00	is on the Nadir at	16 N 18
	is on the Zenith at		16 S 18	never sets at	73 S 42
	is on the lower Meridian at		42 W 00	never rises at	73 N 42
At	60 S 00	Ju rises at long.	17 E 34	and sets at long.	101 W 35
At	55 S 00	Ju rises at long.	23 E 19	and sets at long.	107 W 20
At	50 S 00	Ju rises at long.	27 E 36	and sets at long.	111 W 37
At	45 S 00	Ju rises at long.	31 E 00	and sets at long.	115 W 00
At	40 S 00	Ju rises at long.	33 E 48	and sets at long.	117 W 48
At	35 S 00	Ju rises at long.	36 E 11	and sets at long.	120 W 12
At	30 S 00	Ju rises at long.	38 E 17	and sets at long.	122 W 17
At	25 S 00	Ju rises at long.	40 E 10	and sets at long.	124 W 10
At	20 S 00	Ju rises at long.	41 E 53	and sets at long.	125 W 54
At	15 S 00	Ju rises at long.	43 E 30	and sets at long.	127 W 31
At	10 S 00	Ju rises at long.	45 E 02	and sets at long.	129 W 03
At	5 S 00	Ju rises at long.	46 E 32	and sets at long.	130 W 32
At	0 N 00	Ju rises at long.	48 E 00	and sets at long.	132 W 00
At	5 N 00	Ju rises at long.	49 E 28	and sets at long.	133 W 28
At	10 N 00	Ju rises at long.	50 E 57	and sets at long.	134 W 58
At	15 N 00	Ju rises at long.	52 E 29	and sets at long.	136 W 30
At	20 N 00	Ju rises at long.	54 E 06	and sets at long.	138 W 07
At	25 N 00	Ju rises at long.	55 E 50	and sets at long.	139 W 50
At	30 N 00	Ju rises at long.	57 E 43	and sets at long.	141 W 43
At	35 N 00	Ju rises at long.	59 E 48	and sets at long.	143 W 49
At	40 N 00	Ju rises at long.	62 E 12	and sets at long.	146 W 12
At	45 N 00	Ju rises at long.	65 E 00	and sets at long.	149 W 00
At	50 N 00	Ju rises at long.	68 E 23	and sets at long.	152 W 24
At	55 N 00	Ju rises at long.	72 E 40	and sets at long.	156 W 41
At	60 N 00	Ju rises at long.	78 E 25	and sets at long.	162 W 26
At	65 N 00	Ju rises at long.	86 E 49	and sets at long.	170 W 50
At	70 N 00	Ju rises at long.	101 E 26	and sets at long.	174 E 33
Saturn	is on the upper Meridian at		42 E 41	is on the Nadir at	18 S 45
	is on the Zenith at		18 N 45	never sets at	71 N 15

	is on the lower Meridian at		137 W 19	never rises at	71 S 15
At	60 S 00	Sa rises at long.	11 W 19	and sets at long.	96 E 41
At	55 S 00	Sa rises at long.	18 W 20	and sets at long.	103 E 42
At	50 S 00	Sa rises at long.	23 W 28	and sets at long.	108 E 50
At	45 S 00	Sa rises at long.	27 W 29	and sets at long.	112 E 51
At	40 S 00	Sa rises at long.	30 W 46	and sets at long.	116 E 08
At	35 S 00	Sa rises at long.	33 W 34	and sets at long.	118 E 56
At	30 S 00	Sa rises at long.	36 W 01	and sets at long.	121 E 23
At	25 S 00	Sa rises at long.	38 W 13	and sets at long.	123 E 35
At	20 S 00	Sa rises at long.	40 W 13	and sets at long.	125 E 35
At	15 S 00	Sa rises at long.	42 W 06	and sets at long.	127 E 28
At	10 S 00	Sa rises at long.	43 W 53	and sets at long.	129 E 15
At	5 S 00	Sa rises at long.	45 W 37	and sets at long.	130 E 59
At	0 N 00	Sa rises at long.	47 W 19	and sets at long.	132 E 41
At	5 N 00	Sa rises at long.	49 W 01	and sets at long.	134 E 23
At	10 N 00	Sa rises at long.	50 W 45	and sets at long.	136 E 07
At	15 N 00	Sa rises at long.	52 W 32	and sets at long.	137 E 54
At	20 N 00	Sa rises at long.	54 W 25	and sets at long.	139 E 47
At	25 N 00	Sa rises at long.	56 W 25	and sets at long.	141 E 47
At	30 N 00	Sa rises at long.	58 W 37	and sets at long.	143 E 59
At	35 N 00	Sa rises at long.	61 W 04	and sets at long.	146 E 26
At	40 N 00	Sa rises at long.	63 W 52	and sets at long.	149 E 14
At	45 N 00	Sa rises at long.	67 W 09	and sets at long.	152 E 31
At	50 N 00	Sa rises at long.	71 W 10	and sets at long.	156 E 32
At	55 N 00	Sa rises at long.	76 W 18	and sets at long.	161 E 40
At	60 N 00	Sa rises at long.	83 W 19	and sets at long.	168 E 41
At	65 N 00	Sa rises at long.	94 W 01	and sets at long.	179 E 23
At	70 N 00	Sa rises at long.	116 W 07	and sets at long.	158 W 30

Uranus	is on the upper Meridian at		9 W 28	is on the Nadir at	23 S 07
	is on the Zenith at		23 N 07	never sets at	66 N 53
	is on the lower Meridian at		170 E 32	never rises at	66 S 53
At	60 S 00	Ur rises at long.	51 W 49	and sets at long.	32 E 53
At	55 S 00	Ur rises at long.	61 W 55	and sets at long.	42 E 59
At	50 S 00	Ur rises at long.	68 W 54	and sets at long.	49 E 58
At	45 S 00	Ur rises at long.	74 W 12	and sets at long.	55 E 16
At	40 S 00	Ur rises at long.	78 W 29	and sets at long.	59 E 33
At	35 S 00	Ur rises at long.	82 W 05	and sets at long.	63 E 09
At	30 S 00	Ur rises at long.	85 W 12	and sets at long.	66 E 16
At	25 S 00	Ur rises at long.	87 W 59	and sets at long.	69 E 03
At	20 S 00	Ur rises at long.	90 W 32	and sets at long.	71 E 36
At	15 S 00	Ur rises at long.	92 W 54	and sets at long.	73 E 58
At	10 S 00	Ur rises at long.	95 W 09	and sets at long.	76 E 13
At	5 S 00	Ur rises at long.	97 W 20	and sets at long.	78 E 24
At	0 N 00	Ur rises at long.	99 W 28	and sets at long.	80 E 32
At	5 N 00	Ur rises at long.	101 W 36	and sets at long.	82 E 40
At	10 N 00	Ur rises at long.	103 W 47	and sets at long.	84 E 51

At	15 N 00	Ur rises at long.	106 W 02	and sets at long.	87 E 06
At	20 N 00	Ur rises at long.	108 W 24	and sets at long.	89 E 28
At	25 N 00	Ur rises at long.	110 W 57	and sets at long.	92 E 01
At	30 N 00	Ur rises at long.	113 W 44	and sets at long.	94 E 48
At	35 N 00	Ur rises at long.	116 W 51	and sets at long.	97 E 55
At	40 N 00	Ur rises at long.	120 W 27	and sets at long.	101 E 31
At	45 N 00	Ur rises at long.	124 W 44	and sets at long.	105 E 48
At	50 N 00	Ur rises at long.	130 W 02	and sets at long.	111 E 06
At	55 N 00	Ur rises at long.	137 W 01	and sets at long.	118 E 05
At	60 N 00	Ur rises at long.	147 W 07	and sets at long.	128 E 11
At	65 N 00	Ur rises at long.	165 W 41	and sets at long.	146 E 45

Neptune	is on the upper Meridian at	102 E 24	is on the Nadir at	2 N51
	is on the Zenith at	2 S 51	never sets at	87 S 09
	is on the lower Meridian at	77 W 36	never rises at	87 N 09

At	60 S 00	Ne rises at long.	7 E 27	and sets at long.	162 W 39
At	55 S 00	Ne rises at long.	8 E 19	and sets at long.	163 W 31
At	50 S 00	Ne rises at long.	8 E 59	and sets at long.	164 W 12
At	45 S 00	Ne rises at long.	9 E 32	and sets at long.	164 W 45
At	40 S 00	Ne rises at long.	10 E 00	and sets at long.	165 W 12
At	35 S 00	Ne rises at long.	10 E 24	and sets at long.	165 W 36
At	30 S 00	Ne rises at long.	10 E 45	and sets at long.	165 W 57
At	25 S 00	Ne rises at long.	11 E 04	and sets at long.	166 W 16
At	20 S 00	Ne rises at long.	11 E 21	and sets at long.	166 W 34
At	15 S 00	Ne rises at long.	11 E 38	and sets at long.	166 W 50
At	10 S 00	Ne rises at long.	11 E 54	and sets at long.	167 W 06
At	5 S 00	Ne rises at long.	12 E 09	and sets at long.	167 W 21
At	0 N 00	Ne rises at long.	12 E 24	and sets at long.	167 W 36
At	5 N 00	Ne rises at long.	12 E 39	and sets at long.	167 W 51
At	10 N 00	Ne rises at long.	12 E 54	and sets at long.	168 W 06
At	15 N 00	Ne rises at long.	13 E 10	and sets at long.	168 W 22
At	20 N 00	Ne rises at long.	13 E 26	and sets at long.	168 W 39
At	25 N 00	Ne rises at long.	13 E 44	and sets at long.	168 W 56
At	30 N 00	Ne rises at long.	14 E 03	and sets at long.	169 W 15
At	35 N 00	Ne rises at long.	14 E 24	and sets at long.	169 W 36
At	40 N 00	Ne rises at long.	14 E 48	and sets at long.	170 W 00
At	45 N 00	Ne rises at long.	15 E 15	and sets at long.	170 W 28
At	50 N 00	Ne rises at long.	15 E 48	and sets at long.	171 W 01
At	55 N 00	Ne rises at long.	16 E 29	and sets at long.	171 W 41
At	60 N 00	Ne rises at long.	17 E 21	and sets at long.	172 W 33
At	65 N 00	Ne rises at long.	18 E 32	and sets at long.	173 W 44
At	70 N 00	Ne rises at long.	20 E 16	and sets at long.	175 W 29

Pluto	is on the upper Meridian at	49 E 42	is on the Nadir at	23 S 26
	is on the Zenith at	23 N 26	never sets at	66 N 34
	is on the lower Meridian at	130 W 18	never rises at	66 S 34

At	60 S 00	Pl rises at long.	8 E 20	and sets at long.	91 E 03
At	55 S 00	Pl rises at long.	2 W 04	and sets at long.	101 E 28
At	50 S 00	Pl rises at long.	9 W 12	and sets at long.	108 E 36
At	45 S 00	Pl rises at long.	14 W 37	and sets at long.	114 E 01
At	40 S 00	Pl rises at long.	18 W 59	and sets at long.	118 E 22
At	35 S 00	Pl rises at long.	22 W 38	and sets at long.	122 E 02
At	30 S 00	Pl rises at long.	25 W 49	and sets at long.	125 E 13
At	25 S 00	Pl rises at long.	28 W 39	and sets at long.	128 E 02
At	20 S 00	Pl rises at long.	31 W 14	and sets at long.	130 E 37
At	15 S 00	Pl rises at long.	33 W 38	and sets at long.	133 E 02
At	10 S 00	Pl rises at long.	35 W 55	and sets at long.	135 E 19
At	5 S 00	Pl rises at long.	38 W 08	and sets at long.	137 E 32
At	0 N 00	Pl rises at long.	40 W 18	and sets at long.	139 E 42
At	5 N 00	Pl rises at long.	42 W 28	and sets at long.	141 E 52
At	10 N 00	Pl rises at long.	44 W 41	and sets at long.	144 E 05
At	15 N 00	Pl rises at long.	46 W 58	and sets at long.	146 E 22
At	20 N 00	Pl rises at long.	49 W 23	and sets at long.	148 E 46
At	25 N 00	Pl rises at long.	51 W 58	and sets at long.	151 E 21
At	30 N 00	Pl rises at long.	54 W 47	and sets at long.	154 E 11
At	35 N 00	Pl rises at long.	57 W 58	and sets at long.	157 E 22
At	40 N 00	Pl rises at long.	61 W 38	and sets at long.	161 E 01
At	45 N 00	Pl rises at long.	65 W 59	and sets at long.	165 E 23
At	50 N 00	Pl rises at long.	71 W 24	and sets at long.	170 E 48
At	55 N 00	Pl rises at long.	78 W 32	and sets at long.	177 E 56
At	60 N 00	Pl rises at long.	88 W 57	and sets at long.	171 W 40
At	65 N 00	Pl rises at long.	108 W 38	and sets at long.	151 W 58

Uri Geller
Dec. 20, 1946
32N04'00"
Tropical

Natal Chart
Tel Aviv
8:10:59 ST
Geocentric

Paran Printout
2:0:0 AM EET
34E46'00"

Pl	Rise			Upper			Set			Lower		
	d	m	h m s	d	m	h m s	d	m	h m s	d	m	h m s
☉	193	01	12:52:06	267	17	17:49:06	341	32	22:46:07	87	17	5:49:06
☽	147	23	9:49:32	227	20	15:09:19	307	16	20:29:05	47	20	3:09:19
☿	171	30	11:26:01	247	44	16:30:55	323	58	21:35:50	67	44	4:30:55
♀	147	29	9:49:56	228	07	15:12:26	308	44	20:34:57	48	07	3:12:26
♂	198	30	13:13:58	272	08	18:08:31	345	46	23:03:03	92	08	6:08:31
♃	146	32	9:46:07	225	58	15:03:54	305	25	20:21:41	45	58	3:03:54
♄	28	23	1:53:33	130	40	8:42:39	232	56	15:31:45	310	40	20:42:39
♅	333	00	22:12:01	78	31	5:14:03	184	01	12:16:04	258	31	17:14:03
♆	102	10	6:48:40	190	23	12:41:30	278	35	18:34:20	10	23	0:41:30
♇	31	55	2:07:42	137	41	9:10:43	243	26	16:13:43	317	41	21:10:43
☊	324	29	21:37:56	69	12	4:36:48	173	55	11:35:39	249	12	16:36:48
As	122	45	8:10:59	205	55	13:43:42	289	06	19:16:24	25	55	1:43:42
MC	19	32	1:18:08	122	45	8:10:59	225	57	15:03:49	302	45	20:10:59
Vx	329	26	21:57:43	74	37	4:58:28	179	48	11:59:13	254	37	16:58:28
EP	131	12	8:44:47	212	45	14:10:59	294	18	19:37:11	32	45	2:10:59
⊗	77	53	5:11:32	170	28	11:21:51	263	03	17:32:10	350	28	23:21:51

♆	lower	10	23	♆	rise	102	10	♂	rise	198	30	♆	set	278	35
MC	rise	19	32	MC	upper	122	45\	As	upper	205	55	As	set	289	06
As	lower	25	55	As	rise	122	45/	EP	upper	212	45	EP	set	294	18
♄	rise	28	23	♄	upper	130	40\	MC	set	225	57\	MC	lower	302	45
♇	rise	31	55\	EP	rise	131	12/	♃	upper	225	58/	♃	set	305	25
EP	lower	32	45/	♇	upper	137	41	☽	upper	227	20\	☽	set	307	16
♃	lower	45	58	♃	rise	146	32\	♀	upper	228	07/	♀	set	308	44
☽	lower	47	20\	☽	rise	147	23/\	♄	set	232	56	♄	lower	310	40
♀	lower	48	07/	♀	rise	147	29/	♇	set	243	26	♇	lower	317	41
☿	lower	67	44	⊗	upper	170	28	☿	upper	247	44	☿	set	323	58\
☊	upper	69	12	☿	rise	171	30	☊	lower	249	12	☊	rise	324	29/
Vx	upper	74	37	☊	set	173	55	Vx	lower	254	37	Vx	rise	329	26
⊗	rise	77	53\	Vx	set	179	48	♅	lower	258	31	♅	rise	333	00
♅	upper	78	31/	♅	set	184	01	⊗	set	263	03	☉	set	341	32
☉	lower	87	17	♆	upper	190	23	☉	upper	267	17	♂	set	345	46
♂	lower	92	08	☉	rise	193	01	♂	upper	272	08	⊗	lower	350	28

Data Sources

Berlin Wall
August 13, 1961
1:11 a.m. CET
East Berlin, East Germany
Source: *The Berlin Wall*, Norman Gelb, pp. 144-145; *Die Mauer*, Jürgen Petschull, p. 47

Charles, Prince
November 14, 1948
9:14 p.m. GMT
London, ENGLAND
Source: Buckingham Palace

Copperfield, David
September 16, 1956
7:02 a.m. EDT
Metuchen, NJ, U.S.A.
Source: Birth certificate (Lois Rodden)

Diana, Princess
July 1, 1961
7:45 p.m. BST
Sandringham, England
Source: Buckingham Palace

First Manned Space Flight
April 12, 1961
9:07 a.m. EET
Baikonur, USSR
Source: *The Book of World Horoscopes*

Geldof, Bob
October 5, 1951
2:20 p.m. BST
Dublin, Ireland
Source: From himself

Geller, Uri
December 20, 1946
2:00 a.m. EET
Tel Aviv, Israel
Source: From himself (*American Book of Charts*)

Germany, East
October 7, 1949
1:45 p.m. CET
East Berlin, East Germany
Source: *The Fate of a German Nation* (Ebertin); data provided by Charles Harvey

Germany, West
May 24, 1949
00:00 a.m. CED
Bonn, West Germany
Source: *The Book of World Horoscopes* (2nd Edition)

Honecker, Erich
August 25, 1912
6:00 p.m. CET
Neunkirchen, Saar, Germany
Source: *Astrologie Heute* (Aug/Sept 1992); DAV

Kohl, Helmut
April 3, 1931
6:30 a.m. CET
Ludwigshafen, Germany
Source: Birth certificate

Schiffer, Claudia
August 25, 1912
12:10 p.m. CET
Düsseldorf, Germany
Source: *Meridian* (Jan/Feb 1995); Shelley Ackerman via "Elle"

Schwarzkopf, Norman
August 22, 1934
4:45 a.m. EDT
Trenton, NJ, U.S.A.
Source: Birth certificate (Lois Rodden)

Thatcher, Margaret
October 13, 1925
9:00 a.m. GMT
Grantham, England
Source: From her press office

Wedding, Charles/Diana
July 29, 1981
11:7 a.m. BST
London, England
Source: *Working with Astrology*; (Michael Harding/Charles Harvey)

Worlds First Powered Flight
December 17, 1903
10:35 a.m. EST
Kill Devil Hills, NC, U.S.A.
Source: *The Book of World Horoscopes*

Recommended Reading

Books

*Astro*Carto*Graphy Book of Maps*, Jim Lewis & Ariel Guttman

*The Psychology of Astro*Carto*Graphy*, Jim Lewis with Kenneth Irving

Navigating by the Stars, Edith Hathaway

Working with Astrology, Michael Harding & Charles Harvey

Forecasting Backward and Forward, Bruce Hammerslough

The Astrology of the Macrocosm, Edited by Joan McEvers

Mundane Astrology, M. Baigent, N. Campion, C. Harvey

The Book of World Horoscopes, Nicholas Campion

*The Astro*Carto*Graphy Source Book of Mundane Maps*, Jim Lewis

Hymns to the Ancient Gods, Michael Harding

Essays in Astrology, Robert Hand

Planets in Composite, Robert Hand

Planets in Locality, Steve Cozzi

Synastry, Ronald Davison

Recent Advances in Natal Astrology, Geoffrey Dean

Astronomy for Astrologers, John Filbey, Peter Filbey

Planets on the Move, Maritha Pottinger and Zipporah Dobyns

What are Astrolocality Maps?, Maritha Pottinger

How to write an Astrological Synthesis, Terry Dwyer

Periodicals

Astrologie Heute—Articles on Astro*Carto*Graphy and mundane astrology are often featured in this bi-monthly periodical published in German.

Astrologie Heute
Albisriedstrasse 232
Postfach
CH 8047 Zurich
Switzerland

Welcome to Planet Earth—This is also a bi-monthly periodical which has an excellent coverage of mundane astrology. The articles take a look at events that have recently been in the news and makes very interesting reading.

Welcome to Planet Earth
The Great Bear
P.O. Box 12007
Eugene
OR 97440
U.S.A.

Considerations—This periodical is published every three months and features at least one article on rectification. There are many different methods to rectify a time of birth, but those used in *Considerations* are very impressive. Back issues of *Considerations* are well worth obtaining as they explain how to approach rectification.

Considerations
Ken Gillman
P.O. Box 491
Mount Kisco
NY 10549
U.S.A.

Index